GEOLOGY OF THE ANDES AND ITS RELATION TO HYDROCARBON AND MINERAL RESOURCES

Circum-Pacific Council for Energy and Mineral Resources

Earth Science Series, Volume 11

GEOLOGY OF THE ANDES AND ITS RELATION TO HYDROCARBON AND MINERAL RESOURCES

Edited By

George E. Ericksen, María Teresa Cañas Pinochet, and John A. Reinemund

Published by the Circum-Pacific Council
for Energy and Mineral Resources
Houston, Texas, U.S.A.
1990

Published February, 1990

ISBN: 0-933687-12-5

Cover Design: Ben Servino, U.S. Geological Survey

Cover Illustration: Top left: Callejón de Huaylas, a north-trending Andean valley in northern Peru flanked by glaciated Cordillera Blanca (background), which has a granitic batholith core flanked by Jurassic and Cretaceous marine sedimentary rocks. *Top right:* Site of Plaza de Armas in the city of Yungay that was buried by the Huascarán debris avalanche triggered by the northern Peru earthquake of May 31, 1970. Blocks near palm tree on right are fragments of cathedral walls. *Bottom left:* Open-pit mine at Cerro de Pasco, central Peruvian Andes, where lead-zinc orebodies in Mesozoic limestones are exploited. *Bottom right:* Recent scarp of Salar de Pintados fault at southwestern margin of salar; light-colored silty rock-salt crust of salar in foreground has been etched by floods discharged from gullies in fault scarp. *Photographs by G. E. Ericksen.*

Circum-Pacific Council for Energy and Mineral Resources Earth Science Series
F. L. Wong and H. G. Greene, Editors

1. Tectonostratigraphic Terranes of the Circum-Pacific Region
   edited by D. G. Howell

2. Geology and Offshore Resources of Pacific Island Arcs—Tonga Region
   compiled and edited by D. W. Scholl and T. L. Vallier

3. Investigations of the northern Melanesian Borderland
   edited by T. M. Brocher

4. Geology and Offshore Resources of Pacific Island Arcs—Central and Western Solomon Islands
   edited by J. G. Vedder, K. S. Pound, and S. Q. Boundy

5A. The Antarctic Continental Margin: Geology and Geophysics of Offshore Wilkes Land
   edited by S. L. Eittreim and M. A. Hampton

5B. The Antarctic Continental Margin: Geology and Geophysics of the Western Ross Sea
   edited by A. K. Cooper and F. J. Davey

6. Geology and Resource Potential of the Continental Margin of Western North American and Adjacent Ocean Basins—Beaufort Sea to Baja California
   edited by D. W. Scholl, A. Grantz, and J. G. Vedder

7. Marine Geology, Geophysics, and Geochemistry of the Woodlark Basin—Solomon Islands
   edited by B. Taylor and N. F. Exon

8. Geology and Offshore Resources of Pacific Island Arcs—Vanuatu Region
   edited by H. G. Greene and F. L. Wong

9. Geology and Offshore Resources of Pacific Island Arcs—New Ireland and Manus Region, Papua New Guinea
   edited by M. S. Marlow, S. V. Dadisman, and N. F. Exon

10. Petroleum Resources of China and Related Subjects
   edited by H. C. Wagner, L. C. Wagner, Frank F. H. Wang, and F. L. Wong

11. Geology of the Andes and its Relation to Hydrocarbon and Mineral Resources
   edited by G. E. Ericksen, M. T. Cañas Pinochet, and J. A. Reinemund

12. Geology and Offshore Resources of Pacific Island Arcs—Solomon Islands and Bougainville, Papua New Guinea Regions
   edited by J. G. Vedder and T. R. Bruns

Circum-Pacific Council publications are available from:
The AAPG Bookstore • P.O. Box 979 • Tulsa, Oklahoma 74101-0979 • USA
and its international distributors.

CIRCUM-PACIFIC COUNCIL FOR ENERGY AND MINERAL RESOURCES

FOREWORD

The Earth Science Series of the Circum-Pacific Council for Energy and Mineral Resources (CPCEMR) is designed to convey the results of geologic research in and around the Pacific Basin. Topics of interest include framework geology, petroleum geology, mineral resources, geothermal energy, environmental geology, volcanology, oceanography, tectonics, geophysics, geochemistry, and applications of renewable energy. The CPCEMR supports and publishes results of scientific research that will advance the knowledge of energy and mineral resources potential in the circum-Pacific region. The Earth Sciences Series is specifically designed to publish papers that include new data and new maps, report on CPCEMR-sponsored symposia and workshops, and describe the results of onshore and marine geological and geophysical investigations.

This symposium, sponsored jointly by the CPCEMR and the Servicio Nacional de Geología y Minería of Chile, was an excellent opportunity to disseminate information on the geology of the Andean region and its importance in investigations for mineral and energy resources. This volume contains 32 of the 42 papers that were presented. A Spanish version of selected papers is being published by the Servicio Nacional de Geologia y Mineria. These two volumes will greatly enhance the scientific literature of this region, and will definitely provide an important reference source for future energy and mineral resource exploration.

In addition to the sponsoring organizations, financial support for the symposium was supplied by the Inter-American Development Bank, the U.S. Agency for International Development, and the U.S. Geological Survey. I thank George E. Ericksen (U.S. Geological Survey), María Teresa Cañas Pinochet (Director, Servicio Nacional de Geología y Minería), and John A. Reinemund (CPCEMR) for their dedicated efforts in bringing this volume to fruition.

Michel T. Halbouty
Chairman and President

PREFACE

The symposium on "Geology of the Andes and Its Relation to Hydrocarbons and Mineral Resources" was held in Santiago, Chile, during November 11-15, 1985. Thirty-two papers from this symposium are presented in this volume of the Circum-Pacific Council for Energy and Mineral Resources (CPCEMR) Earth Science Series, and 25 of these papers, chiefly those dealing with the central and southern Andes, are being published in Spanish by the Servicio Nacional de Geología y Minería of Chile. These papers present a broad spectrum of geoscience research in the Andes by geologists with a wide range of backgrounds, interests, and responsibilities. Because of the diversity in backgrounds of the authors and the fact that some manuscripts were prepared in Spanish and others in English, the task of editorial review and author revision has been formidable and time consuming. As a consequence, publication has been delayed.

We believe that all the papers included in this volume contain significant new information that provides new insight to Andean geology and resources. However, we recognize that many subjects relevant to the study of Andean resources are not covered as fully as would be desirable in this volume. Among these are the distribution and characteristics of precious metal deposits, the distribution and potential of energy resources other than hydrocarbons, and industrial mineral resources.

Finally, we believe that both the symposium itself and the preparation of this volume have brought about an interchange of information that will promote better coordination between geoscientists interested in Andean geology and resources and between national geoscience institutions of the Andean countries. In view of the increasing interest in exploration and development of Andean energy and mineral resources, which provide some unique opportunities in a world of declining available resources, this volume provides information that will be of wide interest and use.

We wish to express our gratitude to the government of Chile and to the CPCEMR for their initiative in promoting the symposium, and for financial assistance for the symposium meetings and preparation and publication of this volume. We wish to thank the authors for their cooperation in preparation of a series of papers that are technically sound and which meet rigorous editorial standards. Our editorial work has been greatly facilitated by Mary Stewart and Harlene Padon of the CPCEMR Secretariat.

George E. Ericksen
María Teresa Cañas Pinochet
John A. Reinemund

TABLE OF CONTENTS

HYDROCARBON RESOURCES

Ericksen, G. E., Cañas Pinochet, M. T., and Reinemund, J. A., editors, 1989, Geology of the Andes and its relation to hydrocarbon and mineral resources: Houston, Texas, Circum-Pacific Council for Energy and Mineral Resources Earth Science Series, v. 11.

OPENING REMARKS

Samuel Lira
Minister of Mining
Chile

On behalf of His Excellency the President and the Government of Chile, I have the honor to open this first meeting of the Symposium on "Geology of the Andes and Its Relation to Hydrocarbon and Mineral Resources," organized by the Circum-Pacific Council for Energy and Mineral Resources and the Servicio Nacional de Geología y Minería.

The Chilean Government, understanding the scientific importance of the event we are opening today, in which scientists from many of the countries bordering the Pacific basin are represented, is also participating as official sponsor. This official support and sponsorship acknowledges the scientific importance of this Symposium; it also demonstrates our gratitude to the experts who, under the Circum-Pacific Council, have chosen to hold this meeting in our country with the purpose of increasing communication among Andean country specialists.

In addition, this Symposium is of special interest because the information about mineral resources and their relation to the geological evolution of the Andes is of special interest to a country such as Chile that is so greatly endowed with minerals that they constitute the base of its economy.

Therefore, while I gratefully acknowledge the kind invitation to open this event, I wish to take advantage of this opportunity to discuss some aspects of our mining activity that should be of interest to this distinguished audience.

The Chilean Government, considering the necessity for rapid economic development that was essential in fulfilling the hopes of a better life for all Chileans, has assigned a preponderant role for mining activity in the economic development of Chile. This was imperative because mining is one of the main sources of government income, and a major contribution to the Gross National Product of the country, utilizing directly or indirectly an important portion of labor.

In order to support increased mining activity, fundamental institutional transformations were carried out, guided by the philosophy of respect and fundamental freedom of the individual. As a consequence, the State was able to redefine the way in which it exerted jurisdiction over mining activity. Within this context and in the area of subjects considered at this Symposium, I would like to mention that in November 1980 the Servicio Nacional de Geología y Minería was created as the legal successor of the Servicio de Minas del Estado and the Instituto de Investigaciones Geológicas, with the main aim of serving the Ministry of Mines as the technical adviser in subjects related to compilation, publication, and distribution of information concerning the existence, development, and conservation of all land and sea-floor mineral resources of Chile.

Among the most important concerns of this Servicio were preparation and execution of a plan to obtain the fundamental geologic information required for the formulation of policies affecting mining activity. We, therefore, are emphasizing a national-level program of geological and mining research in order to obtain, during the 80s, reliable information about our mineral resources, which will assist both potential future and present mining enterprise in preparation of effective plans for mineral exploration and exploitation as well as assist in the quantification of mineral resources. This program includes a series of investigations, which at present are at different levels of completion, as follows: (1) Carta Geológica de Chile (40% complete in 1985); (2) Carta Metalogénica de Chile (60% complete in 1985); and (3) Carta Magéntica de Chile (30% complete in 1985). In addition, other maps such as the Hydrological Map of Chile are being prepared. The Servicio has also taken part in the investigation and solution of geotechnical problems related to public works construction.

The Servicio has the responsibility of maintaining and updating the Catastro Minero Nacional, a permanent, up-to-date record of all existing mining claims. This is one of the main goals of the Government, and by late 1985 was about 70% complete. In addition, the Government is carrying out a project for computerized information on more than 234,000 claims. I would like to emphasize that the Catastro Minero Nacional is in response to a long-standing—more than 50 years—need of the mining sector. This computerized Catastro will provide legal protection for the right to explore and exploit mineral deposits, activities that may be affected by lack of precise location and other data about mining claims.

Inasmuch as the Servicio Nacional de Geología y Minería is in charge of the compilation and preservation of earth-science information of all types, it has initiated a program for systematizing of all earth-science information existing in Chile, which has been scattered in the many different organizations. The National Geological and Mining Data Center has been created to carry out this work. This center has financial support of the Banco Exterior de Expaña.

It should be mentioned that the geological investigations carried out in some areas have served as a basis for attracting foreign investment; such has been the case for the copper deposits of Copaquire, Quebrada Blanca, Cerro Colorado, and Mocha projects, and the lithium brines in Salar de Atacama. The Servicio also has participated in the determination of ore reserves in the Andina-Disputada area.

The Government, being convinced that it was necessary to eliminate any doubt concerning the basic concept of the permanence of mine ownership, which is essential to a vigorous mining industry, included in the Constitution of 1980 statutes oriented at the reestablishment of confidence of the private sector in regard to mine investment. These statutes were established by the Organic Law of Mining Claims and, later, by a new Mining Code of Chile, which introduced important improvements and modernized the former code. One of the points most emphasized in the new code is the legal protection for mineral exploration. In this context, legislation places an extraordinary value on exploration, providing improvement in obtaining and protecting exploration claims. This new legislation, with clear and simple standards, guarantees the right of the investor to fruits of his efforts in the case of success, and assures compensation for the assumed risks.

But any mining development policy would not be complete if it were not for a real potential for discovery of new resources large enough to compensate the financial risk of mineral exploration and development. Private foreign investment was encouraged by Decree No. 600 of the Foreign Investment Ordinance; this law gives assurance to the foreign investor of fair and equal treatment with respect to the national investor. As a result of this, and the great competitive advantages presented by the Chilean mining projects, an ever increasing number of new investments, both foreign and national, have been made.

In the late 1970s and early 1980s, mining investments of about 1,100 million U.S. dollars were made in Chile. Moreover, at present, projects of development such as Los Bronces, La Escondida, Collahuasi, Cerro Colorado, Tambo, Choquelimpie, Sancarrón, La Copia, and El Hueso are in different stages of exploration and development. During this same period, other mine development projects have become reality. Included are the opening of the sulphuric acid plant of the Cia. Minera Mantos Blancos, the El Indio gold-silver operation, and major expansion of the El Soldado operation by the Cia. Minera Disputada de las Condes.

During 1984 and the first half of 1985, foreign investments in mineral exploration were more than US$ 40 million, which, added to the previous investments, give a sum of US$ 140 million foreign capital investment during the first half of the 1980 decade.

In conclusion, I would like to emphasize that the Government considers the subject of this Symposium to be of great importance to Chile and other South American countries. Your deliberations will enhance our knowledge concerning orogenic processes and their relationship to origin and distribution of mineral deposits. We believe that the exchange of information in a Symposium of this type will provide guides to future geological research in the Andes, and eventually, the discovery of new energy and mineral resources.

On behalf of the Government I wish to acknowledge your participation in this Symposium, to express gratitude for your efforts, and to wish you a great success in your deliberations.

GEOLOGICAL FRAMEWORK

Ericksen, G. E., Cañas Pinochet, M. T., and
Reinemund, J. A., editors, 1989, Geology of the Andes
and its relation to hydrocarbon and mineral resources:
Houston, Texas, Circum-Pacific Council for Energy
and Mineral Resources Earth Science Series, v. 11.

CHAPTER 1

GEOLOGIC-TECTONIC FRAMEWORK OF THE ANDEAN REGION

José Corvalán
Department of Geology and Geophysics
University of Chile, Santiago, Chile

ABSTRACT

The two main tectonic boundaries postulated for the subdivision of the modern Andean Belt into northern, central, and southern Andes, plus another—the Arica-Santa Cruz line—appear to have been present during the Mesozoic-Cenozoic evolution of the chain. However, tectonic activity along these boundaries, probably deep-seated fault movement, seems to have changed through time and caused changes in the longitudinal extent and depositional history of different paleogeographic domains.

The present central Andes correspond to a pericratonic domain in which only arc-backarc lithologic associations are recognized; important participation of oceanic crust occurred in the northern and southern Andes, where trench-forearc sequences also are evident. Paleogeographic reconstructions made for the Triassic, Jurassic, Cretaceous, and Tertiary suggest that it is possible to separate tectonic segments of different extent in successive periods, limited by transverse structures that were differentially reactivated in some epochs but remained relatively inactive in others. These changing tectonic boundaries, which may reflect corresponding subduction zone boundaries, have controlled uplift and subsidence in different segments of the belt, which did not always coincide with the longitudinal extent of the present northern, central, and southern Andes.

INTRODUCTION

The fundamental work of Gerth (1955) on South American geology showed the Andes Mountains as a complex mountain system. In their present configuration, which is the result of an active Plio-Pleistocene tectonism that affected the whole western border of the South American continent, several major morphostructural units can be recognized (Figure 1). Whether continuous or not, these units are generally parallel to the main axis of the Andes, but parallel units may differ markedly in tectonic style and lithology. Individual units also show important longitudinal geologic and structural changes, even where the continuity of the morphostructural units can be established. For example, such changes are well displayed along the Cordillera Occidental of Perú, Ecuador, and Colombia, which continues southward into the Main Andean Cordillera between Chile and Argentina and on into the Patagonian Cordillera. Pioneer work of Steinmann on the Andean belt, in particular that in Perú (Steinmann, 1930), led to the discovery of important stratigraphic and structural units that he designated as Peruandes, Ecuadorandes, and Chimuandes to distinguish the particular lithologic and structural configurations of the Andes in Perú and Ecuador, respectively (Figure 2).

Major subdivisions of the Andes were redefined by Aubouin and Borrello (1966), Aubouin (1972), and Vicente (1970) who recognized that during the evolution of the Andes in the Mesozoic and Cenozoic there existed pericontinental mountain chains of Andean type and intercontinental chains of alpine or geosynclinal type. The Caribbean Mountain system (Figure 1), which is typical of the latter type, is characterized by abundant ophiolites, flysch deposits, strong metamorphism, and overthrusting. The Andean-type Cordillera of Perú, Bolivia, Chile, and Argentina (Aubouin et al., 1973) extends from the Huancabamba deflection in northern Perú to the Bariloche transversal in southern Chile and Argentina (Figure 3). The Andean Cordillera of Ecuador and Colombia (Septentrional Andes) is interpreted as being transitional to the Caribbean system and the Patagonian Cordillera transitional to the Magallanean (Aubouin et al., 1973).

Gansser (1973) proposed division of the Andes into three main longitudinal segments: northern, central, and southern Andes (Figure 3). The northern Andes extend northward from the Huancabamba deflection, which is characterized by transverse structures that project westward toward the Carnegie Ridge-Galapagos Islands and eastward into an important lineament following the Amazon River. The central Andes end at the latitude of Golfo de Penas, facing the Chile Ridge.

Sillitoe (1974) proposed further segmentation of the Andes

based on morphology, structure, stratigraphy, and distribution of recent volcanic activity, which he believed reflected discontinuities in the underlying subduction zone.

The present review of the geologic and tectonic framework of the Andean region is concerned primarily with (a) the characterization of the major tectonic segments that can be differentiated during the Mesozoic-Cenozoic evolution, (b) a discussion of the lithologic and paleogeographic development, and (c) an evaluation of the major tectonic segments which appear to have controlled relative uplift and subsidence in different segments through time.

MAJOR TECTONIC SEGMENTS

Gansser's (1973) subdivision of the Andes into three main segments is a valid scheme on a regional basis. However, it is probable that deep-seated transverse structures in the central Andes were responsible for the important longitudinal changes in lithofacies and paleogeography during the Mesozoic and Cenozoic. Some of the tectonic boundaries discussed by Sillitoe (1974) may be related to these long-term deep-seated structures.

From north to south, the first important boundary is the Huancabamba deflection, northern Perú (Gansser, 1973), which marks the junction between the northern and central Andes (Figure 3). Changes in structural style across this deflection were reported by Ham and Herrera (1963), who characterized the sub-Andean fault system along the eastern flank of the Andes and western half of the sub-Andean basin in Perú and Ecuador. The system consists of subparallel faults striking northeastward in Ecuador and northwestward in Perú. However, at the Huancabamba deflection itself the main faults strike eastward, and had previously been recognized by Steinmann (1930) in the northern part of the Cordillera Occidental of Perú (north of the Otusco-Huamachuco region) where the main faults and folds trend predominantly east-west. This is the region that Steinmann (1930) considered to be a branch of the Andes extending westward into the Pacific Ocean, and which he named the Chimuandes and considered to be a structurally distinct unit of the main Peruvian Andes (Peruandes). The Andean basin, as Steinmann defined it, was south of the Chimuandes, which would explain the almost complete absence of pre-Cretaceous marine deposits in Ecuador and Colombia.

The next well-documented boundary to the south is the Pisco or Abancay deflection (Ham and Herrera, 1963, Marocco, 1971) where the predominant northwest-trending faults of the Peruvian Andes change to east-trending faults, and which in turn give way to northwest-trending faults to the south, including the belt of reverse faults in the sub-Andean region of Bolivia. At the latitude of this deflection, the marine Mesozoic sedimentary sequence, chiefly of Cretaceous age, grades into more continental facies. Marocco (1971) reported the presence of several east-trending subparallel faults along the Abancay deflection, such as those in the region of Andahuaylas and the major Abancay fault near the city of Abancay. Marocco believed that these faults resulted from Hercynian (late Paleozoic) tectogenesis, and that faulting along the Abancay deflection was reactivated during the Mesozoic. However, Mégard et al. (1971) suggested that the Abancay deflection

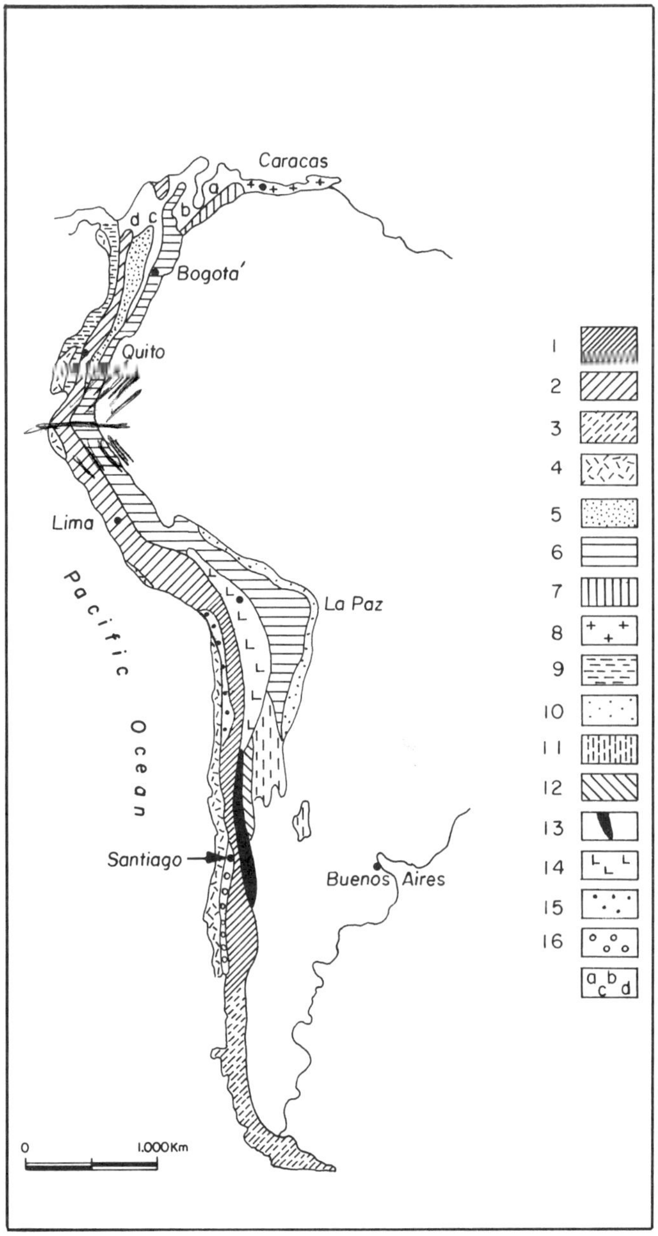

Figure 1. Main morphostructural units of the Andean belt: (1) Main Cordillera; (2) Cordillera Occidental; (3) Patagonian Cordillera; (4) Coastal Range; (5) Cordillera Central; (6) Cordillera Oriental; (7) Venezuelan Andes; (8) Caribbean Mountain System; (9) Pacific Coastal Province; (10) sub-Andean Ranges; (11) Pampean Ranges; (12) pre-Cordillera; (13) Frontal Cordillera; (14) Altiplano; (15) Pampa del Tamarugal; (16) Central Valley; (a) Falcon basin, (b) Maracaibo basin; (c) Lower Magdalena-Cesar Depression; (d) San Juan-Atrato Depression.

may correspond to a younger structure.

Another important tectonic boundary is the Arica-Santa Cruz line (Sonnenberg, 1963), which coincides with an evident break in the Paleozoic structures. Structural trends are northward south of the Santa Cruz line and northwestward to the north of it (Figure 3). Stille (1958) considered this transversal line to correspond to an old structure, possibly even Precam-

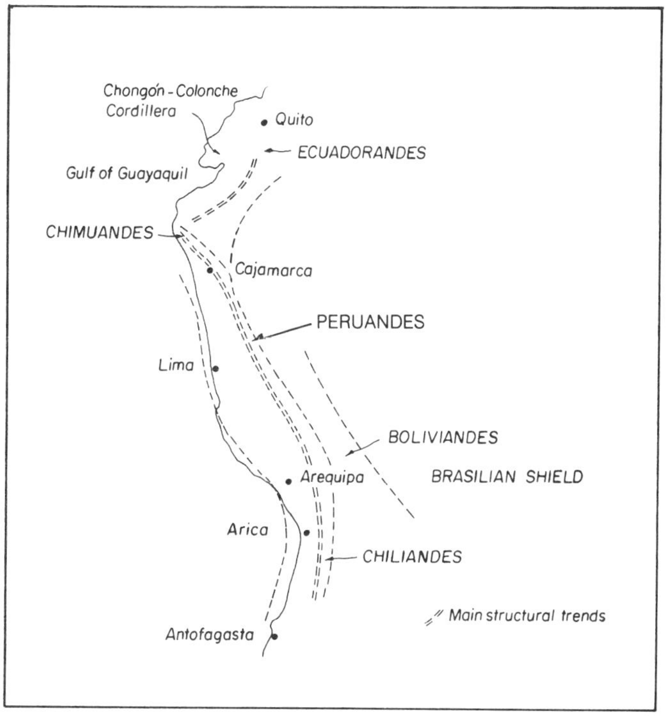

Figure 2. Distribution of structural units of the Peruvian and Ecuadoran Andes (from Steinmann, 1930). Not to scale.

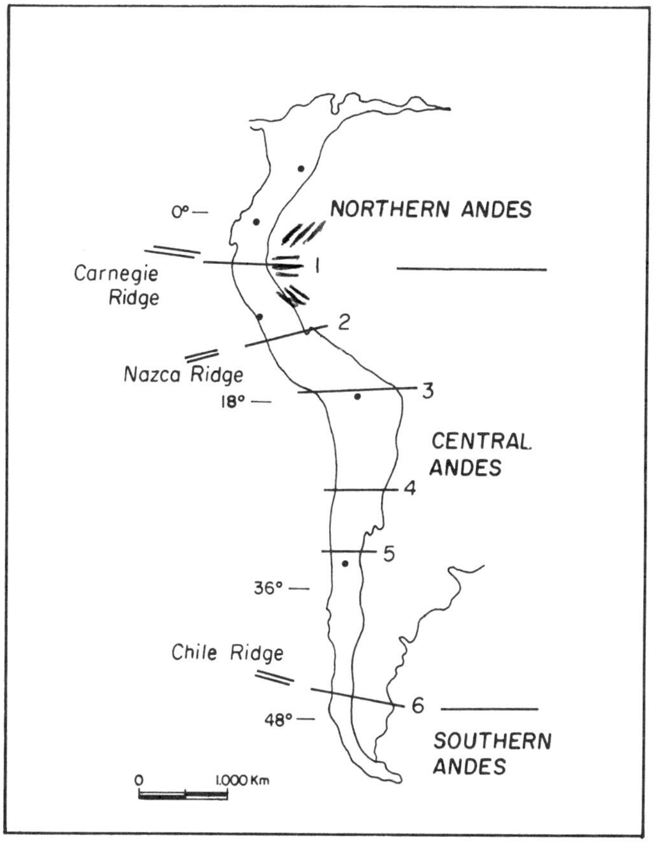

Figure 3. Subdivision of the Andean region and tectonic boundaries referred to in this report (after Gansser, 1973, and Sillitoe, 1974): (1) Huancabamba deflection; (2) Abancay (Pisco) deflection; (3) Arica-Santa Cruz line; (4) 27 °S latitude; (5) 33 °S latitude; (6) Bariloche transversal.

brian, reactivated during the Cenozoic. Although the Arica-Santa Cruz line might be related to a transcurrent fault zone, Sonnenberg (1963) suggested that during Andean orogenesis the movement was predominantly vertical, with uplift of the northern block. The structural continuity of this line into the Arica bend has been postulated (Sonnenberg, 1963) but not demonstrated. The line may correspond to an old megastructure having a complex history of reactivation. If, for example, the Jurassic marine sedimentary sequences exposed south of Arica are compared with those of southwestern Perú, it can be concluded that they probably accumulated under a continuous and common environment. Clearly, the Arica-Santa Cruz line was not an active fault zone during most of the Jurassic. In contrast, the Cretaceous marine sequences, which are well developed in central and southern Perú, become almost entirely continental south of the line, indicating that it was an active tectonic boundary at that time.

Farther to the south in the Andean belt, two other transverse tectonic boundaries can be recognized, one at about lat. 27 °S, and the other at lat. 33 °S (Figure 3). Between them is an Andean segment called the Norte Chico tectonic province (Lomnitz, 1962; Perez and Aguirre, 1968). This segment is characterized by transverse ranges in Chile. In contrast with the Andes northward and southward, it lacks a longitudinal central valley, lacks Quaternary volcanism, and structurally is characterized by reverse and thrust faulting, especially in the Main Cordillera.

The southernmost important tectonic boundary lies at approximately the latitude of Golfo de Penas, facing the Chile Ridge (Figure 3). This coincides with the limit between the central and southern Andes, whose depositional and tectonic histories are markedly different.

LITHOFACIES AND STRATIGRAPHIC DEVELOPMENT OF THE MESOZOIC-LOWER TERTIARY IN THE ANDEAN BELT

The various Andean segments show different evolutions (Steinmann, 1930; Gerth, 1955; Aubouin and Borrello, 1966; Aubouin, 1972; Aubouin et al., 1973; Gansser, 1973). Among the sedimentary rocks exposed in the Andes, those which accumulated during the Mesozoic and Cenozoic are probably the most important because of their great thicknesses and widespread distribution. Clearly, information about these rocks is essential to understanding the geologic history and tectonic development of each of the Andean segments. Present knowledge permits corroboration of some earlier interpretations of Andean development and suggestion of others. From an overall review of the lithofacies and stratigraphic development of the deposits accumulated since the Triassic, the following more important regional features of the northern Andes may be noted.

In Venezuela, the Triassic and Jurassic are represented by continental sedimentary sequences (red beds) and extrusive and intrusive rocks of silicic to intermediate composition. In

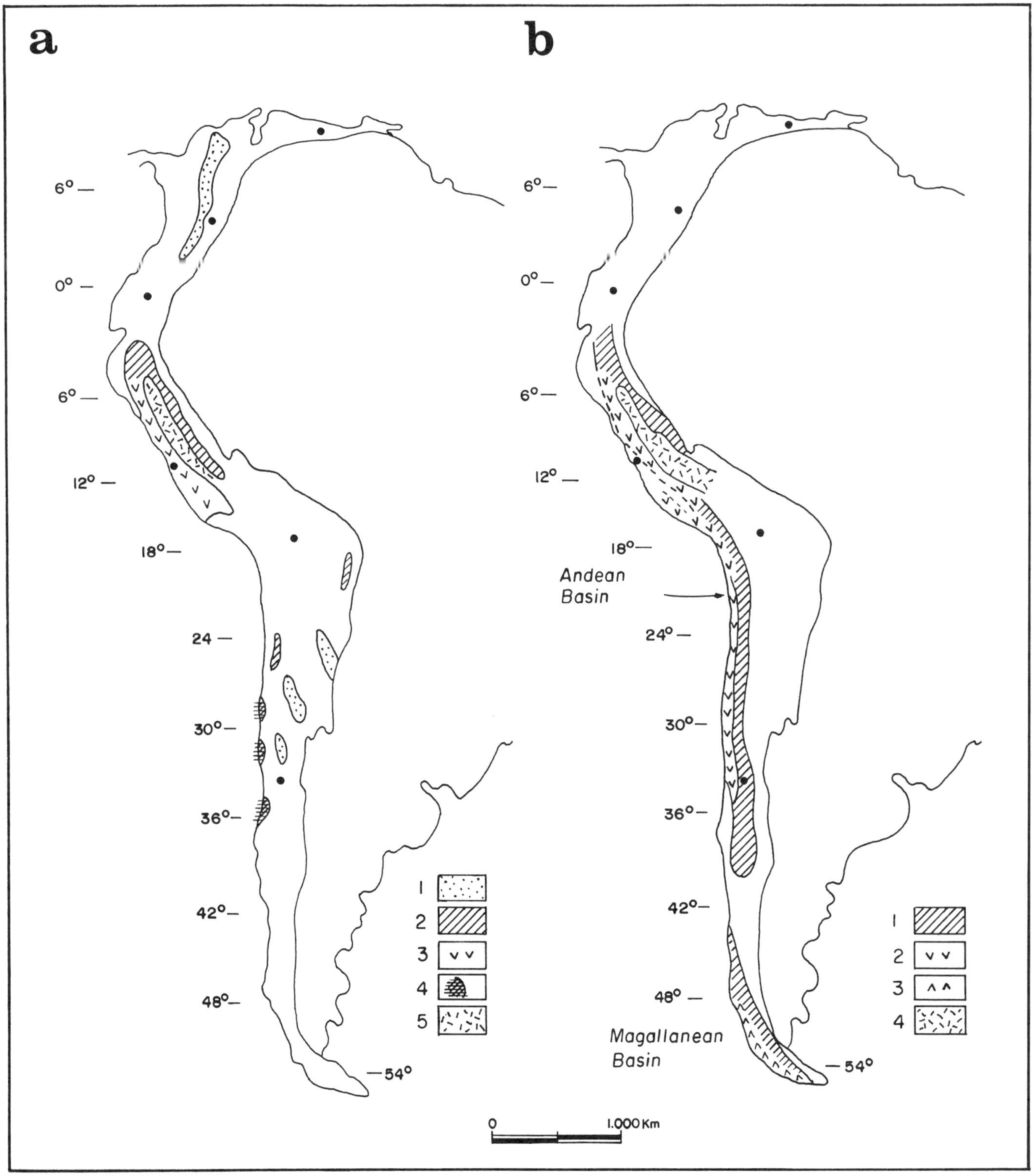

Figure 4a. Distribution of lithofacies of the Triassic system: (1) continental; (2) mostly marine-sedimentary, nonvolcanic; (3) andesitic volcanic-marine sedimentary; (4) platform marine facies; (5) Marañón geanticline.

Figure 4b. Distribution of lithofacies of the Jurassic system: (1) mostly marine sedimentary, nonvolcanic; (2) andesitic volcanic-marine sedimentary; (3) mafic volcanic-marine sedimentary; (4) Marañón geanticline.

Colombia, the Triassic and Jurassic consist of continental volcaniclastic sequences, well exposed in the Serra de Perija and on the western flank of the Santander massif. Included are red conglomerates, sandstones and lutites, and silicic to intermediate pyroclastic and flow rocks. In the upper reaches of the Magdalena River, Colombia, minor intercalations of shallow-water marine lutites and limestones are present. These rocks have a Late Triassic and Early Jurassic fauna, and they represent short-lived marine episodes within a predominantly continental environment. It can be inferred, therefore, that Colombia and the Andean sector of Venezuela were mostly emerged regions, with active continental erosion, deposition, and volcanism during most of the Triassic and Jurassic (Figure 4).

In Ecuador, rocks of Triassic age have been recognized. In contrast with the Jurassic continental facies in Colombia and Venezuela, the Jurassic sequences in the sub-Andean region of Ecuador contain, at least in part, well-developed marine facies. For example, in the Sierra de Cutucú (approximately between lat. 2° and 3° S) are exposed marine limestones and calcareous sandstones with intercalated lutites, which to the west include basaltic volcanic rocks interpreted as submarine flows (Faucher and Savoyat, 1973). Locally, the marine sedimentary rocks contain *Arietites*, which indicate a Liassic (Sinemurian) age. The marine sequence is overlain by continental clastic sedimentary rocks with some intercalations of evaporites in the lower part and with andesitic to basaltic rocks in the upper part.

Along the coastal region of Ecuador, north of the Cordillera Chongón-Colonche, the Basin Igneous Complex (Goossens and Rose, 1973) or Piñón Formation (Tschopp, 1948) includes basaltic lavas, diabase, and mafic intrusive rocks, considered to be of Late Jurassic to Early Cretaceous age. Similar rocks are exposed on the western flank of the Cordillerra Occidental (Formación Diabásica-Porfirítica of Sauer, 1965), where a mafic to ultramafic unit contains intercalated nonfossiliferous lutites and silicsedimentary rocks. According to Goossens and Rose (1973), the Basin Igneous Complex constitutes a continuous unit of Mesozoic rocks in western Colombia.

In northern and central Perú, marine transgressive sedimentary rocks were deposited during the Late Triassic and Jurassic. In the Central and Oriental Cordilleras, sequences of this age consist of limestones, dolomitic limestones, and sandstones (Pucará Group; Bellido, 1969), which apparently extend southward to near Abancay (approximately lat. 13°S).

In western Perú, the chronostratigraphic equivalent of the above Triassic-Jurassic rocks is a volcaniclastic sequence with marine intercalations, the Zaña Group (Bellido, 1969). This sequence presumably accumulated within a trough located in the proximity of the present coast, separated from the eastern marine domain by a rather stable emerged land in the position of the northwest-trending belt of Precambrian and Paleozoic rocks in central and southern Perú. Upper Triassic-Lower Jurassic units are also known in southwestern Perú, where they are predominantly volcanic-sedimentary rocks of continental origin (Yamayo Group), but locally include intercalations of marine lutites and reef limestones referred to as the Volcanico Chocolate (Bellido, 1969).

In the Andean belt to the south of Perú, the Triassic is mostly continental and consists chiefly of thick sequences of clastic sedimentary rocks having abundant interbedded volcanic rocks of silicic to intermediate composition. However, Upper Triassic marine formations are known in the sub-Andean belt of southern Bolivia (between lat. 19° and 23°S). These consist of shallow-water clastic sediments, limestones, and some evaporites. Also, Middle to Upper Triassic marine deposits are known in the pre-Cordillera of Antofagasta, Chile, between lat. 23°30′ and 26°00′ S (Chong and Hillebrandt, 1985), in the Coastal Range of central Chile (Brüggen, 1950; Muñoz Cristi, 1938, 1950; Fuenzalida, 1937; Cecioni and Westermann, 1968; Thiele, 1967; Corvalán, 1965, 1976), and at Alto del Carmen in the Main Cordillera (Brüggen, 1913, 1950; Zeil and Ichikawa, 1958; Reutter, 1974). The sedimentary sequence in the Coastal Range has been interpreted as being a shallow-water marine platform deposit (Corvalán, 1965). It is conformably overlain by marine sediments of Early Jurassic age. In places a small hiatus indicates a break in sedimentation (Corvalán, 1965).

The Jurassic, which is mostly continental in Colombia, Venezuela, and Ecuador (except for the marine Liassic sediments in southeastern Ecuador), is predominantly marine in Perú, Chile, and Argentina, the complete system being represented in the latter two countries. In Perú, the Upper Triassic-Lower Jurassic marine sequence is overlain by Middle and Upper Jurassic sequences that accumulated in well-defined basins having heteropic facies. In the eastern region of Perú, east of the Marañón geanticline (Benavides, 1956), the sub-Andean basin contains continental red beds (Upper Jurassic) and marine sediments (Middle Jurassic). In a western basin, to the west of the Marañón geanticline and extending westward to the present-day coast, the Middle and Upper Jurassic sequences consist of andesitic volcanic rocks and sedimentary rocks of marine origin (Figure 4b). The same facies are found in the Coastal Range of Chile, beginning with transgressive deposits of Hettangian or, more commonly, of Sinemurian age, and having abundant interbedded, predominantly andesitic volcanic rocks that are chiefly submarine. In the Main Cordillera of Chile the chronostratigraphic equivalent (early Liassic to Oxfordian) is a marine sedimentary, volcanic-free sequence. Both lithofacies characterize a western and eastern domain of the Andean basin, which is a backarc basin. The marine conditions in this basin persisted until Oxfordian time and were succeeded by continental clastic deposition and andesitic volcanism during most of the Kimmeridgian and locally Tithonian time.

The Jurassic in the southern Andes, as exposed in the Magallanean basin (Figure 4b), consists of mafic volcanic and plutonic rocks and volcaniclastic flysch (Yaghan Formation) of Late Jurassic-Early Cretaceous age. The mafic rocks are interpreted as being equivalent to the upper part of an ophiolitic complex (Suarez and Pettigrew, 1976; Saunders et al., 1979), which is exposed in a belt extending from lat. 56°S to 49°S (Dalziel, De Wit, and Palmer, 1974).

In the northern Andes, and typically in Colombia, the Cretaceous sediments accumulated in eastern and western basins (Figures 5, 6a), separated by a structural high, now the Central Cordillera. In the eastern basin, particularly in the Surco de Bogota (Julivert, 1973), the Lower Cretaceous (Berriasian to Aptian) consists mainly of black lutites having intercalations of limestone and sandstone, whereas the Upper Cretaceous consists of black lutite, sandstone in which are interbedded evaporites, and phosphatic sediments. In the western basin, extending over parts of the region of the Cordillera Occidental and western flank of the Cordillera Central, the Cretaceous sequence consists of volcaniclastic rocks having abundant basaltic lavas, including pillow lavas (Diabasic Group) of Barremian to Late Cretaceous age (Julivert, 1973).

A similar stratigraphic sequence is found in western Ecuador where the Piñón Formation, exposed west of the Guayaquil suture (Figure 6b), consists of basaltic breccias and flows, with tholeithic affinities, some with pillow structures, and interbedded turbiditic volcaniclastic sediments and thin-bedded pelagic sediments (Baldock, 1982). Overlying this complex, which is considered of Aptian to Turonian or even early Senonian age, is a sequence of tuffites, siliceous tuffs with abundant glauconite, chert, sandstone, and pyroclastic rocks of the Senonian-Maestrichtian Cayo Formation (Olsson, 1942). This formation is interpreted as having accumulated in a fore-arc basin, separated from a nonvolcanic backarc basin to the east by an island arc and the emergent Cordillera Real (Baldock, 1982). In this backarc basin clastic continental sediments accumulated during the Early Cretaceous followed by carbonate rocks of Albian to Campanian age. This sequence of predominantly continental Lower Cretaceous and marine Upper Cretaceous is also found in the Cordilleras Oriental and Occidental of Perú. During the Early Cretaceous, mainly during the Neocomian, these regions were emerged (Benavides, 1956). They were invaded by the sea during Albian transgression, which affected the region from central Perú to Venezuela.

In the northern and central parts of the Cordillera Occidental of Perú, the Lower Cretaceous deposits are composed of predominantly marine facies to the west and of deltaic to continental facies to the east. They contain abundant intercalations of andesitic tuffs and flows, especially at the southwest and northeast margins of the western basin (Audebaud et al., 1973). At the beginning of the Late Cretaceous this western region of Perú began to be uplifted and gradually became a positive element, and volcanic, carbonate, and clastic sediments became progressively more abundant in the basin to the east.

The Upper Cretaceous marine facies of Perú become progressively more continental from north to south. This is evident south of lat. 13°S (Lake Titicaca region) where the uppermost Cretaceous is predominantly continental (Bellido, 1969), having only a few marine intercalations, such as the Vilquechico Formation (Campanian-Maestrichtian), that were deposited during short-lived marine transgressions (Audebaud et al., 1973).

The Cretaceous sequences of northern Chile and Argentina, between lat. 26° and 41°S, strongly contrast with those described above (Figures 5, 6a, 6b). They consist of backarc marine sequences of Neocomian and Tithonian age (Klohn, 1956, 1960; Corvalán, 1956). Included are an eastern sedimentary, nonvolcanic facies and a western volcanic-sedimentary facies. The volcanic rocks are of rhyolitic to andesitic composition. These marine rocks are overlain by Lower (Aptian-Albian) and Upper Cretaceous continental, volcanic, and sedimentary formations that consist of epiclastic red beds overlain by thick sequences of rhyolitic, andesitic, and basaltic volcanic rocks and volcaniclastic sedimentary rocks (Klohn, 1956, 1960; Thomas, 1958).

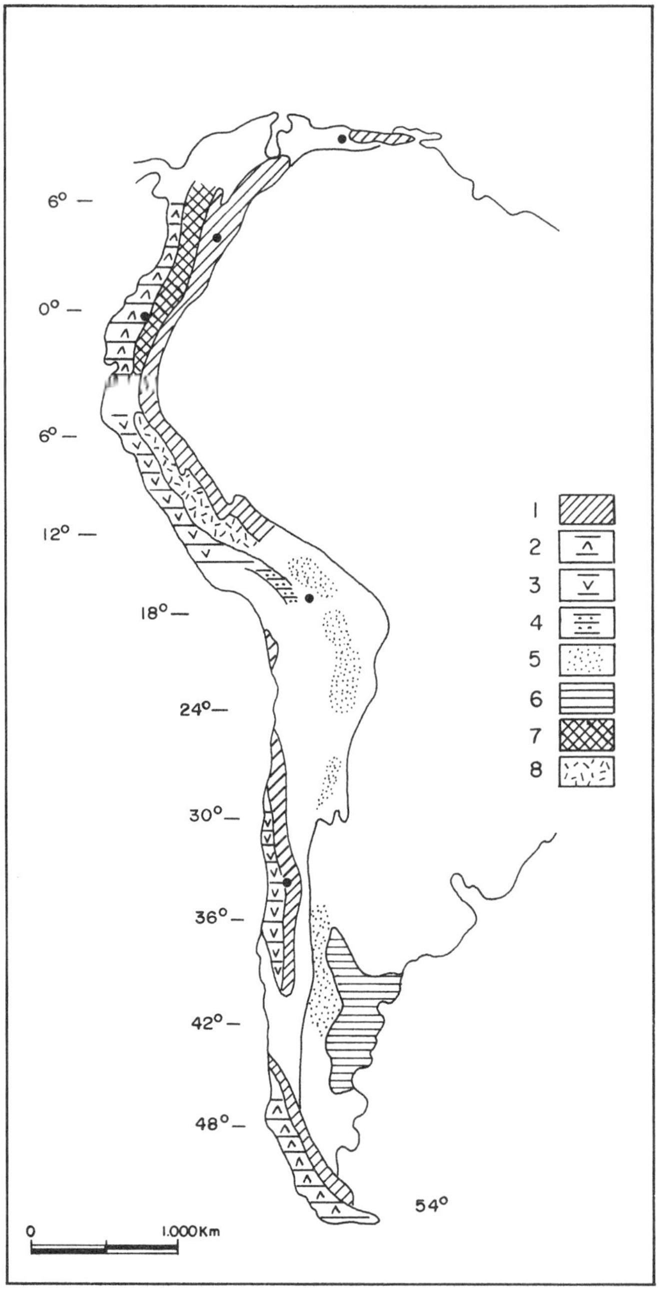

Figure 5. Distribution of lithofacies of the Cretaceous system: (1) marine sedimentary, nonvolcanic; (2) marine sedimentary and mafic to ultramafic volcanic rocks; (3) marine sedimentary and volcanic rocks of intermediate composition; (4) continental sediments with marine intercalations; (5) continental sediments; (6) platform marine deposits; (7) pre-Cretaceous belt of Colombia-Ecuador; (8) Marañón geanticline.

In the southern Andes, the domain of the Magallanean basin, the Lower and Upper Cretaceous rocks are essentially marine. From west to east, the lithologic assemblages are interpreted to be island-arc, marginal basin with ophiolitic associations, and shelf accumulations (Suarez and Pettigrew, 1976). The Upper Cretaceous marine sediments of this region contain important molasse and flysch deposits (Cecioni, 1957).

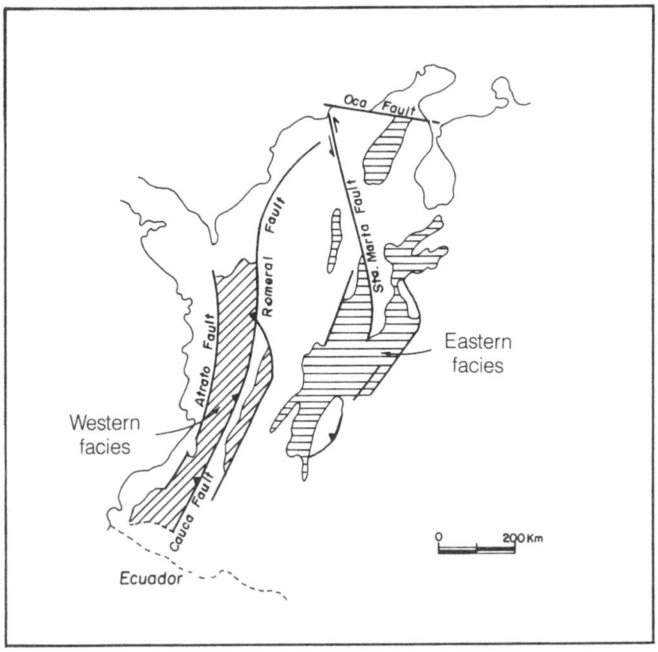

Figure 6a. Distribution of Cretaceous lithofacies in Colombia, and principal faults (after Julivert, 1973).

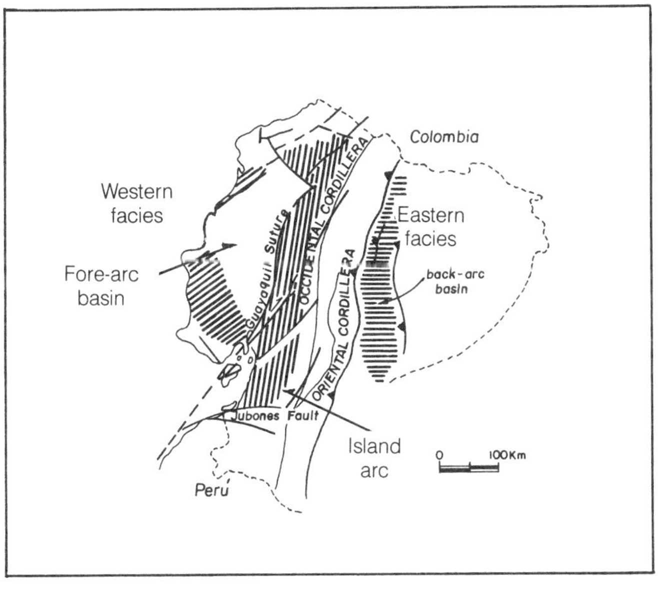

Figure 6b. Distribution of lithofacies in Ecuador, and principal faults (after Baldock, 1982).

The lower Tertiary sequences also show longitudinal changes in lithofacies and stratigraphy along the Andean belt. For example, the Paleocene-Eocene marine sequences in the northern Andes consist chiefly of basin-type, deep-water marine sedimentary rocks having interbedded turbidites. Such sequences are well exposed in northwestern Colombia. Similar sequences are known in the Oriental and Falcon basins of Venezuela and in the Pacific coastal province of western Ecuador, where the Eocene is represented by clastic and calcareous rocks, including carbonates, olistostromes, and turbidites. In contrast, the lower Tertiary marine formations in the central Andes are chiefly shallow-water (littoral to infralittoral) sedimentary rocks, such as those exposed in the Paracas Peninsula in the coastal region of central Perú (Bellido, 1969) and in the Coastal Cordillera of central southern Chile. In southern Chile, the lower Tertiary sequence contains important coal beds that were deposited in a continental environment. The lower Tertiary marine sediments in the Magallanean basin, southernmost South America, accumulated in basins that probably had a depositional and paleogeographic development more similar to the basins of the northern Andean region than to those of the central Andes. During the early Tertiary the central Andean region was subjected to widespread and intense volcanic activity (Figure 7).

The late Tertiary and Quaternary also show longitudinal changes in geology and tectonic style. As has been noted, Sillitoe (1974) recognized these changes as being related to a series of segments, the boundaries of which are deep-seated transverse structures related to late Tertiary-Quaternary plate convergence and subduction. For example, the extensive late Tertiary and Quaternary volcanism south of lat. 33 °S, chiefly andesites and basalts, differs from the predominantly silicic volcanism in northernmost Chile and southern Perú (north of lat. 27 °S). The abrupt onset of more mafic volcanism in this southern region suggests eruption along deep-seated transverse structures related to subduction. The predominance of reverse and thrust faults and absence of recent volcanism and seismic activity between lat. 27° and 33 °S characterize this distinct tectonic province, the Norte Chico province of Chile (Lomnitz, 1962; Pérez and Aguirre, 1968).

PALEOGEOGRAPHIC AND TECTONIC IMPLICATIONS

The two main tectonic boundaries that separate the northern, central, and southern Andes (Gansser, 1973) plus another, the Arica-Santa Cruz line, appear to have been present during the Triassic-early Tertiary interval. However, tectonic activity along these boundaries, probably deep-seated fault movement, seems to have changed through time and caused changes in the longitudinal extent of different paloegeographic domains. For example, the Triassic is represented by distinctly different lithofacies assemblages in three Andean segments. North of the Huancabamba deflection continental sediments and volcanic rocks accumulated in a relatively stable, nonsubsiding or weakly subsiding cratonic area. These Triassic rocks are well exposed in Colombia and Venezuela where they have sparse intercalations of marine sediments deposited during a few short-lived transgressions (Figures 4a, 8).

In northern and central Perú, south of the Huancabamba deflection, a marine environment prevailed during much of the Late Triassic. This region had subsiding basins separated by stable emerged highs such as the Marañón geanticline (Figure 4a). West and east of the Marañón, marine sedimentary-volcanic and exclusively marine sedimentary facies, respectively, accumulated. These sequences are of Late Triassic to Early Jurassic age, and represent an early development of the Andean basin. The Triassic becomes progressively more continental southward in Perú, being almost completely continental near the Arica-Santa Cruz line (Figures 3, 8). This line appears as a possible active structure controlling subsidence and marine deposition in the segment bounded by it and the Huancabamba deflection. Still farther south in the Andean region the Triassic is predominantly continental, consisting of sediments deposited within local, rather stable basins, and widespread volcanic rocks of silicic to intermediate composition. Shallow-water marine basins developed in the southern part of the sub-Andean region of Bolivia and in the pre-Cordillera of Chile between lat. 23°30′ and 26°00′S. Shallow-water marine embayments existed in the coastal region of central Chile during the Middle and Late Triassic (Figure 4a). Rocks of Triassic age are absent farther south in the southern Andes, where the Paleozoic basement is overlain by sedimentary rocks of Late Jurassic age. Evidently, this region was also emergent during the Triassic (Figures 3, 8).

The Triassic paleogeographic environment of the northern Andean segment continued into the Jurassic, and continental clastic deposition, volcanism, and a short-lived marine transgression took place during the Lias. The Huancabamba deflection probably continued as an active tectonic boundary during this time, because marine sedimentation to the south of this boundary, initiated in the Late Triassic, continued in the Early Jurassic (Figures 4b, 8).

The marine depositional environment of the eastern and western basins of northern and central Perú (separated by the Marañón geanticline) also persisted from the Triassic into the Jurassic. Marine sediments were deposited in the eastern basin during the Early and Middle Jurassic, followed by continental sediments in the Late Jurassic. Volcanism accompanied marine deposition in the western basin.

The depositional environment in the Andean basin (Figures 4b, 8) of Argentina and Chile south of the Arica-Santa Cruz line and extending southward to the Bariloche High at about lat. 41°S (Aubouin et al., 1973) was also chiefly marine. This basin, considered to be a backarc basin, seems to have been a single basin trough without an intervening high such as the Marañón geanticline. Marine sedimentation persisted in the eastern part of this basin during the Early and Middle Jurassic, followed by continental sedimentation during the Late Jurassic. In the western part of the basin, marine sedimentation was accompanied by widespread volcanism.

The similar lithostratigraphic development of the Jurassic and continuity of facies from Perú into the Chilean-Argentinian basin implies that the Arica-Santa Cruz line, which was an active structure during the Triassic, was not active during the Jurassic. On the other hand, the prevalence of marine conditions during most of the Jurassic in the central Andean segment, from the Huancabamba deflection in the north to the Bariloche transversal in the south, while continental conditions prevailed to the north and south, indicates that these boundaries were active structures (Figure 8) controlling an actively subsiding central segment.

During the Cretaceous the evolution and paleogeographic relationships of the different segments were more complex, and the distribution of the marine and continental domains suggests subdivision into four segments (Figure 8). North of the Arica-Santa Cruz line marine conditions prevailed during

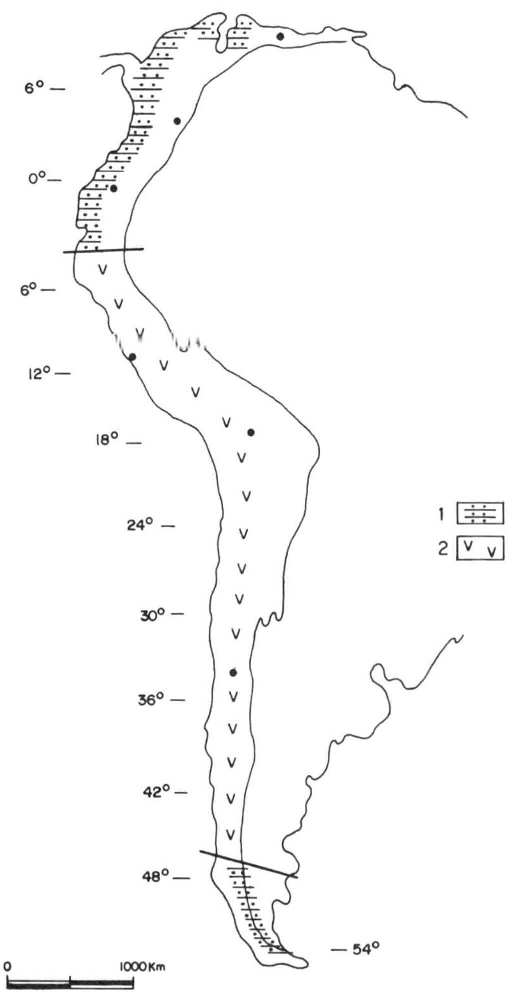

Figure 7. Distribution of early Tertiary (1) basin-type, deep-water marine sedimentary rocks, and (2) continental volcanic rocks and clastic sedimentary rocks.

and after Albian time, suggesting continuous subsidence of this segment. In contrast, to the south, the Lower and Upper Cretaceous are entirely continental. The paleogeographic conditions of the whole segment north of the Arica-Santa Cruz line during the Cretaceous were such that eastern and western facies accumulated in basins separated by structural highs such as the Marañón geanticline in Perú and a land area of Precambrian and Paleozoic rocks in Colombia (Figure 5).

The Huancabamba deflection was evidently an active structural boundary during the Cretaceous. To the north of the Huancabamba deflection the Cretaceous western facies is represented by ophiolitic complexes having associated radiolarites and turbidites. This sequence is broken by thrust faults that form at least three main nappes (Bourgois et al., 1985). In contrast, in Perú to the south of the deflection, ophiolites are absent and volcanic rocks are chiefly of intermediate composition. In the segment between the Huancabamba deflection and the Arica-Santa Cruz line, marine conditions prevailed during much of the Cretaceous, but in the southernmost part

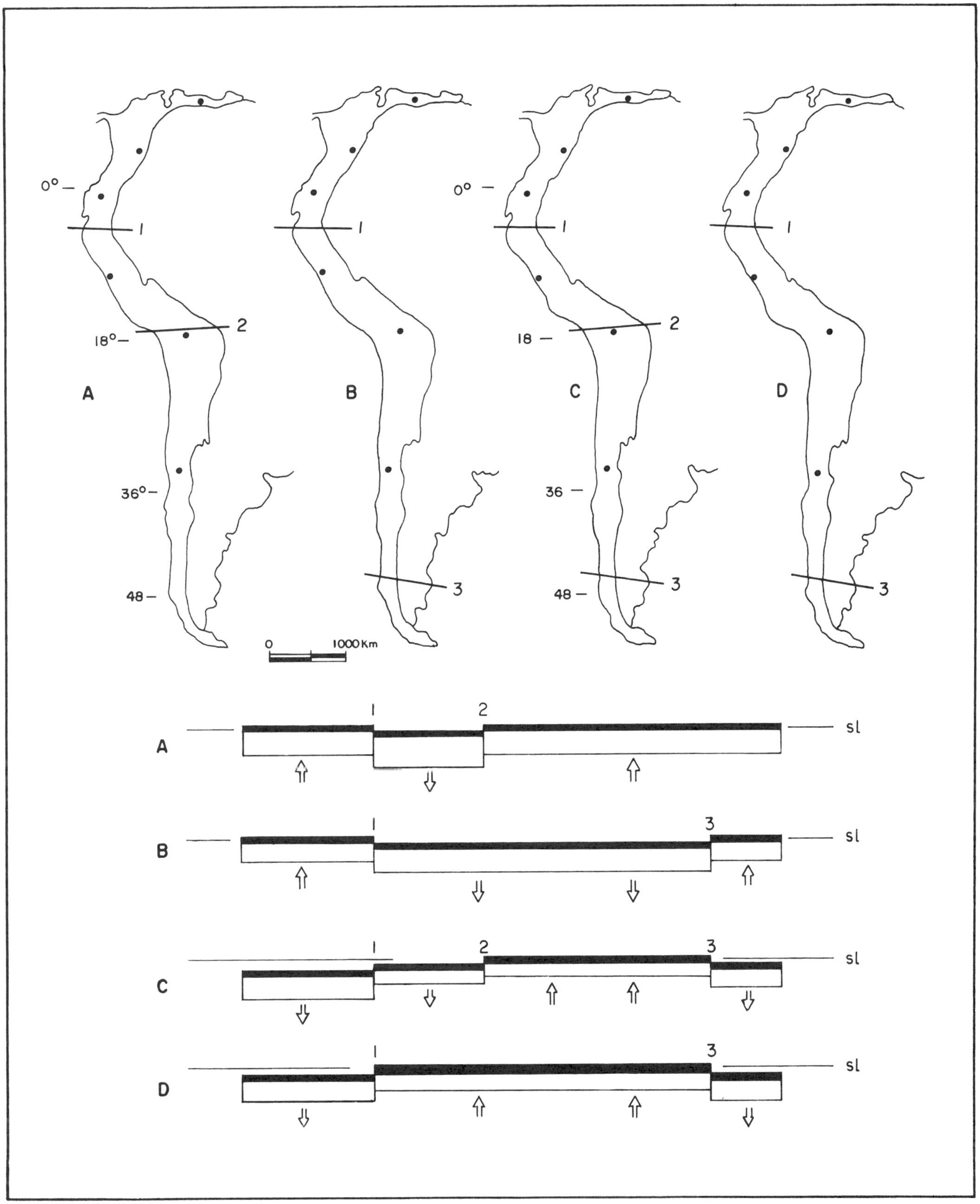

Figure 8. Segments of the Andean region showing relative subsidence during: (A) Triassic; (B) Jurassic; (C) Cretaceous, and (D) early Tertiary. Segment boundaries are: (1) Huancabamba deflection; (2) Arica-Santa Cruz line; (3) Bariloche transversal. sl: subsidence relative to sea level.

of this region the sedimentary environment changes from marine to predominantly continental. In contrast, with the exception of marine sediments deposited in small, shallow-water coastal embayments in central Chile, Lower and Upper Cretaceous marine sedimentary rocks are not known to exist in the Andean region between the Arica-Santa Cruz line and about lat. 42-43 °S. A marine backarc basin existed in this area only during the Neocomian. The Lower and Upper Cretaceous deposits of this region are molasse facies, and include both clastic sediments and volcanic rocks. In the Magallanean Basin, south of the Bariloche High, the Cretaceous is represented by marine facies (Figures 4b, 8). Ophiolitic complexes are present in the lower part of the western sequences. Reverse and thrust faults are the predominant structures. This segment resembles the segment north of the Huancabamba deflection.

During the early Tertiary, the northern and southern Andean regions had basin-type marine environments, whereas the central Andes had a continental environment characterized by widespread volcanism and only minor marine sedimentation in the western part of the basin. The northern and southern regions were actively subsiding segments (Figures 7, 8).

CONCLUSIONS

The geologic history of the modern Andes mountains is recorded in the rocks and structures of Triassic to early Tertiary age. The distribution and nature of these rocks and structures were determined, in large part, by deep-seated transverse structures that divided the region into segments. These structures were active at different times, and, as a consequence, the dimensions and environments of individual segments, as well as their subsidence-uplift conditions, changed with time. For example, three major segments existed during the Triassic: (1) a northern segment to the north of the Huancabamba deflection near the Perú-Ecuador boundary; (2) a central segment between the Huancabamba deflection and the Arica-Santa Cruz line near the Perú-Chile border; and (3) a southern segment south of the Arica-Santa Cruz line. Three major segments also existed during the Jurassic: (1) a northern segment to the north of the Huancabamba deflection; (2) a central segment between the Huancabamba deflection and the Bariloche High, which faces the Chile Ridge off the coast of southern Chile; and (3) a segment south of the Bariloche High. Four segments existed during the Cretaceous, which were separated by the Huancabamba deflection, Arica-Santa Cruz line, and the Bariloche High, all of which were active structures at this time. The Arica-Santa Cruz line probably became inactive at the end of the Cretaceous, and three segments separated by the Huancabamba deflection and the Bariloche High existed during the Tertiary. These segments correspond to the present-day northern, central, and southern Andes of Gansser (1973).

REFERENCES CITED

Aubouin, J., 1972, Chaines liminaires (Andines) et chaines geosynclinales (Alpines): Montreal, International Geological Congress, Montreal, Section 3, p. 438-461.

Aubouin, J., and A. V. Borrello, 1966, Chaînes andines et chaînes alpines: Regard sur la geólogie de la Cordillére des Andes au parallèle de l'Argentine Moyenne: Société Géologique de France, Bulletin 7e Sér, v. VIII, no. 7, p. 1050-1070.

Aubouin, J., A. V. Borello, G. Cecioni, R. Charrier, P. Chotin, J. Frutos, R. Thiele, and J.-C. Vicente, 1973, Esquisse paleógeographique et estructurale des Andes Meridionales: Revue de Géographie Physique et de Géologia Dynamique, 2 Sér., v. XV, fasc. 1-2, p. 5-72.

Audebaud, E., R. Capdevila, B. Dalmayrac, J. Debelmas, G. Laubacher, C. Lefevre, R. Marocco, C. Martinez, M. Mattauer, F. Mégard, J. Paredes, and P. Tomasi, 1973, Les traits géologiques essentieles des Andes Centrales (Pérou-Bolivie): Revue de Géographie Physique et de Géologia Dynamique, 2 Sér., v. XV, fasc. 1-2, p. 73-111.

Baldock, J. W., 1982, Geologia del Ecuador: Quito, Dirección General de Geología y Minas, 62 p.

Bellido, E., 1969, Sinópsis de la Geología del Perú: Lima, Servicio de Geología y Minería, Boletín, no. 22, 54 p.

Benavides, C. V., 1956, Cretaceous system in northern Perú: American Museum of Natural History Bulletin, v. 108, Art. 4, p. 357-493.

Brüggen, J., 1913, Contribución a la jeología del valle del Huasco y del Departamento de La Serena: Santiago, Boletín Sociedad Nacional de Mineria, Boletín 25, nos. 199-200, p. 447-458.

Brüggen, J., 1950, Fundamentos de la geología de Chile: Santiago, Instituto Geográfico Militar, 374 p.

Bourgois, J., J.-F. Toussaint, H. Gonzales, A. Orrego, J. Azema, B. Calle, A. Sesmet, A. Murcia, A. Pablo, E. Parra, and J. Tournon, 1985, Les ophiolites des Andes de Colombie; evolution structurale et signification geodynamique: Memorias VI Congreso Latinoamerica de Geología, v. 1, p. 93.

Cecioni, G., 1957, Cretaceous flysch and molasse in Departamento Ultima Esperanza, Magallanes Province, Chile: American Association of Petroleum Geologists Bulletin, v. 41, p. 538-564.

Cecioni, G., and G. E. Westermann, 1968, The Triassic/Jurassic marine transition of coastal central Chile: Pacific Geology, v. 1, p. 41-75.

Chong, G., and A. Hillebrandt, 1985, El Triásico preandino de Chole entre los 23 °23 ′ y 26 °00 ′ de lat. Sur.: IV Congreso Geologico Chileno Actas, v. 1, p. 1.162-1.210.

Corvalán, J., 1956. Über Marine Sedimente des Tithon und Neocom der Gegend von Santiago. Geol. Rundsch., v. 45, p. 919-926.

Corvalán, J., 1965, Geología General, *in* Geografía económica de Chile. Texto Refundido: Corpoacion de Fomento de la Produccion (Chile), p. 35-82.

Corvalán, J., 1976, El Triásico y Jurásico de Vichuqúen-Tilicura y de Hualañe, Prov. de Curió. Implicaciones paleogeográficas: Congreso Geológico Chileno Actas, v.I, p. A138-A154.

Dalziel, I., M. De Wit, and K. Palmer, 1974, A fossil marginal basin in the Southern Andes: Nature, v. 250, p. 291-294.

Faucher B., and E. Savoyat, 1973, Esquisse géologique des Andes de l'Equateur: Revue de Géographie Physique et de Géologia Dynamique, 2 Sér., v. XV, fasc. 1-2, p. 115-142.

Fuenzalida, H., 1937, Las capas de Los Molles: Santiago, Boletín Huseo Nacional de Historia Natural, v. 16, p. 67-92.

Gansser, A., 1973, Facts and theories on the Andes: Journal of the Geological Society of London, v. 29, p. 93-131.

Gerth, H., 1955, Der geologischer Bau der Südamerikanischen Kordillere: Gebrüder Borntraeger Berlin, 264 p.

Goossens, P. J., and W. I. Rose, 1973, Chemical composition and age determination of tholeiitic rocks in the Basin Igneous Complex, Ecuador: GSA Bulletin, v. 84, p. 1043-1052.

Ham, C., and L. Herrera, 1963, Role of sub-Andean fault system in tectonics of eastern Perú and Ecuador, *in* Backbone of the Americas; Tectonic history from pole: American Association of Petroleum Geologists Memoir 2, p. 47-61.

Julivert, M., 1973, Les traits structuraux et l'evolution des Andes

colombiennes: Revue de Géographie Physique et de Géologia Dynamique, 2 Sér., v. XV, fasc. 1-2, p. 143-146.
Klohn, C., 1956, Estado actual del estudio geológico de la "Formacíon Porfirítica": Santiago, Instituto Ingenieros Minas de Chile, Minerales, año XI no. 55, p. 49-60.
Klohn, C., 1960, Geología del la Cordillera de los Andes de Chile Central, Provincias de Santiago, O'Higgins, Colchagua y Curicó: Santiago, Instituto Investigaciones Geológicas Boletín no. 8, 96 p.
Lomnitz, C., 1962, On Andean structure: Journal of Geophysical Research, v. 67, p. 351-363.
Marocco, R., 1971, Estude géologique de la Chaîne Andine au niveau de la Déflexion d'Abancay (Pérou): Cahiers ORSTOM, Sér. Geol., v. III, no. 1, p. 45-48.
Mégard, F., B. Dalmayrac, G. Laubacher, R. Marocco, C. Martinez, P. J. Paredes, and P. Tomasi, 1971, La Chaîne Hercynienne au Pérou et en Bolivie, Premiers resultats: Cahiers ORSTOM, Sér. Geol., v. III, no. 1, p. 5-44.
Muñoz Cristi, J., 1938, Geología de la región de Longotoma y Guaquén en la provincia de Aconcagua: Santiago, Boletín de Minas y Petróleos, t. 8, no. 81.
Muñoz Cristi, J., 1950, Geología, *in* Geografía económica de Chile: CORFO. t. 1, p. 55-127.
Olsson, A., 1942, Tertiary deposits of north-western South America and Panama: 8th Am. Soc. Congr. Washington, Proceedings, v. 4, p. 231-287.
Perez d'A., E. and L. Aguirre, 1968, Relación entre estructura y volcanismo cuaternario andino en Chile: Simposio Panamericano del Manto Superior, Mexico, v. 2, p. 39-46.
Reutter, K. J., 1974, Entwicklung un Bauplan der Chilenischen Hoch Kordillere in Bereich 29° südlicher Breite: Neves Jahrbuch fur Geologie und Paläontologie Abhandlungen, v. 146, no. 2, p. 153-178.
Sauer, W., 1965, Geología del Ecuador: Quito, Editorial del Mininsterio de Educacion, 383 p.
Saunders, A. D., J. Tarney, R. Stern, and I. W. Salziel, 1979, Geochemistry of Mesozoic marginal basin floor igneous rocks from Southern Chile: GSA Bulletin, v. 90, no. 3, p. 237-258.
Sillitoe, R., 1974, Tectonic segmentation of the Andes: Implications for magmatism and metallogeny: Nature, v. 250, p. 542-545.
Sonnenberg, F., 1963, Bolivia and the Andes, *in* Backbone of the Americas; Tectonic history from pole to pole: American Association of Petroleum Geologists Memoir 2, p. 36-46.
Steinmann, G., 1930, Geologia del Perú: Heildelberg, Carl Winters Universitätsbuchandlung, 448 p.
Stille, H., 1958, Die assyntische Tektonik in geologischen Erdbild: Geologisches Jahrbuch, heft 22.
Suarez, M., and T. Pettigrew, 1976, An upper Mesozoic island arc-back-arc system in the southern Andes and South Georgia: Geological Magazine, v. 113, no. 4, p. 305-328.
Thiele, R., 1967, El Triásico y Jurásico del Departamento de Curepto en la provincia de Talca: Universidad de Chile, Departamento de Geologia, Publicación no. 28, 46 p.
Thomas, H., 1958, Geología de la Cordillera de la Costa entre el valle de La Ligua y la Cuesta de Barriga: Santiago, Instituto de Investigaciones Geológicas, Boletín 2, 86 p.
Tschopp, H., 1948, Geologische Skisse von Ekuador: Swiss Association of Petroleum Geologists and Engineers Bulletin, v. 15, no. 48, p. 14-45.
Vicente, J. C., 1970, Reflexiones sobre la porcíon meridional del Sistema Peripacífico Oriental: Buenos Aires, Conferencia Sobre Problemas de la Tierra Solida, v. 37-1, p. 158-184.
Zeil, W., and K. Ichikawa, 1958, Marine Mittel-Trias in der Hochikordillere der Provinz Atacama (Chile): Neves Jahrbuch fur Geologie und Paläontologie Abhandlungen, v. 106, p. 339-351.

Ericksen, G. E., Cañas Pinochet, M. T., and Reinemund, J. A., editors, 1989, Geology of the Andes and its relation to hydrocarbon and mineral resources: Houston, Texas, Circum-Pacific Council for Energy and Mineral Resources Earth Science Series, v. 11.

CHAPTER 2

CIRCUM-PACIFIC OROGENIC PROCESSES: A VIEW FROM THE ANDES AND THE ANTARCTANDES

I. W. D. Dalziel
Institute for Geophysics
University of Texas
Austin, Texas, U.S.A.

ABSTRACT

Tectonic accretion along the Pacific margin of Gondwana was a widespread and important process prior to initiation of breakup in the Late Jurassic to Early Cretaceous. Since that time, there apparently have been no major additions to the South American and Antarctic continental margins. Rather, the Andes and Antarctandes owe their existence to basin formation and inversion, magmatism, and small-block rotation and migration along a relatively simple ocean-continent convergent boundary.

Several different types of tectonic activity characterize the circum-Pacific mountain ranges. No one model explains them all. Study of cordilleran orogenic processes would be greatly enhanced by acquisition of deep seismic-reflection profiles from several carefully selected transects.

INTRODUCTION

This contribution is a brief overview of work in progress on the evolution of the Andes and the Antarctandes in the context of Gondwana fragmentation. Most of the recent work cited is part of collaborative efforts currently in press. In recent years much attention has been paid in the literature to the identification of so-called "suspect terranes" in mountain belts (Coney, Jones, and Monger, 1980; Jones et al., 1982; Howell, 1985). The importance to orogenesis of collision ("docking") of allochthonous terranes, such as island arcs and oceanic plateaus, with a major continent at a continent-ocean convergent boundary has been emphasized (Nur and Ben Avraham, 1982). Indeed, it has been maintained that high mountain chains such as the Andes can only be formed by collisional processes (Nur and Ben Avraham, 1983).

Yet despite abundant evidence of accretion to the South American part of the Gondwana supercontinent during the Paleozoic and early Mesozoic (Dalziel and Forsythe, 1985; Ramos et al., 1986), there is no evidence of such accretion during the late Mesozoic and Cenozoic building of the Andean Cordillera (Dalziel, 1985). Rotation and migration of small crustal blocks along the Pacific margin probably were more important than accretion (Dalziel and Forsythe, 1985; Beck, in press). Major "suspect terranes" such as the Arequipa massif of southern Perú and the metamorphic complexes of the Chilean coastal cordillera were clearly part of South America before the initiation of Andean orogenesis in the late Mesozoic.

What processes, then, have been most active in the building of the Andes? Classically, subduction-related magmatism has been recognized as playing a major role. Undoubtedly this is correct. More recently, however, it has become apparent that the Late Triassic to Early Cretaceous formation and mid-Cretaceous inversion of a composite backarc basin extending along most of the length of the Cordillera was probably critical in the initiation of tectonic uplift. Similar processes continued on a more local scale during the latest Cretaceous and Cenozoic (Bussell, 1983), controlled by such factors as segmentation of the downgoing lithospheric slab, differences in dip of the segments, and convergence rate (Jordan et al., 1983; Suarez, Molnar, and Burchfiel, 1983; Dalziel, 1985). Overall, tectonic shortening and thickening of the continental lithosphere along the western margin of the South American continent is now believed to have played a major role in Andean orogenesis (Suarez, Molnar, and Burchfiel, 1983; Dalziel, 1985; Molnar, 1986). The relative importance of deformation in relation to magmatism in this regard remains uncertain.

Examination of the Antarctic Peninsula Cordillera to the south of Drake Passage, originally known as the Antarctandes, provides further insights into the processes of cordilleran orogenesis. The peninsula was a magmatic arc during the entire Mesozoic and Cenozoic, displaying virtually no signs of compressional deformation during that time (Thomson et al.,

1983). Recent (1-2 Ma) opening of a composite extensional trough along the Pacific margin of the peninsula (Barker and Dalziel, 1983) suggests the possibility of future basin inversion and associated tectonic uplift resulting from shortening and thickening of the continental lithosphere (see Storey and Garrett, 1985).

The state of stress in the overriding plate and its strength appear to be critical factors in the development of "Andean-type" orogens. This proposition is explored here.

JURASSIC AND EARLY CRETACEOUS EXTENSIONAL PHASE

It is widely recognized that the Andean Cordillera was initiated with subsidence of the elongate west Peruvian trough, Andean basin, and Magellan basin during the Late Triassic, Jurassic, and Early Cretaceous (Aubouin et al., 1973). Although originally described in geosynclinal terms, it now seems clear that these basins represent a composite backarc basin extending over at least 7,500 km of the Andean margin from the Gulf of Guayaquil to the North Scotia Ridge (Figure 1; Chotin, 1976; Dalziel, 1985). Indeed, backarc extension along the Pacific margin during the Jurassic and Early Cretaceous was coeval with widespread extension within Gondwana that led to breakup of the supercontinent (Dalziel et al., in press, a). It seems likely that by Aptian-Albian times the Andean mountain range consisted of an active volcanic arc at the continental margin, perhaps similar to the North American Cascade Mountains of today, bordering a backarc basin having various physiographic features. In its southern (Magellan) part, at least, it was a marginal ocean basin probably comparable to the Bransfield trough on the Pacific side of the Antarctic Peninsula, separated from the ocean basin by an island arc like the South Shetland Islands (Figures 2, 3). At places it was a seaway floored by stretched continental crust, and at other places it was a landmass rising above sea level. The picture presented here of extension in the period from Late Triassic through Early Cretaceous is a generalized one. Clearly, variations in the locus of arc magmatism, and at least local faulting and tilting, occurred throughout this interval.

From its southern extremity on the North Scotia Ridge, it seems likely that the former backarc basin extended into the Weddell Sea in Antarctica (Figure 2; Barker, Dalziel, and Storey, in press; Dalziel et al., in press, a). Any such connection was subsequently disrupted by the formation of Cenozoic oceanic lithosphere in Drake Passage and the Scotia Sea (see Tectonic Map of the Scotia Arc, 1:3,000,000, BAS [Misc.] 3, British Antarctic Survey, Cambridge, 1985). However, the Weddell Sea floor is Mesozoic (LaBrecque and Barker, 1981), and paleomagnetic data indicate rotation of the Antarctic Peninsula clockwise about a rotation pole, in a location that corresponds to that of the present western Drake Passage, during the Cretaceous opening of the Weddell Sea (Figure 4) (Longshaw and Griffiths, 1983; Grunow, Dalziel, and Kent, in press). Moreover, the Antarctic Peninsula, essentially a Mesozoic-Cenozoic magmatic arc (Thomson et al., 1983), appears to be a continuation of the Andean arc on the Pacific side of the Mesozoic backarc basin (Dalziel, 1983). The backarc basin was the site of "burial" metamorphism reflecting a high thermal gradient during the extensional phase (Aquirre, Levi, and

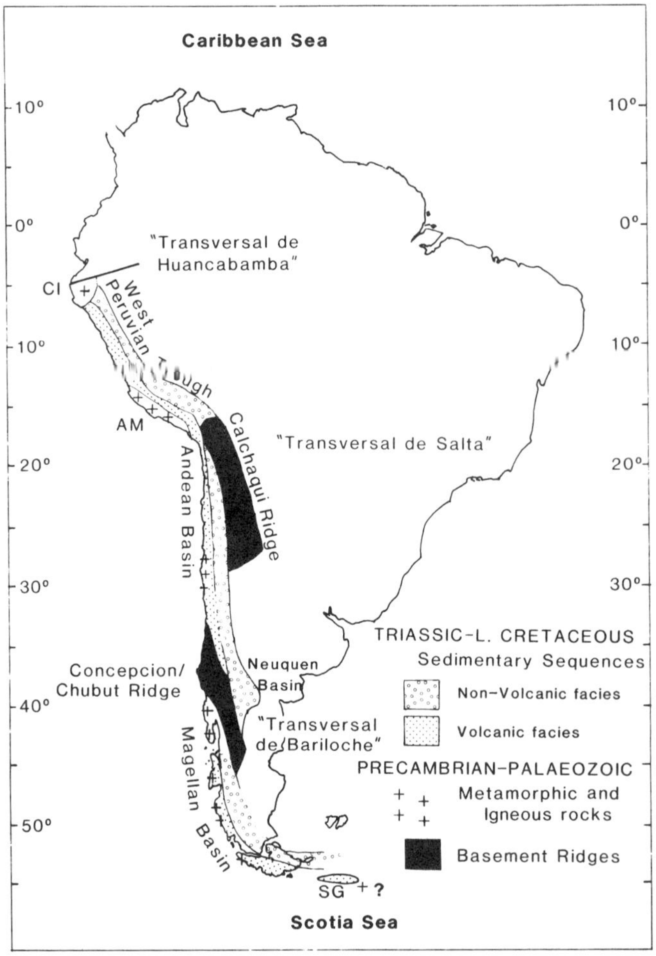

Figure 1. Main elements of the Andean margin of South America in the Late Triassic to Early Cretaceous (after Dalziel, 1985).

Offler, 1978; Levi and Aquirre, 1981; Aberg et al., 1984). During the Middle to Late Jurassic a suite of peraluminous granitic plutons, with relatively high Sr^{87}/Sr^{86} ratios, was emplaced along the basin from northern Chile to Tierra del Fuego as well as in the Ellsworth-Whitmore Mountains crustal block of West Antarctica (Dalziel et al., in press, a and b; M. Suarez, personal communication, 1985). These plutons are contemporaneous with the dominantly silicic Tobifera (or Porfiritica) Formation of southernmost South America and with the enormous mafic igneous field represented by the Karroo, Ferrar, and Tasman gabbros, dolerites, and related basalts (Figure 5). Hence the extensional phase of Andean evolution coincides with the development of a huge thermal anomaly throughout much of the Gondwana supercontinent (Dalziel et al., in press, a and b). Whether a relationship exists between Pacific margin subduction and this anomaly remains unproven at this time.

MID-CRETACEOUS COMPRESSIONAL PHASE

The first significant compression of Jurassic and Cretaceous rocks along the western margin of the South American continent occurred during the mid-Cretaceous. Along at least 7500

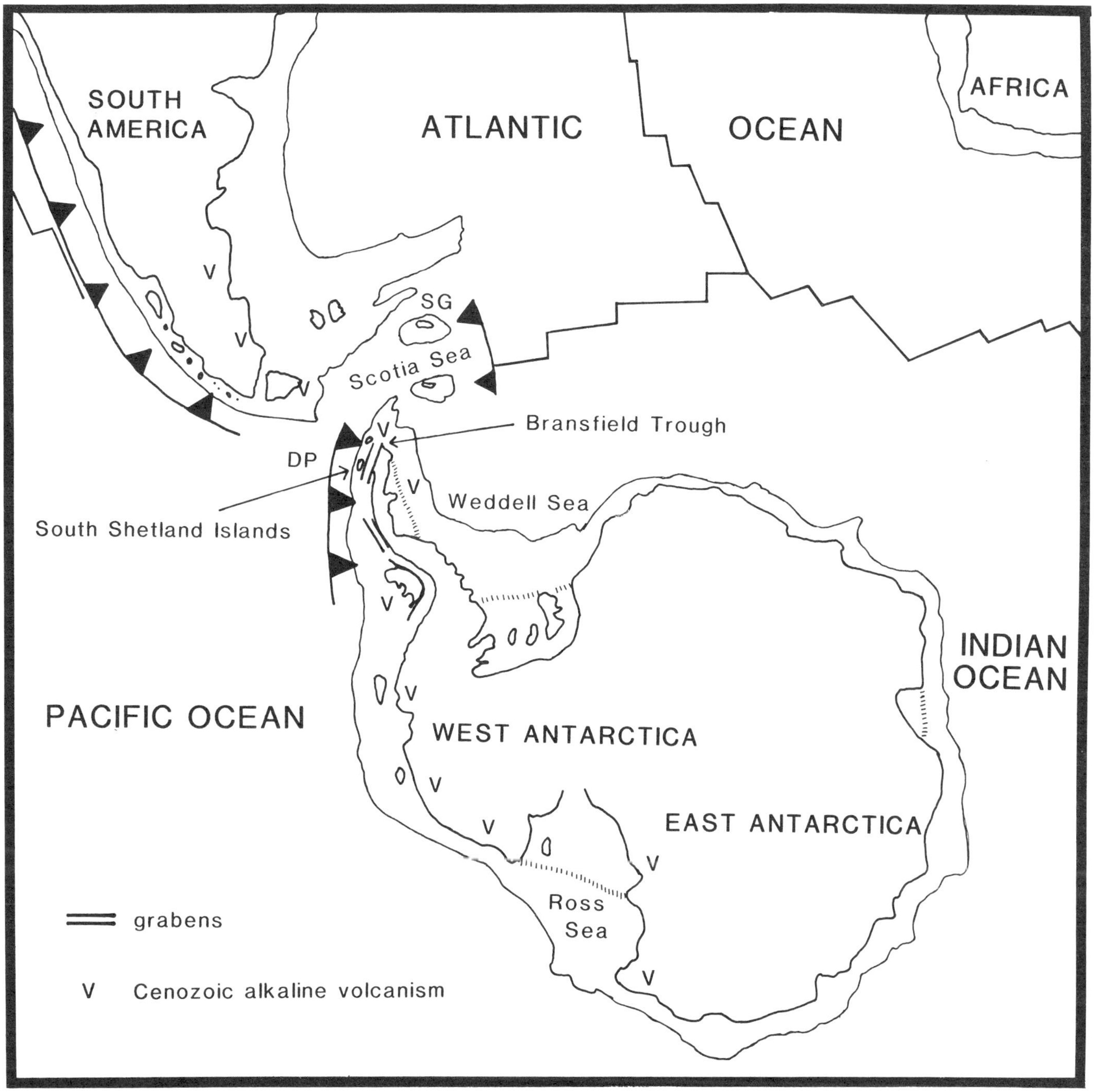

Figure 2. Geotectonic setting of the Andes and Antarctandes in the Cenozoic.

km of the Andean Cordillera, this event (variously referred to as the Subhercynian, Peruvian, Andean, or Mirano phase) affected rocks as young as Aptian prior to the intrusion of calc-alkaline plutons as old as 108 Ma (for review, see Dalziel, 1985). As in the case of earlier extension, the amount and effects of the compression varied considerably. Open folding with little penetrative deformation is common in the West Peruvian trough and the Andean basin, although local recumbent folding and related thrusting has been reported (Vicente et al., 1979). Intense penetrative deformation, even of the pre-Late Jurassic crystalline basement, prevails in Tierra del Fuego. The locus of the deformation is along the line of the composite backarc basin, presumably the major zone of structural weakness in the continental lithosphere in the mid-Cretaceous. Critical in the development of the Cordillera was underthrusting of the backarc basin floor beneath the magmatic arc during this mid-Cretaceous inversion of the backarc basin that resulted in structures verging away from the Pacific Ocean. This is particularly apparent in the Magellan basin (Bruhn and Dalziel, 1977). The compression that caused this underthrusting seems most likely to be related to a global increase in sea-floor spreading rates at this time and related

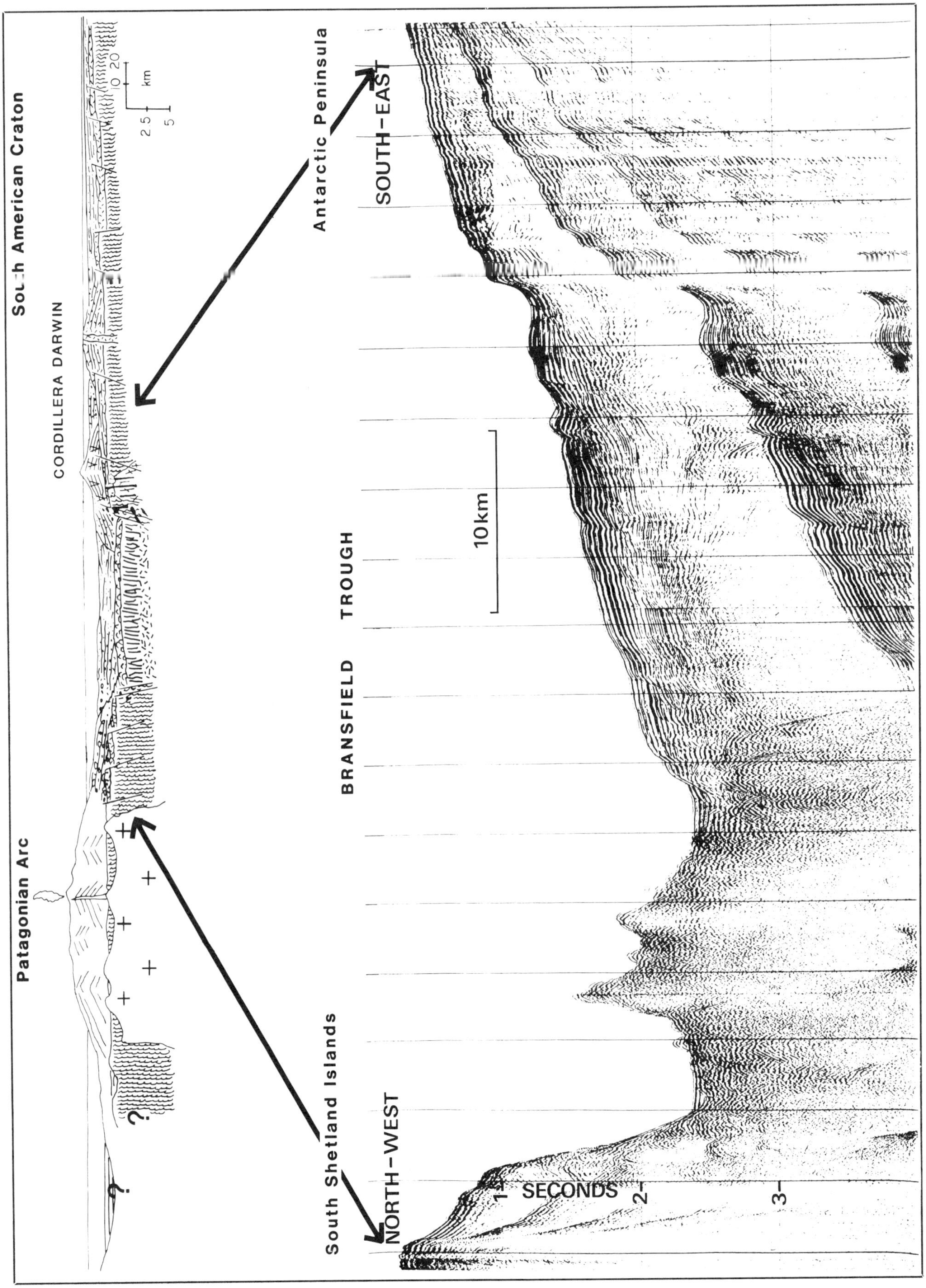

Figure 3. Reconstruction of Magellan basin in Tierra del Fuego during the Early Cretaceous (Dalziel, 1981) compared with a recent single-channel seismic profile across the Bransfield trough (Barker, Dalziel, and Storey, in press).

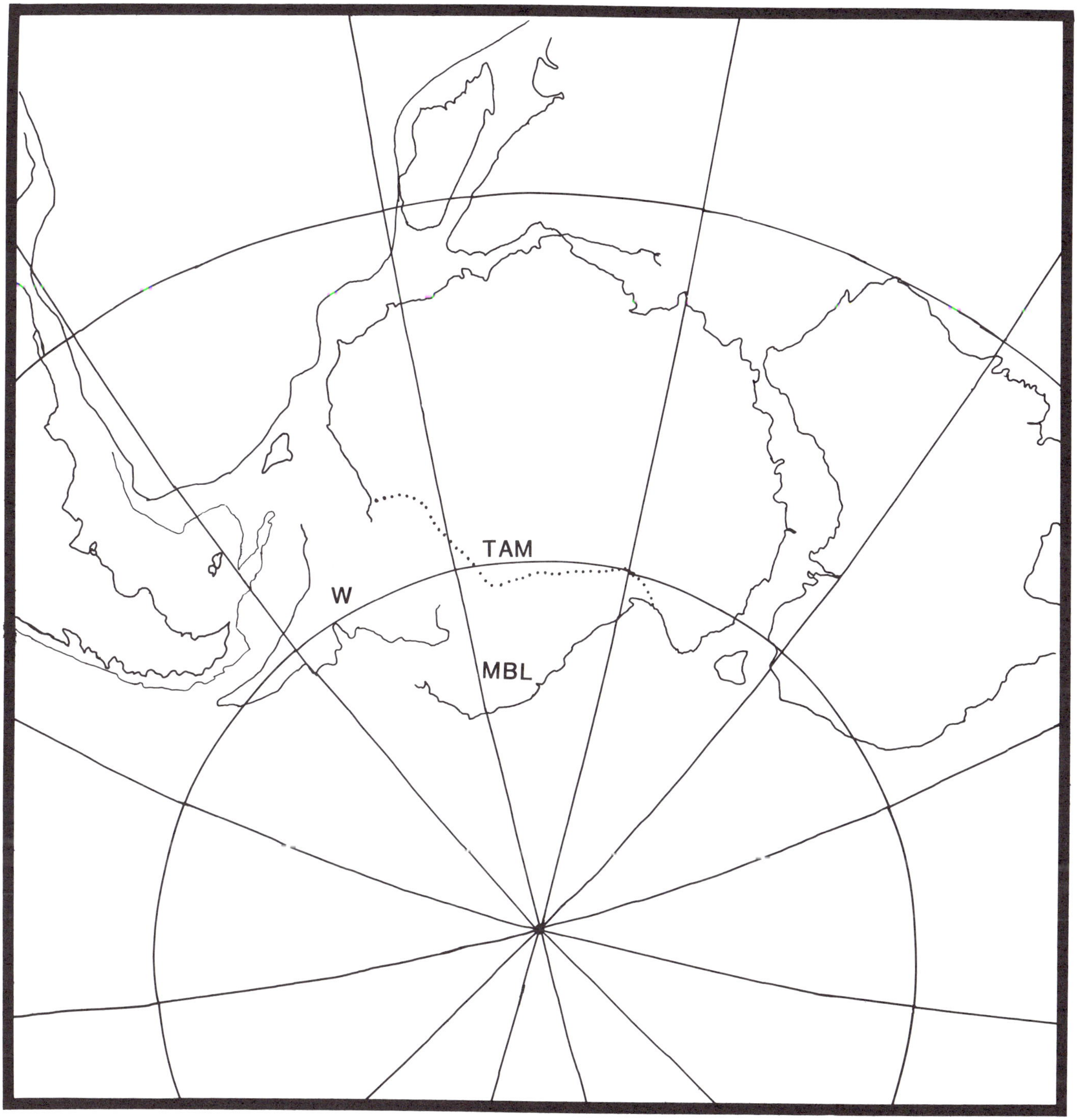

Figure 4. Reconstruction of the Antarctic Peninsula and adjacent crustal blocks of West Antarctica (W—Weddellia) in Gondwana based on recently acquired paleomagnetic data (Grunow, Kent, and Dalziel, 1985; Grunow, Dalziel, and Kent, in press). MBL—Marie Byrd Land. TAM—Transantarctic Mountains.

rapid northward propagation of the South Atlantic ocean basin (see Kominz, 1984). Not only, therefore, did the lithospheric plate convergence rate increase sharply along the western margin of South America, but the continent must also have moved westward relative to the mantle, effectively changing the convergent margin from the "Marianas-type" of Uyeda and Kanamori (1979) to their "Chilean-type." As pointed out by many workers, the main concentration of radiometric dates for the Pacific margin batholiths are in the Late Cretaceous. The simplest explanation is that the increase in sea-floor spreading rates and opening of the South Atlantic ocean basin caused a geologically instantaneous increase in compressive stress within the South American Plate, resulting in collapse of the weak backarc basin along its leading edge. Concomitant increase in the rate of subduction of Pacific Ocean lithosphere beneath the continent then generated a

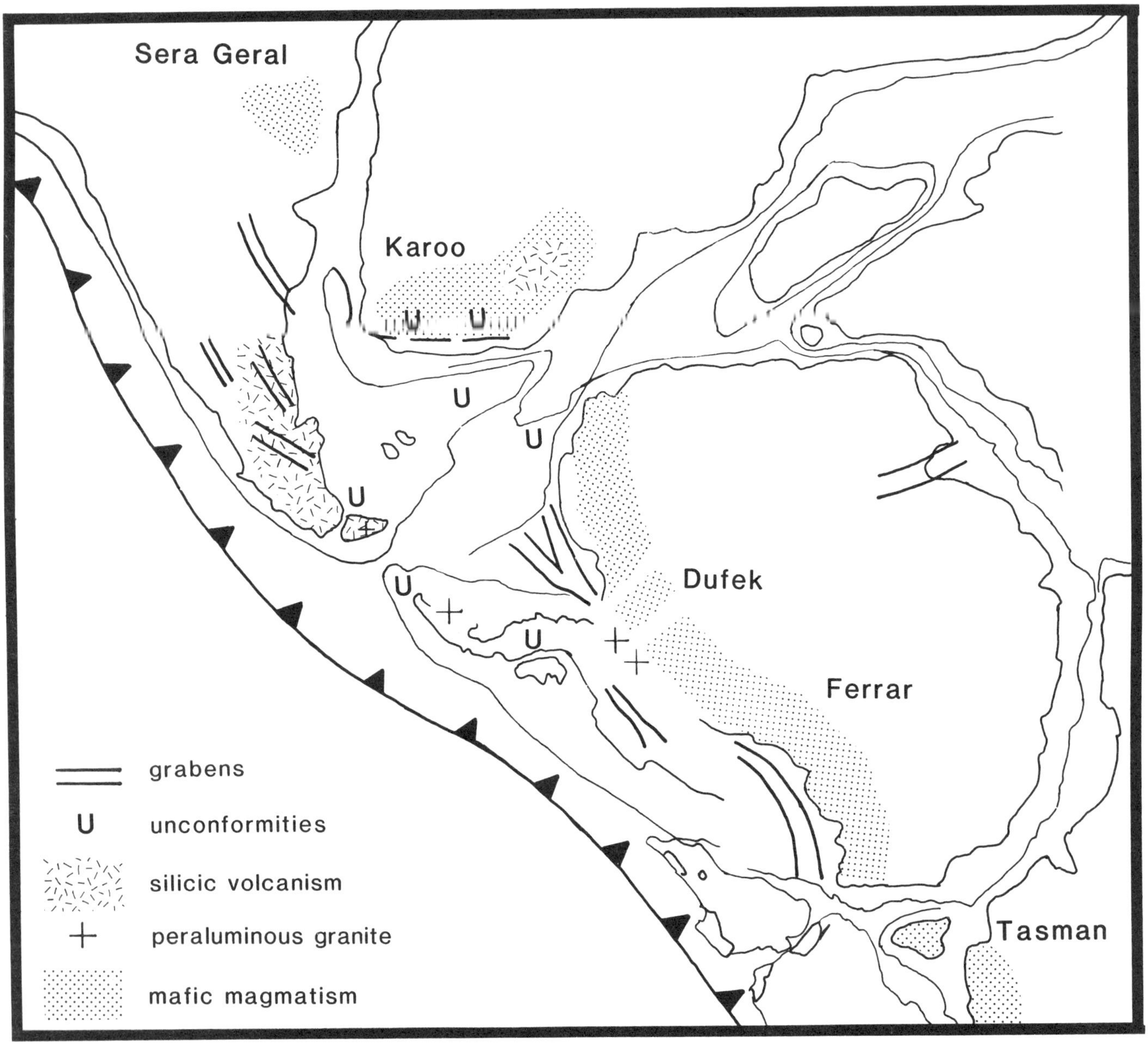

Figure 5. Igneous and tectonic phenomena preceding Gondwana breakup (Dalziel et al, in press, a).

greatly increased volume of calc-alkaline magma that was emplaced as the main batholiths within older arc rocks as well as within the deformed marginal basin floor rocks (Herve, Suarez, and Puig, 1984; Pitcher et al., 1985).

Mid-Cretaceous deformation is absent from the Antarctic Peninsula despite the fact that the Weddell Sea basin to the east was probably opening at that time (Dalziel and Elliot, 1973; Thomson et al., 1983). It is suggested that there was no zone of weakness within the peninsula magmatic arc to collapse at that time, the marginal basin in South America being represented in Antarctica by the then still actively spreading Weddell Sea. Early Cretaceous deformation at the southern extremity of the Antarctic Peninsula may rather be related to relative motion between the peninsula and the East Antarctic craton during separation of east and west Gondwana (DeWit, 1977; Grunow, Dalziel, and Kent, in press).

LATE CRETACEOUS TO PRESENT-DAY ARC AND FOLD-THRUST BELT MIGRATION; SMALL BLOCK MOVEMENTS

Cycles of extension and compression have continued along the length of the Andean Cordillera since mid-Cretaceous (Charrier, 1973), although none appears to have been geographically as extensive as that culminating in the mid-Cretaceous backarc basin inversion. They have, for example, been documented in the Peruvian Andes (Bussell, 1983; Mégard, 1984), and in the northern Patagonian Cordillera (Bartholomew and Tarney, 1984). Nonetheless, each cycle probably contributed, even if only locally, to shortening and thickening of the continental lithosphere along the mountain range. Over the same period, calc-alkaline magmatism has continued to the present (Herve, Suarez, and Puig, 1984; Pilger, 1984; Pitcher et al., 1985), as has foreland folding and thrusting east of the Cordillera (Jordan et al., 1983; Suarez, Molnar, and Burchfiel, 1983). In general, the locus of both processes has migrated eastward with time, but the pattern is complex in detail.

Although paleomagnetic data of high quality are as yet limited, comparison of those from the Cordillera and those from "stable" cratonic South America does not indicate major latitudinal shifts but mainly rotation of small crustal blocks inwards toward the Arica embayment in the continental margin of southernmost Perú and northernmost Chile (see Beck, in press, for review). If such a process occurred over a considerable duration of late Mesozoic and Cenozoic time, it could have contributed significantly to orogenic uplift in the region of the Altiplano, the second largest high-altitude plateau on earth. Whereas some collapse of the high Cordillera is indicated by the predominance of present-day normal faulting, active thrust faulting along the eastern margin of the Cordillera suggests that its elevation is being maintained, at least in part, by active underthrusting on both sides (Suarez, Molnar, and Burchfiel, 1983; Molnar, 1986). It should be noted in this regard that the elevation of the Cordillera diminishes markedly south of the Chile Rise-Chile Trench triple junction at approximately lat. 46°S, where the convergence rate of the Antarctic and South American Plates is significantly lower (~ 2 cm/yr) than that of the Nazca and South American plates to the north (~ 9 cm/yr). In the south, at least one major crustal block, the South Georgia microcontinent, has been removed from the Cordillera to become an allochthonous terrane (Dalziel et al., 1975). It is difficult to see that this will ever be reunited with its parent continent (Dalziel and Forsythe, 1985; Dalziel and Grunow, 1985). In contrast, the South Shetland Islands magmatic-arc terrane, which has separated from the Antarctic Peninsula over the past 2 Ma or so (Barker and Dalziel, 1983) by the opening of the Bransfield Strait trough, could readily be reunited with the Antarctandes by a basin inversion event comparable to that of the mid-Cretaceous in the Andes to the north. Such an inversion probably would result in substantial orogenic uplift of the West Antarctic Cordillera in the northern Antarctic Peninsula region. Other rifts along the West Antarctic margin, which could suffer a similar fate, have been recognized farther south (Storey and Garrett, 1985). Thus the processes that appear to have been important in the formation of the Andean Cordillera are still active at the present time along the southeastern margin of the Pacific Ocean basin.

DISCUSSION: CIRCUM-PACIFIC OROGENESIS

Many different types of tectonic activity are currently taking place around the margins of the Pacific Ocean. Greatly simplifying, one can recognize the following four main situations:

NW Sector (Kamchatka Peninsula to Philippines)—ocean-continent convergence zone with several backarc basins separating magmatic arcs founded on Asian continental crust from their parent continent.

SW Sector (Celebes to New Zealand)—oblique convergence of Pacific and Indo-Australia Plates with complex zone of interaction including island arcs, backarc basins, microcontinents, oceanic plateaus, and seamounts.

SE Sector (Andean margin)—comparatively simple ocean-continent convergence zone characterized by subduction of sea floor and remarkedly uncluttered by oceanic plateaus, island arcs, and seamounts.

NE Sector (Baja California to Alaska)—oblique convergence of oceanic and continental lithosphere including translation of terranes along the continental margin and stretching of continental lithosphere; short zones of arc magmatism.

Several authors (e.g., Silver and Smith, 1983) have pointed out that the SW Sector may be an analog of the Mesozoic margin of western North America. Continuation of the present-day oblique convergence of the Indo-Australia and Pacific Plates probably will result in the "docking" of a collage of terranes of different characteristics with the Australian and Southeast Asian continental margins to form a complex mountain chain of substantial proportions. Conversely, a major change in plate motions could result in closure of the complex of marginal basins from the South China Sea to the Sea of Okhotsk to reunite the Philippine, Japanese and Kuril-Kamchatka arcs with their parent continents. Indeed, thrust-fault solutions for major earthquakes along the arc side of the Sea of Japan indicate that such closure is taking place, at least locally. Initiation of the Andean Cordillera has been ascribed here to a similar event in the mid-Cretaceous. As noted in a previous section, the best modern analog for the embryonic Andes in the Early Cretaceous is, perhaps, the South Shetland Islands arc-Bransfield backarc trough-Antarctic Peninsula continental margin with its more limited backarc-basin development. The SE sector appears to be characterized by what might be described as "accordion tectonics," namely, repeated development and collapse of backarc basins.

Hence, in considering the causes and processes of circum-Pacific orogenesis, one must acknowledge differences with time at any given location. Perhaps the best example of such differences is the contrast between the substantial accretion along the Pacific margin of the South America segment of Gondwana before the Late Jurassic and its subsequent absence. In studying the processes of cordilleran orogenesis it would be of great value to obtain deep seismic-reflection profiles at different localities representing different stages in the evolution of a cordilleran orogen.

ACKNOWLEDGMENTS

I am grateful to my colleagues in Lamont-Doherty Geological Observatory of Columbia University, the British Antarctic Survey, The University of Texas, and several other institutions, who are contributing to the understanding of this fascinating part of the world. Responsibility for this brief overview, however, is mine. Support for my work comes largely from the Division of Polar Programs of the National Science Foundation through grants DPP 86-43404 and DPP 86-43441.

REFERENCES

Aberg, G., L. Aquirre, B. Levi, and J. O. Nystrom, 1984, Spreading-subsidence and generation of ensialic marginal basins: an example from the early Cretaceous of central Chile, *in* B. Kokelaar and M. F. Howells, eds., Marginal basin geology: Geological Society of London Special Publication 16, p. 185-194.

Aquirre, L., B. Levi, and R. Offler, 1978, Unconformities as mineralogical breaks in the burial metamorphism of the Andes: Contribution to Miner. Petrol., v. 66, p. 361-366.

Aubouin, J., et al., 1973, Esquisse paleogeographique et structurale des Andes Meridionales: Review Geogr. Phys. Geol. dyn. Paris, v. 15, p. 11-72.

Barker, P. F., and I. W. D. Dalziel, 1983, Progress in geodynamics of the Scotia Arc region, *in* R. J. Cabré, ed., Geodynamics of the eastern Pacific region, Caribbean and Scotia arcs: American Geophysical Union Geodynamic Series, v. 9, p. 137-170.

Barker, P. F., I. W. D. Dalziel, and B. C. Storey, in press, *in* R. Tingey, ed., The geology of Antarctica: Oxford University Press.

Bartholomew, D. S., and J. Tarney, 1984, Crustal extension in the southern Andes (45-46 °S), *in* B. Kokelaar and M. F. Howells, eds., Marginal basin geology: Geological Society of London Special Publication 16, p. 195-205.

Beck, M., in press, A review of paleomagnetic data from South America, IGCP Project 120, Andean Magmatism.

Bruhn, R. L., and I. W. D. Dalziel, 1977, Destruction of the Early Cretaceous marginal basin in the Andes of Tierra del Fuego, *in* M. Talwani and W. C. Pitman, III, eds., Maurice Ewing Series, Vol. I: Island arcs, deep sea trenches, and back-arc basins: Washington, D.C., American Geophysical Union, p. 395-405.

Bussell, A. M., 1983, Timing of tectonic and magmatic events in the Central Andes of Perú: Journal of the Geological Society of London, v. 140, p. 279-286.

Charrier, R., 1973, Interruptions of spreading and the compressive tectonic phases of the meridional Andes: Earth and Planetary Science Letters, v. 20, p. 242-249.

Chotin, P., 1976, Essai d'interpretation du Bassin Andin chileno-argentin en tant que bassin marginal: Ann. Soc. Geol. Nord., v. 96, p. 177-184.

Coney, P. J., D. L. Jones, and J. W. H. Monger, 1980, Cordilleran suspect terranes: Nature, v. 288, p. 329-333.

Dalziel, I. W. D., 1981, Back-arc extension in the southern Andes: a review and critical reappraisal: Phil. Transactions Royal Society London, v. A300, p. 319-335.

Dalziel, I. W. D., 1983, Evolution of the Scotia arc: a review, *in* R. L. Oliver, P. R. James, and J. B. Jago, eds., Antarctic earth science: Canberra, Australian Academy of Science, p. 283-288.

Dalziel, I. W. D., 1985, Collision and cordilleran orogenesis: an Andean perspective, *in* M. P. Coward and A. C. Ries, eds., Collision tectonics: London, Geological Society Special Publication 19, p. 389-404.

Dalziel, I. W. D., and D. H. Elliot, 1973, The Scotia arc and Antarctic margin, *in* A. E. M. Nairn and F. G. Stehli, eds., The ocean basins and margins: 1. The South Atlantic: New York, Plenum Press, p. 171-245.

Dalziel, I. W. D., and R. F. Forsythe, 1985, Andean evolution and the terrane concept, *in* D. G. Howell, ed., Tectonostratigraphic terranes of the Circum-Pacific region, Circum-Pacific Council for Energy and Mineral Resources Earth Science Series, Proceedings of the Circum-Pacific Terranes Conference: American Association of Petroleum Geologists, v. 1, p. 565-581.

Dalziel, I. W. D., and A. M. Grunow, 1985, The Pacific margin of Antarctica: Terranes within terranes, *in* D. G. Howell, ed., Tectonostratigraphic terranes of the Circum-Pacific region, Circum-Pacific Council for Energy and Mineral Resources Earth Science Series, v. 1, p. 555-564.

Dalziel, I. W. D., et al., 1975, Tectonic relations of South Georgia Island to the southernmost Andes: Geological Society of America Bulletin, v. 86, p. 1034-1040.

Dalziel, I. W. D., et al., in press, a, Extensional tectonics and the fragmentation of Gondwana, *in* M. Coward, J. Dewey, and P. Hancock, eds., Extensional tectonics: The Geological Society of London Special Publication.

Dalziel, I. W. D., et al., in press, b, The Ellsworth-Whitmore crustal block: its role in the tectonic evolution of West Antarctica, *in* Proceedings of the Sixth Gondwana Symposium, Columbus, Ohio, August 1985: American Geophysical Union, Antarctic Research Series.

DeWit, M. J., 1977, The evolution of the Scotia arc as a key to the reconstruction of southwestern Gondwana: Tectonophysics, v. 37, p. 58-81.

Grunow, A. M., D. V. Kent, and I. W. D. Dalziel, 1985, West Antarctica: new paleomagnetic data and tectonic implications: Eos (American Geophysical Union Transactions), v. 66, no. 46, p. 866.

Grunow, A. M., I. W. D. Dalziel, and D. V. Kent, in press, Ellsworth-Whitmore Mountains crustal block Western Antarctica: new paleomagnetic data and their tectonic significance, *in* Proceedings of the Sixth Gondwana Symposium, Columbus, Ohio, August 1985: American Geophysical Union, Antarctic Research Series.

Herve, M., M. Suarez, and A. Puig, 1984, The Patagonian batholith S of Tierra del Fuego, Chile: timing and tectonic implications: The Journal of the Geological Society, v. 141, p. 909-917.

Howell, D. G., ed., 1985, Tectonostratigraphic terranes of the Circum-Pacific region: Circum-Pacific Council for Energy and Mineral Resources, Earth Science Series, vol. 1.

Jones, D. L., A. Cox, P. J. Coney, and M. Beck, 1982, The growth of western North America: Scientific American, v. 247, p. 70-84.

Jordan, T. E., et al., 1983, Andean tectonics related to geometry of subducted Nazca Plate: Geological Society of America Bulletin, v. 94, p. 341-361.

Kominz, M. A., 1984, Oceanic ridge volumes and sea level change—an error analysis, *in* J. Schlee, ed., Inter-regional unconformities and hydrocarbon accumulations: American Association of Petroleum Geologists Memoir 34, p. 109-127.

LaBrecque, J. L., and P. F. Barker, 1981, The age of the Weddell basin: Nature, v. 290, p. 489-492.

Levi, B., and L. Aquirre, 1981, Ensialic spreading-subsidence in the Mesozoic and Paleogene Andes of central Chile: Journal of the Geological Society of London, v. 138, p. 75-81.

Longshaw, S. K., and D. H. Griffiths, 1983, A paleomagnetic study of Jurassic rocks from the Antarctic Peninsula and its implications: Journal of the Geological Society of London, v. 140, p. 945-954.

Mégard, F., 1984, The Andean orogenic period and its major structures in central and northern Perú: The Journal of the Geological Society, v. 141, p. 893-900.

Molnar, P., 1986, The structure of mountain ranges: Scientific American, v. 255, p. 70-79.

Nur, A., and Z. Ben-Avraham, 1982, Oceanic plateaus, the fragmentation of continents, and mountain building: Journal of Geophysical Research, v. 87, p. 3644-3661.

Nur, A., and Z. Ben-Avraham, 1983, Displaced terranes and mountain building, *in* K. J. Hsü, ed., Mountain building processes: London, Academic Press, p. 73-84.

Pilger, R. H., 1984, Cenozoic plate kinematics, subduction and magmatism: South American Andes: The Journal of the Geological Society, v. 141, p. 793-802.

Pitcher, W. S., et al., 1985, Magmatism at a plate edge: The Peruvian Andes: Glasgow and London, Blackie and Sons, Ltd., 328 p.

Ramos, et al., 1986, Paleozoic terranes of the central Argentine-Chilean Andes: Tectonics, v. 5, p. 855-880.

Silver, E. A., and R. B. Smith, 1983, Comparison of terrane accre-

tion in modern southeast Asia and the Mesozoic of the North American Cordillera: Geology, v. 11, p. 198-202.
Storey, B. C., and S. W. Garrett, 1985, Crustal growth of the Antarctic Peninsula by accretion, magmatism and extension: Geological Magazine, v. 122, p. 5-14.
Suarez, G., P. Molnar, and B. C. Burchfiel, 1983, Seismicity, fault plane solutions, and active tectonics of the Andes of Perú, Ecuador, and southern Colombia: Journal of Geophysical Research, v. 88, p. 403-428.
Thomson, M. R. A., et al., 1983, The Antarctic Peninsula—a Late Mesozoic-Cenozoic arc, *in* R. L. Oliver, P. R. James, and J. B. Jago, eds., Antarctic earth science: Canberra, Australian Academy of Science, p. 289-294.
Uyeda, S., and Kanamori, 1979, Back-arc opening and the mode of subduction: Journal of Geophysical Research, v. 84, p. 1049-1061.
Vicente, J. C., et al., 1979, El sobre-escurrimiento de Cincha-Lluta: Elemento del accidente major Andino al NW de Arequipa: Boletin de la Sociedad Geologica de Perú, v. 61, p. 67-100.

Ericksen, G. E., Cañas Pinochet, M. T., and Reinemund, J. A., editors, 1989, Geology of the Andes and its relation to hydrocarbon and mineral resources: Houston, Texas, Circum-Pacific Council for Energy and Mineral Resources Earth Science Series, v. 11.

CHAPTER 3

ANDEAN CHRONOTECTONICS

John J. Gallagher[1]
Occidental Exploration and Production Company
Bakersfield, California, U.S.A.

ABSTRACT

State-of-the-art plate-tectonic information provides a global-scale framework for understanding the principal tectonic events of the Andean region. Decipherable tectonic history begins with Precambrian supercontinents that consist of collages of sutured cratons or microplates. Late Precambrian South America was attached to Africa on the east and may have been attached to Europe on the north and to North American on the west. Later Precambrian and early Paleozoic rift and drift events separated South America from Europe and from North America. These extensional events, combined with basement geology, created a structural fabric which strongly influenced trends of younger, major tectonic elements, including basins and uplifts. Phanerozoic tectonics and tectonostratigraphy of the Andean region were controlled by plate-edge events along the western margin of South America. These events relate to the velocity and direction of adjacent plates of all South America plate boundaries. Some notable examples of tectonism related to plate-edge events occurred during latest Proterozoic, Devonian, Late Carboniferous, Middle Jurassic, Cretaceous, Miocene, and Pliocene. Andean uplifts and basins developed as plate-edge displacements which were transmitted along appropriately oriented basement fabric trends. These effects of tectonism—together with sea-level changes, oceanic upwelling, paleoclimates, and other factors—produced basins with significant amounts of hydrocarbons.

INTRODUCTION

Andean chronotectonics synthesizes geological data over many scales of observation, providing constraints for geological models which can be used to evaluate energy or other resources in the Andean area. This chapter investigates the dynamic processes or cause-effect, action-reaction relationships between plate scale and regional scale in the Andean region. The chapter sketches the general plate-tectonic setting or sequence of events for western[2] South America during the Archean, Proterozoic, and Phanerozoic. The plate-tectonic events are tied to regional tectonics for the northern, central, and southern Andes. The plate and regional tectonics are shown to provide constraints on time and process for sub-Andean basins.

PLATE-TECTONIC SETTING

The chronotectonics of the Andean region shows, within the limitations of present knowledge, how global-scale events were related in time and space to regional tectonic events. These relationships provide a framework for understanding the subsidence and structuring of sub-Andean basins.

The Andean region of South America consists of many tectonic terranes which were accreted or deformed during the Phanerozoic. Accretion and deformation occurred in a mobile belt around the Proterozoic continent built in turn around Archean cratonic cores (Figure 1). The western limit of the Proterozoic rocks probably was set by Paleozoic rifting and drifting. The Phanerozoic mobile belt, representing about 25% of the present-day continent, consists of allochthonous terranes that were accreted to the western margin through convergence, collision, and suture. The Andean belt grew within these accreted terranes.

Pre-Phanerozoic

Worldwide, Archean cratons seem to have been formed by plate-tectonic processes very similar to those operative today but with significant differences in detail (Windley, 1984). The Archean regions of South America are characterized as strong, granitic, cratonic cores or microplates (Figure 1a) around which younger continental masses accumulated. They have clearly influenced the change of shape of western South America through time. They seem to have acted as buttresses during collisional events and to have influenced the distribution of strike-slip faulting and rifting at their margins. The positions of cratonic masses during the Archean are not known, but lithologies and structure suggest their positions during the Proterozoic, as shown in Figure 1b.

Present-day continents are predominately Proterozoic rock added to Archean cratons through plate-tectonic accretion

---

[1]Now at ARCO Oil and Gas Company, 2300 West Plano Parkway, Plano, Texas, U.S.A.

[2]Present-day continental shapes, geographic directions, scales, and political boundaries are used in discussing paleogeography.

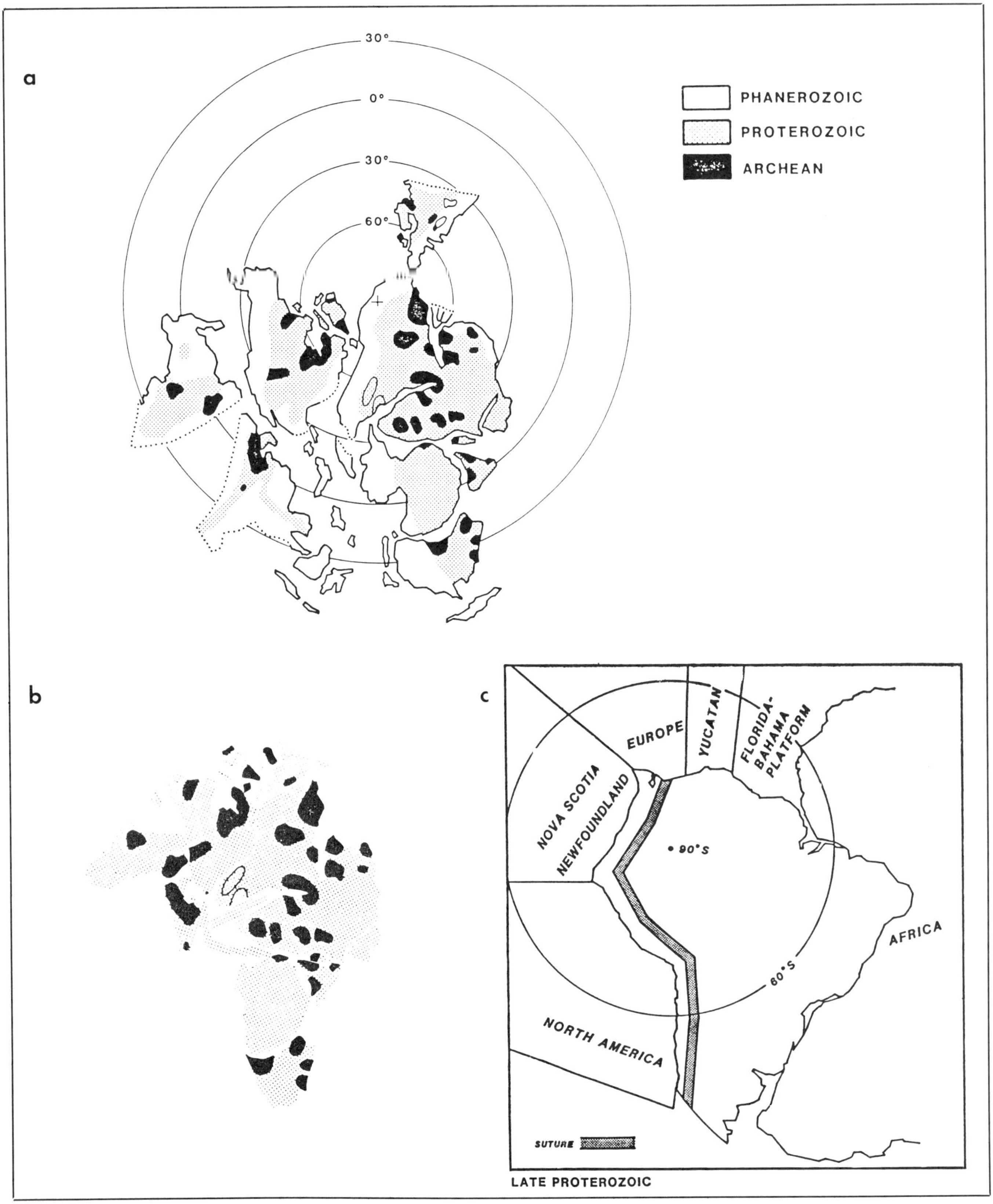

Figure 1. Hypothetical plate-tectonic reconstruction 600 Ma. Proterozoic cratons in southern South America are the Sierras Pampeanas to the west and the Uruguay-Argentina craton to the east (after Burke et al., 1978). Archean and Proterozoic modified from Windley (1984). Continental positions modified from Scotese et al. (1979); Ziegler, Scotese, and Barrett (1983); and Bond, Nickeson, and Kominz (1984). (a) Overview. (b) Phanerozoic fold belts removed and remaining cratons moved closer together. (c) Schematic for positions of continents adjacent to South America.

processes (Windley, 1984; Howell et al., 1985). Present-day continental outlines shown in Figure 1a identify the Proterozoic components in the supercontinent. This reconstruction (Figure 1) combines data from many sources. The plate positions, except for Asia, roughly combine positions shown by Scotese et al. (1979), Ziegler, Scotese, and Barrett (1983) and Bond, Nickeson, and Kominz (1984). The Archean and Proterozoic crustal ages are modified from Windley (1984). The positioning of Asia west of North America provides a means of explaining the well-known late Precambrian rift sequences of sedimentary rocks found on both continents, and provides a western neighbor for North America with suspect terranes removed (Coney, Jones, and Monger, 1980). Furthermore, Johnson, Rong, and Yang (1985) give faunal and stratigraphic evidence that China and North America were linked in the early Paleozoic. Howell et al. (1985) cite glacial evidence for such a linkage. During the late Proterozoic, South America abutted against various parts of North and Central America and Europe (Figure 1c).

The structural trends of Proterozoic rock developed during its 2000-m.y. history and are indeed complex. The trends that most strongly influence Phanerozoic structures may be those related to the last Precambrian extensional event which continued into the Paleozoic. The rift-sequence Proterozoic sedimentary rocks on the various continents suggest pulsating extensional tectonic activity, which led ultimately to the breakup of the last Precambrian supercontinent. Bond, Nickeson, and Kominz (1984) used tectonic subsidence curves to estimate a date of 625 to 555 Ma for the breakup of western South America. North and South America may have separated at this time but probably remained close together, for reasons to be discussed in a later section.

About 600 Ma the northern pre-Andes were at the South Pole and the southern pre-Andes were at about lat. 45 °S, so that the continental polarity was reversed from the present-day continental polarity. South America seems to have moved northward, retaining the reversed continental polarity, until the end of the Proterozoic (570 Ma), when the continent was centered on the equator (Ziegler, Parrish, and Scotese, 1981). Smith, Hurley, and Briden (1981), Ziegler, Scotese, and Barrett (1983) and Bond, Nickeson, and Kominz et al. (1984) agree on the Precambrian continental polarity of South America (that is, reversed), but seem to differ widely on latitudinal positioning. Their range of latitudes for the northern pre-Andes is from lat. 10 °S and for the southern pre-Andes is from lat. 30 °N to lat. 45 °S during the period from 560-600 Ma. Irving (1981) agrees with Ziegler, Scotese, and Barrett (1983) on general latitudinal positioning (that is, between lat. 30 °N and lat. 10 °S) but disagrees on continental polarity. For the purposes of this chapter, the consensus reported by Bond, Nickeson, and Kominz (1984) is used as a working model for the Precambrian to earliest Cambrian. Longitudinal placement is totally arbitrary, because definitive biostratigraphic and other data which might link continents together are lacking. Whatever the details might have been, the Precambrian supercontinent began to rift, drift, and break up as Phanerozoic time began.

Phanerozoic

The plate-tectonic setting of South America and its interactions with adjacent continental plates during the Phanerozoic can be summarized and interpreted with the aid of a series of maps (Figures 2, 3, 4) modified from Scotese et al. (1979) and Ziegler, Scotese, and Barrett (1983). The earliest Paleozoic history was dominated by extension, first on the northern margin during the Cambrian (Figure 2a), then on the western margin during the Silurian (Figure 2b). The remainder of the Phanerozoic plate-tectonic history of western South America was dominated by convergence (see, for example, Cobbing, 1985). During the Devonian the active margins had subduction zones (Figure 2c). By the Late Carboniferous, collisions were under way to the north and south (Figure 2d). The southernmost tip of South America accreted by the Middle Jurassic, and the Permian-Triassic subduction zones jumped offshore (Figures 2e, 3a). Rifting began in the South and Central Atlantic during the Middle Jurassic, or possibly earlier (Figure 3b). By the Early Cretaceous, seafloor spreading was active over much of the South and Central Atlantic and the Caribbean, and microcontinents began accreting to northwestern South America (Figure 3c). Accretion continued throughout the Cretaceous, and culminated in latest Cretaceous or early Paleocene (Figure 3d). Plate convergence and tectonic accretion continued through the Miocene (Figure 4a) and the Pliocene (Figure 4b) to the Present (Figure 4c).

Episodic extensional phases such as the one in the Devonian (Figure 2c) and another at the Paleozoic-Mesozoic boundary were subordinate to convergence and subduction (Figure 2e). Strike-slip movement occurred in the northern pre-Andes from the Miocene to the Present as a consequence of oblique subduction and accretion (Figure 4a).

Paleolatitude

The South American Plate underwent extreme and sometimes rapid latitudinal changes during the Paleozoic. The polarity of the continent was reversed (Figure 1c) until Late Ordovician-Early Silurian, when it changed to normal (Figure 2b)—that is, the northern pre-Andes lying to the north of the southern pre-Andes (Scotese et al., 1979; Irving, 1981). The continental polarity changed because South America passed over the South Pole. South America was over or nearly over the South Pole from the Silurian through the Early Carboniferous. The continental polarity remained normal throughout the Mesozoic and Cenozoic. Latitudinal changes were slow and slight during the Mesozoic and even more so during the Cenozoic (Ziegler, Scotese, and Barrett, 1983; Zoneshain et al., 1985).

Paleoclimate

During the Cambrian and Ordovician, most of South America was in the equatorial wet zone, whereas during the Silurian, Devonian, and Early Carboniferous most of the continent was in the temperate zone. From Late Carboniferous to the Present, segments of the pre-Andean and Andean regions were within the following climatic zones: (1) northern region was in the equatorial wet zone; (2) central region was in the subtropical dry zone; and (3) southern region was in the temperate zone. Glaciation affected the northern pre-Andes in the latest Precambrian to earliest Cambrian and the southern pre-Andes in the Late Ordovician to Early Silurian, and again in the Late Carboniferous through Permian.

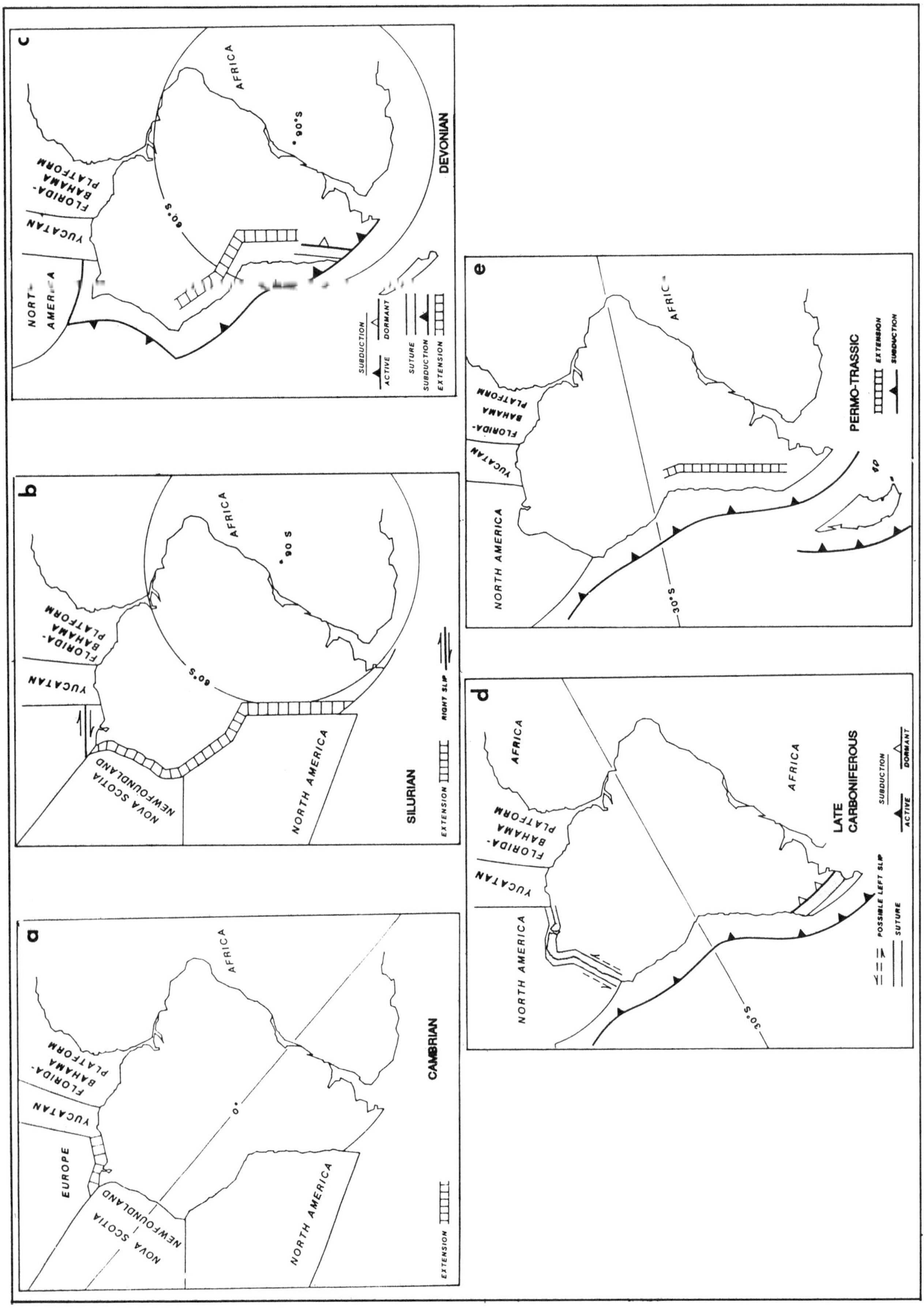

Figure 2. Hypothetical plate-tectonic reconstructions for South America from Cambrian to Permian-Triassic: (a) 570 Ma; (b) 420 Ma; (c) 380 Ma; (d) 320 Ma; (e) 245 Ma.

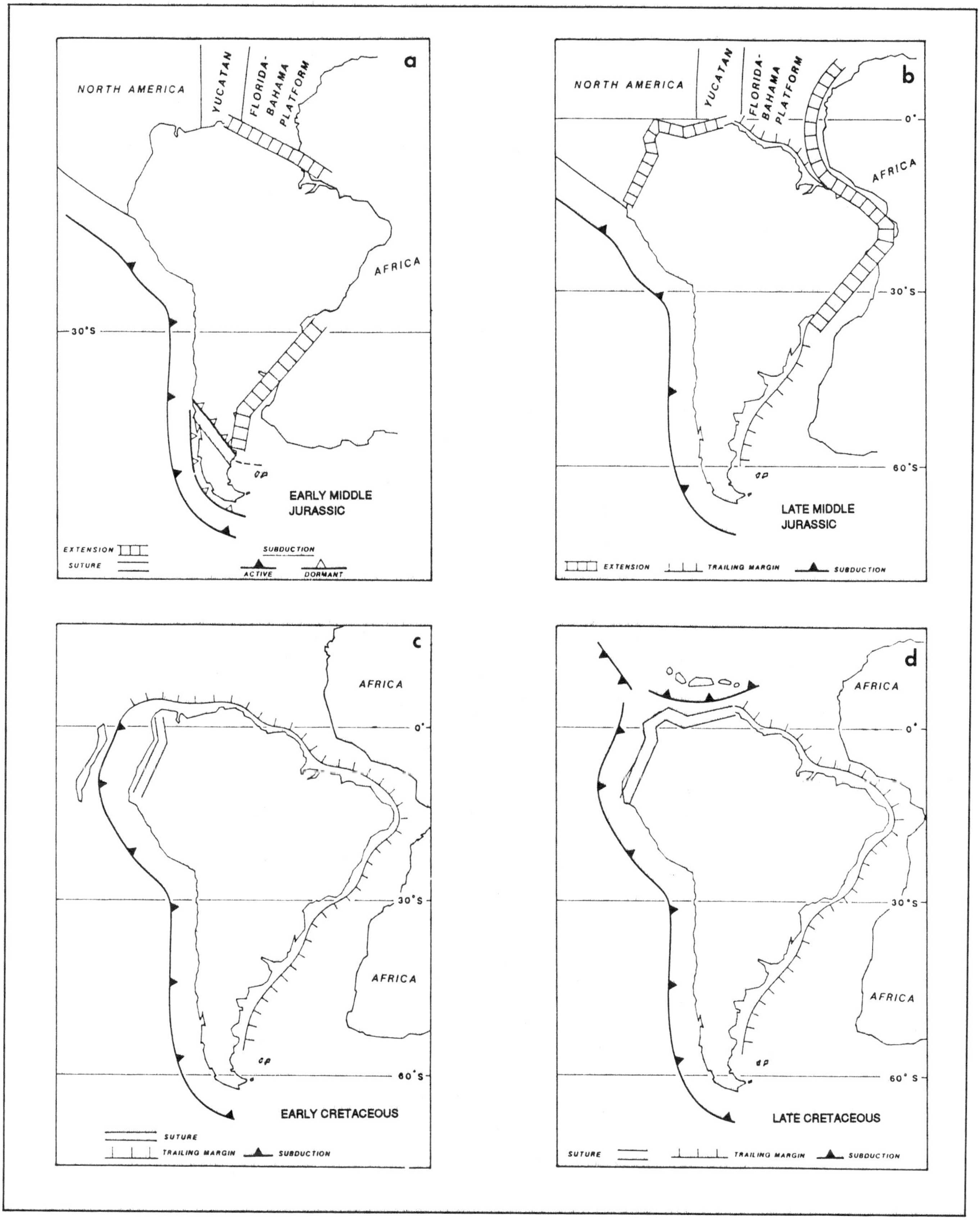

Figure 3. Hypothetical plate-tectonic reconstructions for South America from Middle Jurassic to Late Cretaceous: (a) 180 Ma; (b) 160 Ma; (c) 125 Ma; (d) 65 Ma.

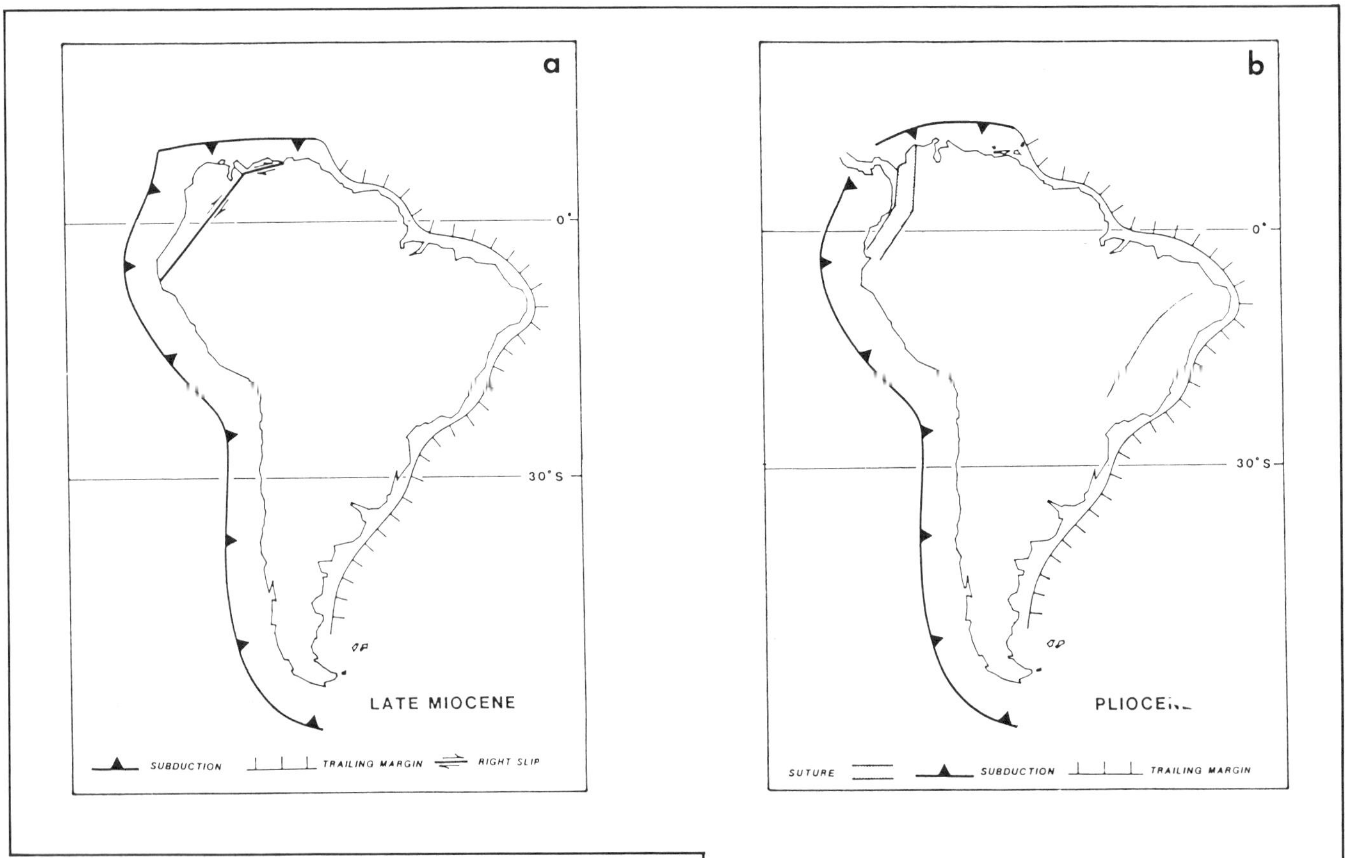

REGIONAL TECTONIC FRAMEWORK

Gansser (1973) proposed a three-fold subdivision of the 9000-km-long Andes. Corvalan (1985) showed that the boundaries of these subdivisions change somewhat with time but that the subdivisions are valid. In the northern and southern subdivision, there are rock assemblages with clear-cut oceanic affinities, which are therefore allochthonous or suspect terranes.

Northern Andes

The northern Andes is made up of suspect and displaced terranes. The present-day plate convergence is exerting compressive, oblique, subduction-related forces on the northwestern corner of the South American continent (Figure 4c). The principal response is right-lateral faulting accompanied by local thrusting along the eastern Andes and Merida-Bocono trends (Figure 5a). Strike-slip movement has taken place on the Santa Marta and Oca faults and thrusting on the Salina, Perija and Villa de Cura fault. Left-slip may have taken place on the Meta zone, as indicated by a map of Martin (1978), but this remains uncertain. An interpretation of development for the tectonic terranes of the northern Andes from the late Proterozoic to the Present, is given in Figures 5 and 6. The corresponding plate-tectonic setting is shown on Figures 2, 3, and 4.

Northwestern South America began in the Precambrian as part of a supercontinent (Figure 5a). Its history was dominated by crustal extension as the supercontinent broke up and separated starting in the Cambrian (Figure 5b) and continuing

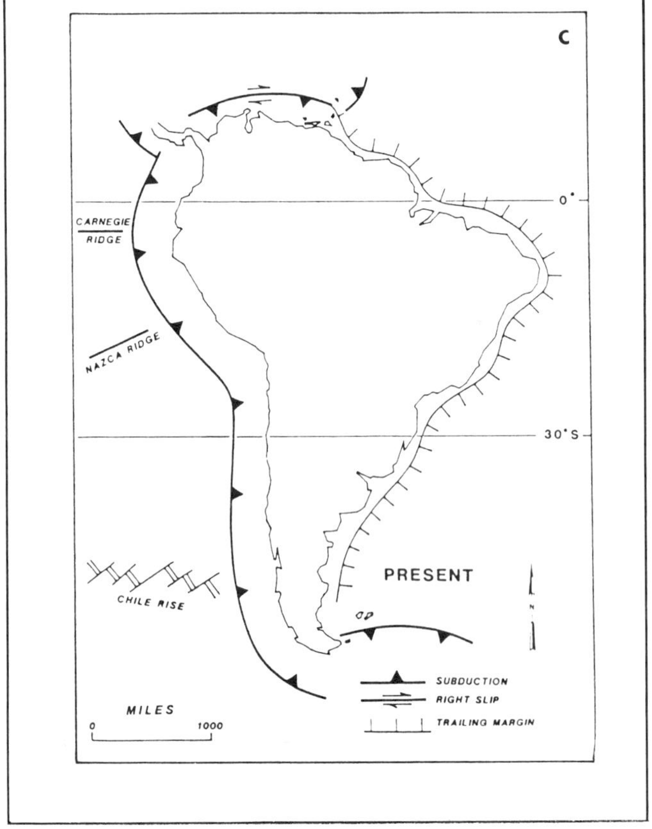

Figure 4. Hypothetical plate-tectonic reconstruction for South America from Miocene to Present: (a) 9 Ma; (b) 3 Ma; (c) 0 Ma.

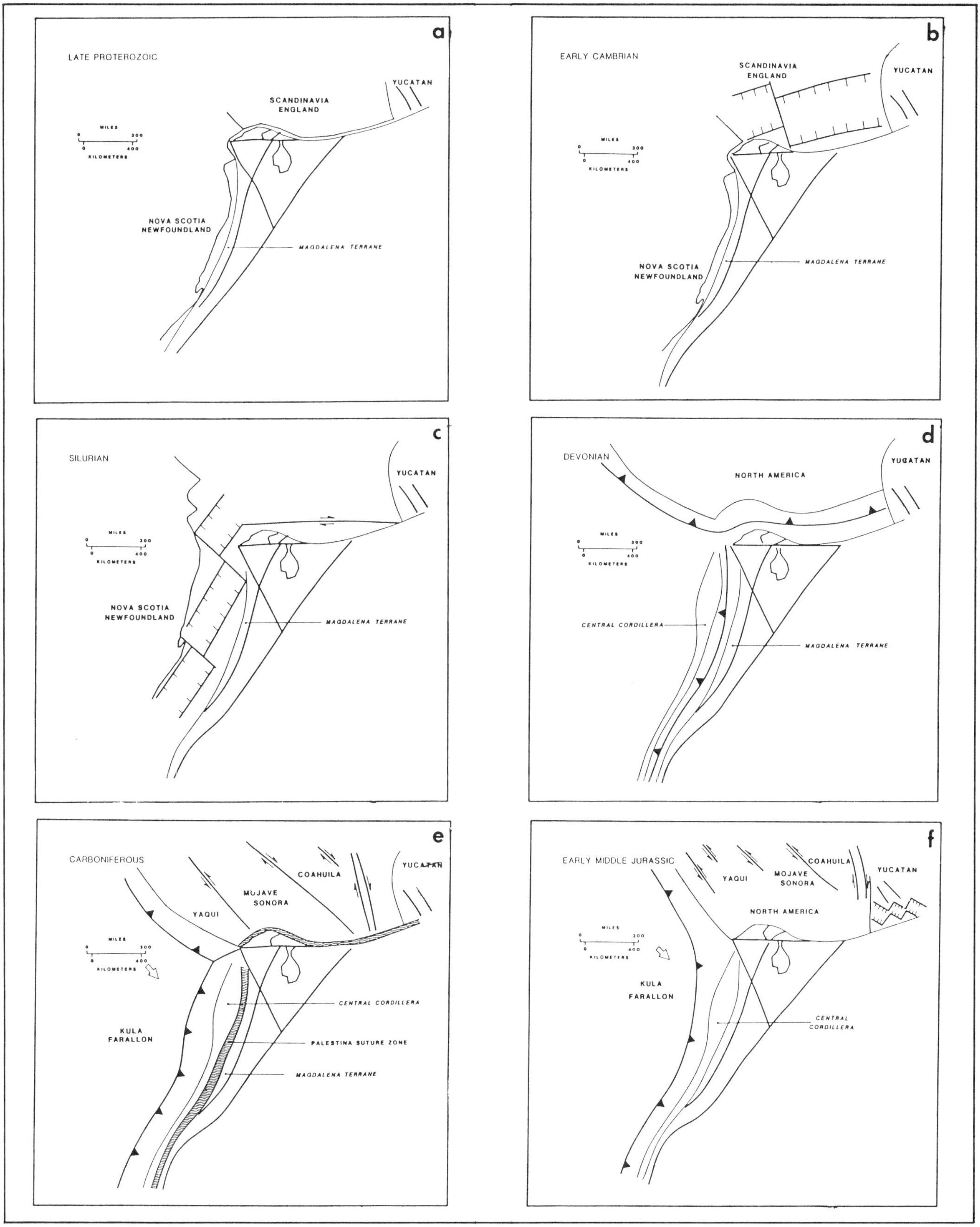

Figure 5. Tectonic framework of northwestern South America from late Proterozoic to Middle Jurassic: (a) 600 Ma; (b) 570 Ma; (c) 420 Ma; (d) 380 Ma; (e) 320 Ma; (f) 180 Ma.

through the Silurian (Figure 5c). These extensional events are thought to have established structural trends for mechanical reasons, in particular because rocks are very weak and fracture easily in extension (Gallagher, 1981). The younger history was dominated by convergence and tectonic accretion of displacement and suspect terranes (Figures 5d, 5e, 5f, 6). Subducting slabs dipped away from northwestern South America in the Devonian (Figure 5d), a time when tectonic activity on the continent was evidently minimal. Major compressional events, like the one in the Carboniferous, are believed to have reactivated existing trends rather than creating new trends (Figure 5e). In the shallow lithosphere, rocks generally slip along existing zones of weaknesses rather than fracture for specific mechanical reasons (Gallagher, 1984, In press a, b).

Until the middle of the Middle Jurassic, the northern Andes was joined with North America on the north and had a continent-ocean subduction complex on the west (Figure 5f). By the end of the Middle Jurassic, the northern margin of South America was under tension, accompanied by the separation of North America from South America (Figure 6a). During the Cretaceous, collision took place on the west margin (Figures 6b, 6c). Collision on the north margin occurred in the Late Cretaceous and early Paleocene (Figure 6c). Both the north and the west margins were subjected to transpression (combined strike-slip and compression) throughout the later Tertiary, with major collisions on the west margin (Figures 6d, 6e).

Suspect Terranes

Northwestern South America is rimmed by suspect terranes (Figure 6f), which have oceanic affinities and may have moved hundreds or even thousands of kilometers from their position of origin. These include a southern group of terranes which extends westward from the Eastern Cordillera overthrust in Colombia and Ecuador and a northern group of terranes in an east-trending zone extending from the Guajira Peninsula in Colombia through the Coastal Cordillera of Venezuela to northern Trinidad.

Northern Terranes

Studies of igneous and metamorphic rocks demonstrate that a zone extending from the Guajira Peninsula to Trinidad collided with northern South America at some time in the Late Cretaceous (Maresch, 1974). Beets et al. (1984) date the island arc-continent collision as Coniacian to Campanian. The suture zone is today marked by dismembered ophiolites which are offset along the Bocono trend in a right-lateral sense as much as 290 km (Dewey and Pindell, 1985; Pindell, 1985). The need for palinspastic restoration along major faults separating northern terranes is supported by work on stratigraphy, structure, and hydrocarbons by Krause and James (1985).

The dip of the precollision subduction slab was northward (Figure 6c). After collision and plate reorganization a new trench formed oceanward, with a slab subducting southward, and the terranes became part of the South America continent for the rest of their history (Figure 6d).

Southern Terranes

Most of western Colombia and western Ecuador are suspect terranes accreted since the Carboniferous (McCourt, Aspden, and Brook, 1984; Etayo, 1985). These terranes are discussed in sequence from west to east, and in terms of time of accretion, from youngest to oldest.

The newest addition to northwestern South America is the coastal terrane which was accreted along the Guayaquil suture (Figures 6c-6f). McCourt, Aspden, and Brook (1984) date the suture as Late Cretaceous to Paleocene, but Dewey and Pindell (1985) illustrate the collision as progressing northward along a right transpressional zone during the late Miocene to Pliocene with convergence continuing at the present (Figures, 6e, 6f). Dewey and Pindell (1985) cite evidence that uplift of the Cordillera Central, Oriental, Occidental, Perija, and Merida is well constrained within the last 9 to 10 m.y. which are a consequence of the coastal terrane accretion.

There is stratigraphic evidence that in the early Pliocene the culmination of the Andean orogeny was accompanied by uplift of accreted oceanic material of the Serrania de Baudo trench-slope (McCourt, Aspden, and Brook, 1984). Rocks in the trench-slope sequence date well back into the Mesozoic; therefore, a trench is shown west of the coastal terrane in Figure 6d.

The present-day coastal plain is made up of a highly complex assemblage of oceanic island arc-trench related rocks. At first glance, the coastal plain might appear to be the forearc portion of the island arc represented by the Western Cordillera. However, there are structures and lithologies typical of sutures in the coastal plain and the Western Cordillera (Martin, 1978; McCourt, Aspden, and Brook, 1984). For these reasons, the coastal plain is illustrated as an allochthonous terrane until the Miocene-Pliocene (Figures 6d, 6e).

The Western Cordillera appears to be a Mesozoic island arc situated above an eastward dipping subducting slab. This arc collided with the Cauca Patia accretionary terrane along the Cauca Patia fault zone during the Late Cretaceous to Paleogene (Figures 6c-6e). Paleontological and radiometric data indicate that the Mesozoic rocks of the Western Cordillera are post-Aptian/Albian to Maestrichtian in age (McCourt, Aspden, and Brook, 1984).

The Cauca Patia accretionary terrane originated as a Mesozoic accretionary prism along a trench associated with a Late Cretaceous volcanic arc in the Central Cordillera (Figures 6a-6c). The zone was compressed and deformed during collision and suture of the Western Cordillera (Figure 6c). Following the collision of the coastal terrane, the Cauca Patia terrane was placed in a backarc position, was subjected to extension, and finally became a graben which today extends from the Gulf of Guayaquil northward to the northern coast of Colombia.

The Central Cordillera had a long and complex history related to subduction (McCourt, Aspden, and Brook, 1984). It may have originated in the Devonian as an island arc above a westward dipping subduction slab (Figure 5d). It collided with the subsiding continental margin of South America along the Palestina fault zone during the Carboniferous (Figure 5e). This was a profound tectonic event marked by regional metamorphism and associated with collision of North and South America. The Central Cordillera was an episodically active arc associated with ocean-continent convergence throughout most of the Mesozoic (Figures 5e, 5f, 6a, 6b). The Central Cordillera suspect terrane is apparently sutured to and in part comingled with the westernmost extension of the Precambrian basement rock which lies beneath the Magdelena displaced terrane.

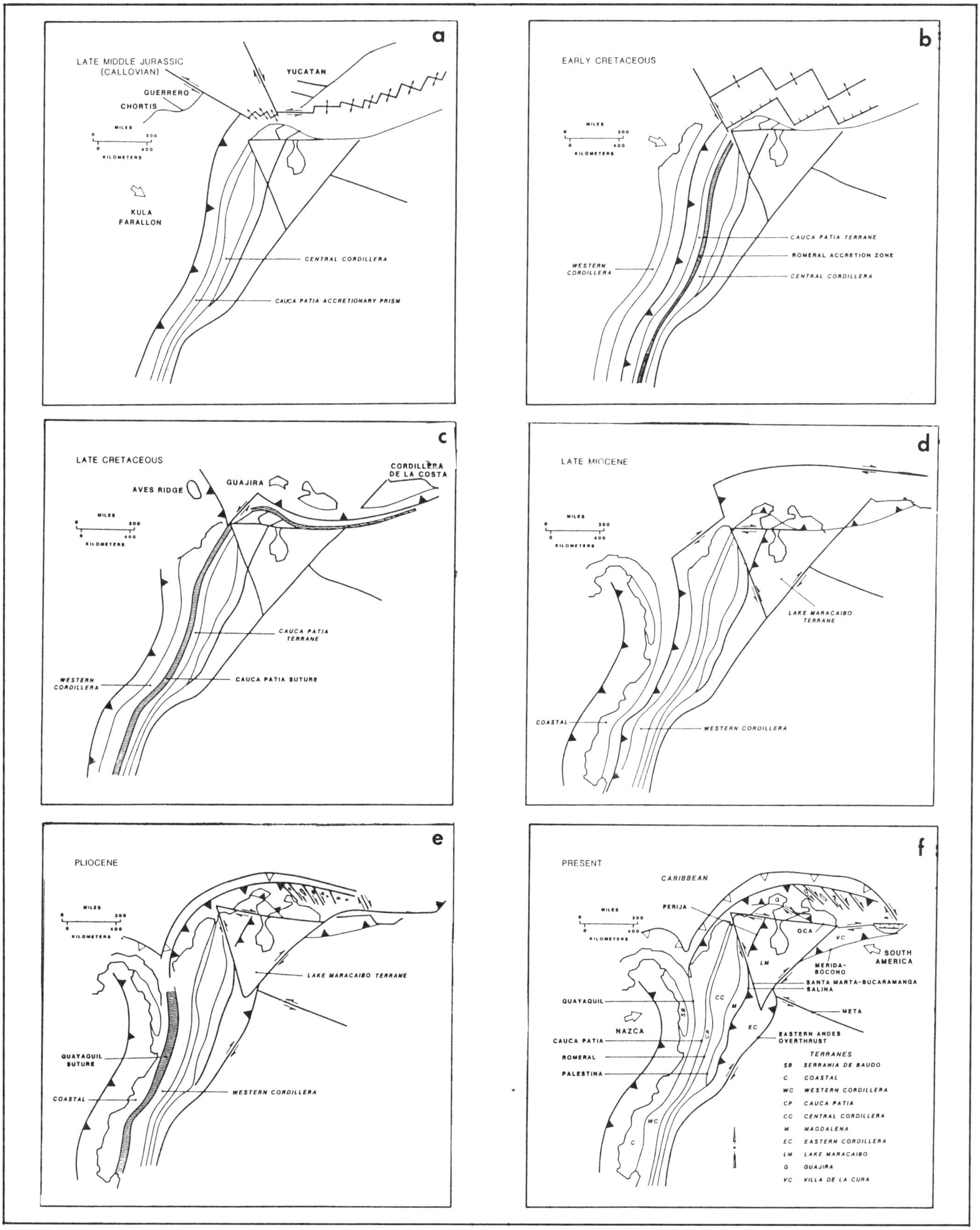

Figure 6. Tectonic framework of northwestern South America from Middle Jurassic to Present: (a) 160 Ma; (b) 125 Ma; (c) 65 Ma; (d) 9 Ma; (e) 3 Ma; (f) 0 Ma.

Displaced Terranes

Displaced terranes are those which are closely associated with and not far separated from Archean and Proterozoic cratons. In contrast to suspect terranes, they have moved only short distances. In the northern Andes, these terranes have been displaced by strike-slip faulting, but (as well as can be determined from available data) remain near their parent Precambrian cratons. Displaced terranes in the northern Andes include several components, all of which moved northeastward during the Neogene. Each component moved as a separate unit but displacements between some of the terranes relative to each other are so small that they may be grouped as two terranes bounded by major faults. These two terranes are the Eastern Cordillera-Magdelena terrane and the Lake Maracaibo terrane, and which are discussed in the next section (Figures 6d-6f).

Eastern Cordillera—Magdelena Terrane

The westernmost projection of the Precambrian craton, which terminates under the Magdelena Valley, may have been formed during late Precambrian and Paleozoic rift events (McCourt, Aspden, and Brook, 1984). Events like these typically result in passive continent trailing margins (Windley, 1984). Scotese et al. (1979) suggest that Nova Scotia and Newfoundland were adjacent to Ecuador and Colombia at that time (Figures 5b, 5c). Their data further suggest that Nova Scotia and Newfoundland separated from the Eastern Cordillera-Magdelena terrane during the latest Ordovician to earliest Silurian, leaving a subsiding, trailing continental margin. This was a time of profound plate-tectonic events. As noted above, at this time South America changed once and for all its continental polarity from reversed to normal as the continent made its final passage over the South Pole.

The passive margin along the western edge of the Eastern Cordillera-Magdalena terrane was destroyed during its collision with the Central Cordillera subduction complex in the Devonian (Figure 5d). The terrane was in a backarc position from the Devonian to the Present.

The Carboniferous collision of North America to the north and west caused regional metamorphism (Figure 5e).

All accreted, displaced, and integral terranes west of the Eastern Cordillera boundary fault system moved northeastward from the Miocene to the Present. The right slip on the Eastern Cordillera overthrust was on the order of 275 km (Dewey and Pindell, 1985). These same tectonic events affected the terranes in the Lake Maracaibo region to the north.

Lake Maracaibo Terrane

The triangular collection of smaller terranes bounded by the Merida-Bocono, Santa Marta-Bucaramanga, and Oca fault systems can be considered to have been displaced as a unit (Dewey and Pindell, 1985). The terranes include the Santa Marta, Perija, Santander, and Merida Cordilleras and the Maracaibo and Cezar basins. During the last 10 m.y., uplift of the ranges has been as much as 12 km. Basins have subsided about 4 km on the average during the Cenozoic (Bonini, Garing, and Kellogg, 1982; Kellogg and Bonini, 1982).

Dewey and Pindell (1985) use published data for fault-slip and vector-triangle diagrams to calculate 290 km of northeastward movement of the Maracaibo terrane relative to the South America autochthon along the Merida-Bocono fault system during the last 9 m.y. (Figure 6c-6f). For the same time frame, the terrane moved 65 km westward along the Oca fault, relative to the Guajira Peninsula, and 110 km northwestward along the Santa Marta-Bucaramanga fault, relative to the Western and Central Cordilleras.

The pre-9 Ma history of the Maracaibo terrain and its components is uncertain. The position shown on Figure 6d and used on Figures 5 and 6a-6c represents a minimum southwestward restoration. The actual position was almost certainly farther to the southwest. This restoration leads to the supposition that the Precambrian basement of the Maracaibo terrane is genetically related to time-equivalent rocks in the British Isles and Scandinavia, or in Nova Scotia and Newfoundland, or perhaps in Virginia, if the terrane was indeed farther south (Figure 5b). The basement fabric would include Precambrian trends as well as roughly east-trending weaknesses developed as a result of the Early Cambrian extensional event which led ultimately to the separation of the British Isles and Scandinavia from the north edge of the terrane (Figure 5b). The Silurian extensional event that moved Nova Scotia and Newfoundland away probably resulted in a northeast-directed trend, roughly parallel to the Salina, Perija, Eastern Cordillera overthrust, and Bocono faults (Figure 5c). The younger history was dominated by compression from the north and west related to subduction and accompanying island-arc, microcontinent, and continent collisions (Figures 5d-5f, 6).

When the displaced terranes are palinspastically restored as, for example, on Figure 6c, the Salina fault is aligned with the Perija fault, and the Guajira-Villa de la Cura suture, shown as offset on Figure 6f, is continuous and unbroken as shown on Figures 6c and 6d. These realignments support the displaced terrane interpretation.

Central Andes

The plate-tectonic development of the central Andes is poorly understood at the present time. Data are sufficient to provide touchstones, as shown in Figures 2 to 4, but they are separated by wide, often confusing, gaps. The central subdivision has eugeosynclinal rocks as well as Precambrian crystalline rocks that have been detached from the South American craton. The latter suggest at least the possibility of displaced terranes which are detached but not far separated from their original craton. Cobbing (1985) maintains that no suspect or displaced terranes can be proven for the central Andes. Accretionary prisms may be considered as suspect terranes, and they occur all along the western margin of South America, including the central and southern margins (Kulm, Schweller, and Masias, 1977; Forsythe, 1982). Nur and Ben-Avraham (1981) suggest more substantial accretionary terranes for the central Andes, and the recent work of Vicente (1985) supports this.

Paleozoic tectonics appear to be dominated by continental collision, whereas Mesozoic and Cenozoic tectonics are dominated by subduction. No major suspect or displaced terranes have been mapped (Cobbing, 1985). However, gravity, seismological, and geological evidence indicates crustal shortening of more than 100 km, suggesting cumulative thrust-fault displacements of that order of magnitude (Lyon-Caen, Molnar, and Suarez, 1985). In addition, Precambrian granitic massifs in the coastal regions in Perú and northern Chile are considered by some workers to be allochthonous (Nur and Ben-Avraham, 1981; Vicente, 1985).

The Precambrian geologic history of the central pre-Andes is little known. At depth, Precambrian granitic rocks of the Proterozoic and Archean craton extend to the western limit of the Eastern Cordillera, Puna, and Altiplano (Breitkreuz and Zeil, 1984; Cobbing, 1985; Vicente, 1985). Similar rocks are found in the Arequipa massif, in the area to the north along the present coast, and as isolated blocks in the region between the Coastal Cordillera and the Western Cordillera (Kulm, Schweller, and Masias, 1977; Thornburg and Kulm, 1981; Breitkreuz and Zeil, 1984; Cobbing, 1985). The massif and isolated blocks may be remnants of the Precambrian or Paleozoic supercontinent mentioned above, or they may be suspect terranes. In any event, Bond, Nickeson, and Kominz (1984), Cobbing (1985), Pitcher and Cobbing (1985) cite data which show that in the Late Precambrian and Early Paleozoic, the central pre-Andes were flanked by the ancestral continent (Figure 1).

One problem with current plate-tectonic reconstructions of the central pre-Andes is that there seem to have been several collisions during the Paleozoic, evidently with North America (Figures 2a, 2b). Data from Bond, Nickeson, and Kominz (1984), and Cobbing (1985) indicate continental collisions in the Late Precambrian, Silurian-Ordovician, Devonian, Early Carboniferous, and Late Carboniferous. Presumably, intervals of extension, during which rifting or strike-slip faulting took place, separate the collisions. Certainly, subduction along one or the other continental margin preceded each collision. The Wilson cycle of repeated opening and closing oceans appears to be valid for the central Andes.

The Devonian to Triassic plate-tectonic history of the central pre-Andes is especially confusing (Figures 2c-2e). Breitkreuz and Zeil (1984) list evidence which suggests continued continental collision. However, North America (if west of South America in the Early Paleozoic) must somehow have moved to a position north of South America, for its southern margin collided with the northern margin of South America in the Late Carboniferous. No other continents are known to be in the appropriate position for such collision (Scotese et al., 1979; Irving, 1981). At least three explanations seem plausible: (1) North America was segmented and one segment collided with western South America while the other segment attached to Europe and collided with South America (at a later time the North American segments sutured together); (2) some yet-to-be-identified non-North American continent(s) or microcontinent(s) collided with South America and later disappeared through drifting, strike-slip faulting, subduction, or another process; and (3) western South America was flanked by a particularly strong subduction zone which caused orogenic effects on the continent similar to effects caused by continental collision.

Breitkreuz and Zeil (1984) document the existence of S-type magmatism which, according to Pitcher (1982), is indicative of "Hercynotype continental collision" and "Encratonic ductile shear belts." Granites of this type have sedimentary affinities which suggest that a substantial volume of sediment was subducted, melted, and injected as a mgma. In light of this, one of the first two of the above explanations seems more likely than the third.

The younger history of the central Andes is dominated by subduction of an oceanic plate beneath a thickening and westward-building continental margin (Breitkreuz and Zeil, 1984). This interpretation is supported by documentation of I-type plutonism, which is indicative of "Andinotype marginal continental arcs" (Pitcher, 1982).

Southern Andes

The southern Andes may be part of two or more suspect terranes, including Patagonia, that may have been accreted to the South American craton during the Silurian-Devonian and the Permian-Triassic (Figures 2c-2e). The younger history of the southern Andes is otherwise much like that of the central Andes, that is, ocean-continent convergence.

Southern South American Allochthons

Mpodozis and Ramos (1985) describe a Silurian-Devonian arc-continent collision which occurred along the Early Paleozoic margin of the South American craton (Figures 2c, 2d). The trench moved oceanward to a new position as the arc, possibly the eastward extension of "Chilenia," accreted to the craton. The region to the south, which is presently southernmost South America, may not have become part of the craton until the Late Permian to Late Jurassic (Figures 2e, 3a, 3b). Gust et al. (1985) describe a northwest-trending forearc basin of late Paleozoic age in southern South America that was flanked on the northeast by a late Paleozoic to early Mesozoic magmatic arc. They present evidence for a Permian-Triassic backarc basin farther to the northeast. Consequently, the region to the southwest of the forearc basin—Patagonia, Tierra del Fuego, and the Falkland Islands (Islas Malvinas)—was allochthonous until the Late Permian to Late Jurassic.

Gust et al. (1985) propose that a subducting slab was located along the late Paleozoic margin of western South America, and that it was steeply dipping until latest Paleozoic, when the dip became more gentle and caused widespread magmatism and compressional deformation. This change in dip may have taken place at the time the southernmost terrane was accreted to South America. After suture closure, the trench moved westward (Figure 3a). They suggest further that by the Late Jurassic the dip of the subducting slab had become steep once again. The Middle to Late Jurassic change in dip was accompanied by extension, which affected both the allochthonous and the autochthonous regions.

The culmination of the extension at about 140 Ma (Dalziel, 1981) was the formation of the Rocas Verdes marginal basin in western South America (Figure 2e). Gust et al. (1985) commented on the broader significance of this event: "the extension and volcanism that began in the Late Triassic and culminated in the mid-Jurassic ignimbritic eruption and the opening of the Rocas Verdes Basin was the initial stages of crustal extension that lead directly to the opening of the South Atlantic." Thus, for southern South America, the plate-tectonic events on the east margin were synchronous with those on the west margin in the middle Mesozoic.

It is not clear where the southernmost accreted terrane originated. It appears to have developed along a convergent segment of the late Paleozoic to early Mesozoic Pacific margin of Gondwana (Nelson, Dalziel, and Milnes, 1980). The southern and western borders have undergone "complex but essentially continuous subduction since mid-Paleozoic" (Dalziel, 1981). If true, this allochthon is one of the largest suspect terranes accreted to South America.

Coastal Suspect Terranes

The coastal suspect terranes of Chile are relatively small (Forsythe, 1982). While the southern South America allochthon was separate from the South American continent, its western margin was a forearc (Figure 2c). Collision of a microcontinent or island arc during the Devonian on the west side of the craton is "Chilenia" of Mpodozis and Ramos (1985). Forsythe (1982) shows that during the late Paleozoic and Triassic subduction, a complex array of small suspect terranes with oceanic affinities accreted to the allochthon edge. As previously mentioned, after the southernmost allochthon accreted to South America in the Permian-Triassic, the subduction zone moved oceanward. The subduction zone along the coast also moved oceanward. As a consequence, during the Jurassic to Early Cretaceous a single, continuous subduction zone came into being along the western margin of South America (Figures 3a, 3b).

The Cretaceous and post-Cretaceous history of the southern Andes is dominated by subduction. Forsythe (1982) describes a sizeable accretionary wedge built up in the arc-trench region.

DISCUSSION

Several highly productive hydrocarbon-bearing basins were developed during tectonic evolution of the Andes. The basins in each of the three major Andean subdivisions are distinctly different in structural style and history. The following discussion shows how plate and regional tectonic events can be used to understand, interpret, or predict broad aspects of basin structure and subsidence for each of the major subdivisions. It provides constraints for basins models but does not give detailed interpretation of structure and stratigraphy of individual basins.

Northern Andean Basins

The plate and regional tectonic analyses suggest two structural grains for the northern sub-Andean basins: (1) a roughly east-west trend resulting from Early Cambrian extension (Figures 2a, 5b) and a roughly north-northeast trend resulting from Paleozoic, possibly Silurian, extension (Figures 2b, 5c). As expected, these trends in the Magdelena, Eastern Andes, Llanos, and Baritas areas were activated by way of the numerous continental collisions, microplate collisions, and island-arc accretions, as well as the subduction events which came later.

The sub-Andean basins of the northern Andes began as extensional basins accompanied by tectonic stretching subsidence. They were closer to the continental edge in the early Paleozoic than they are today and should have been sensitive to even subtle changes in marine conditions. The basins were on continental margins subsiding due to cooling and to sediment and water loading, as described by Watts and Ryan (1976), McKenzie (1978), and Steckler and Watts (1978). This condition seems to have persisted through the Devonian (Figures 2c, 5d).

The most profound event that affected the northern margin was the Carboniferous collision (Figures 2d, 5e) with North America and more or less synchronous eastward movement of the Central Cordillera block to the west. The stress intensity or mean stress was high enough to cause regional metamorphism. In sub-Andean basins floored by Precambrian crystalline rock, the expected result is major compressive, basement-involved structures with a variety of directions. Possibly the most active compressional structures would be associated with northeast or east-west basement trends. The exact nature of these structures is difficult to predict with confidence from the First Principles of regional and plate tectonics. The main problem is that the stress was not strongly anisotropic, so structural trends and their characteristics are difficult to discern and evaluate.

The basins became flexural as a result of this collision. Their subsidence was driven by the tectonic loading of the advancing wedge in the collision zone over the downflexing edge of the buried Precambrian craton. Later, sediment load further downflexed the basins.

A second significant but less profound event was the combined Late Cretaceous collisions to the north with the Guajira-Cordillera de la Costa terrane and to the west with the Western Cordillera (Figures 3d, 6c). The effects of these collisions were similar to, but weaker than, the earlier collision; the intensity of the collision itself was less and the sub-Andean basins were protected in part because the Central Cordillera and Cauca Patia terranes absorbed some of the collisional forces.

The trends associated with the third peak event affecting the northern Andes are pronounced and decipherable. The collision with the coastal terrane in the Neogene led not only to the uplift of the Andes but also to effects in the adjacent basins.

A block-tectonic model can be used to help visualize the tectonic events. Summarized on Figure 7 are the basement-inherited zones of weaknesses or structural trends resulting from Paleozoic extensional tectonic events (Figure 7a). The zones were reactivated by collision of the coastal terrane combined with motions of the Nazca, Caribbean, and South American plates (Figure 7b). Predicted fault activity along structural trends would be as follows:

North-northeast (Paleozoic)	Right slip
East-west (Paleozoic)	Left slip
East-northeast (Andean)	Extension (normal faults)
North-northwest (Andean)	Compression (reverse faults)

Andean fault trends in the region of stress concentration (Figure 7b) would be deflected, depending on the orientation of the local stress trajectories. Structures would be intensified, depending on the relative magnitude of the stress intensity. The dominant style of deformation in the eastern part of the stress concentration is thrusting, and this is verifiable on geologic and tectonic maps (e.g. Martin, 1978).

Central and Southern Andean Basins

The origin of the central sub-Andean basins is enigmatic, because suspect terranes are not known but may exist in the central Andes. If suspect terranes are present, the central

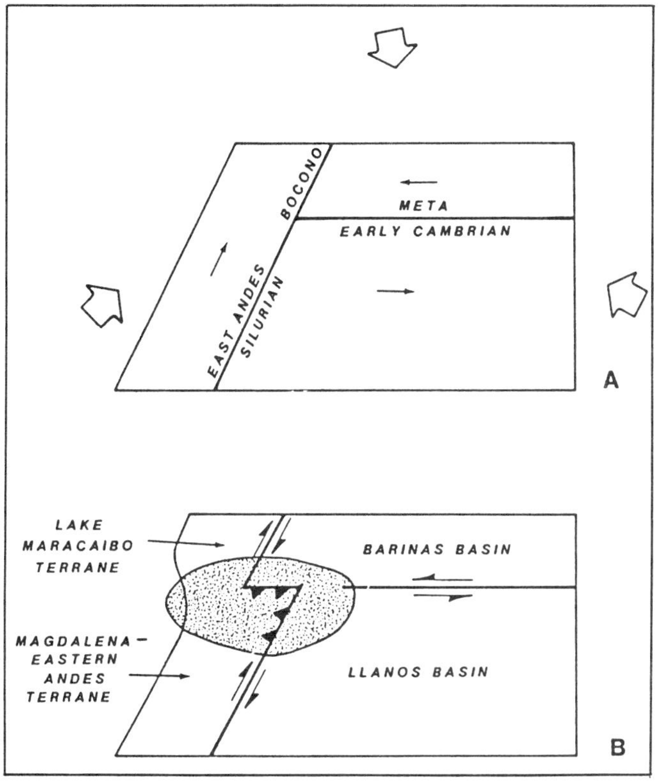

Figure 7. Simplified schematic of the tectonic constraints on the western Llanos and Barinas basins: (a) before 9 Ma and (b) Present. Shading shows area of stress concentration.

Andes fits the East Pacific-suspect-terrane mold and the enigma is explicable. If suspect terranes are not present in the central Andes, another explanation is needed. In this case, four possibilities exist: (1) plate-tectonic denudation, whereby suspect terranes that were accreted to the continental margin were subsequently moved away along strike-slip faults; (2) subducted terranes which were former suspect terranes later subducted; (3) shadow zone such that suspect terranes were somehow intercepted during their journey on the Kula, Farallon, Pacific, and other plates converging on the central Andean margin; and (4) barren spawning ground, whereby no suspect terranes were spawned in the appropriate location to accrete against the central Andean margin.

Whatever the case, the implications for the structural and subsidence history of the central sub-Andean basins are profound. Plate-tectonic reconstruction suggest that at some time in the latest Precambrian or in the early Paleozoic western South America separated from an adjacent continent, possibly North America (Figure 2). Basement trends resulting from these extensional events should follow roughly along the cratonic edge, which seems to be north-northwest for Perú and north-south for northern Chile. The well-known east-west "deflection" trends may be related to this extensional event, or they may be related to the Amazon depression which separated the exposures of Archean cratons. Initial subsidence, then, should be Paleozoic extensional tectonic subsidence followed by cooling, sediment and water loading subsidence. The sub-Andean basins should have been closer to the continental margin because enormous additions of later igneous rock began to separate the basins from the ocean margin. The sub-Andean basins were in a backarc position for the remainder of the Phanerozoic. There were major compressional pulses in the Paleozoic which may have resulted in some flexural subsidence. For example, there was folding, intrusion, and low-grade metamorphism in the Eastern Cordillera in the Late Devonian and Early Carboniferous (Cobbing, 1985). The plate-tectonic cause of this event remains a mystery.

During the compressional episodes, the north-northwest and east-west trends would have been roughly perpendicular to the direction of greatest compression and would have been activated as reverse faults. The various deflections would have been in the correct position to be activated as strike-slip faults.

As backarc basins, subsidence due to crustal extension is expected in the sub-Andean basins. The Wyoming foreland of the western United States could be used as a typical model, which could be generalized using techniques discussed by Beaumont (1981). Sea-level transgressions and regressions should be strong influences on the stratigraphy if the model is valid (see, for example, Vail, Mitchum, and Thompson, 1977; Van Hinte, 1978). However, Withjack and Gallagher (1983) show that sea-level fluctuations alone may not be sufficient to explain extensional basin stratigraphy. They document unconformities which can be related to plate-tectonic events.

During the uplift of the Andes a major episode of flexural downwarp of the basins adjacent to and east of the Andes began and thick sediment piles enhanced the downflexing. The tectonic wedge of the central Andes may have encroached as much as 100 km over the Precambrian basement slab (Lyon-Caen, Molnar, and Suarez, 1985). This encroachment resulted in a fold and thrust belt much like the well-known Montana-Idaho-Wyoming-Utah thrust belt. The sub-Andean basins in the northern part of the central Andes should be analogous to the western United States fold and thrust belt and Wyoming foreland.

The sub-Andean basins in the southern Andes and the southern part of the central Andes are predominantly backarc and forearc basins. The sub-Andean basins of the southern part of the central Andes were backarc basins formed during development of the ancestral Andes. Extensional subsidence, thermal cooling, and sediment load were responsible for basin subsidence.

The sub-Andean basins on the southern Andes had histories identical to those of the southern part of the central Andes except for the collision event of the latest Permian to earliest Triassic. This collision must have had at least local effect on basin structuring and subsidence.

CONCLUSIONS

The synthesis presented in this chapter links plate tectonics to regional structures, which in turn have been used to explain some features of basin history in the Andean region. The following conclusions can be made:

1. Differences in plate-tectonic evolution of the northern, central, and southern Andes are reflected in the regional tectonic patterns. This linking of global and regional processes enhances our understanding of the geology of the region.

2. The northern and western margins of the Precambrian craton were established during the latest Precambrian to early Paleozoic by extensional plate-tectonic events. Structural trends on the adjacent craton roughly parallel the margins. These trends were reactivated during later collisional events, especially Devonian, early Carboniferous, and Andean. This explains the origin of structures with these trends.
3. Palinspastic restoration of northwestern South America for Tertiary time shows that rich Mesozoic, marine hydrocarbon source beds now in the Lake Maracaibo basin were downdip from the Llanos and Barinas basins during the Miocene. This suggests the possibility that some Llanos oil reservoirs were filled by Maracaibo-sourced oil.
4. The eastern Cordillera fault appears to be a strike-slip fault in Andean time, thus challenging previous overthrust interpretations.
5. The stress concentration at the junction of the Eastern Cordillera fault and the Meta zone contributed to the development of faults in the Mesozoic and Cenozoic rocks in adjacent basins. The stress concentration concept can be used to explain the history of development of these faults.

ACKNOWLEDGMENTS

I am grateful to Occidental International Exploration and Production Company for granting permission to publish this work.

REFERENCES CITED

Baldis, B. A. J., and O. L. Bordonaro, 1982, Comparación entre el Cámbrico de la "Great Basin" norteamericana y la precordillera de San Juan, Argentina, su implicancia intercontinental: Quinto congreso latinoamercano de geología, Argentina, ACTASI p. 97-108.

Beaumont, C., 1981, Foreland basins: Geophysical Journal of the Royal Astronomical Society, v. 65, p. 291-329.

Beets, D. J., W. V. Maresch, G. T. Klover, A. Mottana, R. Bocchio, F. F. Beunk, and H. P. Monen, 1984, Magnetic rock series and high-pressure metamorphism as constraints on the tectonic history of the southern Caribbean: Geological Society of America Monograph 162, p. 95-130.

Bond, G. C., P. A. Nickeson, and M. A. Kominz, 1984, Breakup of a supercontinent between 625 Ma and 555 Ma: new evidence and implications for continental histories: Earth and Planetary Science Letters, v. 70, p. 325-345.

Bonini, W. E., J. D. Garing, and J. M. Kellogg, 1982, Late Cenozoic uplifts of the Maracaibo-Santa Marta block, slow subduction of the Caribbean Plate, and results from a gravity study: Ninth Caribbean Geological Conference Transactions, p. 99-105.

Breitkreuz, C., and W. Zeil, 1984, Geodynamic and magmatic stages on a traverse through the Andes between 20° and 24°S (N. Chile, S. Bolivia, N.W. Argentina): Journal of the Geological Society of London, v. 141, p. 861-868.

Burke, K., L. Delano, J. Dewey, A. Edelstein, W. Kidd, K. Nelson, A. Sengor, and J. Stroup, 1978, Rifts and sutures of the world: NASA Contract #NA55-24094, 238 p.

Cobbing, E. J., 1985, The tectonic setting of the Peruvian Andes, *in* Magmatism at a plate edge, the Peruvian Andes: New York, John Wiley and Sons, p. 3-12.

Coney, P. J., D. L. Jones, and J. W. H. Monger, 1980, Cordilleran suspect terranes: Nature, v. 288, p. 329-333.

Corvalan, J., 1985, Geologic-tectonic framework of the Andean region: This conference.

Dalziel, I. W. D., 1981, Back-arc extension in the southern Andes: overview and critical appraisal: Philosophical Transactions, Royal Society of London, v. A300, p. 319-335.

Dewey, J. F., and J. L. Pindell, 1985, Neogene block tectonics of eastern Turkey and northern South America: continental applications of the finite difference method: Tectonics, v. 4, p. 71-83.

Etayo, S. F., 1985, El armazón estratigráfico tectónico de los andes colombianos: This conference.

Forsythe, R., 1982, The late Paleozoic to early Mesozoic evolution of southern South America: a plate tectonic interpretation: Journal of the Geological Society of London, v. 139, p. 671-682.

Gansser, A., 1973, Facts and theories on the Andes: Journal of the Geological Society of London, v. 129, p. 3-131.

Gallagher, J. J., 1981, Tectonics of China: continental scale cataclastic flow: American Geophysical Union Geophysical Monograph 24, p. 259-273.

Gallagher, J. J., 1984, Large-scale faults and the tectonics of China: Fourth International Conference on Basement Tectonics Proceedings, Oslo, Norway, p. 173-185.

Gallagher, J. J., in press, a, Philippine Islands: a tectonic railroad siding: American Association of Petroleum Geologists, Pratt Memorial Conference Proceedings, Tucson, Arizona, 23 p.

Gallagher, J. J., in press, b, Strength of basement rock in Egypt: Fifth International Conference on Basement Tectonics Proceedings, Cairo, Egypt, 35 p.

Gust, D. A., K. T. Biddle, D. W. Phelps, and M. A. Uliana, 1985, Associated middle to late Jurassic volcanism and extension in southern South America: Tectonophysics, v. 116, p. 223-253.

Howell, D. G., E. R. Schermer, D. L. Jones, Z. Ben-Avraham, and E. Scheibner, 1985, Preliminary tectonostratigraphic terrane map of the Circum-Pacific Region: American Association of Petroleum Geologists, Tulsa, Oklahoma, scale 1:17,000,000.

Irving, E., 1981, Phanerozoic continental drift: Physics of the Earth and Planetary Interiors, v. 24, p. 197-204.

Johnson, M. E., J.-Y. Rong, and X.-C. Yang, 1985, Intercontinental correlation by sea-level events in the Early Silurian of North America and China: Geological Society of America Bulletin, v. 96, p. 1384-1397.

Kellogg, J. N., and W. E. Bonini, 1982, Subduction of the Caribbean Plate and basement uplifts in the overriding South American Plate: Tectonics, v. 1, p. 251-276.

Krause, H. H., and K. H. James, 1985, Subandian hydrocarbon resources of Venezuela: This conference.

Kulm, L. D., W. J. Schweller, and A. Masias, 1977, A preliminary analysis of the subduction process along the Andean continental margin, 6° to 45°S: American Geophysical Union Ewing Series I, p. 285-301.

Lyon-Caen, H., P. Molnar, and G. Suarez, 1985, Gravity anomalies and flexure of the Brazilian shield beneath the Bolivian Andes: Earth and Planetary Science Letters, v. 75, p. 81-92.

Maresch, W. V., 1974, Plate tectonic origin of the Caribbean mountains system of northern South America: Geological Society of America Bulletin, v. 85, p. 669-682.

Martin, F. C., 1978, Mapa tectónico Norte de America del sur: Ministerio de Energía y Mines, República de Venezuela, 2 sheets.

McCourt, W. J., J. A. Aspden, and M. Brook, 1984, New geological and geochronological data from the Colombian Andes: continental growth by multiple accretion: Journal of the Geological Society of London, v. 141, p. 831-845.

Mpodozis, C., and V. Ramos, 1985, The Andes of Chile and Argentina: This conference.

McKenzie, D., 1978, Some remarks on the development of sedimentary basins: Earth and Planetary Science Letters, v. 40, p. 25-32.

Nelson, E. P., I. W. D. Dalziel, and A. G. Milnes, 1980, Structural geological of the Cordillera Darwin-collisional-style orogenesis in the southernmost Chilean Andes: Eclogae geolog. helv., v. 73, p. 727-751.

Nur, A., and Z. Ben-Avraham, 1981, Volcanic gaps and the consumption of aseismic ridges in South America: Geological Society of America Memoir 154, p. 729-740.

Pindell, J. L., 1985, Alleghenian reconstruction and subsequent evolution of the Gulf of Mexico, Bahamas and Proto-Caribbean: Tectonics, v. 4, p. 1-39.

Pitcher, W. S., 1982, Granite type and tectonic environment, *in* K. J. Hsu, ed., Mountain building processes: New York, Academic Press, p. 19-40.

Pitcher, W. S., and E. J. Cobbing, 1985, Epilog, *in* Magmatism at a plate edge, the Peruvian Andes: New York, John Wiley and Sons, p. 285-289.

Scotese, C. R., R. K. Bambach, C. Barton, R. Van Der Voo, and A. M. Ziegler, 1979, Paleozoic basemaps: Journal of Geology, v. 87, p. 217-277.

Smith, A. G., A. M. Hurley, and J. C. Briden, 1981, Phanerozoic paleocontinental world maps: Cambridge University Press, 102 p.

Steckler, M. S., and A. B. Watts, 1978, Subsidence of the Atlantic-type continental margin off New York: Earth and Planetary Science Letters, v. 41, p. 1-13.

Thornburg, T., and L. D. Kulm, 1981, Sedimentary basins of the Perú continental margin: Structure, stratigraphy, and Cenozoic tectonics from 6 °S to 16 °S latitude: Geological Society of America Memoir 154, p. 393-422.

Vail, P. R., R. M. Mitchum, Jr., and S. Thompson, 1977, Seismic stratigraphy and global changes of sea level, Part 4: Global cycles of relative changes of sea level: American Association of Petroleum Geologists Memoir 26, p. 84-97.

Van Hinte, J. E., 1978, Geohistory analysis—application of micropaleontology in exploration geology: American Association of Petroleum Geologists Bulletin, v. 62, p. 201-222.

Vicente, J. C., 1985, Early Late Cretaceous overthrusting in the Western Cordillera of central Perú: This conference.

Watts, A. B., and W. B. F. Ryan, 1976, Flexure of the lithosphere and continental margin basins: Tectonophysics, v. 36, p. 25-44.

Windley, B. F., 1984, The evolving continents: New York, John Wiley & Sons, 399 p.

Withjack, M. O., and J. J. Gallagher, 1983, The rifted basins of eastern India: SEAPEX Proceedings v. VI, p. 41-57.

Ziegler, A. M., J. T. Parrish, and C. R. Scotese, 1981, Cambrian world paleogeography: U.S. Geological Survey Open-File Report 81-743, 252 p.

Ziegler, A. M., C. R. Scotese, and S. F. Barrett, 1983, Mesozoic and Cenozoic paleogeographic maps, *in* P. Brosche and J. Sundermann, eds., Tidal friction and the Earth's rotation: Berlin, Springer-Verlag, p. 240-252.

Zoneshain, L. P., L. A. Savostin, A. P. Sedov, and L. P. Voloditina, 1985, Paleogeodynamics world base maps and paleobathymetry for the last 70 Ma: an explanatory note: Tectonophysics, v. 116, p. 189-207.

Ericksen, G. E., Cañas Pinochet, M. T., and
Reinemund, J. A., editors, 1989, Geology of the Andes
and its relation to hydrocarbon and mineral resources:
Houston, Texas, Circum-Pacific Council for Energy
and Mineral Resources Earth Science Series, v. 11.

CHAPTER 4

RIDGE COLLISIONS ALONG THE PLATE MARGINS OF SOUTH AMERICA COMPARED WITH THOSE IN THE SOUTHWEST PACIFIC

H. Gary Greene and Florence L. Wong
U.S. Geological Survey
Menlo Park, California, U.S.A

ABSTRACT

The ocean-continent collision along the Andean margin of western South America has brought ridges and spreading centers into contact with the continent. Several ridges and ridge complexes on the Nazca plate are being subducted beneath the South American plate. Current publicly available data reveal neither present-day continental margin fragmentation nor arc fragmentation such as that observed in the southwest Pacific. However, more detailed study in the future may reveal such fragmentation along the South American margin, which may have played a role in the accumulation of hydrocarbons as it has in the southwest Pacific.

When large bathymetric highs—such as ridges and spreading-ridge systems—collide with continents and island arcs at subduction zones, some structural trends may be disrupted. Collisions of ridges with arcs in the southwest Pacific commonly fragment the arcs into small blocks (50-100 km on a side) between arc-transverse faults. This effect is well demonstrated in the New Hebrides arc in the southwest Pacific and to a lesser degree in the adjoining Solomon Islands and Tonga arcs. The collision of the D'Entrecasteaux zone with the central New Hebrides arc, the Woodlark spreading-ridge system with the central Solomon Islands arc, and the Louisville Ridge with the Tonga arc have produced various fragmentation features. Minor effects of ridge-arc collision include an increase in displacement along preexisting arc-transverse faults. Major effects of ridge-arc collision include disruption of arc-parallel faults, folds, or basins, fault rupture to tap primitive magma sources at mantle depths, and development of local wrench tectonics.

In the island arcs of the southwest Pacific, where large sediment packages have accumulated, arc fragmentation improves the possibilities for hydrocarbon generation and accumulation by deepening basins and creating more structural traps. If such arc environments prove to be good prospects in the southwest Pacific, then similar ridge subduction zones may be hydrocarbon-producing candidates along the Andean margin.

INTRODUCTION

The Cocos, Nazca, and Antarctic plates are converging with the South American plate along the western (Andean) margin of Central and South America (Figure 1) at the relative rate of 9 cm/yr. This convergence has produced the greatest relief anywhere in the world with the formation of the Perú-Chile Trench (as deep as 8000 m) and the Andean mountain range (maximum height slightly more than 7000 m). Nazca plate is being subducted throughout the region extending from southern Colombia, through Ecuador and Perú to southern Chile (Figure 1).

Collision and subduction of oceanic ridges along convergent plate margins have long been associated with established or suspected vertical movements of arc and forearc areas and with the disruption of magmatic activity, seismicity, and trench continuity (e.g., Barazangi and Isacks, 1976; Vogt et al., 1976; Marsh, 1979; Nur and Ben-Avraham, 1981; see also summary by Taylor and Exon, 1987). In addition to these effects, ridge-oceanic arc collisions cause small-scale structural disruption by arc-transverse faults that can progress to include wrench tectonic features.

Unlike the ocean-island arc converging plate boundaries in the southwest Pacific, the Andean margin is an ocean-continent margin (Chinn and Isacks, 1983). The Benioff zone under the Andean margin has relatively shallow (< 30 °) dips,

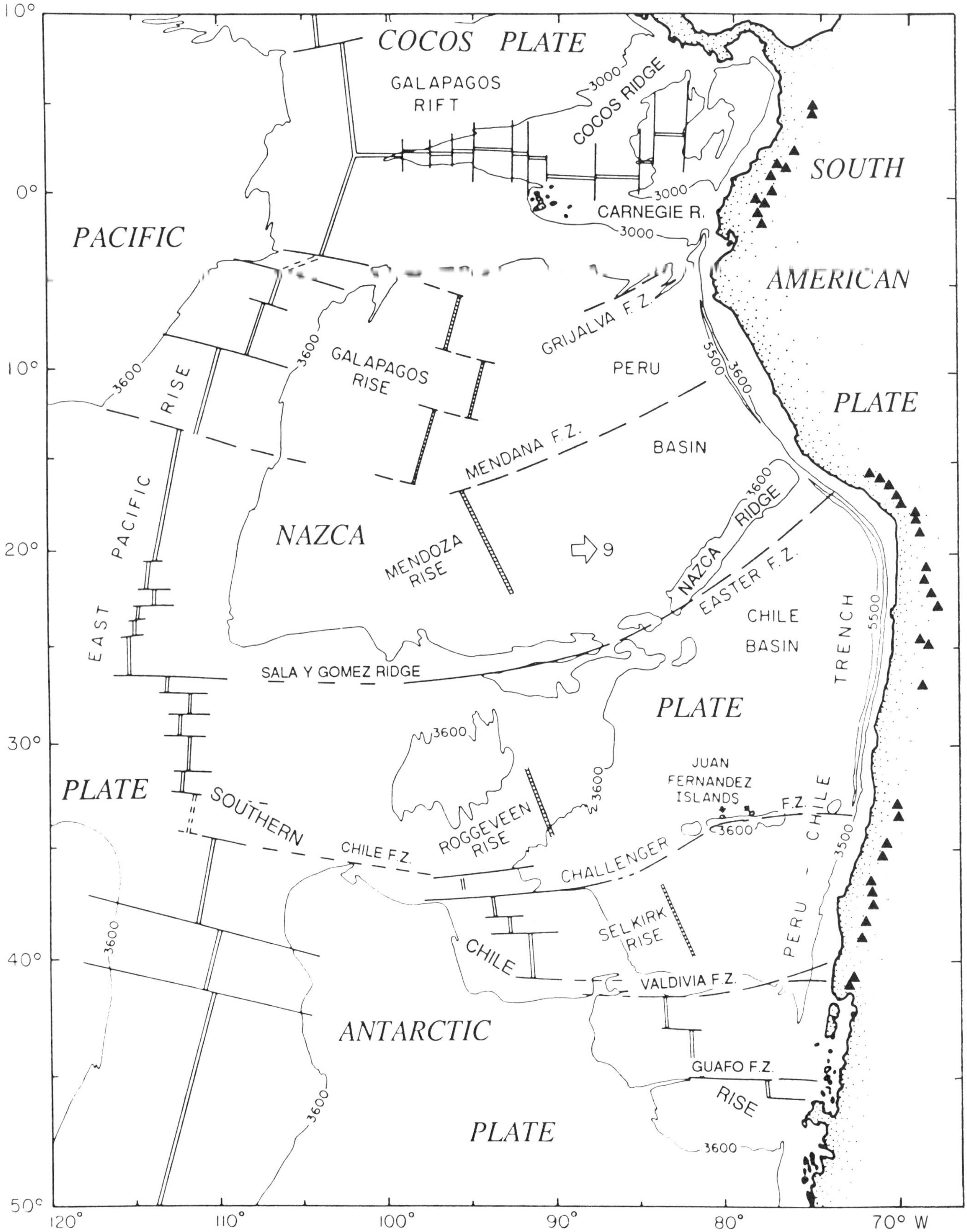

Figure 1. Regional tectonic features of the Nazca plate (modified from Mammerickx, Herron, and Dorman, 1980; Schweller, Kulm and Prince, 1981; Drummond et al., 1984). Solid triangles locate active volcanoes (from Nur and Ben-Avraham, 1981).

which steepen at depth across a broad downdip width (200 km vs. 80 km in the New Hebrides) (Chinn and Isacks, 1983). Rates of plate convergence are comparable to those in the southwest Pacific, about 9 to 10 cm/yr (Minster et al., 1974; Chase, 1978; Dubois et al., 1978; Drummond et al., 1984). Large-scale (~ 1000 km) segmentation of magmatic arc activity has been defined by gaps in lines of active volcanoes (Figure 1; Hall and Wood, 1985; Nur and Ben-Avraham, 1981). Cretaceous and Cenozoic sedimentary basins have been mapped off Ecuador, Perú, and Chile (Figure 2; Thornburg and Kulm, 1981; Rosania-S., this volume). Many of the basins formed between the Cretaceous and Miocene and were subsequently structurally deformed during the Neogene and Quaternary. Some of these basins have produced commercial quantities of oil and gas (Travis, Gonzales, and Pardo, 1976; Shepherd and Moberly, 1981). Major ridges that are presently colliding with the Andean margin and which might contribute to margin fragmentation are, from north to south, the Carnegie Ridge off Ecuador, the Nazca Ridge off Perú, and the Juan Fernandez Ridge and the southern Chile rise off Chile (Figure 1). If continental margin fragmentation enhances hydrocarbon prospects, as we believe it does, then exploration of such areas should be encouraged.

Detailed investigations along the summit platforms of the Solomon Islands and New Hebrides arcs show that the sediment-filled intra-arc basins were deepened and fragmented as the result of collisions with the Woodlark spreading-ridge system and the D'Entrecasteaux zone of ridges. Similar effects resulted from the collision of the Louisville Ridge with the Tonga arc. On the basis of the sparse seismic-reflection data currently available, most collisions of ridges or ridge complexes with continents either do not seem to result in margin fragmentation or the degree of fragmentation is considerably less than we observe in the southwest Pacific. Nevertheless, the eastern margin of the Pacific, where the Nazca Ridge and the southern Chile rise are colliding with the Andean margin, is a region where fragmentation may have taken place.

In 1982 and 1984, the U.S. Geological Survey in cooperation with the Australian Bureau of Mineral Resources, the New Zealand Geological Survey (known as the ANZUS Tripartite partners) and the Committee for Coordination of Joint Prospecting for Mineral Resources in South Pacific Offshore Areas (CCOP/SOPAC) undertook investigations for hydrocarbon resources in the offshore areas of Tonga, Vanuatu, Solomon Islands, and Papua New Guinea. These investigations provided new data that have enhanced our understanding of the geologic framework and tectonic evolution of island arcs and arc-ridge collisional processes.

The purpose of this paper is to describe and compare ridge-plate margin collision zones along the Andean margin and in the southwest Pacific. Where the ridges are colliding, the trench is obscure or nonexistent, distinct seismic and volcanic gaps occur along the arc, and the forearc is uplifted. In the southwest Pacific, but generally absent along the Andean margin, there are disrupted plate margin structures or arc fragmentation that may be related to ridge collision. The factors affecting the magnitude of arc fragmentation or whether or not it occurs are unclear. The rate of plate convergence, dip of the subducting slab, impingement angle[1], displacement of the impact point[2], seismic vs. aseismic ridges, and size of the subducting ridge or ridge complex are possible factors. The following descriptions of these features, which reveal diverse characteristics for each example, serve as a basis for future inquiry.

This paper draws heavily upon the results and conclusions of Couch and Whitsett (1981), Nur and Ben-Avraham (1981), Thornburg and Kulm (1981), Forsythe and Nelson (1985), Hall and Wood (1985), and Cande and Leslie (1986) for the Andean margin. The discussion of the geology of the southwest Pacific is drawn from the results of the Tripartite-CCOP/SOPAC investigations (Scholl and Vallier, 1985; Vedder, Pound, and Boundy, 1986; Taylor and Exon, 1987; Greene and Wong, 1988).

WESTERN SOUTH AMERICA

Carnegie Ridge

The Carnegie Ridge is a broad (~ 250 km) ridge that stands about 1500 m above the surrounding seafloor (Figure 1). The ridge is a continuous feature with little variation in width or height. When the Carnegie Ridge first collided with Ecuador is unknown, but probably this was no earlier than ~24 Ma when the Nazca plate formed. Its impingement angle off Ecuador is ~ 10°. The Guayaquil Peninsula onshore of the Carnegie Ridge is uplifted. Hall and Wood (1985) attribute the increased uplift of the Andes in this region to the subduction of the Carnegie Ridge. This uplift is similar to the that of Espiritu Santo Island in Vanuatu (see below) which resulted from the collision of the D'Entrecasteaux zone of ridges (Taylor, Jouannic and Bloom, 1985). The Progreso basin (Figure 2), opposite the Carnegie Ridge, is a pull-apart basin filled with 6000 m of sediment and seems unaffected by recent ocean ridge subduction (Shepherd and Moberly, 1981).

Nazca Ridge

The Nazca Ridge (Figure 1) is a broad (~ 250-km-wide) continuous aseismic ridge that rises to as much as 4000 m above the adjacent seafloor. It first collided with the South American plate at ~15 Ma (Nur and Ben Avraham, 1981). The Nazca Ridge has an impingement angle of nearly 40° (Drummond et al., 1984), and the eastward plate motion causes the impact point (which is presently located at about 15°S) to sweep southward with time. The ridge is being subducted along a downgoing slab having a shallow (~ 12°-20°) dip (Couch and Whitsett, 1981; Chinn and Isacks, 1983; Suarez, Molnar, and Burchfiel, 1983). Old sedimentary forearc shelf and slope basins in the area (Thornburg and Kulm, 1981) have been little affected by the subduction of the Nazca Ridge.

Juan Fernandez Ridge

Two major areas of ridge collisions occur along the Pacific margin of Chile, one is the aseismic Juan Fernandez Ridge and the other is the southern Chile rise (Figure 1). The Juan Fernandez Ridge is colliding with an impingement angle of

[1]We consider the impingement angle to be the acute angle between the strike of the subducting ridge and the plate convergence direction.

[2]The impact point is the point of intersection of the subducting ridge and the plate margin.

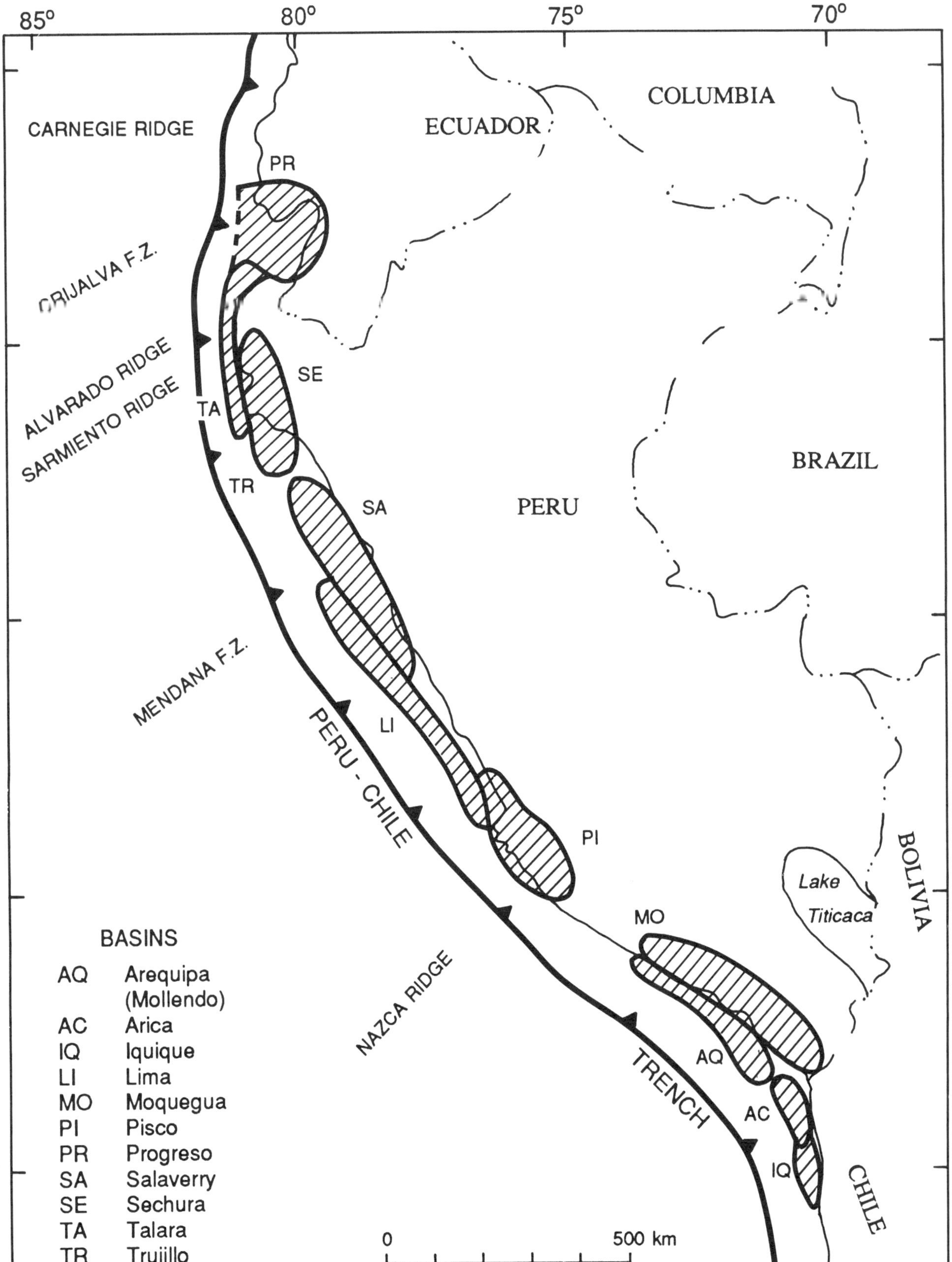

Figure 2. Major sedimentary basins along the Andean margin and ridges and fracture zones on the Nazca plate (modified after Coulbourn, 1981; Kulm et al., 1981; Shepherd and Moberly, 1981).

about 5°, which causes a slight northward migration of the impact point at the Perú-Chile Trench, just offshore from Valparaiso (Drummond et al., 1984). The ridge is discontinuous, averages nearly 50 km in width, and is composed of irregular seamounts having heights ranging from 1000 to 1500 m above the surrounding seafloor. The dip of the downgoing slab is shallow, about 15° (Chinn and Isacks, 1983; Hall and Wood, 1985). A sedimentary basin, the Valparaiso basin, is perched on the landward trench slope at a water depth of 2.5 km a little south of the impact point (Mordojovich, 1981).

Southern Chile Rise

The southern Chile rise, a spreading-ridge system, has been colliding with the Andean margin since 14 Ma (Cande and Leslie, 1986). The impingement angle of the spreading system is about 40° but that of the ridge segment that is currently being subducted is about 90°. Nevertheless, because of the general trend of the ridge, the impact point has migrated and will continue to migrate northward with time (Figure 1; Drummond et al., 1984; Cande and Leslie, 1986). The spreading system is a continuous, broad (>900 km wide) ridge with an average height of 1000 m above the adjoining seafloor. However, its offset spreading centers on the ridge are much narrower and stand higher above the seafloor (Cande and Leslie, 1986). The southern Chile rise is being subducted along with the shallowly dipping (12°-15°) oceanic plate (Chinn and Isacks, 1983; Hall and Wood, 1985).

The collision of the the southern Chile rise postdates formation of the major faults and basins in the vicinity of the Golfo de Penas (Figure 3) where seismic-reflection profiles indicate a complex pattern of differential uplift and subsidence that took place when the impact point was far south of its current position (Cande and Leslie, 1986). However, Cande and Leslie (1986) note that the southern Chile rise collision did cause uplift and plutonism on the continental margin adjacent to the impact point in late Neogene and Quaternary times. They conclude that tectonic activity on the continental margin is much greater when short oceanic ridge segments, offset by large transform faults, collide with the margin. These observations indicate the degree of tectonic impact an active colliding ridge or other bathymetric high might have on the margin of an island arc or continent.

A few sedimentary basins are strung out along this part of the Chilean margin. Although the Chiloe basin to the north of the Golfo de Penas has not been affected by the southern Chile rise collision, it may yet be found that the Madre de Dios and Golfo de Penas basins (Figure 3) were affected as the impact point migrated northward. Petroleum exploration is currently (mid-1980s) underway in this region because the basins contain thick (>2 sec, Cande and Leslie, 1986) sediments in a region of high heatflow.

SOUTHWEST PACIFIC

The Solomon Islands, New Hebrides, and Tonga arcs lie along a complex system of active volcanic arcs extending from Papua New Guinea southward through the Solomon Islands to Vanuatu (Figure 4) where the system is displaced eastward along the Hunter fracture zone through Fiji to Tonga. From Tonga, the arc system continues southward through the Kermadec Islands to New Zealand (Figure 4). This arc system defines a major tectonic boundary associated with the collision zone of the Australia-India and Pacific plates.

The basin and forearc sedimentary sequences of New Hebrides, Solomon Islands, and Tonga arcs are believed to have been part of a set of forearc depositional aprons or lenses associated with a late Oligocene (or earlier) to middle Miocene volcanic island arc system known as the outer Melanesian arc (Figure 5; Karig, 1970; Chase, 1971; Gill and Gorton, 1973; Falvey, 1975, 1978; James and Falvey, 1978). This island arc formed by westward subduction of the Pacific plate beneath the Australia-India plate at the Vityaz Trench. The early(?) Miocene collision and docking of the Ontong Java Plateau initiated rifting of the outer Melanesian arc that divided the region into the Solomon Islands arc, the New Hebrides arc, the Fiji platform, and the Tonga-Lau arc (Figure 6; Falvey and Greene, 1988; Wells, 1987).

In early late Miocene time, the arc system associated with the Vityaz subduction zone may have included one long linear forearc basin which consisted of (1) the northern island belt and Central Solomons Trough of the Solomon Islands, (2) the Eastern (island) Belt, Vanikolo basin, and Central Basin of Vanuatu, (3) Vanua Levu and Bligh Water basin of Fiji, and (4) the sedimentary Tonga platform (Figure 6; Falvey, 1978; Falvey and Greene, 1988). The volcanic arc of this system was made up of the southern island belt of the Solomon Islands, the Western (island) Belt of the New Hebrides arc, Viti Levu of Fiji, and the Lau Ridge. A reversal to eastward subduction in late Miocene time further modified these regional components. Active spreading in the North Fiji and Lau Basins continues to fragment the region (Figure 7).

Central New Hebrides Arc-D'Entrecasteaux Zone of Ridges Collision

The New Hebrides arc extends for a distance of 1700 km from the Santa Cruz Islands in the north to Matthew and Hunter Islands in the south (Figure 8). Along this arc the Australia-India plate is being subducted beneath the North Fiji Basin, thought to be a microplate that lies between the New Hebrides arc and the Pacific plate to the northeast (Figure 7; Chase, 1971; Kroenke, 1984). Structures in the central part of the arc (~14.5°-17°S) are of local rather than regional extent (Figure 8). For example, the New Hebrides Trench to the west shallows and loses much of its physiographic expression, similar to what occurs where the Nazca Ridge is colliding with South America. The arc platform consists of two rather than one island-bearing horst ridges (Western and Eastern Belts) with an intervening graben that underlies the deep (>3000 m) Central Basin, which contains as much as 5 or 6 km of sediment. Absent in this central area are backarc grabens such as the Vot Tande and Coriolis Troughs, which are continuous to the north and south (Collot, Daniel, and Burne, 1988). All of these features in the central arc coincide with the zone where the D'Entrecasteaux zone (DEZ) is colliding with the New Hebrides arc.

The DEZ is a broad (~100 km wide) aseismic zone composed of two fairly regular and continuous ridges, separated by intervening basins, that rise about 3000 m above the adjacent seafloor. For about the past 8-6 m.y., the ridges of the DEZ have been colliding with the central New Hebrides arc

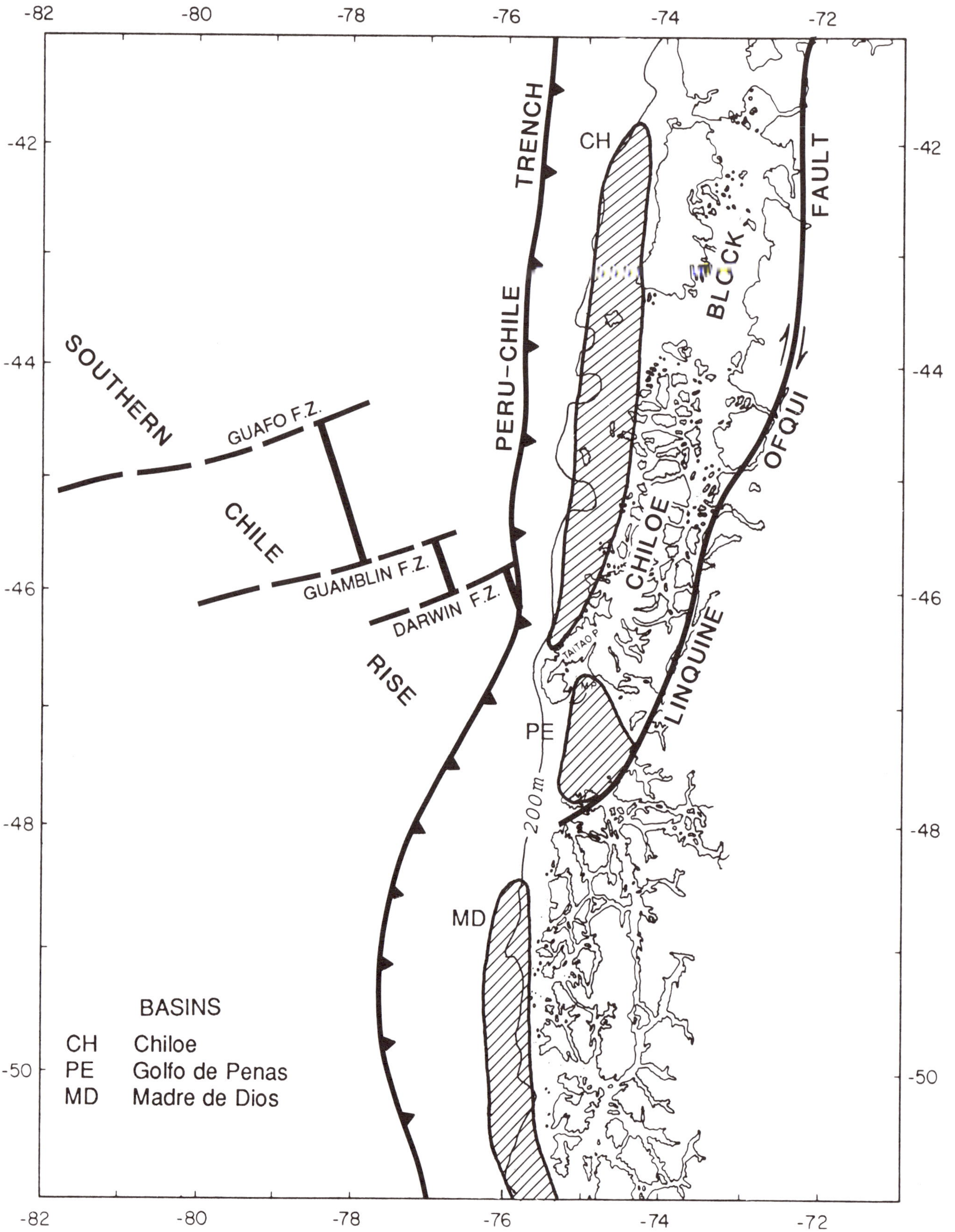

Figure 3. Structural elements in the region of collision of the southern Chile rise with the Perú-Chile Trench (modified from Cande and Leslie, 1986; Gonzales, this volume).

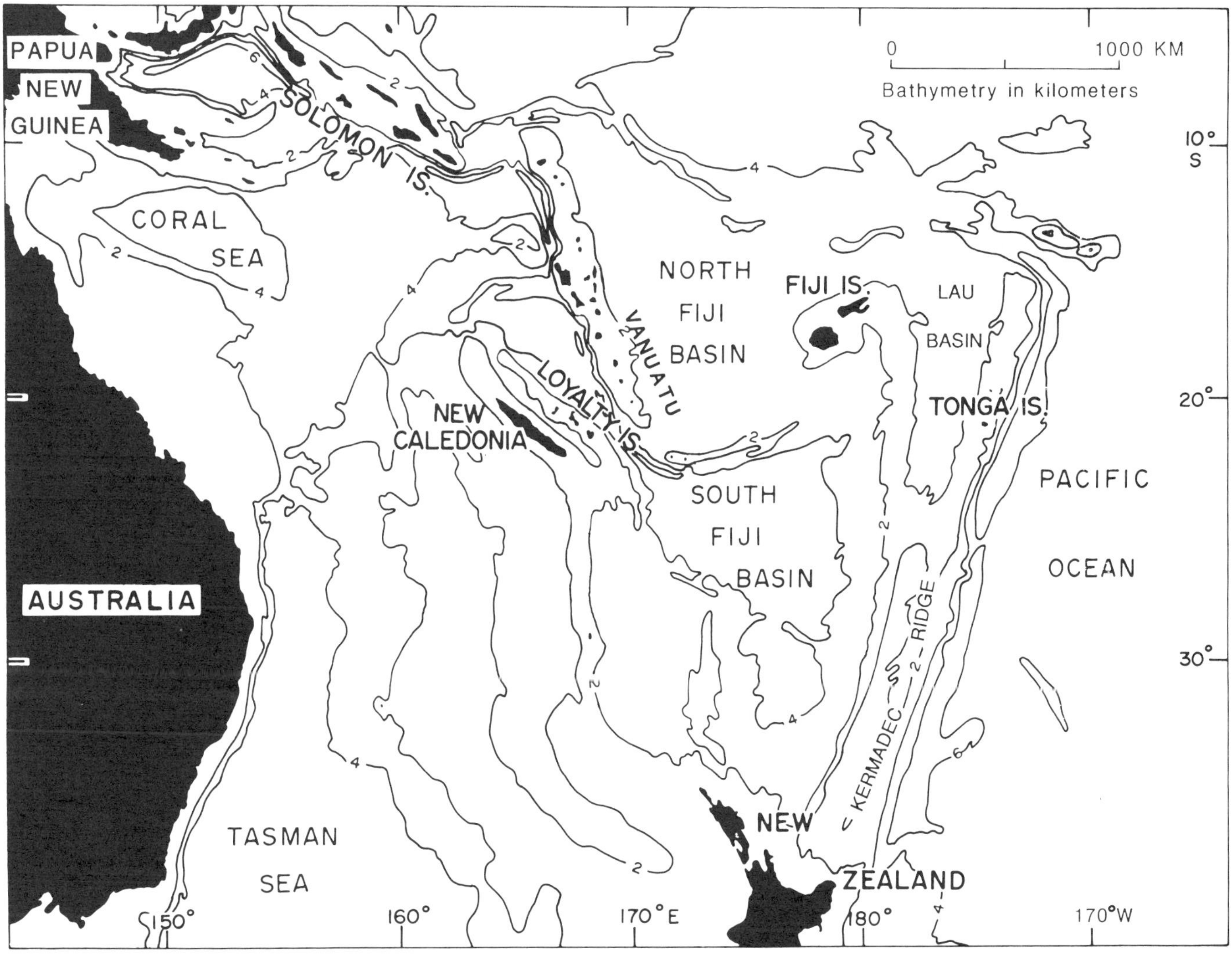

Figure 4. Map showing location of island arcs and basins in the southwest Pacific.

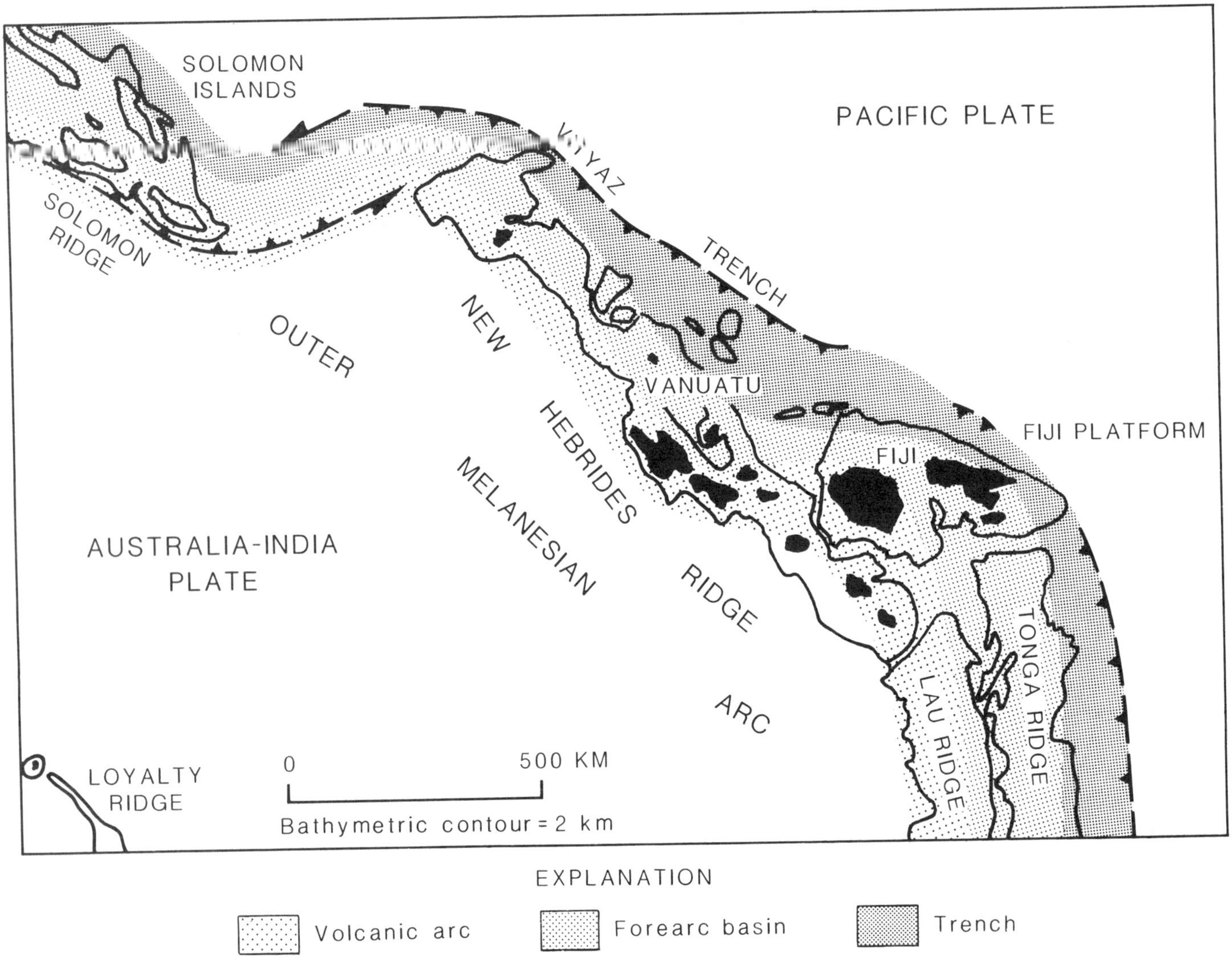

Figure 5. Middle and late Miocene tectonic reconstruction of the outer Melanesian arc showing the pre-fragmented Solomon-New Hebrides-Fiji-Tonga region (modified after Falvey, 1978; Falvey and Greene, 1988).

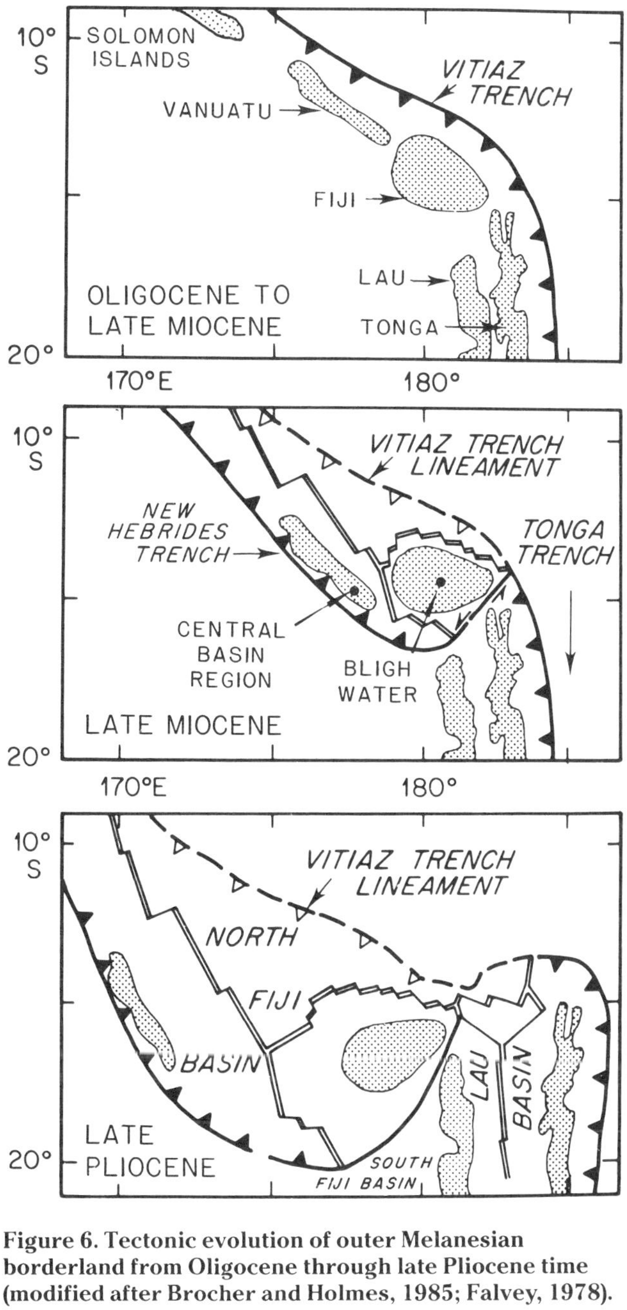

Figure 6. Tectonic evolution of outer Melanesian borderland from Oligocene through late Pliocene time (modified after Brocher and Holmes, 1985; Falvey, 1978).

(Greene et al., 1988). The relative motion between the two plates here is 10 cm/yr (Minster and Jordan, 1978) and the dip of the downgoing slab is steep (70 °-60 °) but variable (Dubois et al., 1978; Isacks et al., 1981; Dunkley, 1983; Louat, Hamburger and Monzier, 1988). The impingement angle is 5 °-20 ° where the DEZ is colliding with the islands of Espiritu Santo and Malakula (Dubois et al., 1978; Isacks et al., 1981). The impact point now migrates northward, but because the DEZ is arcuate with the westward segment striking northeast, the present-day direction of plate convergence will ultimately cause the impact point to migrate southward.

Although the structure of the New Hebrides arc represents a tectonic history that has been influenced by a shifting stress field for the past 30 m.y., the collision of the DEZ with the New Hebrides arc appears to have produced the most recent structures and is in the process of fragmenting the arc (Greene et al., 1988).

Detailed mapping by Greene and Johnson (1988) and Macfarlane et al. (1988) shows that in the Central Basin region, the summit platform of the arc is cut by major faults oriented parallel and oblique (arc-parallel and arc-oblique faults) to the arc axis, which are in turn offset by faults that cut transversely (arc-transverse faults) across the arc and exhibit transcurrent movement (Figure 9). The arc-parallel and arc-oblique faults are normal faults that bound horsts and grabens. For example, the North and South Aoba basins (subdivisions of the Central Basin) are grabens downdropped along major arc-parallel faults, which are truncated along the northern boundary of the North Aoba basin by the Santa Maria fault zone and along the southern boundary by the Aoba fault zone (both arc-transverse faults). The Aoba fault zone appears to converge with the Santa Maria fault zone in the backarc region east of Maewo Island. These converging fault zones bound a structural block known as the Aoba structural wedge (Figure 9; Greene et al., 1988).

Greene et al. (1988) defined other similar structural wedges in this region, which from south to north are: the Efate wedge bound by the Epi and Vulcan fault zones, the Malakula wedge bound by the Ambrym and Epi fault zones, and the Ambrym wedge bound by the Aoba and Ambrym fault zones. These wedges are 100-200 km on a side and may indicate wrench tectonics (Moody and Hill, 1956; Wilcox, Harding and Seely, 1973). The Ambrym wedge has rotated counterclockwise as a result of left-lateral strike-slip along the Ambrym fault zone (Greene et al, 1988). Right lateral movement along the Aoba fault zone has offset structural features of the North Aoba basin about 25 km to the east in relation to analogous structures of the South Aoba basin (Figure 9; Greene et al., 1988).

Greene et al. (1988) compared the structural evolution of the arc with the strain ellipsoid and concluded that subduction of the DEZ caused wrenching that displaced major structural elements. The preexisting arc-parallel normal faults continued to be active and probably reflect tension over the buoyant upbowing that results from subduction of lower-density bathymetric features such as the ridges and seamounts of the DEZ. Northward sliding of the DEZ down the subduction plane clogged the physiographic trench, and although its lateral collision appears to have caused the major transcurrent rifts across the arc platform that fragment the arc from south to north, the steepness of the subducting slab eliminates this possibility (Greene et al., 1988).

The two largest volcanic edifices in the arc, Aoba and Ambrym, lie along two major faults (Aoba and Ambrym fault zones) overlying that part of the downgoing slab carrying the DEZ (Greene and Johnson, 1988; Greene et al., 1988). Both Ambrym and Aoba contain voluminous primitive picritic arc lavas (Gorton, 1977; Crawford, Greene, and Exon, 1988). Greene et al. (1988) speculated that impact of the DEZ, which stands 3 km higher than the surrounding ocean crust, caused uplift of Espiritu Santo Island (Taylor, Jouannic, and Bloom, 1985) and mantle-deep fracturing of the brittle crust of the overriding plate. Arc magmas, generated by "normal" arc-magma processes above the subducted slab found easy access to the surface along these deep faults, and erupted after only minimal fractionation in transit to the surface.

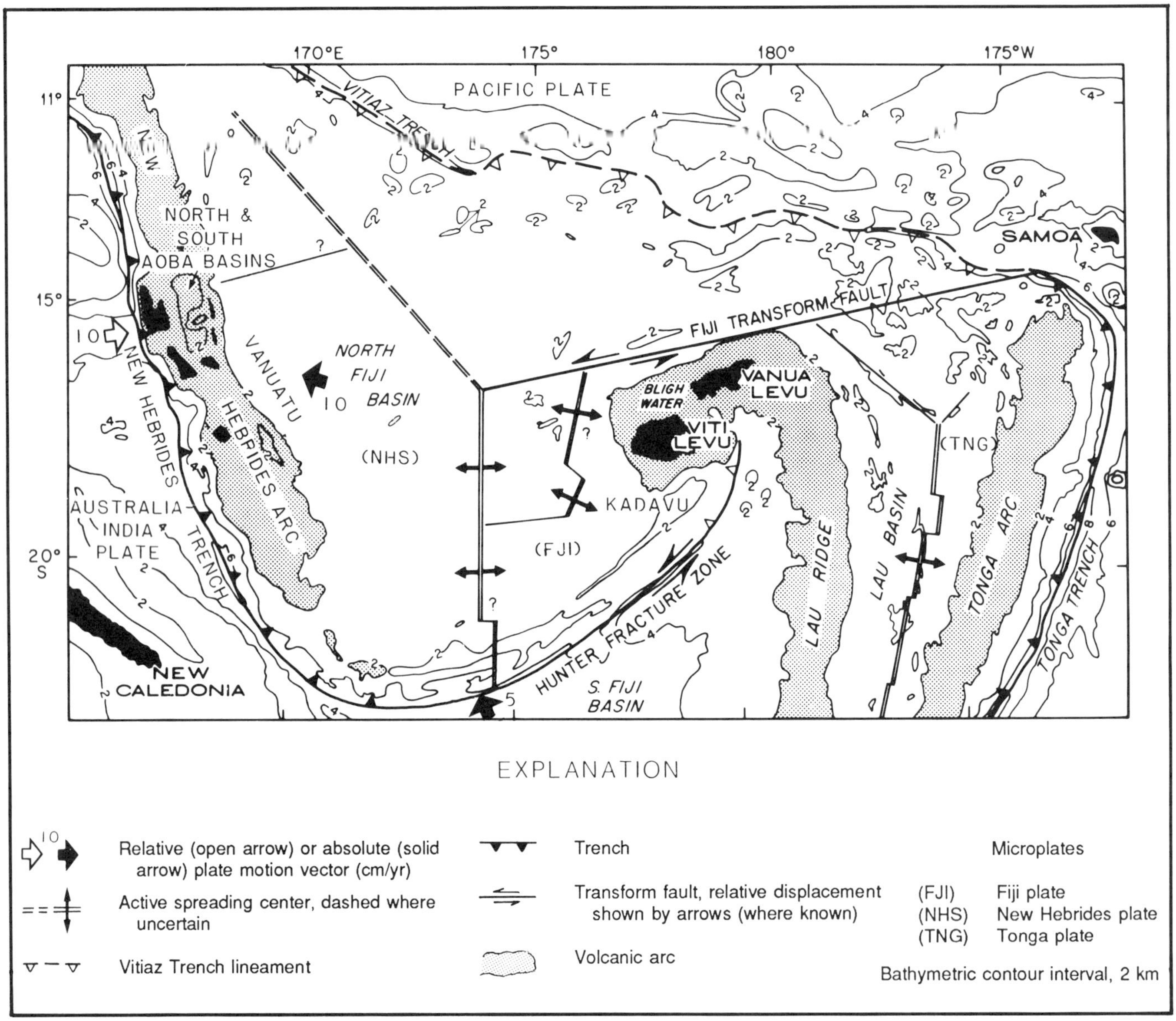

Figure 7. Present-day tectonic setting of the North Fiji Basin region (modified after Brocher and Holmes, 1985; Falvey, 1978). Plate motion vectors from Drummond et al. (1984).

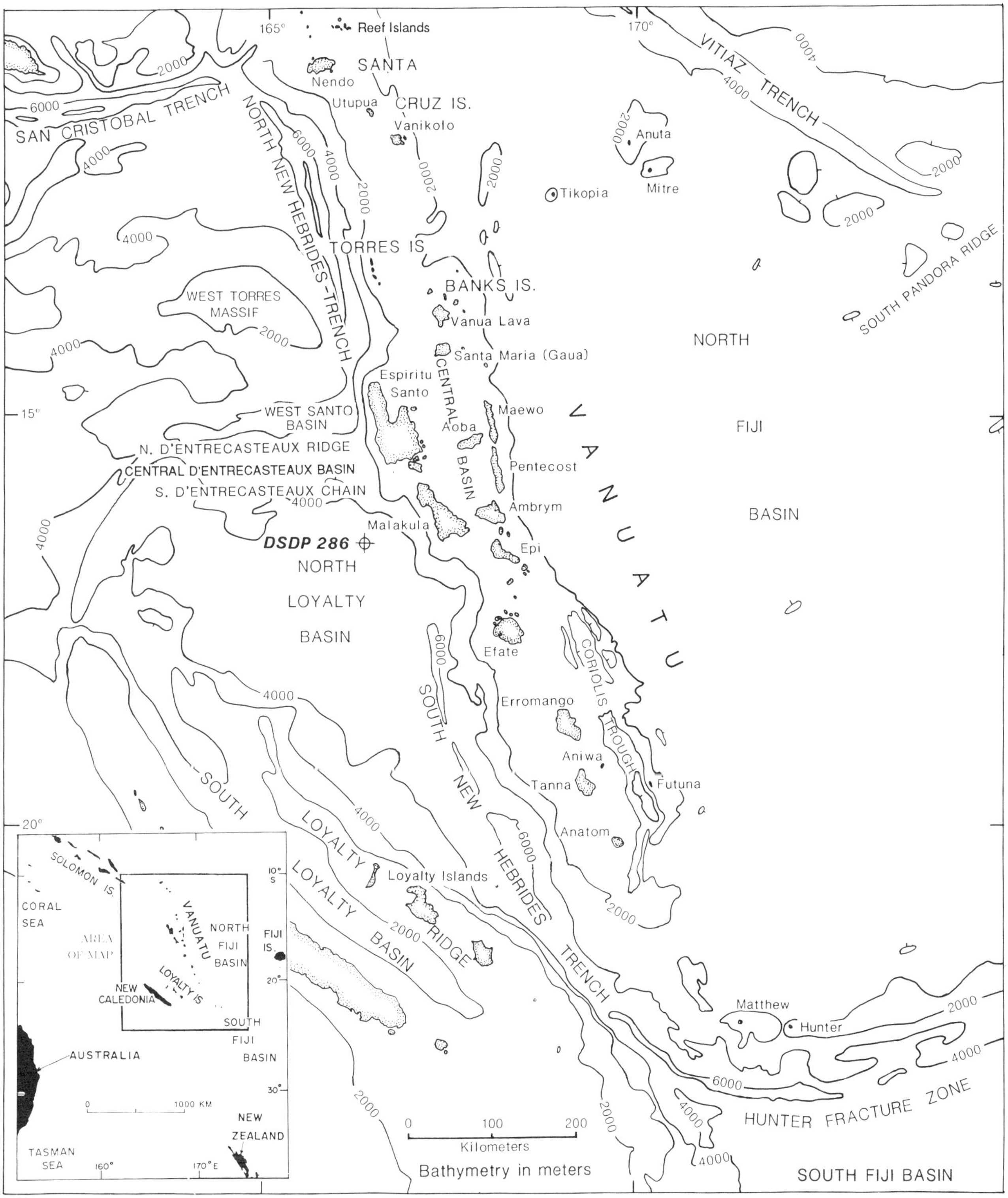

Figure 8. Physiography of the New Hebrides island arc region. DSDP=Deep Sea Drilling Project site (Andrews, Packham et al., 1975).

Arc fragmentation is clearly evident where the DEZ has collided with the central New Hebrides arc. Here the impingement of the DEZ on the arc pushed the arc eastward and fractured the brittle rocks in the process. Arc-parallel faults, sedimentary basins, backarc troughs, and other structures have been disrupted and rotated in small blocks in front of the currently northward-sweeping oceanic-ridge zone.

Central Solomon Islands-Woodlark Basin Collision

The Solomon Islands arc, which is similar tectonically to the New Hebrides arc, extends for 1100 km from Bougainville in the north to San Cristobal in the south (Figure 10). The arc marks the eastward subduction of the Solomon Sea/Australia-India plates beneath the Pacific plate at a rate of 10-11 cm/yr since late Miocene time (Figure 11; Luyendyk, Macdonald and Bryan, 1973; Weissel, Taylor, and Karner, 1982; Taylor, 1984; Drummond et al., 1984). The dip of the downgoing slab is steep at the extreme northwest and southeast ends of the arc (Dunkley, 1983; Vedder and Coulson, 1986), but nearly horizontal in the central part of the arc (Cooper, Kroenke and Resig, 1986).

The Solomon Islands arc is composed of a northern and southern belt of islands with an intervening intra-arc basin, the Central Solomons Trough (Figure 10). The trough appears to be a structural downwarp in which are several sedimentary basins that contain at least 5 km of sediment (Bruns et al., 1986). Many normal, arc-parallel faults have been mapped along the central part of the arc (Kroenke, Resig, and Cooper, 1986), but are not discernible in the data reported by Bruns et al. (1986).

The seismic Woodlark spreading-ridge system is colliding with the islands of the New Georgia Group (Figure 10). The ridge segment currently being subducted rises approximately 1000 m above the adjoining seafloor and its impingement angle is about 75°. The progressive subduction of the spreading ridge system has moved the impact point back and forth along a limited stretch of the trench as the end of one ridge segment is consumed and another arrives at the trench (Taylor and Exon, 1987). This relative movement is unlike the unidirectional sweep of the DEZ past the central New Hebrides arc.

Opposite the Woodlark spreading-ridge system, an arc-transverse, left-lateral fault zone, mapped by de Broin, Aubertin, and Ravenne (1977) off southern Choiseul and Manning Strait, coincides with a longer fault that we speculate continues across the Central Solomons Trough to Kolombangara and Gizo Islands (Figure 10). Kolombangara could have been built by magma tapped from the mantle via this fault zone. Coulson and Vedder (1986) report that the rocks on the New Georgia Group, including Kolombangara, are composed mainly of highly porphyritic olivine basalt and picritic basaltic lava and breccia (Ramsay, Crawford, and Foden, 1984). Chemically unlike the calc-alkaline suites found elsewhere on the arc, this suite of rocks is attributed to the subduction of the active Woodlark spreading-ridge system beneath the New Georgia Group (Dunkley, 1983; Taylor, 1984; Johnson et al., 1987).

Northwest of the New Georgia Group the narrow (~30 km wide) but regular and continuous, aseismic Woodlark Rise

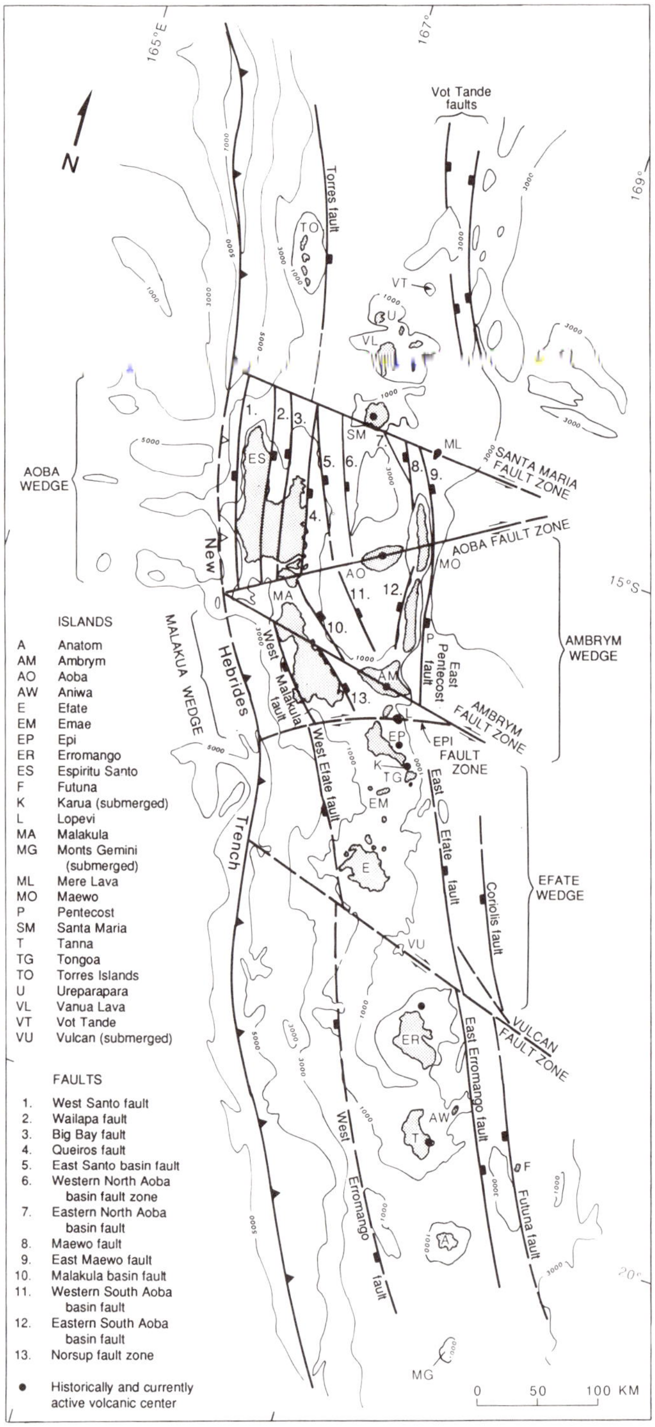

Figure 9. Generalized structure map showing major faults and fault zones of the central New Hebrides arc, Vanuatu (from Greene et al., 1988).

which has an impingement angle of ~10°, is colliding with the arc at the Shortland Islands. The impact point may be sweeping slightly northwestward with spreading from the Woodlark spreading ridge (Figure 10). De Broin, Aubertin, and Ravenne (1977) show another transverse fault zone here extending northeast across the Bougainville Strait and offsetting Bougainville from Choiseul (Figure 10). From the offset of

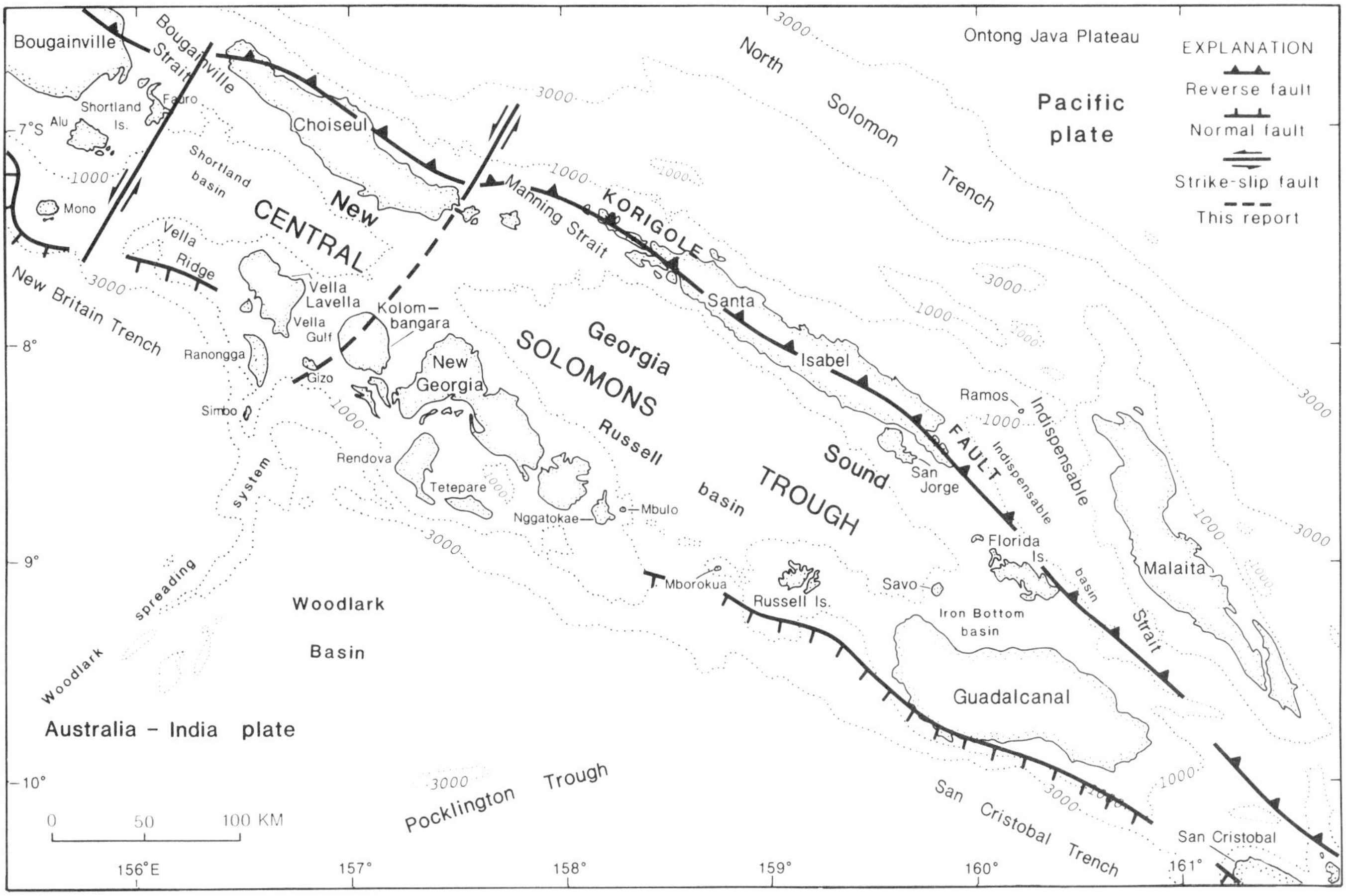

Figure 10. Structural elements (modified after de Broin, Aubertin, and Ravenne, 1977) and sedimentary basins (Bruns et al., 1986) of the Solomon Islands.

the lines of islands and faults, we interpret this arc-transverse fault as having a left-lateral displacement, though it was originally mapped as having right-lateral displacement.

The central Solomon Islands arc has been fragmented: the axis of the Shortland basin is displaced from that of the Russell basin, and Choiseul is left-laterally offset from Bougainville (Figure 10; de Broin, Aubertin and Ravenne, 1977). A block bounded by the two strike-slip faults, the Korigole fault, and the New Britain Trench is about 150 km on a side, the same order of magnitude as the New Hebrides wedges. Whether arc-transverse structures are the result of current subduction processes or earlier pre-reversal ones is not known.

Tonga Arc-Louisville Ridge Collision

The Tonga Ridge extends for nearly 800 km from Tafahi Island in the north to the Louisville collision zone in the south where there is a distinct kink in the trend (Figure 12; Dupont, 1982). The arc marks the westward subduction of the Pacific plate beneath the Australia-India plate at approximately 15 cm/yr (Lonsdale, 1986). The dip of the Benioff zone beneath the Tonga-Kermadec arc varies from 43° to 45° (Isacks, Sykes, and Oliver, 1968; Ewart, Brothers, and Mateen, 1977; Isacks et al., 1981). The active volcanic chain is located along the Tofua arc. Behind the arc, spreading has formed the Lau basin and caused the Lau Ridge to separate from the Tonga Ridge (Figure 12; Karig, 1970). Except where the Louisville Ridge obscures it, the deep (in places as deep as 10,800 m) Tonga Trench lies east of the Tonga Ridge. The Tonga Trench becomes the Kermadec Trench south of the Louisville Ridge and continues to offshore New Zealand (Figure 12).

The islands of Vava'u, Ha'apai, Nomuka and of the Tongatapu-'Eua group are the principal inactive volcanic islands, which consist of elevated masses of limestone and volcanic rocks capped with Quaternary reef limestone (Scholl, Vallier, and Maung, 1985). A broad shallow (average depth of about 500 m) westward dipping platform flanks the islands. Beneath the southern Tonga platform (south of Tongatapu to the Louisville collision zone) lie 3 km of forearc-basin sediment (Scholl, Vallier, and Packham, 1985). Oil seeps on Tongatapu indicate that hydrocarbon generation and accumulation occur in the region.

The aseismic Louisville Ridge is approximately 100 km wide and supports a narrower central band of 40-50-km wide seamounts. Locally, the Louisville Ridge rises about 3000 m above the adjacent seafloor and has been colliding with the Tonga arc during late Cenozoic time. Its impingement angle (~45°) has caused a 900-km southward migration of the impact point to its present position near the southern end of the southern Tonga platform during the last 3 m.y. (Figure 12;

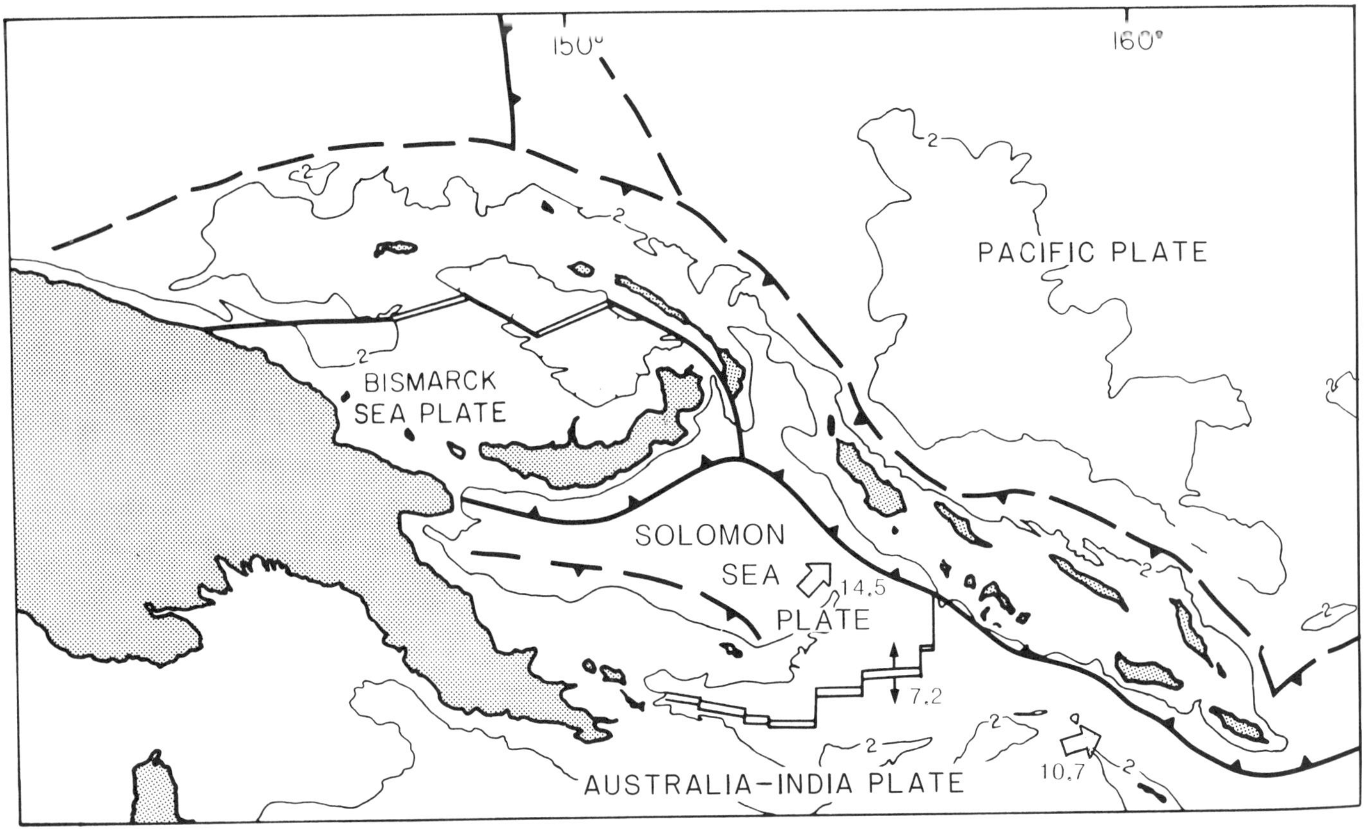

Figure 11. Sketch map showing inferred plate boundaries in Papua New Guinea-Solomon Islands region (from Cooper, Kroenke and Resig, 1986). Relative plate motion vectors from Taylor (1987).

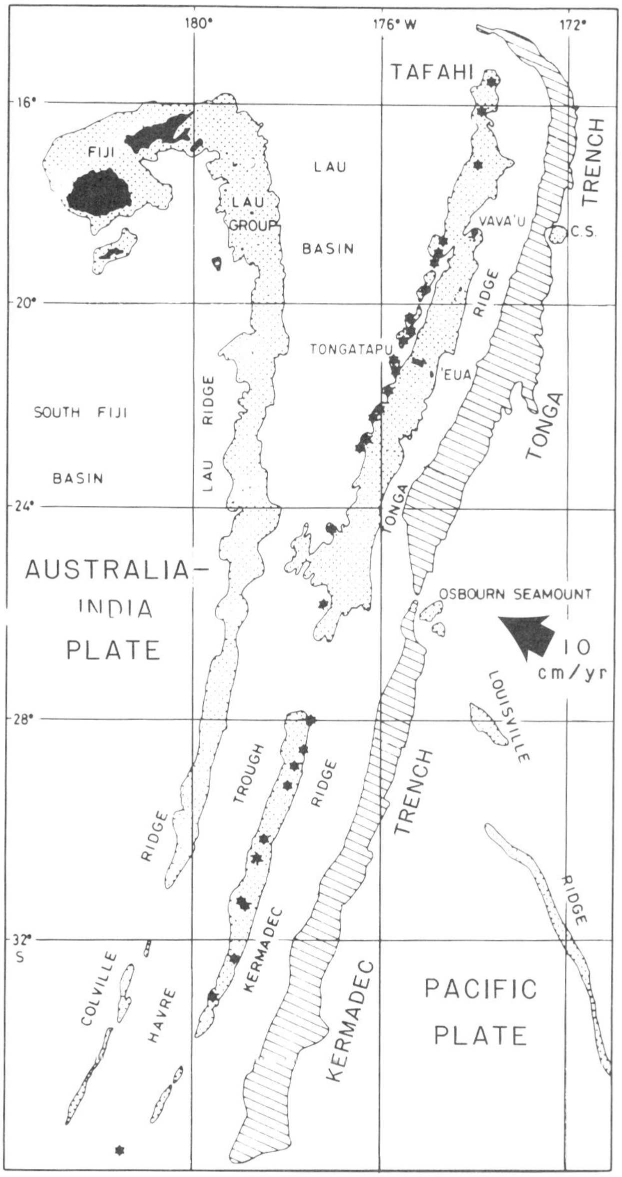

Figure 12. Tonga and Kermadec island arcs and associated trenches, basins, and ridges. Ridges are shaded, trenches are hachured, and volcanoes are shown by stars. CS=Capricorn seamount (modified after Dupont and Herzer, 1985). Arrow indicates absolute plate motion of the Pacific plate (Drummond et al., 1984).

Dupont and Herzer, 1985). The eastward displacement of the active volcanic arc coincides with the Tonga arc segment already traversed by the Louisville Ridge (Figure 12).

Arc-transverse and arc-parallel faults probably formed before the Louisville Ridge first collided with the Tonga arc, but these structures may have been reactivated by the collision of the ridge (Figure 13; Herzer and Exon, 1985). The western and eastern margins of the southern Tonga platform are bounded by arc-parallel normal faults, downthrown to the west and east, respectively, that are truncated and displaced along arc-transverse faults. Herzer and Exon (1985) interpreted the movement on these transverse faults as being predominantly vertical. Based on the map pattern, we see small (< 100 km on a side) blocks that might be rotated structural wedges. Herzer and Exon (1985) indicate that some of these faults may have formed prior to the collision of the Louisville Ridge. Pelletier and Louat's (in press) analysis of recent shallow seismicity shows that the strike-slip first motions occur along the entire length of the Tonga-Kermadec arc, suggesting that movement on the arc-transverse faults may be predominantly strike-slip. Rather than or in addition to the collision of the Louisville Ridge, these faults may have formed in response to the opening of the Lau Basin, which would introduce arc-transverse forces from the west (D.W. Scholl, oral commun., 1988).

Arc fragmentation appears to be an inherited and ongoing process in the Tonga arc. Perhaps because of its relatively rapid traverse from north to south, the current subduction of the Louisville Ridge seems to be causing little or no uplift of the fragmented arc.

SUMMARY

The foregoing examples of ridge collisions at plate margins show the great diversity of the effects of such collisions along plate margins. Some ridges are subducted entirely in a short stretch of plate margin; others are swept along unidirectionally or back and forth along a margin segment. Large structural disruptions can be credited to ridge collisions at some convergent margins, whereas the effect is negligible at others. Active spreading systems along the subducting ridges introduce another order of complexity.

Our limited observations of colliding ridges in the South Pacific suggest that the more oblique impingement angles are less effective at disrupting structure than the more orthogonal ones. Consider a ridge that makes a 45 angle with a north-south trending plate margin. If the subducting plate is moving due east, the impact point will probably migrate along the plate margin; successive segments of the ridge will contact the plate margin at different places. If the subducting plate is moving parallel to the ridge trend, successive ridge segments will collide with the same interval of plate margin, compounding the collision effects at one point. The extreme structural deformation at the central New Hebrides arc, for example, appears to result from the collision of a ridge (the DEZ) that trends almost parallel to the convergence direction of the host Australia-India plate. Impingement angles and structural disruptions of other ridge systems discussed are shown in Table 1.

To date, little detailed investigation of the tectonic deformation associated with ridge collision has been done along the convergent margin of South America where the collisional processes described for the southwest Pacific may apply. Deformation of the forearc region of Perú where the Nazca Ridge is being obliquely subducted may be comparable to the DEZ-New Hebrides arc or Louisville Ridge-Tonga arc collisions. The southern Chile rise may be deforming the southern Chile margin much as the collision of the Woodlark spreading system deforms the Solomon Islands arc. The hydrocarbon accumulation and mineralization of southwest Pacific basins opposite ridge collision zones compels efforts to understand ridge-margin collisional processes along the South American

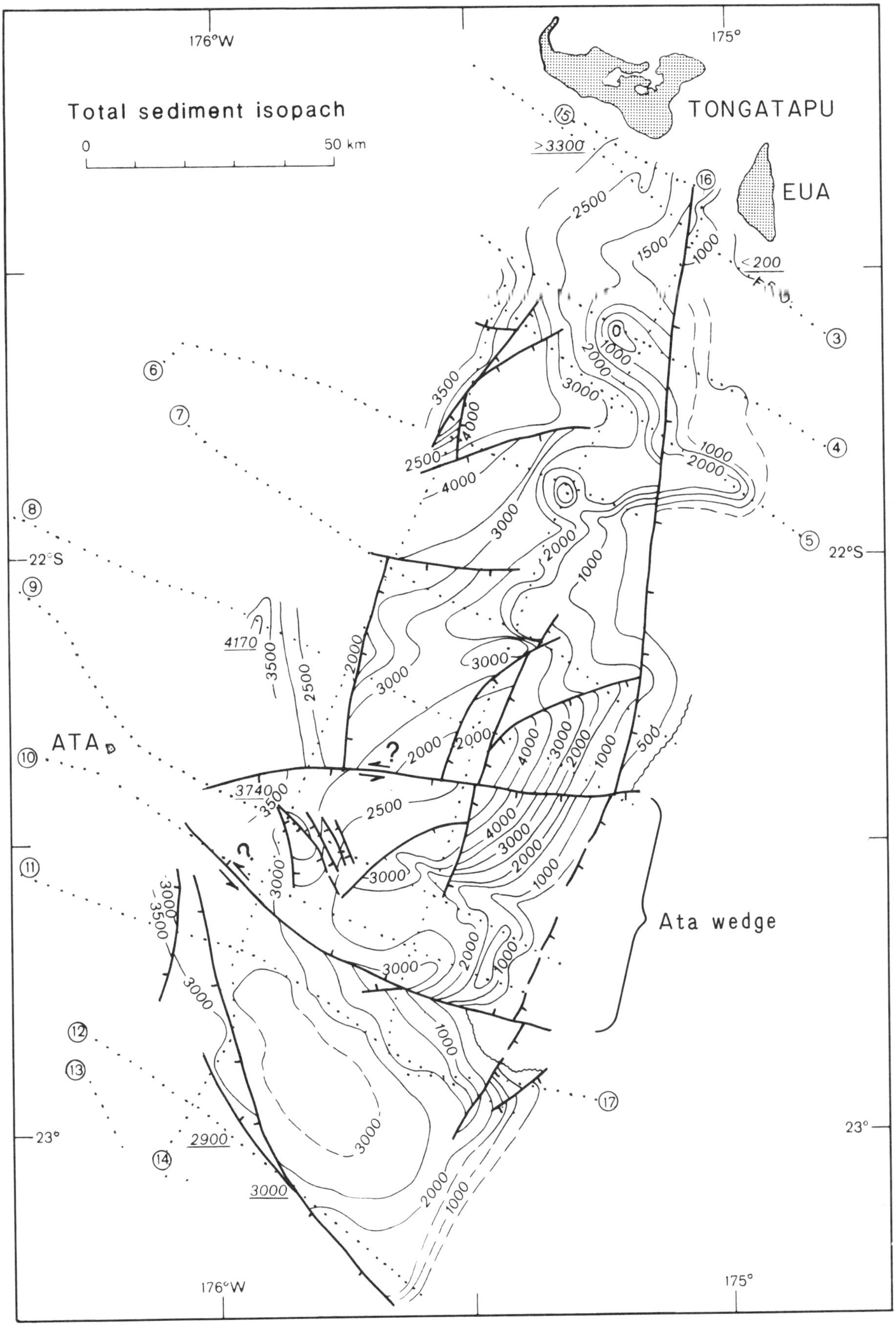

Figure 13. Isopach map of total sediment and faults of the southern Tonga platform. Question marks indicate faults that may have had transcurrent movement. Contours in meters; contours and faults dashed where inferred. Dotted lines and circled numbers indicate R/V *S.P. Lee* tracklines (modified after Herzer and Exon, 1985).

Table 1. Impingement angles of subducting ridges or ridge segments. (Angles measured from Drummond et al., 1984.)

Subducting Ridge	Overlying Plate	Impingement Angle (°)	Structural Disruption
Carnegie Ridge	Ecuador	10	some
Nazca Ridge	Peru	40	little
Juan Fernandez	Chile	5	some
Southern Chile Rise (SCR)	Chile	40	some
SCR ridge segment		90	
D'Entrecasteaux zone	New Hebrides	5-20	much
Woodlark spreading system (WSS)	Solomon Islands	10	little
WSS ridge segment		75	
Louisville Ridge	Tonga	45	little

margin that may be important to energy and mineral development.

ACKNOWLEDGMENTS

We thank G.E. Ericksen, S. Lewis, J.A. Reinemund, D.W. Scholl, J.G. Vedder, and R. von Huene for reviewing this paper.

REFERENCES

Andrews, J.E., Packham, G.H., et al., 1975, Site 286, *in* Andrews, J.E., Packham, G.H., et al., Initial Reports of the Deep Sea Drilling Project, v. 30: Washington, D.C., U.S. Government Printing Office, p. 69.

Barazangi, M. and Isacks, B., 1976, Spatial distribution of earthquakes and subduction of the Nazca plate beneath South America: Geology, v. 4, p. 686-692.

Brocher, T.M. and Holmes, R., 1985, Tectonic and geochemical framework of the northern Melanesian borderland-An overview of the KK820316 leg 2 objectives and results, *in* Brocher, T.M., ed., Geological investigations of the northern Melanesian borderland: Houston, Texas, Circum-Pacific Council for Energy and Mineral Resources Earth Science Series, v. 3, p. 1-11.

Bruns, T.R., Cooper, A.K., Mann, D.M., and Vedder, J.G., 1986, Seismic stratigraphy and structure of sedimentary basins in the Solomon Islands region, *in* Vedder, J.G., Pound, K.S., and Boundy, S.Q., eds., Geology and offshore resources of Pacific island arcs-Central and western Solomon Islands: Houston, Texas, Circum-Pacific Council for Energy and Mineral Resources Earth Science Series, v. 4, p. 99-108.

Cande, S.C. and Leslie, R.B., 1986, Late Cenozoic tectonics of the southern Chile trench, Journal of Geophysical Research, v. 91, no. B1, p. 471-496.

Chase, C.G., 1971, Tectonic history of the Fiji Plateau: Geological Society of America Bulletin, v. 82, p. 3087-3110.

Chase, C.G., 1978, Plate kinematics—the Americas, east Africa, and the rest of the world: Earth and Planetary Science Letters, v. 37, p. 355-368.

Chinn, D.S. and Isacks, B.L., 1983, Accurate source depths and focal mechanisms of shallow earthquakes in western South America and in the New Hebrides island arc: Tectonics, v. 2, no. 6, p. 529-563.

Collot, J.Y., Daniel, J., and Burne, R.V., 1985, Recent tectonics associated with the subduction/collision of the D'Entrecasteaux zone in the central New Hebrides: Tectonophysics, v. 112, p. 325-356.

Cooper, P.A., Kroenke, L.W., and Resig, J.M., 1986, Tectonic implications of seismicity northeast of the Solomon Islands, *in* Vedder, J.G., Pound, K.S., and Boundy, S.Q., eds., Geology and offshore resources of the Pacific island arcs—Central and western Solomon Islands: Houston, Texas, Circum-Pacific Council for Energy and Mineral Resources Earth Science Series, v. 4, p. 1-18.

Couch, R. and Whitsett, R.M., 1981, Structures of the Nazca Ridge and the continental shelf and slope of southern Perú, *in* Kulm, L.D., Dymond, J., Dasch, E.J., Hussong, D.M., and Roderick, R., eds., Nazca plate—Crustal formation and Andean convergence: Geological Society of America Memoir 154, p. 569-586.

Coulbourn, W.T., 1981, Tectonics of the Nazca plate and the continental margin of western South America, 18 °S to 23 °S, *in* Kulm, L.D., Dymond, J., Dasch, E.J., Hussong, D.M., and Roderick R., eds., Nazca plate—Crustal formation and Andean convergence: Geological Society of America Memoir 154, p. 587-618.

Coulson, F.I. and Vedder, J.G., 1986, Geology of the central and western Solomon Islands, *in* Vedder, J.G., Pound, K.S., and Boundy, S.Q., eds., Geology and offshore resources of the Pacific island arcs—Central and western Solomon Islands: Houston, Texas, Circum-Pacific Council for Energy and Mineral Resources Earth Science Series, v. 4, p. 59-87.

Crawford, A.J., Greene, H.G., and Exon, N.F., 1988, Geology, petrology and geochemistry of submarine volcanoes in the area around Epi Island, New Hebrides island arc, *in* Greene, H.G. and Wong, F.L., eds., Geology and offshore resources of Pacific island arcs—Vanuatu region: Houston, Texas, Circum-Pacific Council for Energy and Mineral Resources Earth Science Series, v. 8, p. 301-327.

de Broin, C.E., Aubertin, F., and Ravenne, C., 1977, Structure and history of the Solomon-New Ireland region, *in* International Symposium on Geodynamics in South-west Pacific, Noumea, New Caledonia, 1976: Paris, Editions Technip, p. 37-49.

Drummond, K.J., Nishiwaki, C., Corvalan, J., Doutch, H.F., and Craddock, C., 1984, Plate tectonic map of the circum-Pacific region, Pacific basin sheet: Tulsa, Oklahoma, American Association of Petroleum Geologists, scale 1:17,000,000.

Dubois, J., Dugas, F., Lapouille, A., and Louat, R., 1978, The troughs at the rear of the New Hebrides island arc—Possible mechanisms of formation: Canadian Journal of Earth Science, v. 15, no. 3, p. 351-360.

Dunkley, P.N., 1983, Volcanism and the evolution of the ensimatic Solomon Islands arc, *in* Shimozuru, D. and Yokoyana, I., eds., Arc volcanism—Physics and tectonics: Tokyo, Terra Scientific Publishing Company (Terra Pub.), p. 225-241.

Dupont, J., 1982, Morphologie et structures superficielles de l'arc insulaire des Tonga-Kermadec, *in* Contribution a l'étude geodynamique du Sud-Ouest Pacifique: Paris, ORSTOM, v. 147, p. 263-282.

Dupont, J. and Herzer, R.H., 1985, Effect of subduction of the Louisville Ridge on the structure and morphology of the Tonga arc, *in* Scholl, D.W. and Vallier, T.L., eds, Geology and offshore resources of Pacific island arcs—Tonga region: Houston, Texas, Circum-Pacific Council for Energy and Mineral Resources Earth Science Series, v. 2, p. 323-334.

Ewart, A., Brothers, R.N., and Mateen, A., 1977, An outline of the geology and geochemistry, and the possible petrogenetic evolution of the volcanic rocks of the Tonga-Kermadec-New Zealand island arc: Journal of Volcanology and Geothermal Research, v. 2, p. 205-250.

Falvey, D.A., 1975, Arc reversals, and a tectonic model for the North Fiji Basin: Australian Society of Exploration Geophysicists Bulletin, v. 6, p. 47-49.

Falvey, D.A., 1978, Analysis of paleomagnetic data from the New Hebrides: Australian Society of Exploration Geophysicists Bulletin, v. 9, p. 117-124.

Falvey, D.A. and Greene, H.G., 1988, Origin and evolution of the sedimentary basins of the New Hebrides arc, *in* Greene, H.G. and Wong, F.L., eds., Geology and offshore resources of Pacific island arcs—Vanuatu region: Houston, Texas, Circum-Pacific Council for Energy and Mineral Resources Earth Science Series, v. 8, p. 413-442.

Forsythe, R. and Nelson, E., 1985, Geological manifestations of ridge collision—Evidence from the Golfo De Penas-Taitao Basin, southern Chile: Tectonics, v. 4, no. 5, p. 477-495.

Gill, J.B. and Gorton, M.P., 1973, A proposed geological and geochemical history of eastern Melanesia, *in* Coleman, P.J., ed., The western Pacific—Island arcs, marginal seas and geochemistry: Perth, University of Western Australia Press, p. 543-566.

Gorton, M.P., 1977, The geochemistry and origin of Quaternary volcanism in the New Hebrides: Geochimica et Cosmochimica Acta, v. 41, p. 1257-1270.

Gonzalez, Eduardo, this volume, Hydrocarbon resources in the coastal zone of Chile.

Greene, H.G. and Johnson, D.P., 1988, Geology of the intra-arc basin of the New Hebrides arc based on single-channel seismic-reflection data, *in* Greene, H.G. and Wong, F.L., eds., Geology and offshore resources of Pacific island arcs—Vanuatu region: Houston, Texas, Circum-Pacific Council for Energy and Mineral Resources Earth Science Series, v. 8, p. 177-199.

Greene, H.G., Macfarlane, A., Johnson, D.P., and Crawford, A.J., 1988, Structure and tectonics of the New Hebrides arc, *in* Greene, H.G. and Wong, F.L., eds., Geology and offshore resources of Pacific island arcs—Vanuatu region: Houston, Texas, Circum-Pacific Council for Energy and Mineral Resources Earth Science Series, v. 8, p. 377-412.

Greene, H.G. and Wong, F.L., eds., 1988, Geology and offshore resources of Pacific island arcs-Vanuatu region: Houston, Texas, Circum-Pacific Council for Energy and Mineral Resources Earth Science Series, v. 8, 442 p.

Hall, M.L. and Wood, C.A., 1985, Volcano-tectonic segmentation of the northern Andes: Geology, v. 13, p. 203-207.

Herzer, R.H. and Exon, N.F., 1985, Structure and basin analysis of the southern Tonga forearc, *in* Scholl, D.W. and Vallier, T.L., eds., Geology and offshore resources of Pacific island arcs—Tonga region: Circum-Pacific Council for Energy and Mineral Resources Earth Science Series, v. 2, p. 55-74.

Isacks, B.L., Cardwell, R.K., Chatelain, J.L., Barazangi, M., Marthelot, J.-M., Chinn, D., and Louat, R., 1981, Seismicity and tectonics of the central New Hebrides island arc, *in* Simpson, D.W. and Richards, P.G., eds., Earthquake prediction—An international review: American Geophysical Union Maurice Ewing Series 4, p. 93-116.

Isacks, B.L., Sykes, L.R., and Oliver, J., 1968, Seismology and the new global tectonics: Journal of Geophysical Research, v. 73, p. 5855-5899.

James, A. and Falvey, D.A., 1978, Analysis of paleomagnetic data from Viti Levu, Fiji: Australian Society of Exploration Geophysicists Bulletin, v. 9, p. 115-123.

Johnson, R.W., Jaques, A.L., Langmuir, C.H., Perfit, M.R., Staudigel, H., Dunkley, P.N., Chappell, B.W., and Taylor, S.R., 1987, Ridge subduction and forearc volcanism—Petrology and geochemistry of rocks dredged from the western Solomon arc and Woodlark Basin, *in* Taylor, B. and Exon, N.F., eds., Marine geology, geophysics, and geochemistry of the Woodlark Basin-Solomon Islands: Houston, Texas, Circum-Pacific Council for Energy and Mineral Resources Earth Science Series, v. 7, p. 155-226.

Karig, D.E., 1970, Ridges and basins of the Tonga-Kermadec island arc system: Journal of Geophysical Research, v. 75, p. 239-254.

Kroenke, L.W., 1984, Vanuatu and the Eastern Outer Solomon Islands—Formation of the New Hebrides and Vitiaz arcs and development of the North Fiji Basin, *in* Kroenke, L.W., ed., Cenozoic tectonic development of the southwest Pacific: UN ESCAP, CCOP/SOPAC Technical Bulletin 6, p. 63-75.

Kroenke, L.W., Resig, J.M., and Cooper, P.A., 1986, Tectonics of the southeastern Solomon Islands—Formation of the Malaita Anticlinorium, *in* Vedder, J.G., Pound, K.S., and Boundy, S.Q., eds., Geology and offshore resources of Pacific island arcs—Central and western Solomon Islands: Houston, Texas, Circum-Pacific Council for Energy and Mineral Resources Earth Science Series, v. 4, p. 109-116.

Kulm, L.D., Schrader, H., Resig, J.M., Thornburg, T.M., Masias, A., and Johnson, L., 1981, Late Cenozoic carbonates on the Perú continental margin—Lithostratigraphy, biostratigraphy, and tectonic history, *in* Kulm, L.D., Dymond, J., Dasch, E.J., Hussong, D.M., and Roderick, R., eds., Nazca plate—Crustal formation and Andean convergence: Geological Society of America Memoir 154, p. 469-508.

Lonsdale, P., 1986, A multibeam reconnaissance of the Tonga trench axis and its intersection with the Louisville guyot chain: Marine Geophysical Researches, v. 8, p. 295-327.

Louat, R., Hamburger, M., and Monzier, M., 1988, Shallow and intermediate-depth seismicity in the New Hebrides arc—Constraints on the subduction process, *in* Greene, H.G. and Wong, F.L., eds., Geology and offshore resources of Pacific island arcs—Vanuatu region: Houston, Texas, Circum-Pacific Council for Energy and Mineral Resources Earth Science Series, v. 8, p. 329-356.

Luyendyk, B.P., Bryan, W.B., and Jezek, P.A., 1974, Shallow structure of the New Hebrides island arc: Geological Society of America Bulletin, v. 85, p. 1287-1300.

Luyendyk, B.P., MacDonald, K.C., and Bryan, W.B., 1973, Rifting history of the Woodlark Basin in the southwest Pacific: Geological Society of America Bulletin, v. 84, p. 1125-1134.

Macfarlane, A., Carney, J.N., Crawford, A.J., and Greene, H.G., 1988, Vanuatu—A review of the onshore geology, *in* Greene, H.G. and Wong, F.L., eds., Geology and offshore resources of Pacific island arcs—Vanuatu region: Houston, Texas, Circum-Pacific Council for Energy and Mineral Resources Earth Science Series, v. 8, p. 45-91.

Mammerickx, J., Herron, E., and Dorman, L., 1980, Evidence for two fossil spreading ridges in the southeast Pacific: Geological Society of America Bulletin, v. 91, p. 263-271.

Marsh, B.D., 1979, Island arc development—Some observations, experiments and speculations: Journal of Geology, v. 87, p. 687-713.

Minster, J.B., Jordan, T.H., Molnar, P., and Haines, E., 1974, Numerical modelling of instantaneous plate tectonics: Geophysical Journal of the Royal Astronomical Society of London, v. 36, no. 3, p. 541-576.

Minster, J.B. and Jordan, T.H., 1978, Present-day plate motions: Journal of Geophysical Research, v. 83, p. 5331-5354.

Moody, J.D. and Hill, M.J., 1956, Wrench-fault tectonics: Geological Society of America Bulletin, v. 67, no. 9, p. 1207-1246.

Mordojovich, Carlos, 1981, Sedimentary basins of Chilean Pacific offshore, *in* Halbouty, M.T., ed., 1981, Energy resources of the Pacific region: American Association of Petroleum Geologists Studies in Geology n. 12, p. 63-82.

Nur, A. and Ben-Avraham, Z., 1981, Volcanic gaps and the consumption of aseismic ridges in South America, *in* Kulm, L.D., Dymond, J., Dasch, E.J., Hussong, D.M., and Roderick, R., eds., Nazca plate—Crustal formation and Andean convergence: Geological Society of America Memoir 154, p. 729-740.

Pelletier, B. and Louat, R., in press, Seismotectonics and present-day relative plate motions in the Tonga-Lau and Kermadec-Havre region: submitted to Tectonophysics, July 1988.

Ramsay, W.R.H., Crawford, A.J., and Foden, J.D., 1984, Field setting, mineralogy, chemistry and genesis of arc picrites, New Georgia, Solomon Islands: Contributions to Mineralogy and Petrology, v. 88, p. 386-402.

Rosania-S., Giovanni, this volume, Petroleum prospects of the sedimentary basins of Ecuador.

Scholl, D.W. and Vallier, T.L., eds., 1985, Geology and offshore resources of Pacific island arcs—Tonga region: Houston, Texas, Circum-Pacific Council for Energy and Mineral Resources Earth Science Series, v. 2, 488 p.

Scholl, D.W., Vallier, T.L., and Maung, T.U., 1985, Introduction, *in* Scholl, D.W. and Vallier, T.L., eds., Geology and offshore resources of Pacific island arcs—Tonga region: Houston, Texas, Circum-Pacific Council for Energy and Mineral Resources Earth Science Series, v. 2, p. 3-16.

Scholl, D.W., Vallier, T.L., and Packham, G., 1985, Framework geology and resource potential of southern Tonga platform and adjacent terranes—A synthesis, *in* Scholl, D.W. and Vallier, T.L., eds., Geology and offshore resources of Pacific island arcs—Tonga region: Houston, Texas, Circum-Pacific Council for Energy and Mineral Resources Earth Science Series, v. 2, p. 457-488.

Schweller, W.J., Kulm, L.D., and Prince, R.A., 1981, Tectonics, structure, and sedimentary framework of the Perú-Chile Trench, *in* Kulm, L.D., Dymond, J., Dasch, E.J., Hussong, D.M., and Roderick, R., eds., Nazca plate—Crustal formation and Andean convergence: Geological Society of America Memoir 154, p. 323-350.

Shepherd, G.L. and Moberly, R., 1981, Coastal structure of the continental margin, northwest Perú and southwest Ecuador, *in* Kulm, L.D., Dymond, J., Dasch, E.J., Hussong, D.M., and Roderick, R., eds., Nazca plate—Crustal formation and Andean convergence: Geological Society of America Memoir 154, p. 351-391.

Suarez, G., Molnar, P., and Burchfiel, B.C., 1983, Seismicity, fault plane solutions, depth of faulting, and active tectonics of the Andes of Perú, Ecuador, and southern Colombia: Journal of Geophysical Research, v. 88, no. B12, p. 10,403-10,428.

Taylor, B., 1987, A geophysical survey of the Woodlark-Solomons region, *in* Taylor, B. and Exon, N.F., eds., Marine geology, geophysics, and geochemistry of the Woodlark Basin-Solomon Islands: Houston, Texas, Circum-Pacific Council for Energy and Mineral Resources Earth Science Series, v. 7, p. 25-48.

Taylor, B. and Exon, N.F., 1987, An investigation of ridge subduction in the Woodlark-Solomons region—Introduction and overview, *in* Taylor, B. and Exon, N.F., eds., Marine geology, geophysics, and geochemistry of the Woodlark Basin-Solomon Islands: Houston, Texas, Circum-Pacific Council for Energy and Mineral Resources Earth Science Series, v. 7, p. 1-24.

Taylor, F.W., Jouannic, C., and Bloom, A.L., 1985, Quaternary uplift of the Torres Islands, northern New Hebrides frontal arc—Comparison with Santo and Malekula Islands, central New Hebrides frontal arc: Journal of Geology, v. 93, p. 419-438.

Thornburg, T. and Kulm, L.D., 1981, Sedimentary basins of the Perú continental margin—Structure, stratigraphy, and Cenozoic tectonics from 6 °S to 16 °S latitude, *in* Kulm, L.D., Dymond, J., Dasch, E.J., Hussong, D.M., and Roderick, R., eds., Nazca plate—Crustal formation and Andean convergence: Geological Society of America 154, p. 393-422.

Travis, R.B., Gonzales, G., and Pardo, A., 1976, Hydrocarbon potential of coastal basins of Perú, *in* Halbouty, M.T., Maher, J.C., and Lian, H.M., eds., Circum-Pacific energy and mineral resources: American Association of Petroleum Geologists Memoir 25, p. 331-338.

Vedder, J.G. and Coulson, F.I., 1986, Introduction to geology and offshore resources of the Solomon Islands, *in* Vedder, J.G., Pound, K.S., and Boundy, S.Q., eds., Geology and offshore resources of Pacific island arcs—Central and western Solomon Islands, Circum-Pacific Council for Energy and Mineral Resources Earth Science Series, v. 4, p. 1-18.

Vedder, J.G., Pound, K.S., and Boundy, S.Q., eds., 1986, Geology and offshore resources of Pacific island arcs—Central and western Solomon Islands: Houston, Texas, Circum-Pacific Council for Energy and Mineral Resources Earth Science Series, v. 4, 306 p.

Vogt, P.R., Lowrie, A., Bracey, D.R., Hey, R.N., 1976, Subduction of aseismic oceanic ridges; effects on shape, seismicity, and other characteristics of consuming plate boundaries: Geological Society of America Special Paper no. 172, 59 p.

Weissel, J.K., Taylor, B., and Karner, G.D., 1982, The opening of the Woodlark Basin, subduction of the Woodlark spreading system and the evolution of northern Melanesia since mid-Pliocene time: Tectonophysics, v. 87, p. 253-277.

Wells, R.E., 1987, Origin of the oceanic basalt basement of the Solomon Islands arc and its relationship to the Ontong Java Plateau—insights from Cenozoic plate motion models, *in* Vedder, J.G., Bruns, T.R., and Colwell, J.B., compilers, Aspects of regional geology, Solomon Islands and Papua New Guinea, joint cruise report, R/V *S.P. Lee*, 1984: U.S. Geological Survey Administrative Report, p. D1-D14.

Wilcox, R.E., Harding, T.P., and Seely, D.R., 1973, Basic wrench tectonics: American Association of Petroleum Geologists Bulletin, v. 57, p. 74-96.

Ericksen, G. E., Cañas Pinochet, M. T., and Reinemund, J. A., editors, 1989, Geology of the Andes and its relation to hydrocarbon and mineral resources: Houston, Texas, Circum-Pacific Council for Energy and Mineral Resources Earth Science Series, v. 11.

CHAPTER 5

THE ANDES OF CHILE AND ARGENTINA

Constantino Mpodozis
Servicio Nacional de Geología y Minería
Santiago, Chile

Victor Ramos
Servicio Geológico Nacional
Buenos Aires, Argentina

ABSTRACT

The Andean Cordillera has been considered a simple model of a mountain chain, but detailed analyses of its geologic evolution show a more complex history. During the early Paleozoic, the southwestern margin of the Gondwana continent was near the western side of the present-day Sierras Pampeanas of Argentina. The site of the present Precordillera was a carbonate platform, facing an ocean to the west, which separated a microcontinental block, the Chilenia terrane, from the rest of Gondwana. This ocean basin was consumed along an eastward-dipping subduction zone below the Sierras Pampeanas during the Late Devonian, when Chilenia moved eastward and was accreted to the Gondwana continental margin. A new subduction zone was then formed at the west side of the accreted terrane, leaving a large accretionary prism at the site of the present Chilean Coastal Range. A late Paleozoic-early Mesozoic magmatic arc, consisting of a belt of granitoids and related rhyolitic volcanic rocks, developed in this accretionary prism and in the adjacent continental terrane.

Lateral growth of the continental margin ended in the Paleozoic, and in the early Mesozoic a magmatic arc and associated back-arc basins were developed on the late Paleozoic basement. Subsequent subduction erosion or strike-slip faulting eliminated large pieces of the Paleozoic forearc assemblages. In northern Chile, these processes reached a maximum, resulting in formation of a Jurassic magmatic arc at the present-day continental margin.

The outstanding feature of the Mesozoic evolution of the Andes of Chile and Argentina is a tectonic segmentation characterized by differences in the geologic evolution, changes in the nature of backarc basins, differences in structural style, and behavior of related magmatic arcs of the various segments. As a consequence of these differences, the following well-defined segments have been recognized:

Segment A (lat. 21°-27°S) had a Jurassic-Early Cretaceous magmatic arc, preserved in the present Coastal Range of Northern Chile, to the east of which was the Tarapacá basin, an ensialic, extensional, backarc basin that became filled with carbonate and clastic sediments.

Segment B (lat. 27°-35°S) also had a Jurassic magmatic arc along the Coastal Range. In the Early Cretaceous the backarc basin evolved into an "aborted" marginal basin where large volumes of andesitic and basaltic volcanic rocks were erupted and which interfinger eastward with the carbonate and clastic sediments of a narrow shelf (Aconcagua platform) along the stable continental edge.

Segment C (lat. 35°-40°S) had a magmatic arc, which remained stationary from the Jurassic to the present, along the main axis of the Cordillera Principal; to the east was the Jurassic-Cretaceous Neuquén basin, an extensive and deep ensialic foreland basin.

Segment D (lat. 41°-49°S) was characterized by an oscillatory magmatic arc and only minor extension behind the arc during the Jurassic and Early Cretaceous.

Segment E (49°-56°S) had an ocean-floored backarc basin (Magallanes marginal basin) developed to the east of a magmatic arc in the Early Cretaceous. Subsequent middle Cretaceous closure caused formation of a metamorphic core complex along the foreland edge of the marginal basin.

The Andean region experienced severe deformation during the middle to Late Cretaceous, and the diverse segments reacted in different ways, showing conspicuous differences in the patterns of folding and faulting from one segment to another. The Andean Cordillera of Chile and Argentina was again segmented during the Miocene, resulting in the present-day morphotectonic pattern that apparently is controlled by processes such as plate convergence velocities and inclination of the Benioff zone. These processes determined the distribution of the late Cenozoic volcanism, caused uplift of the Altiplano and the Sierras Pampeanas of Argentina, and brought about the final tectonic changes of the geological and morphological provinces now recognized in the Andean region. The boundaries of the Tertiary segments differ slightly from those of the Mesozoic, but undoubtedly the features of each were determined by the previous segmentation.

INTRODUCTION

The large number of modern geological provinces that can be distinguished in the Andes of Chile and Argentina (Figure 1) owe their distribution to segmentation of the subducted Nazca Plate (Jordan et al., 1983a). In some of the current geological literature (e.g., James, 1971; Smith, 1980) the Andes have been considered as a type of mountain range in which paleogeography, magmatism, and tectonics are related to the evolution of a simple active margin, and which is controlled by the subduction of Pacific oceanic crust under the South American continent. However, analysis of the Paleozoic geology and structure indicate a much more complex history characterized by collision and accretion of continental and oceanic terranes. Many authors believe that collisional tectonics is a process essential to formation of mountain ranges (see Nur and Ben-Avraham, 1983), and some even doubt that normal subduction of oceanic crust alone could produce the deformation within a mountain system. In contrast with the Paleozoic, the Mesozoic-Cenozoic Andes lack evidence of collision of major terranes (Dalziel, 1986; Ramos et al., 1984), and recent paleomagnetic data (Forsythe et al., 1986b) for the northern and central coastal zone of Chile do not indicate the existence of large allochthonous terranes younger than the Triassic. Instead of collisional processes, the Mesozoic-Cenozoic Andean orogenesis seems to have been related to tectonic erosion of the continental margin. The absence of accretionary prisms together with the short distance between the extinct Jurassic magmatic arc and the present trench in northern Chile could be explained by subduction erosion (Rutland, 1971; Coira et al., 1982). Such a process (Kulm, Schweller, and Masias, 1977; Scholl et al., 1980; von Heune, 1984) could have destroyed large volumes of the Pacific margin of South America.

Three stages of Andean evolution can be recognized: (1) a Cambrian to Late Devonian stage, corresponding to the Famatinian tectonic cycle; (2) the Early Carboniferous to Middle Triassic Gondwana tectonic stage; and (3) the Late Triassic to Quaternary Andean tectonic cycle.

EARLY PALEOZOIC (FAMATINIAN) TECTONIC CYCLE

An outstanding feature of the early Paleozoic in western Argentina is the presence of a disrupted sequence of basaltic pillow lavas and associated ultramafic rocks exposed in a belt more than 1500 km long in the region between lat. 23°-3°S (Borello, 1969; Haller and Ramos, 1984). These rocks have been interpreted as remnants of an oceanic realm (Kay, Ramos, and Kay, 1984; Ramos et al., 1986) that separated Gondwana from a series of unconnected "Pacific" microcontinental blocks to the west. These blocks were finally accreted to the South American continental margin during middle to late Paleozoic. Terrane accretion is documented by a belt of calc-alkaline plutonic rocks, presently exposed within the continent, 500 km or more east of the present trench, and by the recently discovered early Paleozoic subduction complexes between the Cordillera Frontal and the Precordillera (Figure 1) (Ramos et al., 1984, 1986). The following early Paleozoic segments are distinguished in the Andean region of Argentina and Chile.

Northern Segment

Northern Chile and Argentina differ from the region to the south in having Precambrian basement exposed in the westernmost part of the Andean chain (Hervé et al., 1981). Isolated outcrops of medium- to high-grade metamorphic rocks of Precambrian age, exposed at Belén and Sierra de Moreno, Chile (Figure 2), probably are the southern extension of the Arequipa massif (Shackleton et al., 1979). Radiometric ages of these rocks range from 1460 to 1000 Ma (Pacci et al., 1980; Breitkreuz and Zeil, 1984; Damm, Pichowiak, and Todt, 1986). Precambrian metamorphic rocks also are exposed in northern Argentina to about lat. 26° (Salar de Antofalla), where they are partly covered by Cenozoic volcanic rocks (Segerstrom and Turner, 1972).

The Arequipa-Belén-Antofalla block (Figure 2) was the site of calc-alkaline plutonic and volcanic activity during the Cambrian and Ordovician; the igneous rocks emplaced at this time show radiometric ages ranging from 487 to 429 Ma (Mpodozis et al., 1983; Palma, Parica, and Ramos, 1986; Damm, Pichowiak, and Todt, 1986).

An Ordovician basin formed to the east of the metamorphic terrain and was filled with a thick sequence of turbidites with volcanic debris that is widely exposed in the Argentine Puna (Aceñolaza, 1982; Turner, 1970). The western part of this basin was floored by mafic volcanic rocks (Allmendinger et al., 1982; Kay, Ramos, and Kay, 1984). Eastward, near the stable Ordovician continental margin, a volcanic belt known as Faja Eruptiva de la Puna (Figure 2), developed. This has been interpreted by some as a possible subduction-related magmatic arc that formed over a thinned continental or "quasioceanic" crust (Coira et al., 1982), although more detailed studies are needed to better understand the tectonic significance of this belt. A stable shelf developed to the east between the above magmatic arc and the Pampean foreland where carbonate and clastic sediments were deposited. Available data suggest the existence of two parallel magmatic arcs during the Ordovician—a western arc in the Arequipa-Belén-Antofalla block and an eastern arc in the Faja Eruptiva de la Puna (Palma, Parica, and Ramos, 1986). Magmatic activity ceased during the latest Ordovician. All these rocks were intensely deformed during the Ocloyic orogeny at the end of the Ordovician (Turner and Méndez, 1979).

The above elements suggest an evolutionary model wherein the Arequipa-Belén-Antofalla block collided with the South American continental margin during Ocloyic deformation (Coira et al., 1982).

Central Segment

The central segment (lat. 26°-38°S) differs from the northern segment as follows: (1) formation of a granitoid belt in the western part of the Sierras Pampeanas (Figure 1) of Argentina, which represents a southern extension of the Faja Eruptiva de la Puna (Figure 2); (2) appearance of a new terrane—the Precordillera Terrane (Figure 2)—west of the Sierras Pampeanas, consisting of an extensive lower Paleozoic carbonate platform for which the tectonic relationships with its bounding provinces are unknown; and (3) presence of an "oceanic" basin between the Precordillera Terrane and the present Cordillera Frontal (Figure 1), which persisted until Middle to Late Devonian. The presence of this oceanic basin is indicated by

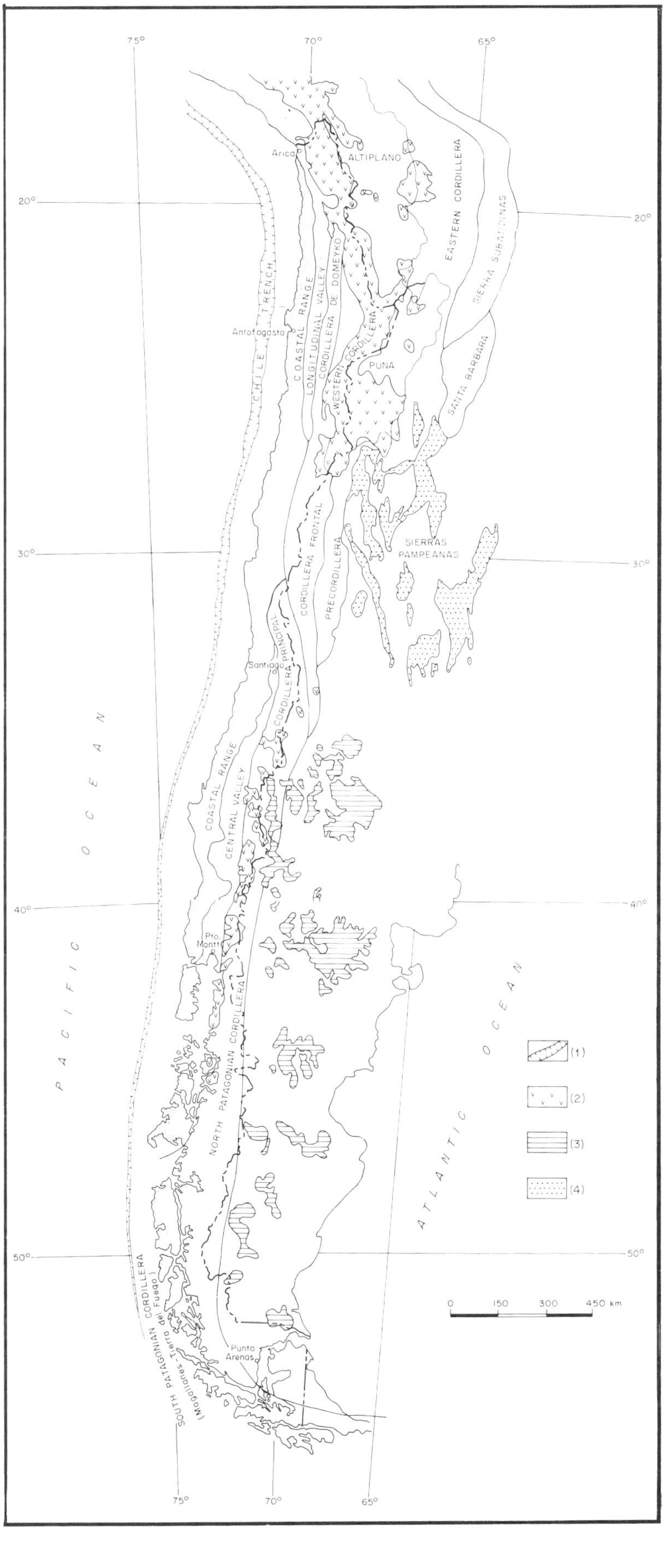

Figure 1. Major physiographic and geologic provinces of the Andes in Chile and Argentina: (1) late Cenozoic "arc" volcanics; (2) late Cenozoic backarc volcanics; and (3) Precambrian crystalline rocks. Modified from Jordan et al. (1983a) and Muñoz and Stern (1986).

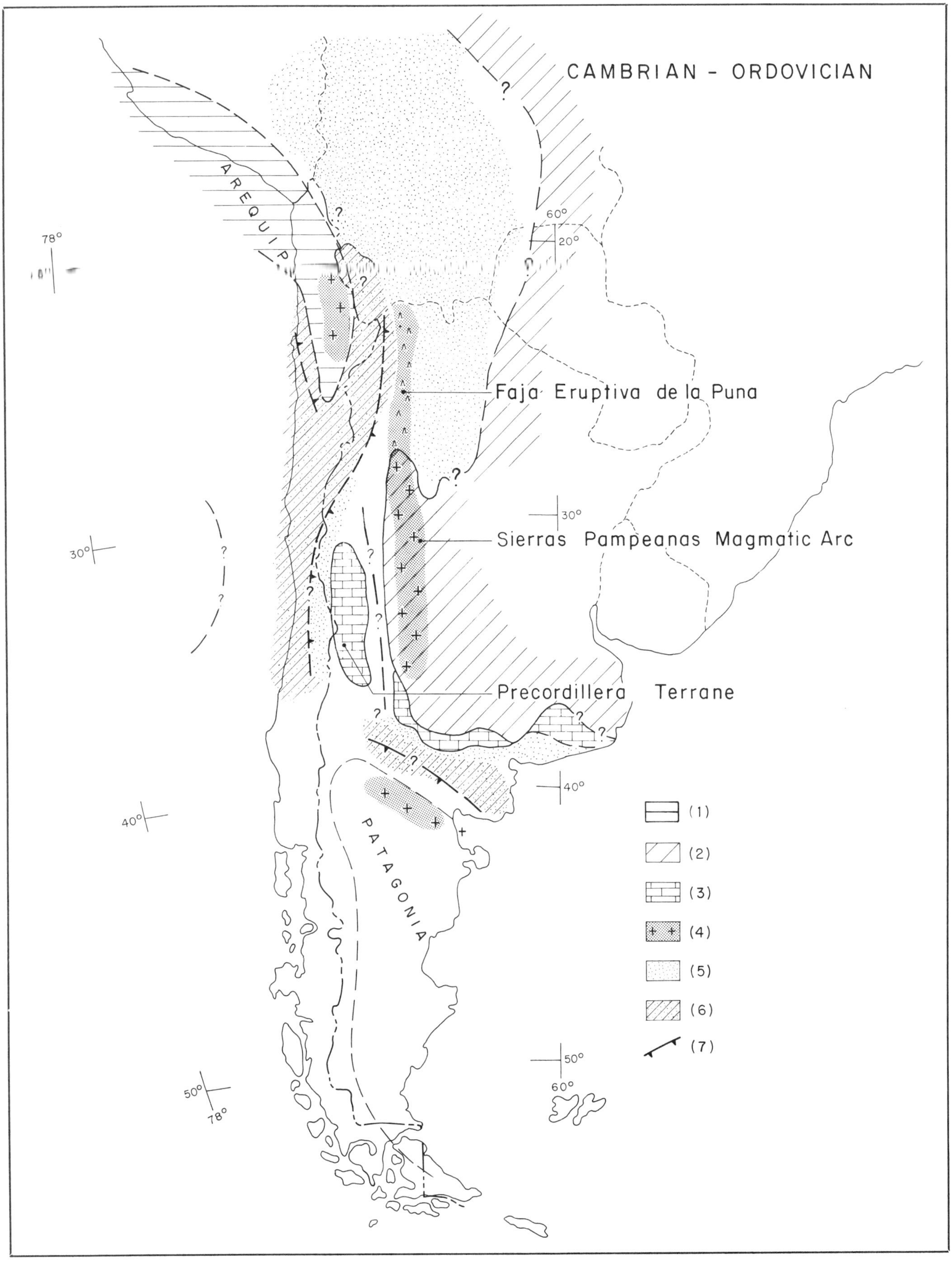
CAMBRIAN - ORDOVICIAN
AREQUIPA
Faja Eruptiva de la Puna
Sierras Pampeanas Magmatic Arc
Precordillera Terrane
PATAGONIA
78°
30°
40°
50°
60°
20°
(1)
(2)
(3)
(4)
(5)
(6)
(7)

the interfingering toward the west of the slope facies of the Precordillera with a sequence of pillow basalts cut by sills and ultramafic intrusive bodies, chemically similar to evolved oceanic tholeiites (Kay, Ramos, and Kay, 1984; Haller and Ramos, 1984). Deformed turbidites interbedded with pillow basalts, such as those exposed in the western part of the Precordillera (Cortaderas and Sandalio ranges, near Uspallata), have been interpreted as part of a Siluro-Devonian subduction complex (Ramos et al., 1986).

The granitoid belt of the western Sierras Pampeanas (Figure 1), described by Hervé et al. (1982), may represent the roots of an early Paleozoic magmatic arc related to an eastward-dipping subduction zone. The main activity of this arc lasted until the Early Carboniferous, when changes in the composition of the igneous rocks indicate a crustal contamination (Rapela, Heaman, and McNutt, 1982) that seems to reflect waning magmatic activity. A complex mélange in the region between the Precordillera and Sierras Pampeanas (Figure 1) (Ramos et al., 1986) indicates that a considerable gap may have existed between the Precordillera and the margin of Gondwana.

The collision of the basement of the Cordillera Frontal, which constitutes the Chilenia terrane (Figure 3), occurred during latest Devonian to earliest Carboniferous. The nature of the Chilenia basement is not well known because Precambrian rocks are, in general, not exposed in the Cordillera Frontal. However, the extensive outcrops of late Paleozoic high-silica rhyolites and granites suggest the presence then of an old sialic basement (Mpodozis et al., 1985). The collision caused uplift of a positive element, the Proto Precordillera (Figure 4) of Rolleri and Baldis (1967). This uplifted block divided the late Paleozoic facies into an eastern realm in the region of the central and eastern Precordillera, where alluvial fan and proximal fluvial sediments were deposited (Lopez-Gamundi Azcuy, and Guerra, 1986), and a western realm, where littoral estuarine to deltaic marine sediments were deposited during the middle Carboniferous to Early Permian (Caminos, 1979).

Southern Segment

Little is known about the early Paleozoic paleogeography of the region south of lat. 36 °S. Precambrian basement rocks, which yielded ages of 1200 Ma (Linares et al., 1985), have been recognized in the Andean foothills of Neuquén Province, Argentina (lat. 38 °-39 °S). The intracontinental fold thrust belt of the Ventania System, south of Buenos Aires, appears to be related to the Cape System of South Africa, and has been traditionally interpreted as an aulacogen (see Harrington, 1970). However, recent information about the Ventania System and the adjacent North Patagonian massif indicates that Patagonia, and perhaps the Antarctic Peninsula and other minor fragments, may have been accreted to Gondwana along a northward-dipping subduction zone during middle to Late Permian (Ramos, 1984). Similarly, Winter (1984) proposed that the Cape System of South Africa resulted from convergence of a continental block with a north-dipping subduction zone.

The Cape System has only minor pyroclastic rocks derived from an unknown source, but coeval magmatic rocks are known in the North Patagonian massif (Llambías and Rapela, 1985). A predrift reconstruction of the Patagonia-South Africa join, such as that of Martin, Hartnady, and Goodland (1981), shows that the eastern extension of the Paleozoic magmatic arc of Patagonia lies adjacent to the Cape System.

Figure 2. Cambrian-Ordovician paleogeography: (1) Precambrian basement of the Arequipa-Belén-Antofalla block; (2) emerged land areas along the ancestral South American continental margin; (3) carbonate platforms; (4) magmatic "arcs"; (5) marine terrigenous sediments accumulated over a continental basement; (6) sediments deposited on "oceanic" basement; and (7) axis of paleotrench. The broken line in the Pacific represents the hypothetical eastern margin of the Chilenia terrane.

LATE PALEOZOIC (GONDWANA) TECTONIC CYCLE

In contrast with the Famatinian tectonic cycle, accretion of large "exotic" terranes, with the possible exception of Patagonia, did not take place during late Paleozoic. Only minor late Paleozoic-earliest Mesozoic allochthonous terranes have been recognized; the only well-documented terrane is that of fusulinid-bearing Tarlton limestone in the Madre de Dios Archipelago (Figure 5), southern Chile (lat. 50 °-52 °S) (Mpodozis and Forsythe, 1983). This limestone is associated with a large composite accretionary prism of Late Devonian to possibly early Mesozoic age that is preserved along the whole Pacific margin of Chile south of lat. 25 °S (Hervé et al., 1981). In addition, paleomagnetic studies indicate that Triassic felsic volcanic rocks near the coastal town of Los Vilos, Chile (lat. 32 °S) also may be allochthonous, as recently proposed by Forsythe et al. (1986b).

Remnants of the late Paleozoic subduction complex have been identified in several regions along the present Chilean coast, including blueschists, and fragments of the oceanic floor (Bell, 1982; Hervé et al., 1982; Godoy et al., 1984; Sepúlveda, 1984).

During the late Paleozoic, complex sequences of granitoids, both of "I" and "S" types, and related volcanic rocks were emplaced in different segments of the Cordillera Frontal and Coastal Range (Figures 1, 5) to the east of the growing accretionary prism (Mpodozis et al., 1985; Hervé et al., 1985). These igneous rocks are dominated by late-stage rhyolitic volcanic rocks such as those of the Choiyoi Group (Figure 6) in the Cordillera Frontal of Chile and Argentina between lat. 27 °-34 °S (Mpodozis et al., 1976; Caminos, 1979). Sequences of rhyolitic lavas and ignimbrites, several thousand meters thick, have been interpreted as evidences of either the products of generalized anatexis during a period of crustal extension (Zeil, 1979) or as the differentiated felsic units of a calc-alkaline subduction-related assemblage (Coira et al., 1982; Ramos et al., 1984).

In Argentina, a series of sedimentary foreland basins known as the Tarija, Paganzo, Rio Blanco-Uspallata, San Rafael, and Sauce Grande-Claromeco basins (Archangelsky, 1986) was developed east of the above-mentioned magmatic arc during the late Paleozoic.

Several phases of compressional late Paleozoic deformation

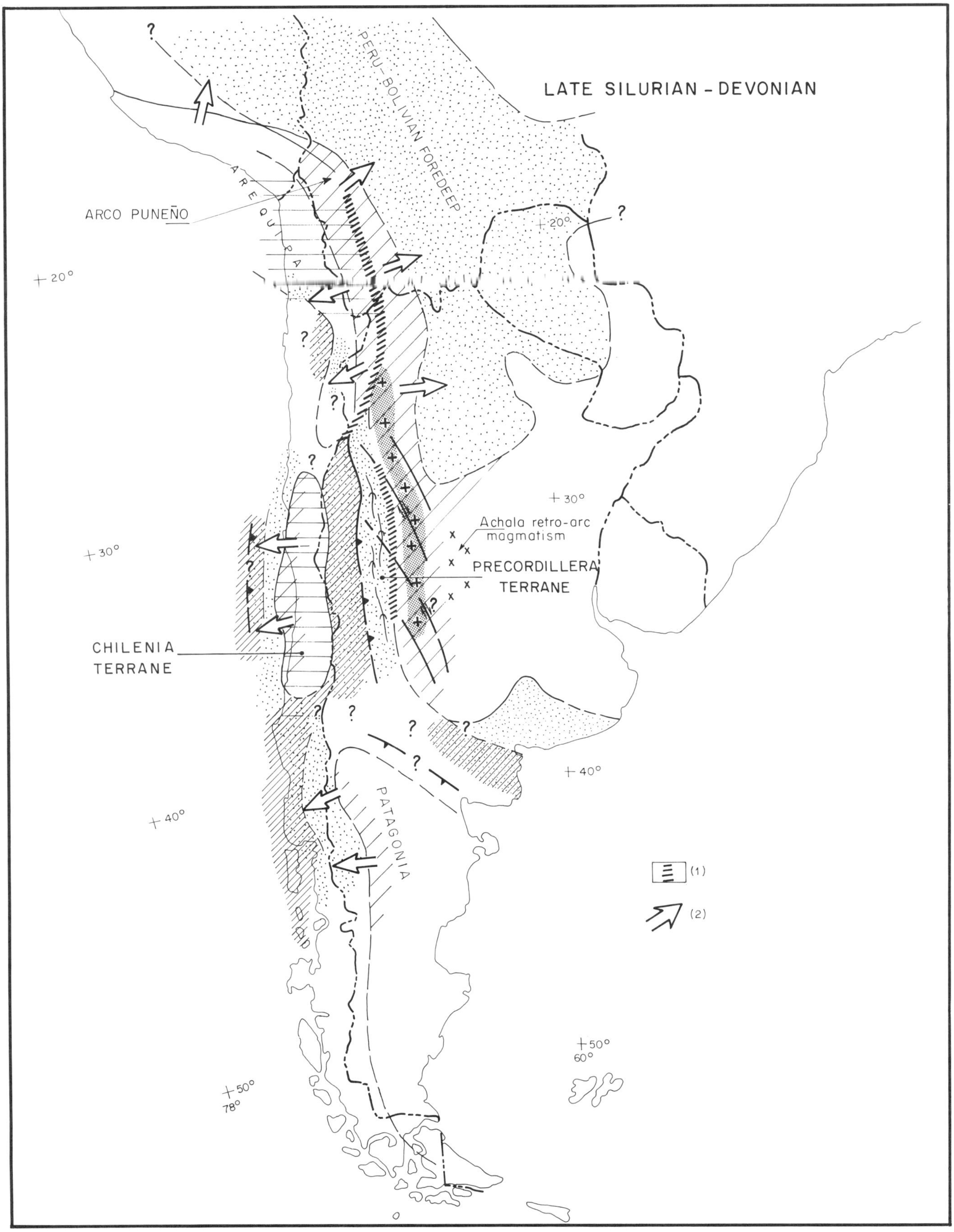

Figure 3. Late Silurian-Devonian paleogeography. Symbols are as in Figure 2, except (1) suture zones and (2) provenance direction of terrigenous sediments.

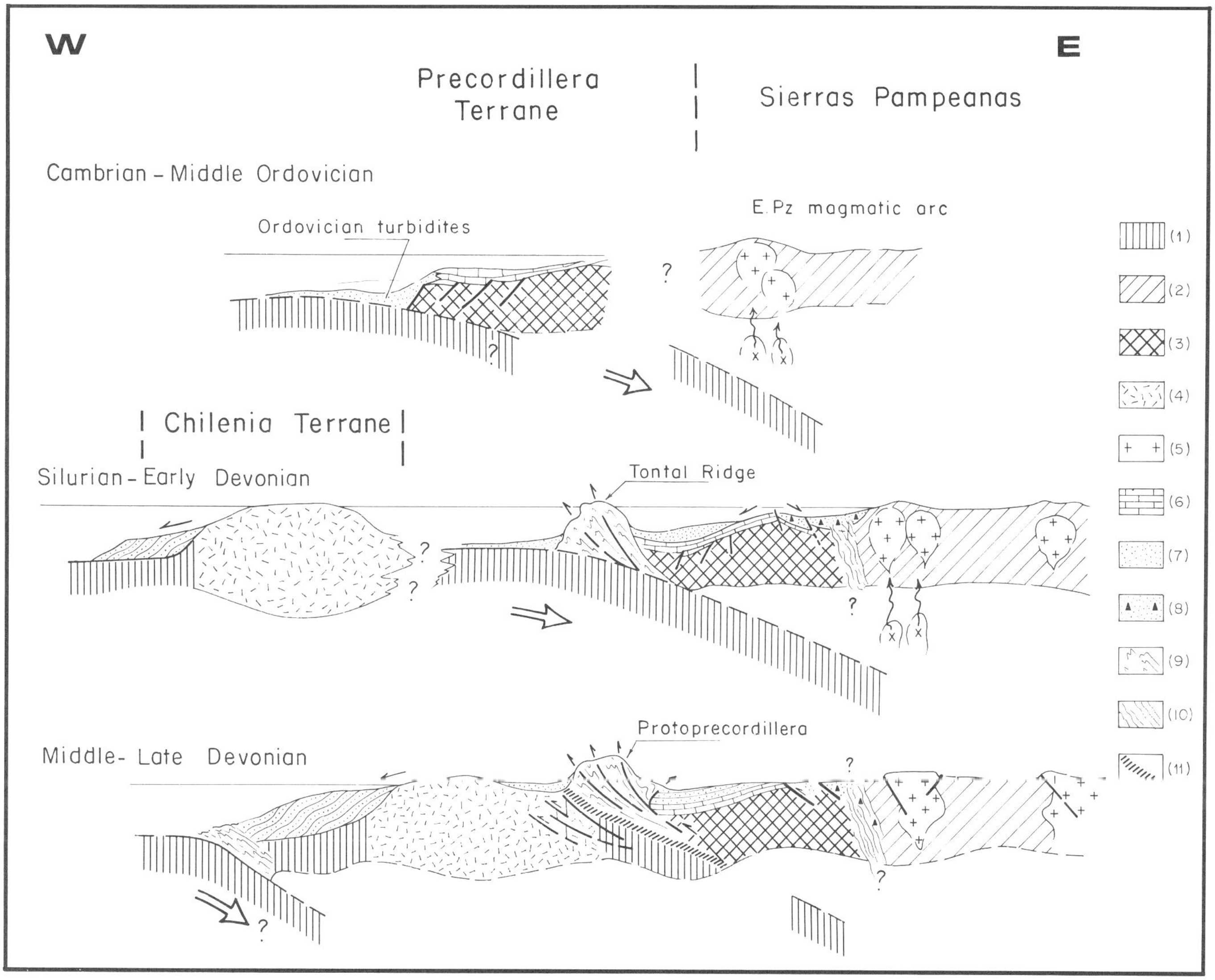

Figure 4. Diagram showing the progressive accretion of Chilenia terrane to the South American continent, between lat. 28°-36°S, during the early Paleozoic: (1) oceanic crust; (2) Precambrian basement of the Sierras Pampeanas terrane; (3) possible Precambrian basement of the Precordillera terrane; (4) "Chilenia" terrane; (5) calc-alkaline granitoids; (6) limestones; (7) turbidites; (8) olistostromes; (9) subduction complexes; (10) pervasively sheared early Paleozoic melange between the Precordillera and the Sierras Pampeanas; and (11) probable Devonian suture between the Precordillera and the Sierras Pampeanas. Also shown is the source region of the Sierras Pampeanas "arc" magmas (x) in the asthenospheric wedge over the early Paleozoic subduction zone.

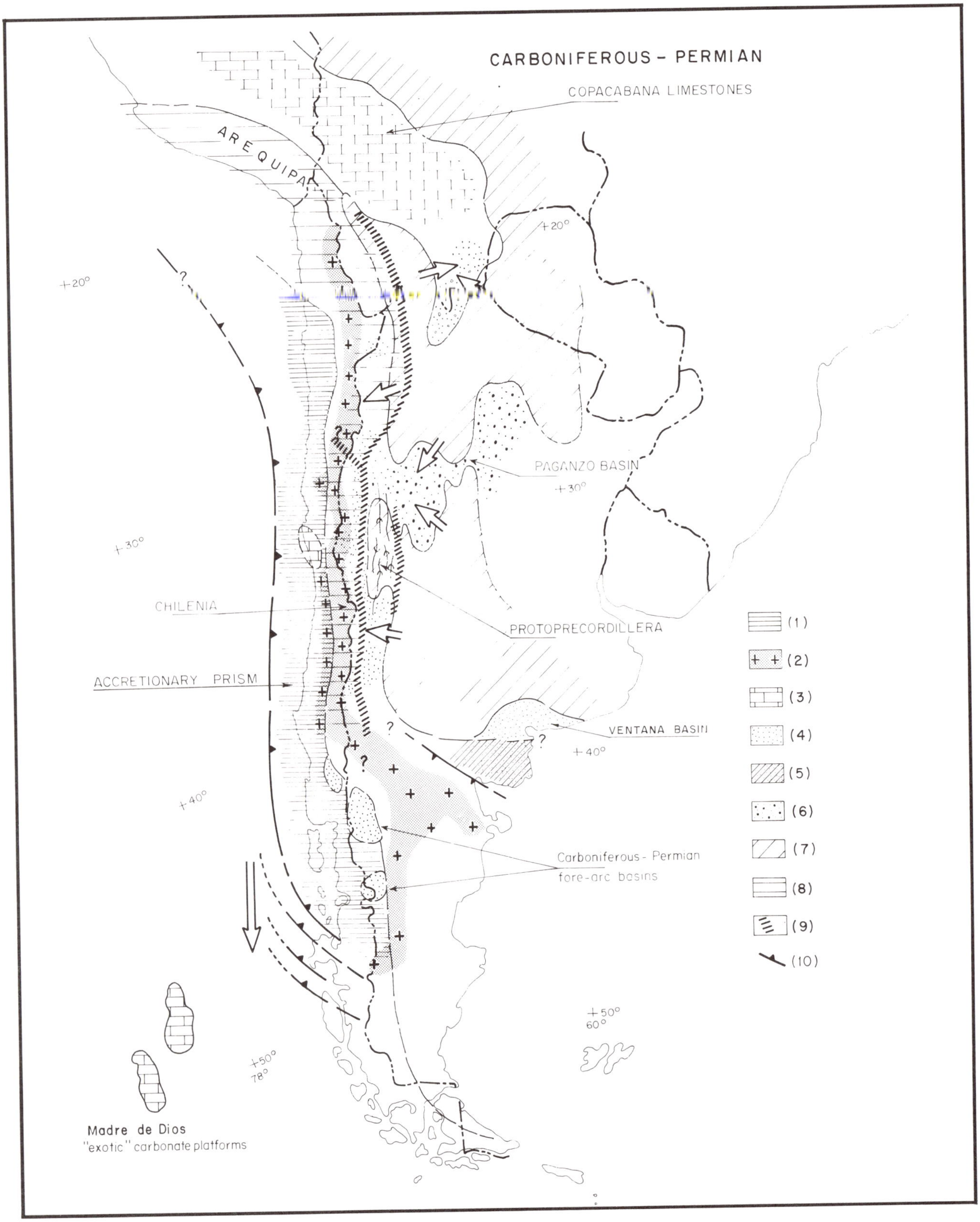

Figure 5. Late Paleozoic paleogeography: (1) accretionary prism and forearc assemblages; (2) magmatic arc; (3) carbonate rocks; (4) shallow-water marine-deltaic-terrigenous sediments; (5) the ocean-floored "basin" between Patagonia and Sierra de La Ventana; (6) terrestrial red beds; (7) uplifted area; (8) "exotic" terranes accreted to the South American margin before the Carboniferous; (9) suture zone; and (10) inferred paleotrench position; note progressive southward displacement.

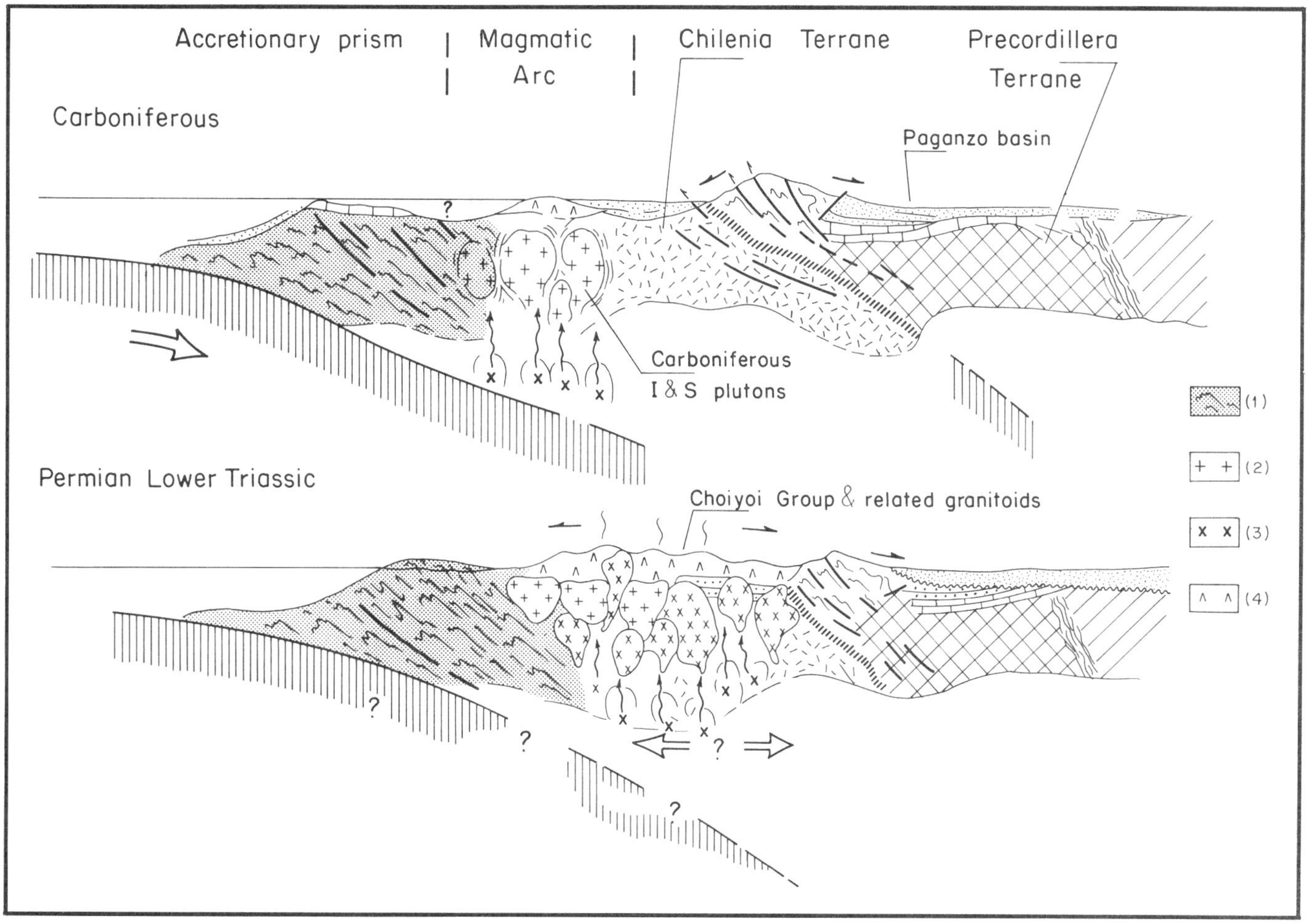

Figure 6. Diagrammatic cross section of western South America at lat. 29°-33°S (central Chile and Argentina) during Carboniferous-Early Triassic: (1) late Paleozoic accretionary prism; (2) Carboniferous granitoids; (3) Permian-Early Triassic intrusives; and (4) late Paleozoic-Early Triassic felsic volcanic rocks. Other symbols as in Figure 4.

are recognized in Argentina. The middle Permian San Rafael phase is the most important (Azcuy and Caminos, 1986); folding and thrusting associated with this phase could have been contemporaneous with the proposed time of accretion of Patagonia (Ramos, 1984).

MESOZOIC TO CENOZOIC (ANDEAN) TECTONIC CYCLE

Although the accretionary prism continued to grow along the Pacific margin of southernmost Patagonia during the Triassic to Early Jurassic, such accretion ended along the coast of north-central Chile in the late Paleozoic. During the earliest Mesozoic, a major paleogeographic change took place as the magmatic arc shifted westward to a new position near the present Chilean coast. A complex series of interconnected basins formed behind the arc during the Jurassic-Early Cretaceous (Figure 7). North of lat. 38°S, the magmatic arc was superimposed on the late Paleozoic accretionary prism, whereas the basement of the backarc basins was the late Paleozoic magmatic belt. South of lat. 40°S, both the Jurassic-Cretaceous arc and the basins developed on the former late Paleozoic forearc zone.

This composite system of arcs and related backarc basins remained active until the middle Cretaceous, when a generalized tectonic event caused major, deformation in the area of the former backarc basins. The basins were destroyed, leaving only a series of continental magmatic arcs as the dominant tectonic element during the Late Cretaceous and early Tertiary. In some areas the magmatic arcs migrated eastward with time, whereas in others they remained stationary, or even shifted westward toward the ocean. Mesozoic and Cenozoic forearc assemblages either never existed, were removed by "subduction erosion," or were shifted by lateral displacement along strike-slip faults at the Pacific margin of the continent.

The Jurassic-Cretaceous backarc basins show a variety of types ranging from ensialic sedimentary basins to oceanic-crust-floored "marginal" basins. These backarc basins reacted differently to deformation and, as a consequence, gave rise to the five major segments shown in Figure 7, each with distinctive characteristics.

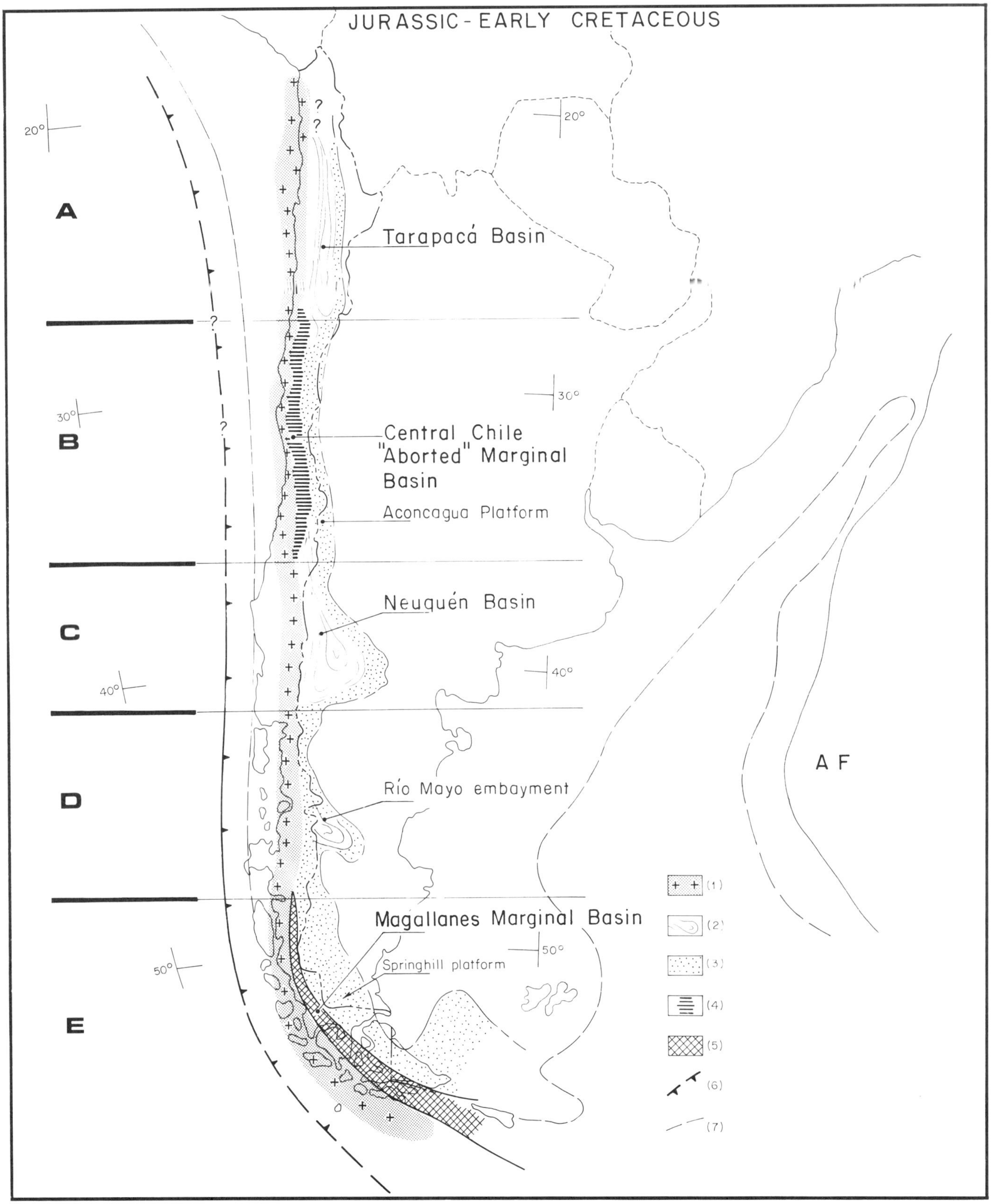

Figure 7. Jurassic-Early Cretaceous arc and backarc basin systems of the Chile-Argentine Andes: (1) magmatic arc; (2) ensialic backarc basin; (3) sedimentary platform; (4) Early Cretaceous volcanic infill of the central Chile "aborted" marginal basin; (5) oceanic crust of the Early Cretaceous Magallanes marginal basin; (6) possible paleotrench position: and (7) continental shelf edge. The characteristics and evolution of these features are discussed in the text by reference to their locations within segments A-E shown on this figure.

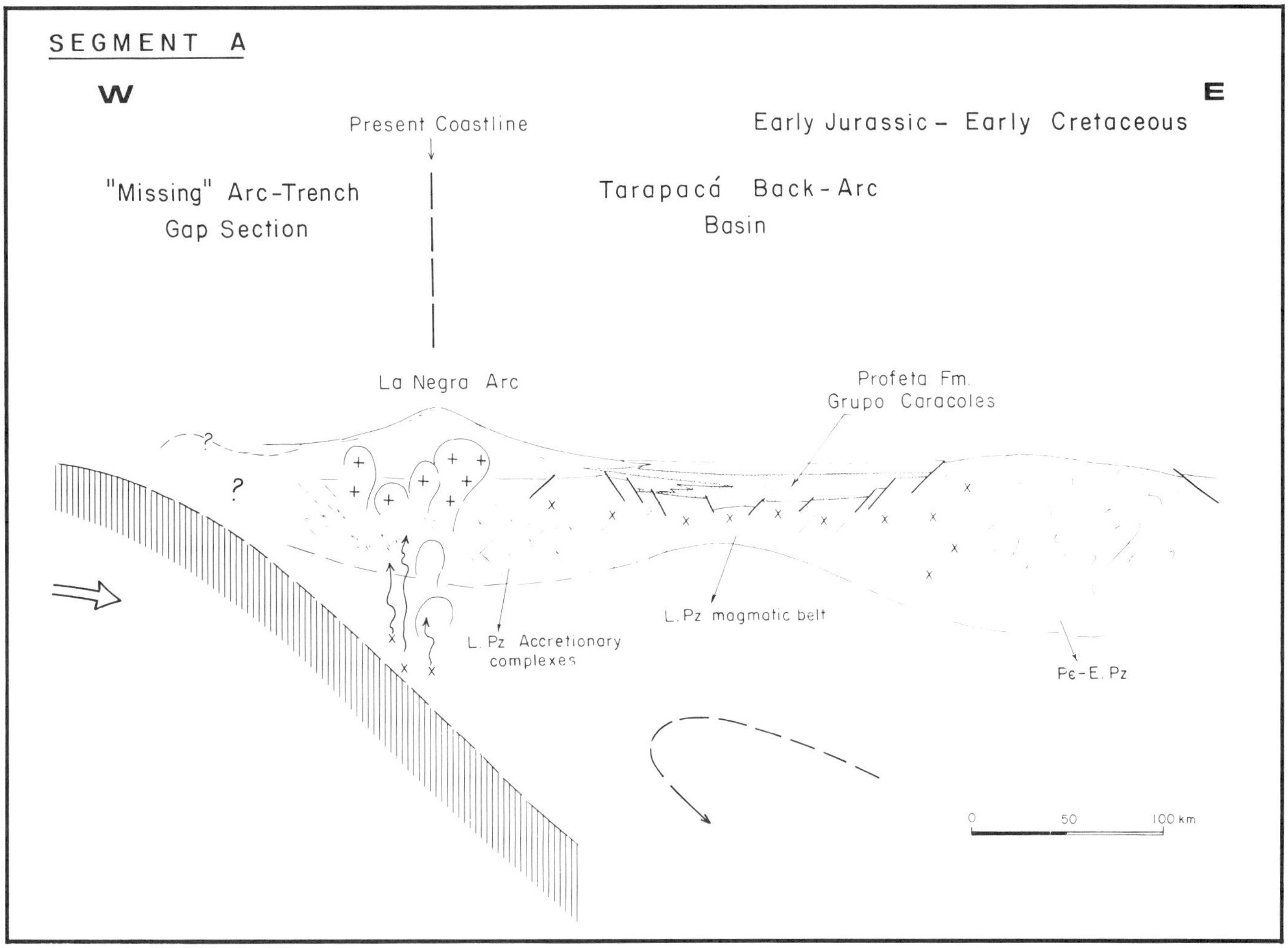

Figure 8. Jurassic-Early Cretaceous magmatic arc and backarc basin in segment A (northern Chile). The La Negra arc formed on late Paleozoic accretionary complexes; the sialic basement of the Tarapacá basin is in the zone of the late Paleozoic magmatic belt. Also shown is the source region of the arc magmas (x). A hypothetical convective roll in the asthenosphere behind the La Negra arc might have been responsible for extension in the backarc basin.

Segment A (Lat. 21°-27°S)

Segment A (northern Chile and northwestern Argentina) shows the following (Figures 8-11)—a magmatic arc in the Coastal Range that was active during the Jurassic-Early Cretaceous; a backarc ensialic basin (Tarapacá basin) in which late Triassic-early Cretaceous marine and terrestrial sediments accumulated; middle Cretaceous deformation, which caused uplift of the sialic backarc basin floor to form the Proto-Cordillera de Domeyko; discontinuous migration of the magmatic foci toward the east from Late Cretaceous to Tertiary; progressive migration of the orogenic front toward the foreland until the still-active thrust-fold belt of the sub-Andean Ranges was formed along the eastern foothills of the Andes; and active Quaternary volcanism.

Triassic-Early Cretaceous

The earliest recorded Mesozoic event in northern Chile was Late Triassic marine transgression that formed a small basin along the site of the present-day Cordillera de Domeyko (Figure 1) (Chong and von Hillebrandt, 1985). This basin was bordered on the west by highlands from which the alluvial fan, red conglomerates and breccias of the Cifuncho Formation, exposed along the present coast south of Taltal, were derived (García, 1967; Suarez, Naranjo, and Puig, 1985).

During the Early Jurassic, the marine domain of Cordillera de Domeyko progressively expanded northward to form the elongate and deep Tarapacá basin (Figure 7) in which subsidence was controlled by extensional tectonics. Several thousand meters of carbonate and terrigenous sediments of Jurassic age (Profeta Formation, Caracoles Group, Figure B) accumulated in this basin (Harrington, 1961; Naranjo and Puig, 1984; Coira et al., 1982). In the Coastal Range, Hettangian-Sinemurian marine sediments having interbedded pyroclastics conformably overlie the late Triassic continental deposits (García, 1967; Naranjo and Puig, 1984). In the Sinemurian, intensive volcanism and plutonism began at the La Negra magmatic arc (Figure 8), which extended along the present-day Coastal Range for more than 1000 km from Arica (lat. 18°S) to Chañaral (lat. 26°S). The La Negra is composed of thousands of meters of andesites and basaltic andesites in which are sporadic intercalations of marine and continental

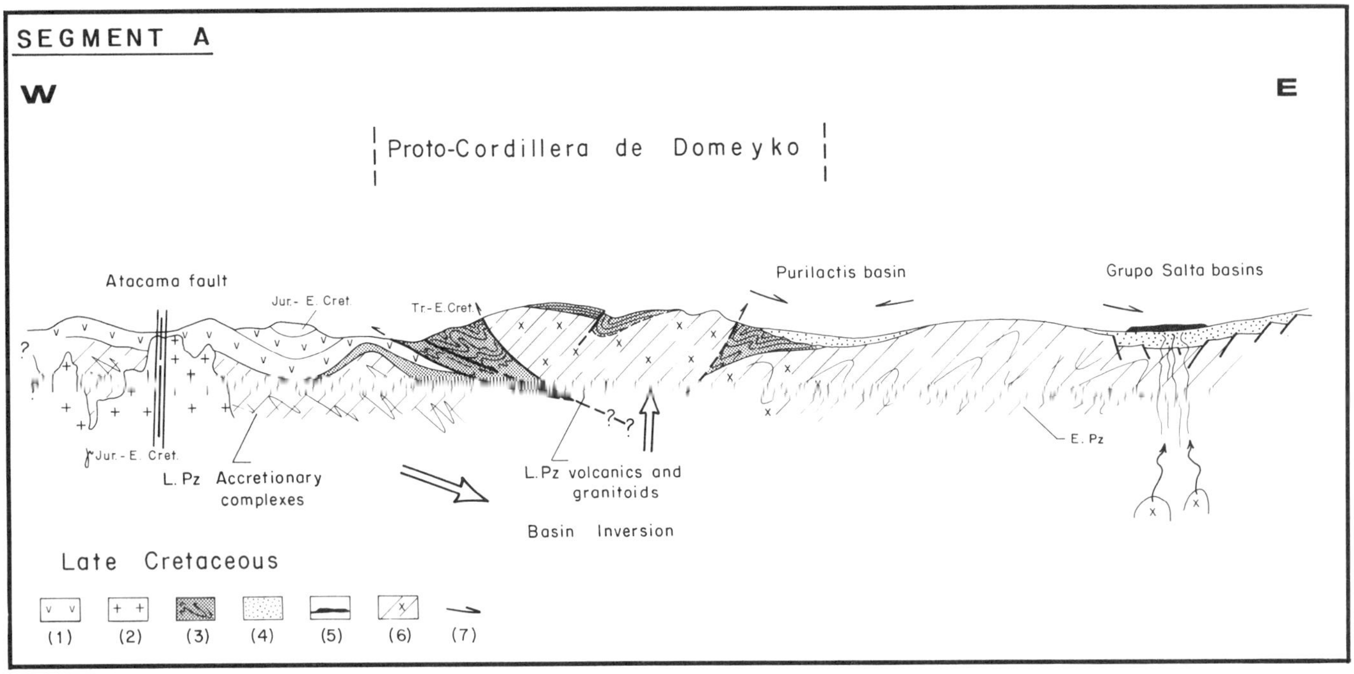

Figure 9. Cross section at lat. 25 °S immediately after middle Cretaceous deformation: (1) lavas from the extinct La Negra arc; (2) Jurassic-Early Cretaceous granitoids; (3) deformed sediments of the Tarapacá backarc basin; (4) Late Cretaceous red beds; (5) alkaline basalts; (6) uplifted sialic floor of the Tarapacá basin to form the Proto-Cordillera de Domeyko (PCD); and (7) provenance directions of the Late Cretaceous sediments. Large arrows indicate direction of middle Cretaceous uplift in the PCD and the east-directed underthrust of the La Negra arc terrane below the PCD.

sediments. Accumulation of this sequence continued until the Early Cretaceous (García, 1967; Ferraris and Di Biase, 1978; Silva, 1977; Naranjo and Puig, 1984). Large gabbroic to granitic intrusive bodies, showing radiometric ages ranging from 190 to 115 Ma (Maksaev, 1984; Boric, Díaz, and Maksaev, 1985), also were emplaced. Stratiform copper deposits are present in lavas in the lower part of the La Negra sequence (Losert, 1972; Boric and Díaz, 1984).

The Atacama fault (Figure 9), a 1000-km-long fault related to the La Negra magmatic arc, became active during Early Cretaceous, suggesting that major strike-slip movement took place at this time (Arabaz, 1971). South of Antofagasta, several massive magnetic bodies were emplaced along the fault during the Early Cretaceous (Bookstrom, 1977; Espinoza, 1979).

The sea progressively retreated from the backarc basin in the north (Sierra de Moreno) during Oxfordian time (Maksaev, 1978; Skarmeta and Marinovic, 1981), whereas in the southern part of the Tarapacá basin (Qebrada Asientos and Cordillera de Domeyko) marine conditions continued until the Tithonian, and in some areas until the Neocomian (Harrington, 1961; Pérez, 1982). The entire backarc basin was filled with continental red beds during the Early Cretaceous, and some interbedded lava flows, derived from the La Negra arc, extended eastward to the Sierra de Moreno region (Arca Formation of Maksaev, 1978).

Middle Cretaceous

A major tectonic reorganization occurred during the middle Cretaceous, when magmatic activity in the La Negra arc ceased and the rocks of the backarc region were deformed. The sialic floor of the Tarapacá basin was uplifted and became a positive topographic element, the Proto-Cordillera de Domeyko (PCD), which separated the coastal region from the present-day Altiplano-Puna (Figure 9). A complex basin developed to the east of the PCD after the middle Cretaceous, in which a sequence of continental red beds, the Purilactis Formation (Figure 9), derived from the erosion of the PCD was deposited (Marinovic and Lahsen, 1984). At the same time, a series of taphrogenic basins (Grupo Salta basins, Figure 9) formed in the Argentine foreland, and thick sequences of interbedded conglomerates, red beds, and alkalic volcanic rocks were deposited in them (Turner, 1960; Salfity, 1979). These volcanic rocks and related subvolcanic intrusions were emplaced during the three cycles of 130-100 Ma, 80-75 Ma, and 65-60 Ma (Salfity, 1982). These cycles were associated with a continuous synrift sedimentation. An ephemeral marine transgression from the north and southwest connected the various basins during Campanian-Maestrichtian time (Salfity, 1982).

Tertiary

Discontinuous migration of magmatism and deformation toward the foreland took place during the Tertiary (Figure 10). Volcanism began during latest Cretaceous and continued until the Eocene, and was widespread in the present-day Chilean Longitudinal Valley, as shown by the abundant andesites and rhyolites of the Chile-Alemania, Icanche, and Augusta Victoria formations (Naranjo and Puig, 1984; Maksaev, 1978, 1984). The volcanic rocks and related subvolcanic porphyries have K-Ar ages ranging from 72 to 39.9 Ma (Boric, Díaz, and Maksaev, 1985). Volcanic activity decreased in intensity dur-

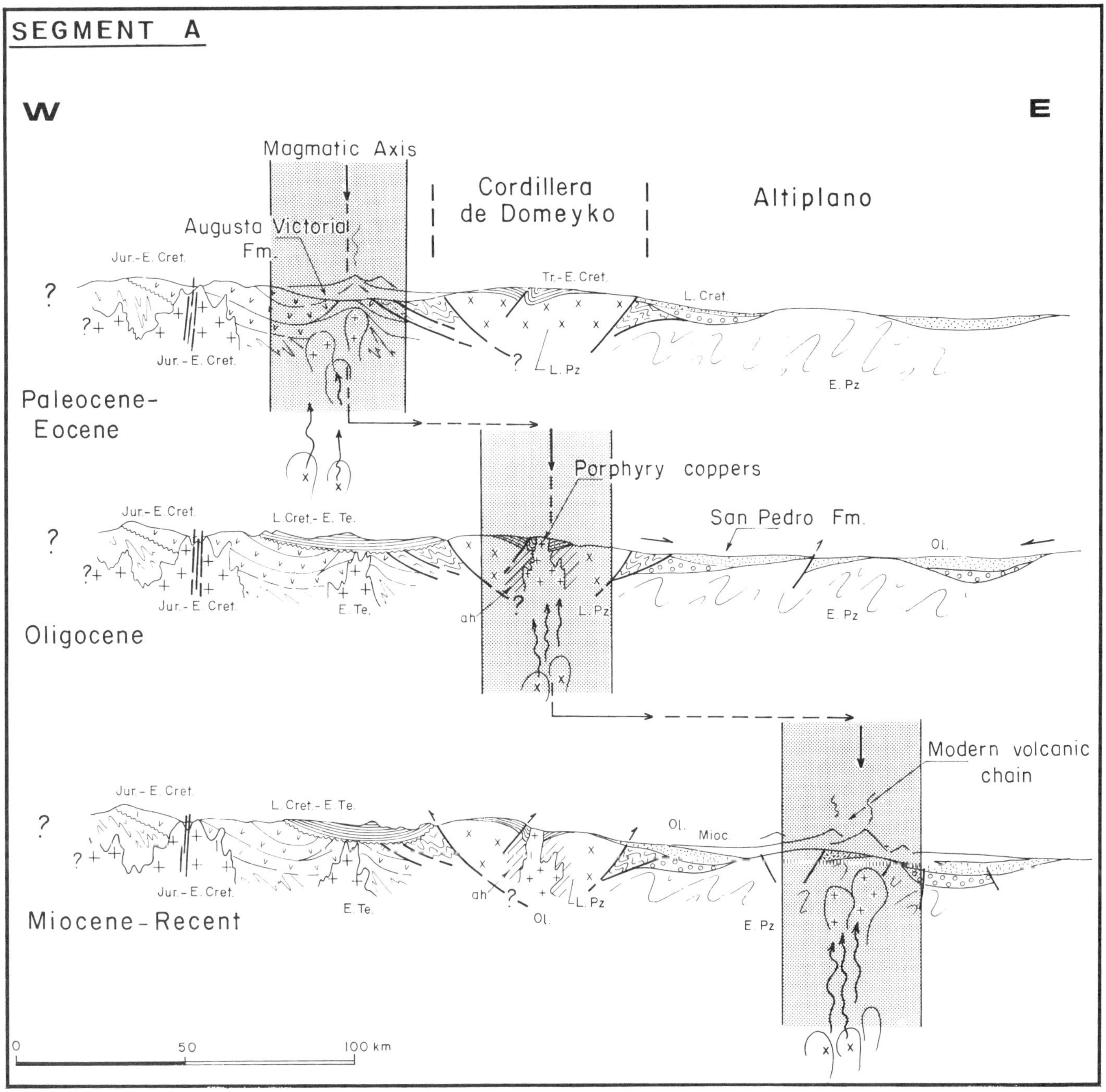

Figure 10. Diagrams illustrating the discontinuous eastward migration of magmatic foci in segment A during the Cenozoic.

ing the Oligocene, and to the east of the PCD, both in Chile and in the Argentine Puna, red continental sediments and interbedded evaporites were deposited (Sichal, San Pedro, and Tambores formations, Pastos Grandes Group; Coira et al., 1982). Igneous intrusive activity took place along the PCD axis during late Eocene and Oligocene (48-28 Ma), and the large porphyry copper deposits of this region are associated with intrusions emplaced at this time. The sparse tuff beds in the Sichal and San Pedro formations (Maksaev, 1978; Travisany, 1978) are the only volcanic rocks of known Oligocene age in this part of Chile. The Nevado de Acay Granite (26 Ma) and associated volcanic rocks constitute the only relatively large Oligocene igneous complex in northwestern Argentina (Mirré, 1974; Llambías, Sato, and Tomsic, 1986). This interval of sparse magmatism coincides with a period of highly oblique convergence between the Nazca and South American plates (Cande, 1983). The large Oligocene porphyry copper deposits were emplaced along the PCD axis, which was a zone of crustal weakness characterized by strike-slip faults such as the "West Fissure" at Chuquicamata. A large part of the horizontal oblique convergence vector could have been absorbed in this zone (see Dewey, 1980).

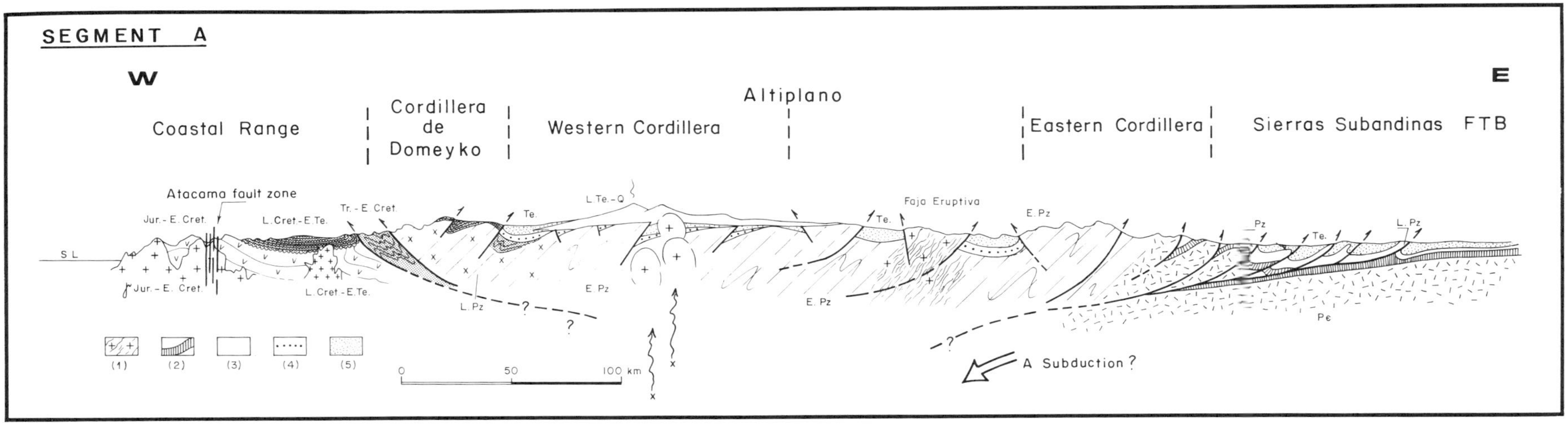

Figure 11. Generalized profile of the Andes at lat. 22 °S (segment A). Symbols as in Figure 9 except (1) Faja Eruptiva de La Puna, and (2-5) sedimentary cover of the Sierras Sub-Andinas region—(2) early Paleozoic, (3) late Paleozoic, (4) Late Cretaceous, (5) Tertiary continental deposits. Simplified structural section of the Sierras sub-Andinas according to Mingram et al. (1979) and Allmendinger et al. (1983).

During the early Miocene, an increase in the rate of convergence of the Nazca and South American plates, as well as a shift toward a more orthogonal convergence vector, coincided with a period of deformation in the Cordillera de Domeyko, which resulted in folding of the late Eocene-Oligocene strata and reactivation of reverse faults. At this time, intense volcanism began in the Western Cordillera (Figure 1). Explosive caldera-type, rhyolitic volcanism was particularly intense to the north of lat. 22 °S, where the widespread ignimbrites of the Altos de Pica and Oxaya formations were emplaced (Salas et al., 1966; Lahsen, 1982; Gardeweg and Ramírez, 1984). Andesitic stratovolcanoes were dominant south of lat. 25 °S (Zentilli, 1974; González Ferrán, Baker, and Rex, 1984). Volcanic activity became much more widespread during the late Miocene after a new phase of deformation (Quechua phase), which coincided with cessation of spreading in the Farallon-Nazca ridge and the beginning of spreading along the East Pacific Rise (Mammerickx, Herron, and Dorman, 1980). The Quechua phase, although evident in the Western Cordillera, seems to have been of greater importance in the Argentine Puna (Coira et al., 1982; Jordan and Gardeweg, 1986). The front of compressional deformation migrated eastward during the late Cenozoic from the Western Cordillera toward the Eastern Cordillera and sub-Andean ranges (Figure 1). Deformation in the easternmost flank of the sub-Andean ranges has continued until the present along an active thrust-fold belt in the Sierras sub-Andinas (Figure 11) that extends from northwestern Argentina through Bolivia into the eastern Andean foothills of Perú (Mingram et al., 1979; Allmendinger et al., 1983; Mégard, 1984).

Structural Geology

A schematic and simplified structural section of the Andes at lat. 23 °S (Figure 11) shows two major zones of crustal shortening—the thrust-fold belt of the Sierras sub-Andinas and the Cordillera de Domeyko. In the Sierras sub-Andinas and adjacent Eastern Cordillera (Figure 11), part of the Precambrian basement of the Andean foreland apparently is currently being thrust below the uplifted Puna plateau as a part of an "A subduction" process similar to that proposed by Bally (1981) for the Canadian Rockies. In contrast, the deep structure of the Cordillera de Domeyko, though still uncertain, has been interpreted as the result of thrusting to the west of the sialic floor of the Tarapacá basin along faults that flatten at depth (Figure 11). Movement on these faults may have started in the middle to Late Cretaceous. If this hypothesis is correct, the abnormally thick crust of the Andes at segment A may have an important component of tectonic thickening related to shortening along both zones of deformation.

It is evident that the distance between the old La Negra arc and the present trench is too short when compared with that of other volcanic fronts such as the late Cenozoic arc in the Western Andean Cordillera. Large pieces of the forearc zone and parts of the Mesozoic magmatic arc must have been removed during the Cretaceous and Tertiary, either by subduction erosion or by longitudinal strike-slip faults such as the Atacama fault (see Rutland, 1971; Coira et al., 1982). These phenomena could also be related, at least in part, to the episodical migration of magmatic foci toward the continental interior during the Tertiary (Figure 8).

Segment B (Lat. 27 °-33 °S)

Segment B (central Argentina and Chile) shows the following (Figures 12-15)—Jurassic magmatic arc in the Coastal Range; backarc sedimentary platform (Aconcagua platform) toward the east; development of an "aborted marginal basin" during the Early Cretaceous in which huge volumes of andesites and basalts were erupted through the thinned and stretched continental crust behind the arc; progressive deformation and thrusting toward the east of the Early Cretaceous "marginal" basin, beginning in middle Cretaceous, to form the Aconcagua thrust-fold belt, eastward migration of magmatism during the Cretaceous-Tertiary; progressive migration of the orogenic front toward the Precordillera and block faulting of the basement of the Sierras Pampeanas during the late Cenozoic; and absence of Quaternary volcanism.

Segment B is characterized by lack of both Quaternary volcanism and a Central Valley in Chile (Figure 1). In contrast with segment A, Andean deformation extended eastward from the trench for more than 1000 km, forming the fault-bounded basement blocks of the Sierras Pampeanas (Jordan et al., 1983a, 1983b). Such a wide zone of deformation is correlated with a shallow-dipping Benioff zone (Barazangi and Isacks, 1976; Bevis and Isacks, 1984). Segment B approximately coincides with the zone of collision of Chilenia during the Paleozoic (Ramos et al., 1984; Ramos et al., 1986). A large subsiding volcanic zone, the central Chile "aborted" basin, formed in this segment during Late Cretaceous. Although the geologic history of the segment as a whole was similar, some important differences can be observed from north to south. The features shown in the south half of the segment (lat. 31 °31 ′-33 °05 ′S) are typical of segment B as a whole.

Jurassic-Early Cretaceous

Marine transgression and volcanism of Middle to Late Triassic age are recorded by the El Quereo Formation, which accumulated on the late Paleozoic terrane of the coastal region of central Chile (Cecioni and Westermann, 1968). The Ajial and Cerro Calera Formations, consisting chiefly of rhyolitic pyroclastic rocks, were emplaced in the Coastal Range (lat. 32 °-34 °S) during an interval of Early Jurassic volcanism (Thomas, 1958; Piracés, 1977). Volcanic eruptions were partly submarine, as indicated by the presence of interbedded Bajocian limestones (Figure 12). At the same time, large granitic to gabbroic plutons of the Mincha Superunit were emplaced in the coastal region (Rivano et al., 1985). These plutons have K-Ar ages ranging from 191 to 138 Ma (Munizaga and Vicente, 1982; Hervé et al., 1985; Rivano et al., 1985). In the backarc region to the east, a Liassic to Dogger marine transgression initiated a clastic-carbonate platform environment, the Aconcagua platform (Figure 12), on the late Paleozoic igneous basement of the Cordillera Frontal (Schiller, 1912; Vicente, 1972; Ramos, 1985a, 1985b). Uplift during Late Jurassic caused a change to subaerial conditions, as indicated by the Horqueta Formation in the Coastal Range (Piracés, 1977). Meanwhile, in the backarc region, an Oxfordian marine regression is indicated by thick evaporite units such as the Auquilco gypsum covered by Upper Jurassic continental red beds. These sediments were derived by erosion of volcanic rocks in the Coastal Range. They thin progressively

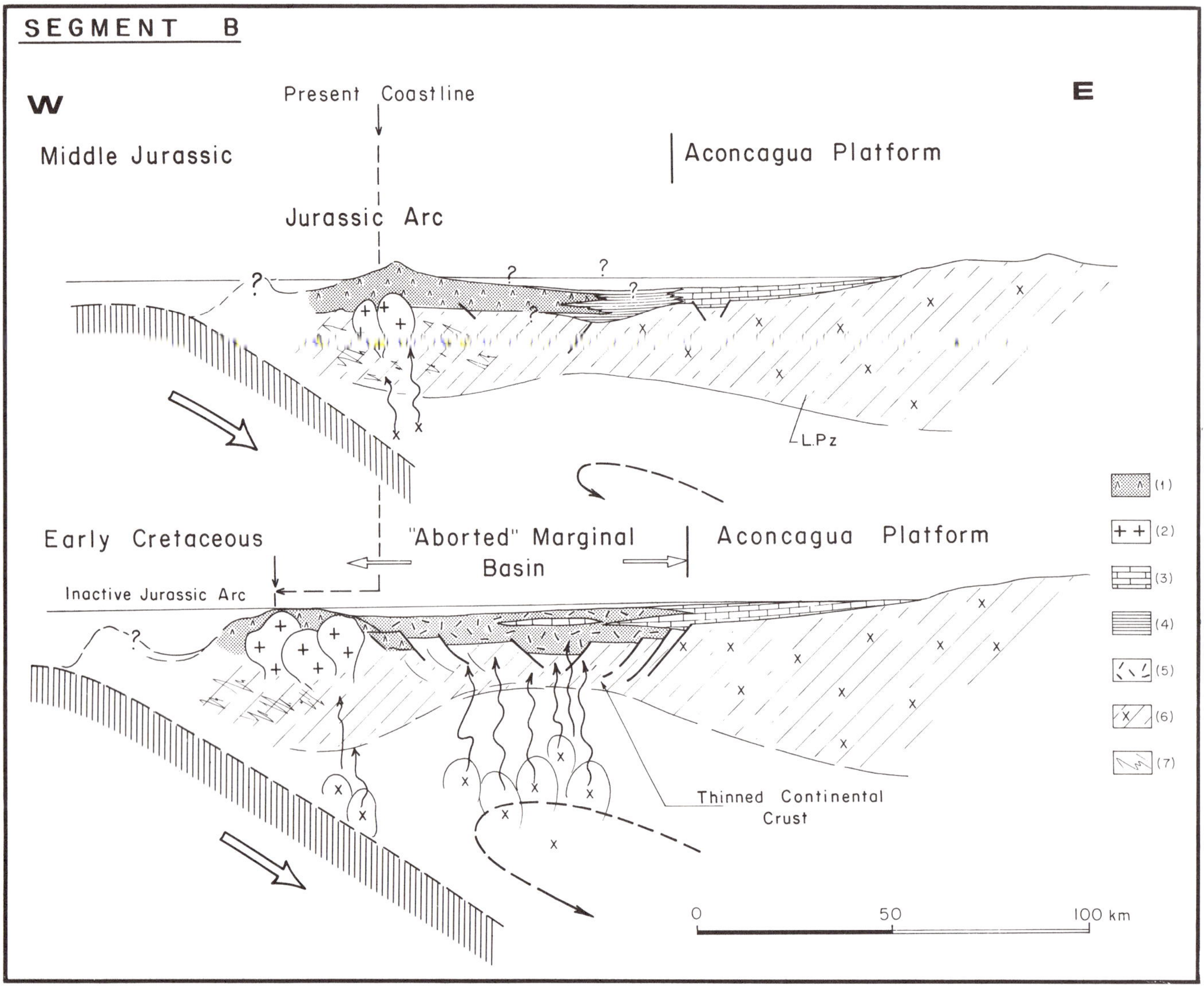

Figure 12. Paleogeographic sections for Jurassic-Early Cretaceous of central Chile. Segment B (lat. 27°-35°S): (1) Jurassic felsic volcanic rocks; (2) granitoid plutons; (3) marine and continental sediments of the Aconcagua platform; (4) deep-basin sedimentary facies; (5) Early Cretaceous volcanic infill (andesites and basalts) of the central Chile "aborted" marginal basin; (6) late Paleozoic plutonic-volcanic basement of the Cordillera Frontal; and (7) late Paleozoic accretionary complexes.

eastward and change from alluvial fan facies having sporadic intercalated volcanics, the Rio Damas Formation in Chile, to fine fluvial sandstones of the Tordillo Formation in Argentina (Groeber, 1953).

The sea again advanced over the Aconcagua platform during Early Cretaceous (Ramos, 1985a, 1985b) and was followed by—(a) a noticeable decrease of plutonism in the Coastal Range, where only small stocks of this age are found just to the east of the Jurassic magmatic belt (Nasi, 1984; Rivano and Sepúlveda, 1985); (b) a progressive change in the composition of volcanism to andesites and flood basalts interbedded with marine carbonate rocks (Lo Prado and Veta Negra formations) having low Sr^{87}/Sr^{86} initial ratios (Piracés, 1976; Levi and Aguirre, 1981; Nasi, 1984; Aberg et al., 1984); and (c) an eastward expansion of volcanism into the Aconcagua platform, where the Early Cretaceous lavas of the Los Pelambres Formation interfinger with Neocomian limestones along the Chile-Argentina border. Aberg et al. (1984) and Levi and Aguirre (1981) interpreted the widespread volcanic activity of the Early Cretaceous as taking place in a short-lived "aborted" marginal basin of Hauterivian-Albian age. The term "aborted" was used by Aberg et al. (1984) to denote a zone where large volumes of basalts and andesites were erupted through a thin, attenuated continental crust which did not evolve into an oceanic-crust-floored basin such as that in segment E (Dalziel, De Witt, and Palmer, 1974). Large volumes of volcanic rocks were erupted during a period of crustal extension, when an elevated thermal gradient associated with subsidence caused pervasive "burial" metamorphism of the volcanic pile. The basin was filled with coarse red conglomerates interbedded with lavas and limestones of the Las Chilcas Formation (Thomas, 1958) at the end of the Early Cretaceous.

Late Cretaceous-Tertiary

The Late Cretaceous-Tertiary was characterized by progressive deformation and eastward migration of the magmatic foci. At the beginning of the Late Cretaceous, the marginal basin rocks were intruded by large granodioritic plutons (Figure 13) of the Illapel Superunit (Rivano et al., 1985), which have ages of 130-85 Ma (Drake et al., 1982; Munizaga and Vicente, 1982; Rivano et al., 1985). The magmatic activity continued to the Miocene with the development of discrete belts that are progressively younger eastward (Rivano et al., 1985). Volcanic activity was reestablished during the Late Cretaceous along the western slopes of the Cordillera Principal (Thomas, 1958; Nasi, 1984). Early Miocene lavas and pyroclastic rocks of andesitic to rhyolitic composition (Farellones Formation), having K-Ar ages of 22 to 11 Ma, are still farther to the east (Munizaga and Vicente, 1982; Vergara et al., 1985; Ramos et al., 1985; Rivano et al., 1985). Both extrusive and intrusive magmatism ceased in the late Miocene, perhaps the result of shallowing of the Benioff zone in these latitudes (Jordan et al., 1983b; Kay et al., 1985) and the disappearance of the asthenospheric wedge between the continental lithosphere and the subducted oceanic plate.

Structural Geology

The structural style of segment B is different from that of segment A, chiefly in having the marginal basin terrain thrust eastward over the Aconcagua platform. The closure of this basin and the related shortening produced the Aconcagua fold-thrust belt (AFTB) along the eastern border of the Cordillera Principal (Figures 14, 15). The AFTB consists of a series of west-dipping thrust sheets (five in the Mendoza River valley; Ramos, 1985b), which are composed of the Upper Jurassic and Necomian sedimentary rocks (Figure 14) of the Aconcagua platform. The Oxfordian gypsum units (Figure 14) acted as the decollement horizon (Schiller, 1912; Vicente, 1972; Yrigoyen, 1976; Ramos, 1985b). The crustal shortening due to thrusting in the AFTB is estimated to be approximately 29 km (Ramos, 1985b).

The Aconcagua fold-thrust belt developed progressively during the interval of Late Cretaceous to Miocene. The Farellones Formation of Miocene age (Figure 16) unconformably overlies the deformed rocks of the inner (western) zones of the AFTB (Yrigoyen, 1976; Rivano and Sepúlveda, 1986). Oligocene-early Miocene alluvial fan sediments, the Santa Maria Conglomerates of Schiller (1912), accumulated along the eastern AFTB front and were later involved in the thrusting, confirming the eastward migration of the thrust front (Polanski, 1972; Ramos, 1986). The shallowing of the Benioff zone during the Miocene (Kay et al., 1985) was followed by deformation in the Cordillera Frontal and formation of high-angle reverse faults. It was also accompanied by tectonic reactivation and related shortening of about 50 km in the Precordillera (Ortiz and Zambrano, 1981; Baldis et al., 1982). Andean thrust faulting began in the Precordillera during middle to late Miocene and is still active in the eastern foothills. The Precordillera is interpreted as being a thin-skinned Andean fold-thrust belt similar to the sub-Andean Ranges of segment A.

Eastward migration of deformation during middle to late Miocene caused block faulting of the Precambrian-early Paleozoic crystalline basement in the Sierras Pampeanas of Argentina (Figure 16). Some of the faults show evidences of Quaternary displacements (Jordan et al., 1983b; Introcaso, Lyon, and Ramos, 1986).

Segment C (Lat. 33 °-41 °S)

Segment C (Andes of south-central Chile and south Mendoza and Neuquén provinces, Argentina) shows the following (Figures 17, 18)—a stationary Jurassic-Quaternary magmatic arc developed along the axis of the present-day Cordillera Principal; toward the east, development of a large backarc ensialic basin (Neuquén basin) that became filled with marine and continental sediments during the Jurassic-Early Cretaceous; deformation of this basin, beginning in middle Cretaceous, which caused shortening and faulting of the basement below a decollement characterized by a complex system of overthrust and underthrust faults associated with wrench faults; and active Quaternary volcanism.

A fundamental feature of segment C is the scarcity of volcanic and intrusive rocks of Mesozoic age in the Coastal Range. The magmatic activity during this interval was concentrated in the Cordillera Principal along an arc that was stationary, showing no evidence of migration with time. To the east of the arc, a wide ensialic and asymmetric foreland basin, the Neuquén backarc basin (Figure 17), developed as a marine embayment. The Neuquén embayment shares some common features with the Aconcagua and Tarapacá basins, but differs in the pattern of subsequent deformation. Although deformation was intense in the inner Cordilleran sector, where the Agrio fold-thrust belt (Figure 18) was formed, it was slight in the outer foreland area, where most of the primary features of sedimentary facies are still in place. The Neuquén basin is one of the major oil producing areas of Argentina and consequently has been studied in great detail.

Magmatic Arc

The igneous rocks exposed between lat. 33° and 38°S consist of late Tertiary volcanic and intrusive rocks. However, to the south of lat. 38°S, Quaternary glaciation has removed the Tertiary volcanic cover and exposed granitic basement rocks in the Cordillera Principal. Young volcanic rocks are represented by symmetrical cones of modern volcanoes. North of lat. 38°S, the only pre-Tertiary magmatic rocks exposed are scattered Late Cretaceous stocks. Extensive Oligocene-Miocene lacustrine sediments of the Lolco Formation (Salinas, 1981) at Lonquimay interfinger to the north with volcanic and volcaniclastic rocks of the Cura-Mallin Formation (Niemeyer and Muñoz, 1983). These two units are cut by Miocene granodioritic stocks and are overlain by the early to middle Miocene volcanic Trapa-Trapa Formation, the late Miocene volcanic Campanario Formation, and the Pliocene volcanic Cola de Zorro Formation (Muñoz and Niemeyer, 1984). Superimposed on these units are the Quaternary stratovolcanoes of the Southern Volcanic Zone of the Andes (SVZ) (Moreno, 1976). The magmatic activity east of the Cordillera Principal resulted in emplacement of andesitic subvolcanic intrusions and lava flows of Miocene to Pliocene age (Groeber, 1946; Haller et al., 1985). Between lat. 34° and 37°S, extensive backarc plateau basalts developed at the same time (Llambías, 1966; González Díaz, 1972; Muñoz and Stern, 1986).

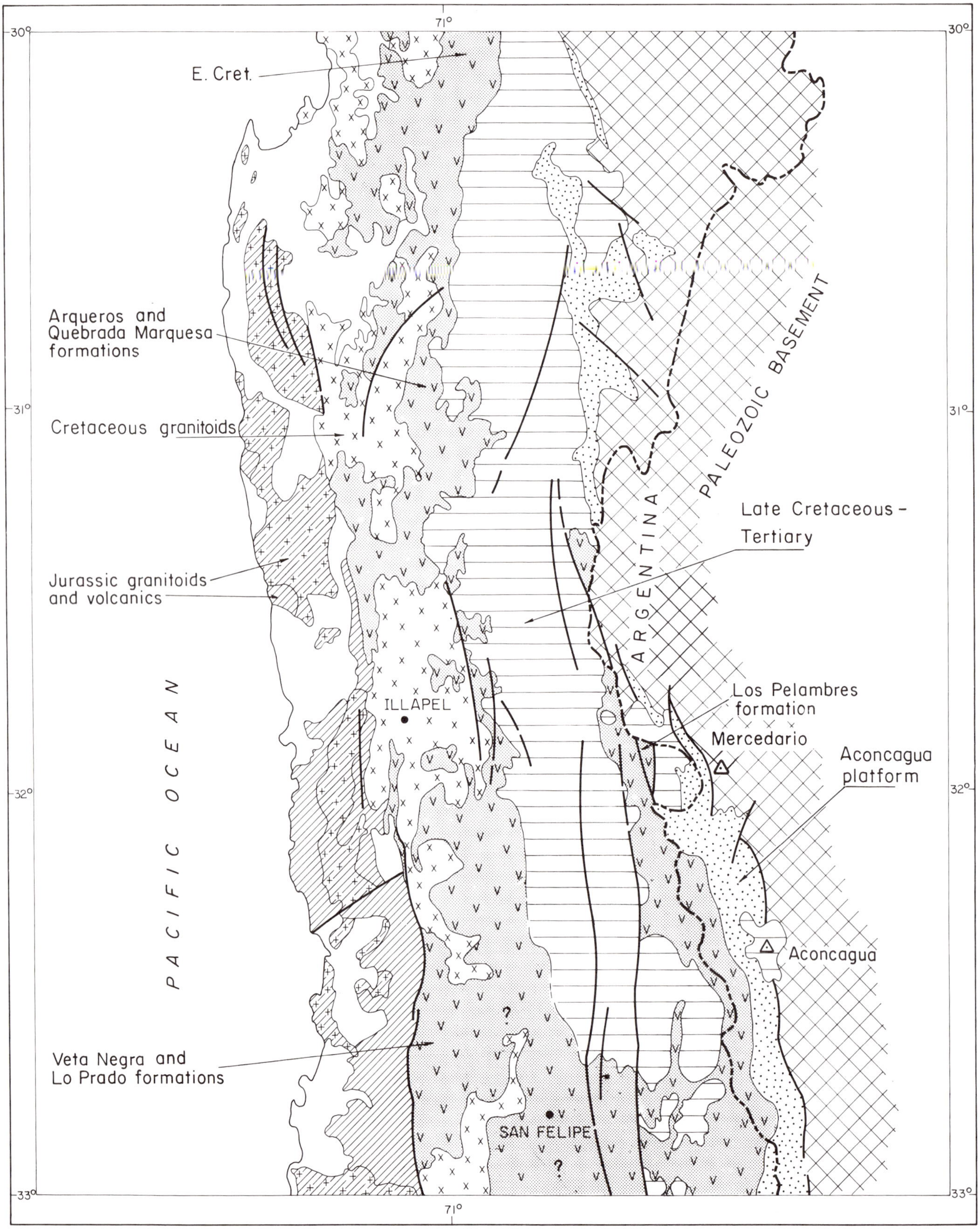

Figure 13. Distribution of the Early Cretaceous "aborted" marginal basin volcanic rocks (shaded area) in central Chile (lat. 30°-33°S). After Rivano and Sepúlveda (1986), Rivano et al. (in preparation), and Ramos (1985a).

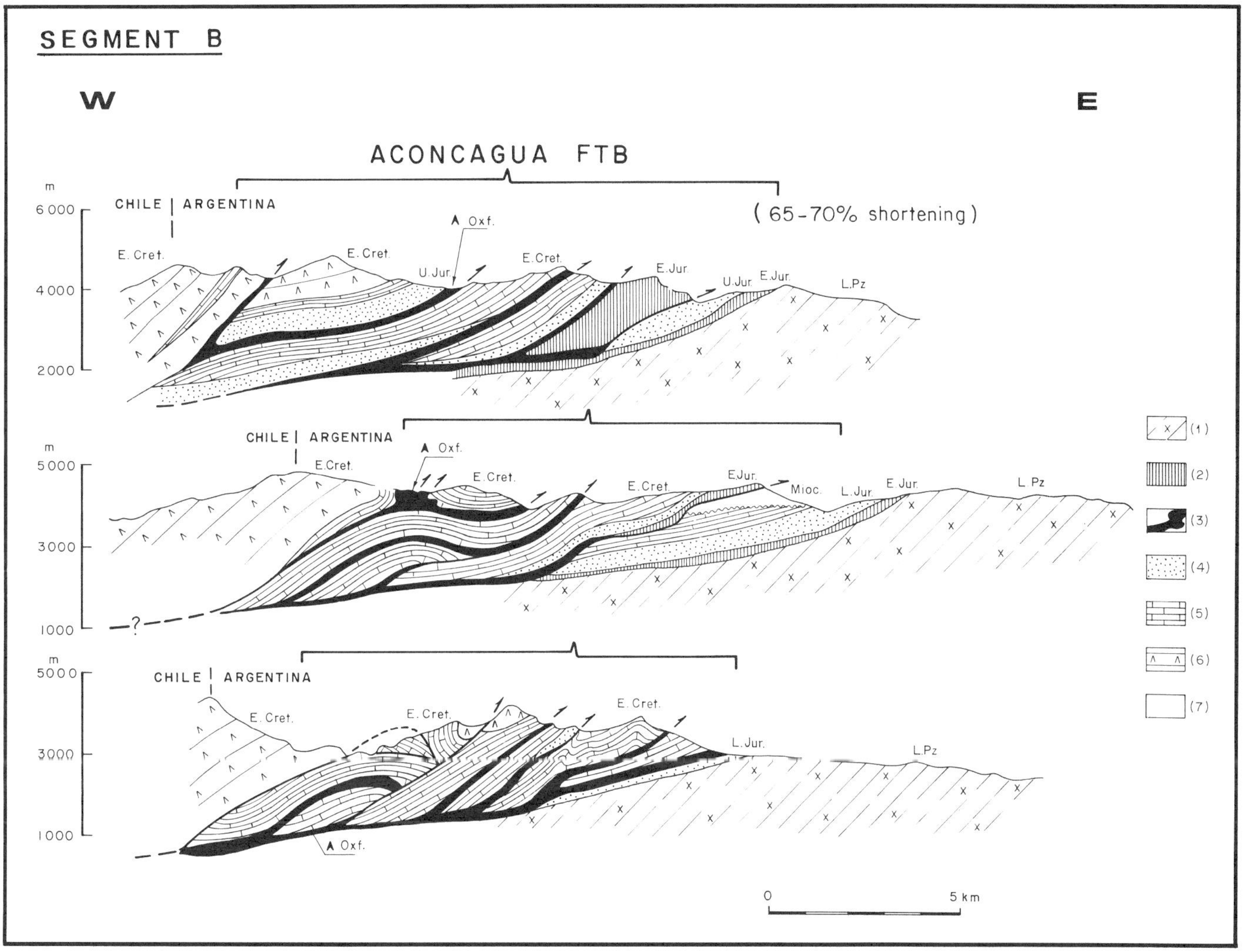

Figure 14. Cross sections of the Aconcagua fold-thrust belt south of Puente del Inca-Las Cuevas area, Mendoza province, Argentina (lat. 33 °S), according to Ramos (1985b): (1) late Paleozoic basement of the Cordillera Frontal; (2) Early Jurassic marine sediments; (3) Late Jurassic (Oxfordian) evaporites; (4) Late Jurassic continental red beds; (5) Early Cretaceous limestones of the Aconcagua platform; (6) Early Cretaceous volcanic rocks (Los Pelambres Formation) of the "aborted" marginal-basin terrane; and (7) Tertiary continental sediments.

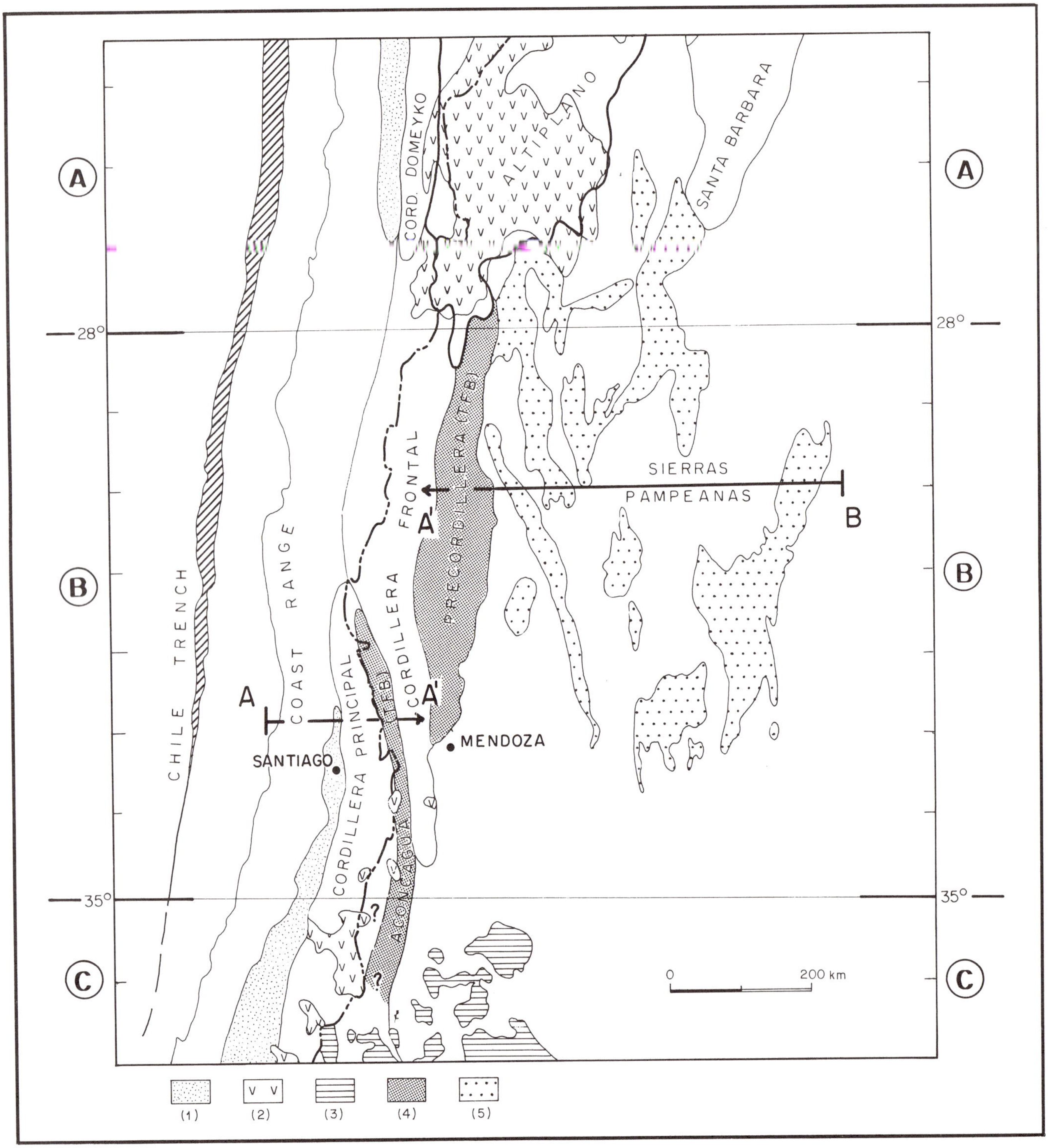

Figure 15. Major morphotectonic elements of the Andes in segment B (modified from Jordan et al., 1983a): (1) Cenozoic infill of the Chilean Central Valley; (2) modern arc volcanic rocks; (3) modern "backarc" lavas; (4) Andean thrust and fold belts (Aconcagua FTB and Precordillera); and (5) uplifted Precambrian-early Paleozoic crystalline massifs (Sierras Pampeanas).

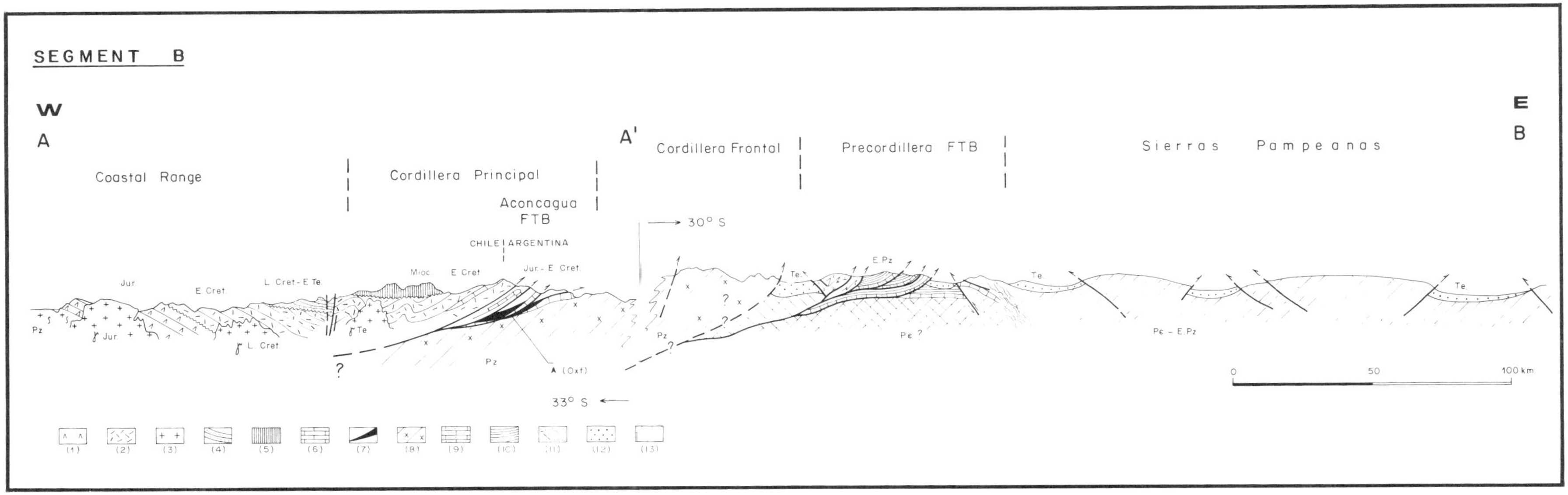

Figure 16. Composite structural section of the Andes in segment B: (1) Jurassic felsic volcanic rocks; (2) Early Cretaceous andesites and basalts of the central Chile "aborted" marginal basin; (3) granitoid plutons; (4) Late Cretaceous(?) volcanic rocks and red beds; (5) Miocene volcanic rocks (Farellones Formation); (6) Jurassic-Early Cretaceous sedimentary rocks of the Aconcagua platform; (7) Late Jurassic gypsum; (8) late Paleozoic crystalline and sedimentary units of the Cordillera Frontal; (9) Ordovician limestones of the Precordillera; (10) early Paleozoic sheared melanges; (11) Tertiary continental sediments; (12) Tertiary sediments (Te); and (13) Precambrian-early Paleozoic basement of the Sierras Pampeanas.

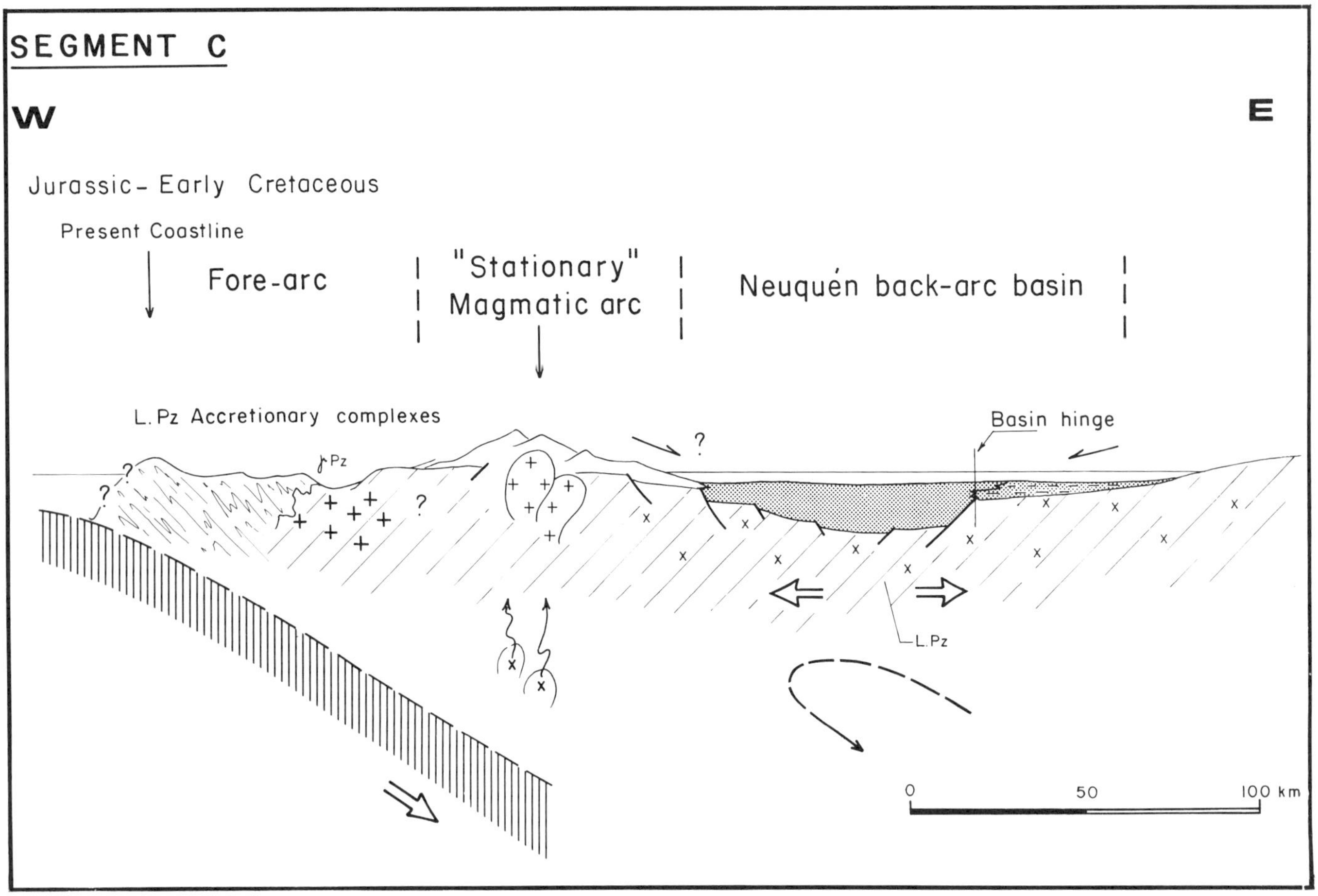

Figure 17. Paleogeographic section of segment C (lat. 38 °S) for Jurassic-Early Cretaceous time. Symbols as in Figure 12.

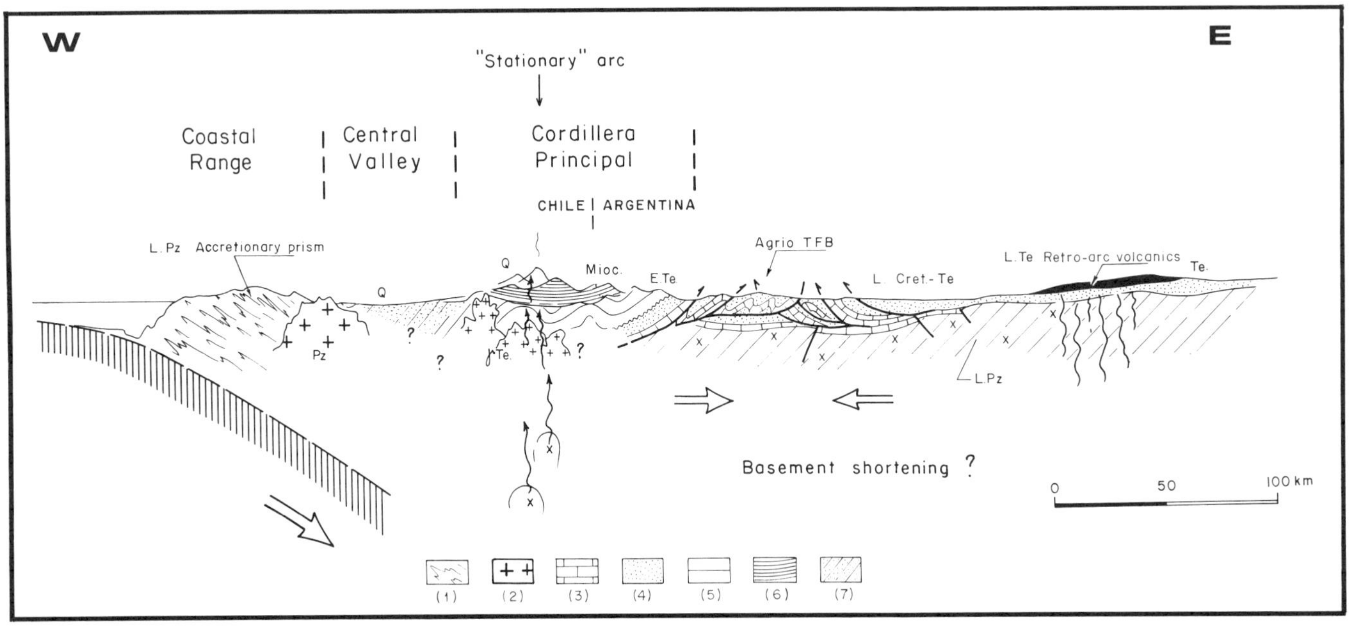

Figure 18. Structural section for segment C: (1) late Paleozoic accretionary prism; (2) late Paleozoic coastal batholith; (3) Jurassic-Early Cretaceous sediments of the Neuquén backarc basin; (4) Late Cretaceous Neuquén Group; (5) early Cenozoic lacustrine sediments; (6) Miocene volcanic rocks; and (7) late Cenozoic sedimentary infill of the Central Valley of Chile.

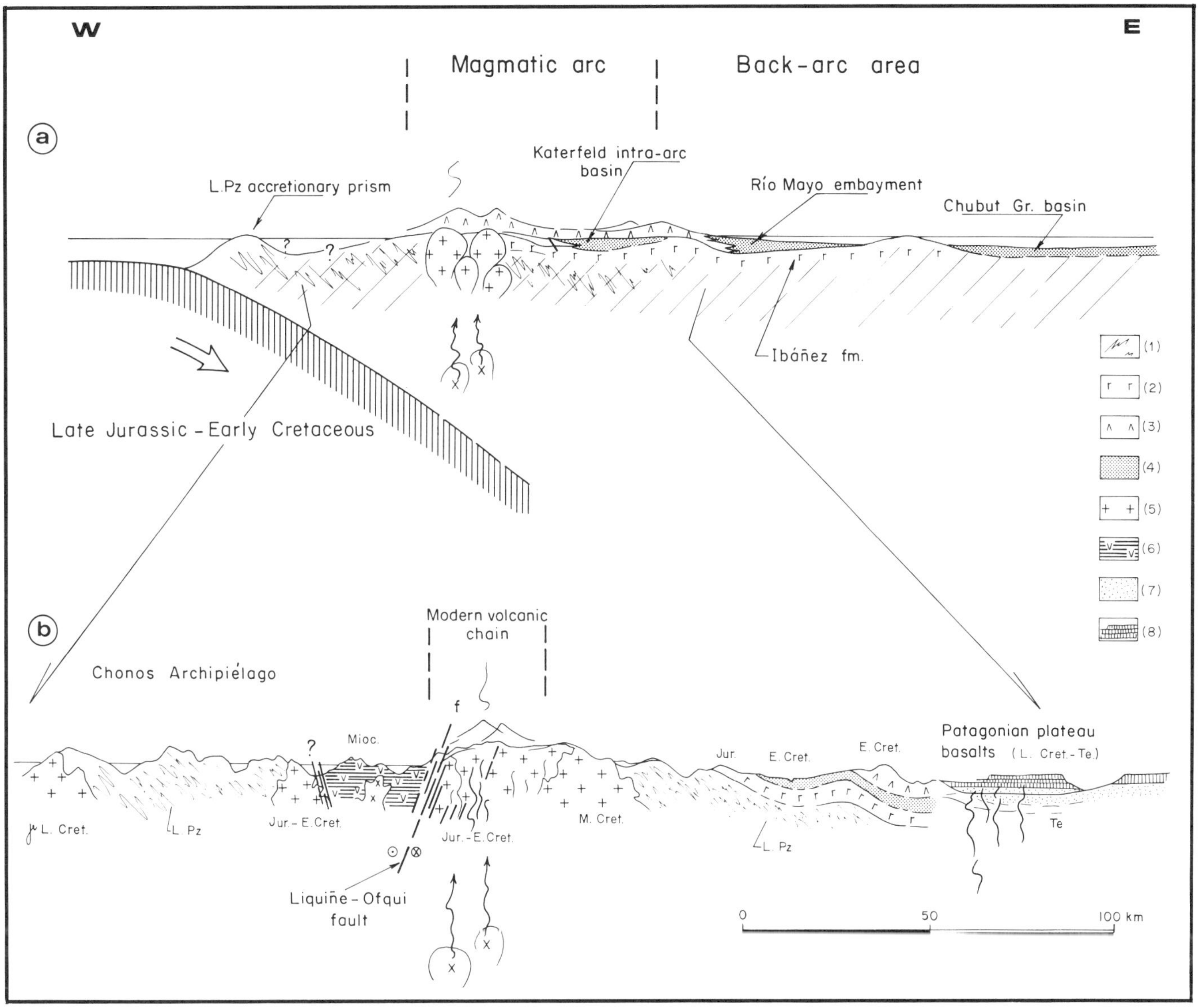

Figure 19. Diagrammatic sections for segment D—(a) Early Cretaceous paleogeography and (b) structural profile (lat. 46 °S). Symbols show: (1) late Paleozoic accretionary prism; (2) Jurassic rhyolitic volcanic rocks; (3) Early Cretaceous "arc" lavas; (4) Early Cretaceous intraarc to backarc basin sediments; (5) Patagonian batholith; (6) Miocene lavas of the Traiguén formation; (7) Tertiary sedimentary rocks; and (8) Late Cretaceous-Tertiary plateau basalts.

South of Temuco (lat. 38 °S), the Cordillera Principal is characterized by late Paleozoic granitic plutons (295-282 Ma) which are associated with metamorphic rocks and late Paleozoic sediments. These rocks are cut by Jurassic (220-160 Ma), Cretaceous (90 Ma), and Miocene (16-10 Ma) granodioritic stocks (Moreno and Parada, 1976; Munizaga et al., 1985). Sparse andesitic roof pendants in the Curarrehue region are considered to be of Cretaceous age. A north-south–trending strike-slip fault, the Liquiñe-Ofqui fault (Figure 19) (Hervé, 1984), cuts across the plutons along the main axis of the Andean Cordillera. As a result of the combined effects of shear heating along this fault zone and the emplacement of the Miocene stocks, the pre-Miocene rocks show progressively older K-Ar ages outward from the main Cordillera axis (Hervé et al., 1982).

Widespread early Tertiary volcanism occurred east of the Andean Cordillera in the region between lat. 39 °S and lat. 40 °30 ′S. Volcanic rocks formed along two distinctive belts, a western belt where 3000 m of andesites, the Ventania Formation, were erupted during the Paleocene-Eocene (González Bonorino, 1973; Ramos, 1982), and an eastern belt where rhyolitic-dacitic volcanic rocks were erupted during the Cretaceous to Eocene (Rapela et al., 1984). A marine to continental basin, the Nirihuao basin, formed between the two belts during the Oligocene and Miocene (Cazau, 1980).

Neuquén Basin

The Neuquén backarc basin (Figure 17), east of the Andean Cordillera between lat. 36 ° and 39 °S, was the site of three sedimentary cycles—the Cuyano of Early Jurassic age, the Chacayano-Loteniano of Middle to Late Jurassic age, and the

Mendociano of Tithonian to Neocomian age (Groeber, 1953; Malumián, Nullo, and Ramos, 1983). These cycles show a common pattern of facies, changing from alluvial facies in the east to marine platform, slope, and basin facies in the west (Gulisano, Gutiérrez Pleiming, and Digregorio, 1984a; Gulisano et al., 1984). The subsidence of the basin was thermally controlled and seems to reflect an important thinning of the continental crust. As the edge of the slope migrated eastward from Jurassic to Early Cretaceous, centers of deposition also shifted eastward, indicating that the extensional tectonic regime persisted during most of the Jurassic-Early Cretaceous interval (Ramos, 1986). The basin fill of thousands of meters of Mesozoic sediments [illegible] evidence of deformation until middle Cretaceous. The western border of the Neuquén basin is not exposed, but its position is indicated by a conglomeratic facies of Valanginian age, the source of which was to the west (Burckhardt, 1900).

Molassic continental deposits accumulated in the Neuquén basin during Late Cretaceous, when the basin was filled by 1300 m of red beds and conglomerates of the Neuquén Group (Cazau and Uliana, 1973). At the end of the Cretaceous, thin, shallow-water marine limestones and shales were deposited by the first Atlantic transgression at this latitude. The Cenozoic is represented by thick alluvial-fan and fluvial deposits that cover most of the extra-Andean region.

Structural Geology

The structural evolution (Figure 18) of the Neuquén embayment began with deformation between the Early Cretaceous and the time when deposition of the Neuquén Group started. A fold-thrust belt was developed, and as a consequence of tectonic loading the Neuquén Group foredeep was formed. Subsequent deformation at the end of the Eocene and during the middle Miocene caused crustal shortening and development of the present structure of the basin. This structure is characterized by a series of east-verging folds and related thrusts, which developed a complex thrust front that is bounded to the east by an area of gentle folds, where most of the present oil fields are located. The basin is crossed by a major wrench fault, the Huincul fault, which has been intermittently active since middle Cretaceous, and along which are exposed a series of flower structures (Ploskiewicz et al., 1984).

Segment D (Lat. 41 °-49 °S)

Segment D (North Patagonian Cordillera) has the following characteristics (Figure 19)—a late Jurassic oscillatory magmatic arc in which magmatic foci shifted eastward during middle Cretaceous, retreating westward during Late Cretaceous and again eastward during the Tertiary; ephemeral marine transgressions east of the magmatic arc in the Middle Jurassic-Early Cretaceous, which resulted in the sedimentary fill of the fault-bounded Río Mayo embayment; weak deformation in the backarc zone represented by long-wavelength open folds and minor normal faults, whereas the magmatic arc was affected by large Cenozoic faults showing strike-slip displacements; and active Quaternary volcanism.

Jurassic

An Early Jurassic marine transgression from the Pacific deposited black shales and carbonate rocks having abundant ammonite fauna in an elongate basin east of the present Patagonian Cordillera. This sequence is exposed in the Pampa de Agnia and Tepuel areas. It has interbedded andesite flows and pyroclastics that indicate the presence of a volcanic zone to the west, somewhere in the extra-Andean region of Chubut, between Bariloche (lat. 41 °S) and Lago Fontana (lat. 44 °30 ′S) (Haller and Lapido, 1982).

The magmatic arc shifted westward in the Middle Jurassic, when andesites, dacites, and rhyolites of the Ibáñez and Lago Fontana formations were erupted along the eastern slopes of the present-day Patagonian Cordillera (Figure 1). The youngest volcanic units have interbedded carbonate rocks (Ramos et al., 1982; Niemeyer et al., 1984). This volcanism was synchronous with the development of thick sequences of rhyolitic lavas and ignimbrites in extra-Andean Patagonia (Bruhn, Stern, and De Witt, 1978; Uliana et al., 1986). These sequences, which make up the Chon Aike and Marifil formations, have been interpreted as the products of extensive crustal anatexis in a volcano-tectonic rift environment predating the opening of the south Atlantic (Malumían and Ramos, 1984).

Cretaceous

The Cretaceous was a time of marine transgressions, intensive volcanism, and emplacement of the main Patagonian batholith. The earliest intrusions of the Patagonian batholith were emplaced during the Jurassic, cutting across the late Paleozoic forearc assemblages of the Los Chonos Archipelago (Godoy et al., 1984), however, most of the granitic bodies of the batholith are of Early to middle Cretaceous age (Ramos et al., 1982; Niemeyer et al., 1984). True granites were intruded along the eastern margin of the batholith during the middle Cretaceous (Wells, 1976). During the Late Cretaceous, intrusive activity shifted to the west, and isolated small stocks of granodiorite having ages ranging from 75 to 67 Ma were emplaced along the western shores of the Taitao Peninsula (Godoy et al., 1984). Granitic rocks of Miocene age have been found along the batholith axis (Bartholomew and Tarney, 1984). The distribution of plutonic rocks in this segment reveals an oscillatory behavior of the magmatic arc that differs from the eastward migration of magmatic foci that is typical of the northern segments of the Chile-Argentine Andes.

A small marine basin, the Río Mayo embayment (Figure 19), developed to the west of the Patagonian batholith during Tithonian time and lasted until early Barremian time (Aguirre Ureta and Ramos, 1981). Black shales and sandstones, containing interbeds of andesites and dacites, were deposited in an intra-arc basin (Katerfeld basin, Figure 19) during the Early Cretaceous in the Coyhaique-Lago Fontana region (Skarmeta, 1976; Ramos and Palma, 1983). During middle Cretaceous, the extensive volcanic rocks of the Divisadero Formation were erupted over most of the eastern foothills of the Patagonian Cordillera and adjacent extra-Andean region (Niemeyer et al., 1984). This sequence of andesitic, dacitic, and rhyolitic lavas is unconformably overlain by backarc plateau basalts of Late Cretaceous age (Franchi and Page, 1980).

Structural Geology

Deformation of the rocks in this segment occurred chiefly during middle Cretaceous, but the crustal shortening was sig-

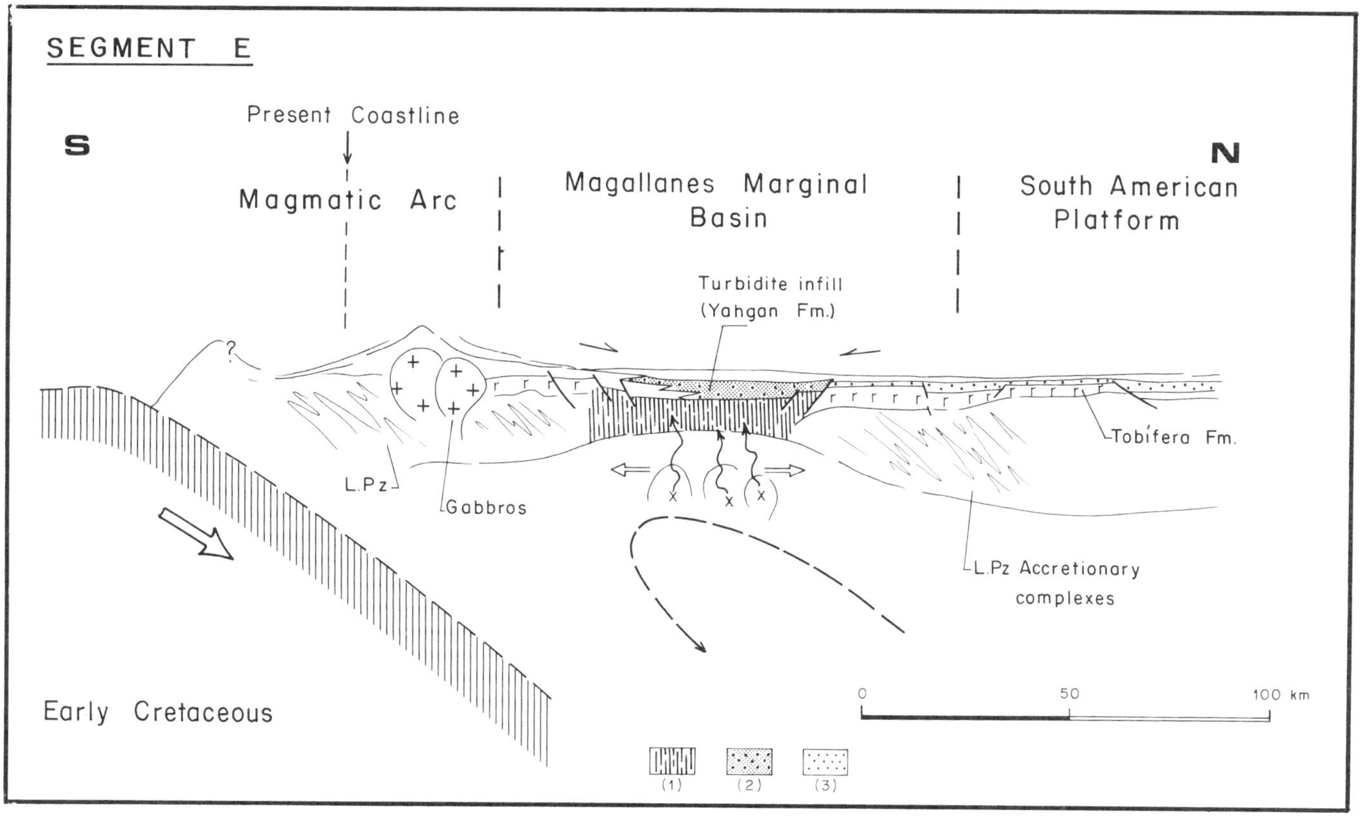

Figure 20. Paleogeographic section for the Early Cretaceous segment E (Tierra del Fuego). Symbols as in Figure 19 except (1) "oceanic" floor of the Magallanes marginal basin; (2) turbidite infill of the marginal basin (Yahgan Formation); and (3) Early Cretaceous marine sedimentary cover of the South American platform.

nificantly smaller than that related to the Andean orogeny in the northernmost Andean segments. The resulting structures are represented only by gentle folds and normal faults with minor displacements in the Mesozoic sequences east of the Patagonian batholith (Ramos, 1983; Niemeyer et al., 1984). However, the batholith itself is cut by the Liquiñe-Ofqui wrench fault (Figure 19) (Hervé et al., 1982; Forsythe and Nelson, 1985), along which most of the coastal area (Chiloe Block) seems to have been displaced northward. Miocene eruptive centers located along this fault zone were the sources of the volcanic rocks in the Traiguén Formation (Figure 19) (Skarmeta, 1976; Céspedes, 1975; Bartholomew and Tarney, 1984).

After deformation, a thin sequence of sandstones and pyroclastic rocks, the Rio Frias Formation (Roth, 1908), was deposited in the eastern foothills of the Cordillera during middle to late Miocene.

During late Cenozoic, isolated stratovolcanoes came into being near the Liquiñe-Ofqui fault zone (Figure 19) (Stern, Skewes, and Durán, 1976; Lopez-Escobar et al., 1985), and extensive alkaline plateau basalts, which carry peridotite xenoliths, were erupted in the extra-Andean Patagonia (Stern et al., 1985).

Segment E (Lat. 49-56 °S)

Segment E (Austral Patagonian Andes) shows the following (Figures 20, 21)—crustal anatexis and extensive rhyolitic volcanism during Middle-Late Jurassic to form a magmatic arc along the Pacific margin of Patagonia; disruption of the continental crust in the backarc zone and formation of a marginal basin floored by oceanic crust (Magallanes marginal basin) in the Early Cretaceous; closure and deformation of the marginal basin during middle Cretaceous due to foreland shift of the magmatic arc; formation of a thrust-load foredeep during Late Cretaceous (Magallanes foredeep), which received detritus from the uplifted arc and marginal basin terrane; retreat of the magmatic arc toward the ocean and progressive deformation of the foredeep during the Tertiary; and minor Quaternary volcanism.

Segment E has been the object of intensive investigation since the mid-1970s, and many aspects of its geological history have been reported elsewhere (Dalziel, De Witt, and Palmer, 1974; Suárez and Petigrew, 1976; Bruhn, Stern, and De Witt, 1978; Caminos et al., 1981). An unusual feature of this segment was the opening of the Magallanes marginal basin (Figure 20), an oceanic-floored backarc basin, during Early Cretaceous. Closure of this basin during the middle to Late Cretaceous resulted in pervasive ductile deformation along the foreland margin of the basin to form the metamorphic core complex of Cordillera Darwin (Figure 21) in southern Tierra del Fuego (Nelson, Dalziel, and Milnes, 1980). Syntectonic plutons (Canal Beagle Plutonic Group) were emplaced along the oceanic side of the marginal basin during the middle Cretaceous (Hervé et al., 1984). Posttectonic granitic plutons were emplaced oceanward during the Late Cretaceous (Hervé et al., 1984).

The stable continental margin to the east and north of the

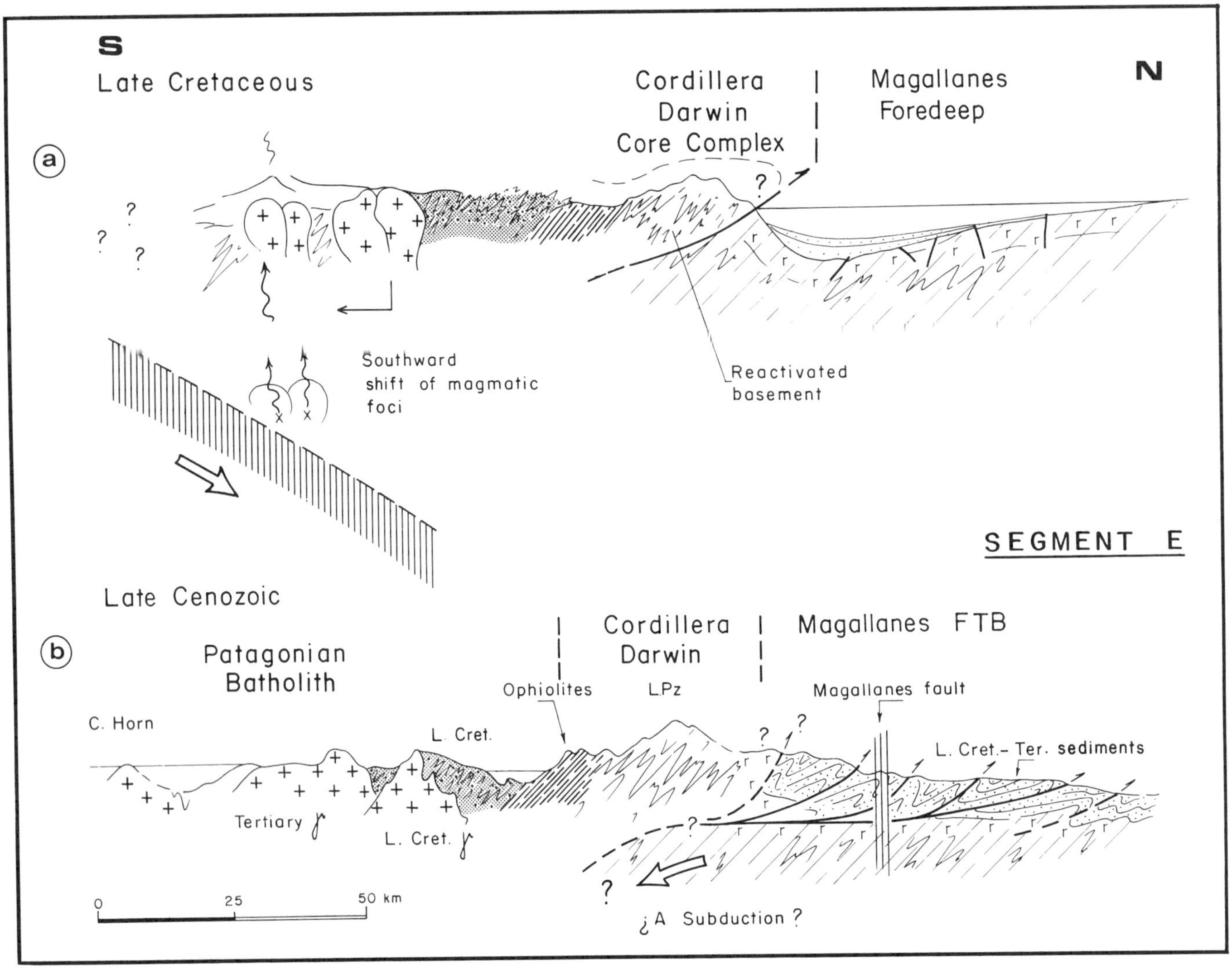

Figure 21. Late Cretaceous (a) and late Cenozoic (b) structural sections of the Andean Cordillera in Tierra del Fuego (segment E, long. 68 °W).

Magallanes marginal basin subsided as a result of the tectonic loading related to the closure and uplift of the basin and formation of the Cordillera Darwin core complex (Figure 21). As a consequence, coarse turbiditic sediments derived in part from the uplifted basin terrane were deposited in a narrow and deep trough, the Magallanes foredeep (Figure 21), the axis of which migrated progressively toward the stable South American foreland during the Late Cretaceous to Miocene (Natland et al., 1974; Winslow, 1979; Uliana et al., 1986). The foothills structure of the Southern Patagonian Cordillera and the Fueguian Andes is characterized by the Magallanes fold-thrust belt (Figure 21), which resulted from shortening and deformation of the foredeep strata. The orogenic front migrated progressively eastward and northward throughout Late Cretaceous and Cenozoic (Ramos, 1979; Winslow, 1979).

A few granitic stocks were emplaced along the eastern foothills of the Andes during the late Miocene (Nullo, Proserpio, and Ramos, 1978). Late Cenozoic strike-slip movements along the Magallanes fault (Fuenzalida, 1976) in Tierra del Fuego are coeval with the opening of the Scotia Sea. Segment E also shows minor Quaternary volcanism related to the slow subduction of the Antarctic Plate underneath the continental border (Stern et al., 1985).

CONCLUSIONS

The Andean Cordillera of Chile and Argentina shows a much more complex history than has previously been recognized. It is a region dominated by Mesozoic and Cenozoic sedimentation, magmatism, and tectonic deformation superimposed on a Paleozoic basement characterized by accretionary tectonics. The early Paleozoic was the time of major terrane accretion, details of which are still obscure. Accretion essentially terminated in late Paleozoic when most of the western margin of Gondwana changed to an "Andean-type" magmatic arc, bounded oceanward by a wide accretionary prism. The Mesozoic-Cenozoic history of this Andean region is characterized by destruction of the continental margin by "subduction erosion" and strike-slip faulting. Evidence for the accretion of large terranes is absent. Subduction erosion or strike-slip faulting is believed to have eliminated most of the late Paleozoic forearc assemblages and, perhaps, resulted in the recycling of large volumes of siliceous crust in the adjacent modern volcanic arc, as has been suggested by Rutland (1971).

During the Jurassic and Early Cretaceous, the Andean region was characterized by a series of magmatic arcs and backarc basins. The distribution of these elements and the patterns of deformation define five major Andean segments which, from north to south, are designated segments A-E. The degree of crustal attenuation determined the nature of the basin floors, which varied from "oceanic" in segment E to ensialic in segments A and C. In contrast, segment B was characterized by subsidence within and behind the magmatic arc, and development of an "aborted" marginal basin that became filled with a thick sequence of andesites and basalts. Segment D differs in showing no evidence of major backarc extension.

The various segments reacted in different ways to closure of backarc basins during the middle Cretaceous. In segment A, the magmatic arc was underthrust beneath the uplifted sialic floor of the Tarapacá backarc basin, whereas in segment B the volcanic fill of the central Chile "aborted" marginal basin was overthrust toward the east onto the adjacent Aconcagua sedimentary platform. Middle Cretaceous marginal-basin closure in segment E resulted in metamorphism of the rocks between the margin of the basin and the foreland, thus forming the Cordillera Darwin core complex. The Mesozoic and Cenozoic history of the magmatic arcs differs from one segment to another, showing either a continuous or episodical eastward migration as in segment B, stationary arc as in segment C, or oscillatory migration of the magmatic foci as in segment D.

The tectonic regime of the Chilean-Argentine Andes during the Jurassic-Early Cretaceous was dominated by backarc extension and subsidence, and resembles the "Mariana-type" of subduction described by Uyeda (1982). However, the subduction mode shows a distinctive change in the middle Cretaceous, when the extensional regime in the backarc basin environment terminated and a new period of high-stress "Chilean-type" subduction began. This subduction mode characterizes the present active margin of southern South America. The tectonic events that affected the Andean region probably reflect the major changes in rates of convergence and subduction within the Pacific realm during the Mesozoic (Larson and Pitman, 1972; Rona and Richardson, 1978; Mammerickx, Herron, and Dorman, 1980; Cande, 1983). The end of the low-stress Mariana-type subduction may be correlated with the beginning of the active drift of the South American continent following its separation from Africa and the opening of the south Atlantic. After middle Cretaceous, the stress regime could have been incompatible with the survival of backarc basins, and the subsidence of the late Cretaceous foreland basins of segments C and E was in response to the tectonic load produced by the adjacent growing Andean Cordillera.

Modern active margins bordering the Pacific basin show a striking tectonic asymmetry (Chotin, 1981). The western Pacific margin is characterized by extensive and complex systems of intraoceanic arcs and related extensional marginal basins, whereas these features are absent at the eastern margin where convergence is now dominated by a high-stress, Chilean-type subduction. These differences seem related to absolute motions of the overriding plates. For example, in the western Pacific, the Asian Plate appears to be stationary or is retreating from the trench, whereas the South American Plate is actively advancing toward the Pacific (Uyeda and Kanamori, 1982). Our information shows that the present configuration of the eastern Pacific margin has existed only since the middle Cretaceous. Prior to this time, from the Paleozoic to the earliest Cretaceous, the eastern Pacific margin was in some respects quite similar to the present-day western Pacific margin.

ACKNOWLEDGMENTS

The authors wish to express their appreciation to the authorities and geologists (too numerous to be cited individually) of the Servicio Geológico Nacional (Argentina) and the Servicio Nacional de Geología y Minería (Chile) who supported the field and laboratory investigations on which this chapter is based. We are grateful to Richard Allmendinger and Suzanne M. Kay for their constructive comments and Oriana Gonzalez for preparation of the illustrations.

REFERENCES CITED

Aberg, G., L. Aguirre, B. Levi, and J.O. Nystrom, 1984, Spreading subsidence and generation of ensialic marginal basin: An example from the Early Cretaceous of central Chile, *in* B.P. Kokelaar and M.F. Howells, eds., Marginal basin geology: Geological Society of London Special Publication 16, p. 185-193.

Aceñolaza, G.F., 1982, El sistema Ordovícico en Sudamérica: Tucumán, Acta Geológica Lilloana, v. 16, p. 7-109.

Aguirre Ureta, M.B., and V.A. Ramos, 1981, Estratigrafía y paleontología de la Alta Cuenca del río Roble, Cordillera Patagónica: Buenos Aires, VIII Congreso Geológico Argentino Actas 3, p. 101-138.

Allmendinger, R., T.E. Jordan, M. Palma, and V.A. Ramos, 1982, Perfil estructural en la Puna Catamarqueña (25-27 °S), Argentina: Buenos Aires, V Congreso Latinoamericano de Geología Actas 1, p. 499-518.

Allmendinger, R., V.A. Ramos, T.E. Jordan, M.A. Palma, and B.I. Isacks, 1983, Paleogeography and Andean structural geometry, northwest Argentina: Tectonics, v. 2, p. 1-16.

Arabaz, W.J., 1971, Geological and geophysical studies of the Atacama fault zone in northern Chile: Pasadena, California Institute of Technology, unpublished Ph.D. dissertation, p. 1-264.

Archangelsky, S., ed., 1986, The Carboniferous System in the Argentine Republic: Córdoba, Subcommittee on Carboniferous Stratigraphy, p. 1-359.

Azcuy, C.L., and R. Caminos, 1986, Diatrofismo, *in* S. Archangelsky, ed., The Carboniferos System in the Argentine Republic: Córdoba, Subcommittee on Carboniferous Stratigraphy, p. 225-237.

Baldis, B., M.S. Beresi, O. Bordonaro, and A. Vaca, 1982, Síntesis evolutiva de la Precordillera Argentina: Buenos Aires, V Congreso Latinoamericano de Geología Actas 4, p. 399-495.

Bally, A.W., 1981, Thoughts on the tectonics of folded belts, *in* K.R. McClay and N.J. Price, eds., Thrust and nappe tectonics: Geological Society of London Special Publication 9, p. 13-22.

Barazangi, M., and B.I. Isacks, 1976, Spatial distribution of earthquakes and subduction of the Nazca Plate beneath South America: Geology, v. 4, p. 686-692.

Bartholomew, D.S., and J. Tarney, 1984, Geochemical characteristics of magmatism in the Southern Andes (45-46 °S), *in* R.S. Harmon and B.A. Barreiro, eds., Andean magmatism: Chemical and isotopic constraints: Shiva Geological Series, p. 220-250.

Bell, C.M., 1982, The lower Paleozoic metasedimentary basement of the Coastal Ranges of Chile between 28°30′ and 27°S: Revista Geológica de Chile, no. 17, p. 21-29.

Bevis, M., and B.L. Isacks, 1984, Hypocentral trend surface analysis: Probing the geometry of Benioff zones: Journal of Geophysical Research, v. 89 (B7), p. 6153-6170.

Bookstrom, A., 1977, The magnetite deposits of El Romeral, Chile: Economic Geology, v. 72, p. 1101-1130.

Borello, A.V., 1969, Los geosinclinales de la Argentina: Buenos Aires, Dirección Nacional Geología y Minería Anales, v. 14, p. 1-36.

Boric, R., and F. Díaz, 1984, Yacimentos metálicos de la región de Antofagasta: Santiago, Servicio Nacional de Geología y Minería, Curso de actualización de la Geología de Chile, Apuntes, p. D1-D19.

Boric, R., F. Díaz, and V. Maksaev, 1985, Magmatic events and related metallogenesis in the Antofagasta region, northern Chile: Santiago, Departamento de Geología, Universidad de Chile, Comunicaciones, v. 35, p. 37-40.

Breitkreuz, C., and W. Zeil, 1984, Geodynamic and magmatic stages on a transverse through the Andes between 20° and 24° S (N Chile, SE Bolivia, NW Argentina): Journal of the Geological Society of London, v. 141, p. 861-868.

Bruhn, R.L., C.R. Stern, and M.J. De Witt, 1978, Field and geochemical data bearing on the development of a Mesozoic volcano-tectonic and back arc basin in southernmost South America: Earth and Planetary Science Letters, v. 41, p. 32-46.

Burckhardt, C., 1900, Coupe geologique de la Cordillera entre las Lajas et Curacautín: La Plata, Museo de La Plata, Anales 2, p. 1-136.

Caminos, R., 1979, Cordillera Frontal: Córdoba, Segundo Symposium Geología Regional Argentina, Academía Nacional de Ciencias, 1, p. 397-454.

Caminos, R., M.A. Haller, O. Lapido, A. Lizuain, R. Page, and V.A. Ramos, 1981, Reconocimieno geológico de los Andes Fueguinos, Territorio Nacional de Tierra del Fuego: Buenos Aires, VIII Congreso Geológico Argentino Actas 3, p. 759-786.

Cande, S., 1983, Nazca-South America Plate interactions, 80 My to present: EOS (Transactions, American Geophysical Union), v. 64, p. 64-65.

Cazáu, L., 1980, Cuenca de Ñirihuao-Ñorquinco-Cushamen, *in* J.C.M. Turner, ed., Geología regional Argentina: Córdoba, Academía Nacional Ciencias, II, p. 1149-1171.

Cázau, L., and M. Uliana, 1973, El Cretácico superior continental de la cuenca neuquina: Buenos Aires, V Congreso Geológico Argentino Actas 3, p. 131-164.

Cecioni, G., and T.E. Westermann, 1968, The Triassic-Jurassic marine transition of Coastal-Central Chile: Tokyo, Pacific Geology, v. 1, p. 41-75.

Céspedes, S., 1975, El Terciario de Isla Traiguén y Golfo de Tres Montes, Provincia de Aysén: Santiago, Empresa Nacional Petróleo (Enap-Chile), unpublished report, p. 1-5.

Chong, G., and A. von Hillebrandt, 1985, El Triásico pre-Andino de Chile entre los 23°30′ y 26°00′ de Lat. Sur.: Antofagasta, IV Congreso Geológico Chileno Actas 1, p. 1-152.

Chotin, P., 1981, Dyssymetrie fondamentale dans l'evolution des zones de subduction de l'Ouest et de l'Est-Pacifique, dupuis le Trias: Bulletin Société Géologique de France, v. 23, p. 245-252.

Coira, B.I., J. Davidson, C. Mpodozis, and V.A. Ramos, 1982, Tectonic and magmatic evolution of the Andes of northern Argentina and Chile: Earth Science Reviews, v. 18, p. 303-332.

Dalziel, I.W.D., 1986, Collision and cordilleran orogenesis: An Andean perspective, *in* Collision tectonics: Geological Society of London Special Publication II, p. 389-404.

Dalziel, I.W.D., M.J. De Witt, and K. Palmer, 1974, Fossil marginal basin in the southern Andes: Nature, v. 250, p. 291-294.

Damm, K.W., S. Pichowiak, and W. Todt, 1986, Geochimie, petrologie un geochronologie der plutonite und des metamorphen grundgebirges in Nordchile: Berlin, Berliner Geowiss, Abhandlung (A), v. 66, p. 73-146.

Dewey, J., 1980, Episodicity, sequence and style at convergent plate boundaries, *in* D.W. Strangway, ed., The continental crust and its mineral deposits: Geological Association of Canada Special Paper 20, p. 553-573.

Drake, R., M. Vergara, F. Munizaga, and J.C. Vicente, 1982, Geochronology of Mesozoic-Cenozoic magmatism in central Chile, Lat 31° 36° S: Earth Science Reviews, v. 18, p. 353-363.

Espinoza, S., 1979, Una hipótesis sobre la metalogenesis de las franja ferrífera chilena: Arica, II Congreso Geológico Chileno Actas 2 (c), p. 1-2.

Ferraris, F., and F. Di Biase, 1978, Hoja Antofagasta: Santiago, Instituto de Investigaciones Geológicas, Carta Geológica de Chile no. 30, p. 1-48.

Forsythe, R.D., and E.P. Nelson, 1985, Geological manifestations of ridge collision, evidence from the Golfo de Penas-Taitao basin, Chile: Tectonics, v. 4, p. 239-246.

Forsythe, R.D., E.P. Nelson, M.J. Carr, M.E. Kaeding, M. Hervé, C. Mpodozis, J.M. Soffia, and S. Harambour, 1986a, Pliocene near trench magmatism in southern Chile: A possible manifestation of ridge collision: Geology, v. 14, p. 23, 27.

Forsythe, R.D., D.V. Kent, C. Mpodozis, and J. Davison, 1986b, Paleomagnetism of Permian and Triassic rocks, central Chilean Andes: Madison, Wisconsin, Proceedings of the Antarctic Geoscience Conference.

Franchi, M., and P. Page, 1980, Los basaltos cretácicos y la evolución magmática del Chubut Occidental: Revista Asociación Geológica Argentina, v. 35, p. 208-229.

Fuenzalida, P.R., 1976, The Magellan fault zone, *in* Proceedings, Symposium of Andean Antarctic problems: Naples, International Association of Volcanology and Geochemistry of the Earth Interior, p. 373-391.

García, F., 1967, Geología del Norte Grande de Chile, *in* Symposium sobre el geosinclinal Andino: Santiago, Sociedad Geológica de Chile, v. 3, p. 1-138.

Gardeweg, M., and C.F. Ramírez, 1984, Volcanismo Cenozoico superior del Altiplano Chileno: Santiago, Servicio Nacional de Geología y Minería Curso de actualización de la geología de Chile, Apuntes, p. E1-E31.

Godoy, E., J. Davidson, F. Hervé, C. Mpodozis, and K. Kawashita, 1984, Deformación sobreimpuesta y metamorfismo progresivo en un prisma de acreción paleozoico, Archipiélago de Los Chonos, Aysen, Chile: Buenos Aires, IX Congreso Geológico Argentino Actas 4, p. 211-232.

González Bonorino, F., 1973, Geología del area entre San Carlos de Bariloche y Llao Llao, Provincia de Rio Negro: San Carlos de Bariloche, Fundación Bariloche, Dep. Rec. Nat. Publication 16, p.1-53.

González Díaz, R., 1972, Descripción geológica de la Hoja 30d "Payún Matrú," Provincia de Mendoza: Buenos Aires, Dirección Nacional Geología y Minería Boletin 130, p. 1-190.

González Ferrán, O., P.E. Baker, and D.C. Rex, 1984, Tectonic-volcanic discontinuity at latitude 27° S Andean Range, associated with Nazca Plate subduction: Tectonophysics, v. 112, p. 423-441.

Groeber, P., 1946, Observaciones geológicas a lo largo del Meridiano 70. 1, Hoja Chos Malal: Sociedad Geológica Argentina, v. 1, p. 177-208.

Groeber, P., 1953, Mesozoico, *in* Geografía de la República Argentina: Buenos Aires, Sociedad Argentina de Estudios Geograficos GAEA, v. 2, p. 1-154.

Gulisano, C.A., A.R. Gutiérrez Pleimling, and R.E. Digregorio, 1984, Esquema estratigráfico de la secuencia jurásica al este de la provincia del Neuquén: IX Congreso Geológico Argentino Actas 1, p. 236-259.

Gulisano, C.A., A.R. Gutiérrez Pleimling, and R.E. Digregorio, 1984, Análisis estratigráfico de intervalo Tithonian-Valanginiano (Formaciones Vaca Muerta, Quintuco y Mulichinco) en el suroeste de la privincia del Neuquén: Buenos Aires, IX Congreso Geológico Argentino Actas 1, p. 221-235.

Haller, M., and O. Lapido, 1982, The Jurassic-Cretaceous volcanism in the Septentrional Patagonian Andes: Earth Science Reviews, v. 18, p. 395-410.

Haller, M., and V.A. Ramos, 1984, Las ofiolitas famatinianas (Eopaleozoico) de las provincias de San Juan y Mendoza: Buenos Aires, IX Congreso Geológico Argentino Actas 2, p. 66-83.

Haller, M., F. Nullo, C. Prosperpio, P. Parica, M. Cagnoni, and J. Walker, 1985, Major elements geochemistry of early Tertiary Andean volcanics (34°-36° S): Santiago, Departamento de Geología, Universidad de Chile, Comunicaciones, v. 35, p. 97-100.

Harrington, H.J., 1961, Geology of parts of Antofagasta and Atacama provinces of northern Chile: AAPG Bulletin, v. 45, p. 169-197.

Harrington, H. J., 1970, Las Sierras Australes de Buenos Aires, República Argentina: Cadena aulacogénica: Revista Asociación Geológica Argentina, v. 25, p. 151-181.

Hervé, F., J. Davidson, E. Godoy, C. Mpodozis, and V. Covacevich, 1981, The late Paleozoic in Chile: Stratigraphy, structure and possible tectonic framework: Anais. de Academía Brasileria de Ciencias, v. 53, p. 361-373.

Hervé, F., K. Kawashita, F. Munizaga, and M. Bassei, 1982, Edades Rb-Sr de los cinturones metamórficos pareados de Chile central: Concepción, III Congreso Geológico Chileno Actas (D), p. 116-135.

Hervé, F., K. Kawashita, and F. Munizaga, 1984, Rb-Sr isotopic age from late Paleozoic metamorphic rocks of central Chile: Journal of the Geological Society of London, v. 141, p. 877-884.

Hervé, F., K. Kawashita, M.A. Parada, M. Brook, R. Pankhurst, N. Snelling, and R. Drake, 1985, Granitoids of the Coast Range of central Chile, geochronology and geological setting: Santiago, Departamento de Geología, Universidad de Chile, Comunicaciones, v. 35, p. 105-108.

Hervé, M., 1984, La zona de falla Liquiñe en Liquiñe, Santiago, Departamento de Geología, Universidad de Chile, Comunicaciones, v. 34, p. 101-105.

Hervé, M., M. Suárez, and A. Puig, 1984, The Patagonian Batholith south of Tierra del Fuego, Chile: Timing and tectonic implications: Journal of the Geological Society of London, v. 14, p. 909-917.

Introcaso, A., A. Lyon, and V.A. Ramos, 1986, La estructura profunda de las Sierras de Córdoba: Revista Asociación Geológica Argentina.

James, D.E., 1971, Plate tectonics model for the evolution of the central Andes: Geological Society of America Bulletin, v. 82, p. 3325-3346.

Jordan, T.E., and M. Gardeweg, 1986, Tectonic evolution of the late Cenozoic Central Andes, *in* Z. Ben-Avraham, ed., Mesozoic and Cenozoic evolution of the Pacific margin: Oxford University Press (in press).

Jordan, T.E., B. Isacks, R. Allmendinger, J. Brewer, V.A. Ramos, and C. Ando, 1983a, Andean tectonics related to geometry of subducted plates: Geological Society of America Bulletin, v. 94, p. 341-361.

Jordan, T.E., B. Isacks, V.A. Ramos, and R.R. Allmendinger, 1983b, Mountain building in the Central Andes: Episodes, v. 3, p. 20-26.

Kay, S.M., V.A. Ramos, and R. Kay, 1984, Elementos mayoritarios y trazas de las vulcanitas ordovícias en la Precordillera Occidental: Basaltos de rift oceánicos tempranos próximos al margen continental: Buenos Aires, IX Congreso Geológico Argentino Actas 2, p. 48-65.

Kay, S.M., V. Maksaev, C. Mpodozis, C. Nasi, and R. Moscoso, 1985, Evolution of mid-late Tertiary igneous rocks in the main Chilean Cordillera (29-31° S): Correlation with changes in the slab geometry: Santiago, Departamento de Geología, Universidad de Chile, Comunicaciones, v. 35, p. 113-118.

Kulm, L.D., W.J. Schweller, and A. Masias, 1977, A preliminary analysis of the geotectonic processes of the Andean continental margin 6° to 45° S, *in* M. Talwani and W.W. Pitman, eds., Problems in the evolution of island arcs, deep sea trenches, and backarc basins: American Geophysical Union, Maurice Ewing Series, I, p. 285-302.

Lahsen, A., 1982, Upper Cenozoic volcanism and tectonism in the Andes of northern Chile: Earth Science Reviews, v. 18, p. 285-302.

Larson, R., and W. Pitman, 1972, World-wide correlation of magnetic anomalies and its implications: Geological Society of America Bulletin, v. 83, p. 3645-3662.

Levi, B., and L. Aguirre, 1981, Ensialic spreading subsidence in the Mesozoic and Paleogene Andes of Central Chile: Journal of the Geological Society of London, 138, p. 75-81.

Linares, E., M. Cagnoni, M. De Campo, and H. Ostera, 1985, Geocronologiá de las rocas metamorficas y eruptivas de la provincia del Neuquén y del Noroeste de la Provincia de Río Negro, República Argentina: Santiago, Departamento de Geología, Universidad de Chile, Comunicaciones, v. 35, p.129-136.

Llambías, E., 1966, Geología y petrografía del Volcan Payún Matrú: Tucumán, Acta Geológica Lilloana 8, p. 265-310.

Llambías, E., and C.W. Rapela, 1985, Geología de los complejos eruptivos de La Esperanza, provincia de Rio Negro: Revista Asociación Geológica Argentina, v. 39, p. 220-243.

Llambías, J.E., A.N. Sato, and J. Tomsic, 1986, Geologia y caracteristicas quimicas del stock terciario del Nevado de Acay y vulcanitas asociadas, provincia de Salta: Revista Asociación Geológica Argentina, v. 40, p. 158-175.

Lopez-Escobar, L., M. Tagiri, K. Notsu, W. Hildreth, R.S. Harmon, and J. Davidson, 1985, Geochemical characteristics of rocks from the Andean SVZ between 41°30′ and 46° S: Santiago, Departamento de Geología, Universidad de Chile, Comunicaciones v. 35, p. 137-141.

Lopez-Gamundi, O., C.L. Azcuy, and A. Guerra, 1986, Cuencas Rio Blanco y Calingasta-Uspallata, *in* S. Archangelsky, ed., The Carboniferous system in the Argentine Republic: Córdoba, Subcommittee on Carboniferous Stratigraphy, p. 93-116.

Losert, J., 1972, Genesis of copper mineralizations and associated alterations in the Jurassic volcanic rocks of the Buena Esperanza mining area (Antofagasta Province, Chile): Santiago, Universidad de Chile, Departamento de Geología Publication 40, 64 p.

Maksaev, V., 1978, Cuadrángulo Chitigua y sector occidental de cuadrángulo Cerro Palpana, Region de Antofagasta: Santiago, Instituto de Investigaciones Geológicas Carta Geológica de Chile, p. 1-31.

Maksaev, V., 1984, Mesozoico a Paleógeno de la region de Antofagasta, in Seminario Actualización de la geología de Chile: Santiago, Servicio Nacional de Geología y Minería, C1-C20.

Malumián, N., and V.A. Ramos, 1984, Magmatic intervals, transgression-regression cycles and oceanic events in the Cretaceous and Tertiary of southern South America: Earth and Planetary Science Letters, v. 67, p. 228-237.

Malumián, N., F.E. Nullo, and V.A. Ramos, 1983, The Cretaceous of Argentina, Chile, Paraguay and Uruguay, *in* M. Moullade and A.E.M. Nairn, eds., The Phanerozoic geology of the world: the Mesozoic, B: Amsterdam, Elsevier Science Publications, p. 265-304.

Mammerickx, J., E. Herron, and L. Dorman, 1980, Evidence for two fossil spreading centers in the southeast Pacific: Geological Society of America Bulletin, v. 34, no. 3, p. 207-218.

Marinovic, N., and A. Lahsen, 1984, Hoja Calama: Santiago, Servicio Nacional de Geología y Minería, Carta Geológica Chile, no. 58, p. 1-140.

Martin, A.K., C.J.N. Hartnady, and S.W. Goodland, 1981, A revised fit of South America and south central Africa: Amsterdam, Earth and Planetary Science Letters, v. 54, p. 293-305.

Mégard, F., 1984, The Andean orogenic period and its major structures in central and northern Perú: Journal of the Geological Society of London, v. 141, p. 893-900.

Mingram, A., A. Russo, A. Pozzo, and L. Cazau, 1979, Sierras Sub-Andinas, *in* J.C.M. Turner, ed., Geología regional Argentina: Córdoba, Academía Nacional Ciencias I, p. 95-138.

Mirré, J.C., 1974, El Granito de Acay, intrusivo de edad Terciaria en el Ambiente de la Puna: Revista Asociación Geológica Argentina, v. 29, p. 205-212.

Moreno, H., 1976, The upper Cenozoic volcanism in the Andes of southern Chile (from 40°00′ to 41°30′ S), *in* Proceedings Symposium on Andean Antarctic Problems: Naples, International Association of Volcanology and Geochemistry of Earth Interior, p. 143-171.

Moreno, H., and M.A. Parada, 1976, Esquema geológico de la Cordillera de los Andes entre los paralelos 39°00′ y 41°30′S: I Congreso Geológico Chileno Actas 1, p. A213-A226.

Mpodozis, C., and R. Forsythe, 1983, Stratigraphy and geochemistry of accreted fragments of the ancestral Pacific floor in southern South America: Paleoceanography, Paleoclimatology, Paleoecology, v. 41, p. 103-124.

Mpodozis, C., M.A. Parada, S. Rivano, and J.C. Vicente, 1976, Acerca del plutonismo tardi-hercínico de la Cordillera Frontal entre los 30° y 33° sur (Provincias de Mendoza y San Juan, Argentina; Colquimbo Chile): Buenos Aires, VI Congreso Geológico Argentino Actas 1, p. 143-172.

Mpodozis, C., F. Hervé, J. Davidson, and S. Rivano, 1983, Los granitoides de Cerros de Lila, manifestaciones de un episodio intrusive y termal del Paleozoico inferior de Los Andes del Norte de Chile: Revista Geológica de Chile, no. 18, p. 3-14.

Mpodozis, C., C. Nasi, R. Moscoso, P. Cornejo, V. Maksaev, and M.A. Parada, 1985, The late Paleozoic-Early Triassic magmatic belt of the Chilean Frontal range (28°-31°S): Igneous "stratigraphy" and tectonic setting: Santiago, Departamento de Geología, Universidad de Chile, Comunicaciones, v. 35, p. 161-166.

Munizaga, F., and J.C. Vicente, 1982, Acerca de la zonación plutónica y del volcanismo miocénico en los Andes de Aconcagua (Lat. 32°-33° S): Datos radiométricos K-Ar: Revista Geológica de Chile, no. 16, p. 3-21.

Munizaga, F., F. Hervé, M. Brook, R. Pankhurst, and N. Snelling, 1985, Geochronology of granitoids of the Lake Region, Chile (39°-42° Lat. S): Santiago, Departamento de Geología, Universidad de Chile, Comunicaciones, v. 35, p. 167-170.

Muñoz, J., and J. Niemeyer, 1984, Hoja Laguna del Maule: Santiago, Servicio Nacional Geología y Minería, Carta Geológica de Chile, no. 64, p. 1-98.

Muñoz, J., and C. Stern, 1986, Transverse structural and petrochemical variations across the segment 38°-39° S of the Plio-Quaternary magmatic belt of the southern South American continental margin.

Naranjo, J.A., and A. Puig, 1984, Hojas Taltal y Chañaral.: Santiago, Servicio Nacional de Geología y Minería, Carta Geológica de Chile no. 62-63, p. 1-140.

Nasi, C., 1984, Geología de la Cordillera de la Costa de Chile central, *in* Seminario de actualización geología de Chile: Santiago, Servicio Nacional de Geología y Minería, L1-L17.

Natland, M.L., E. González, A. Cañón, and M. Ernst, 1974, A system of stages for correlation of Magallanes Basin sediments: Geological Society of America Memoir 139, p. 1-126.

Nelson, E.P., I.W.D. Dalziel, and A.G. Milnes, 1980, Structural geology of Cordillera Darwin: Collision style orogenesis in the southernmost Chilean Andes: Ecologae Geologicae Helveticae, v. 73, p. 727-751.

Niemeyer, H., and J. Muñoz, 1983, Hoja laguna del Maule, region del Bío-Bío: Santiago, Servicio Nacional de Geología y Minería, Carta Geológica de Chile, no. 57, p. 1-52.

Niemeyer, H., J. Skarmeta, R. Fuenzalida, and W. Espinoza, 1984, Hoja Península de Taitao y Puerto Aysén: Santiago, Servicio Nacional de Geología y Minería, Carta Geológica de Chile, no. 60-61, p. 1-80.

Nullo, F., C. Proserpio, and V.A. Ramos, 1978, Estratigrafía y tectónica de la vertiente este del Hielo Continental Patagónico, Argentina-Chile: VII Congreso Geológico Argentino Actas 1, p. 455-470.

Nur, A., and Z. Ben-Avraham, 1983, Oceanic plateaus, the fragmentation of continents and mountain building: Journal of Geophysical Research, v. 87, p. 3644-3661.

Ortiz, A., and J.J. Zambrano, 1981, La provincia geológica Precordillera Oriental: VII Congreso Geológico Argentino Actas 3, p. 59-74.

Pacci, D., F. Munizaga, F. Hervé, and K. Kawashita, 1980, Acerca de la edad precámbrica de la formación Esquistos de Belén, Departamento de Parinacota, Chile: Revista Geológica de Chile, no. 11, p. 43-50.

Palma, M.A., P. Parica, and V.A. Ramos, 1986, El Granito Archibarca: Su edd y significado tectónico, provincia de Catamarca: Revista Asociación Geológica Argentina.

Pérez, E., 1982, Bioestratigrafía del Jurásico de Qda, Asientos, norte de Potrerillos, Región de Atacama: Santiago, Servicio Nacional Geología y Minería Boletín 37, p. 1-149.

Piracés, R., 1976, Estratigrafía de la Cordillera de la Costa entra la cuesta El Melón y Limache, Provincia de Valparaiso, Chile: I Congreso Geológico Chileno Actas 1, A65-A82.

Piracés, R. 1977, Geología de la Cordillera de la Costa entre Catapilco y Limache, region de Aconcagua: Santiago, Unpublished thesis, Departamento de Geología, Universidad de Chile, p. 1-188.

Ploszkiewicz, J.V., I.A. Orchuela, J.C. Vaillard, and R.F. Vinez, 1984, Compresión y desplazamiento lateral en la Zona de Falla Huincul: Estructuras asociadas, provincia de Neuquén: Buenos Aires, XI Congreso Geológico Argentino Actas 2, p. 163-169.

Polanski, J., 1972, Descripción geológica de la Hoja 24 a-b, Cerro Tupungato, Provincia de Mendoza: Buenos Aires, Dirección Nacional Geología y Minería Boletín 98, p. 127-342.

Ramos, V.A., 1979, Tectónica de la region de Rio y Lago Belgrano, Cordillera Patagonica, Argentina: Arica, II Congreso Geológico Chileno Actas 1(B), p. 1-32.

Ramos, V.A., 1982, Las ingresiones pacíficas del Terciario en el Norte de la Patagonia: Concepción, III Congreso Geológico Chileno Actas 1(A), p. 262-288.

Ramos, V.A., 1983, Evolucion tectónica y metalogénesis de la Cordillera Patagónica: San Juan, II Congreso Nacional de Geología Económica Actas 1, p. 108-124.

Ramos, V.A., 1984, Patagonia: un continente paleozoico a la deriva?: IX Congreso Geológico Argentino Actas 2, p. 311-325.

Ramos, V.A., 1985a, El Mesozoico de la Alta Cordillera de Mendoza: facies y desarrollo estratigráfico, Argentina: Buenos Aires, IV Congreso Geológico Chileno Actas 1, p. 491-513.

Ramos, V.A., 1985b, El Mesozoico de la Alta Cordillera de Mendoza: Reconstruccion tectónica de sus facies, Argentina: Antofagasta, IV Congreso Geológico Chileno Actas 1, p. 104-118.

Ramos, V.A., 1986, The tectonics of the central Andes: 30°-33° S latitude, *in* S. Clark and D. Burchfiel, eds., Processes in continental lithospheric deformation: Geological Society of America Special Publication.

Ramos, V.A., and M.A. Palma, 1983, Las lutitas pizarreñas del Cerro Dedo Lago La Plata, provincia del Chubut: Buenos Aires, Revista Asociación Geológica Argentina, v. 38, p. 148-160.

Ramos, V.A., H. Niemeyer, I. Skarmeta, and J. Muñoz, 1982, The magmatic evolution of the Austral Patagonian Andes: Earth Science Reviews, v. 18, p. 411-443.

Ramos, V.A., T.E. Jordan, R.W. Allmendinger, S.M. Kay, J.M. Cortes, and M.A. Palma, 1984, Chilenia: Un terreno alóctono en la evolucion paleozoica de los Andes Centrales: Buenos Aires, IX Congreso Geológico Argentino Actas 2, p. 84-106.

Ramos, V.A., S.M. Kay, C. Cingolani, and K. Kawashita, 1985, The volcanic rocks of Cerro Aconcagua, Cordillera Principal (32° S), Argentina: Santiago, Departamento de Geología, Universidad de Chile, Comunicaciones, v. 35, p. 191-194.

Ramos, V.A., T.E. Jordan, R.W. Allmendinger, C. Mpodozis, S.J. Kay, L. Cortes, and M.A. Palma, 1986, Paleozoic terranes of the Central Andes: Tectonics, v. 5, p. 855-880.

Rapela, C.W., L.M. Heaman, and R.J. McNutt, 1982, Rb-Sr geochronology of granitoid rocks from the Pampean Ranges, Argentina: Journal of Geology, v. 90, p. 574-582.

Rapela, C.W., L.S. Spalletti, J.C. Merodio, and E. Aragón, 1984, El vulcanismo paleoceno-eoceno de la provincia volcánia andinopatagónica: Buenos Aires, IX Congreso Geológico Argentino Relatorio, p. 189-214.

Rivano, S., and P. Sepúlveda, 1985, Las calizas de la Formacion Huentelauquén: Depósitos de aguas templadas a frías en el Carbonífero Superior-Pérmico Inferior: Revista Geológica de Chile, no. 25-26, p. 29-38.

Rivano, S., and P. Sepúlveda, 1986, Hoja Illapel, Region de Coquimbo: Santiago, Servicio Nacional de Geologia y Minería, Carta Geológica de Chile.

Rivano, S., and P. Sepúlveda, M. Hervé, and A. Puig, 1985, Geocronología K-Ar de las rocas intrusivas entre los 31°-32° latitud Sur, Chile: Revista Geológica de Chile, no. 24, p. 63-74.

Rolleri, E., and B. Baldis, 1967, Paleogeography and distribution of Carboniferous deposits in the Argentine Precordillera, Coloquio de la IUGS: Amsterdam, La estratigrafia del Gondwana, UNESCO, Ciencias de la Terrra 40, p. 1-11.

Rona, P.A., and E.S. Richardson, 1978, Early Cenozoic global plate tectonic reorganization: Earth and Planetary Science Letters, v. 40, p. 1-11.

Roth, B., 1908, Beiträge zur Gliederung der Sedimentablagerungen in Patagonien und der Pampas region: Stuttgart, Neues Jahrbuch Mineralogie Geologie und Palaeontologie, v. 26, p. 92-150.

Rutland, R.W., 1971, Andean orogeny and sea floor spreading: Nature, v. 233, p. 252-255.

Salas, R., R. Kast, F. Montecinos, and I. Salas, 1966, Geología y recursos minerales del Departamento de Arica: Santiago, Instituto de Investigaciones Geológicas, Boletín 21, 114 p.

Salfity, J., 1979, Paleogeología de la cuenca del Grupo Salta (Cretácico-Eogénico) del Norte de Argentina: Buenos Aires, VII Congreso Geológico Argentino Actas 1, p. 505-515.

Salfity, J., 1982, Evolución paleogeográfica del Grupo Salta (Cretácico-Eogénico), Argentina: Buenos Aires, V Congreso Latinamericano Geología Actas 1, p. 11-26.

Salinas, P., 1981, El Terciario lacustre del Valle del Lolco (Formacion Lolco), IX Region, Chile: San Luis, VIII Congreso Geológico Argentino Actas 3, p. 471-482.

Schiller, W., 1912, La Alta Cordillera de San Juan y Mendoza y parte de la Provincia de San Juan: Buenos Aires, Anales Ministerio de Agricultura, Sección de Geología y Minería, v. 7, 68 p.

Scholl, D.W., R. von Huene, T.L. Vallier, and D.G. Howell, 1980, Sedimentary masses and concepts about tectonic process at underthrust ocean margins: Geology, v. 28, p. 271-291.

Segerstrom, K., and J.C.M. Turner, 1972, A conspicuous flexture in regional structural trend in the Puna of northwestern Argentina: U.S. Geological Survey Professional Paper 800-B, p. 205-209.

Sepúlveda, P., 1984, Geología del Paleozoico en la zona de la coste del Norte Chico, *in* Seminario actualización de la geología de Chile: Santiago, Apuntes, Servicio Nacional Geología y Minería, J1-J7.

Shackleton, R.M., A.C. Ries, M.P. Coward, and P.R. Cobbold, 1979, Structure, metamorphism and geochronology of the Arequipa Massif of coastal Perú: Journal of the Geological Society of London, v. 136, p. 195-214.

Silva, L.I., 1977, Hojas Pisagua y Zapiga: Santiago, Instituto Investigaciones Geológica, Carta Geológica de Chile, no. 24, p. 1-8.

Skarmeta, J., 1976, Evolucion tectónica y paleográfica de los Andes Patagónicos de Aysen (Chile) durante el Neocomiano: Santiago, I Congreso Geológico Chileno Actas 1, B1-B15.

Skarmeta, J., and N. Marinovic, 1981, Hoja Quillagua, Región de Antofagasta: Santiago, Instituto Investigaciones Geológica, Carta Geológica de Chile, no. 51, p. 1-63.

Smith, A.G., 1980, Subduction and coeval thrust belts with particular reference to North America, *in* K.R. McClay and N.J. Price, eds., Thrust and nappe tectonics: Geological Society of London Special Publication 9, p. 111-124.

Stern, C.H., K. Futa, S. Saul, and M.A. Skewes, 1985, Evolution of the subcontinental mantle lithosphere below southernmost South America: Santiago, Departamento de Geología, Universidad de Chile, Comunicaciones, v. 35, p. 227-231.

Stern, E., M. Skewes, and A.M. Durán, 1976, Volcanismo orogénico en Chile Austral: Santiago, I Congreso Geológico Chileno Actas 2, F195-F212.

Suárez, M., and T.H. Pettigrew, 1976, An upper Mesozoic island-arc and basin system in the southern Andes and South Georgia: Geological Magazine, 113, p. 305-328.

Suárez, M., J.A. Naranjo, and A. Puig, 1985, Estratigrafía de la Cordillera de la Costa, al sur de Taltal, Chile, Etapas iniciales de la evolucion Andina: Revista Geológica de Chile, no. 24, p. 19-28.

Thomas, H., 1958, Geología de la Cordillera de la Costa entre el valle de La Ligua y la cuesta de Barriga: Santiago, Instituto de Investigaciones Geológicas Boletín 2, p. 1-80.

Travisany, V.J., 1978, Mineralizacion cuprífera en areniscas de la Formacion San Pedro en el distrito San Bartolo: Santiago, Departamento de Geología, Universidad de Chile, (unpublished thesis) p. 1-70.

Turner, J.C.M., 1960, Estratigrafía de la Sierra de Santa Victoria y adyacencias: Córdoba, Academia Nacional de Ciencias Boletín, v. 41, p. 163-196.

Turner, J.C.M., 1970, The Andes of northwestern Argentina: Stuttgart, Geologische Rundschau LIX, p. 1028-1063.

Turner, J.C.M., and V. Méndez, 1979, Puna, *in* Segundo Simposio Geol. Reg. Arg: Córdoba, Academia Nacional Ciencias, I, p. 13-56.

Uliana, M., K. Biddle, V.W. Phelps, and D.A. Gust, 1986, Significado del vulcanismo y extensión mesojurásica en el extremo meridional de Sudamérica: Buenos Aires, Revista Asociación Geológica Argentina, v. 40, p. 231-253.

Uyeda, S., 1982, Subduction zones: An introduction to comparative subductology: Tectonophysics, v. 81, p. 133-159.

Uyeda, S., and H. Kanamori, 1982, Back-arc opening and the mode of subduction: Journal of Geophysical Research, v. 84, p. 1049-1061.

Vergara, M., R. Charrier, F. Munizaga, S. Rivano, R. Thiele, and R. Drake, 1985, Miocenic volcanism in the central Chilean Andes, Cordillera Principal (31°30′-34°30′S lat.): Santiago, Departamento de Geología, Universidad de Chile, Comunicaciones, v. 35, p. 49-252.

Vicente, J.C., 1972, Apercu sur l'organization el l'evolution des Andes argentino-chiliennes centrales au parallele de Aconcagua: Montreal, XXIV International Geological Congress, Section 3, p. 42-436.

von Heune, R., 1984, Structural diversity along modern convergent margins and the role of overpressured pore fluids in subduction zones: Société Géologique de France Bulletin, v. 26, p. 207-219.

Wells, P., 1976, The geochemistry of the Patagonian Batholith between 45°-46° S: University of Birmingham, England, M.S. thesis, p. 1-65.

Winslow, M., 1979, Mechanism for basement shortening in the Andean foreland fold belt of southern South America, *in* K.R. McClay and N.J. Price, eds., Thrust and nappe tectonics: Geological Society of London Special Publication 9, p. 513-526.

Winter, H. de la R., 1984, Tectonostratigraphy as applied to analysis of South African Phanerozoic basins: Pretoria, Transactions of the Geological Society of South Africa, v. 87, p. 169-179.

Yrigoyen, M., 1976, Observaciones geológicas alrededor del Aconcagua: Santiago, I Congreso Geológico Chileno Actas 1, p. A169-190.

Zeil, W., 1979, Vulkanismus und geodynamik an der wende Palüozoikum-Mesozoikum in den zentralen und südlichen Anden (Chile-Argentinien): Zentralblatt Geologie Palaeontologie, v. 3-4, p. 298-318.

Zentilli, M., 1974, Geological evolution and metallogenetic relationship in the Andes of northern Chile between 26° and 29° S: Queens University, Ph.D. thesis, p. 1-446.

Ericksen, G. E., Cañas Pinochet, M. T., and
Reinemund, J. A., editors, 1989, Geology of the Andes
and its relation to hydrocarbon and mineral resources:
Houston, Texas, Circum-Pacific Council for Energy
and Mineral Resources Earth Science Series, v. 11.

CHAPTER 6

EARLY LATE CRETACEOUS OVERTHRUSTING IN THE WESTERN CORDILLERA OF SOUTHERN PERÚ

Jean-Claude Vicente
Département de Géotectonique
Université Pierre et Marie Curie
Paris, France

ABSTRACT

The importance of early Late Cretaceous overthrusting in the Western Cordillera of southern Perú has been demonstrated by geologic mapping in the region northwest of Arequipa. Three superposed tectonic units include (1) the lower, autochthonous Huambo unit; (2) the middle, parautochthonous Yura unit, which is widely detached and thrust over the Huambo unit; and (3) the upper, allochthonous Lluta unit, which is an overturned segment of the Precambrian Arequipa massif thrust over the Yura unit. Hence, the Arequipa massif and its Mesozoic cover consist of a huge, crystalline-cored recumbent anticline (Lluta fold nappe) overthrust northeastward for at least 30 km along a low-angle fault. This overthrust is marked by a series of tectonic windows and klippen that extend northwestward for at least 120 km. The magnitude of this structure indicates that tectonism of the Western Cordillera of southern Perú is much greater than had been previously known.

A tectonic window having a style similar to that observed in the Arequipa area exists at Palca, near the Chilean border. This suggests the presence of large overthrust faults in southernmost Perú, which may extend into northernmost Chile.

In central Perú, the tectonic style, as shown by exposures along the Huara River, is a typical thin-skinned, easterly wedging fold and thrust belt, regionally detached and having an imbricate zone at its front and a penetrative cleavage and schistosity zone at its rear. The general tectonic polarity implies compression from the west. The abrupt change in the Churin area, from eastern cratonic-derived sediments to western are-derived volcanics, implies the lack of a large marine basin. This zone of change is the inferred Churin overthrust, which involves the Coast Range allochthon and the destruction of a former backarc "marginal" basin by a typical arc-continent collision. Consequently, the Late Cretaceous tectonics are now viewed as a direct result of large-scale crustal convergence (more than 100 km) and collision between the sialic basement of the arc (Paracas-Arequipa block) and the Brazilian craton.

INTRODUCTION

The purpose of this chapter is to emphasize the importance of early Late Cretaceous overthrusting in the Western Cordillera of southern Perú (Figures 1, 2) and to discuss its significance within the Andean context. The discussion is based on systematic mapping (1:50,000 scale) by a team of postgraduate students guided by me during my stay as visiting professor at the University of Arequipa (1977-1983). The study began with a field survey of the relationship between the Precambrian basement and the Mesozoic cover (Figure 3) as first shown by Vargas (1970) in his geologic map of the Arequipa quadrangle. This led to a differentiation of two tectonic units (Vicente et al., 1979): (1) the Yura unit, a lower unit regarded as relatively autochthonous; and (2) the Lluta unit, an upper allochthonous unit stratigraphically linked to the Precambrian Arequipa massif, which is overturned and thrust over the Yura unit. This relationship is illustrated in Figures 4, 5, and 6, which show that the contacts between the units delineate a series of tectonic windows (Liquiña, Siguas, Lluta, Gloriahuasí) and klippes (Pitchil, Paccpaco). The structure of the area northwest of Arequipa is exceptionally well exposed in the deep Siguas and Colca canyons (Figure 6B and 6G). The so-called Arequipa massif of this region and associated younger rocks constitute an enormous recumbent Precambrian-cored anticline which was overthrust northeastward along a low-angle fault plane. Moreover, the root of this structure is extensively intruded by the Arequipa batholith, which was

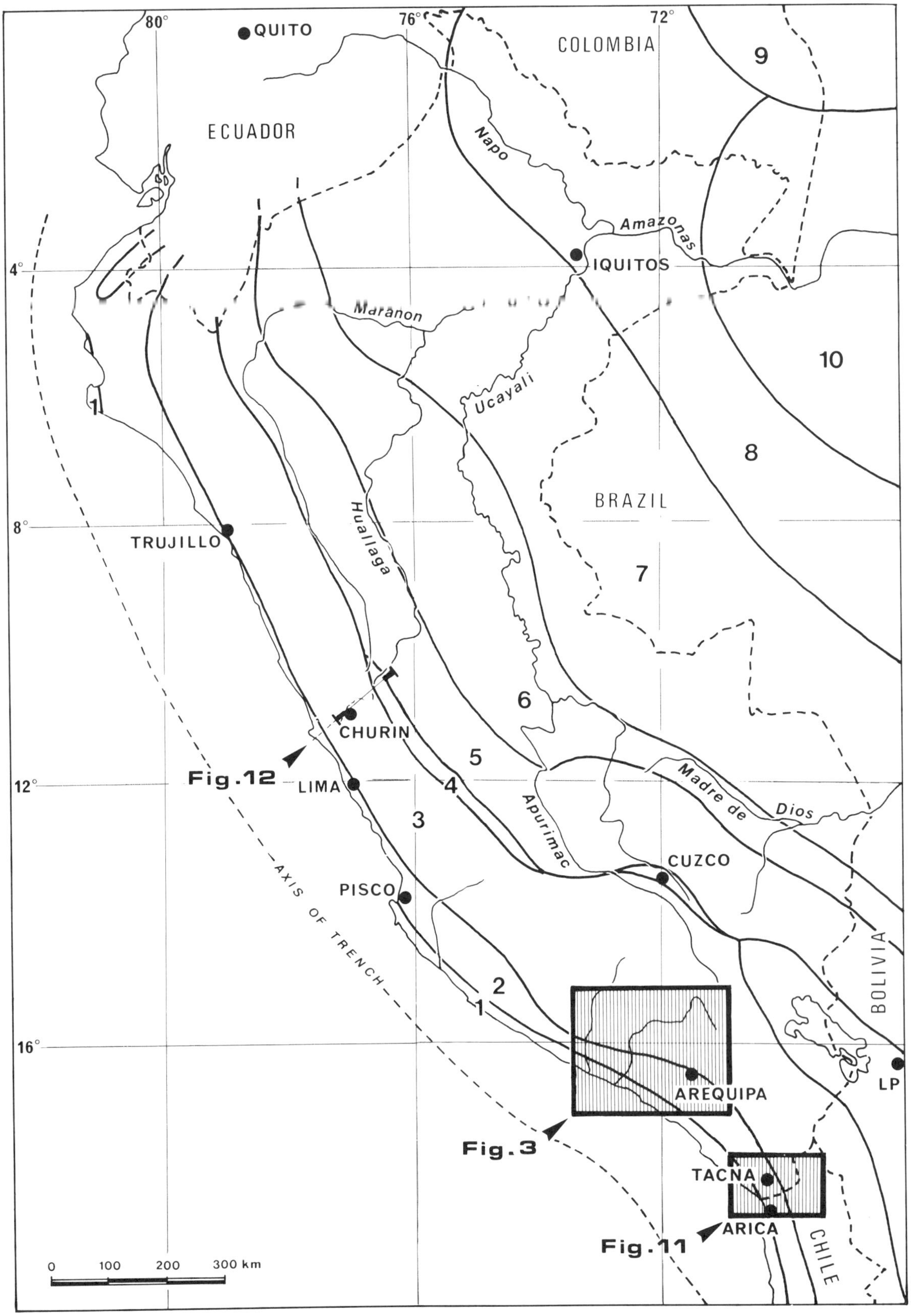

Figure 1. Main physiographic provinces of the Peruvian Andes and the adjacent foreland, showing principal areas discussed in the text. (1) Coast Range; (2) Coastal plains; (3) Western Cordillera; (4) Inter-Andean basins or Altiplano; (5) Eastern Cordillera; (6) sub-Andean zone; (7) Peri-Andean basins (Marañon, Ucayali, Madre de Dios); (8) Iquitos arch; (9) Guyana shield; and (10) Amazon basin.

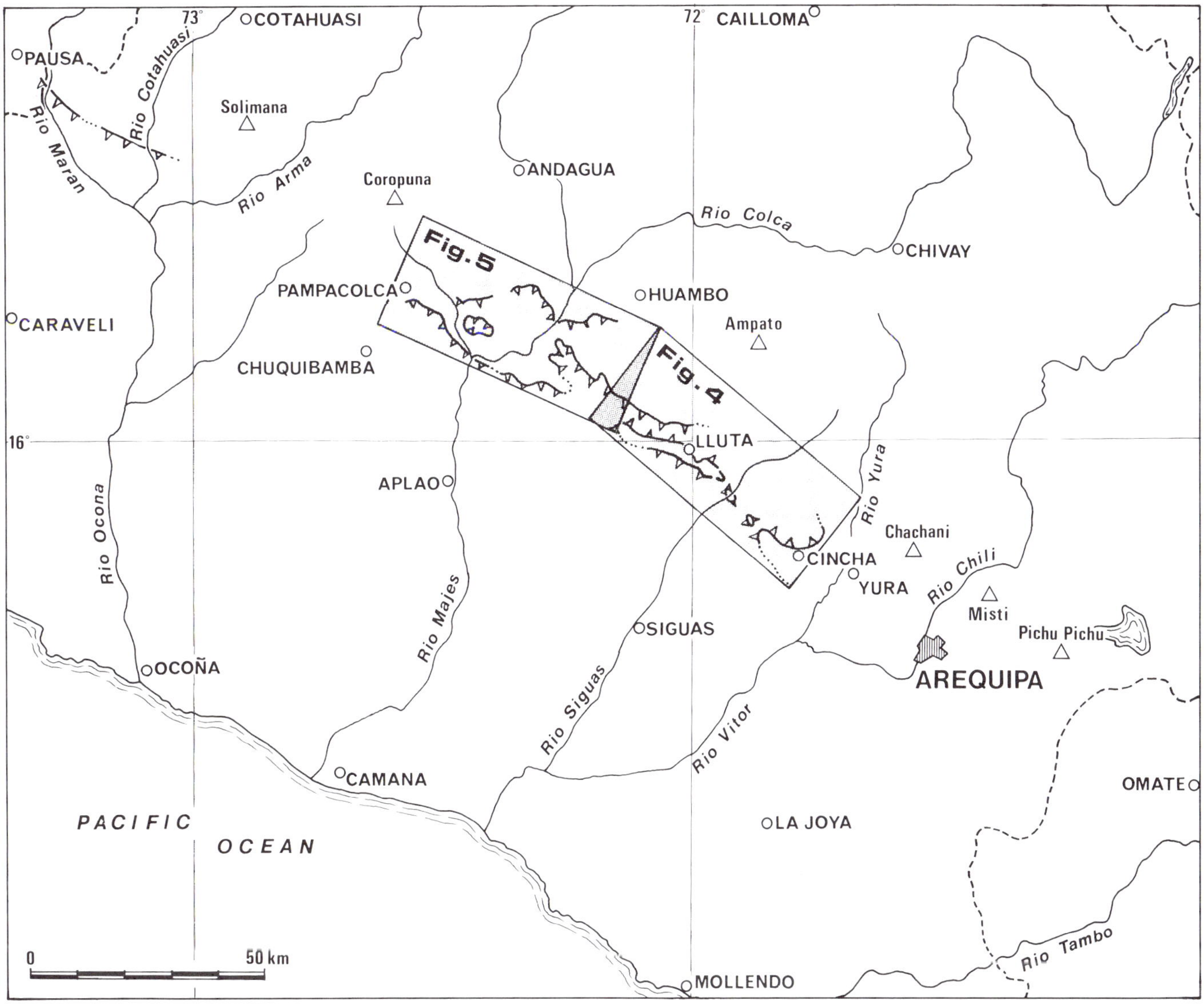

Figure 2. Sketch map of the Arequipa region, showing location of Cincha-Lluta and Rio Colca areas (Figures 4, 5).

emplaced as a series of intrusive pulses between 97 and 78 Ma (Beckinsale et al., 1985; Mukasa and Tilton, 1985). This indicates that the principal tectogenesis occurred in the early Late Cretaceous.

CINCHA-LLUTA OVERTHRUST

The Cincha-Lluta overthrust is the major fault between the allochthonous and autochthonous masses northwest of Arequipa. It can be traced discontinuously for more than 120 km from Cincha to Pampacolca (Figure 2). Surveys across the Cotahuasi and Maran Rivers imply that it may continue still farther northwest to near Pausa.

The tectonic style of the region is illustrated by cross sections on Figure 6. Figure 6 also shows the lithostratigraphic column of the Lluta unit and the standard nomenclature of the formations of this unit. Note the dominance of margin sediments throughout the column, except for the fluvial red beds of the Murco Formation. The dynamic evolution of the Arequipa basin is indicated by the various formations shown in Figure 3 (Vicente et al., 1982). Except for the turbiditic facies, which is the more distal facies of the allochthonous unit, there do not appear to be significant differences in facies and thickness between the Yura and Lluta units. This attests to the width of the Arequipa basin.

The Lluta unit is more than 3000 m thick (not including the 1000 m of the Lower Jurassic Chocolate Formation). It shows contrasting lithologies of competent sequences (Labra, Hualhuani, and Murco quartzites) and incompetent sequences (Socosani and Cachios shales, Murco and Chilcane pelites and evaporites, Puente turbidites), which have affected the tectonic behavior and the morphologic expression of the Lluta unit.

Rio Siguas Profile

A profile exposed along the upper course of Rio Siguas (Figure 6B) shows two major features: (1) an overturned section in which the Precambrian rocks of the Arequipa massif overlie

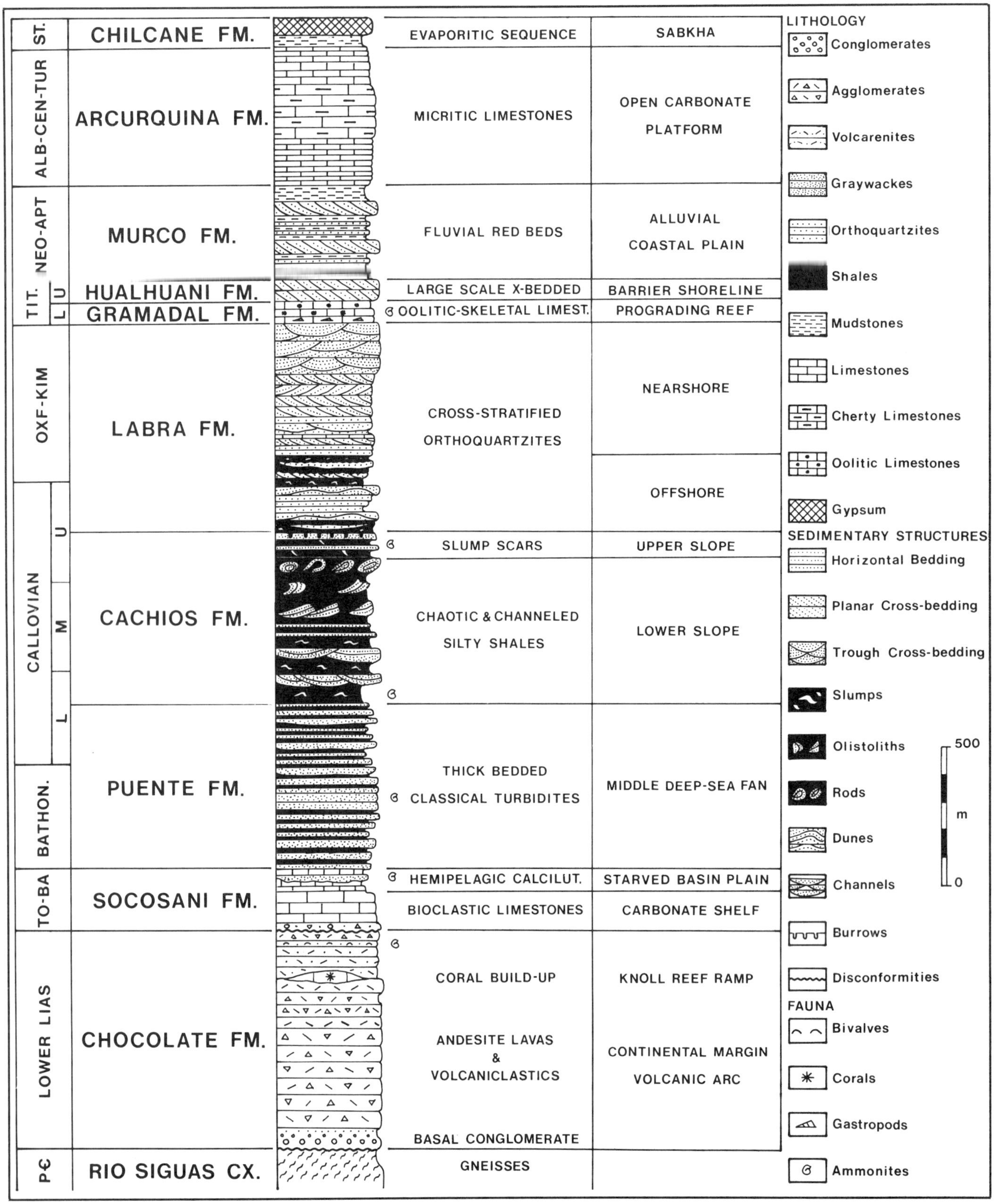

Figure 3. Lithostratigraphic column of the Mesozoic sedimentary cover of the Arequipa massif (Lluta unit) showing the main facies and environments of deposition.

the Mesozoic Lluta unit, and (2) the "Rio Siguas tectonic window," in which the Cretaceous Murco Formation is exposed beneath Precambrian rocks.

Downstream from Tingo (Figure 6B), the lower part of the Chocolate Formation is overturned and lies beneath Precambrian gneiss. The contact is an angular unconformity. The gneissic foliation dips 80 °SSW, and the Chocolate bedding is inclined 45 °SW. The base of this formation is marked by a transgressive unit consisting of a 70-m-thick coarse conglomerate consisting of reworked angular and subangular fragments of Precambrian basement rocks (Zavala, 1981).

With the exception of the drag folds, all the sequence upstream from this unconformable contact shows a nearly homoclinal style with a progressive steepening to the northeast, which outlines the reverse limb of a huge Precambrian-cored anticline. Only in the front does the Arcurquina Formation exhibit a comparatively independent and tight folding. Stratigraphic succession and standard polarity (sole marks, graded bedding, cross-bedding) show that the entire Lluta unit is inverted.

In the Rio Siguas window a sequence of quartzites and variegated pelites (lithologically indistinguishable from the typical Murco Formation) is exposed at the bottom of the canyon beneath Precambrian gneiss (Figure 6B). This sequence shows tight asymmetrical folds and imbricate faults facing northeast. At the fault contact, the Murco quartzites show a strong cataclasis with shear cleavage, boudins, and cleavage mullions. The thrust plane is well exposed on both sides of the canyon, and outlines a buttonhole-shaped window (Figure 4).

Cerro Liquiña Half-Window

The tectonic relationships between the Yura and Lluta units (Vicente et al., 1979; Valdivia, 1979) are exceptionally well shown in the Cerro Liquiña area, 12 km southeast of the Rio Siguas window (Figure 4). In this area it was first recognized that the lower para-autochthonous unit corresponded to the Yura of the type area (Jenks, 1948; Benavides, 1962; Vargas, 1970), which accounts for its designation as the Yura unit. Figure 6A shows that the relatively autochthonous unit, with its large open folds, is overlain mainly by overturned strata of the allochthonous unit. The latter shows the particularly imbricate style of the Socosani Formation (Figure 7B), on the northeast side of Cerro Liquiña; the hectometric folding of the Puente turbidites, with a complete series of synformal anticlines and antiformal synclines; and the strongly folded Arcurquina Formation "floating" on gypsum diapirs (Figure 6B).

The structure of Cerro Liquiña (Figure 7) shows a nappe anticline having a core of Murco quartzites and limbs of older allochthonous Chocolate Formation. Imbricate faulting and striations along secondary shear planes (Figure 7) indicate a thrusting toward the northeast (Sequeiros, 1979). Gypsum diapirs along the contact in the Cincha sector (Figure 2) and along Quebrada Liquiña represent tectonic accumulations along the sole thrust.

Rio Lluta Profile

The autochthonous Yura unit, exposed in the Lluta window (Figure 6C) has tight, chevronlike folds 1-2 km wide and pinched synclines. The allochthonous Lluta unit is remarkably well exposed in this area. Moreover, the deep notch of Rio Lluta allows one an appreciation of the size of the thrust sheet. The plunging frontal lobe or arch-bend of the nappe is observed upstream from Lluta, at the very bottom of the canyon. Also, upstream from the overturned limb, the transition to the normal limb accounts for the general outward steepening of dips already noted on the Rio Siguas cross section.

The Lluta unit displays a large recumbent fold (Figure 6C) in whose overturned limb more than 6 km of strata are exposed. The Precambrian Arequipa massif is the basement core of this spectacular Lluta fold nappe. Consequently, it is estimated that the real thickness and amplitude of this structure is comparable to that in the classic Helvetic nappes. In many ways, the style and size of the Lluta nappe with respect to the Arequipa massif is reminiscent of the Morcles nappe with respect to Mont Blanc (Ramsay, 1981).

Reconstruction of the eroded normal limb and the truncated strata in the overturned limb indicate that the fold developed prior to thrusting. Similarly, the truncated folds in the Yura unit show that the overthrust sheet moved over an already folded autochthon. This accounts for the numerous exotic blocks and lenses of quartzites observed along the thrust plane (Vicente et al., 1979).

Among minor structures, it must be pointed out that the leading imbricate faults in the Chocolate-Socosani sequence are associated with minor thrust faults involving disharmonic mesoscopic folding of the Puente Formation and local slaty cleavage development, and the left-handed pattern of drag folds in the Cachios Formation supports its overturned position (Figure 6C). At the front, the tightly folded Arcurquina Formation, almost chevron in style, overrides and pushes forward large masses of Upper Cretaceous evaporites (Chilcane Formation). The intense folding of the Arcurquina Formation is in marked contrast with the rest of the unit and involves the frictional locking and crumpling of the detached part of the inverted limb at the front of the recumbent fold.

As a consequence of a general culmination toward the northwest, we can judge the real dimension of the inverted limb. In effect, if we follow the Murco Formation as a marker horizon, we observe that is becomes progressively overturned, and in the Cerro Tintayquisma area (Figure 6D) it exhibits a horizontal and inverted flat-lying position over the Arcurquina Formation. The differential detachment between Murco and Arcurquina formations can be interpreted as a typical "strut-thrust" (according to the terminology of Badgley, 1965, p. 165) with a net slip of about 2 km, where the movement of the competent Murco quartzites was compensated by folding in the underlying, also inverted, Arcurquina Formation. The "Lluta window" can be traced 15 km to the northwest (Figure 4), where the Murco quartzites of the Yura unit are below the overturned Socosani and Puente formations of the Lluta unit in the transverse range between Cerro Tororunca and Cerro Pucaguada.

Rio Colca Profile

The Yura unit is well exposed in the 4000-m-deep gorge of the Rio Colca (Figure 6G). It shows a series of large, boxlike anticlines at Cerros Huanaquituyoc and Accarota and pinched synclines at Satay, Huacombra, and Pichilhuay. The important Jasmin thrust fault marks the northeastern margin of the Yura

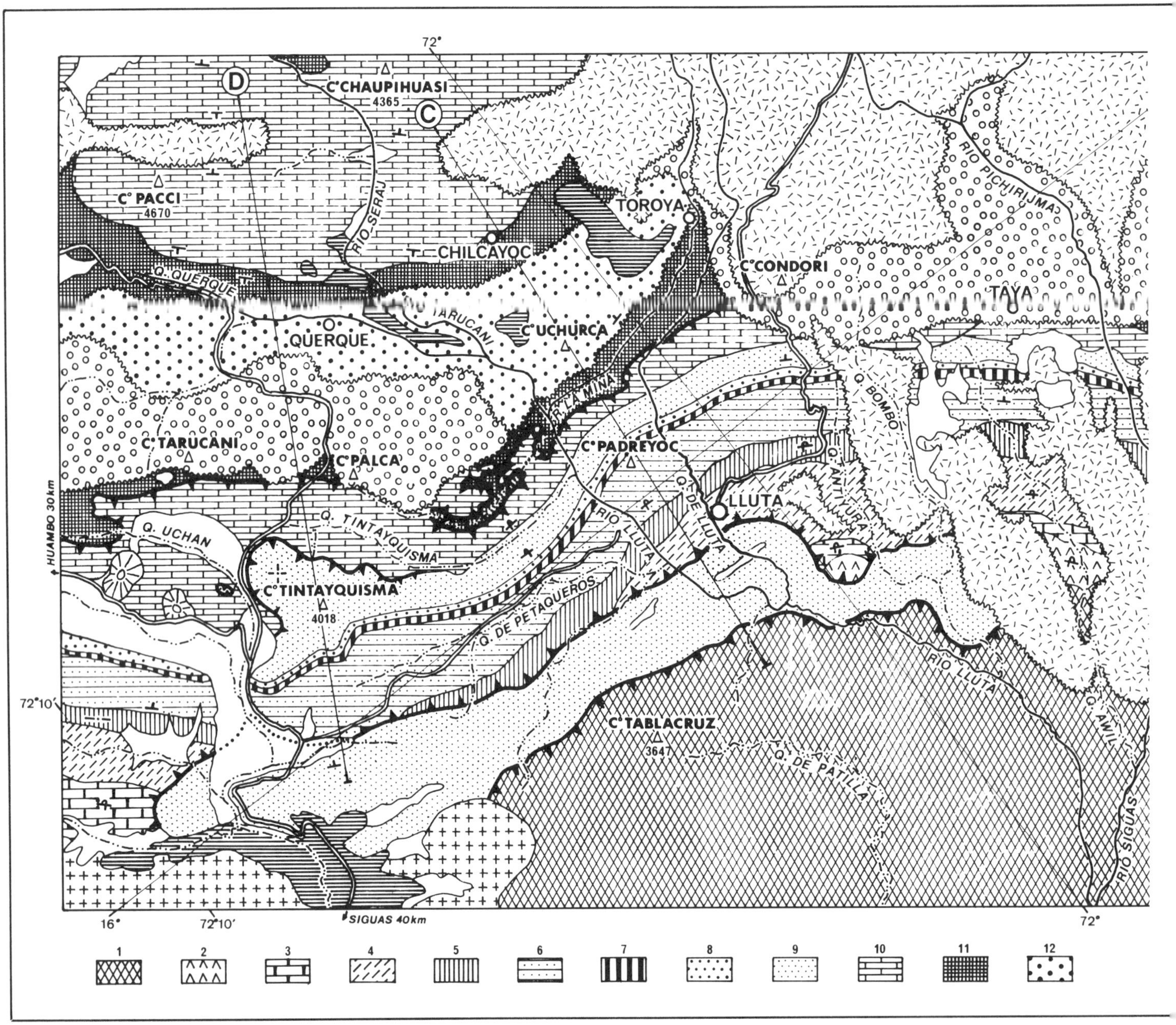

Figure 4. Geological map of Cincha-Lluta area (see Figure 2),showing locations of cross sections A, B, C, and D (Figure 6). After Vicente et al., 1979. (1) Rio Siguas metamorphic complex (Precambrian); (2) Chocolate Formation (lower Lias); (3) Socosani Formation (Toarcian-Bajocian); (4) Puente Formation (Bathonian-lower Callovian); (5) Cachios Formation (middle to upper Callovian); (6) Labra Formation (Oxfordian-Kimmeridgian); (7) Gramadal Formation (lower Tithonian); (8) Hualhuani Formation (upper Tithonian); (9) Murco Formation (Neocomian-Aptian); (10)

unit and the beginning of the autochthonous Huambo unit (Huaman, 1980), which implies that the Yura unit is para-autochthonous. Furthermore, the implied 4-5 km of net slip along the thrust and the box-fold geometry suggest a major detachment of the Yura unit that probably developed at the level of the Cachios shales. In that case, the Jasmin thrust fault must flatten downward into a bedding-plane fault. The back thrust, which dips to the northeast and developed on the southwest flank of the Accarota box fold, appears to have formed in response to compression in the core of the structure because of the detachment (Ramsay, 1967). With regard to the allochthon itself, at that latitude it is structurally high and therefore is poorly preserved. However, it can be recognized southeast of Gloriahuasi at Timar (Figure 6F) as an inverted Precambrian basement to Chocolate-Socosani formations sequence, and at Cerro San Cristobal, as a klippe of tightly folded Arcurquina Formation resting on masses of Upper Cretaceous evaporites (Huaman, 1980). Also, the allochthon is found west of Capiza at Cerro Mamas (Figure 6I) with the same sequence as at Timar and at Cerro Paccpaco (Figure 6H), as a klippe consisting of an overturned sequence of the Labra-Gramadal-Hualhuani-Murco formations (Perez, 1981). In this

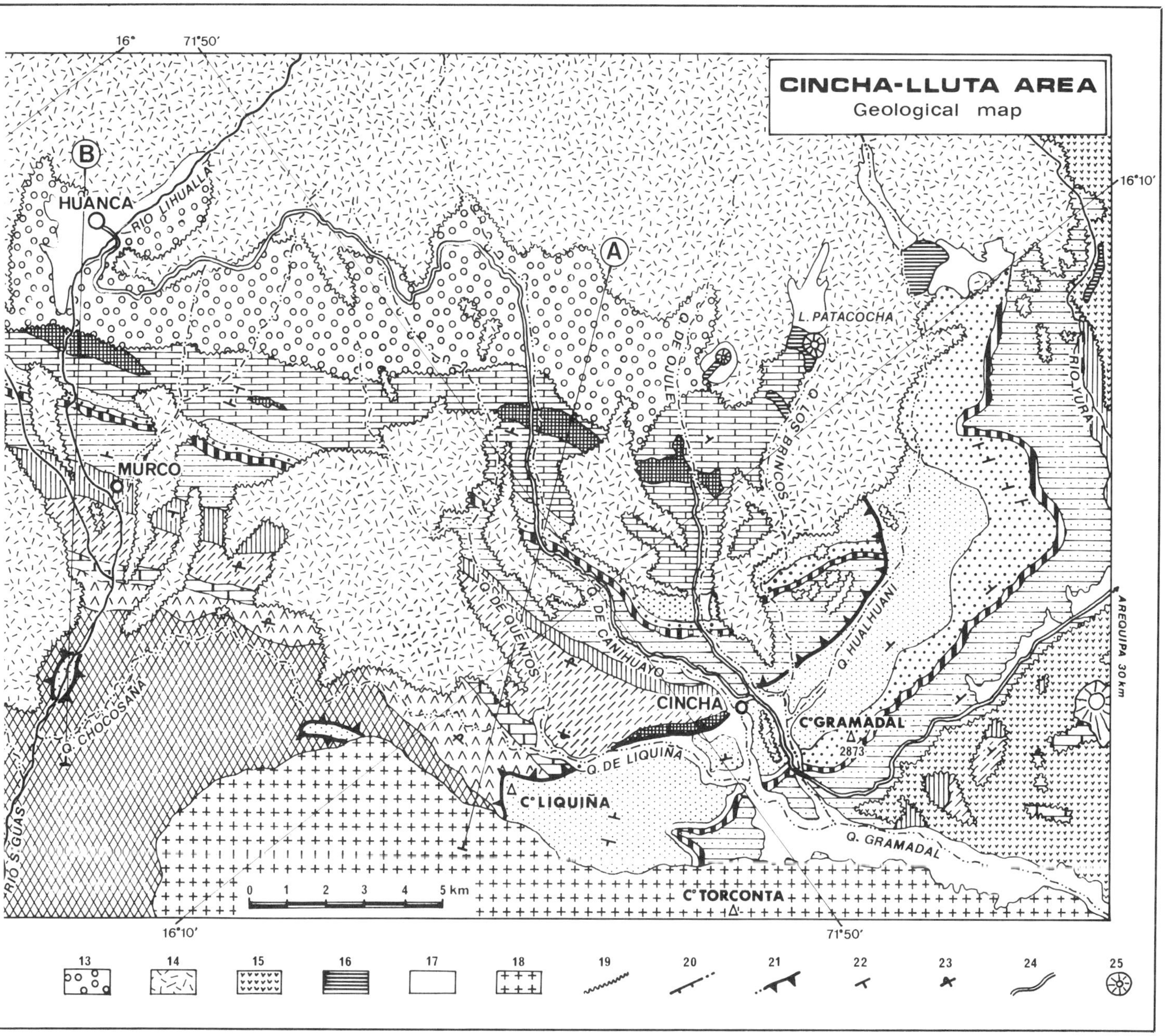

Arcurquina Formation (Albian to Coniacian); (11) Chilcane Formation (Santonian); (12) Uchurca Formation (Eocene); (13) Huanca Formation (Oligocene); (14) Tacaza Formation (Miocene); (15) Sencca Formation (upper Pliocene); (16) basalt flow (Quaternary); (17) alluvium; (18) Arequipa batholith (Upper Cretaceous); (19) late Andean unconformities; (20) normal fault; (21) thrust fault (teeth on overthrust block); (22) strike and dip of beds; (23) overturned beds; (24) road; and (25) volcanic cone.

manner, the characteristic feature of the Lluta unit is strong folding of the Socosani Formation with development of axial-plane slaty cleavage and late shear planes. This situation suggests an increase in the intensity of deformation northward, perhaps related to a more important overthrust.

Main Features of the Lluta Nappe

In summary, the Lluta nappe is a large crystalline-cored recumbent fold having a minimum width of 10 km, computed between the frontal hinge with deformed gypsum masses and the inner edge of the Rio Siguas window. That amount represents a minimum displacement of the nappe. The net slip may be estimated with reference to the position of the key horizon given by the basement-cover contact below the plane of the overthrust. Because the thrust was formed by stretching of the overturned limb of the recumbent fold, outcrops in the Rio Siguas window and the tectonic style of the Yura unit may be used to restore the missing structure of the inverted limb of the stretched syncline at the root. Thus, the reconstructed section (Figure 8) shows that the shortening associated with the Satay syncline, Huanaquituyoc anticline, and a missing syn-

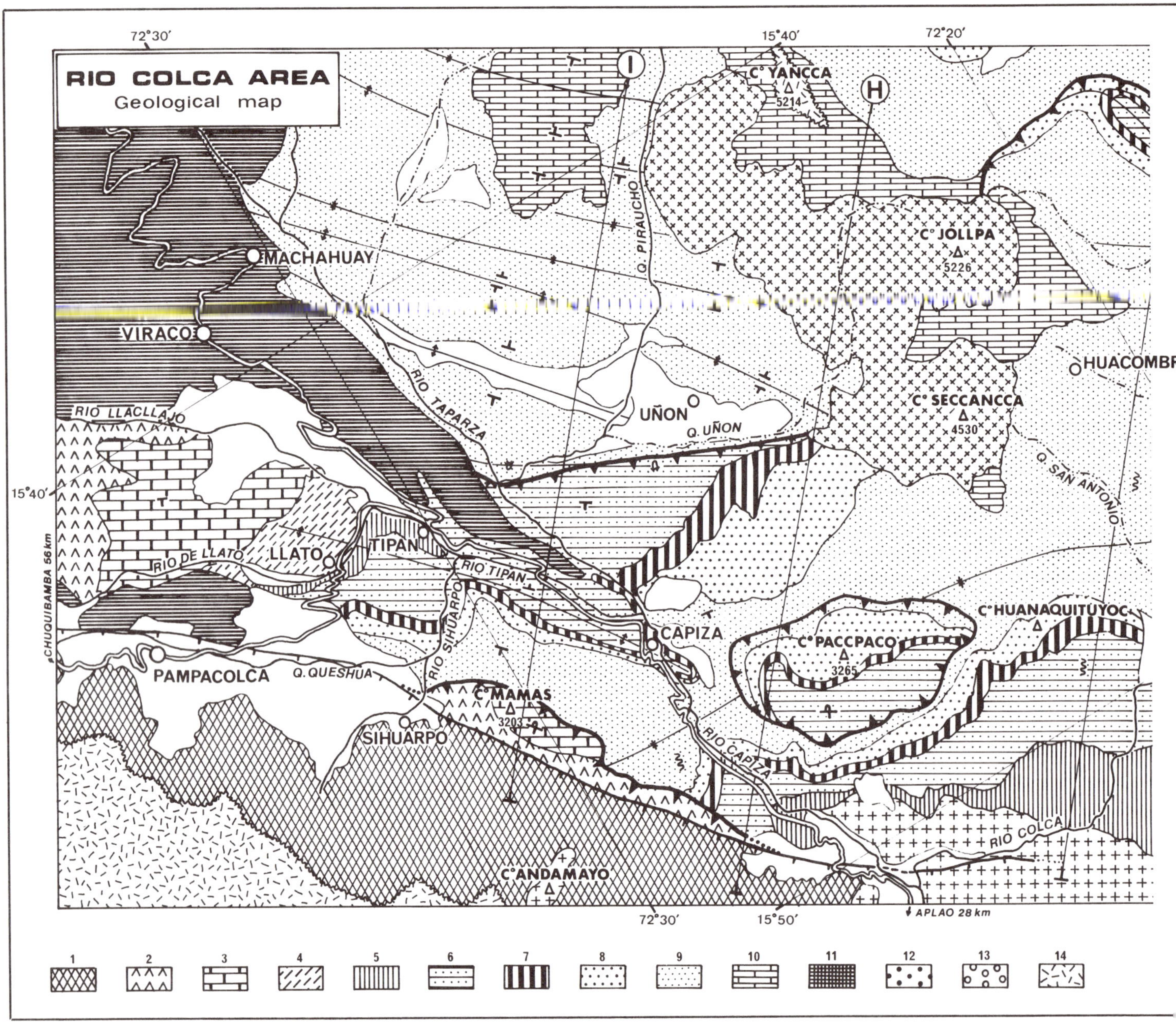

Figure 5. Geological map of Rio Colca area (see Figure 2), showing location of sections E, F, G, H, I (Figure 6). After Huaman, 1980; Perez, 1981. (1) Rio Siguas metamorphic complex (Precambrian); (2) Chocolate Formation (lower Lias); (3) Socosani Formation (Toarcian-Bajocian); (4) Puente Formation (Bathonian-lower Callovian); (5) Cachios Formation (middle to upper Callovian); (6) Labra Formation (Oxfordian-Kimmeridgian); (7) Gramadal Formation (lower Tithonian); (8) Hualhuani Formation (upper Tithonian); (9) Murco Formation (Neocomian-Aptian); (10) Arcurquina

cline, suggests a minimum displacement of 15 km along the overlying Cerro Mamas overthrust. Then, the total displacement of the nappe, which includes this 15 km displacement plus the 10 km minimum width, is at least 25 km. But we also have to consider the shortening across the detached Yura unit, which is about 40% of its original length. Therefore, 30 km of minimum displacement for the Lluta nappe does not seem at all excessive.

The nappe had a pre-erosion thickness of at least 5.5 km, which includes 2.5 km of crystalline core overlain by 3 km of sedimentary cover. Together with the width of the structure, this provides the conditions for the dynamic metamorphism that developed within and below the thrust sheet. Especially noteworthy in the Andean context is the ductile deformation exhibited by the basement core, possibly related to the thermal effects of intrusion of the Arequipa batholith. This would explain (1) the general retrograde metamorphism to greenschist facies suffered by the katazonal gneisses, which can be connected to widespread cataclasis (Perez, 1981); (2) the mid-Cretaceous thermal event, clearly registered by K-Ar ages for gneisses from Cerro Verde (Stewart, Everden, and Snelling, 1974); and (3) the foliation exhibited by the Torconta tonalite

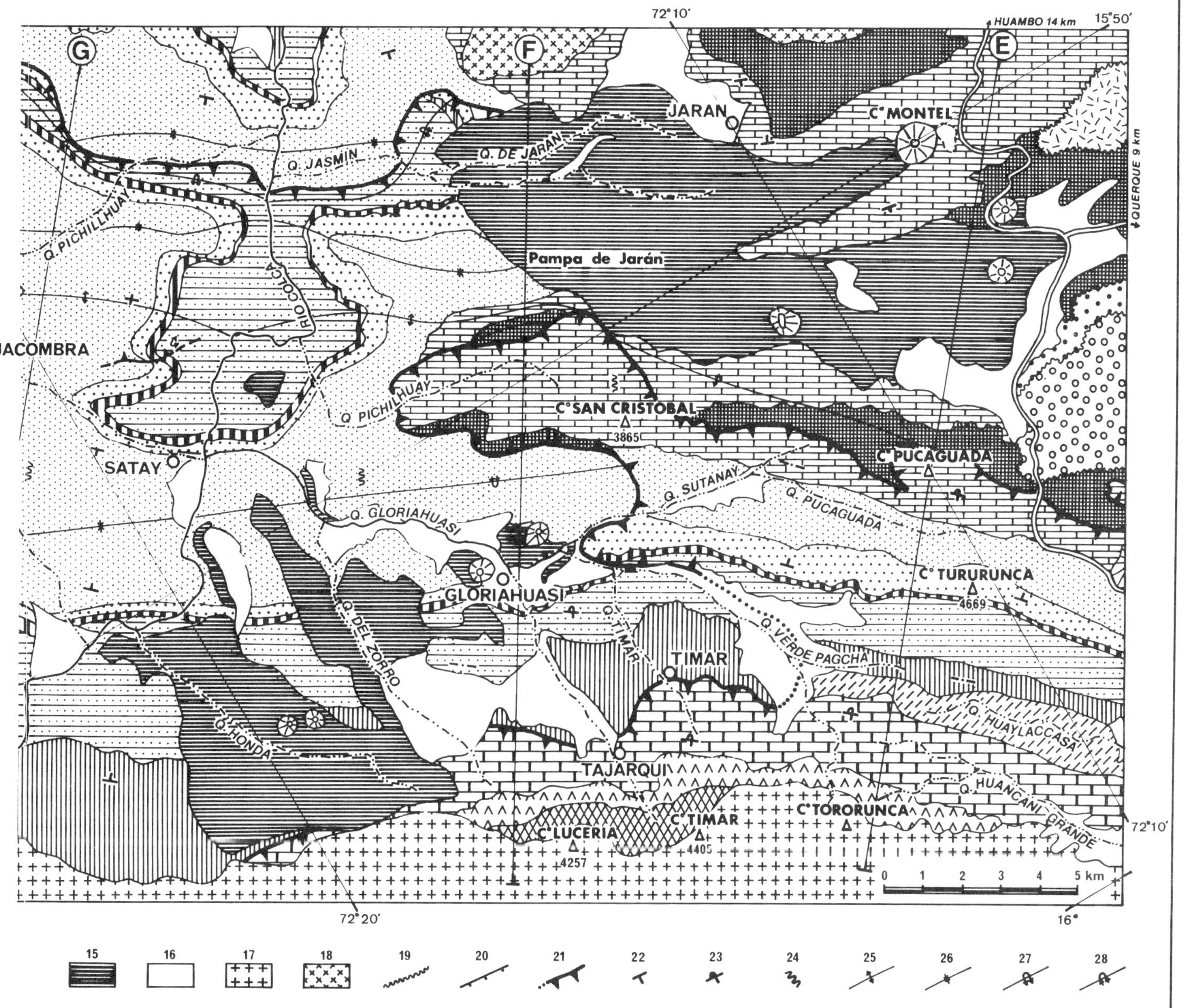

Formation (Albian to Coniacian); (11) Chilcane Formation (Santonian); (12) Uchurca Formation (Eocene); (13) Huanca Formation (Oligocene); (14) Tacaza Formation (Miocene); (15) basalt flows (Quaternary); (16) alluvium; (17) Arequipa batholith (Upper Cretaceous); (18) Cerro Jollpa microdiorite (upper Miocene); (19) late Andean unconformities; (20) normal fault; (21) thrust fault (teeth on overthrust block); (22) strike and dip of beds; (23) overturned beds; (24) small-scale folding; (25) anticline; (26) syncline; (27) overturned anticline; and (28) overturned syncline.

(Incahuasi Group) associated with crystal deformation and recrystallization under greenschist conditions, suggesting syntectonic emplacement (Johan, Le Bel, and McMillan, 1980).

A two-stage development of the Lluta overthrust involved, as typical "stretch" thrust, overextended recumbent folding and a long period of ductile shearing in the gneissic basement prior to thrust faulting. Such ductile deformation of the crystalline core at shallow depths (< 5 km) implies a high geothermal gradient. This again implies a close relationship between the ascent of the batholith, the rise of the fold, and the retrograde metamorphism of the basement.

At the root of the thrust, the plutons, of the Arequipa segment were emplaced between 100 and 80 Ma (late Albian to early Campanian) (Beckinsale et al., 1985; Mukasa and Tilton, 1985), whereas at the front, the red beds involved in the overthrust are Santonian (87.5-84 Ma, DNAG time scale, Palmer, 1983), as indicated by the presence of *Tissotia steinmanni* (Pardo, 1967, unpublished report). This implies that the deformation developed progressively, taking an interval of almost 20 m.y. to be completed, for which 30 km of minimum displacement indicates an advance of at least 1.5 mm per year.

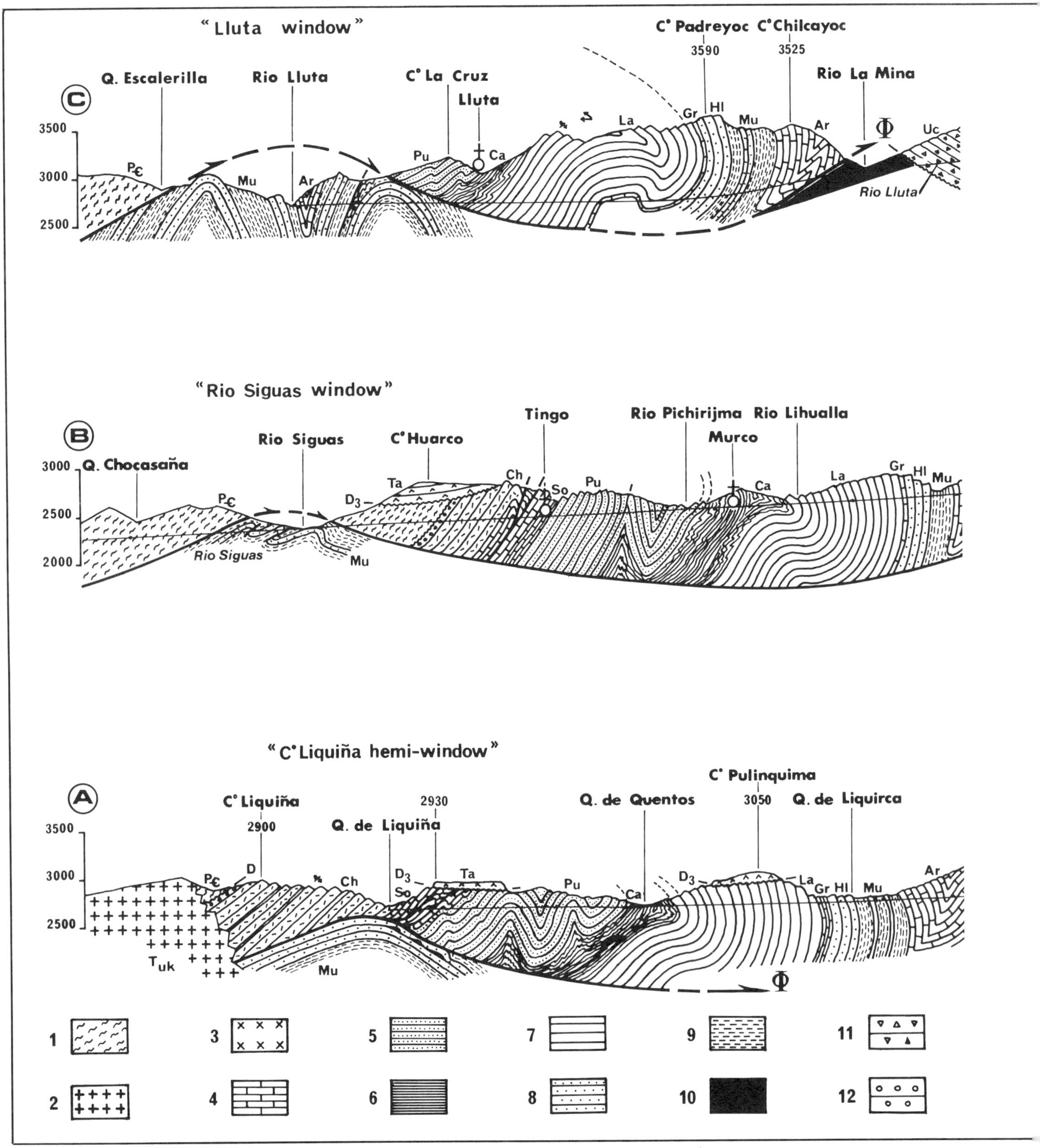

Figure 6. Cross sections of Cincha-Lluta and Rio Colca areas, southern Perú (see Figure 4 and 5 for locations of sections). Based in part on sections by Vicente et al., 1979; Huaman, 1980; Perez, 1981. No vertical exaggeration. (1) gneiss; (2) tonalite; (3) microdiorite; (4) limestone; (5) graywacke; (6) silty shale; (7) subarkose; (8) orthoquartzite; (9) red pelite; (10) gypsum; (11) pyroclastic breccias and agglomerates; (12) conglomerate; (13) andesite; (14) rhyolite; (15) basalt; (16) gravel; (17) unconformity; (18) Lluta unit overthrust; (19) Jasmin thrust fault; (20) fault; (21) overturned beds; (22) anticline axis; and (23) syncline axis. (Pe) Rio Siguas metamorphic complex (Precambrian); (Ch) Chocolate

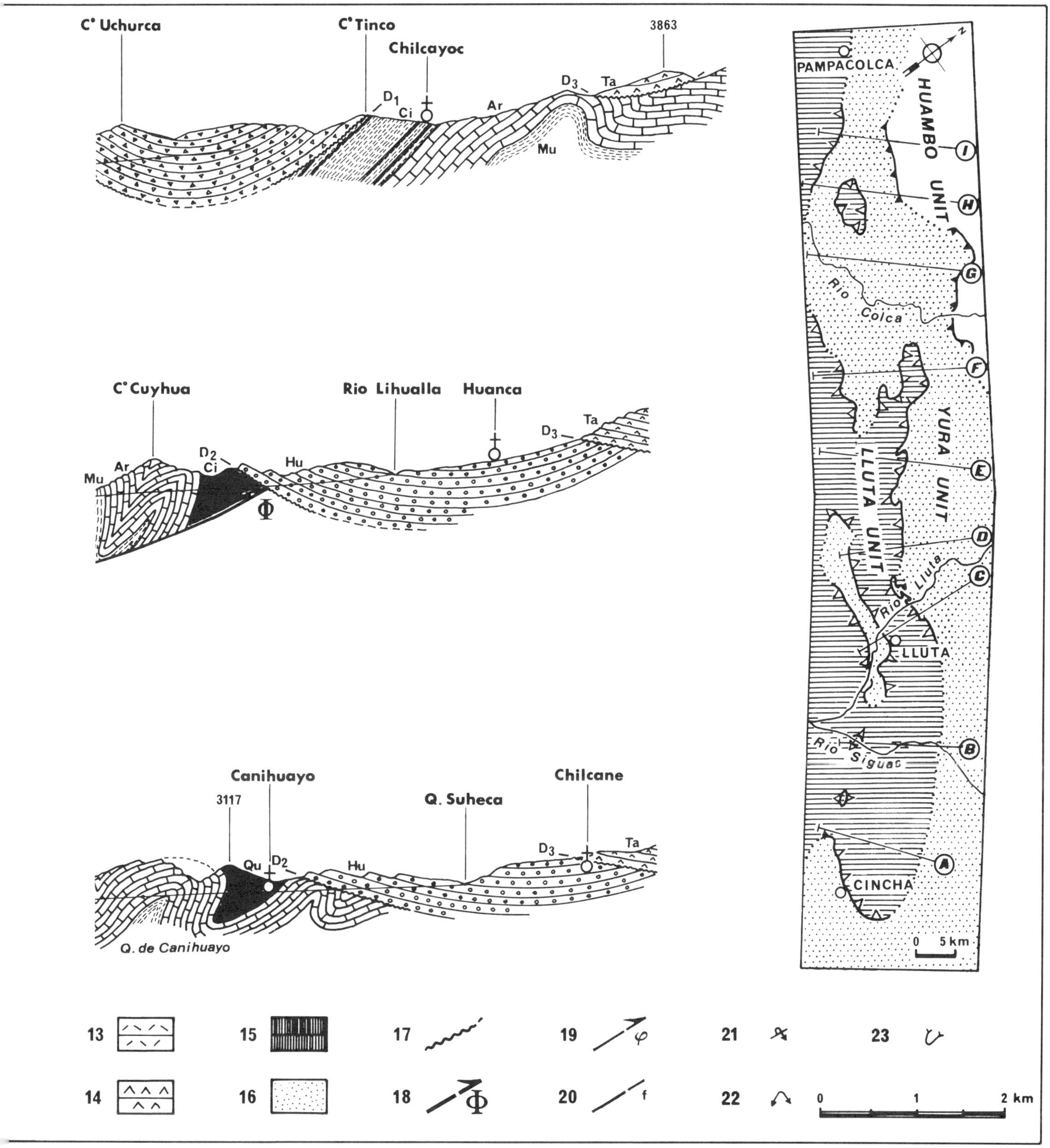

Formation (Lower Jurassic); (So) Socosani Formation (Toarcian-Bajocian); (Pu) Puente Formation (Bathonian-lower Callovian); (Ca) Cachios Formation (middle to upper Callovian); (La) Labra Formation (Oxfordian-Kimmeridgian); (Gr) Gramadal Formation (lower Tithonian); (Hl) Hualhuani Formation (upper Tithonian); (Mu) Murco Formation (Neocomian-Aptian); (Ar) Arcurquina Formation (Albian to Coniacian); (Ci) Chilcane Formation (Santonian); (Uc) Uchurca Formation (Eocene); (Hu) Huanca Formation (Oligocene); (Ta) Tacaza Formation (Miocene); and (S_1) slaty cleavage.

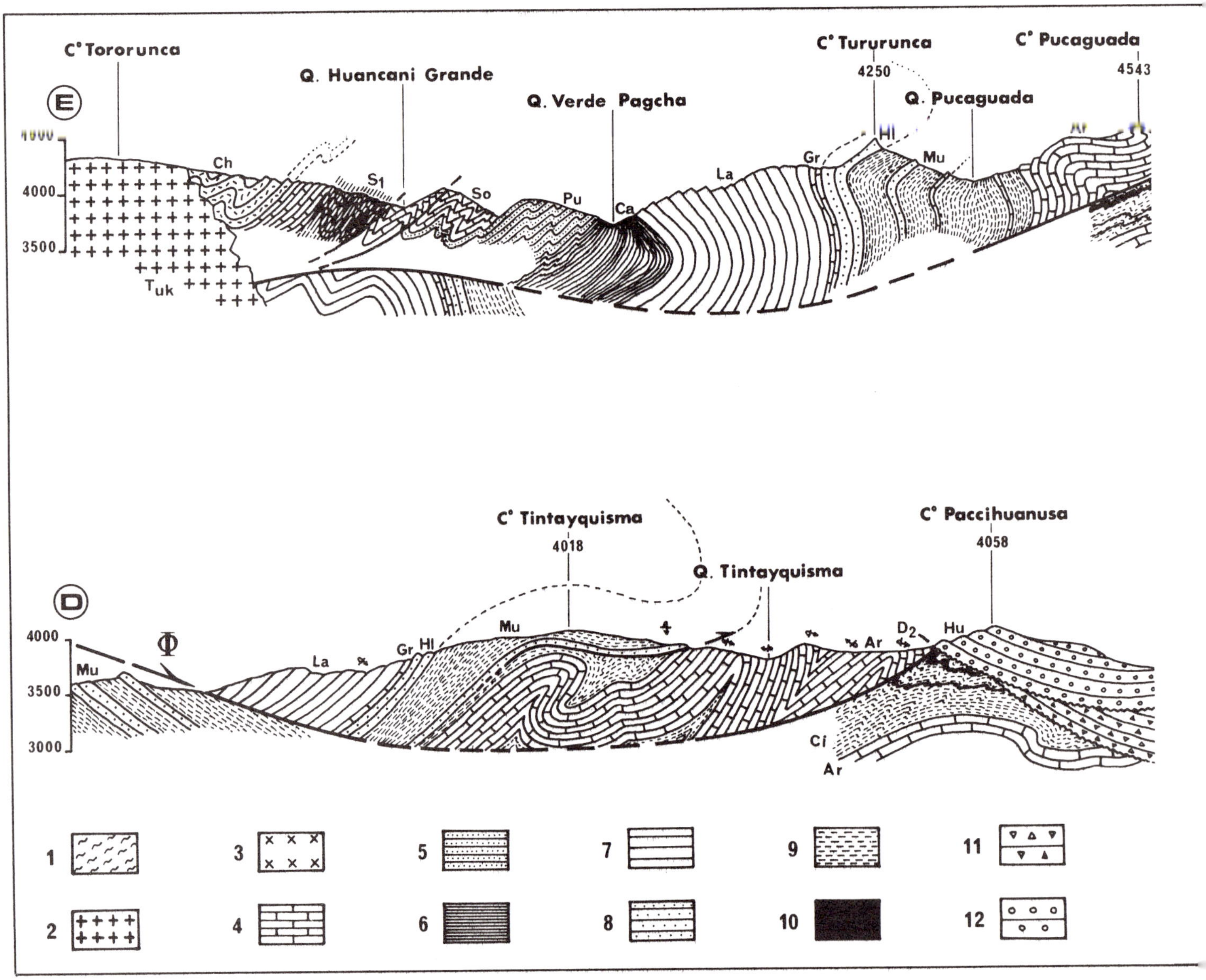

Figure 6 continued.

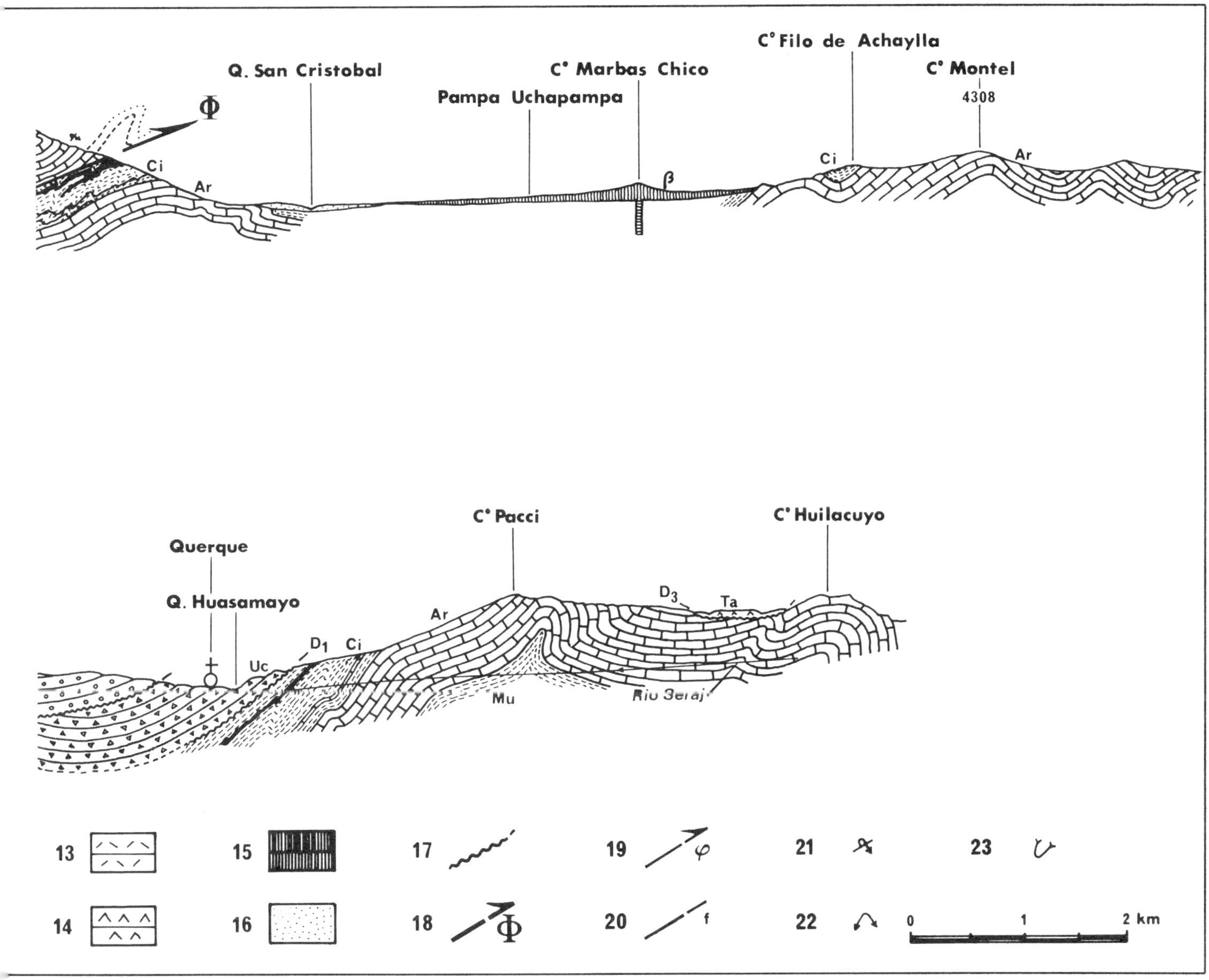
C° Filo de Achaylla
Q. San Cristobal
C° Marbas Chico
C° Montel
4308
Pampa Uchapampa
Φ
Ci
Ar
β
C° Pacci
C° Huilacuyo
Querque
Q. Huasamayo
D3
Ta
D1
Uc
Mu
Rio Seraj
13
14
15
16
17
18
19
20
21
22
23
0
1
2 km

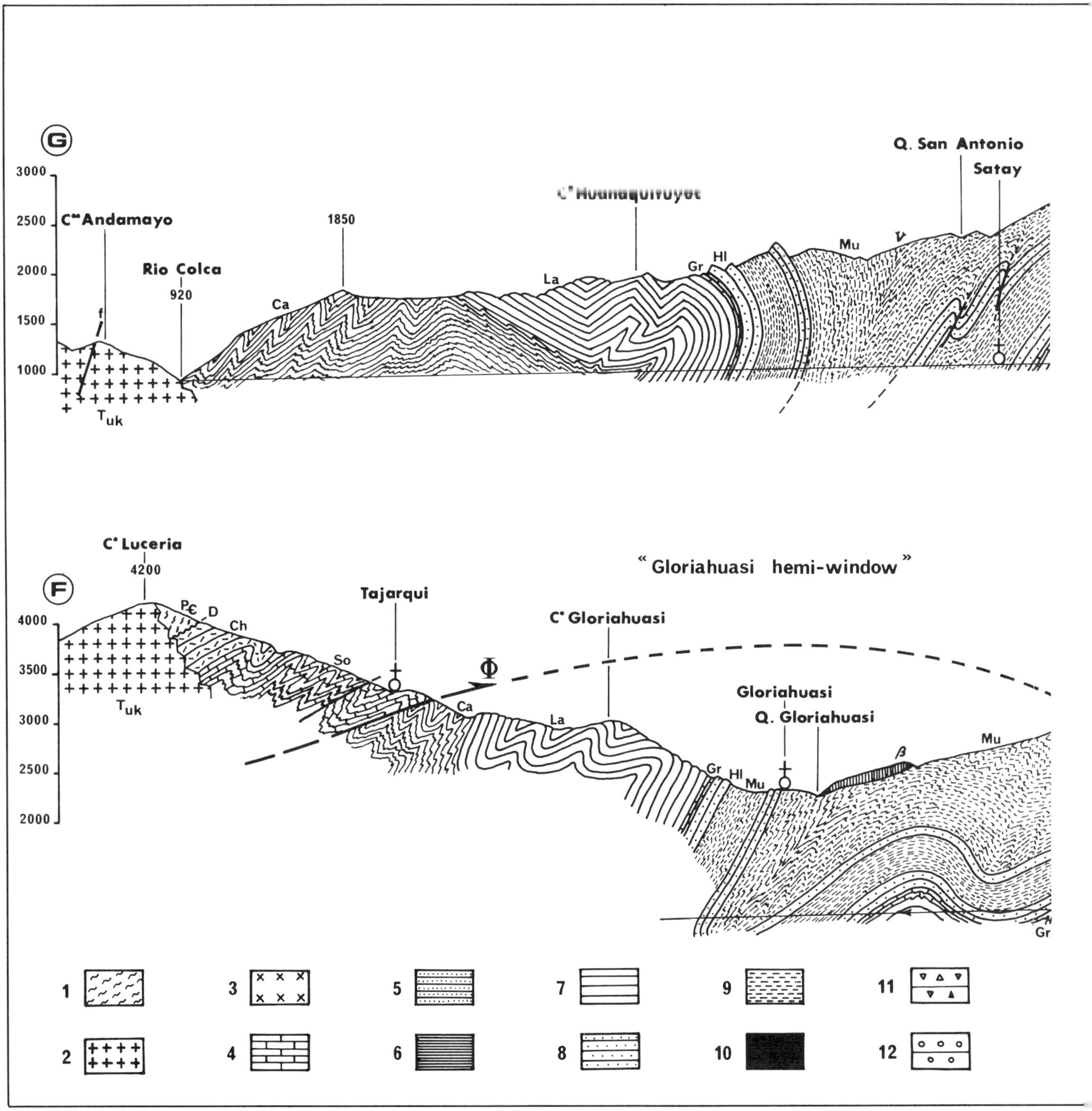

Figure 6 continued.

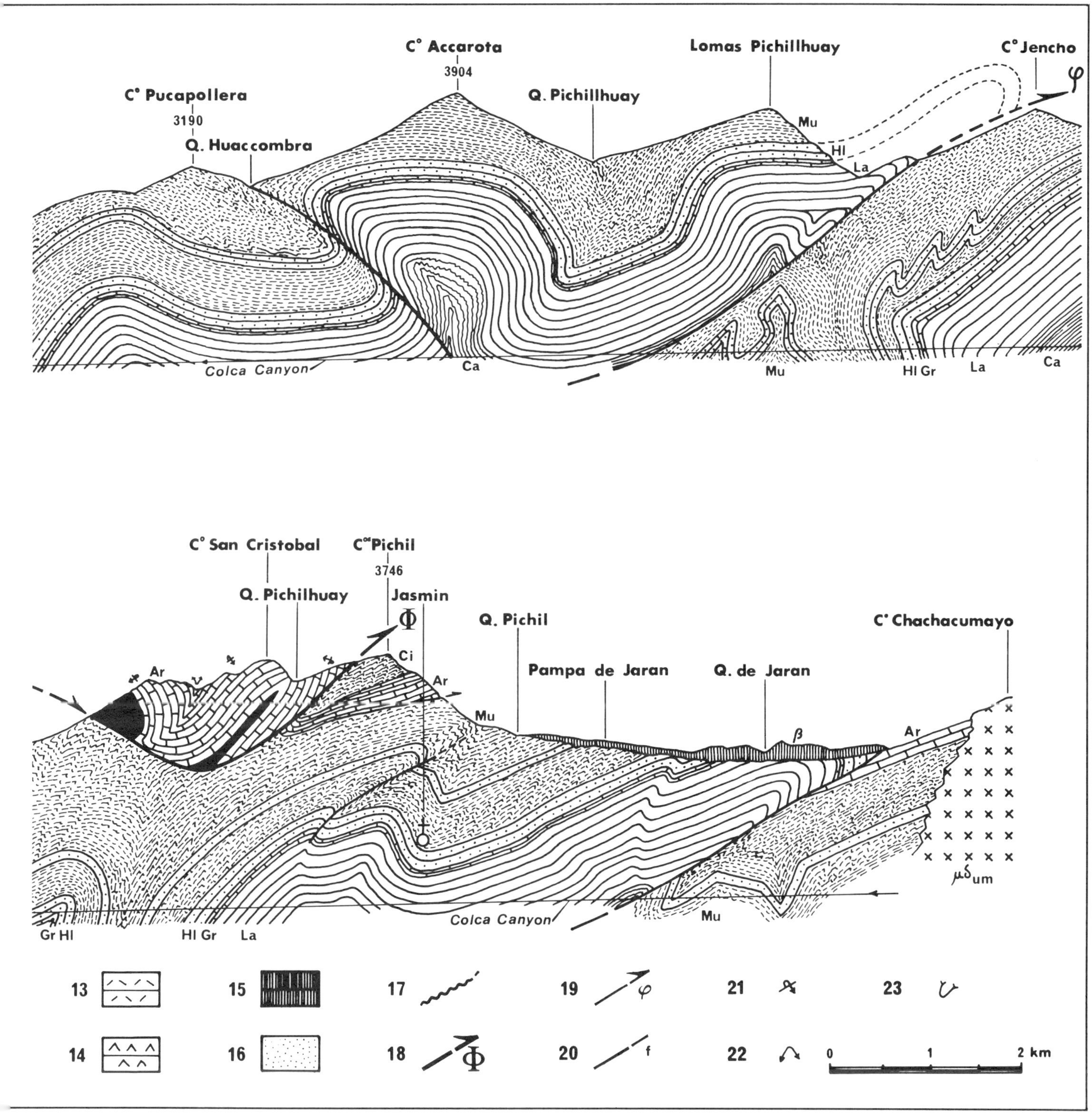
C° Pucapollera
3190
Q. Huaccombra
C° Accarota
3904
Q. Pichillhuay
Lomas Pichillhuay
C° Jencho
Mu
Hl
La
Colca Canyon
Ca
Mu
Hl Gr
La
Ca
C° San Cristobal
C° Pichil
3746
Q. Pichilhuay
Jasmin
Ar
Ci
Ar
Mu
Q. Pichil
Pampa de Jaran
Q. de Jaran
C° Chachacumayo
β
Ar
μδum
Colca Canyon
Mu
Gr Hl
Hl Gr
La
13
14
15
16
17
18
19
20
21
22
23
0
1
2 km

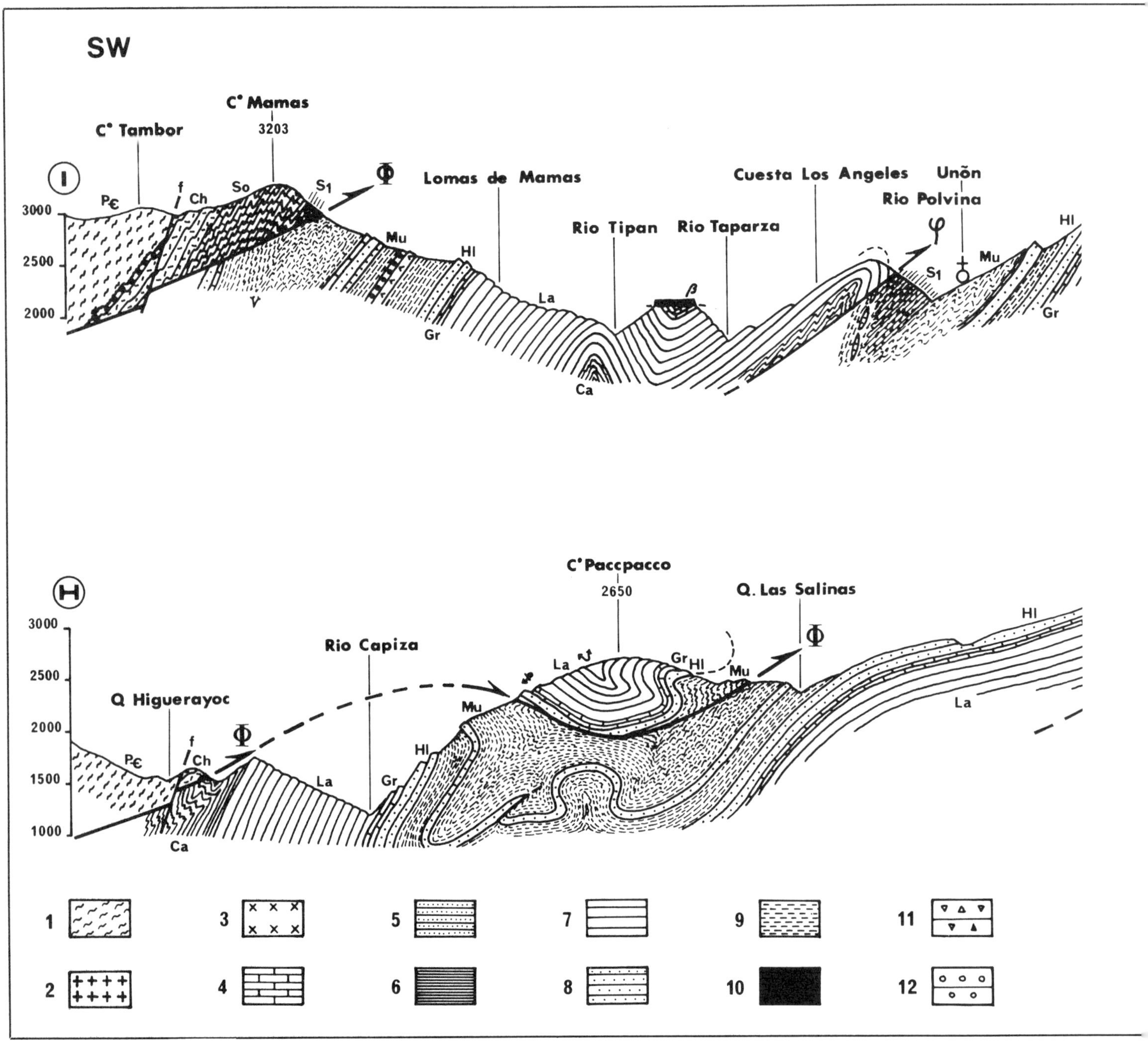

Figure 6 continued.

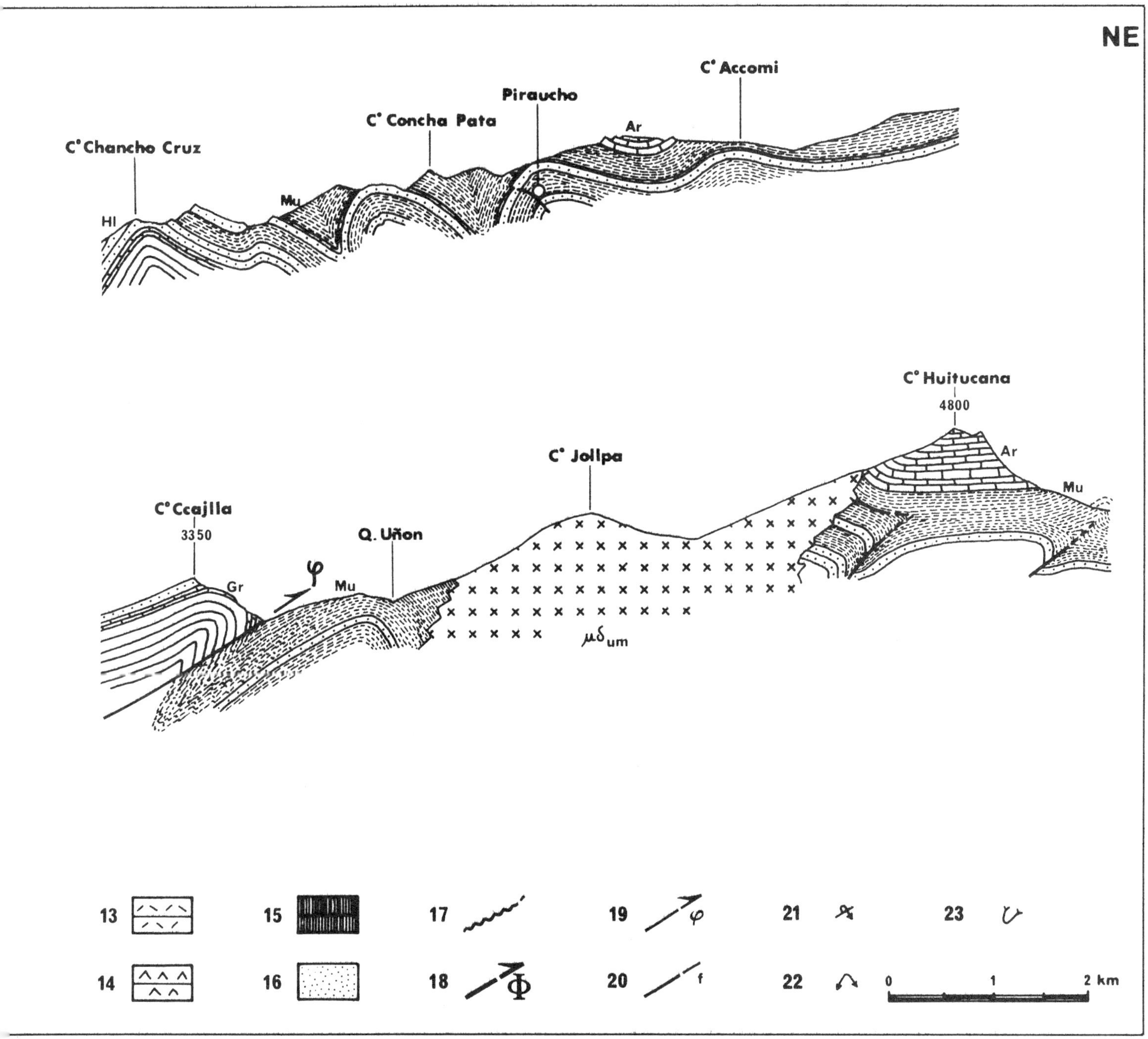
NE
C° Accomi
Piraucho
C° Concha Pata
C° Chancho Cruz
Ar
Mu
Hl
C° Huitucana
4800
C° Jollpa
Ar
Mu
C° Ccajlla
3350
Q. Uñon
Gr
Mu
φ
μδum
13
14
15
16
17
18 Φ
19 φ
20 f
21
22
23
0 1 2 km

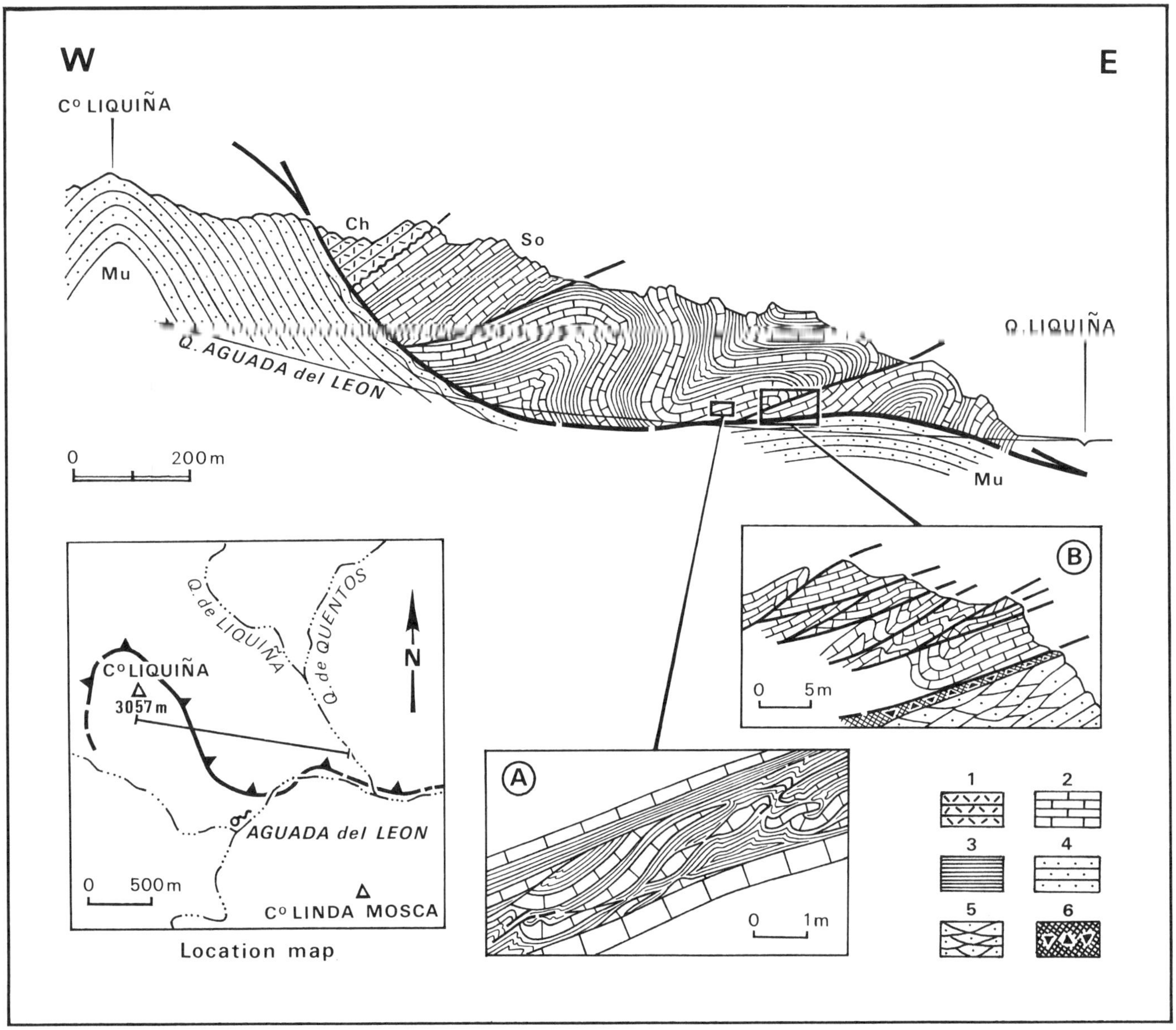

Figure 7. Cross section along Quebrada Aguada del Leon, east side of Cerro Liquiña, showing structural style of the overthrust Lluta unit (after Vicente et al., 1979). (A) Diagram illustrating shear planes and boudinage developed within incompetent beds of the Socosoni Formation; and (B) small imbrications developed within incompetent beds. (1) andesitic flows and volcaniclastics (Chocolate Formation); (2) massive limestone and (3) thin-bedded calcilutite (Socosani Formation); (4) orthoquartzite (Murco Formation); (5) shear slices; and (6) tectonic breccia and mylonite.

Because of the magnitude of Late Cretaceous thrusting in the Arequipa region, we can no longer speak of weak deformations for the Western Cordillera of southern Perú (Audebaud, Laubacher, and Marocco, 1976). Rather, the overthrusting of the Arequipa massif and its cover toward the northeast appears to be one of the major tectonic features in this part of the Andes. It implies a crustal shortening without precedent in the Andean context.

PALCA WINDOW

The Palca window is about 40 km northeast of Tacna, near the border with Chile (Figure 9). Interest in the Palca locality stems from the study of Wilson and Garcia's (1962) map of this region, which shows a curious faulted open anticline having a core of marine Bajocian rocks separated from an envelope of lowermost Jurassic volcanic rocks by peripheral reverse faults. The faults are arcuate and bound a circular area partly obscured by Quaternary deposits. These relationships suggest the possible existence of a tectonic window (Vicente, 1981) and warrant detailed study of the area. That a tectonic window exists, as is shown in Figure 9, also is indicated by the asymmetry of drag folds and shear planes in the San Francisco Formation near the thrust plane, and by right-lateral monoclinal kink folds in the volcanic rocks of the Junerata Formation, an obvious indicator of left-lateral shear near the basal thrust (Johnson, 1977). The sense of movement is northeastward thrusting of the upper unit.

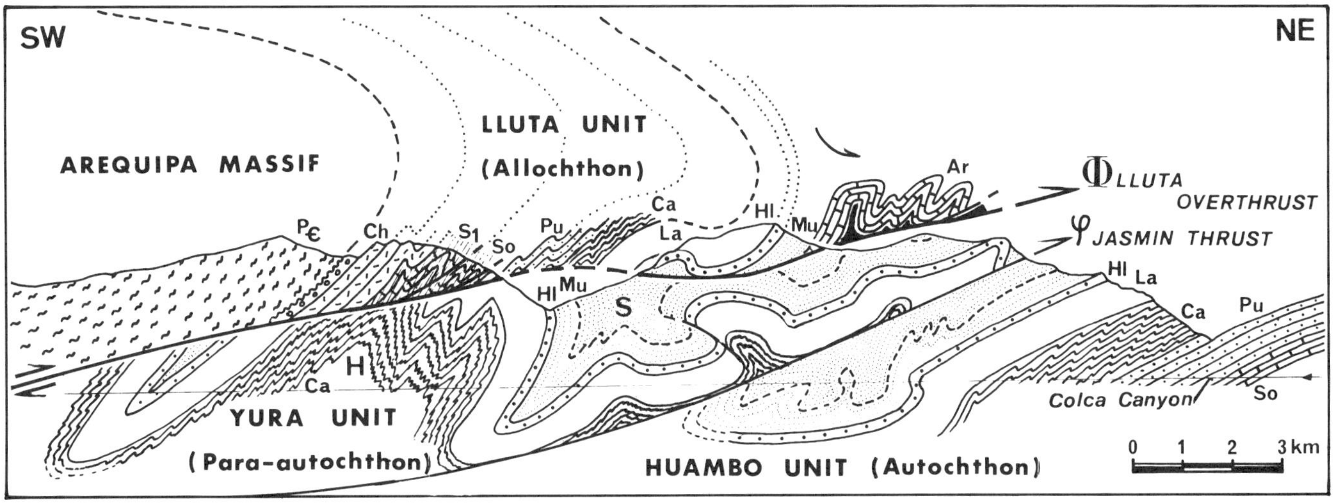

Figure 8. Schematic cross section illustrating relationship and style of main tectonic units of the Rio Colca section. (H) Huanaquituyoc anticline; (S) Satay syncline. Abbreviations of formation names as in Figure 6.

The Pelado Formation (Sinemurian to Bathonian) and the Yura Group (Callovian to Tithonian) belong to the allochthon. They are different from the rocks to the west, beyond the Lluta Tertiary intrusive and near San Francisco Pampa (Figure 9), that probably belong to the autochthon. Consequently, the Pelado and San Francisco formations are viewed as correlative but of different facies and tectonic settings. The cross section suggests a preliminary delineation of thrusts in the southernmost Western Cordillera of Perú that may continue into northernmost Chile.

CHURIN OVERTHRUST

The Churin overthrust is considered to be a fundamental structure separating (just north of the latitude of Lima) the internal and external zones (the western eugeosynclinal and eastern miogeosynclinal zones, according to the terminology of Cobbing, 1973, 1974, and 1978). Evidence for the overthrust is based on tectonic and paleogeographic evidence from critical analysis of the tectonic profile of the Rio Huara above the Coastal batholith, specifically the cross section between Churin and Yanahuanca via Oyon, Uchucchacua, and Chinche (Figure 10). This section shows that the Continental Divide separates an eastern zone of broad folds and widely spaced reverse faults (Figure 10), involving up to Upper Cretaceous to lower Tertiary red beds (Casapalca Formation), from a western zone of tight folds and imbricate faults involving beds no younger than the Coniacian Celendin Formation (Romani, 1982), and where Upper Cretaceous red beds are absent. This suggests that the two zones were deformed at different times. Indeed, the conspicuous lack of red beds within the western zone cannot be attributed merely to erosion. In short, contrary to what is usually accepted (Cobbing, 1973; Mégard, 1973, 1984; Cobbing et al., 1981), we consider that the main episode of deformation in the western zone occurred immediately before the deposition of Upper Cretaceous red beds and thus was during the early Late Cretaceous (Santonian) Peruvian orogeny. The deformation of the eastern zone took place during the late Eocene Incaic phase. Thus, as in southern Perú, the Upper Cretaceous red beds of central Perú seem to be a direct sedimentary consequence of the Late Cretaceous tectogenesis of the Western Cordillera.

Imbricate Subzone

The imbricate subzone (Figure 10) has been well described on a regional scale by Harrison (1953), Harrison and Wilson (1960), Wilson and Reyes (1964), Wilson, Reyes, and Garayar (1967), Szekely (1967, 1969), Cobbing (1973), Romani (1982), Nuñez del Prado (1983) and Romani, Nuñez del Prado, and Angeles (1983). This distinctive belt, about 4-5 km wide, exhibits a well-exposed stack of imbricate slices of Upper Cretaceous massive white limestones (Jumasha Formation). By virtue of its tectonic style and its position between the western flank of the Uchucchacua boxlike anticline and the first main thrust to the west, the imbricate subzone can be described (according to the terminology of Boyer and Elliot, 1982) as a typical trailing imbricate fan. This system of imbricate structures in the footwall of a major thrust was caused by the "bulldozer" action of the overlying thrust sheet. Consequently, the overturned higher wedge of Cerro Cashe Cusmar (Figure 10) suggests a typical "footwall horse" cut from the inverted limb of the syncline. The outer listric thrusts and their characteristics "thin-skinned" style, involving solely the upper Jumasha limestones (Romani, 1982), can be interpreted as obvious "out-of-the syncline thrusting" (Dahlstrom, 1970) that moved the upper beds away from the syncline axis and against the western back limb of the Uchucchacua anticline.

Concentrically Folded Subzone

The concentrically folded and extensively overthrust subzone (Figure 10) has undergone regional detachment above the coal-bearing shales of the Upper Jurassic Oyon Formation (Cobbing, 1973, Figure 23). This is suggested by the geometry of folds, since "in the cores of the anticlines it is difficult to include any considerable volume of rock greater than the expected from the Oyon shale" and "the fact that the basement is never brought to surface exposure" (Coney, 1971) in the Western Cordillera. This 22-km-wide subzone consists of

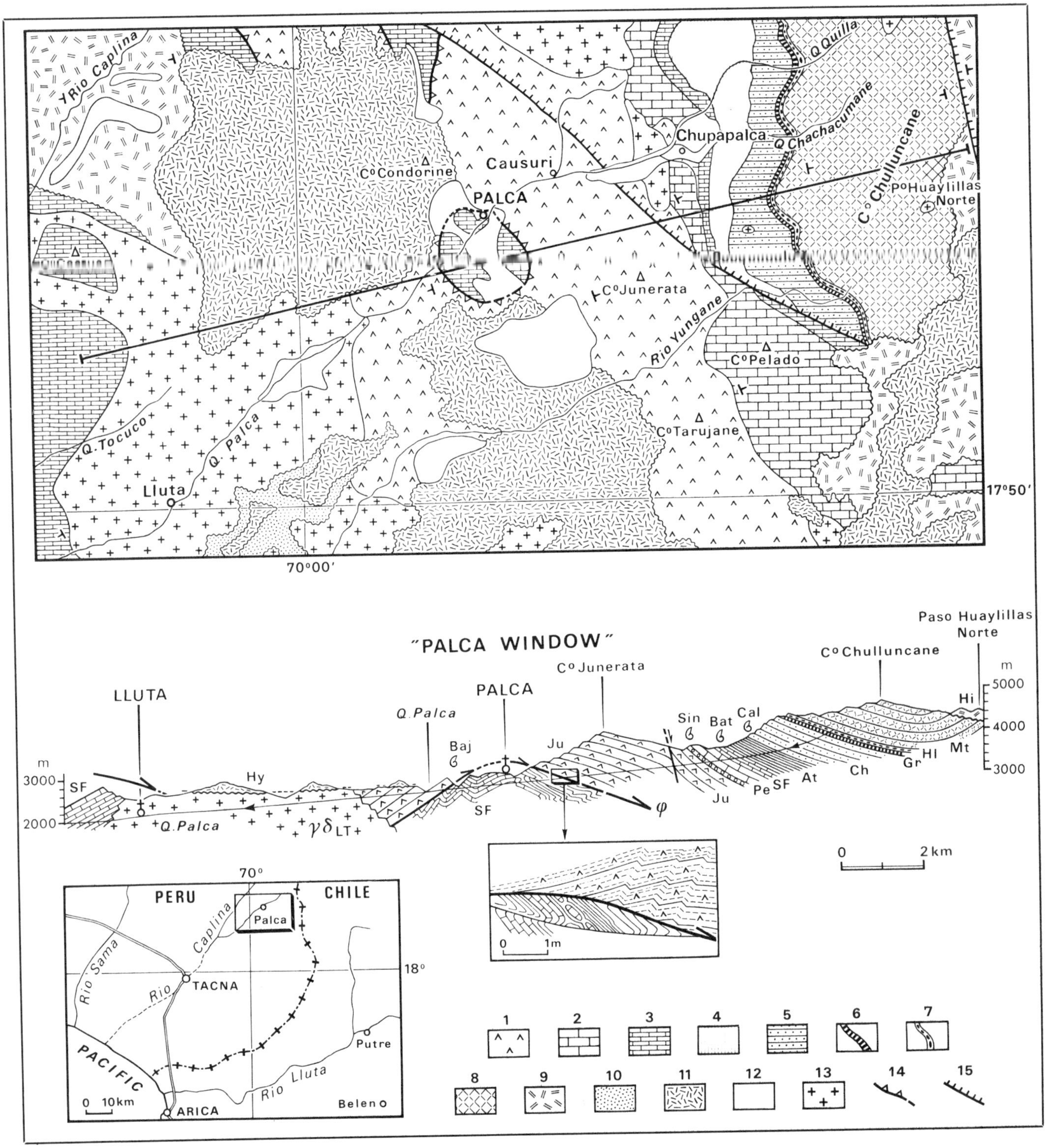

Figure 9. Geology and structure of Palca area (see Figure 1). Geological map modified from Wilson and Garcia (1962). (1) Junerata Formation (Ju) (= Chocolate Formation, Lower Jurassic); (2) Pelado Formation (Pe) (Sinemurian to Toarcian); (3) San Francisco Formation (SF) (Aalenian-Bajocian-Bathonian); (4) Ataspaca Formation (At) (Callovian); (5) Chachacumane Formation (Ch) (Oxfordian-Kimmeridgian); (6) Gramadal Formation (Gr) (lower Tithonian); (7) Hualhuani Formation (Hl) (upper Tithonian); (8) Matalaque Formation (Mt) (Neocomian); (9) Huilacollo Formation (Hi) (= Toquepala Formation, Eocene); (10) Moquegua Formation (Oligocene); (11) Huaylillas Formation (Hy) (Miocene); (12) alluvium; (13) Andean granitoids (early Tertiary) (LT) Lluta granodiorite; (14) overthrust (refines shape); and (15) normal fault.

four distinctive thrust-faulted anticline-syncline pairs. As can be seen in Figure 10, the folds tend to be overturned toward the northeast, and the thrust faults are on their east limbs. Also, the open synclines in massive limestones of the Jumasha Formation and much tighter anticlines in quartzites of the Chimu Formation reveal a disharmonic fold geometry. This implies an upper zone of decollement along the Pariatambo shale horizon, upon which the Jumasha carbonates have become detached from the underlying rocks and ejected out of the cores of the synclines. The chevron anticlines in massive quartzites of the Chimu Formation become progressively tighter to the southwest, so that the Viroc anticline (Figure 10) is almost isoclinal. The eastern thrust faults agree with the fore-limb pattern and may be referred to as typical "break thrusts" that cut diagonally across the strata of the overturned fore-limb before passing downward into a bedding-plane thrust. The thrust in the Viroc anticline (Figure 10) belongs to the class of "axial plane thrusts," which is in harmony with the tightness and lack of space in the core of that competent fold. This also explains the strong chevron mesoscopic folding in the incompetent Oyon Formation of the anticlinal core and the drag against the fault plane.

Similar-Type Folded Subzone

The similar-type folded subzone, which is well exposed near Churin (Figure 10), exhibits very complex mesoscopic folds, tens of meters wide, marked by a strong flattening with thick hinges and thin limbs, and a general subvertical axial-plane slaty cleavage. In detail, the shape of the folds varies considerably. S to M folds have been observed, as well as isoclinal and even sheath folds, such as those visible near Churin. The penetrative deformation coincides with a pronounced change of facies, indicated by a much higher proportion of shales and siltstones and a decrease in the thickness and proportion of sandstones; this suggests a transition to slope facies compared with the platform facies to the east. Calculations applied to some examples of isoclinal folds according to Ramsay's (1962) method have shown consistent values of nearly 65% flattening. It is not surprising that the geometry of folds closely approaches the ideal "similar" pattern. On the other hand, it is difficult to estimate the true amount of shortening of this zone without restoration of the folds to their early parallel shape. However, a rough estimate, based on De Sitter's (1958) limit of 36% as the maximum shortening possible by flexural-slip folding alone, indicates a total shortening of nearly 80% in the Churin area. Inasmuch as the axial plane cleavage is subvertical, with a preferential inclination to the southwest, the direction of regional compression responsible for the shortening, flattening, and metamorphism was initially subhorizontal and was oriented southwest-northeast. Thus, in view of the general tectonic polarity shown on the cross section, the imposed compression was from the southwest. In this context, the Churin subzone must be considered as closely related to the previous zones and therefore also detached, which suggests that a major tectonic contact may occur toward the southwest.

Casma Unit

At the southwest end of the section (Figure 10) is a thick volcanogenic sequence, consisting mainly of thin-bedded pyroclastics, that is referred to as the Churin facies of the Casma Group (Webb, 1976). This sequence represents the eastern facies, developed east of the batholith, of the thick Albian volcanogenic unit that extends up to the Coast, and which was intruded by the Coastal batholith. The most obvious tectonic feature in this sequence is its strong subvertical schistosity and greenschist metamorphism, which so markedly obscures bedding that an accurate assessment of its tectonic style is difficult. However, detailed observations on outcrop and in thin sections show a fracture cleavage oblique to bedding that could be an axial plane cleavage of folds whose hinges remain to be found. The style of folding may be much more complicated than appears at first sight. Very likely it is comparable with the isoclinal style described by Myers (1974) for the Señal Cochapunta Formation of the Tapacocha fold belt, just to the north, and as confirmed by the few observations recently carried out at Paccho Tingo, 17 km downstream from Churin.

Churin Overthrust Concept

On the basis of paleogeography, it would appear that the Casma unit has been thrust eastward over rocks of the "similar-type folded subzone" (Figure 10). It is not possible to explain the abrupt contrast along the Rio Churin by what Cobbing (1978) referred to as miogeosyncline-eugeosyncline by a simple change in facies. Thus, the so-called "miogeosynclinal sequence" is, in fact, typical of a subsiding open marine platform receiving first Lower Cretaceous siliciclastics, and then middle Cretaceous carbonates, but always at shallow depth. Because the term "basin" used by Cobbing (1978) appears to be inexact, we prefer to speak of the "Chavin platform." We saw that the westernmost facies near Churin represents the edge of the platform margin and indicates the westward transition to the basin proper. Furthermore, as was shown by Wilson (1963), clastic sediments were derived exclusively from the cratonic foreland or Brazilian shield, to the east, and appear to be free of volcanic material. On the other hand, the "eugeosynclinal sequence," chiefly represented by the Casma Group, reflects volcanic-arc activity to the west (Webb, 1976), and accumulation of volcanic materials on an emergent basement after westerly derived siliciclastics were deposited (Wilson, 1963). The general relationships indicate a thickening of the volcanic sequence and an increase of pillow lavas toward the west, and a higher proportion of fine-grained pyroclastics and intercalated sediments toward the east in the Casma unit (Cobbing et al., 1981). The absence of volcanic rocks in the platform facies indicates the presence of a large marine basin that is equivalent to a typical backarc marginal basin between the active volcanic arc and the platform.

The Casma Group corresponds to what Karig and Moore (1975) described as a "volcaniclastic apron," a submarine fan complex built into deep water from the rear flank of a volcanic arc. In this scheme (Figure 11), the Churin facies represents the distal end of the apron, whereas facies along the coast belong to the proximal part. Consequently, the bulk of the starved basin developed to the east, beyond the end of the apron, and all its eastern side would have disappeared beneath the Churin overthrust. This implies that the Casma Group is allochthonous and is thrust over rocks of a marginal basin. This thrust would correspond to a major suture recording the destruction of the previous Andean marginal basin (to retain the terminology of Cobbing, 1978, we shall use the term

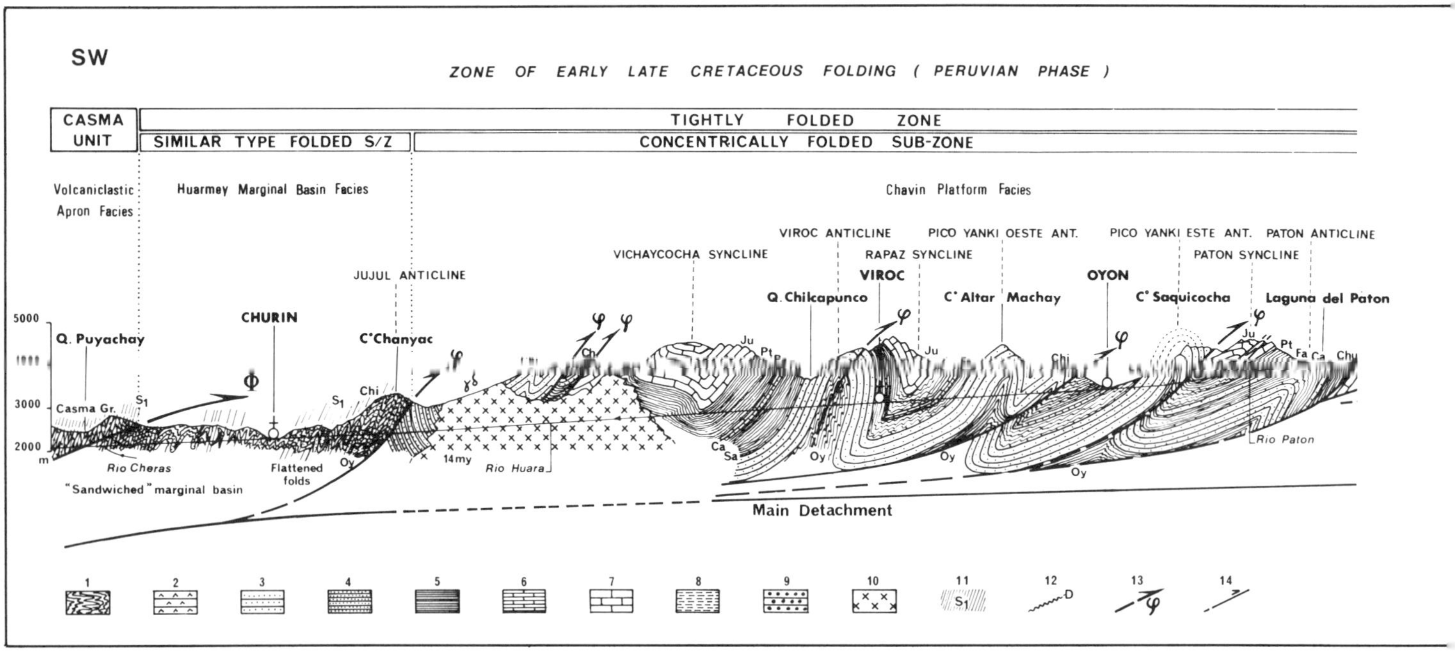

Figure 10. **Cross section along the Churin-Yanahuanca line, Western Cordillera of central Perú, east of the Andean batholith (see Figure 1). (1) phyllite and schist; (2) volcanic and volcaniclastic rocks; (3) quartz arenite; (4) arkosic arenite; (5) shale; (6) thin-bedded limestone; (7) massive limestone; (8) red pelites; (9) conglomerate; (10) granitic rocks; (11) slaty cleavage; (12) unconformity; (13) thrust fault; (14) reverse fault. (Pz) Yanahuanca Group (lower Paleozoic); (Mi) Mitu Formation (Upper Permian-Lower Triassic); (Pu) Purcara Group (Upper Triassic-Lower Jurassic);**

"Huarmey marginal basin") during a typical arc-continent collision. Contrary to prevailing opinion (Mégard, 1978; Cobbing et al., 1981; Bally, 1981), the opening and closing of the Andean backarc basin as a "marginal basin" appears to have played a major role in the Mesozoic history of the central Andean Cordillera. It was an especially weak feature with regard to arc-continent collision, which accounts for the similarity between the structures developed here and those of a typical foreland thrust and fold belt (Suarez, Molnar, and Burchfiel, 1983; Mégard, 1984).

Backarc Main Thrust

In the tectonic framework described above, the Casma unit represents solely the thrust volcaniclastic apron of the arc. The style of folding of this unit, whether the isoclinal style of the eastern front or the "style dejectif" with pinched synclines (e.g., the Canoas syncline described by Myers, 1974) of the coastal region, implies an underlying detachment zone. In addition, its strong subvertical cleavage and differences in lithology on either side of the batholith reflect a strong horizontal compression with a relative easterly translation. This indicates the presence of a still more important major thrust along the present coastline, which forms the tectonic contact between the Lower Cretaceous magmatic arc and its volcaniclastic apron. If we accept an essentially undeformed basement under the Western Cordillera and a detached sedimentary cover, which represents a situation similar to that in the Canadian Rocky Mountains (Bally, Godoy, and Stewart, 1966; Price, 1981) or in the overthrust belt of western Wyoming or central Utah (Gries, 1983), it is on this detachment that the tectonic shortening across the cover must be balanced by an equivalent net convergence between the sialic basement of the arc and the South American craton (Figure 12). Unfortunately, at this latitude the zone in question is now below sea level; it crops out only in the coastal range south of Pisco (Figure 1), as the "Paracas block or geanticline" (Myers, 1975), a unit consisting of remnants of the eroded arc massif in the form of a Precambrian basement with a volcanic cover and an impressive swarm of dikes (Caldas, 1978). Only in that area can its tectonic relationship to the backarc thrust belt be studied, and can its possible link with the previously described overthrust of the Arequipa massif be substantiated.

Tectonic Model

A paleogeographic and structural model is proposed that differs considerably from the usually accepted model involving vertical movements and block faulting (Myers, 1975; Pitcher and Bussell, 1977; Cobbing, 1978, 1985; Cobbing et al., 1981). The conspicuous thrust and fold belt in the Western Cordillera of central Perú displays substantial supracrustal shortening on equivalent large horizontal convergence of the basement at a deeper crustal level. Shortening between the Churin overthrust and the front of the imbricate fan alone is about 50 km; the amount of shortening of the original length is 60% for the imbricate subzone, 45% for the concentrically folded subzone, and 80% for the similar-type folded subzone of the Churin area (Figure 10). In the Casma unit to the west, the current width of 85 km and a minimum shortening of 25% (based on deformed ammonites on the coast measured by Myers, 1974) imply at least another 30 km of shortening. Therefore, total shortening is more than 80 km, without including a minimum value for the net slip of the Churin over-

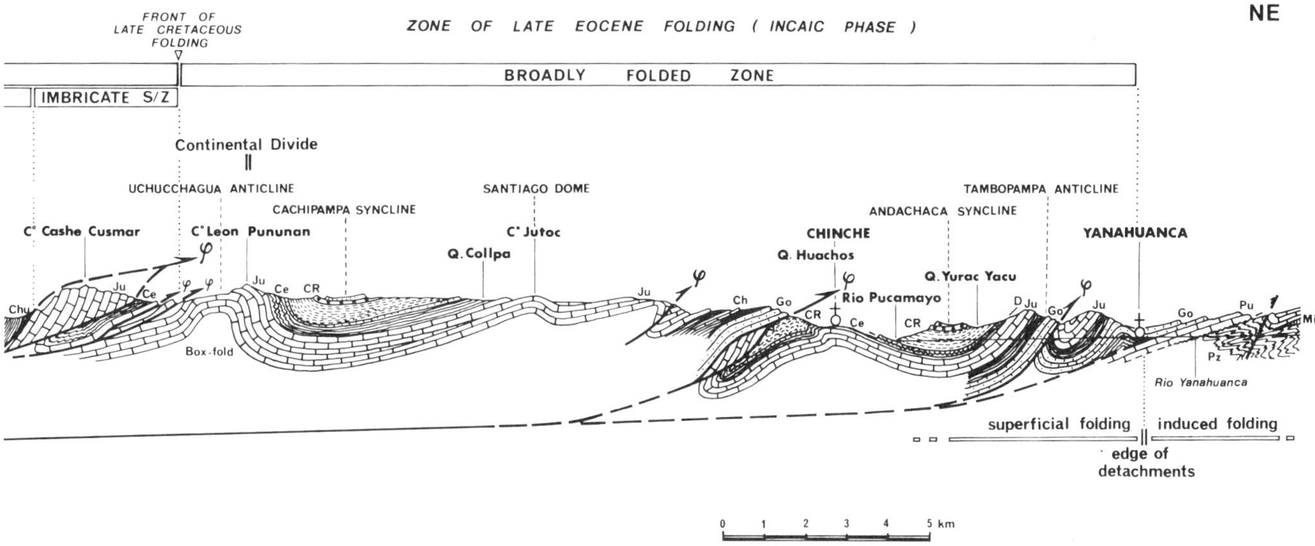

(Oy) Oyon Formation (Upper Jurassic); (Go) Goyllarisquisga Formation (Neocomian); (Chi) Chimu Formation (Lower Cretaceous); (Sa) Santa Formation (Valanginian); (Ca) Carhuaz Formation (upper Valanginian-Aptian); (Fa) Farrat Formation (Aptian); (Ph) Pariahuanca Formation (lower Albian); (Chu) Chulec Formation (middle Albian); (Pt) Pariatambo Formation (upper middle Albian); (Ju) Jumasha Formation (upper Albian-Turonian); (Ce) Celendin Formation (Coniacian); and (CR) Capas Rojas or Casapalca Formation (Cretaceous-Eocene).

thrust, nor that of the Paracas block. Thus, it is estimated that a minimum of 100 km of shortening occurred during Late Cretaceous tectogenesis. Such large-scale shortening cannot be accounted for by mere shearing of steep deep-seated fractures in the basement as proposed by Mégard (1984). A large, net crustal convergence evidently took place between the eastern edge of the Paracas block and the Brazilian craton, which resulted in destruction of the "marginal" Huarmey backarc basin. This resembles Price's (1981) tectonic model for the foreland thrust and fold belt of the southern Canadian Rockies. Price (1981) viewed this as "an example of intra-plate convergence involving the collapse of a back-arc or marginal basin, behind the main easterly dipping Cordilleran subduction zone." Such crustal convergence has been compared with a type of subduction involving an arrangement of masses of continental crust that Bally (1975) has called "A- (or Ampferer) subduction." Though that notion has been applied to the Andes of northern Perú (Bourgois and Janjou, 1981), the same reservations made by Price (1981) can be made about the amount of continental crust really involved. Moreover, the thin-skinned deformation style and the basal detachment indicate that the frontal imbricate fan cannot be considered as representing the direct upward emergence due to subduction. If a crustal shear zone representing the root zone of the Late Cretaceous collision is present, it probably would be along the present coast at the margin of the Paracas block (Figure 12). In this structural context, the Andean synkinematic metamorphism is explained in the framework of the tectonic model of Mattauer, Collot, and Driessche (1981) for the Canadian Rockies. The necessary conditions are provided by the tectonic thickening of the sedimentary prism by folding and thrusting and the subduction below the overriding arc. Also, Bussell's (1983) estimate of the amount of unroofing over the Coastal batholith indicates the importance of later uplift (at least 4-5 km) and erosion.

Tertiary deformation (except a slight detachment caused by the Incaic phase, which marks the end of the progressive forward deformation of the sedimentary wedge of the High Plateau, Figure 10) is characterized by block faulting, wherein the basement of the Eastern Cordillera and the sub-Andean region was affected by steep, mainly west-dipping, reverse faults. This suggests two stages in the tectonic development of the central Andes: (1) a Cretaceous stage, characterized by convergence of the island arc and the continental craton, which ended in the collapse of the marginal basin; and (2) a Tertiary stage where, after collision, deformation continued toward the east, but by a progressive high-angle thrust faulting of the foreland basement, which is equivalent to an underthrusting by the Brazilian shield beneath the eastern margin of the incipient belt (Suarez, Molnar, and Burchfiel, 1983). Altogether, this contrast in style suggests the model involving the structures of the thin-skinned overthrust belt of the North American Cordillera and the high-angle foreland thrusts of the Laramide basement uplifts of Colorado and Wyoming (Smithson, et al., 1979; Gries, 1983).

Andean Marginal Basin

The floor of the Andean "marginal basin" evidently was continental, although strongly attenuated (Atherton, Pitcher, and Warden, 1983; Atherton, Warden, and Sanderson, 1985), as indicated by the abnormally high geothermal gradients

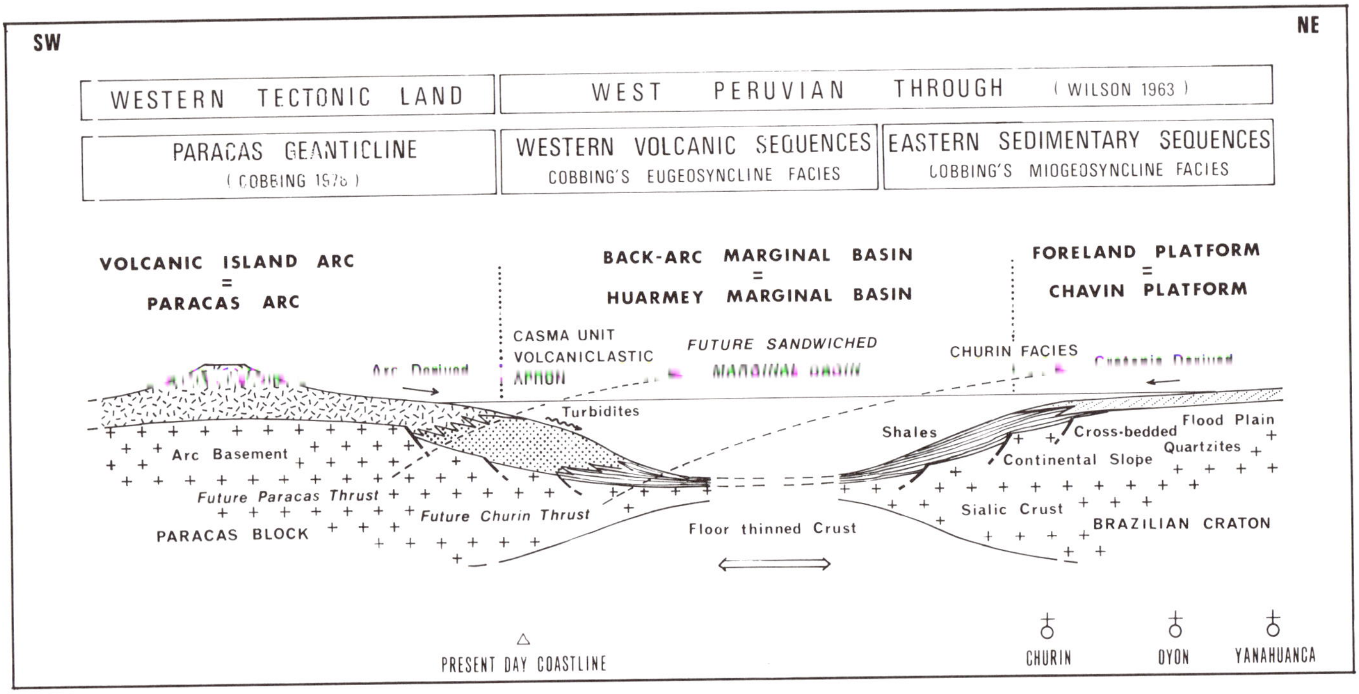

Figure 11. Diagrammatic restored section across the West Peruvian trough during the Early Cretaceous. Note asymmetry of terrigenous supplies (arc and cratonic derived), ensialic backarc marginal basin of aborted type (floor thinned crust) and relation to Cobbing's terminology.

shown by the hydrothermal metamorphic facies of the Casma Group (Offler et al., 1980; Aguirre and Offler, 1985) and the gabbroic intrusions and associated basic dikes and sills present in the western facies of the group (Regan, 1985). The basin exhibits many features intermediate between a simple, extensional, continental-floored backarc basin and an ocean-floored marginal basin. Its designation by Aguirre and Offler (1985) as an "ensialic marginal basin of aborted type" is apt. From a geotectonic point of view, these transitional features make sense if compared with features to the south and north. In comparison with the Aconcagua-Neuquen and Tarapacá basins of the Meridional Andes (Jensen et al., 1976; Jensen and Vicente, 1979), where relatively little rifting took place, the Arequipa basin appears to be far more subsident and active (Vicente et al., 1982). The contrast with the Huarmey basin to the north is still more striking because the oceanic character of Lower to middle Cretaceous sediments becomes more and more evident northward. It appears that the amount of stretching and thinning of the continental crust of the backarc basin increased progressively northward; true ocean-floored ophiolite assemblages were developed in Ecuador and Colombia, and, as tholeiitic volcanic associations, are considered to be formed in marginal basins (Millward, Marriner, and Saunders, 1984; Bourgois et al., 1987).

CONCLUSIONS

Early Late Cretaceous tectogenesis appears to have been the major phase in the tectonic development of southern Perú, a geodynamic scenario similar to the proposed for the Meridional Andes (Vicente et al., 1973). The structures in this region are similar in style and scale to Alpine structures if one accepts the thesis (1) that the basement nappe of the Arequipa massif is an immense recumbent, Precambrian-cored anticline thrust eastward at least 30 km, and (2) that the thin-skinned easterly tapering fold and thrust belt of central Perú is regionally detached and draped over the edge of the cratonic foreland. These structures are considered to be characteristic features of the Western Cordillera of Perú. Moreover, there is a clear and progressive northward increase in the intensity of deformation, especially in regional metamorphism, which suggests that still more spectacular structures will be found in northernmost Perú. The Arequipa massif overthrust probably is continuous with that of the Paracas block. The style of the fold and thrust belt persists along the Andes of central Perú (Szekely, 1967, 1969; Hillebrandt and Mirwald, 1969; Coney, 1971; Cobbing, 1974) and northern Perú (Wilson and Reyes, 1964; Wilson, Reyes, and Garayar, 1967). In the simple tectonic model of collision between the Andean magmatic arc and the South American craton, referred to above, the Arequipa massif overthrust seems to be the backarc main thrust, whereas the early Late Cretaceous foreland fold and thrust belt can be viewed as the subduction-related complex, tectonically prograded over the western edge of the continental craton at the time of convergence and obliteration of the marginal basin. The structural style of the Western Cordillera of southern Perú is remarkably similar to that of the Canadian Rocky Mountains and the overthrust belt of western North America.

The above view of intense horizontal displacements on shallow-dipping detachments contrasts with previously published cross sections showing thrust faults steepening with depth into the basement. The latter erroneously extrapolate the tectonic style of the Eastern Cordillera to the Western Cor-

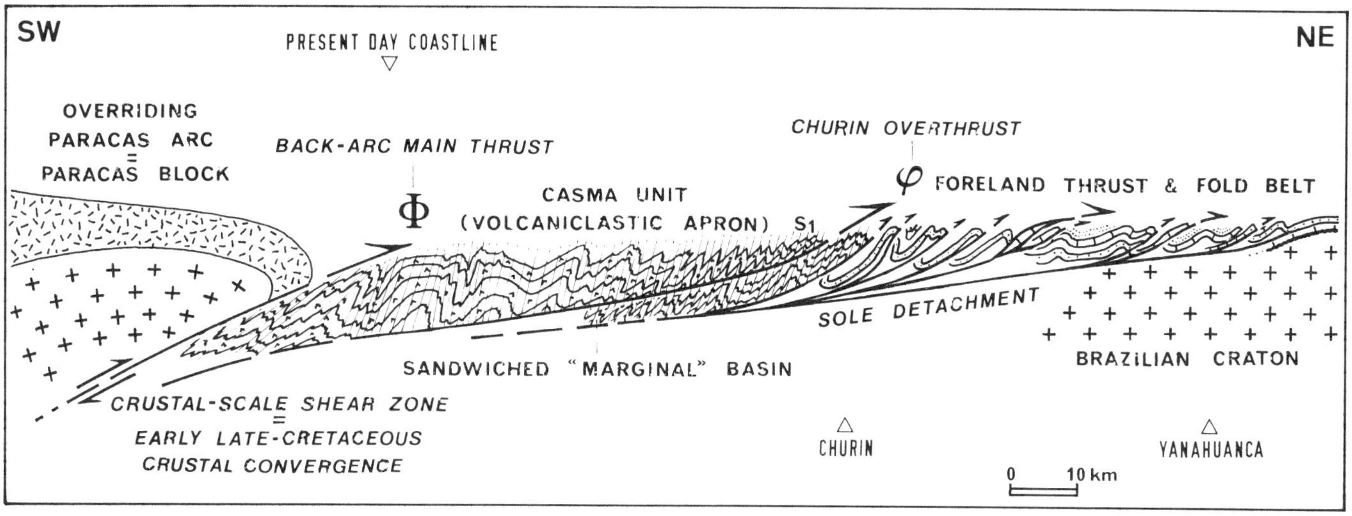

Figure 12. Schematic regional cross section illustrating tectonic style of the Western Cordillera of central Perú at the latitude of Churin. Coastal batholith not shown.

dillera; in reality, the two completely different styles succeed one another, both in time and space. There needs to be a reappraisal of structural cross sections of the Western Cordillera of Perú, where tangential structures are widely underestimated. Deep seismic-reflection profiles across the Andes are also required to corroborate the basal décollement of Mesozoic thrusts.

Finally, the importance of the nature of the floor of the Andean backarc or marginal basin on the tectonic style and amount of shortening involved must be viewed in the context of arc-continent collision and the progressive deformation of the sedimentary wedge. Such deformation probably took place over an interval of nearly 20 m.y., from the end of late Albian time until the beginning of late Santonian time.

ACKNOWLEDGMENTS

The work was undertaken in the Department of Geology and Geophysics, University San Agustin of Arequipa (Perú), during a stay as visiting Professor (1977-1983) in accordance with a cooperative agreement between the French and Peruvian Foreign Offices. The writer wishes to express his thanks to the University of Arequipa and the National Research Council of France (CNRS), which, under contract LA 215, supported the research on which this report is based. He also wishes to thank the students who carried out the field investigation: Fortunato Sequerios and Mario Valdivia (1979), David Huaman (1980), Victoriano Perez, and Javier Zavala (1981). J. Aubouin, H. Charlesworth, O. Jensen, R. Shagan, and J. Skarmeta contributed helpful reviews of the manuscript.

REFERENCES CITED

Aguirre, L., and R. Offler, 1985, Burial metamorphism in the western Peruvian trough: Its relation to Andean magmatism and tectonics, *in* W.S. Pitcher et al., eds. Magmatism at a plate edge, the Peruvian Andes: Glasgow, Blackie and Son Ltd., p. 59-71.

Atherton, M.P., W.S. Pitcher, and V. Warden, 1983, The Mesozoic marginal basin of central Perú: Nature, v. 305, p. 303-306.

Atherton, M.P., V. Warden, and L.M. Sanderson, 1985, The Mesozoic marginal basin of Central Perú: A geochemical study of within-plate-edge volcanism, *in* W.S. Pitcher et al., eds., Magmatism at a plate edge, the Peruvian Andes: Glasgow, Blackie and Son Ltd., p. 47-58.

Audebaud, E., G. Laubacher, and R. Marocco, 1976, Coupe géoligique des Andes du sud du Pérou de l'Océan Pacifique au Bouclier Brésilien: Geologische Rundschau, v. 65, p. 223-264.

Badgley, P.C., 1965, Structural and tectonic principles: New York, Harper and Row Publishers, Inc. 521 p.

Bally, A.W., 1975, A geodynamic scenario for hydrocarbon occurrences (abs.): Tokyo, Ninth World Petroleum Congress Proceedings, v. 2, Applied Science, p. 33-34.

Bally, A.W., 1981, Thoughts on the tectonics of folded belts, *in* McClay and Price, eds., Thrust and nappe tectonics: Geological Society of London Special Publication 9, p. 13-32.

Bally, A.W., P.L. Godoy, and G.A. Stewart, 1966, Structure, seismic data and orogenic evolution of southern Canadian Rocky Mountains: Canadian Journal of Earth Science, v. 3, p. 713-723.

Beckinsale, R.D., A.W. Sanchez-Fernandez, M. Brook, E.J. Cobbing, W.P. Taylor, and N.D. Moore, 1985, Rb-Sr whole-rock isochron and K-Ar age determination for the Coastal batholith of Perú, *in* W.S. Pitcher et al., eds., Magmatism at a plate edge, the Peruvian Andes: Glasgow, Blackie and Son Ltd., p. 177-202.

Benavides, V., 1962, Estratigrafía pre-Terciaria de la región de Arequipa: Sociedad Geológica del Perú Boletín, v. 38, p. 5-63.

Bourgois, J., and D. Janjou, 1981, Subduction océanique, subduction continentale et surrection andine: l'example du Pérou septentrional: Paris, Comptos Rendus Acad. Sciences, v. 293, Série II, p. 859-864.

Bourgois, J., J.F. Toussaint, H. Gonzalez, J. Azema, B. Calle, A. Desmet, L.A. Murcia, A.P. Acevedo, E. Parra, and J. Tournon, 1987, Geological history of the Cretaceous ophiolitic complexes of northwestern South America (Colombian Andes): Tectonophysics, v. 143, p. 307-327.

Boyer, S.E., and D. Elliot, 1982, Thrust systems: American Association of Petroleum Geologists Bulletin, v. 66, p. 1196-1230.

Bussell, M.A., 1983, Timing of tectonic and magmatic events in the central Andes of Perú: Journal of the Geological Society of London, v. 145, p. 279-286.

Caldas, J., 1978, Geología de los cuadrangulos de San Juan, Acari y Yauca: Instituto de Geología y Minería del Perú Boletín, v. 30, 153 p.

Cobbing, E.J., 1973, Geología de los cuadrangulos de Barranca, Ambar, Oyon, Huacho, Huaral y Canta: Servicio de Geología y Minería del Perú Boletín, v. 26, 172 p.

Cobbing, E.J., 1974, The tectonic framework of Perú as a setting for batholithic emplacement: Pacific Geology, v. 8, p. 63-65.

Cobbing, E.J., 1978, The Andean geosyncline in Perú, and its distinction from Alpine geosynclines: Journal of the Geological Society of London, v. 135, p. 207-218.

Cobbing, E.J., 1985, The tectonic setting of the Peruvian Andes, *in* W.S. Pitcher et al., eds., Magmatism at a plate edge, the Peruvian Andes: Glasgow, Blackie and Son Ltd., p. 3-12.

Cobbing, E.J., W.S. Pitcher, J.J. Wilson, J.W. Baldock, W.P. Taylor, W. McCourt, and N.J. Snelling, 1981, The geology of the Western Cordillera of northern Perú: Inst. Geol. Sc. London, Overseas Mem. 5, 143 p.

Coney, P.J., 1971, Structural geology of the Cordillera Huayhuash, Andes of Perú: Geological Society of America Bulletin, v. 82, p. 1863-1884.

Dahlstrom, C.D.A., 1970, Structural geology in the eastern margin of the Canadian Rocky Mountains: Bulletin of Canadian Petroleum Geology, v. 18, p. 332-406.

De Sitter, L.U., 1958, Boudins and parasitic folds in relation to cleavage and folding: Geologie en Mijnbouw, v. 20, p. 277-286.

Gries, R., 1983, Oil and gas prospecting beneath Precambrian of foreland thrust plates in Rocky Mountains: American Association of Petroleum Geologists Bulletin, v. 67, p. 1-28.

Harrison, J.V., 1953, Geologiá del Camino entre Canta y Huayllay: Instituto Nacional de Investigación y Fomento Mineros Perú Boletín, v. 9, 36 p.

Harrison, J.V., and J. Wilson, 1960, Geología de la región comprendida entre Huacho y Vinchos: Sociedad Geológica del Perú Boletín, v. 35, p. 5-62.

Hillebrandt, A.V., and P. Mirwald, 1969, Zur geologie der Yarumario-Gruppe: Munchner Anden-Kundfahrt 1967, p. 11-40.

Huaman, D., 1980, Geologia de los sectores de Andahua-Ayo-Gloriahuasi (15°25′-15°57′ S. lat., Departmento Arequipa, Perú): University San Agustin, Arequipa, thesis, 115 p.

Jenks, W.F., 1948, Geología de la Hoja de Arequipa al 1/200,000: Instituto Geológico del Perú Boletín, v. 9, 204 p.

Jensen, O., and J.C. Vicente, 1979, La fase infra-neocomiana a la latitude de Copiapó (Chile): argumentos indirectos: Santiago, Comunicaciones, v. 27, p. 19-40.

Jensen, O., J.C. Vicente, J. Davidson, and E. Godoy, 1976, Etapas de la evolución marina jurásica de la cuenca andina externa (mioliminar) entre los paralelos 26° y 29°30′ sur: Santiago, Primer Congreso Geologico Chileno, v. 1, p. A273-A295.

Johan, A., L. Le Bel, and W.J. McMillan, 1980, Etude comparative des Batholithes de La Caldera (Pérou) et de Guichon Creek (Canada), deux exemples de plutonisme a minéralisations associées de la Cordillère péripacifique, *in* Minéralisations liées aux granitoides: Mem. B.R.G.M., Orléans, no. 99, p. 21-70.

Johnson, A.M., 1977, Styles of folding, *in* Developments in geotectonics 11: Amsterdam, Elsevier Publishing Company, 406 p.

Karig, D.E., and G.F. Moore, 1975, Tectonically controlled sedimentation in marginal basins: Earth and Planetary Science Letters, v. 26, p. 233-238.

Mattauer, M., B. Collot, and V.D. Driessche, 1981, Interpretation des linéations transversales de l'infrastructure de la Cordillere sud-canadienne: une nouvelle hypothèse sur la formation de la chaîne: Paris, C.R. Acad. Sc., v. 293, Série II, p. 183-185.

Mégard, F., 1973, Etude géologique d'une transversale des Andes au niveau du Pérou central: University Montpellier, thesis, 263 p.

Mégard, F., 1978, Etude géologique des Andes du Pérou central: Paris, Mem. ORTSTOM, no. 86, 310 p.

Mégard, F., 1984, The Andean orogenic period and its major structures in central and northern Perú: Journal of the Geological Society of London, v. 141, p. 893-900.

Millward, D., G. Marriner, and A.D. Saunders, 1984, Cretaceous tholeiitic volcanism in the Western Cordillera of Colombia, *in* Geological evolution of the Andes: Journal of the Geological Society of London, v. 141, p. 857-860.

Mukasa, S.B., and G.R. Tilton, 1985, Zircon U-Pb ages of super-units in the Coastal Batholith, Perú, *in* W.S. Pitcher et al., eds., Magmatism at a plate edge, the Peruvian Andes: Glasgow, Blackie and Son Ltd., p. 203-207.

Myers, J.S., 1974, Cretaceous stratigraphy and structure, western Andes of Perú between latitudes 10-10°30′: American Association of Petroleum Geologists Bulletin, v. [illegible], p. [illegible].

Myers, J.S., 1975, Vertical crustal movements of the Andes in Perú: Nature, v. 254, p. 672-674.

Nuñez del Prado, H., 1983, Estudio geológico del sector de Uchucchacua-Hacienda Otuto: University San Agustin, Arequipa, thesis, 224 p.

Offler, R., L. Aguirre, B. Levi, and S. Child, 1980, Burial metamorphism in rocks of the western Andes of Perú: Lithos, v. 13, p. 31-42.

Palmer, A.R., 1983, The decade of North American geology, 1983 geologic time scale: Geology, v. 11, p. 504-505.

Perez, V., 1981, Geología del sector de Pampacolca-Tipan-Union (Prov. Castill, Departmento Arequipa): University San Agustin, Arequipa, thesis, 129 p.

Pitcher, W.S., and M.A. Bussell, 1977, Structural control of batholithic emplacement in Perú: a review: Journal of the Geological Society of London, v. 133, p. 249-256.

Price, R.A., 1981, The Cordilleran foreland thrust and fold belt in the southern Canadian Rocky Mountains, *in* McClay and Price, eds., Thrust and nappe tectonics: Geological Society of London Special Publication 9, p. 427-448.

Ramsay, J.G., 1962, The geometry and mechanics of formation of "similar" type folds: Journal of Geology, v. 70, p. 309-327.

Ramsay, J.G., 1967, Folding and fracturing rocks: New York, McGraw-Hill Book Co., 568 p.

Ramsay, J.G., 1981, Tectonics of the Helvetic Nappes, *in* McClay and Price, eds., Thrust and nappe tectonics; Geological Society of London Special Publication 9, p. 293-309.

Regan, P., 1985, The early basic intrusions *in* W.S. Pitcher et al., eds., Magmatism at a plate edge, the Peruvian Andes: Glasgow, Blackie and Sons Ltd., p. 72-83.

Romani, M., 1982, Géologie de la région miniere Uchucchacua-Hacienda Otuto, Pérou: University Grenoble, thesis, 175 p.

Romani, M., H. Nuñez del Prado, and C. Angeles, 1983, Estudio tectónico detallado de la zona de Uchucchacua-Hda Ututo: Sociedad Geológica del Perú Boletín, v. 71, p. 307-320.

Sequeiros, F., 1979, Geología del area Cerro Redondo-Ojule (Departmento Arequipa): University San Agustin, Arequipa, thesis, 125 p.

Smithson, S.B., J.A. Brewer, S. Kaufman, J.E. Oliver, and C.A. Hurich, 1979, Structure of the Laramide Wind River uplift, Wyoming, from COCORP deep reflection data and from gravity data: Journal of Geophysical Research, v. 84, p. 5955-5972.

Stewart, J.W., J.F. Everden, and N.J. Snelling, 1974, Age determinations from Andean Perú: a reconnaissance survey: Geological Society of America Bulletin, v. 85, p. 1107-1116.

Suarez, G., P. Molnar, and B.C. Burchfiel, 1983, Seismicity, fault plane solutions, depth of faulting, and active tectonics of the Andes of Perú, Ecuador, and southern Colombia: Journal of Geophysical Research, v. 88, p. 10403-10428.

Szekely, T.S., 1967, Geology near Huallacocha lakes, central high Andes, Perú: American Association of Petroleum Geologists Bulletin, v. 51, p. 1346-1353.

Szekely, T.S., 1969, Structural geology, Cochas to Yauricocha, central high Andes, Perú: American Association of Petroleum Geologists Bulletin, v. 53, p. 553-567.

Valdivia, M.A., 1979, Geología del area Cincha-Mina Torconta (Departmento Arequipa): University San Agustin, Arequipa, thesis, 117 p.

Vargas, P., 1970, Geología del cuadrangulo de Arequipa: Serv. Geol. Min., Lima, Boletín, v. 24, 64 p.

Vicente, J.C., 1981, Elementos de la estratigrafía mesozoica surperuana, *in* Volkheimer and Musacchio, eds., Cuencas sedimentarias del Jurásico y Cretácio de America del sur: Buenos Aires, v. 1, p. 319-351.

Vicente, J.C., R. Charrier, J. Davidson, A. Mpodozis, and S. Rivano, 1973, La orogenesis sub-hercínica: fase mayor de la evolución paleogeográfica y estructural de los Andes arentino-chilenos centrales: Buenos Aires, V Congreso Geológico Argentino, v. 5, p. 81-98.

Vicente, J.C., F. Sequeiros, M.A. Valdivia, and J. Zavala, 1979, El sobre-escurrimiento de Cincha-Lluta; elemento del accidente mayor Andino al NW de Arequipa: Sociedad Geológica del Perú Boletín, v. 61, p. 67-99.

Vicente, J.C., B. Beaudoin, A. Chavez, and I. Leon, 1982, La Cuenca de Arequipa (Sur Perú) durante el Jurásico-Cretácico inferior: V Congreso Latinoamericano Geol., v. 1, p. 121-153.

Webb, S., 1976, The volcanic envelope of the Coastal batholith in Lima and Ancash, Perú: University of Liverpool, Ph.D. thesis.

Wilson, J.J., 1963, Cretaceous stratigraphy of central Andes of Perú: American Association of Petroleum Geologists Bulletin, v. 47, p. 1-34.

Wilson, J.J., and W. Garcia, 1962, Geología de los cuadrangulos de Pachia y Palca (Hojas 36v y 36x): Lima, Boletín Com. Carta Nacional Perú, 82 p.

Wilson, J.J., and L. Reyes, 1964, Geología del cuadrangulo de Pataz (Hoja 16-h): Lima, Boletín Com. Carta Nacional Perú, v. 9, 91 p.

Wilson, J.J., L. Reyes, and J. Garayar, 1967, Geología de los cuadrangulos de Mollenbamba, Tayabamba, Huaylas, Pamabamba, Carhuaz y Hauri: Lima, Boletín Serv. Geol. Min. Perú, v. 16, 95 p.

Zavala, J., 1981, Geología del sector Murco-Taya (Departmento Arequipa): University San Agustin, Arequipa, thesis 143 p.

Ericksen, G. E., Cañas Pinochet, M. T., and
Reinemund, J. A., editors, 1989, Geology of the Andes
and its relation to hydrocarbon and mineral resources:
Houston, Texas, Circum-Pacific Council for Energy
and Mineral Resources Earth Science Series, v. 11.

CHAPTER 7

STRUCTURE OF THE ANDEAN CONVERGENT MARGIN AND SOME IMPLICATIONS FOR HYDROCARBON RESOURCES

Roland von Huene
U.S. Geological Survey
Menlo Park, California, U.S.A.

ABSTRACT

The structure of the Perú margin is known from good-quality multichannel seismic data and from the study of cores recovered during Offshore Drilling Project (ODP) leg 112. A 15-to-20-km-wide accretionary complex has developed along the front of the margin. This complex was built against the truncated edge of continental crust that has been part of the Peruvian margin since the early Eocene. This transition from accreted Eocene to Neogene crustal elements is also recognized in older single-channel seismic records along much of the Chilean margin. The width of the accretionary complex appears to be related to the amount of sediment in the trench axis. Mesozoic and Cenozoic sedimentary sequences on the continental shelf extend tens of kilometers seaward from the present shelf edge to the midslope area. These sedimentary sequences are sufficiently old to contain mature hydrocarbons. Thus the anticlinal structures along the edge of the shelf are possible places where such hydrocarbons might migrate and be trapped.

INTRODUCTION

Structures at the leading edge of some convergent margins have recently been clarified by carefully reprocessing multichannel seismic data (e.g., Nankai Trough, Aoki, Tamano, and Kato, 1982; Central America, Shipley et al., 1980; Central Aleutian Trench, McCarthy and Scholl, 1984; and Eastern Aleutian Trench, von Huene, Miller, and Fisher 1986). John Miller and I have reprocessed four multichannel seismic records from offshore Perú (Kulm, Miller, and von Huene, 1986; Miller, Hussong, and von Huene, 1986; von Huene and Miller, 1988). The original records were acquired during the Nazca Plate Project, a decade-long multi-investigator effort summarized in Memoir 154 of the Geological Society of America. The seismic data appeared in Kulm et al. (1981a), Hussong and Wipperman (1981), and Shepherd and Moberly (1981). Although 24 channels were recorded, only 12 channels were originally processed. Our reprocessing of all 24 channels produced an improved image that shows accretion at the foot of the trench slope and the truncated continental margin buried beneath sediment on the middle slope (Figure 1).

The conclusions based on geophysical data and conventional sampling (dredging and coring) were tested by drilling during the Ocean Drilling Project (ODP) leg 112 (von Huene, Suess et al., 1987). Study of the drill cores showed that the continental crust across the shelf and upper slope consists of post-Oligocene slope deposits unconformably overlying Eocene shelf and upper slope deposits. Regional tectonic subsidence during the Miocene depressed the Eocene rocks as much as 5 km below original depths of deposition. Late Miocene trench sediment accreted against the truncated Eocene rock and was covered with Pliocene to Quaternary slope sediment. These results confirmed the interpretation of geophysical data. They also indicated a correspondence of tectonic history on land with major plate tectonic events in the Pacific basin (von Huene, Suess et al., 1987).

Along the Chilean margin, the results of the study of multichannel seismic records was applied to a reinterpretation of single-channel records. The structures of accretionary and continental terranes, as seen clearly in the multichannel records, were first studied in a record where only a single channel of the multichannel records was printed, with a 5:1 vertical exaggeration like that of the single-channel data off Chile. The accretionary and continental terranes could be recognized in the single-channel record, but with less certainty. Continental and accreted terranes were then sought in existing single-channel seismic records off Chile (von Huene, Kulm, and Miller, 1985). The single-channel data were augmented by multichannel seismic records across the continental shelf obtained during petroleum exploration off central

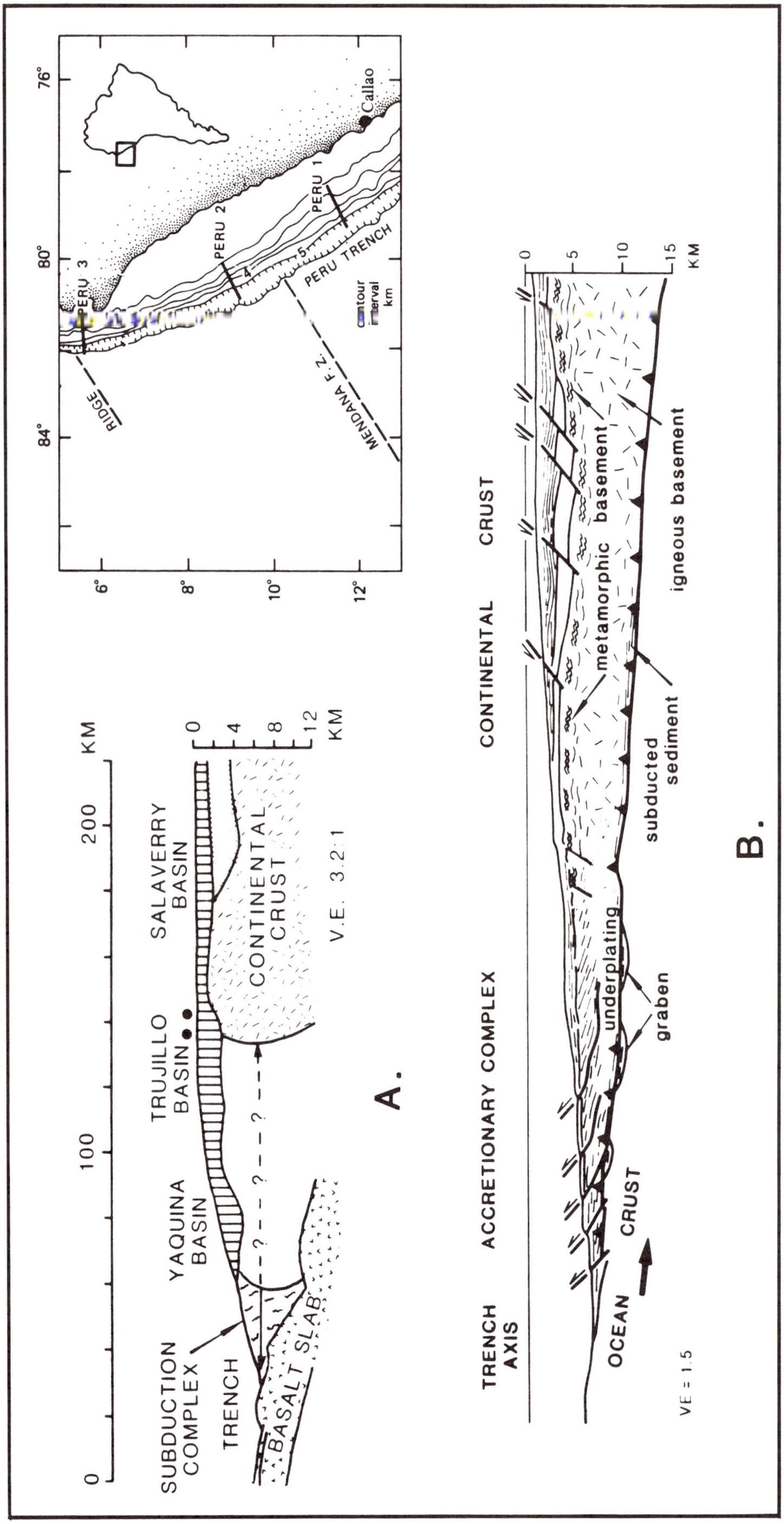

Figure 1. Diagrammatic structure in record Perú 2. (A) Interpretation based on original 12-channel (from Kulm et al., 1981a). (B) Interpretation based on reprocessed 24-channel record (from von Huene, Kulm, and Miller, 1985).

Chile, as well as by exploratory drilling to the metamorphic basement (Mordojovich, 1974). The structure of the Chilean margin appeared similar to that of the Peruvian margin.

The juxtaposition of the accretionary complex against the truncated edge of the continent indicates that erosion of the continental crust occurred before the present accretion. The time of this erosion and the time of the accretion, as well as their relationship to subsidence of the seaward edge of the continent and the tectonic history of the Andes, are only known north of the Nazca Ridge. Nonetheless, rock of continental affinity is truncated some 20 to 30 km seaward of the edge of the continental shelf along central Chile. These continental-margin sedimentary rocks locally contain hydrocarbons along the coast. In this chapter, I review the results of earlier studies and briefly describe the reprocessed multichannel seismic records. Then, I comment on the implications for hydrocarbon exploration along the Perú-Chile continental shelf.

RESULTS OF STUDIES DURING NAZCA PLATE PROJECT

Studies of the Andean margin made during the Nazca Plate Project (Kulm et al., 1981b; Hussong and Kulm, 1985) indicated that off Perú, the Andean continental margin consists of an accretionary complex stacked against the seaward edge of the continental crust (Figure 1A). The complex presumably consists of slices of tholeiitic basalt and the overlying metalliferous sediment, carbonate, and clay (Yeats, Hart, et al., 1976) that were detached from the descending Nazca Plate (Kulm et al., 1981a). The Nazca Plate currently converges with the South American continent at about 10 km/m.y. (Minster and Jordan, 1978; Pardo-Casas and Molinar, 1987) and is broken on the seaward side of the trench by a large horst and graben resulting from the downward flexure of the plate. In the trench axis, the Nazca Plate sediment is overlain by ponded sediment (Rosato and Kulm, 1981; Bourgois et al., 1986).

In the seismic records of the Nazca Plate Project, the structure of the landward slope of the trench is obscured by diffractions that mask all but the top of the oceanic crust; however, the oceanic crust can be followed 25 to 35 km beneath the continental margin (Kulm et al., 1981a; Hussong and Wipperman, 1981). The lateral extent of the accretionary complex was interpreted differently by various authors (Jones, 1981, 85 km; Kulm et al., 1981a, 15 to 85 km; Hussong and Wipperman, 1981, 15 km). Seismic velocities within the first 15 km landward of the trench axis— clearly part of the accretionary complex—range from 1.7 to 2.0 km/s. Investigators interpreted the adjacent crust, which has a velocity of 5.6 km/s, as older accreted sediment (Jones, 1981), or as fractured continental crust (Hussong and Wipperman, 1981). The reprocessed seismic records off Perú showed that the latter interpretation was correct (von Huene, Kulm, and Miller, 1985).

The seaward edge of the continental crust off Perú is probably composed of metamorphic rocks similar to those recovered from two offshore drill sites, from exposures on the upper slope, from islands, and from coastal outcrops (Kulm et al., 1982; Kulm, Thornburg, and Dang, 1984). These metamorphic rocks consist of phyllite and mica schist and are presumed to be of Paleozoic age similar to the metamorphic rocks on land (Thornburg and Kulm, 1981; Thornburg, 1984). These rocks have a velocity of about 6.0 km/s beneath the upper slope and shelf and perhaps 5 km/s beneath the middle slope where the fracturing is suspected (Hussong et al., 1976; Hussong and Wipperman, 1981).

A series of forearc basins occupies the Peruvian continental shelf and upper slope. Sediment of Eocene to Holocene age fills similar basins onshore (Kulm, Thornburg, and Dang, 1984); sedimentary rocks of these ages have been recovered during industrial drilling offshore (Shepherd and Moberly, 1981). Forearc basins developed on the continental shelf during the Cenozoic and were filled with detritus eroded from the uplifting Andean orogenic belt. The Lima basin—an upper slope basin off southern Perú—subsided 600 m during the past 2 to 3 m.y., and the bulk of its sediment fill appears to be of Neogene age (Kulm et al., 1981c; Thornburg and Kulm, 1981; Kulm, Thornburg, and Dang, 1984).

The juxtaposition of metamorphic rock against accreted trench sediment indicates that a large volume of the continent is missing. The inferred tectonic erosion of the seaward edge of the continent and the subsequent change to a dominantly accretionary regime was a fundamental idea relating to the evolution of the Andean margin that was discussed during the Nazca Plate Project.

RESULTS FROM REPROCESSING SEISMIC DATA

Three seismic lines (Figures 1, 2) were shot in 1973 by Seiscom-Delta for the Nazca Plate Project. Two 300 in.^{3} and two 1000 in.^{3} airguns were the sound source, which was fired every 35 m. The resulting signal was detected by a hydrophone streamer with a 1600-m-long active array composed of 24 hydrophone groups. Data were recorded on a DFS III system at a 4 msec sampling rate with record lengths up to 13 sec. The original field data were first processed with the same sequence of steps that was used during the Nazca Plate Project (demultiplexed, deconvolved, and bandpass filtered, 24-fold stacking), but using 24 rather than 12 channels. This was followed by research level processing (Miller, Hussong, and von Huene, 1986) to remove diffractions. Diffractions were removed by migration after stack using velocities determined from constant-velocity migration; particularly difficult areas were also migrated before stack.

The line drawings (Figure 2) show most of the features of the upper and lower landward slopes of the trench. Depth sections were constructed by using the refraction velocities in each area, because stacking velocities converted to interval velocities were not sufficiently representative of the rock velocities in deep-water areas. The velocities of a particular geologic unit were applied by using the reflection-time intercepts. Thus the depth to a horizon is uncertain, but the overall geometric distortion of a time section is minimized. The original records on which Figure 2 is based are published elsewhere (von Huene, Kulm, and Miller, 1985; Kulm, Miller, and von Huene, 1986; Miller, Hussong, and von Huene, 1986; von Huene and Miller, in press).

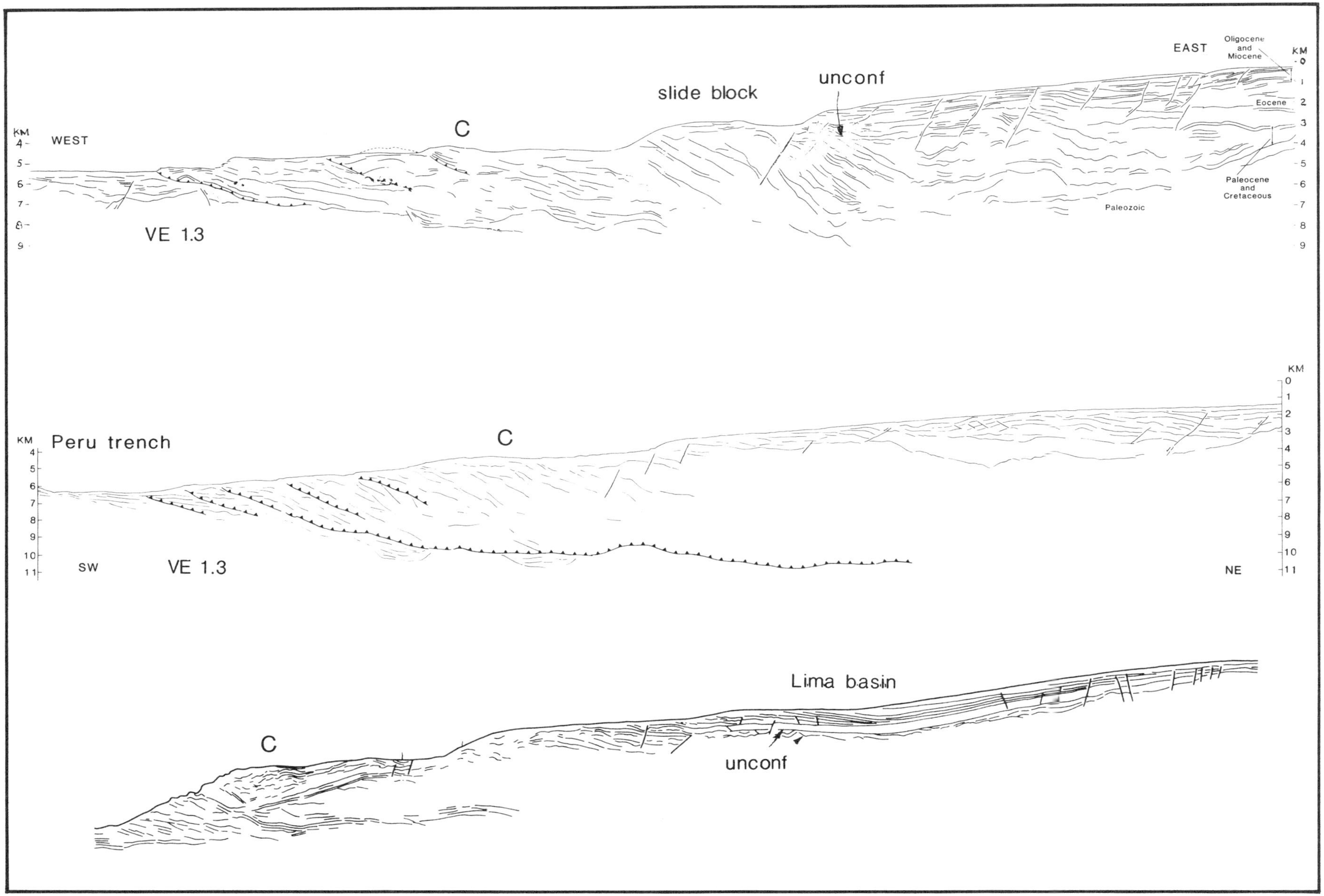

Figure 2. Interpreted depth sections of Perú 2 and Perú 3 and a time section of Perú 1. C marks westernmost extent of recognizable continental crust.

FEATURES IN REPROCESSED RECORDS

Record Perú 3, just south of the Talara oil field, is the northernmost of the Nazca Plate multichannel seismic lines (Figure 1) and was initially studied by Shepherd and Moberly (1981). The first reprocessed version was described by Miller, Hussong, and von Huene (1986). Subsequent interpretations were made after a Seabeam survey (Bourgois et al., 1986) and further acquisition of seismic data in the area (von Huene et al., in press). Here the stratigraphy of the continental shelf was extended from the onshore information around the Talara oil field and in offshore wells (Shepherd and Moberly, 1981). A sediment sequence of Cretaceous to Quaternary age, deposited in shallow and intermediate water depths, extends to the midslope area (Figure 2). A Neogene slope sediment was deposited above an unconformity on the Eocene rocks, and erosion has removed much of the Oligocene section. A locally inferred Cretaceous and Paleocene sequence exceeds 2 km in thickness; also it is probably thickened tectonically in the midslope area where beds have been tilted landward. Cretaceous rock has not been sampled offshore. However, its presence was inferred from the gravity field and the stratigraphy onshore, and it is thought to be underlain by crystalline continental basement. The dip of the Cretaceous to Quaternary section is parallel to the sea floor across most of the upper and middle slope. The section is cut by numerous small normal faults.

The angular unconformity that cuts the Eocene section also extends across landward-tilted strata and is an upper Eocene and Oligocene time horizon. Since the present phase of convergent-margin tectonism began at the end of the Eocene and the beginning of the Oligocene (McKee and Noble, 1989), any rock cut by the unconformity was part of the continent before the present phase of convergent tectonism began. This unconformity helps separate continental crust from the accreted sediment associated with the present arc-trench system. The unconformity can be traced on seismic records to the lower slope (Figure 2), restricting the accretionary complex to a 15-km-wide area (better illustrated in von Huene et al., in press).

During the last half million years the accretionary complex was covered by a huge debris avalanche (Bourgois et al., 1986; von Huene et al., in press) that probably resulted from gravity failure of the truncated continental crust against which the accretionary complex was stacked. Seabeam mapping revealed a large curvilinear scarp that marks a slip surface seaward of which a 15- by 33-km block detached and slid. As the block moved downslope and rotated backward, its forward part was oversteepened and became unstable, thus forming the debris avalanche. Until the debris avalanche was recognized, the transition from accretionary to continental crust was unclear and was interpreted incorrectly.

Record Perú 2 is near lat. 9 °S off central Perú where the shelf is about 120 km wide (Figure 1). Here, the transition from the continental to the accretionary regime was located at the juncture of the middle and upper slope, as based on the initial interpretation of the reprocessed seismic records (von Huene, Kulm, and Miller, 1985; Kulm, Miller, and von Huene, 1986). However, based on later reprocessing, the transition was moved to the juncture of the middle and lower slope (von Huene, Suess, et al., 1987). The 5-km difference between the two interpretations represents a transitional area containing displaced blocks produced during truncation of the continental crust. The blocks were then covered with slope sediment and subsequently became the buttress against which trench sediment was accreted (Figure 2). The stratigraphy of the upper slope consists of three units. The upper unit has a sedimentary structure similar to a delta and overlies a massive unit of uniform thickness that rests on the acoustic basement. These units were difficult to correlate with units on the shelf in the adjacent Trujillo basin on the shelf, where industry drill holes provided some temporal control, because seismic records and the sedimentary strata are not continuous. The section is cut by numerous small normal faults, which are growth faults formed simultaneously with the sedimentation.

Record Perú 1 (Figure 2) is near lat. 11 °S, close to Lima, where the shelf is relatively narrow and the middle slope is wide. The record includes the Lima basin on the upper and middle slope (Hussong and Wipperman, 1981; Kulm, Thornburg, and Dang, 1984). The strata of the Lima basin are underlain by an angular unconformity that cuts across 1-km-thick tilted and stratified sections near the basin center, such as those seen on many other lines in this area. A second unconformity overlies the crystalline basement. The Lima basin is also cut by many small normal growth faults. The slope becomes increasingly deformed seaward of the Lima basin, and the structure is unclear toward the transition to the accretionary complex. The transition to continental crust appears to consist of a seaward-pointing wedge covered by irregular strata. The accretionary complex, consisting of the landward-dipping strata at the front of the margin, is only 9 km wide.

The three reprocessed seismic records all show a small accretionary complex stacked against the truncated edge of continental crust. The continental nature of the crust can be inferred from the crystalline metamorphic rocks recovered from exposures at the edge of the shelf, and less commonly from the slope (Kulm, Thornburg, and Dang, 1984). This structure was tested by drilling during leg 112 of the ODP (von Huene et al., in press). Basement was not reached, but the overlying sedimentary rocks consist of Eocene shelf- and upper-slope deposits that were deposited in water as much as 5 km shallower than present depths. The Eocene is cut by an unconformity and all or most of the Oligocene and lower Miocene sediments are missing. The slope deposits are Neogene mudstones that locally contain primary upwelling facies and the transported products from upwelling centers nearshore. The study of drill cores confirmed that the upper slope of the Peruvian margin consists of continental crust and the lower slope is an accretionary complex. The tectonic evolution involved considerable subsidence beneath the present continental slope subsequent to middle Eocene time and tectonic erosion (von Huene, Suess, et al., 1988).

STRUCTURE ALONG OTHER PARTS OF ANDEAN MARGIN

Just how representative this Peruvian cross section is of the rest of the Andean continental margin has been questioned (von Huene, Kulm, and Miller, 1985). An early study of subduction processes between lat. 6 °S and 45 °S (Kulm, Schweller, and Masias, 1977) related the presence or absence of benches

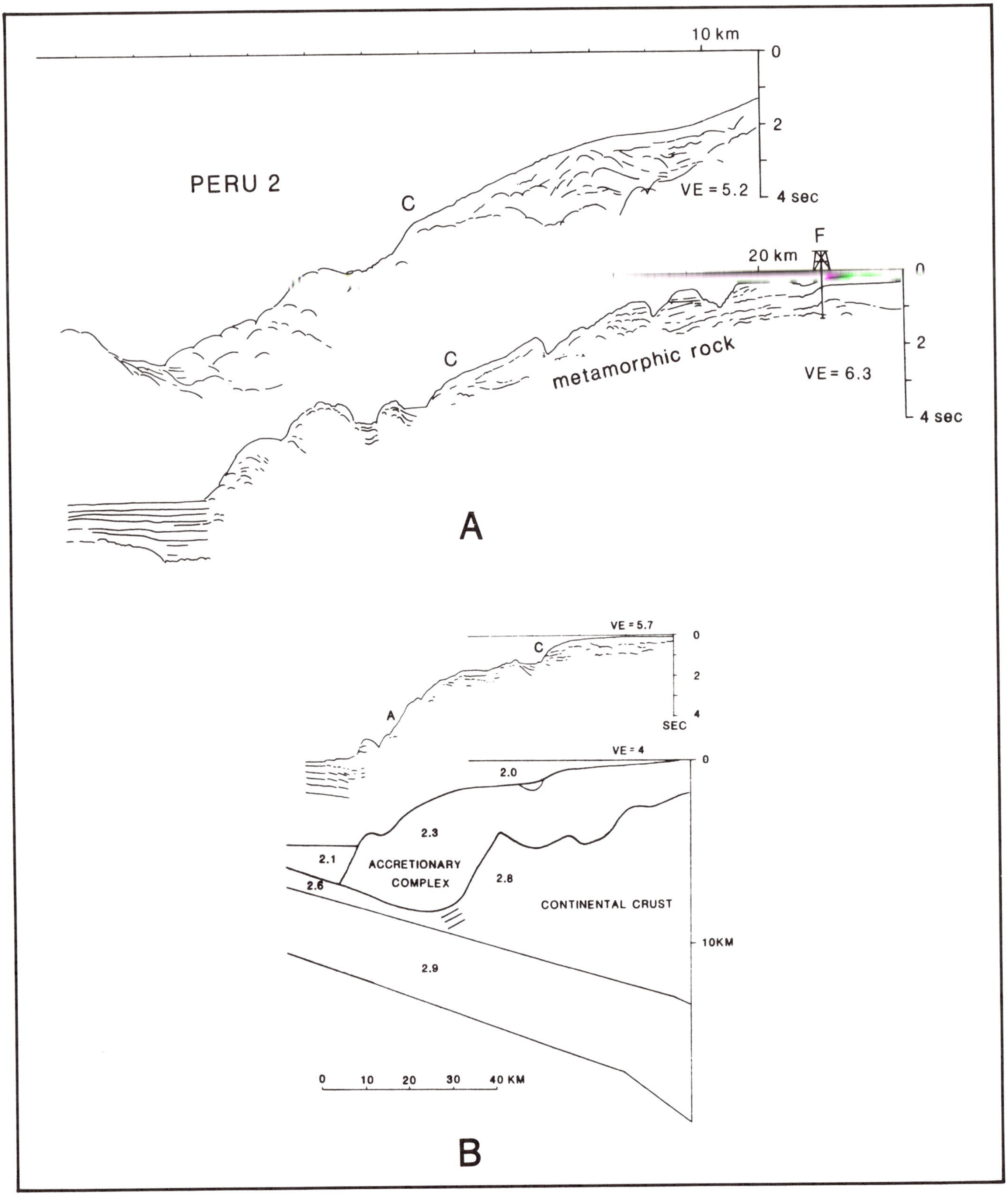

Figure 3. (A) Comparison of single-channel image of record Perú 2, at high vertical exaggeration, with a similar image along line 39, off southern Chile. The upper slope profile is notched by canyons, but at its westernmost margin (C) the continental crust and the accretionary complex have similar seismic characteristics and topography in both records (see von Huene, Kulm, and Miller, 1985, for more complete examples and discussion). (B) Gravity model across southern Chile margin (Couch et al., 1981) showing continental crustal rock densities (2.3 g/cc) and accretionary rock densities. The seismic profile is the corresponding single-channel seismic record.

and basins (assumed to be morphological expressions of thrust faults) to the tectonic process of accretion. If benches were present, the margin was considered accretionary, and if they were not it was considered erosional. The Perú-Chile Trench was divided into three provinces, and from their general morphology, a corresponding tectonic process was inferred. The Peruvian margin and a province off southern Chile were considered to be moderately accretionary, whereas another province along northern Chile was believed to be dominantly erosional (Kulm, Schweller, and Masias, 1977). Subsequent studies have shown that benches on trench slopes are not exclusively compressional features but may be large slumps (for example, Nasu et al., 1981; Aubouin, Bourgois, and Azema, 1984), diapirs, or diapirically cored ridges (Biju-Duval et al., 1982), and perhaps even normal faults as in the records shown here. However, the conclusions of Kulm, Schweller, and Masias (1977) have been little modified by new evidence and subsequent studies.

In the accretionary province off south-central Chile, sediment floods the trench axis because abundant precipitation and a mountainous terrain on land provided large volumes of detritus. Major rivers, which continue as large submarine canyons across the shelf and slope, carry considerable sediment directly to the trench axis. The volume of trench sediment in the area between lat. 33 °S and 35 °S is more than an order of magnitude greater than the sediment in the Perú Trench, and the total volume of accreted sediment is probably correspondingly greater.

Although there are no multichannel seismic records across the Chile margin from the shelf to the trench axis, a study of single-channel records, multichannel records, and the gravity and magnetic fields across the shelf, together with the results of exploratory drilling for oil and gas (Mordojovich, 1974), indicate a continuation of the truncated continental crust southward to at least the south Chile rise (Figure 3). The seaward extent of this basement was studied in single-channel records and gravity data (von Huene, Kulm, and Miller, 1985). When the 24-channel data of Perú 2 are displayed at a vertical exaggeration comparable to that of most of the older single-channel records (from 5x to 10x), they show a characteristic pattern in the accreted and continental parts (Figure 3). There is also a morphological break where the seismic records suggest the change from accreted to continental regimes (Figure 4).

Along central and southern Chile, where the trench contains significant volumes of ponded sediment, the tectonic history of the margin is similar to that of Perú. South of Valparaiso, Paleozoic metamorphic rock, lithologically the same as that making up the core of the Chilean Coast Range, was recovered at the edge of the shelf by drilling (Mordojovich, 1974). This rock is less magnetic and exhibits a higher seismic velocity than the overlying Cretaceous to Quaternary sand and shale. A unit with low magnetic expression and high seismic velocities was followed as seismic basement across the shelf and onto the continental slope, indicating that the metamorphic basement of the Chilean continent extends regionally to the upper continental slope (Mordojovich, 1974). Its extent downslope is less certain. Single-channel seismic records indicate that the structural regime found on the shelf continues to about the middle slope (von Huene, Kulm, and Miller, 1985). A steepened middle-slope area and a local midslope terrace are similar to those that are found in the area of transition from continental crust to the accretion complex off Perú (Figure 3 and 4). Similarities in the Eocene to Quaternary stratigraphy are consistent with the structural similarities. The regional hiatus between Eocene and Miocene strata found onshore and offshore across the Peruvian shelf (von Huene, Suess, et al., 1988) is observed on land in the region of the Arauco Peninsula south of Concepcion, Chile (Pineda, 1986). Offshore, the Oligocene was determined to be missing in four exploratory drill holes (Mordojovich, 1974). These structural and stratigraphic similarities suggest a parallel tectonic history, and I point them out to support the inference that, like the Peruvian margin, the Chilean upper slope could consist of a depressed Cretaceous to Oligocene continental shelf section extending to the middle slope.

Off northern Chile (Valparaiso to Antofagasta), the unique morphology of the Chile Trench was apparent in the earliest continuous echo-sounding profiles of Fisher and Raitt (1962), and was a matter of continuing study by Hayes (1966), Scholl et al. (1970), Kulm, Schweller, and Masias (1977), and Schweller, Kulm, and Prince (1981). Kulm Schweller, and Masias (1977) noted that here the continental slope is steeper than elsewhere along the trench, particularly in the middle and upper slope area; the continental shelf is narrow or absent; and sediment along the trench axis is sparse.

Single-channel records between Antofagasta and Valparaiso (lat. 23 °S to 33 °S) indicate that the slope sediment is about 100 m thick or less, except in local basins that are more numerous to the south (von Huene, Kulm, and Miller, 1985). Seismic refraction and gravity measurements also indicate a thin sediment cover in this region. Fisher and Raitt (1962) reported rocks of 4.7 km/sec velocity near the surface of the landward slope of the trench. Couch et al. (1981) modeled the slope using a density of 2.5 g/cc—a density corresponding to that of a well-consolidated rock—from the coast to the trench axis. Thus, morphology, sample data, shallow seismic reflection, crustal seismic refraction, and gravity interpretation all indicate sparse sediment on the slope, little accretion adjacent to the trench axis, and mostly consolidated rock. The consolidated rock offshore is probably a seaward continuation of the metamorphic and volcanic rocks onshore. Thus, the truncation of the continent off Perú appears to continue south along northern Chile.

IMPLICATIONS FOR HYDROCARBON EXPLORATION

The juxtaposition of accreted complex and a truncated continental crust, revealed in seismic records off Perú, appear common to at least parts of the Andean convergent margin. As Scholl and Vallier (1983) have noted, the Perú-Chile margin has one of the longest (more than 5,000 km) truncation scars bordering any continent. The size of the frontal accretionary complex is largest where the trench is filled with ponded sediment, and appears to be absent off northern Chile where the trench axis has only local small ponds of sediment.

The extension of Paleozoic continental crust, with its cover the Mesozoic and Cenozoic sedimentary rocks far seaward from the coast, is inconsistent with earlier notions concerning the plate tectonics of convergent margins. Most "classical"

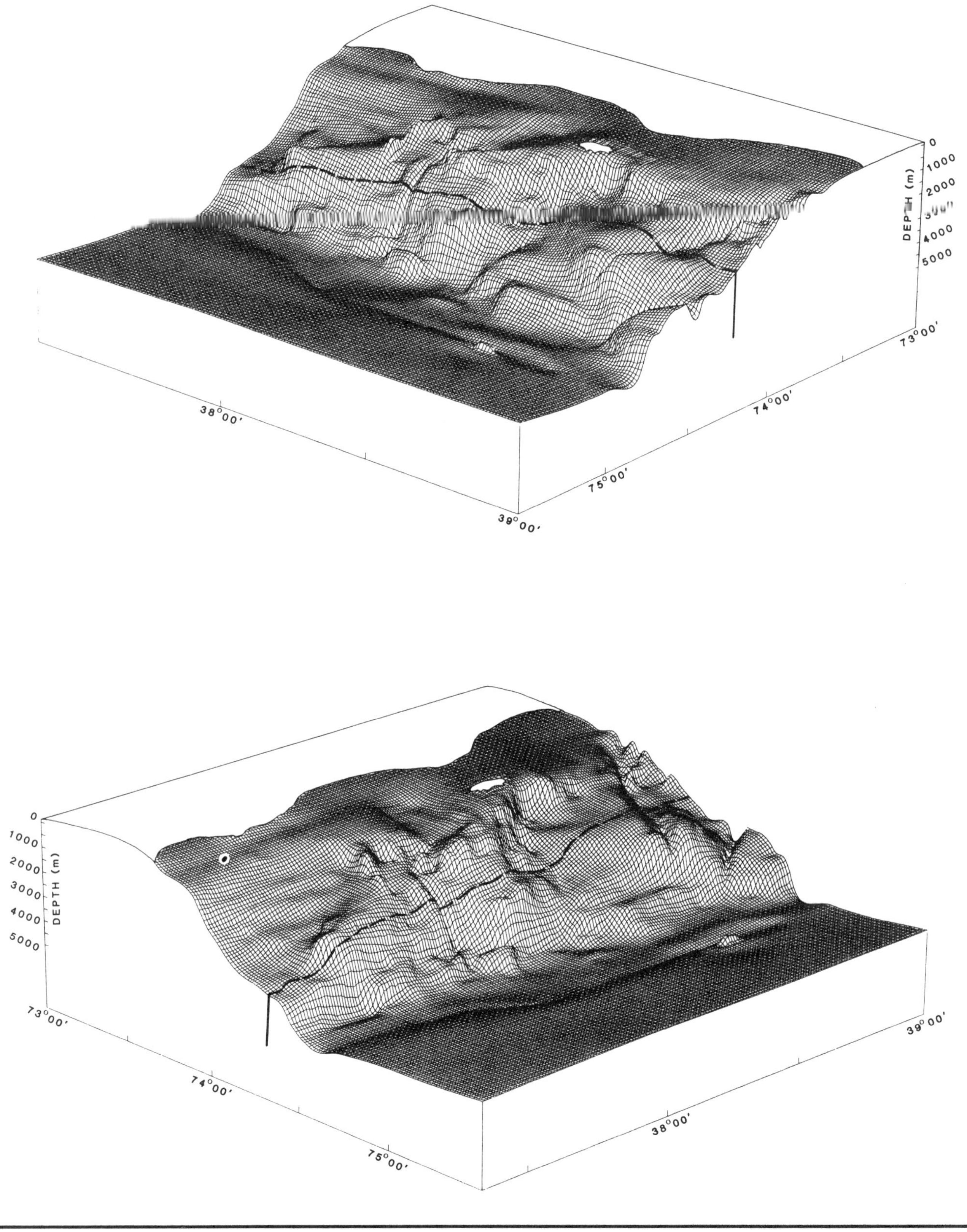

Figure 4. Mesh perspective diagrams off southern Chile. Heavy dashed line shows estimated westernmost extent of continental crust as interpreted from morphology and single-channel seismic records (after von Huene, Kulm, and Miller, 1985).

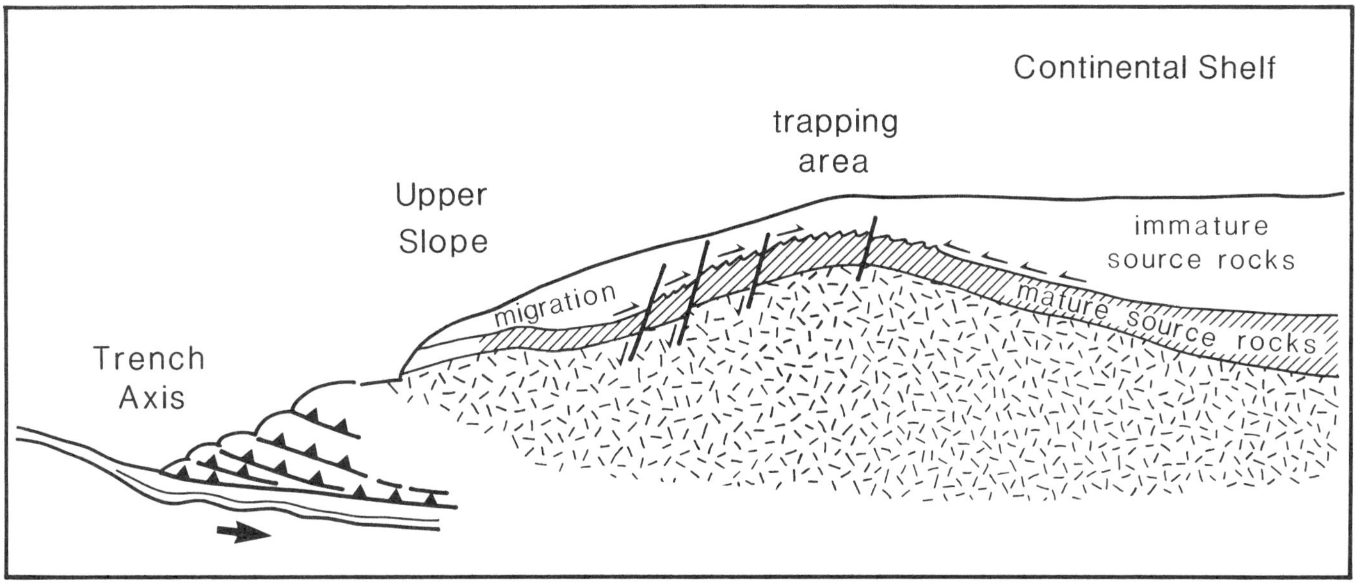

Figure 5. Diagram showing structural configurations for possible hydrocarbon entrapment at the edge of the shelf off western South America.

convergent-margin models show systematically younger tectonic units in a progression of thrust slices extending seaward. Such a progression of younger thrust slices seaward would make the edge of the shelf and upper slope less appealing than the shelf as a hydrocarbon prospect for several reasons. First, the thermal gradients along convergent margins are generally low near the trench and high at the volcanic arc. Thus, the thermal maturity of hydrocarbons would be expected to increase landward. Second, the classical concept of a progressively younger age of sediment seaward would virtually dictate immaturity in a seaward direction and maximum maturity in a landward direction. The presence of unmetamorphosed sedimentary sequences as old as Cretaceous along the upper continental slope of the Andean margin increases the possibility of finding mature hydrocarbons in sedimentary sequences far seaward of the edge of the shelf. Furthermore, the structural highs common along shelf edges might provide traps for a source terrane on the seaward side as well as on the landward side (Figure 5).

The few exploratory wells drilled on structural highs at the edge of the shelf have failed to show major hydrocarbon accumulations, but they did not test the concept of potential hydrocarbon migration up the continental slope. Although the wells drilled off southern Chile were mostly in midshelf positions, significant shows of gas were encountered (Mordojovich, 1974). Published reports of hydrocarbon exploration near the shelf edge on Isla Mocha are not found in the general literature, although some studies there were encouraging (Florial Garcia, oral communication, 1967). Drill holes off central Perú are located on basement ridges in midshelf positions and are covered mainly by Neogene sediment, so they did not test potential trapping structures at the edge of the shelf (Kulm, Thornburg, and Dang, 1984). Obviously, more information is needed to evaluate the hydrocarbon potential of the thousands of kilometers of the unexplored shelf edge.

Two areas where investigations of the edge of the shelf might be particularly rewarding are off Talara in northern Perú and off southern Chile. The source terrane off Talara might extend seaward as suggested by the thick Cretaceous and Eocene sections in seismic record Perú 3. If a suitable structural trap is found by making geophysical surveys along the edge of the shelf, some geochemical testing of shallow samples could be informative. Off southern Chile, the Neogene(?) sedimentary sequence across and down the slope thickens and is unbroken. If that sequence has zones of favorable porosity, similar to the porosity found during drilling on the shelf (Mordojovich, 1974), migration of gas from the underlying older rock is possible. Perhaps the area south of the Chile rise at lat. 45 °S has good potential. Here, the spreading oceanic ridge is being subducted, imparting a higher-than-normal heatflow, and thus a greater-than-normal thermal maturity is produced in the overlying sedimentary sequence of the margin (Cande et al., 1987).

REFERENCES CITED

Aoki, Y., T. Tamano, and S. Kato, 1982, Detailed structure of the Nankai Trough from migrated seismic sections: American Association of Petroleum Geologists Memoir 34, p. 309-322.

Aubouin, J., J. Bourgois, and J. Azema, 1984, A new type of active margin: the convergent extensional margin, as exemplified by the Middle America trench off Guatemala: Earth and Planetary Science Letters, v. 67, p. 211-218.

Biju-Duval, B., P. LeQuellec, A. Mascle, V. Renard, and P. Valery, 1982, Multibeam bathymetric survey and high resolution seismic investigations on the Barbados Ridge complex (Eastern Caribbean): a key to knowledge and interpretation of an accretionary wedge: Tectonophysics, v. 86, p. 275-304.

Bourgois, J., G. Pautot, W. Bandy, T. Boinet, P. Chotin, P. Huchon, B. Lepanay, F. Monge, J. Monlay, B. Pelletier, M. Sosson, and R. von Huene, 1986, Tectonic regime of the Andean convergent margin off Perú (SEAPERC cruise of the R/V *Jean Charcot,* July 1986): C.R. Academy Sci. Paris, p. 1599-1604.

Cande, S.C., R.B. Leslie, J.C. Parra, and M. Hobart, 1987, Interaction between the Chile Ridge and Chile Trench; geophysical and geothermal evidence: Journal of Geophysical Research, v. 92, p. 495-520.

Couch, R., R. Whitsett, B. Huehn, and L. Briceno-Guarupe, 1981, Structures of the continental margin of Perú and Chile, *in* L.D. Kulm et al.., eds., Nazca Plate: crustal formation and Andean convergence: Geological Society of America Memoir 154, p. 703-726.

Fisher, R.L., and R.W. Raitt, 1962, Topography and structure of the Perú-Chile Trench: Deep-Sea Research, v. 9, p. 423-443.

Hayes, D.E., 1966, A geophysical investigation of the Perú-Chile Trench: Marine Geology, v. 4, p. 309-351.

Hussong, D.M., and L.D. Kulm, 1985, Perú-Chile continental margin and adjacent ocean floor, *in* D.M. Hussong and L.D. Kulm, eds., Ocean Margin Drilling Program, Regional Atlas Series, Atlas 6, Marine Science International: Woods Hole Oceanographic Institute.

Hussong, D.M., and L.K. Wipperman, 1981, Vertical movement and tectonic erosion of the continental wall of the Perú-Chile Trench near 11°30'S latitude, *in* L.D. Kulm et al., eds., Nazca Plate: crustal formation and Andean convergence: Geological Society of America Memoir 154, p. 509-524.

Hussong, D.M., et al., 1976, Crustal structure of the Perú-Chile Trench: 8-12° latitude, *in* G.H. Sutton et al., eds., Geophysics of the Pacific Ocean basin and its margin: American Geophysical Union Monograph 19, p. 71-86.

Jones, P.R., III, 1981, Crustal structures of the Perú continental margin and adjacent Nazca Plate, 9°S latitude, *in* L.D. Kulm et al., eds., Nazca Plate: crustal formation and Andean convergence: Geological Society of America Memoir 154, p. 423-443.

Kulm, L.D., J. Miller, and R. von Huene, 1986, The Perú continental margin, record section 2, *in* R. von Huene, ed., Seismic images of convergent margin tectonic structure: American Association of Petroleum Geologists Studies in Geology 26, p. 40-41.

Kulm, L.D., W.J. Schweller, and A. Masias, 1977, A preliminary analysis of the subduction processes along the Andean continental margin, 6° to 45°S, *in* M. Talwani and W.C. Pitman, III, eds., Island arcs, deep sea trenches, and back-arc basins: American Geophysical Union, Series 1, p. 285-302.

Kulm, L.D., T. Thornburg, and S. Dang, 1984, Drill hole stratigraphy, dredge lithologies, and sample locations, *in* D.M. Hussong and L.D. Kulm, eds., Ocean Margin Drilling Program, Regional Atlas Series, Atlas 6, Marine Science International: Woods Hole Oceanographic Institute.

Kulm, L.D., R.A. Prince, W. French, S. Johnson, and A. Masias, 1981a, Crustal structure and tectonics of the central Perú continental margin and trench, *in* L.D. Kulm et al., eds., Nazca Plate: crustal formation and Andean convergence: Geological Society of America Memoir 154, p. 445-468.

Kulm, L.D., J. Dymond, E.J. Dasch, and D.M. Hussong (eds.), 1981b, Nazca Plate: crustal formation and Andean convergence: Geological Society of America Memoir 154, 824 p.

Kulm, L.D., et al., 1981c, Late Cenzoic carbonates on the Perú continental margin: lithostratigraphy, biostratigraphy, and tectonic history, *in* L.D. Kulm et al., eds., Nazca Plate: crustal formation and Andean convergence: Geological Society of America Memoir 154, p. 469-508.

Kulm, L.D., J.M. Resig, T.M. Thornburg, and H.J. Schrader, 1982, Cenozoic structure, stratigraphy, and tectonics of the central Perú forearc, *in* J.K. Legget, ed., Trench-forearc geology: sedimentation and tectonics on modern and ancient active plate margins: London, Blackwell Scientific Publications, p. 151-170.

McCarthy, J., and D.W. Scholl, 1984, Mechanisms of subduction accretion along the central Aleutian Trench: Geological Society of America Bulletin, v. 96, p. 691-701.

McKee, E.H., and D.C. Noble, 1989, Precious-metal mineralization of Cenozoic age in the Andes of Perú, *in* M.T. Canas P., G.E. Ericksen, and J.A. Reinemund, eds., Geology of the Andes and its relation to hydrocarbons and mineral resources.

Miller, J., D.M. Hussong, and R. von Huene, 1986, Perú continental margin record section 3, *in* R. von Huene, ed., Seismic images of convergent margin tectonic structure: American Association of Petroleum Geologists Studies in Geology 26, p. 42-43.

Minster, J.B., and T.H. Jordan, 1978, Present-day plate motions: Journal of Geophysical Research, v. 83, p. 5331-5354.

Mordojovich, K.C., 1974, Geology of a part of the Pacific margin of Chile, *in* C.A. Burk and C.L. Drake, eds., The geology of continental margins: New York, Springer-Verlag, p. 591-598.

Nasu, N., R. von Huene, Y. Ishiwada, M. Langseth, T. Bruns, and E. Honza, 1981, Interpretation of multichannel seismic reflection data, legs 56 and 57, Japan trench transect, Deep Sea Drilling Project, *in* Initial reports of the Deep Sea Drilling Project, 56-57: Washington, U.S. Government Printing Office, v. 7, p. 7.

Pardo-Casas, F., and P. Molinar, 1987, Relative motion of the Nazca (Farallon) and South American Plates since Late Cretaceous time: Tectonics, v. 6, p. 233-248.

Pineda, V., 1986, Evolucion paleogeografica de la cuenca sedimentaria Cretacico-Terciaria de Arauco, *in* J. Frutos, R. Oyarzun, and M. Pinchaera, eds., Geologia y recursos minerales de Chile: Universidad de Concepcion, p. 375-390.

Rosato, V.J., and L.D. Kulm, 1981, Clay mineralogy of the Perú continental margin and adjacent Nazca Plate: implications for provenance, sea level changes, and continental accretion, *in* L.D. Kulm et al., eds., Nazca Plate: crustal formation and Andean convergence: Geological Society of America Memoir 154, p. 545-568.

Scholl, D.W., and T.L. Vallier, 1983, Subduction and the rock record of Pacific margins, *in* S.W. Carey, ed., Expanding earth symposium: Sydney, 1981, University of Tasmania, p. 235-245.

Scholl, D.W., M.N. Christensen, R. von Huene, and M.S. Marlow, 1970, Perú-Chile Trench sediments and sea-floor spreading: Geological Society of America Bulletin, v. 81, p. 1339-1360.

Schweller, W.J., L.D. Kulm, and R.A. Prince, 1981, Tectonics, structure, and sedimentary framework of the Perú-Chile Trench, *in* L.D. Kulm et al., eds., Nazca Plate: crustal formation and Andean convergence: Geological Society of America Memoir 154, p. 323-349.

Shepherd, G.L., and R. Moberly, 1981, Coastal structure of the continental margin, northwest Perú and southwest Ecuador, *in* L.D. Kulm et al., eds., Nazca Plate: crustal formation and Andean convergence: Geological Society of America Memoir 154, p. 351-391.

Shipley, T.H., K.J. McMillen, J.S. Watkins, J.C. Moore, H. Sandoval-Ochoa, and J.L. Worzel, 1980, Continental margin and lower slope structures of the Middle America Trench near Acapulco, Mexico: Marine Geology, v. 35, p. 65-82.

Thornburg, T., 1984, Multichannel seismic interpretations—shelf and upper slope basins, *in* D.M. Hussong and L.D. Kulm, eds., Ocean Margin Drilling Program, Regional Atlas Series, Atlas 6, Marine Science International: Woods Hole Oceanographic Institute.

Thornburg, T., and L.D. Kulm, 1981, Sedimentary basins of the Perú continental margin: structure, stratigraphy, and Cenozoic tectonics from 6°S to 16°S latitude, *in* L.D. Kulm et al., eds., Nazca Plate: crustal formation and Andean convergence: Geological Society of America Memoir 154, p. 393-422.

von Huene, R., and J. Miller, 1988, Migrated multichannel seismic-reflection records across the Perú continental margin, *in* E. Suess, R. von Huene, et al., eds., Initial reports of the Ocean Drilling Project (Part A), v. 112, p.

von Huene, R., L.D. Kulm, and J. Miller, 1985, Structure of the frontal part of the Andean convergent margin: Journal of Geophysical Research, v. 90, p. 5429-5442.

von Huene, R., R. Miller, and M. Fisher, 1986, The eastern Aleutian continental margin record section 111, *in* R. von Huene, ed., Seismic images of convergent margin tectonic structure: American Association of Petroleum Geologists Studies in Geology 26, p. 23-25.

von Huene, R., E. Suess, et al., 1987, Convergent tectonics and coastal upwelling; a history of the Perú continental margin: Episodes, v. 10, p. 87-93.

von Huene, R., J. Bourgois, J. Miller, and G. Pautot, in press, Massive collapse along the front of the Andean convergent margin off Perú: Journal of Geophysical Research.

von Huene, R., E. Suess, et al., 1988, Results of leg 112 drilling; Perú continental margin; Part 1, Tectonic history: Geology, v. 16.

Yeats, R.S., and S.R. Hart, et al., 1976, Initial reports of the Deep Sea Drilling Project, 34: Washington: U.S. Government Printing Office, 803 p.

MINERAL AND COAL RESOURCES

Ericksen, G. E., Cañas Pinochet, M. T., and Reinemund, J. A., editors, 1989, Geology of the Andes and its relation to hydrocarbon and mineral resources: Houston, Texas, Circum-Pacific Council for Energy and Mineral Resources Earth Science Series, v. 11.

CHAPTER 8

COAL RESOURCES OF ARGENTINA

Federico A.J. Bergmann and Alfredo N. Xicoy
Yacimientos Carboniferos Fiscales
Buenos Aires, Argentina

ABSTRACT

The Argentinian coal deposits are in sequences of molassic sediments of Paleozoic to late Tertiary age. They are particularly widespread in the western part of the Argentine, w here 250 coal occurrences are known in four major coal basins. Of the known coal fields, 15 have total reserves (measured, indicated, and inferred) of 585 million tonnes, of which 580 million tonnes are in the Río Turbio field in southern Argentina. Subbituminous coals are now being mined in this field, chiefly for use by thermoelectric plants of Buenos Aires and San Nicolas. In addition, potential resources of 7,350 million tonnes of deeply buried lignite occur in the Río Turbio region. Because of their depth, about 500 m, these coals are not currently economic.

Pilot-plant tests have shown that the Río Turbio coals can be mixed with imported coking coals, in amounts up to about 15%, to make conventional metallurgical coke.

INTRODUCTION

As a consequence of the energy crisis created by World War II, Argentina initiated studies for the evaluation and utilization of its energy resources in 1941. The following information was considered essential: (1) evaluation of the effects of the energy crisis; (2) evaluation of the nonrenewable fuel potential of Argentina; (3) evaluation of the possibility of substituting coal for some fuel-oil uses so that a greater portion of petroleum could be utilized for production of light derivatives (gasoline and others); and (4) determination of availability and resources of coking coal that might serve as a basis for establishing a steel industry in Argentina. Yacimientos Petroliferos Fiscales (YPF) began these studies in 1941. Later, the studies were taken over by organizations such as Yacimientos Carboniferos Fiscales (YCF), created specifically for the exploration, exploitation, industrialization, and commercialization of coals, as well as peat, asphaltites, and oil shale.

The coals of Argentina are found in four major coal basins as shown in Figure 1 and are associated with Carbo-Triassic, Carbo-Jurassic, Tertiary, and Neogene sedimentary sequences. These coals were formed in paralic and limnic environments. The Neogene coals are high-volatile subbituminous coals and lignites, whereas the older coals are low- to high-volatile bituminous coals (Table 1).

EXPLORATION AND ASSESSMENT

Exploration of coals and other solid fuels of Argentina, from 1941 to 1985, included the following: (1) general geologic reconnaissance of formations having a potential for coal and other solid fuels; (2) topographic and geologic survey of selected areas of particular interest; (3) exploratory drilling (including down-hole geophysical measurements) and/or exploratory mining; (4) interpretation of field data; (5) analyses of chemical and physical properties; and (6) calculation of reserves and analysis of economic potential. Coal exploration was largely suspended during the late 1970s and early 1980s, when international oil prices were decreasing, but began again in 1982. In 1985, exploration was being carried out in the Río Turbio field, Province of Santa Cruz (Figure 1).

A total of 250 occurrences of coal have been identified, and of this total, 17 areas were selected for detailed study. Reserves of coal in these areas were determined according to criteria established by the U.S. Geological Survey and U.S. Bureau of Mines (1980, p. 2), as follows:

"*Measured*—Quantity is computed from dimensions revealed in outcrops, trenches, workings, or drill holes; grade and/or quality are computed from the results of detailed sampling. The sites for inspection, sampling, and measurements are spaced so closely and the geologic character is so well defined that size, shape, depth, and mineral content of the resource are well established.

"*Indicated*—Quantity and grade and/or quality are computed from information similar to that used for measured resources, but the sites for inspection, sampling, and measurement are farther apart or are otherwise less adequately spaced. The degree of assurance, although lower than that for measured resources, is high enough to assume continuity between points of observation.

"*Inferred*—Estimates are based on an assumed continuity beyond measured and/or indicated resources, for which there is geologic evidence. Inferred resources may or may not be supported by samples or measurements."

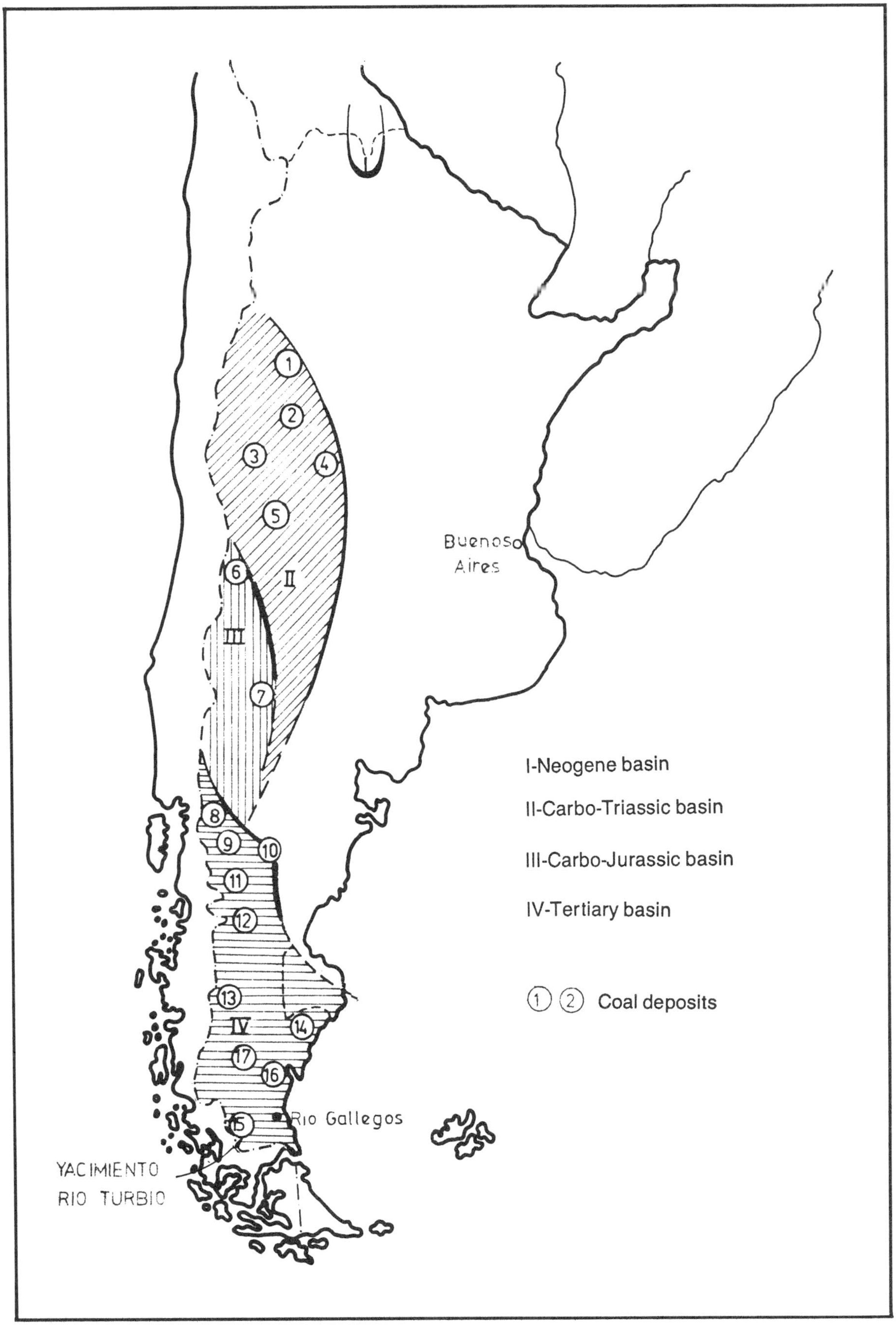

Figure 1. Coal basins of Argentina.

Table 1. Coal resources of Argentina, as determined up to 1985, grouped according to ASTM (1982) classification.[1]
[Parentheses after coal field name show location number in Figure 1, and name of province in which field occurs.]

Type of Coal	Field	Resources (tonnes × 10^3)
IDENTIFIED RESOURCES		
Semianthracite	Cervantes (6, Mendoza)	100
Low-volatile bituminous	La Negra (2, La Ríoja)	55
Mid-volatile bituminous	Río Tambillos (1, La Ríoja) Santa Máxima-El Saltito (5, Mendoza) Burgos (7, Neuquén)	655
High-volatile bituminous A	La Delfina (3, San Juan)	50
High-volatile bituminous B	Richard (4, San Juan) J. Newbery (8, Neuquén) Pico Quemado (9, Río Negro) Lepá (12, Chubut)	3100
High-volatile bituminous C	La Criolla (13, Santa Cruz)	70
Subbituminous A (some high-volatile bituminous C)	Río Turbio (15, Santa Cruz)	580,000
Subbituminous A-C	Indio (10, Chubut) Santa Ana (11, Chubut) Cabo Curioso (14, Santa Cruz)	915
TOTAL		584,945
POTENTIAL RESOURCES		
Lignite	Río Coyle (16, Santa Cruz)	5,000,000
Lignite	Río Santa Cruz (17, Santa Cruz)	2,350,000
TOTAL		7,350,000

[1]Tonnages for the Río Tambillos, Pico Quemado, Río Turbio, Indio, and Cabo Curioso fields include measured, indicated, and inferred reserves. Only inferred reserves are represented for the other fields.

RESOURCES OF COAL AND OTHER SOLID FUELS

The coal resources of Argentina, as of 1985, are summarized in Table 1. The tonnages cited in this table for the Río Tambillos, Pico Quemado, Río Turbio, Indio, and Cabo Curioso fields include measured, indicated, and inferred reserves. Only inferred resources are cited for the other fields. Table 1 also show that the Río Turbio field contains about 99% of the identified coal reserves of Argentina.

Asphaltite, peat, and bituminous shales also were investigated along with the coals (Borrello, 1956). The major peat deposits of Argentina are on the island of Tierra del Fuego, the southern tip of the country. Reserves are more than 90,000,000 tonnes of equivalent air-dried peat having an average heating value of 3,800 Kcal/Kg (16,865 KJ/Kg; 6,840 Btu/lb). Asphaltite veins are found in the southern part of the province of Mendoza and in the northern part of the province of Neuquen. These areas have identified resources of 960,000 tonnes of asphaltite, of which 420,000 tonnes are orthoasphaltites (soluble in organic solvents) and 540,000 tonnes are meta-asphaltites or asphaltic pyrobitumens. The asphaltites have an average heating value of 8,200 Kcal/Kg (34,235 KJ/Kg; 14,760 Btu/lb).

The potential resources of lignite (Table 1) could be exploited only if new technology were developed. The Río Coyle lignites are in a 500-km^2 belt that crosses the Río Coyle some 20-25 km from the Atlantic coast. Three main lignite beds, which have a total thickness of 5-6 m, are present at a depth of 680 m (Luna, 1978). A 180-km^2 lignite field along the Río Santa Cruz (Table 1) has as many as three lignite beds at depths of about 470 m (Luna, 1979). The lignites in both fields have average heating values of 2,000 Kcal/Kg (8350 KJ/Kg; 3600 Btu/lb).

Bituminous shales of Triassic age are present at Rincon Blanco, approximately 70 km west of the city of San Juan. The shales contain 3-4% kerogen, which through distillation would give an estimated 45,000,000 tonnes of oil with a heating value of 10,000 Kcal/Kg (41,750 KJ/Kg; 18,000 Btu/lb).

Rio Turbio Coal Field

The Río Turbio coal field is located at long. 72°17′W, lat. 51°35′S, about 260 km from the port of Río Gallegos (Figure 1). The field is along the Argentine-Chile border and is about 45 km long and 5-10 km wide (Figure 2). Coal beds of this field are in a 1,000-m-thick sequence of lower Tertiary marine and continental sedimentary rock overlying the Cretaceous Cerro Cazodor Group. The vegetable matter from which the coals were formed accumulated in paralic environments. *Nothofagus* sp. has been identified in these coals (Borello, 1956). The coals occur in two zones in the middle part of the Group, separated stratigraphically by a 280-m coal-free sequence. The upper coal zone contains three coal beds having total reserves of 415,000,000 tonnes, which are distributed (from top to bottom) as follows: Dorthea bed (70% of total reserve), bed "A"

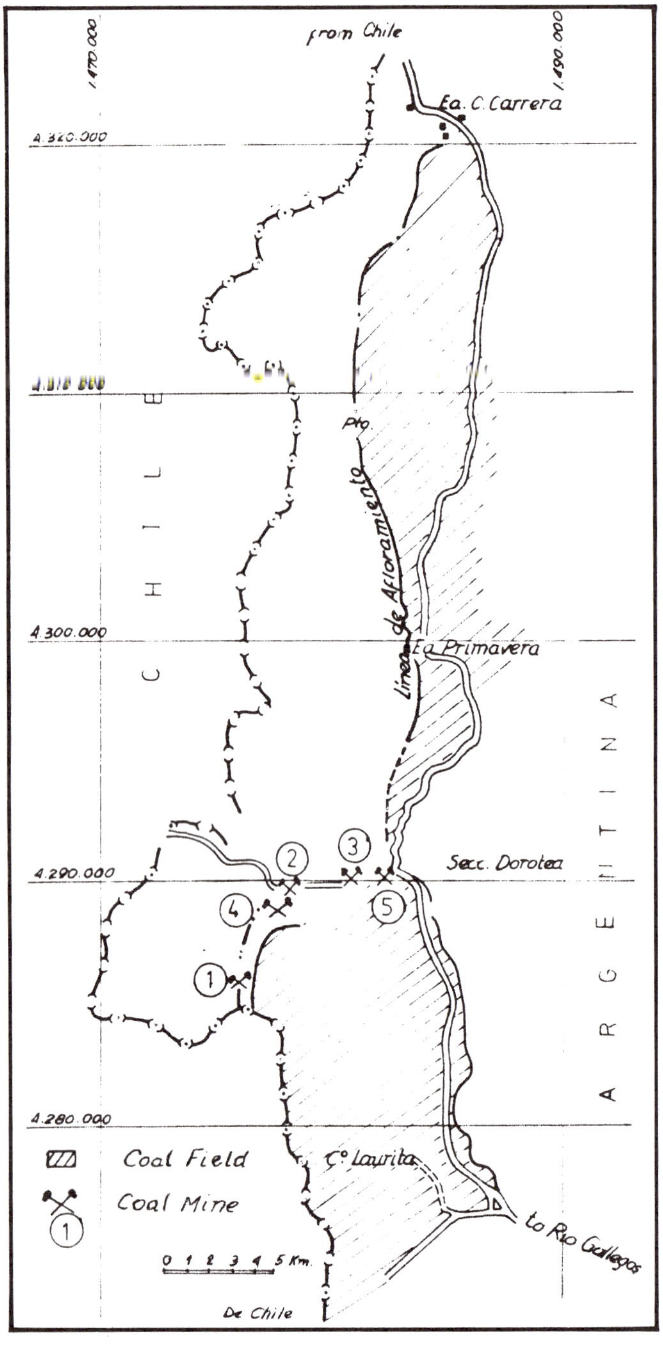

Figure 2. Río Turbio coal field.

(27% of reserve), and bed "B" (3% of reserve). The Dorthea bed is 1.2-2.0 m thick, the "A" bed 0.7-2.0 m thick, and the "B" bed 0.5-1.0 m thick. The stratigraphic interval between the Dorthea and "A" bed is 8 m, and that between the "A" and "B" beds is 15 m. The lower coal-bearing zone has total reserves of 165,000,000 tonnes of which 60% is in an upper bed and 40% is in a lower bed. The upper bed ranges from 1.8 to 2.6 m in thickness, whereas the lower bed is 1.0-7.0 m thick. The average exploitable thickness is not more than 1.5 m for each bed. The stratigraphic distance between the two beds is about 12 m. The coal-bearing sequence dips gently (about 10°) eastward. Physical and chemical characteristics of coals from the Río Turbio field are given in Table 2.

Production from the Río Turbio field, entirely by Yacimientos Carboniferos Fiscales, has been chiefly from the Dorthea bed at mines 3 and 5 (Figure 2). These mines are underground operations in which continuous mining machines are used to extract the coal. Shield-type hydraulic supports and wood or metallic stulls are utilized for roof support at active mine faces. Slushers, conveyer belts, and electric-traction mine cars are used to transport the coal. The 1984 production from these mines was 500,000 tonnes of marketable coal. The mine-run coal is purified by flotation in a plant having a capacity for treatment of 700 t/h. During treatment the coal is prepurified and the 50 × 20 mm material is treated by flotation using magnetite in a Wemco suspension drum; the finer material is purified in cyclones. Recovery is 50%. Electricity for the operation is produced by a 6,000-kw and two 7,500-kw steam-powered generators. Capacity is sufficient to also supply the needs of a local town having about 10,000 inhabitants, most of whom are directly or indirectly concerned with the coal-mining operations.

A large part of the marketable coal is transported to Río Gallegos by a 75-cm-gauge train. According to the circumstances, this coal is loaded directly into ships using a mine-car dumping system and transporting belts; or is transferred to a 30,000 tonne capacity stock yard. Of the stock yard coal, a small amount is sold locally; the remainder is loaded into ships (as they become available), utilizing a mechanical loading system with a 1,000 t/h capacity. The ships are both state-owned and privately owned. Because of high tides (as much as 10 m) and lack of appropriate port facilities, the maximum capacity of these ships is 8,500 tonnes (Yacimientos Carboniferos Fiscales, 1984).

Pico Quemado Coal Field

The Pico Quemado coal field in the province of Río Negro, 70 km south of the Nirihuao train station, near Barriloche, has the second largest known coal reserves in Argentina. It is owned by a private company. Coal production is small, and most is sold locally. The coal is suitable for use by the steel industry, but lack of accurate information and the estimated small reserves have discouraged large-scale development (Bergmann and Xicoy, 1981). The field consists of four coal beds of Tertiary age exposed in a small syncline. Estimated reserves are 2,640,000 tonnes.

The average composition of Pico Quemado coal is:

Humidity	4%
Volatile matter	31%
Fixed carbon	42%
Ash	23%
Heating value (KJ/Kg)	25,490
(Kcal/Kg)	(6,106)
(Btu/lb)	(10,990)
Total sulfur	0.5%

POSSIBLE USES OF ARGENTINE COALS

Of the Argentina coals, only those of the Río Turbio field have been adequately tested (Yacimientos Carboniferos Fiscales, 1984) to determine possible uses, and only the coals of

Table 2. Physical and chemical characteristics of Río Turbio coals.

		Coarse coal (purified; 17% ash)	Fine coal (purified; 17% ash)	Fine purified coal (11% ash)
I. IMMEDIATE ANALYSIS				
on humid base		%	%	%
Humidity		10.1	13.8	13.4
Volatile Matter		35.3	33.9	36.4
Fixed Carbon		37.9	38.1	40.3
Ash		16.7	14.2	9.9
Heating Value (Higher)	KJ/Kg	23,897	23,392	24,836
Heating Value (Higher)	(Kcal/Kg)	5,724	5,633	5,980
Heating Value (Higher)	(Btu/lb)	10,303	10,139	10,764
Heating Value (Lower)	KJ/Kg)	22,820	22,250	23,700
Heating Value (Lower)	(Kcal/Kg)	5,465	5,328	5,675
Heating Value (Lower)	(Btu/lb)	9,837	9,590	10.215
Combustible Sulfur		0.7	0.6	0.6
Total Sulfur		1.0	1.0	0.9
II. ELEMENT ANALYSES				
on pure base				
Fixed Carbon		75.8	76.5	77.9
Hydrogen		6.2	6.2	7.0
Nitrogen		1.0	0.8	0.9
Combustible Sulfur		1.0	0.8	0.8
Oxygen		16.0	15.7	13.2
III. FUSIBILITY OF ASH				
Softening Point C°		1290	1280	1280
Melting Point		1440	1420	1390
Liquefaction Point		1490	1480	1440
IV. PULVERIZATION INDEX				
Hardgrove		36.1	37.2	39.2
V. WEIGHT PER M^3				
Normal Kg/m^3		855	808	827
Compacted Kg/m^3		917	835	n.d.
VI. SWELLING INDEX (FSI)				0-0.5

this field are discussed here. The Río Turbio coal is suitable for thermal applications and has the advantage of being a low-sulfur coal. Pulverized coal from the Río Turbio field can be used in dry-bottom furnaces and boilers of classic design because of relatively high ash-fusion temperature (Table 2). Coarse purified coal (20 mm size) has little granulometric stability, but this defect could be resolved by size reclassification of the coal or preparation of coal briquets with asphaltic agglutinants.

Table 3 gives a summary of gasification tests and products of hypothetical gasification plants. These tests were made during the years 1949-1954 by Lurgi-Rhurchemic Germany and Hydrocarbon Research, Inc., of the United States, who investigated high-pressure synthesis (Fischer-Tropch) and medium-pressure synthesis, respectively. In addition, Hydrocarbon Research made hydrogenation tests of the light oils and tars produced by gasification (Table 3) by treating these subproducts with 98% H_2 gas obtained from synthesis gas. Various end-products were obtained, including 24,000 $Kcal/m^3$ gas and 62/64 octane gasoline.

Hydropyrolysis tests of coal from the Dorthea bed were made in the Tulsa, Oklahoma, laboratories of the Cities Service Co. (R.S. Matyas and W.J. Peterson, unpublished report, 1981), with the following results:

Working temperature (°C)	800	800	800	700
Working pressure (Kg/c^2)	105	55	105	55
Converted coal percentage:				
CO_x	2.3	5.3	2.0	1.7
CH_4	48.8	38.0	17.9	12.8
C_2H_4	3.4	1.9	13.4	11.2
Total gas	54.5	45.2	33.3	25.7
BTX	3.2	5.0	7.8	12.1
Light oil	1.0	1.6	8.8	10.0
Heavy oil	2.6	1.5	6.4	3.6
Total oil	6.8	8.1	23.0	25.7
Total conversion	61.3	53.3	56.3	51.4

Coking tests were made on purified coal from the Dorthea bed, with the following results:

Temperature of distillation	Semicoke* Coke**	Anhydrous tar	Liquid	Gas
Low (500°)	64%*	19%	8%	9%
High (1000°)	60%**	7%	14%	19%

Because the semicoke and coke residues are pulverulent, additional tests were undertaken to determine the possibility

Table 3. Tests for Río Turbio Results of Coal Gasification.

(A) Lurgi-Rhurchemie (unpublished reports, 1954)

Coal consumption for gas production	100 t Coal/Dorthea/hr:850,000 t/yr
Gas production for synthesis	11,099 Nm^3/hour

Synthetic hydrocarbons (including alcohol)

Solid wax	32,800	t/yr
Slack wax	26,600	
Heavy oil	26,600	
Medium oil	26,600	
Light oil	13,600	
Liquid gas	4,000	
Alcohol	2,800	
Total synthetic hydrocarbons	133,000	t/yr

Tar Derivates: remaining from gasification

Tar-oil	73,500	t/yr
Gross benzene	18,400	
Gross phenic oil	4,850	
Total remaining gasified product	96,750	t/yr

(B) Hydrocarbon Research Inc. (unpublished report, 1949)

X1 First alternative (5000 t/day of gross coal treated)

Product	Value	Unit	Amount	Unit
Combustible gas for furnace	3,420	Cal/m^3	716,400	m^3/day
Combustible gas for furnace	18,500		19,600	
Catalytic polymer, gasoline	8,330	Cal/l	314,800	l/day
Treated synthetic gasoline	8,330		686,000	
Tar from polymerization	9,660		22,100	m^3/day
Light tar and oils from coal distillation	9,660		467	
Gas oil	9,660		152,300	l/day
Residual paraffin	8,660		16,700	
Oxygenized chem. org. prod.	6,900	Cal/kg	5,800	Kg/hr

X2 Second alternative (4500 t/d gross carbon treated)

Product	Value	Unit	Amount	Unit
Gas for furnace	3,400	Cal/m^3	716,400	m^3/day
Gas for furnace	18,500		18,970	
Enriched gas	24,000		69,370	
Catalytic polymer gasoline	8,330	Cal/1	314,800	1/day
Treated synthetic gasoline	8,330		686,500	
Gasoline for engines	8,330		779,000	
Oxygenized chem. org. prod.	6,900	Cal/kg	5,800	Kg/hr

of using this coal for making metallurgical coke. These tests involved both one- and two-stage distillations (Yacimientos Carboniferos Fiscales, 1984). The tests for Río Turbio coal were developed in France as a pilot project, and then continued in Argentina. They show that it is possible to make conventional metallurgic coke by adding Río Turbio coal to a typical coke mixture consisting, for example, of 70% high-volatile coal and 30% low-volatile coal. It was found that coke could be made by adding up to 15% of Río Turbio coal to such a coking coal mixture. Likewise, it was found that an excellent metallurgical coke could be made, in bee-hive coke ovens, of mixtures consisting of 80% petroleum coke and 20% tar. Additional tests showed that this petroleum coke could be substituted for the semicoke made from Dorthea purified coal. One-stage coking tests also were made to determine the feasibility of producing coke from briquets consisting of a mixture of semicoke from Río Turbio coal and coking coal, with and without agglutinants. These tests showed that suitable metallurgical coke could be made of such mixtures containing 25-50% Río Turbio coal.

Two-stage coking tests, which involve degasification of the coal to produce pulverulent semicoke or coke for making metallurgical coke from briquets, were made on Río Turbio coal. It was found that good metallurgical coke could be made from Río Turbio semicoke if 20-30% of coking coal was added to the briquets of semicoke. Furthermore, it was found that metallurgical coke could be made from briquets with pulverulent coke even without addition of coking coal. Unfortunately, two-stage coking was found to be a relatively high-cost process that might be suitable only for special needs or in a closed economy.

CONCLUSIONS

Bituminous coals of Carboniferous to Tertiary age are widespread in western and southern Argentina. Although studies of these coals have been carried out since World War II, the coals are still imperfectly known, and 99% of the identified reserves are in the Río Turbio field, southern Argentina. In addition, relatively large amounts of peat and subeconomic lignite are known to exist in Tierra del Fuego and the province of Santa Cruz, southern Argentina. The Río Turbio coals do not make a satisfactory conventional metallurgic coke, but such coke can be made by mixing Río Turbio coal with imported coking coal, thereby decreasing the need for imported coal. Gasification tests of Río Turbio coal have shown the feasibility of producing a variety of gaseous and liquid hydrocarbons suitable for diverse uses.

REFERENCES

American Society of Testing and Materials, 1982, Standard classification of coals by rank: ASTM Annual Book of ASTM Standards, Part 26, D388-82, p. 238-242.

Bergmann, F.A.J., 1983, Incidencia de la facies metamórfica en la versatilidad carbo-química del mineral de Río Turbio (Provincia de Santa Cruz): Revista de la Asociación Geológicia Argentina, v. 38, p. 201-202.

Bergmann, F.A.J., and A.N. Xicoy, 1981, Reseña sobre el potentical carbonífero Argentino: Buenos Aires, Ciencia Energética, v. 12, p. 3-10.

Borrello, A.V., 1956, Recursos minerales de la República Argentina, III, Combustibles sólidos minerales: Buenos Aires, Revista del Instituto Nacional de Ciencias Naturales, 665 p.

Borrello, A.V., 1969, Los geosinclinales de la Argentina: Buenos Aires, Direccion Nacional de Geología y Minería, Annals 14, 187 p.

Luna, P., 1978, Informe sobre los resultados de pozos ejecutados en la zona de Río Coyle, Santa Cruz: Yacimientos Carboníferos Fiscales (Argentina), unpublished report, 35 p.

Luna, P., 1979, Informe sobre los resultados de 6 pozos ejecutados en la zona del curso medio del Río Santa Cruz, Provincia de Santa Cruz: Yacimientos Carboníferos Fiscales (Argentina), unpublished report No. 11111, 30 p.

Suizer, H., 1981, Cubicación de reservas del Yacimiento Río Turbio (Province of Santa Cruz): Yacimientos Carboníferos Fiscales (Argentina), unpublished report, 114 p.

U.S. Geological Survey and U.S. Bureau of Mines, 1980, Principles of resource/reserve classification for minerals: U.S. Geological Survey Circular 831, 5 p.

Vera, R., J.C. Orlandino, and A. Ferrari, 1962, Elaboración de coque metalúrgico, en escala piloto, incorporando carbon de Pico Quemado, en las mezclas que actualment coquiza SOMISA en la Planta General Savio de San Nicolás: Argentina, Yacimientos Carboníferos Fiscales, unpublished laboratory report No. 49, 6 p.

Wood, G.H., Jr., T.M. Kehn, M.D. Carter, and W.C. Culbertson, 1983, Coal resource classification systems of the U.S. Geological Survey: U.S. Geological Survey Circular 891, 65 p.

Yacimientos Carboníferos Fiscales, 1984, Sintesis panorámica del carbón, con referencias al carbón argentino: Argentina, Yacimientos Carboníferos Fiscales, published report, 100 p.

Ericksen, G. E., Cañas Pinochet, M. T., and Reinemund, J. A., editors, 1989, Geology of the Andes and its relation to hydrocarbon and mineral resources: Houston, Texas, Circum-Pacific Council for Energy and Mineral Resources Earth Science Series, v. 11.

CHAPTER 9

COAL ZONES OF COLOMBIA

Raul Duran Rodriguez and Carlos Ariel Arboleda Otalora
Instituto Nacional de Investigaciones Geológico-Mineras, INGEOMINAS
Bogotá, Colombia

ABSTRACT

The many and various coals distributed throughout the Andean region of Colombia represent the largest reserves of all the Latin American countries. These coals differ in quality and potential use according to the basins in which they were deposited. The coals are of Late Cretaceous age in the central and eastern parts of the Colombian Andes, early Tertiary in the northern Andes, and later Tertiary in the western Colombian Andes.

Most of the coals were formed in deltaic and transitional environments, some with clear continental influence, and range from lignite to bituminous and anthracite. Despite the physical and chemical differences from one basin to another, the main characteristics of Colombian coals are as follows:

(1) Northern Andes: Approximately 55 coal beds are known in the Cerrejon area, 40 of which are thicker than 1 m. Some beds are locally as thick as 10, 20, and 26 m. Analyses of coals from this area show the following averages: 8% ash, 11,850 Btu/lb calorific power, 0.6% sulfur, and 9% moisture. These coals have excellent thermal qualities, high calorific power, and low sulfur and ash content, which make them highly competitive in the international market.

(2) Central and Eastern Andes (Cundinamarca-Boyaca and Norte de Santander): The coals with the best coking properties are found in the areas of Checua-Lenguazaque, Department of Cundinamarca; analyses show 4.13% ash, 21 to 30% volatile material, 13,045 to 15,200 Btu/lb calorific power, and 0.4 to 2.4% sulfur. Coal beds range from 0.80 to 3.0 m in thickness, and have average dips of 30° to 35°. Some coals in the Department of Boyaca have coking properties, whereas others are agglomerating, but most are thermal coals having 6 to 19% ash, 31 to 45% volatile material, 0.6 to 1.8% sulfur, and 11,700 to 14,200 Btu/lb calorific power. The coal beds dip 10° to 30° and range in thicknesses from 0.80 to 2.70 m. Some coals in the Norte de Santander area are suitable for semicoke but are used chiefly as thermal coals.

(3) Western Andes (Antioquia and Valle del Cauca): The Antioquia coals have the following characteristics: 1.5 to 26% ash, 7,250 to 14,000 Btu/lb calorific power, and 0.30 to 3.6% sulfur. Thicknesses range from 0.50 to 2.9 m, and dips from 30° to 60°. These are used as thermal coals. The coals in the Valle del Cauca are folded and commonly have steep to overturned dips. Analyses show the following: 2.8 to 4.8% ash, 11 to 49% volatile material, 8,900 to 13,750 Btu/lb calorific power, and 0.9 to 6% sulfur. Thicknesses range from 2.4 to 6.6 m, and dips from 30° to 90°.

The latest study conducted by the Instituto Nacional de Investigaciones Geológico-Mineras in 1982 shows the following reserve estimates for seven regions that might be exploited within the next 20 years; these represent 90% of the total reserves of Colombia.

All coals	Million t	Percentage of total
Measured reserves	3,898	22.8
Indicated reserves	399	2.3
Potential reserves	12,796	74.9
TOTAL	17,093	100.0
Thermal coals only		
Measured reserves	3,759	33.6
Potential reserves	7,167	64.1
TOTAL	11,187	100.0

These reserves are estimated for mine depths of not more than 600 m, which is considered feasible for commercial exploitation in Colombia. However, such depths have not been reached in any of the active mines.

The mining industry in the country is still operated with simple hand-mining methods, except for the underground operation of the Chapa mine, which supplies Acerias Paz de Rico, and the open-pit operation that recently began in the Cerrejon zone. In the 1980s, coal production has been about 5 million t per year, which will increase when new mines are opened in the Cerrejon zone. Production in the Cerrejon area is expected to increase to 15 million t per year by 1990.

INTRODUCTION

Status of Information on Colombian Coal

This report is based chiefly on unpublished reports by the technical personnel of the Instituto Nacional de Investigaciones Geológico-Mineras (INGEOMINAS), by the Colombia Coal Company (CARBOCOL), and by other coal-producing companies. Geological information available for the coal zones of Colombia is limited; only the detailed studies carried out by CARBOCOL in the Cerrejon area and in some specific zones in the central part of the country give adequate geological data. Measured, indicated, and potential coal reserves of Colombia total 17.1 billion t, which consist of measured reserves, 18.2%; indicated reserves, 5.2%; and potential reserves, 76.6%. These reserves represent about 42% of the known coal resources of Latin America.

Distribution and Age of Coal Deposits

Coal deposits are widespread in the Andean region of Colombia, occurring in the three main Andean ranges—Cordillera Oriental, Cordillera Central, and Cordillera Occidental. All the exploitable coals of this region are of Cretaceous to middle Tertiary age. The principal coal deposits are those in the Cordillera Oriental, which are of Late Cretaceous (Maestrichtian) to Miocene age. Coals in the Cordilleras Central and Occidental are of Oligocene-Miocene age. The coals can be further subdivided according to age: (1) those in the central area of the Cordillera Oriental are in the Guaduas Formation of Maestrichtian to early Paleocene age; (2) coals in the northern part of Colombia (Cerrejon, La Jaqua, and Norte de Santander [Figure 1]) are of late Paleocene-middle Miocene age; and (3) coals in the western part of Colombia (Valle del Cauca and Antioquia [Figure 1]) are of Miocene age.

In the Late Cretaceous, sedimentation in the region of the Colombian Andes became more continental, and favorable coal deposition environments began to migrate northward in the region east of the ancestral Central Cordillera. Later, the favorable environments of shallow deltaic plains migrated westward. The first significant coal beds were deposited during the Maestrichtian and continued, almost without interruption, until the middle Miocene. Deposition in the Valle del Cauca and Antioquia basins (Figure 1) began during the Oligocene. The Antioquia basin is essentially continental, whereas the Valle del Cauca basin shows cyclic marine transgressions, as recorded by fossils. The coals in the Guaduas Formation were deposited in a transitional deltaic plain environment resulting from marine regression. El Cerrejon and La Jagua-La Loma coals (Figure 1) also formed in a transitional to continental, subaerial deltaic environment marked by rapid subsidence.

Characteristics and Classification

The coals of Colombia are tentatively divided into two groups on the basis of petrographic constituents: (1) those in the western and northern parts of the country, which show a predominance of exinites over inertinites; and (2) those in the Cordillera Oriental, which show a predominance of inertinites over exinites. Petrographic studies have been limited, however, and definitive information is available for only a few localities. For example, in the Amaga area of Antioquia the most abundant maceral constituent is vitrinite, which represents approximately 70 to 95% of the total. In the Checua-Lenguazaque area of Cundinamarca, where the largest number of petrographic analyses have been made, the coals show a high content of vitrinite and 10 to 30% inertinites.

The distribution of types of coals in Colombia is shown in Figure 1. Coking coals are found only in the central part of the country (Checua-Samaca and Rio Frio areas), and the coals with the best coking characteristics are those of the Guacheta-Samaca area. Studies conducted to date show that only 4% of the coal reserves of Colombia have coking characteristics. The coals from Norte de Santander produce coke for siderurgical use in spite of their relatively high content of volatile matter. Most of the Colombian coals are highly bituminous, having only average contents of volatile matter. In general, the coals of the Cordillera Oriental are agglomerating, with high calorific value. The Valle del Cauca, Risaralda, and Antioquia coals are generally subbituminous high-volatile coals. These coals were locally affected by Tertiary intrusions, which have changed them to semianthracite. Scant information about coals in the northern part of Colombia indicates that they range in rank from lignite to high-volatile bituminous types. The coals from El Cerrejon, which have been studied most intensively of all Colombian coals, are classified as high-volatile bituminous coals having excellent thermal quality, high calorific value, and low contents of ash and sulfur.

RESERVES

Reserve estimates given in the following table for the 7 areas shown in Figure 1 are based on Duran et al. (1979) and unpublished data supplied by governmental agencies and private coal companies (figures are in million tonnes).

Area	Reserves Measured	Reserves Indicated	Potential Resources	Total
1. Cerrejon	3,800	—	—	3,800
2. La Jagua	300*	300*	1,250*	1,850
3. San Jorge	5	35*	5,100*	5,140
4. Cundinamarca-Boyaca	328	695	11,550	12,573
5. Norte de Santander	40	142	874	1,056
6. Antioquia	21	120	—	141
7. Valle del Cauca	19	12	110	141
TOTAL	4,513	1,304	18,884	24,701
PERCENT OF TOTAL	18.2	5.2	76.6	100

*Re-evaluated figures.

DESCRIPTIONS OF MAIN COAL-BEARING AREAS

More than 40 coal-bearing areas have been identified in Colombian territory, most of which are poorly known geologically. Seven areas having the greatest amount of geological

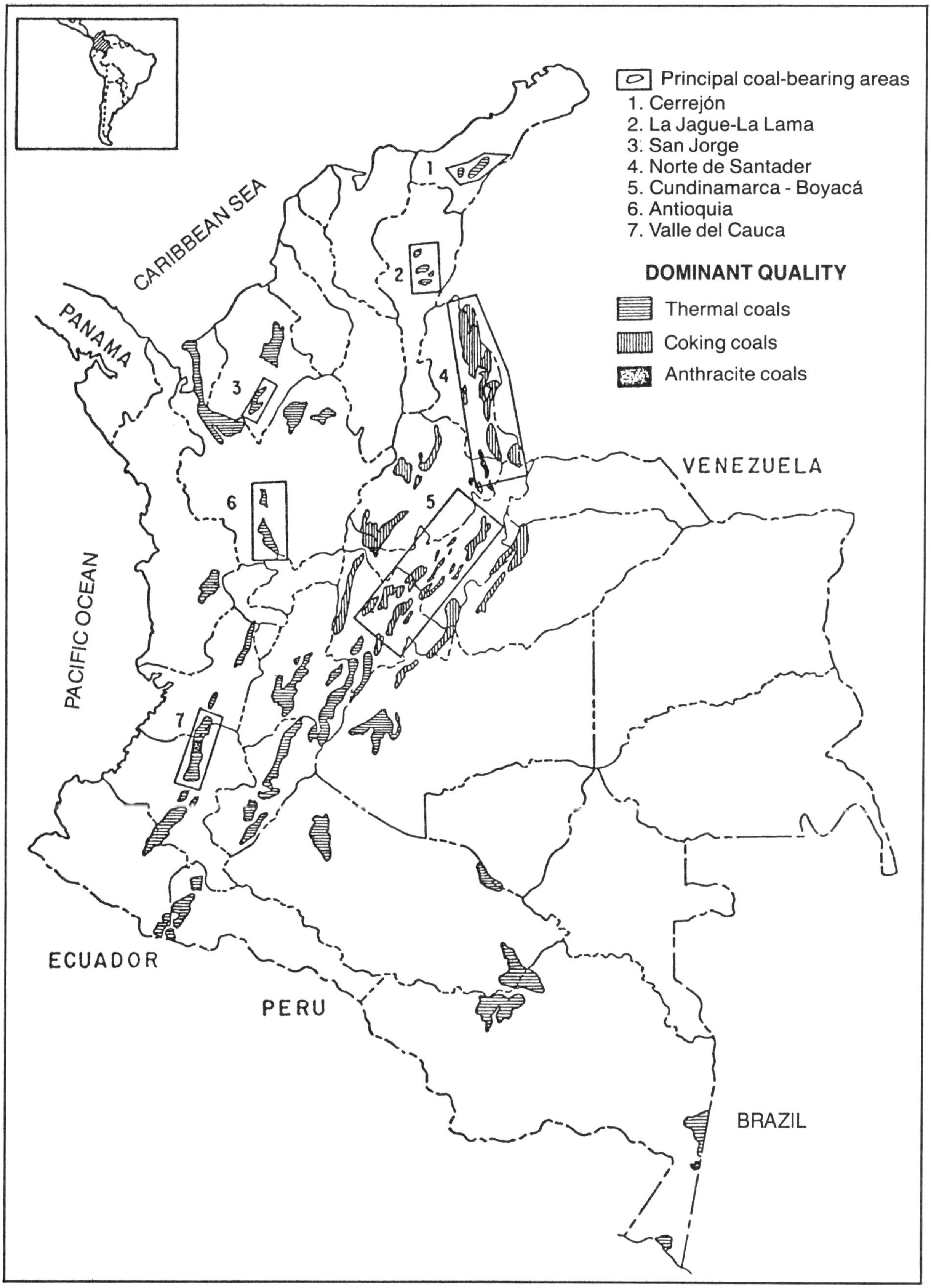

Figure 1. Index map showing the distribution of coal in Colombia.

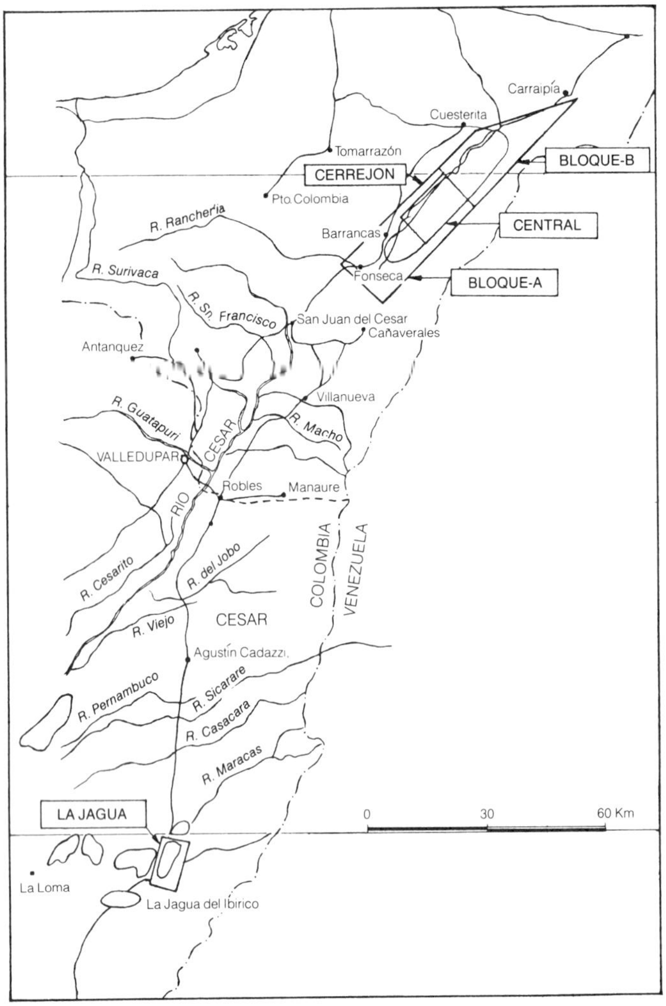

Figure 2. Distribution of coal in the El Cerrejon and La Jagua areas.

and other information, and the greatest potential for development, were selected for discussion. These areas (Figure 1) are El Cerrejon, La Jagua-La Loma, Alto de San Jorge, Norte de Santander, Cundinamarca-Boyaca, Antioquia, and Valle del Cauca. For the purpose of discussion, these areas are grouped into three regions—Atlantic, Central, and Pacific.

Atlantic Region

El Cerrejon area

The El Cerrejon area (Figure 2), in the basin of the Cesar and Rancheria rivers, Department of Guajira, has been known since the 19th century. Systematic studies of this area have been carried out since 1945. INGEOMINAS began systematic study in 1970, and, subsequently, detailed studies were made by the Peabody Company on a concession granted by the Colombian government. In 1974, Intercor, a subsidiary of Exxon, made studies of the coal-bearing Cerrejon Formation.

The El Cerrejon coal deposits (Figure 3) extend through an area of 500 km^2, where Jurassic and younger rocks crop out in a Tertiary synclinorium. The coals are in the Cerrejon Formation, which consists of gray- to yellowish-brown sandstone that is locally calcareous and contains interstratified gray carbonaceous shale. The formation contains 30 to 40 coal beds ranging in thickness from a few centimeters to 10 meters or more. The thickest coal bed locally is as much as 26 m thick.

The El Cerrejon coal field has been subdivided into three blocks (Figure 3). The central block, covering an area of 10,000 hectares, has been studied most intensively. Coal was discovered in this block in the 19th century, but detailed studies to define the field did not begin until the 1960s. INGEOMINAS carried out the first diamond drilling in 1969, but a more complete exploration program was conducted in 1977 by CARBOCOL to obtain more information about the field, particularly in [illegible]. It was determined that approximately 200 million t of coal could be exploited in open pits, and the remaining 250 million t could be extracted in underground mines. Drilling confirmed that the block has 38 coal beds, 28 of which range in thicknesses from 1 to 12 m. Measured reserves, which are approximately 450 million t, extend to a depth of 150 m. Analyses of samples from this block show the following range of values (in percent, except as noted):

Moisture	5.60	8.50
Ash	1.16	12.49
Volatile material	33.70	43.40
Fixed carbon	44.00	59.00
Sulfur	0.40	0.70
Calorific value (cal/g)	6,158	8,055
Swelling index	0	1.59

The analyses and coking tests indicate that El Cerrejon coals do not coke directly, but because of their low ash and sulfur contents they can be mixed with coking coals to produce a satisfactory coke for a variety of uses. The analyses also show that the El Cerrejon coals have excellent thermal qualities because of their high calorific value and low ash and sulfur content, making these coals highly competitive in the international market.

The reserves, classified according to international nomenclature, are as follows:

Depth of 0 to 100 m	950,000,000 t
Depth of 100 to 200 m	1,050,000,000 t
Depth of 200 to 300 m	1,000,000,000 t

La Jagua area

La Jagua area (Figure 1) is 100 km south of Valledupar, capital of the Department of Cesar, and 5 km east of the town of La Jagua (Figure 4). The coals are in the Los Cuervos Formation of Paleocene age exposed in the Jagua de Ibirico syncline, which has a strike (NE) length of about 15 km. The Los Cuervos Formation has been divided into three units, of which the lowest unit, containing 6 to 9 coal beds having thicknesses ranging from 0.80 to 6.90 m, is the most important. Analyses

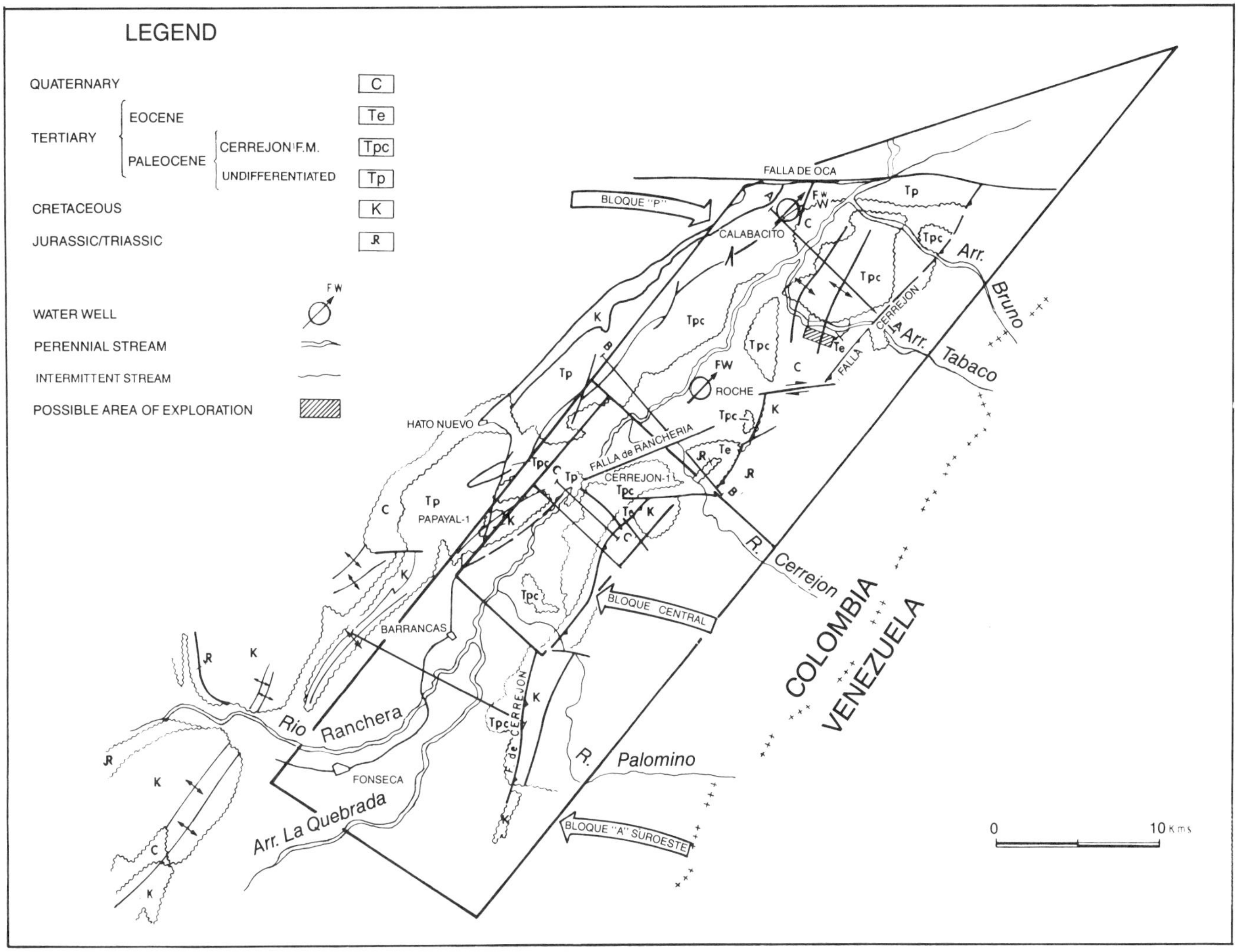

Figure 3. Geologic map of the El Cerrejon coal field.

of coals from this unit show the following ranges (in percent, except as noted):

Moisture	3.2	to	10.0
Ash	1.4	to	17.0
Volatile material	18.5	to	39.8
Fixed carbon	25.9	to	56.7
Sulfur	0.63	to	1.71
Calorific value (cal/g)	6,300	to	8,000
Swelling index	0	to	1.5

These are high-volatile bituminous coals. Inferred reserves in the La Jagua area are approximately 470 million t.

Alto San Jorge area

The coals in the Alto San Jorge area (Figure 1) are in the Cienega de Oro Formation of Tertiary age (Oligocene) in an area dominated by two separate, faulted, north-trending synclines.

Studies in the Alto San Jorge area began in 1984 in two blocks: (1) the 513 km^2 San Jorge block; and (2) the 265 km^2 San Pedro Ure block. Phase I of the study included surface exploration in which it was estimated that approximately 5.1 billion t of coal existed to a depth of 500 m. Phase II of the study began in June 1985 and included subsurface exploration, determination of indicated reserves, and definition of quality. Analyses of coals from the Alto San Jorge area are as follows (in percent, except as noted):

Moisture	8.33	to	15.10
Ash	8.95	to	16.11
Volatile material	30.25	to	55.62
Fixed carbon	27.73	to	50.95
Sulfur	0.31	to	3.03
Calorific value (cal/g)	6,500	to	7,650
Swelling index	0	to	1.5

Central Region

Coals are abundant in the Cundinamarca-Boyaca area in the Cordillera Oriental north of Bogota (Figures 1, 5). These are the best coking coals in Colombia, and have been studied

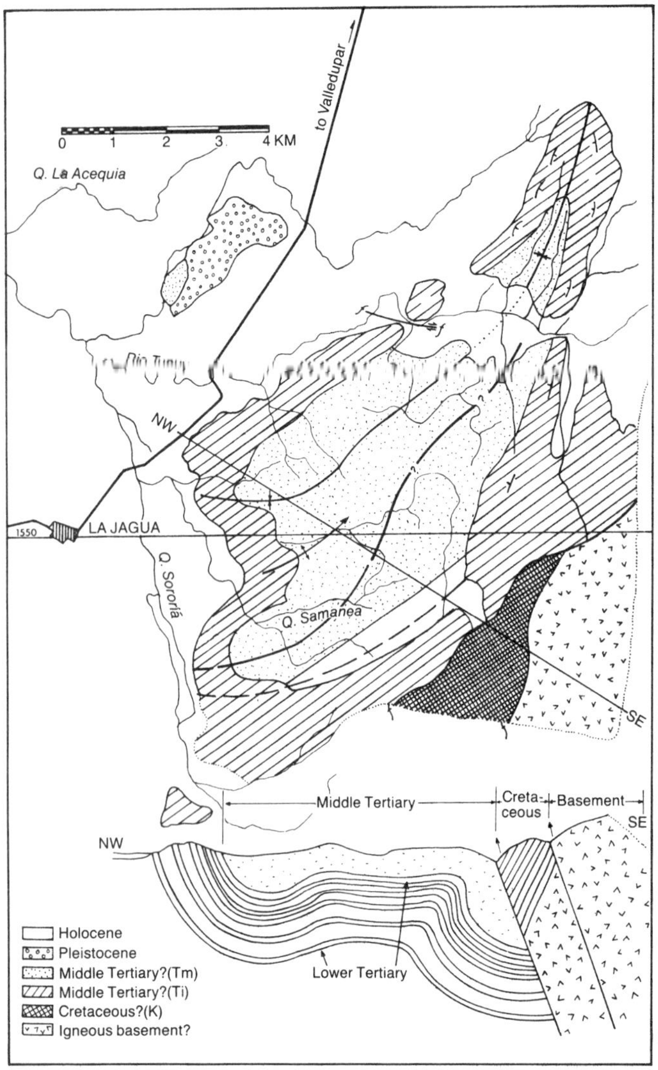

Figure 4. Photogeologic map of the La Jagua coal area.

intensively. The coals are in the Guaduas Formation of Maestrichtian to Paleocene age, which is underlain by the Guadalupe and Ermitano formations and overlain by the Arenisca del Cacho and Picacho formations.

The Guaduas Formation consists of compact claystone, carbonaceous claystone, sandstone layers, and many coal beds associated with claystone in the lower and middle parts of the formation. The formation ranges from 250 m to 1,200 m in thickness, and, on the basis of thick sandstone layers, has been subdivided into three or five lithostratigraphic units. The coal-bearing region extends along the axis of the Cordillera Oriental, where the sedimentary rocks have been folded into a series of synclines and anticlines that strike northeast. Large longitudinal and transverse faults also are present. In general, the anticlines make up the sierras bordering the Cundinamarca and Boyaca tablelands; synclines, in which the Guaduas Formation and its coal beds are preserved, are exposed in the intermountain valleys. Analyses of the coals from the Cundinamarca-Boyaca area show the following (in percent, except as noted):

Moisture	0.3	to	8.0
Ash	2.0	to	19.0
Volatile material	17.2	to	45.0
Fixed carbon	35.2	to	78.0
Sulfur	0.2	to	4.0
Calorific value (cal/g)	5,300	to	8,500
Swelling index	0	to	9.0

These are high-volatile bituminous coals appropriate for thermal and industrial use; bituminous coals with average content of volatile matter appropriate for mixtures; and bituminous coals with low content of volatile matter appropriate for siderolgical use.

As many as 12 coal beds have been recognized in the area, and their thickness ranges from 0.8 to 3 m. Measured reserves are 330,000,000 t (of which 283,000,000 t are coking coals), and indicated reserves are 695,000,000 t.

Norte de Santander area

The Norte de Santander area, in northeastern Colombia near the border with Venezuela (Figures 1, 6) includes the Catatumbo, Zulia, Pamplona, Mutiscua, Toledo, Salazar, and Cerro Tasajero areas, which are part of the Maracaibo basin. The coals are present in the Catatumbo Formation (Maestrichtian), Los Cuervos Formation (Paleocene-Lower Eocene), and Carbonera Formation (Upper Eocene-Lower Oligocene). Two to five coal beds that have been recognized have thicknesses ranging from 0.6 to 3 m and dips of 30° to 50°. Analyses of coals from the three formations are as follows (in percent, except as noted):

	Catatumbo	Los Cuervos	Carbonera
Moisture	0.6-9.1	0.6-7	0.4-1.8
Ash	1.6-23.5	2.7-25.3	3.9-14.8
Volatile material	34.2-43.5	28.3-57.3	38.0-48.3
Fixed carbon	33.4-55.5	38.1-63.2	42.9-49.3
Sulfur	0.37-1.91	0.47-3.67	0.42-3.33
Calorific value (cal/g)	5,058-7,591	5,788-8,339	6,746-8,067
Swelling index	0-3.5	0-6	1-1.5

According to these analyses, the coals may be classified as follows:

Catatumbo Formation: High-volatile bituminous
Los Cuervos Formation: Medium- to high-volatile bituminous
Carbonera Formation: High-volatile bituminous

Some of the coal beds in the Cuervos Formation have coking properties.

Pacific Region

Antioquia area

The Antioquia area is in the valley of the Rio Cauca (Figure 7), which is underlain by Paleocene metamorphic rocks, Mesozoic igneous and volcanic rocks, and Cenozoic igneous and sedimentary rocks. The coals are chiefly in the middle

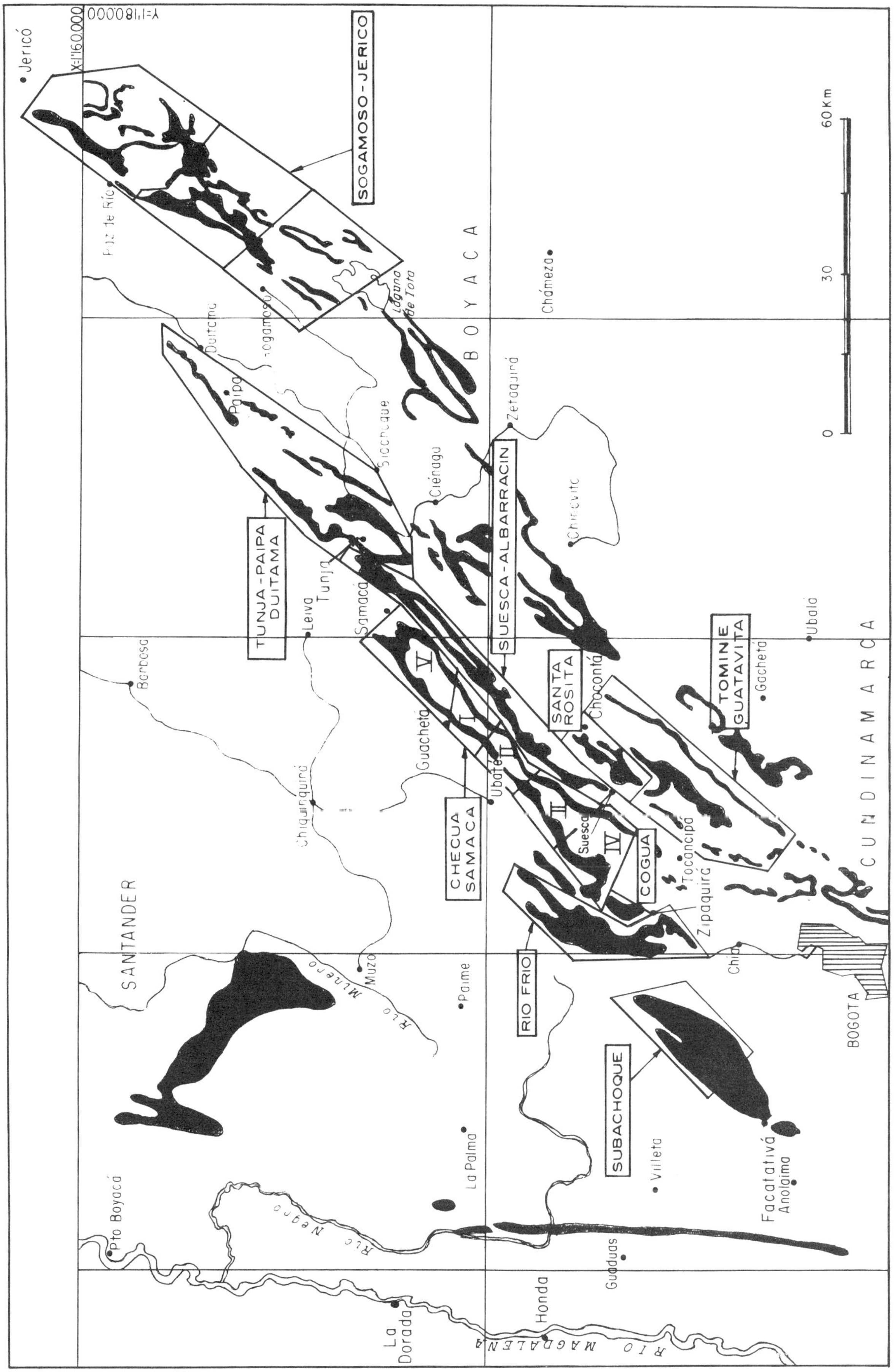

Figure 5. Distribution of coal in the Cundinamarca-Boyaca area.

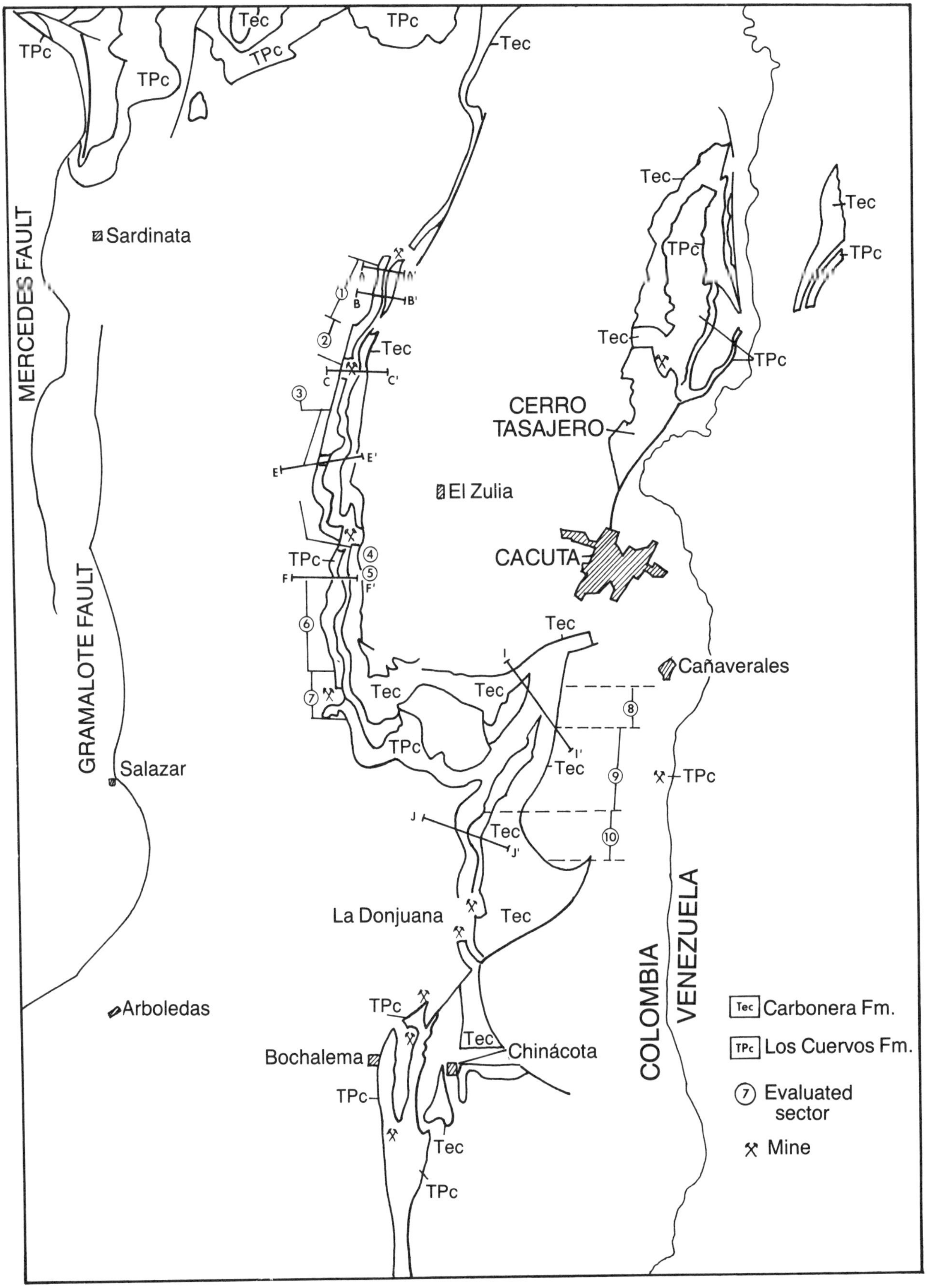

Figure 6. Distribution of coal in the Norte de Santander area.

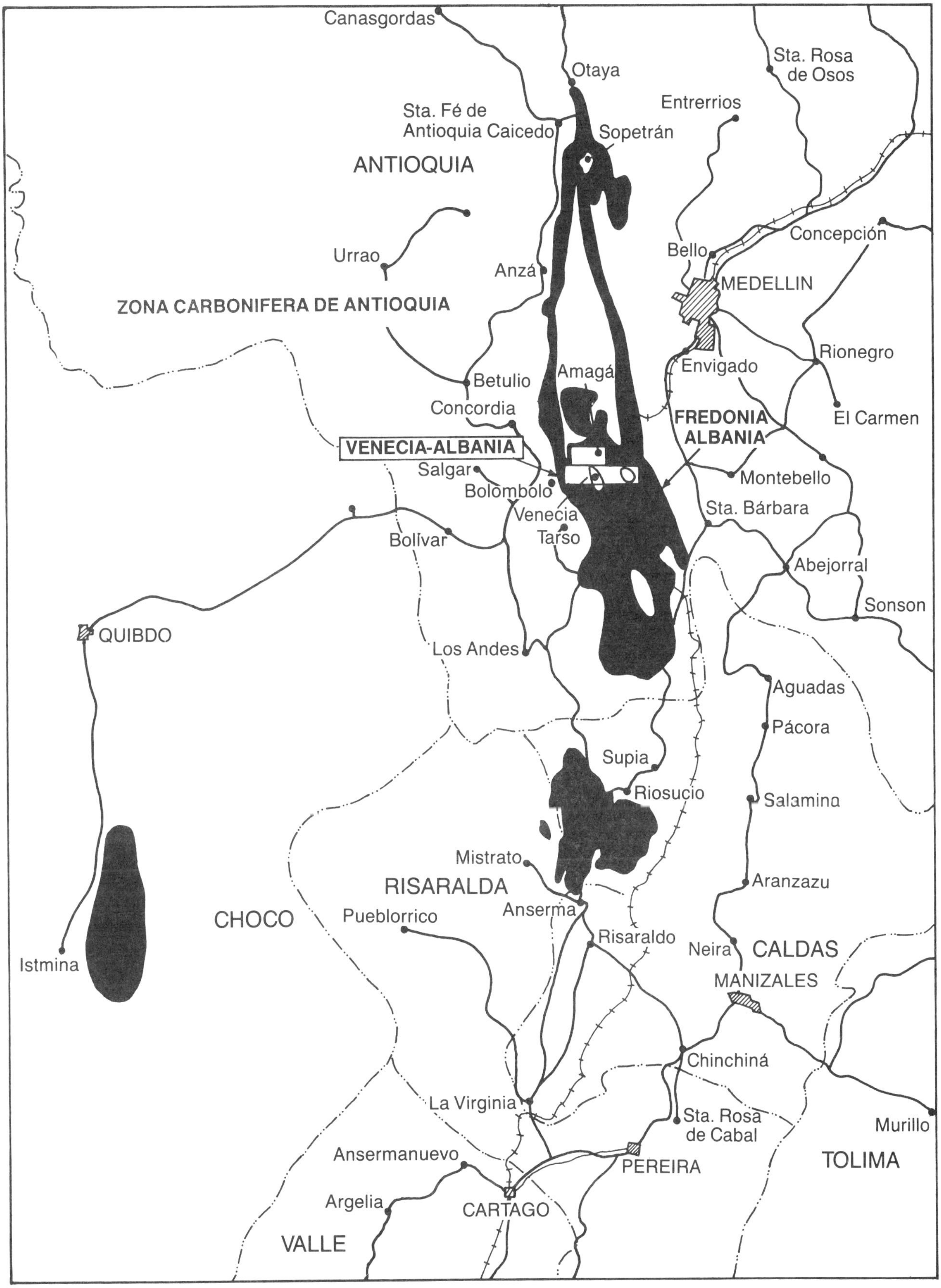

Figure 7. Distribution of coal in the Antioquia area.

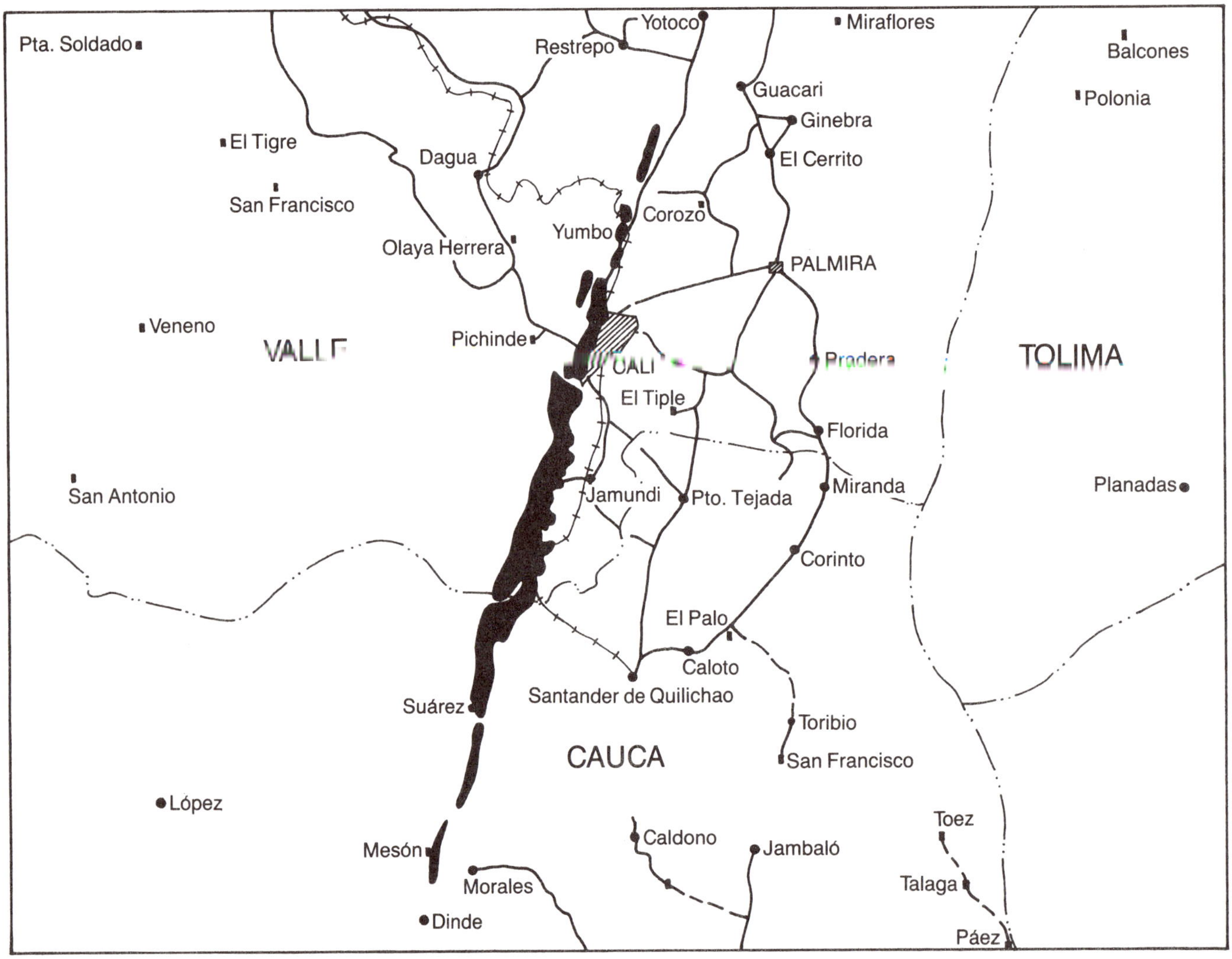

Figure 8. Distribution of coal in the Valle de Cauca area.

member of the upper Oligocene to lower Miocene Amaga Formation; the upper and lower members contain a few coal beds of little economic importance. The middle member, which consists of alternating sandstones, siltstones, and claystones, extends over an area of 30 km^2. The coal beds range in thickness from 0.60 to 3.70 m. Locally, the coals are affected by andesitic intrusions, some in the form of dikes that locally transformed the bituminous coal to anthracite.

Coals in the Antioquia area are exploited for use in the city of Medellin and the Valle del Cauca. The Anitioquia area has 21 million t of measured coal reserves and an additional 120 million t of indicated reserves. Analyses of these coals are as follows (in percent, except as noted):

Ash	0.8	to	15.2
Volatile material	1.5	to	25.0
Fixed carbon	18.0	to	74.6
Sulfur	0.32	to	3.5
Calorific value (cal/g)	4,031	to	7,934

Valle del Cauca area

The Valle del Cauca coal-bearing area is on the eastern flank of the Cordillera Occidental. It ranges in width from 1 to 5 km and extends for 75 km, from north of the city of Cali southward to the south of Suarez, Department of Cauca (Figure 8). The coals are in the Cauca Group of Paleocene to late Oligocene age, which comprises the Chimborazo, Guachiente, and Ferreira formations. The Chimborazo Formation, also called Los Confites Member, overlies rocks of the Diabasic Group. It is predominantly conglomeratic and has a thickness of 575 m. The Guachiente Formation, which is 400 m thick, consists of sandstone, claystone, coal beds, and conglomerate, and includes the Leona fossiliferous horizon. The Ferreira Formation,which is 500 m thick, consists of conglomerate, claystone, coal beds, and sandstone, and includes the San Francisco fossiliferous horizon. The rocks of the Cauca Group are cut by quartz-diorite to dacite-porphyry intrusions, which locally caused formation of anthracite. The coal-bearing rocks are exposed in a broad synclinorium whose western flank is exposed west of the Cauca River. The central part of this syn-

Table 1. Coal Consumption in Colombia
1980-1984
(thousands of tonnes)

Year	Coking Coal	%	Thermal Coal	%	Total	%
1980	651	16	3395	84	4046	100
1981	569	14	3605	86	4174	100
1982	444	10	3853	90	4297	100
1983	472	10	4041	90	5412	100
1984	497	11	4095	89	4591	100

Table 2. Consumption of Coal, by Uses
1980-1984
(thousands of tonnes)

Year	Thermal Power*	Industrial	Residential	Transportation	Total
1980	1103	2002	283	7	3395
1981	1263	2050	286	6	3605
1982	1295	2264	288	6	3853
1983	1442	2302	291	6	4041
1984	1418	2378	293	6	4095

*Includes industrial self-generation.

clinorium is covered by the Popayan Formation (Pliocene-Pleistocene) and alluvium of the Cauca River. The many subsidiary synclines and anticlines in the synclinorium strike north, and are broken by normal and reverse faults. Analyses of coals from this area are as follows (in percent, except as noted):

Ash	1.4	to	2.6
Volatile material	2.9	to	22.0
Fixed carbon	13.5	to	44.6
Sulfur	0.33	to	6.32
Calorific value (cal/g)	6,050	to	7,610

From 4 to 19 coal beds, ranging in thickness from 0.40 to 2.50 m, have been recognized. Measured reserves are approximately 19 million t; indicated reserves 12 million t, and inferred reserves 110 million t.

Borde Llanero and Amazon Regions

The information available regarding the Borde Llanero and Amazon regions is incomplete. In the Leticia (Amazon) area, in the southernmost part of Colombia, at least two coal beds ranging in thickness from 0.80 to 1.20 m have been identified. The coals are high-volatile subbituminous rank. Because of the potential importance as an energy source in this remote area, detailed exploration of these coals is planned for the near future.

MINING

Seventy percent of coal mining activities in Colombia are hand-mining operations in shallow underground mines. The remaining 30% is by intermediate to semimechanized mining methods. The underground mines are operated with room and pillar systems. Long-wall mining is carried out at the Chapa de Paz del Rio mine, and inverse stope mining at the Colombian West (Valley).

Open-pit mining began in 1987 at El Cerrejon as an up-to-date mechanized operation. Beginning in 1990 this mining operation is expected to produce as much as 15 million t per year, and production is expected to continue at this level for at least 23 years.

PRODUCTION

From 1980-1984 the average annual rate of production increased 13.7%. The large Cerrejon projects (Central and North zones) and the Caribbean coal project have shown increasing production since 1982. In 1984, the total production from these projects was 1,809,000 t, of which approximately 500,000 t was exported. In1984, the total coal production of Colombia reached 6.9 million t, of which 6.4 million t (93%) was thermal coals and 0.5 million coking coals. Most of this production came from the Atlantic region and the Cundinamarca-Boyaca area, which together produced 74% of the total; Antioquia produced 11% and Valle del Cauca 9%. The remaining production came mainly from Norte de Santander, with lesser amounts from other areas.

CONSUMPTION

Annual consumption of coal (Table 1) increased 545,000 t (12%) from 1980 to 1984. The annual consumption of coking coals decreased by more than 150,000 t (34%) between 1980 and 1984; annual consumption of thermal coal increased 700,000 t (17%). Eighty-five percent of the coking coal consumption is for high-oven coking, and the remainder is used for other industrial processes. The high-oven consumption decreased at an annual rate of 7.4%, equivalent to a decrease of 40,000 t per year. Most thermal coal was used by the industrial sector; in 1984 this amounted to 2.4 million t, or 58% of the total. The remainder of the thermal coal was used in making cement (37%) and brick (21%); additional small amounts were used in the manufacture of textiles, paper, food, and other products.

Consumption of thermal coal (Table 2) by thermal power plants was 1.4 million t in 1984, or 35% of the total national consumption. From 1980 to 1984, thermal coal consumption showed an average annual increase of 6.5%. The residential sector and transportation consumption in this period was about 7.3% of annual consumption of thermal coal (the only transportation use was by the Acerias Paz de Rio railroad).

CONCLUSIONS

Colombia is capable of exporting approximately 50 million t of thermal coal and 3 million t of coking coal per year. The fields in the Atlantic region, of which the Cerrejon field is most important, have started production for export. The El Cerrejon coal has been exported to countries in Europe, and to Israel, several Central American countries, and the United States. Studies of areas having coking coals, now under way, should define the resources of coking coal available for export, chiefly to countries such as Brazil, Argentina, and Mexico.

REFERENCES

Angarita, L., and C. Sanchez, 1981, Ensayo de los carbones de Checua-Lenguazaque: Medellín, Ponencia ante el III Congreso Colombiano de Geología: Instituto Nacional de Investigaciones Geológico-Mineras, unpublished report.

Arboleda, C., 1985, Exploracíon y evaluación de los carbones de Boyacá: Instituto Nacional de Investigaciones Geológico-Mineras, unpublished report.

Beltran, E., 1974, Carbones Colombianos: Bogota, Ministerio de Minas y Petróleo, unpublished report.

Blanco, G., and A. Granados, 1975, Reconocimiento de los carbones de la cuenca Lebrija-San Vicente de Chucuri (Santander): Instituto Nacional de Investigaciones Geológico-Mineras, unpublished report.

Blanco, G., et al., 1977, Clasificación internacional de carbones del sinclinal Checua-Lenguazaque: Instituto Nacional de Investigaciones Geológico-Mineras, unpublished report.

Blanco, G., 1981, Estudio petrográfico de los carbones Checua-Lenguazaque: Medellín, Ponencia ante el III Congreso Colombiano de Geologia: Instituto Nacional de Investigaciones Geológico-Mineras, unpublished report.

Camacho, R., and G. Guarin, 1981, Exploracíon y geología, Bloque B de El Cerrejón: Medellín, III Congreso Colombiano de Geologia: Instituto Nacional de Investigaciones Geológico-Mineras, unpublished report.

Duran, R., et al., 1976, Geología y recursos carboníferos del área de San Luis, Santander: Instituto Nacional de Investigaciones Geológico-Mineras, unpublished report.

Duran, R., et al., 1978a, Terminología propuesta para la descripción de los recursos reservas y clasificación de los carbones Colombianos: Instituto Nacional de Investigaciones Geológico-Mineras, unpublished report.

Duran, R., et al., 1978b, Carbón (separata) recursos minerales de Colombia: Bogotá, Instituto Nacional de Investigaciones Geológico-Mineras, Publicación Geológico Especial 1, p. 95-129.

Duran, R., et al., 1979a, Zonas carboniferas de Colombia: Bogota, Instituto Nacional de Investigaciones Geológico-Mineras, Publicación Geológico Especial 3, 29 p.

Duran, R., et al., 1979b, Evaluación de reservas de carbón en siete zonas de Colombia: Bogotá, Instituto Nacional de Investigaciones Geológico-Mienras, Publicación Geológico Especial 6, 134 p.

Economic Associated Ltd., 1974, Explotación de carbón coquizable de la cuenca carbonífera de Checua-Lenguazaque: Instituto Nacional de Investigaciones Geológico-Mineras, Biblioteca 553, p. 2-154.

Empresa Nacional Adaro (España-Ingeominas), 1975, Examen y perspectiva de los yacimientos Colombianos de carbón: Instituto Nacional de Investigaciones Geológico-Mineras, Biblioteca 553, p. 1-54.

Franco, F., 1977, Cuencas carboniferas de Colombia: Instituto Nacional de Investigaciones Geológico-Mineras, Compilación actualizada, unpublished report.

Franco, F., 1981, Centro de minas de carbón de la sub-zona Tunja-Duitama: Instituto Nacional de Investigaciones Geológico-Mineras, unpublished report.

Gaitan, E., 1978, Los recursos energéticos de Colombia, una politica para el futuro: Bol. de Mineria y Energia, v. 2, p. 14-16.

Garces, H., 1974, Evaluacion de las cuencas carboníferas de Colombia: Instituto Nacional de Investigaciones Geológico-Mineras,unpublished report.

Japan International Cooperation Acency (JICA), 1971, Informe sobre los yacimientos de la Jagua, Caucasia y Urabá: Instituto Nacional de Investigaciones Geológico-Mineras, Bib. L-229.

Mateus, E., and E. Ospina, 1966, Informe de la cuenca Hullera Tunjua-Paipa-Duitama: Instituto Nacional de Investigaciones Geológico-Mineras, unpublished report.

Mejia, J., 1981, Análisis petrográfico de muestras de carbones Colombianos: Ponencia III Congreso Colombiano de Geología, unpublished report.

Mejia, J., and L. Mateus, 1977, Recursos carboníferos de La Jagua de Ibirico: Instituto Nacional de Investigaciones Geológico-Mineras, unpublished report.

Mejia, J., and E. Ospina, 1977, Reseña geológica y minera de los carbones del Valle del Cauca: Instituto Nacional de Investigaciones Geológico-Mineras, unpublished report.

Mejia, J., et al., 1980, Areas carboníferas del Departamento de Norte de Santander: Instituto Nacional de Investigaciones Geológico-Mineras, unpublished report.

Mojica, P., et al., Informe sobre la prospección geológica para carbon en el área de Subachoque, Departamento de Cundinamarca: Instituto Nacional de Investigaciones Geológico-Mineras, unpublished report.

Ministerio de Minas y Energía, 1977, Proyectos de generación eléctrica 1977-1966: Bol. Min. Ener. 1(2):27 p.

Pulido, O., et al., 1980, Geología y recursos carboníferos del area Amagá-Venecia-La Albania (Antioquia): Instituto Nacional de Investigaciones Geológico-Mineras, unpublished report.

Ramirez, I., 1977, Interpretación y evaluación preliminar de la cuenca carboníferas de Checua-Lenguazaque: Instituto Nacional de Investigaciones Geológico-Mineras, unpublished report.

Rodriguez, M., 1974, Ensayos de Coquización de los carbones vallecaucanos: Instituto Nacional de Investigaciones Geológico-Mineras, unpublished report.

Sanchez, C., 1981, Panorama de los carbones Colombianos: Instituto Nacional de Investigaciones Geológico-Mineras, unpublished report.

Ericksen, G. E., Cañas Pinochet, M. T., and Reinemund, J. A., editors, 1989, Geology of the Andes and its relation to hydrocarbon and mineral resources: Houston, Texas, Circum-Pacific Council for Energy and Mineral Resources Earth Science Series, v. 11.

CHAPTER 10

GEOLOGY AND RESOURCES OF SALARS IN THE CENTRAL ANDES

George E. Ericksen
U.S. Geological Survey
Reston, Virginia, U.S.A.

Raul Salas O.
Servicio Nacional de Geología y Minería
Santiago, Chile

ABSTRACT

Salars of the central Andes constitute one of the world's major continental evaporite complexes that contain significant portions of the known resources of lithium and boron. At least 100 salars and saline lakes ranging in area from a few square kilometers to about 9000 km^2 are found in individual closed basins within a 4000-km^2 area of internal drainage in northwestern Argentina, western Bolivia, northern Chile, and southern Perú. An arid climate persisted in this region during most of the late Tertiary and Quaternary. Compared with glaciation in the Andes Mountains to the north and south, Pleistocene glaciation in this region was scanty. Major Andean uplift took place during the late Tertiary and Quaternary, which was a time of intense volcanism. The basins in which the salars and saline lakes are found were formed both by faulting and by volcanic eruptions. The volcanic rocks are the major sources of the saline materials in the salars. Most of the salars formed by desiccation of saline lakes during the Holocene and by subsequent deposition of saline materials from evaporation of spring water, ground water, and annual or intermittent flood water. Some closed basins have had complex histories dating back to the late Tertiary or even to the Late Cretaceous.

The salars consist of hard to soft saline crusts that overlie sequences of lacustrine sediments. The saline crusts, generally not more than a few meters thick, contain a variety of saline minerals, of which halite and gypsum are most abundant. Mirabilite, thenardite, and ulexite are found in moderate to minor amounts in nearly all salars. Several other saline minerals are present in minor amounts in many salars, and some of these are moderately abundant in a few salars. Five general types of salars can be recognized: (1) zoned salars, consisting of an interior halite zone surrounded by a sulfate zone; (2) gypseous salars, consisting chiefly of gypsum and lesser amounts of other sulfate minerals, halite, and ulexite; (3) salars having crusts of silty halite that formed chiefly by capillary evaporation of near-surface ground water or brine in lacustrine sediments; (4) salars consisting chiefly of mud flats subject to annual flooding and having thin ephemeral crusts of halite or sulfate minerals during dry seasons; and (5) Salar Grande, in the coastal region of northern Chile, which is a 40-km–long valley filled with coarsely crystalline halite to a depth of as much as 160 m.

Although the salars probably have been exploited for rock salt since pre-Columbian time, systematic exploitation did not start until the 19th century. At that time, several salt quarries were opened in Salar Grande, Chile, which, prior to the 1880s, was in Peruvian territory. Ulexite was mined in several salars before the end of the 19th century. Small amounts of $NaNO_3$ were recovered from salars in the coastal region of northern Chile during the late 19th century and early 20th century. In the early 1940s, several thousand tonnes of KCl and KNO_3 were produced from the K-rich saline crust of Salar de Bellavista, in the coastal region of northern Chile. Since the 1950s, the salars have been exploited as sources of rock salt, borate minerals, and Na_2SO_4. Production of Li_2O_3, using Li-rich brines from Salar de Atacama, Chile, began in 1984.

INTRODUCTION

The central Andean region of northwestern Argentina, western Bolivia, northern Chile, and southern Perú is an area characterized by an arid climate and predominantly internal drainage, where many salars (salt-encrusted playas) and saline lakes now exist (Figure 1, Table 1). The region includes the Andean Highlands, as well as the western Andean front ranges and coastal lowlands of northern Chile. This region of internal drainage, which extends over an area of about 400,000 km^2, is subdivided into about 150 closed basins. At least 100 of the basins contain salars or saline lakes having areas ranging from a few square kilometers to 9000 km^2. An arid climate has persisted in this region throughout late Tertiary and Quaternary time, and the coastal region of northern Chile may have had an arid climate since early Tertiary (Mor-

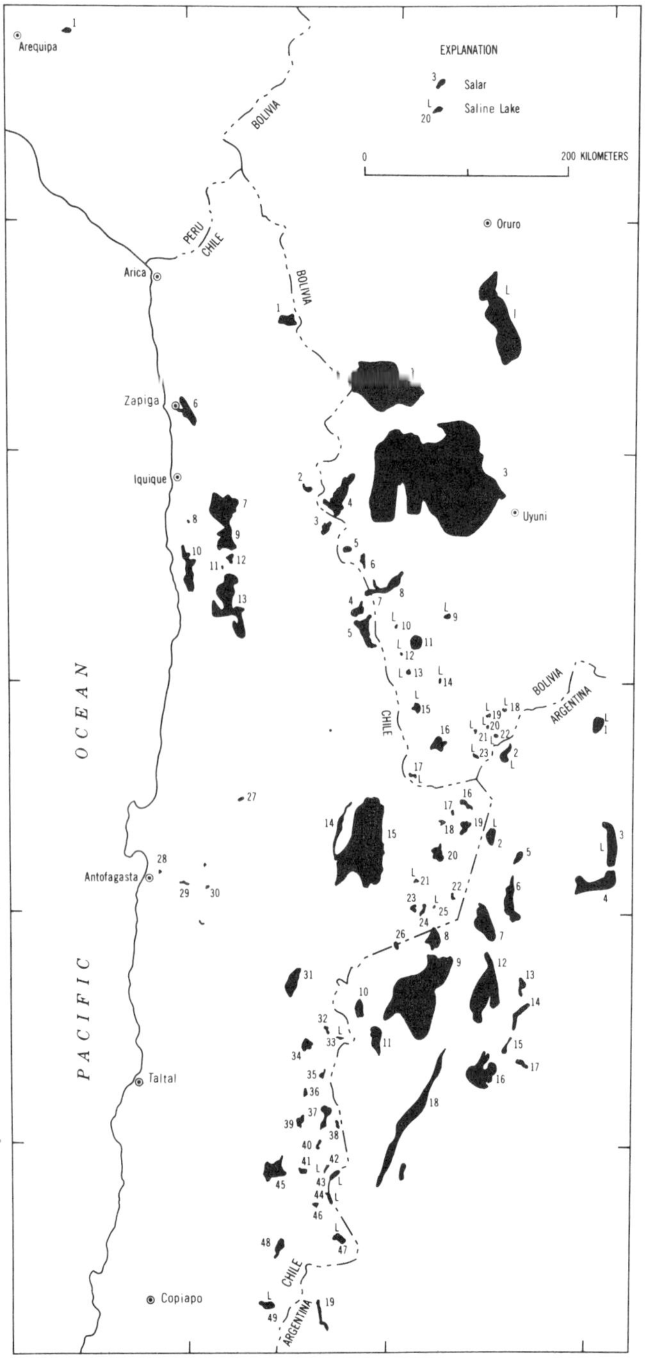

Figure 1. Index map of the central Andean region showing locations of principal salars and saline lakes. Numbers refer to Table 1.

Table 1. Principal salars and saline lakes in the central Andean region (locations shown on Figure 1).

ARGENTINA			
1.	Laguna Pozuelos	10.	Salar de Lullaillaco
2.	Laguna Vilama	11.	Salar de Rio Grande
3.	Laguna Guayatáyoc	12.	Salar Pocitos
4.	Salinas Grandes	13.	Salar de Pastos Grandes
5.	Salar Olaroz	14.	Salar Centenario
6.	Salar de Cauchari	15.	Salar Ratones
7.	Salar del Rincón	16.	Salar del Hombre Muerto
8.	Salar de Incahausi	17.	Salar Diabillos
9.	Salar de Arizaro	18.	Salar de Antofalla
BOLIVIA			
1.	Lago Poopo	12.	Laguna Cachi
2.	Salar de Coipasa	13.	Laguna Khara
3.	Salar de Uyuni	14.	Laguna Capína
4.	Salar de Empexa	15.	Laguna Colorada
5.	Salar de La Laguna	16.	Salar de Challviri
6.	Salar Laguani	17.	Laguna Verde
7.	Salar de Ollagüe	18.	Laguna Corante
8.	Salar de Chiguana	19.	Laguna Mama Khumu
9.	Laguna Tarija	20.	Laguna Chojllas
10.	Laguna Cañapa	21.	Laguna Loromayu
11.	Lagunas Pastos Grandes	22.	Laguna Coruto
		23.	Laguna Kalina
CHILE			
1.	Salar de Surire	25.	Laguna Tuyacto
2.	Salar de Huasco	26.	Salar de Pular
3.	Salar de Coposa	27.	Salar de Pampa Blanca
4.	Salar de San Martín	28.	Salar del Carmen
5.	Salar de Ascotán	29.	Salar de Navidad
6.	Salar de Obispo	30.	Salar Mar Muerto
7.	Salar de Pintados	31.	Salar de Punta Negra
8.	Salar de Soronal	32.	Salar de Aguas Calientes III
9.	Salar de Bellavista	33.	Laguna Azufrera
10.	Salar Grande	34.	Salar de Pajonales
11.	Salar de Lagunas	35.	Salar de Gorbea
12.	Salar de Sur Viejo	36.	Salar de Agua Amarga
13.	Salar de Llamara	37.	Salar de La Isla
14.	Llano de Paciencia	38.	Salar de Las Pariãs
15.	Salar de Atacama	39.	Salar de Aguilar
16.	Salar de Tara	40.	Salar Grande (II)
17.	Salar de Aguas Calientes I	41.	Salar de Piedra Parada
18.	Salar de Pujsa	42.	Lagunas Bravas
19.	Salar de Quisquiro	43.	Laguna Wheelright
20.	Salar de Aguas Calientes II	44.	Laguna Escondida
21.	Laguna Miscanti	45.	Salar de Pedernales
22.	Salar de Laco	46.	Salar Wheelright
23.	Salar de Talar	47.	Laguna Verde
24.	Salar de Purisunchi	48.	Salar de Maricunga
		49.	Laguna Negro Francisco
PERU			
1.	Laguna Salinas		

timer, 1973; Mortimer and Saric, 1972, 1975). At present, the Andean Highlands have a climate transitional between that of the Peruvian Andes, which have a summer (December-March) season of rain, and the southern Andes of Chile and Argentina, which have a winter (June-September) season of rain and snow. The central Andean region is an area of sparse Pleistocene glaciation in comparison with the Andean Highlands to the north and south. At altitudes above 3000 m, present-day annual rainfall is about 100-200 mm, but in places it ranges from as little as 50 mm to slightly ore than 300 mm. Annual rainfall is greatest in the Altiplano of southern Perú and northern Bolivia, where it exceeds 300 mm, and decreases southward to less than 100 mm in southwestern Bolivia and nearby Argentina and Chile (Montes de Oca, 1982). Rainfall decreases with altitude along the western Andean front to a minimum of only a few millimeters per year

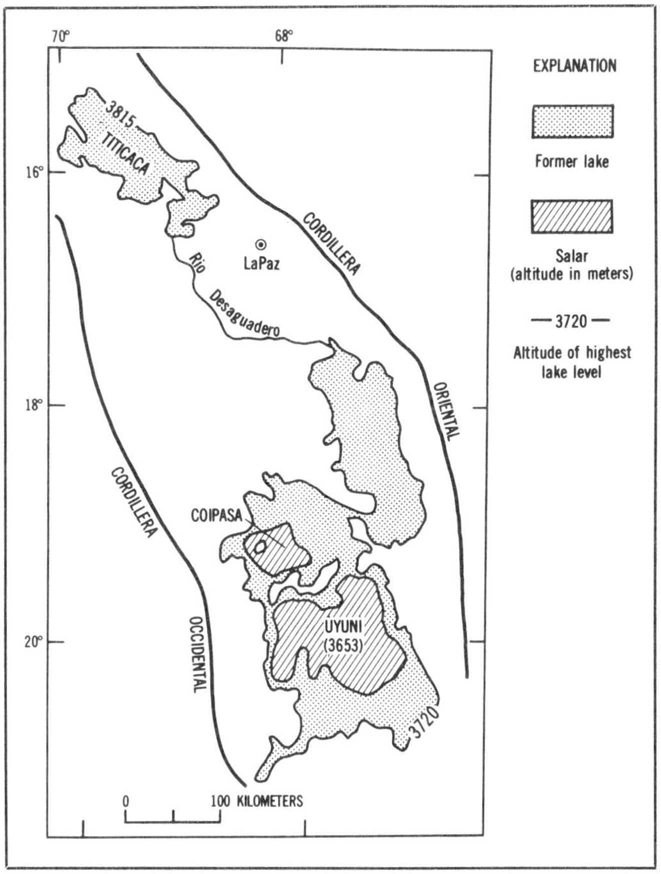

Figure 2. Maximum extensions of former large lakes in the Bolivian Altiplano, 12,360 ± 120 to 10,640 ± 280 years B.P., showing the salars Uyuni and Coipasa. These salars are the evaporite remnants of Lago Tauca, the large lake in the southern part of the Altiplano. Modified from Servant and Fontes, 1978. Present-day Lago Titicaca is also a remnant of the larger lake shown here.

in the Atacama Desert of the coastal region of northern Chile.

Many lakes formed in the central Andes during the Quaternary in response to periods of slightly increased rainfall or of slightly increased temperatures that caused melting of glacial ice and snow. As a result, many of the basins that now contain salars have multiple shorelines and other features of former lakes (Stoertz and Ericksen, 1974). The largest lakes were in the Bolivian Altiplano during the late Pleistocene (Figure 2), when they covered a maximum area of about 50,000 km^2 (Servant and Fontes, 1978). Lake Titicaca, on the Perú-Bolivian border, is a remnant of a larger lake shown in Figure 2, and Lago Poopo and the salars Coipasa, Uyuni, and Empexa are remnants of another, which Servant and Fontes named Lago Tauca.

Systematic study of Andean salars began in the early 1960s. Many reports about them have been published since that time. General features of salars in northern Chile were described by Ericksen (1963), Stoertz and Ericksen (1974), Vila (1974, 1975), and Chong (1984). Moraga et al. (1974) made a detailed study of Salar de Atacama, the largest salar in northern Chile, to evaluate the lithium resources. Vila (1976) described salars in the coastal region of northern Chile. Ericksen, Chong, and Vila (1976) discussed the lithium resources of Andean salars, and Ericksen, Vine, and Ballon (1977), Ballivian and Risacher (1981), and Davis et al. (1982) described salars in southwestern Bolivia. Risacher (1978) and Retting, Jones, and Risacher (1980) discussed the geochemistry of salars in southwestern Bolivia. Muessig (1966) described borate deposits in northwestern Argentina. Nicolli et al. (1982a, 1982b) described the salars in northwestern Argentina, including a discussion of the results of systematic exploration for lithium in Salar de Hombre Muerto. Catalano (1964a, 1964b) discussed Salar de Hombre Muerto and thermal-spring deposits of ulexite in northwestern Argentina. Alonso and Viramonte (1985) also discussed thermal-spring borate deposits in northwestern Argentina. Earlier reports about Andean salars are by Reichert (1907), Bárnabe (1915), and Catalano (1926, 1927, 1930) on borate deposits in northwestern Argentina; Leiding (1942) on borate deposits in northern Chile; and Jochamowitz (1917) on borate deposits in southern Perú. Brüggen (1918) and Gale (1918) discussed potash resources of Salar de Pintados, Chile.

This chapter is based on our investigations of Andean salars that began in 1961 and on published and unpublished reports of other investigators. Work in Chile was carried out in cooperation with the Instituto de investigaciones Geológicas, now the Servicio Nacional de Geología y Minería; work in Bolivia was with the Servicio Geológico de Bolivia; and work in Argentina was with the Servicio Minero Nacional and the Dirección General de Fabricaciones Militares.

PHYSIOGRAPHIC AND GEOLOGIC SETTING

The many closed basins within the area of internal drainage in the central Andes owe their existence to intensive block faulting and volcanism during the late Tertiary and Quaternary. The three major longitudinal physiographic units—the Coastal Range, and Central Valley, of northern Chile, and the Andean Highlands to the east—developed during differential uplift that began in the early Miocene (Mortimer and Saric, 1972, 1975). The Central Valley became largely filled with debris from the uplifted Andes by mid-Miocene time. The Coastal Range is a Miocene or even earlier Tertiary landscape (Mortimer and Saric, 1972) that has since been modified, chiefly by block faulting and by wind erosion. The Andean Highlands of this region consist of a broad longitudinal depression, called the Altiplano, which centers in western Bolivia but which extends from southern Perú to northern Argentina. This depression is filled with continental clastic sediments ranging in age from Cretaceous to Quaternary. It is flanked on the east and west by uplands having discontinuous, longitudinal mountain ranges consisting predominantly of lower Paleozoic sedimentary rocks, Mesozoic and Tertiary plutonic rocks, and Mesozoic sedimentary and volcanic rocks. To the west of the Altiplano, the Andean Highlands have an extensive cover of late Tertiary ash-flow tuffs, clastic continental sediments, and hundreds of stratovolcanoes.

Base altitude in the Andean Highlands is 3500-4000 m, and many mountain peaks and volcanic cones extend to altitudes of 5000-6000 m. The highest peaks are volcanic cones of about 6500-6800 m. The Altiplano slopes gently southward from southern Perú (Lake Titicaca near the north end of the

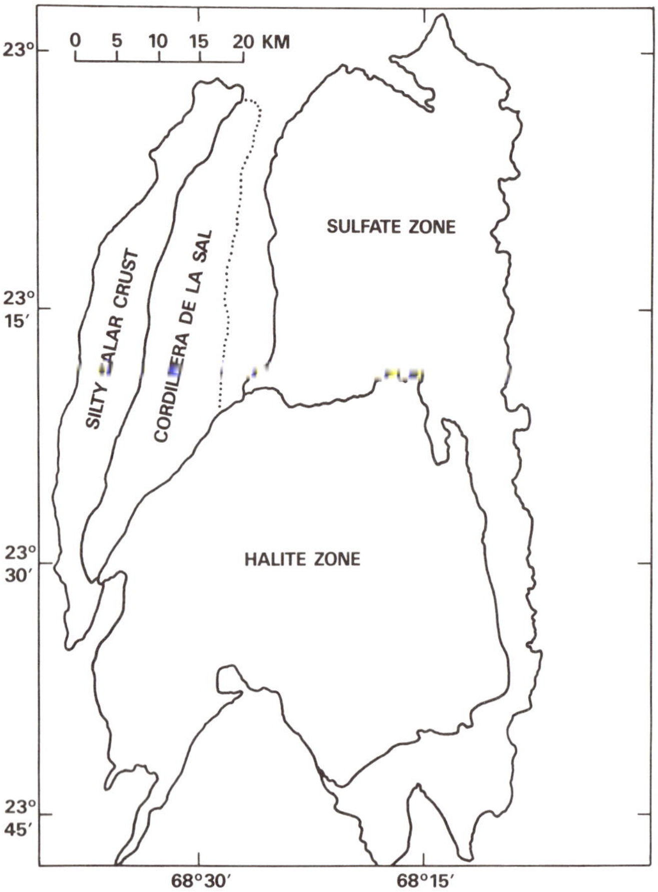

Figure 3. Sketch map of Salar de Atacama, Chile, showing asymmetrical zoning in which the halite crust is crowded against the southwestern margin of the salar. The Cordillera de la Sal consists of an intensely folded sequence of continental saline clastic sediments of Tertiary age; several domal structures have cores of coarsely crystalline halite. Modified from Morago et al. (1974).

Altiplano is at an altitude of 3815 m) to Salar de Uyuni, which, at an altitude of 3653 m, is the lowest part. The Altiplano again rises southward into Argentina, where it is characterized by many individual closed basins, rather than by the broad basins and open valleys of the Bolivian Altiplano to the north.

The late Tertiary and Quaternary volcanic rocks are of particular importance, because they have been the major source of the saline constituents of the lakes and salars of the region. Two distinctly different types of volcanism can be recognized. The earliest volcanic phase, during the Miocene and early Pliocene, was dominated by explosive, caldera-type eruptions of dacitic to rhyolitic ash-flow tuffs that now blanket extensive areas of the Andean Highlands. Superimposed on the ash-flow tuff terrane are hundreds of andesitic stratovolcanoes of Pliocene and Quaternary age. Several of the volcanoes now show fumarolic activity, but only a few have had historic eruptions. The ash-flow tuffs contain considerable amounts of water-soluble saline material that has been leached and carried into the lakes and salars by meteoric waters.

In contrast with the Andean Highlands, the geology of the coastal region of northern Chile is dominated by late Tertiary and Quaternary block faulting of pre-Tertiary rocks. In the Coastal Range, salars and former saline lakes formed in basins isolated by late Tertiary and Quaternary faults, whereas those in the Central Valley formed in depressions resulting from differential uplift as well as from faulting and in basins isolated by alluvial fans extending outward from the Andean front. The Central Valley is a structural depression filled largely with continental sediments, of Miocene to Holocene age, to a depth of as much as 1000 m (Mordójovich, 1965). In northernmost Chile, this infilled depression was dissected by deep transverse valleys that formed after mid-Miocene, when the fill overtopped the Coastal Range in this area (Mortimer and Saric, 1972). From Zapiga southward to near Taltal (Figure 1), only one stream, the Rio Loa, now crosses the Central Valley and cuts through the Coastal Range.

Tectonic movements associated with Andean uplift have affected the closed basins and salars and saline lakes in several ways. Basins have been tilted, causing migration of lakes and salars toward the depressed sides of the basins (Stoertz and Ericksen, 1974). Such salars tend to show an asymmetrical zoning in which the halite crust is crowded against the depressed side and the sulfate zone toward the upraised side (Figure 3). This asymmetrical zoning contrasts with symmetrically zoned crusts of salars in basins that have not been tilted, which consequently consist of a central halite zone surrounded by a sulfate zone. Shorelines of former saline lakes and clastic sediments deposited in such lakes in titled basins may be gently inclined toward the depressed basin sides. Some salars abut active faults, whereas others are in sag basins on active faults. Still others have been displaced by faults. The best example of recent fault displacement is at Salar Grande, in the coastal range of northern Chile. This salar is crossed by the Atacama fault (Figure 4), a longitudinal fault that shows recent displacement throughout its strike length of nearly 700 km. Lacustrine sediments of former lakes in northwestern Argentina have been folded and faulted, and the Tincalayu borate deposit is thought to be a folded remnant of a buried salar crust of late Quaternary or Holocene age (Pratt, 1961).

TYPES OF SALARS AND STRUCTURAL FEATURES OF SALAR CRUSTS

Different types of salars may be distinguished on the basis of the relative abundances of the various saline minerals present and on the physical characteristics of the salar surface. As noted in the previous section, many salars are zoned, having a hard, dense to porous, relatively thick interior halite zone surrounded by a sulfate zone consisting chiefly of gypsum, but having various amounts of admixed halite, mirabilite, thenardite, and ulexite. Typical zoned salars are Atacama (Figure 3) and Pedernales in Chile, Empexa in Bolivia, and Hombre Muerto in Argentina (Figure 1). Salar de Uyuni, Bolivia, is an unusual zoned salar, because most of its 9000-km^2 expanse is covered by a hard halite crust, and because the sulfate zone is confined to narrow muddy or marshy marginal areas. The upper part of the lacustrine sediments beneath the halite crust of this salar contain gypsum, as probably do the sediments beneath the halite crusts of other zoned salars. Many salar crusts consist chiefly of moist, relatively soft, granular gypsum or crystal meshworks of gypsum, admixed halite and mirabilite, and local layers and nodules of ulexite. Such salars are subject to widespread flooding during the rainy season and have perennial ponds within the salar, as well as along margins (Figure 5). Typical examples of salars of this type are Ascotan and Surire, Chile, and Lagunas Pastos Grandes, Bolivia (Figure 5). Some salars or parts of salars consist chiefly of a surface of moist lacustrine sediments that is subject to seasonal flooding; thin ephemeral crusts or scattered crystals of saline minerals form on such surfaces only during the dry season. The near-surface lacustrine sediments may contain layers of gypseous muds and impure to relatively pure ulexite. Lagunas Salinas in southern Perú is a typical salar of this type.

Salars in the coastal region of northern Chile, and some salars in the Andean Highlands, have hard, silty, halite-rich crusts that were formed chiefly by capillary evaporation of near-surface saline ground water in lacustrine sediments. These salar crusts, which generally are not more than a meter thick, are built up by saline material deposited by capillary evaporation of ground-water brine at the base of the otherwise dry crust. Typical salars are Bellavista and Obispo, Chile, and Arizaro, Argentina. Salar del Carmen and Salar de Pampa Blanca in the coastal region of northern Chile have silty crusts that contain considerable amounts of nitratite.

Salar Grande, also in the coastal region of northern Chile, is unique in that it consists of a longitudinal basin, about 40 km long and 5 to 8 km wide, filled with dry, coarsely crystalline, high-purity rock salt to a depth of as much as 160 m. Although its origin is somewhat uncertain, we believe that the halite was deposited from residual chloride-rich brines draining from a huge Pliocene-Pleistocene saline lake in the Central Valley to the east. This lake, of which Salar de Llamara (Figure 1, Table 1) is a desiccation remnant, was named Lago Soledad by Brüggen (1950). Marginal areas of this salar have remnants of a thick (5 m or more), eroded anhydrite crust that evidently was deposited as the Lago Soledad brines became progressively enriched in NaCl. Lago Soledad either periodically overflowed a low divide between the two basins or flowed into the Salar Grande through fault channelways.

Salar crusts exhibit diverse structures that formed during primary deposition and desiccation and during subsequent modification by rainwater and springwater, fog, wind, regional tectonic deformation and local faulting, and diurnal temperature changes (Table 2). Most of these features have been described by Stoertz and Ericksen (1974) and are discussed only briefly here. Of the primary structures, zoning of crusts of Andean salars results chiefly from progressive concentration of brines in saline lakes and deposition of the saline minerals according to their solubilities. Such crusts, however, may be modified by subsequent deposition of saline minerals from annual flood waters or by capillary evaporation of ground water. The typical zoned salar consists of an interior halite crust underlain by gypsum-rich lacustrine sediments and surrounded by a zone of gypsum mixed with other sulfate minerals and halite, and, in places, with layers and nodules of

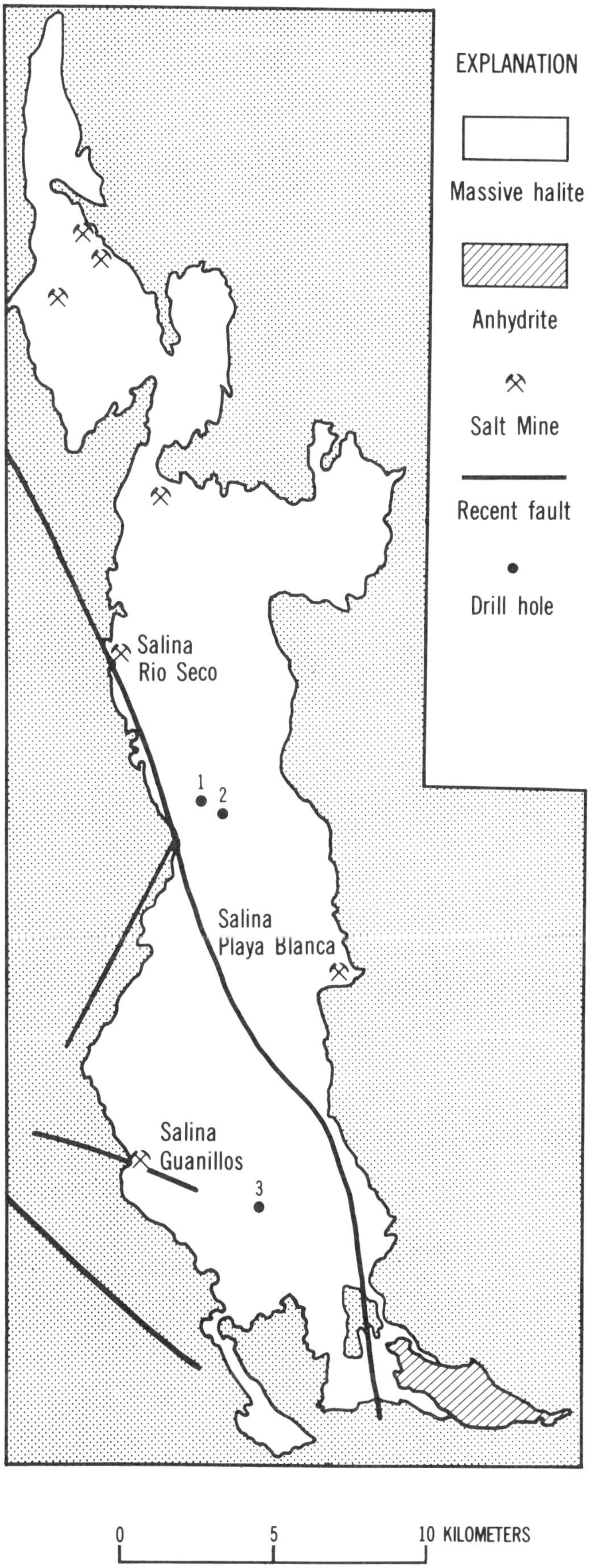

Figure 4. Sketch map of Salar Grande, a salt-filled basin near the coast in northern Chile, showing trace of the Atacama fault, which cuts diagonally across the salar. This fault and others shown have had recent movement. From Stoertz and Ericksen (1974).

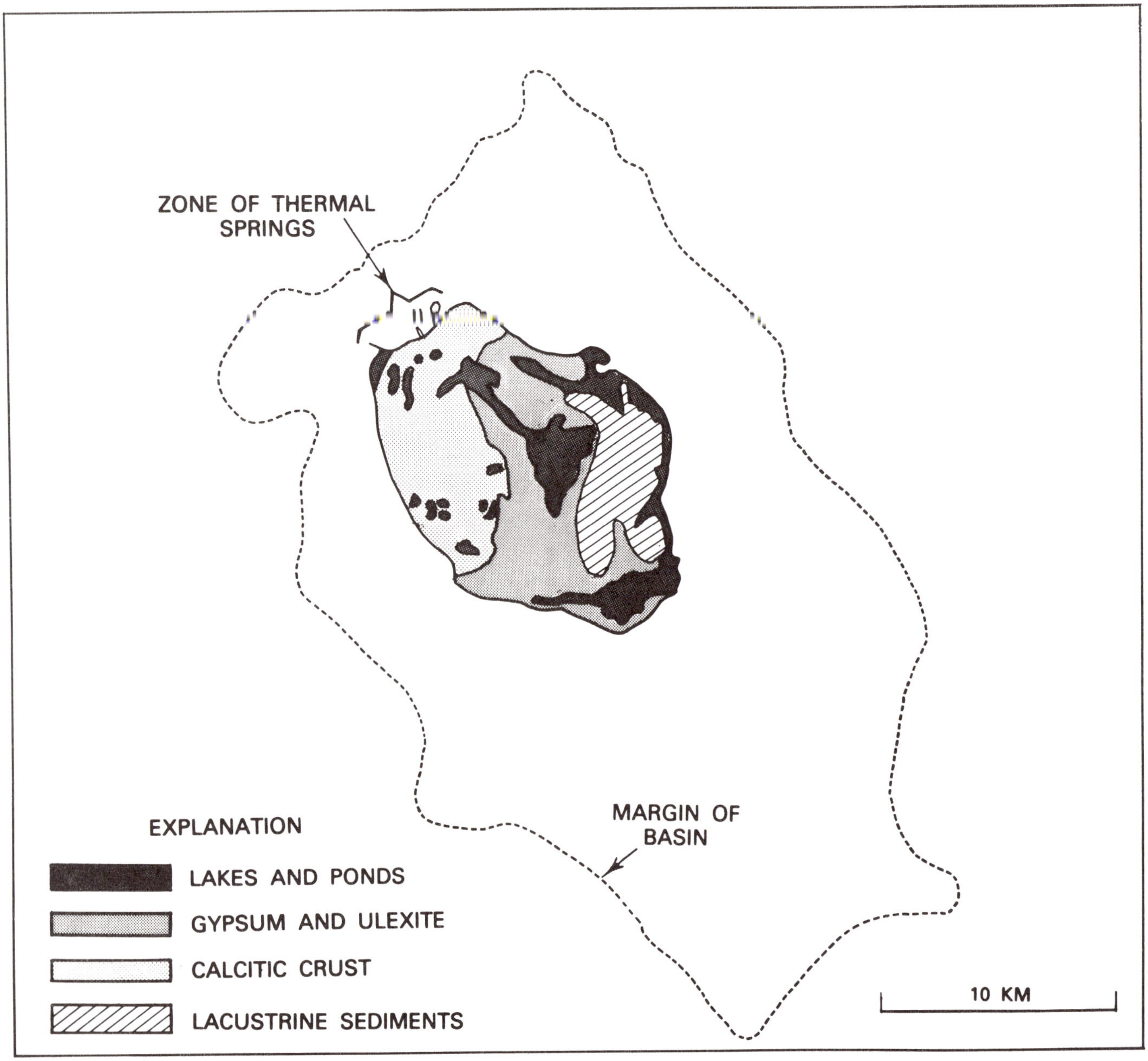

Figure 5. Lagunas Pastos Grandes, a salar in southwestern Bolivia, showing an internal gypsum-ulexite crust and abundant lakes and ponds. The calcitic crust in the western part of the salar was deposited by thermal-spring waters. After Ballivian and Risacher (1981).

ulexite. Depositional layering, best displayed by thick rock-salt crusts in the Andean Highlands, is the result of sporadic flooding and desiccation of surface brines. In Salar de Uyuni, distinct layers of brine-saturated salt, ranging from a centimeter or two to 10 to 20 cm thick, are separated by laminae of organic-rich mud. Such muds accumulated during annual floods. Desiccation polygons are characteristic of halite crusts and generally are about a meter in diameter. However, some thick halite crusts also have giant desiccation polygons, as much as 100 m in diameter, which evidently formed subsequently to the normal-size polygons. Salt blisters and pressure ridges formed by desiccation and expansion of thin (not more than a few centimeters thick), newly deposited halite crusts. Such crusts do not show desiccation polygons and evidently formed by rapid drying of shallow surface brines on muddy flats. In addition to these primary features, most salars in the Andean Highlands have marginal ponds and marshes that are fed by local springs and streams. Others consist chiefly of a perennial saline lake having marginal, perennial, or ephemeral saline crusts.

Massive rock-salt crusts have been extensively modified subsequent to their formation. Where not subject to annual flooding, they tend to have extremely rugged, blocky, pinnacled surfaces due to rainwater leaching. In contrast, rock-salt crusts subject to annual flooding tend to be smooth. Such flooding is local at many salars, and the transition between the

Table 2. Structural features of Andean salars (from Stoertz and Ericksen, 1974).

PRIMARY STRUCTURES	
Mineral zoning	Desiccation polygons
Depositional layering	Salt blisters and pressure ridges
SECONDARY STRUCTURES	
Surface-water flooding	Wind
Salt cusps and crenulate margins	Gypsum dunes
Smooth rock-salt crusts	Gypseous ramparts
Salt channels	Diurnal temperature changes
Springs	Chaotic blocks and pressure ridges in silty halite crusts
Salt channels	Salt-solution tubes
Brine pools	Salt cones
Rainwater and fog	Salt veins
Rugged rock-salt crusts	Salt sheet
Salt pinnacles	Tectonic deformation
Salt nodules	Asymmetric zoning
Salt saucers	Giant oriented polygons
Salt stalacities	Faults
	Fault sag basins

smooth zones and the blocky zones above flood level may be gradational, as, for example, around brine pools in Salar de Pedernales, or abrupt, with the unflooded crust marked by a cuspate or crenulate margin, as in the northeastern part of Salar de Atacama. Salar de Uyuni, which is flooded annually, consists of a smooth, flat, hard, rock-salt crust surrounded by a narrow zone of perennial marshes. The rock-salt crust of this salar, which extends over an area of nearly 9000 km^2, is probably the most extensive hard, flat surface on earth. The average relief is less than a centimeter, which is the average size of scattered halite crystals and upraised or depressed margins of desiccation polygons. This salar also has local mounds of dried salt froth, as much as 10 m in diameter and 20 cm high, that consist of wind-blown salty foam accumulated on the wet salar surface as the surface dried after annual flooding.

At some salars, wind-blown water from fresh-water springs on the windward sides has etched multiple narrow channels extending for hundreds of meters out onto halite crusts. Similar channels occur in upraised halite crusts at the margins of zones subject to annual flooding. Renewed spring activity in lacustrine sediments beneath thick rock-salt crusts dissolve the salt, forming steep-sided "brine pools" in the crust, which stabilize at a size commensurate with the rate of spring-water flow. These pools range from irregular tubelike openings along margins of desiccation polygons, as is well displayed in the north-central part of Salar de Uyuni, to circular pools many meters in diameter, as in Salar de Pedernales and Salar de Atacama.

Fog, rain, and wind cause additional modification of salar crusts. Slow leaching of blocky, silty salar crusts in the coastal region of northern Chile by fog has reduced irregular blocky fragments to rounded "salt nodules" and has changed surfaces of flat polygonal slabs to shallow, concave "salt saucers." Undersides of loose, upraised slabs of halite crusts in the Andes have "salt stalactites" formed by dripping rainwater solutions. Gypseous ramparts, consisting of 50-cm- to 1-m-high walls of nodular gypsum, have been formed by wind-blown spray along the lee sides of spring-fed ponds at the western side of Salar de San Martín. Gypsum dunes are found on the lee sides of several salars in the Andean Highlands. They consist of wind-blown gypsum sand, eroded from newly deposited saline material on the salar surfaces, from which the more soluble saline minerals have been leached by rainwater.

Several features of salar crusts are attributed to expansion and contraction of the crusts due to diurnal temperature changes. The effects are the greatest in the silty crusts of salars of coastal Chile, where the crusts are dry and diurnal temperature changes are great. On a quiet day, a person standing in a salar can hear an almost continuous sound of popping and cracking as the crust adjusts to temperature changes. Addition of new salt (by capillary rise and evaporation of saline ground water) in cracks formed by this movement, as well as new saline crust deposited beneath the older crust, causes heaving and expansion which forms sharp pressure ridges 50-75 cm high and rugged surfaces of loose polygonal slabs. "Salt veins" are deposited in expansion cracks within polygonal blocks of hard rock-salt crusts of salars in the Andean Highlands. Where rock-salt crusts are relatively thin, and brine levels are at depths of not more than a few tens of centimeters, lacy "salt sheets" may be extruded at margins of desiccation polygons.

One of the most unusual effects of diurnal temperature changes is the formation of "salt-solution tubes" along fractures in hard rock-salt crusts where brine levels are at depths of not more than a meter. These tubes, which are as much as 10 cm in diameter and a meter long (extending from brine level to the salar surface), evidently form by diurnal pumping of moist air out of the crust and concomitant capillary rise of brine to form a "salt cone" over the tube at the salar surface. Such cones form more rarely on dry salt crusts where moisture is furnished by rain and fog.

Tectonic deformation, as discussed in the section Physiographic and Geologic Setting, has affected the symmetry of closed basins and of salars within them. Recent faults define the margins of some salars and have displaced the crusts of others. In addition, stresses related to faulting have caused development of giant oriented polygons as much as 150 m across, which are outlined by intersecting sets of fractures. Recent fault movement has initiated spring activity within several salars, of which one manifestation is formation of "brine pools" in halite crusts. Sag basins several meters deep and many tens of meters in diameter are found along the recently active Atacama fault where it cuts across Salar Grande (Figure 4).

MINERALOGY

The minerals listed in Table 3 include those for which we have positive identification, chiefly by x-ray diffraction techniques, and minerals reported by others but not identified in our studies. We believe that some of the latter minerals, as noted in the table, may have been misidentified and that additional study is needed to confirm their presence.

Relatively few of the minerals shown in Table 3 are abundant and widespread. Others are found in minor amounts in

Table 3. Saline minerals in salars of the central Andean region. (Italicized minerals are believed to need additional study to confirm their presence. Numbers in parentheses after mineral name refer to publications in which occurrence is reported; other minerals have been identified by authors of present report.)

HALIDES	
Halite, NaCl	*Bischofite* (3), $MgCl_2 \cdot 6H_2O$
Sylvite (1,2), KCl	*Carnallite* (3), $KMgCl_3 \cdot 6H_2O$
Tachyhydrite, $CaMg_2Cl_6 \cdot 12H_2O$	
SULFATES	
Gypsum, $CaSO_4 \cdot 2H_2O$	Aphyhitalite (Glaserite) (1, 4, 5), $K_3Na(SO_4)_2$
Anhydrite, $CaSO_4$	*Syngenite* (5), $K_2Ca(SO_4)_2 \cdot H_2O$
Mirabilite, $Na_2SO_4 \cdot 10H_2O$	*Polyhalite* (5), $K_2Ca_2Mg(SO_4)_4 \cdot 2H_2O$
Thenardite, Na_2SO_4	
Glauberite (4, 5), $Na_2Ca(SO_4)_2$	
Hydroglauberite, $Na_4Ca(CO_4)_3 \cdot 2H_2O$	
Leonite, $K_2Mg(SO_4)_2 \cdot 4H_2O$	
Bloedite (5), $Na_2Mg(SO_4)_2 \cdot 4H_2O$	
BORATES	
Ulexite, $NaCaB_5O_6(OH)_6 \cdot 5H_2O$	*Gowerite* (9), $CaB_6O_{10} \cdot 5H_2O$
Borax (2, 6, 7), $Na_2B_4O_5(OH)_4 \cdot 8H_2O$	*Probertite* (2), $NaCaB_5O_7(OH)_4 \cdot 3H_2O$
Inyoite (7, 8), $Ca_2B_6O_6(OH)_{10} \cdot 8H_2O$	*Colemanite* (10), $Ca_2B_6O_{11} \cdot 5H_2O$
Tincalconite (6), $Na_2B_4O_5(OH)_4 \cdot 3H_2O$	*Hydroboracite* (11), $CaMgB_6O_8(OH)_6 \cdot 3H_2O$
Kernite (6), $Na_2B_4O_6(OH)_2 \cdot 3H_2O$	
Inderite (6), $MgB_3O_3(OH)_5 \cdot 5H_2O$	
Escurrite (6), $Na_4B_{10}O_{17} \cdot 7H_2O$	
CARBONATES	
Calcite, $CaCO_3$	
Dolomite (12), $CaMg(CO_3)_2$	
Natron (2), $Na_2CO_3 \cdot 10H_2O$	
Thermonatrite (2), $Na_2CO_3 \cdot H_2O$	
NITRATES	
Nitratite, $NaNO_3$	

[1] Ericksen (1963)
[2] Ballivian and Risacher (1981)
[3] Runge et al. (unpublished data, Instituto de Investigaciones Geologicas, 1974)
[4] Gale (1918)
[5] Vila (1976)
[6] Muessig and Allen (1957)
[7] Muessig (1966)
[8] Muessig (1958)
[9] Salas (1975)
[10] Cadima and Lafuente (unpublished data, Servicio Geologico de Bolivia, 1967)
[11] Catalano (1926)
[12] Muessig (1958)

most salars but are abundant in only a few, and still others are found in only minor amounts in any salar. Halite and gypsum are by far the most abundant saline minerals, occurring in all salars. The typical halite crust is hard and white, has a sugary texture, and is highly porous. The unique, coarsely crystalline halite in Salar Grande has an average grain size of 1 to 2 cm, and local coarsely crystalline zones contain anhedral crystals as much as 75 cm in diameter. Gypseous salar crusts consist of medium-grained gypsum and coherent meshworks of gypsum crystals. Nodules and lenticular layers of feathery, white ulexite are found in gypseous crusts, in lacustrine sediments having ephemeral crusts, and in perennial marshes at salar margins. Nodules of ulexite are generally less than 10 cm in diameter, and ulexite layers generally have maximum thicknesses of 30 to 40 cm, but layers as much as a meter thick have been found in some salars.

Borate minerals other than ulexite are found in only a few of the salars. Small amounts of borax are associated with ulexite in several salars and have been found in a borate-rich thermal-spring tufa in northwestern Argentina (Muessig, 1966). Borax is the dominant mineral in the Tincalayu borate deposit, Argentina, which also contains minor amounts of tincalconite (the desiccation product of borax), kernite, inderite, and escurrite (Muessig and Allen, 1957). This is the only known occurrence of these minerals in the Andes. Small amounts of inyoite have been found in a thermal-spring deposit at the east side of Laguna Salinas, Perú (Muessig, 1958). Other reported borate minerals (Table 3) require confirmation.

Mirabilite and its desiccation product, thenardite, are widespread and moderately abundant in the salars. Mirabilite is found as prismatic crystals, as icelike crystalline masses, and as impregnating material in marginal sulfate zones of zoned salars, in gypseous crusts, and on exposed surfaces of brine-saturated lacustrine sediments in the Andean Highlands. Mirabilite is stable in these areas because maximum surface temperatures are below the mirabilite-thenardite transition temperature of 32.4 °C. However, mirabilite does slowly desiccate in the dry air of this region, and consequently, salars in the Andean Highlands also contain powdery, secondary thenardite. Thenardite also is moderately abundant and widespread in salars of the coastal region of northern Chile, where it occurs both as powdery material resulting from desiccation of mirabilite and as large prismatic and bipyramidal crystals. Lenticular layers of icelike mirabilite, as much as 30 cm thick, occur at the base of a 50-cm- to 1-m-thick silty halite crust in the northeastern part of Salar de Pintados (Ericksen, Mrose, and Fahey, 1970). This mirabilite contains nodular masses of tan hydroglauberite, the only known occurrence of this mineral in the central Andes. This area also has local zones in which the salar crust consists chiefly of a dry, coherent meshwork of sand-size thenardite grains and scattered prismatic crystals of translucent, white thenardite as much as 20 cm long. Clusters of tan, bipyramidal crystals of thenardite, 2 to 8 cm long, are found locally in the coarsely crystalline rocksalt of Salar Grande.

Of the other sulfate minerals listed in Table 3, only anhydrite is abundant; others are either widespread minor constituents, local minor constituents, or mineralogical rarities in Andean salars. Anhydrite is widespread in salars and soils in the Atacama Desert of northern Chile, where it is principally a desiccation product of gypsum and is stable because of the extreme dryness of this desert. It is the dominant mineral in the northern part of Salar de Llamara, in a zone that stretches across the northern side of this salar to the southern part of Salar Grande (Figure 4). This anhydrite is as much as 10 m thick in places and has a rugged surface showing a relief of several meters, which is characterized by solution and collapse structures. Bloedite and glauberite have been identified only in

Salar de Pintados, Chile, but these minerals probably occur in minor amounts in most salars in the central Andes. We have found leonite in newly deposited K-rich salt from experimental evaporation pits in the southern part of Salar de Bellavista. Aphthitalite reportedly was found in this same locality (Ericksen, 1963), but we have been unable to confirm its presence here or in other salars of the region.

Halides other than halite have been found in a few salars. Sylvite in small amounts has been found in the rock-salt crust of Salar de Uyuni (Ballivian and Risacher, 1981) and has been reported from the K-rich crust in the southern part of Salar de Bellavista (Ericksen, 1963). We have been unable to confirm the presence of sylvite in either of these salars. We found efflorescences of feathery, white tachyhydrite at the site of abandoned borate workings in the southwestern part of Salar de Pedernales, Chile. This mineral is stable only at low temperatures and probably forms here and in sulfate zones of other salars in the Andean Highlands only during winter months, when average temperatures approach 0 °C. Bischofite and carnallite, reported to occur in Salar de Bellavista, need confirmation.

Except for calcite, carbonate minerals have been found in only a few salars of the central Andes. Calcitic marls and ostracod-rich beds occur in lacustrine sediments of several salars. Algal limestones, which were deposited in former Lago Tauca, are widespread in the southern part of the Bolivian Altiplano. Calcareous tufa is a characteristic thermal-spring deposit of the region, and $CaCO_3$-rich crusts were deposited by thermal-spring waters at Lagunas Pastos Grandes (Figure 5) and Salar de Empexa (Ballivian and Risacher, 1981). Dolomite has been identified in lacustrine sediments of one salar in Argentina (Muessig, 1958) and may occur in such sediments in others. Natron and thermonatrite, the only sodium carbonate minerals identified, have been found in saline evaporites at Laguna Collpa and nearby saline lakes in southern Bolivia.

Nitratite occurs in several salars in the coastal region of northern Chile, where it is associated with halite and sulfate minerals in silty halite crusts. Nitratite is not a normal mineral of saline complexes, and its presence in this region reflects formation (probably by microbial activity in moist soils of the salars) and slow accumulation under conditions of the long-term, nearly rainless climate of the Atacama Desert. This climate also has been a major factor in the accumulation of the well-known nitrate caliche deposits of this region. Some salars have been contaminated with waste waters from nearby nitrate beneficiation plants, and, in addition to nitratite, contain concentrations of iodate and perchlorate, both of which are constituents of nitrate ore.

BRINE CHEMISTRY

The brines of the central Andean salars are chiefly NaCl types in which the relative abundances of the constituents are $Na^+ >> K^+ > Mg^{2+} > Li^+ Ca^{2+}$ and $Cl^- >> SO_4^{2-} >> B_2O_3 > HCO_3^-$ (Table 4). Typical saturated brines have about 350,000 mg/L dissolved solids, of which 80% or more is NaCl. Sulfate is present in various concentrations, but generally from about 5000 mg/L to 20,000 mg/L. Chloride-rich brines of a few salars have SO_4^{2-} concentrations ranging from about 25,000 mg/L to nearly 90,000 mg/L. These brines would be classed as chloride-sulfate brines. A few saline lakes in southwestern Bolivia have relatively high concentrations of Na_2CO_3. Concentrations of K^+, Li^+, and B_2O_3 are relatively high (Table 4) in most of the brines.

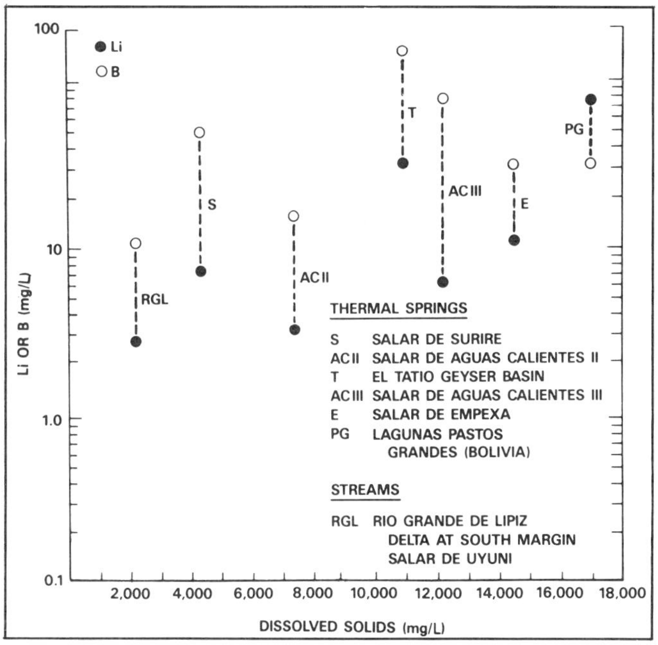

Figure 6. Li and B concentrations plotted against total dissolved solids for selected thermal springs and streams in the central Andes. Based on analyses from Vila (1975), Salas (unpublished data, 1975), Rettig, Jones, and Risacher (1980), Cusicanqui, Majon, and Ellis (1975), and Ballivian and Risacher (1981).

The brines of Andean salars formed by evaporative concentration of meteoric waters in which the relative amounts of Cl^-, SO_4^{2-}, and HCO_3^- show a considerable range, but which have relative cation abundances of $Na^+ > Ca^{2+} > K^+ > Mg^{2+} Li^+$. These waters generally contain several hundred to several thousand mg/L dissolved solids and have relatively high concentrations of Li and B (Figure 6). Thermal springs tend to be more saline and have higher values for Li and B than do streams and ground water, and they show concentrations of these elements that differ somewhat independently from total dissolved solids (Figure 6).

Evolution of the NaCl-rich brines in the central Andes involves major depletion of Ca^{2+}, HCO_3^-, SO_4^{2-}, SiO_2, and B_2O_3 during evaporative concentration of the meteoric waters and enrichment of K^+, Mg^{2+}; and removal of Li^+, Ca^{2+}, and HCO_3^- from the solutions as thermal-spring tufa, as algal limestone such as that deposited in Lago Tauca in the southern Bolivian Altiplano, and as $CaCO_3$-rich marls and ostracod-rich lacustrine sediments. SiO_2 is removed chiefly by diatoms to form the widespread diatomaceous lacustrine sediments. SO_4^{2-} is removed from the solutions chiefly as gypsum, but small amounts of SO_4^{2-} are reduced to H_2S by bacteria in organic-rich muds in some halite crusts and in lacustrine sediments. Rettig, Jones, and Risacher (1980) have shown that major depletion of HCO_3^- and SO_4^{2-} takes place in saline waters having less than 35,000 mg/L dissolved solids. B_2O_3 is removed as ulexite, also at relatively low concentrations of dissolved solids.

Table 4. Chemical analyses of selected brines from salars in the central Andes (measurements in mg/L; n.d. = not determined; do. = ditto).

Salar	Cl	SO_4	HCO_3	Na	K	Ca	Mg	Li	B_2O_3	Dissolved solids
				Argentina[1]						
Hombre Muerto	194,800	11,100	n.d.	121,900	9,340	1,000	268	914	1,455	340,920
Rincon	190,500	15,990	n.d.	122,200	6,570	280	2,120	350	1,609	338,510
Pocitos	190,600	7,440	n.d.	123,100	3,400	600	1,290	97	708	326,747
Pastos Grandes	178,700	26,080	n.d.	118,200	4,730	740	2,980	440	2,220	332,560
Centenario	192,700	19,980	n.d.	112,300	8,170	320	7,550	1,020	3,765	343,210
Rio Grande	148,900	10,610	n.d.	92,400	3,710	800	2,600	420	1,673	259,960
Arizaro	190,700	8,260	n.d.	119,500	160	760	1,840	160	138	321,423
				Bolivia						
Uyuni[2]	191,800	13,200	592	94,900	13,500	461	11,800	700	1,136	351,000
Empexa[2]	120,000	34,100	430	67,200	3,400	259	8,480	213	702	239,000
Coipasa[2]	186,000	25,300	785	100,400	9,080	253	12,120	338	2,208	368,000
Pastos Grandes[3]	194,000	2,460	n.d.	101,000	14,200	3,100	3,480	1,640	3,041	320,825
Cáñapa[3]	126,000	17,900	n.d.	68,900	12,000	600	1,670	712	2,011	228,407
				Chile						
Atacama[4]	183,100	16,140	560	103,000	12,900	520	6,130	760	1,705	370,000
do.	182,800	25,500	220	98,000	19,500	300	8,500	1,200	1,673	385,000
do.	189,500	15,900	230	91,100	23,600	450	9,650	1,570	1,416	370,000
Surire[5]	131,380	11,430	n.d.	73,200	13,200	890	3,830	540	3,700	235,620
Azufrera[6]	172,130	87,990	0	60,000	14,960	88	48,640	86	740	400,000
Laco[6]	109,630	15,360	620	62,200	4,800	820	6,251	101	1,078	210,000
Huasco[7]	112,980	26,700	n.d.	65,000	13,500	610	5,880	480	2,333	277,270
San Martín[8]	68,000	1,570	415	30,100	2,410	8,469	1,646	170	985	112,960
Ascotán[9]	70,000	25,000	2,900	45,000	3,530	920	5,125	186	2,520	153,600

[1]All analyses from Nicolli et al. (1982b)
[2]Analyses by Shirley L. Rettig (unpublished data, U.S. Geological Survey, 1976)
[3]Ballivian and Risacher (1981)
[4]Moraga et al. (1974)
[5]Salas (unpublished data, Instituto de Investigaciones Geológicas, 1975)
[6]Vila (1975)
[7]Corporación de Fomento de la Producción (unpublished data, 1981)
[8]Instituto de Investigaciones Geológicas (unpublished data, 1967)
[9]Analysis by Shirley L. Rettig (unpublished data, U.S. Geological Survey, 1978)

Up to the point of NaCl saturation, the amounts of K^+ and Li^+ are more or less those that would be expected from evaporative concentration of the meteoric waters of this region, as shown by the Li and K curves (Figure 7) for the average concentrations in several salars in the Chilean Andes. These curves are approximately parallel to the curve for normal evaporative concentration. Mg^{2+} behaves similarly. In contrast, the concentrations of Li and K relative to Cl for Salar de Uyuni (Figure 7) are less than expected from the curve for simple or straightforward evaporative concentration; the lower concentrations indicate progressive loss of these ions. This loss of Li and K in Salar de Uyuni may be due to adsorption of these ions by montmorillonitic clays in the lacustrine sediments (Rettig, Jones, and Risacher, 1980). Salar de Atacama (Figure 7) differs in showing an increase of Li content relative to Cl but normal evaporative concentration of K. Mg, which is not shown in Figure 7, also shows normal evaporative concentration in Salar de Atacama. The apparent enrichment of Li relative to K and Mg in Salar de Atacama is not easily explained. Perhaps K^+ has been depleted in the brines by formation of K-bearing saline minerals that have not yet been recognized in this salar. In addition, this is the oldest basin in the Andes, having been in existence since the Late Cretaceous or early Tertiary, and has had many cyclic changes in climate, tectonic deformation, and volcanism, all of which might have affected the sources of the saline constituents and, consequently, the composition of the brines.

Although the relative concentrations of Li^+, K^+, and Mg^{2+} tend to be relatively constant in individual salars, concentrations may differ considerably from one salar to another (Figures 8, 9). The ratios for Li:K indicated by the curves in Figure 8 range from about 1:9 to 1:24, which is probably near the maximum range for salars in the central Andes, and those for Mg:K are 1:1.2 to 1:8, also near the maximum range.

Once the brines in the salars reach NaCl saturation, further changes in relative concentrations of the solute constituents take place during crystallization of halite. The principal result is a progressive increase in the concentrations of Li^+, K^+, and Mg^{2+} and a decrease in Na^+, relative to Cl^-. As a consequence, salars having extensive halite crusts tend to show the greatest concentrations of Li, K, and Mg, and the amounts of these ions may be much greater than would be expected from evaporative concentration. This is shown in Figure 7 by the sharp deflections of the curves for salars Uyuni and Atacama

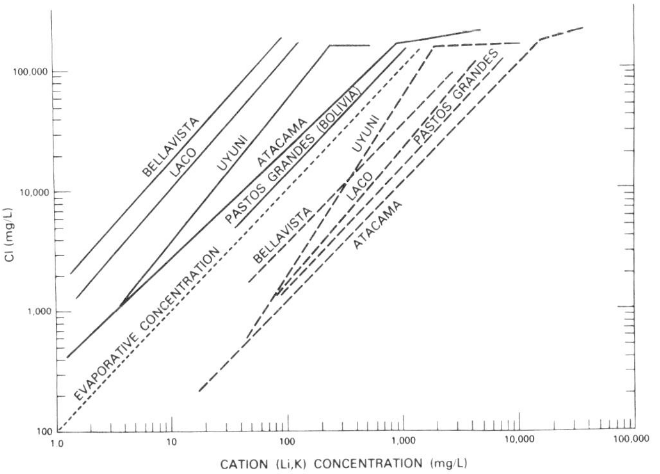

Figure 7. Li and K concentrations (solid lines) plotted against Cl concentrations for brines from selected Andean salars, compared with curves for normal evaporative concentrations below NaCl saturation (dashed line). Deflected upper parts of curves indicate evaporative concentrations at Cl saturation. Based on chemical analyses in Chong (1984), Moraga et al. (1974), and Rettig, Jones, and Risacher (1980).

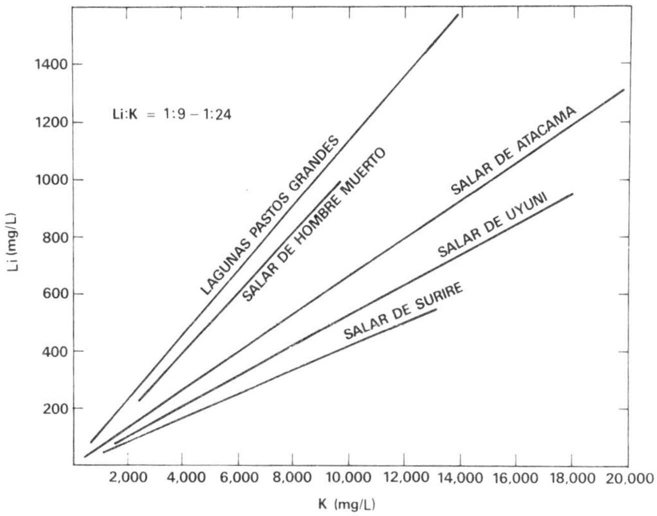

Figure 8. Li:K ratios of brines from selected Andean salars. Plotted from analyses by Nicolli et al. (1982a), Ballivian and Risacher (1981), Ericksen, Vine, and Ballon (1977), Moraga et al. (1974), and Salas (unpublished data, 1975).

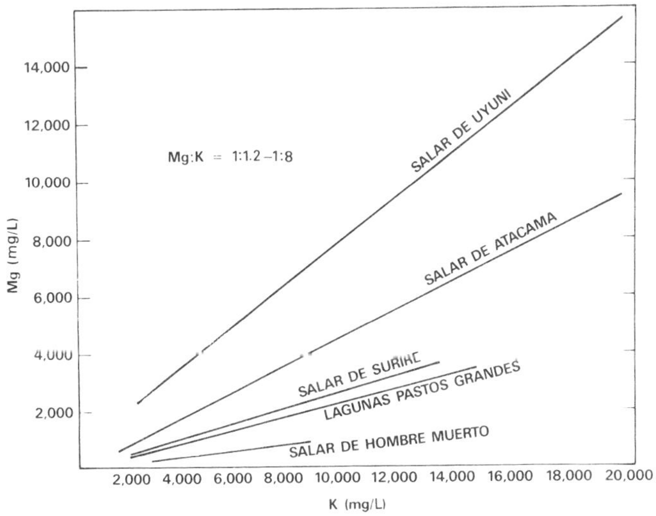

Figure 9. Mg:K ratios of brines from selected Andean salars (see Figure 6 for sources).

at or near chloride saturation of 115,000 to 120,000 mg/L. Salar de Hombre Muerto, Argentina, shows Li-K-Mg:Cl ratios in saturated brines similar to those of Salar de Uyuni (Nicolli et al., 1982b). As indicated by the slopes of the deflected parts of the curves, rates of increase in the concentration ratios of Li:Cl and K:Cl at Cl saturation may differ from one salar to another. Rettig, Jones, and Risacher (1980) compared Li and K concentrations in saturated brines of Salar de Uyuni with bromine concentrations (which presumably are affected only be evaporative concentration) to show that the relative amounts of Li and K were less than would be expected from evaporative concentration. These authors suggested that the brines, as well as the saline waters from which they formed, lost Li, K, and Mg by adsorption on montmorillonitic clay in the lacustrine sediments beneath the salar crusts. The curves for Salar de Atacama (Figure 7) show greater rates of increase for Li^+ and K^+ at Cl saturation than do those for Salar de Uyuni. Unfortunately, Br was not analyzed for the Atacama brines (Moraga et al., 1974), so that comparison of Li-K:Br ratios of the two salars cannot be made.

SOURCES OF SALINE CONSTITUENTS

The distribution of the salars and of accumulations of saline materials along present-day streams and marshes clearly shows that late Tertiary and Quaternary volcanic rocks have been the principal sources of the saline materials in Andean salars. Additional, probably small, amounts of saline material were leached from older rocks or transported by wind from distant sources. Saline material in the volcanic rocks includes water-soluble salts that were incorporated at the time of eruption, and saline constituents in volcanic glass, which are readily leached during hydration of the glass. Widespread saline encrustations along streams and in marshes in areas of abundant volcanic rocks attest to the presence of readily leachable saline material in these rocks. Furthermore, many salars are in basins where volcanic rocks and atmospheric fallout are the only possible sources of the saline materials. Thermal springs in the region tend to be saline, commonly containing several thousand mg/L dissolved solids, and have relatively high concentrations of Li and B, both of which are characteristic of meteoric waters draining rhyolitic volcanic terranes. For example, thermal springs in the Tatio Geyser basin, near the Chile-Bolivia border about 60 km north of Salar de Atacama (Figure 1, Table 1), contain 10,000-15,000 mg/L NaCl and as much as 47 mg/L Li and 183 mg/L B (Cusicanqui, Majon, and Ellis, 1975). Another NaCl-rich thermal spring at Salar de Surire (Figure 6) contains 4200 mg/L dissolved solids and 8.3 mg/L Li and 47 mg/L B. Among the most unusual thermal springs are those in northwestern Argentina that have aprons of

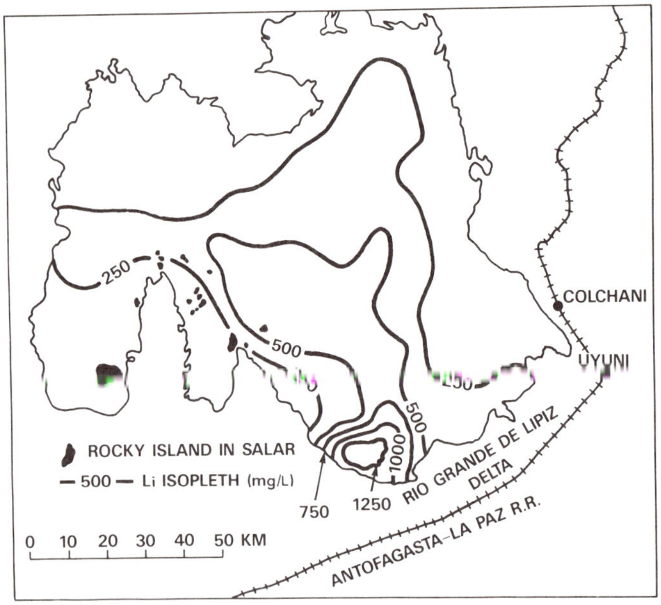

Figure 10. Distribution of Li in near-surface interstitial brines of the porous halite crust of Salar de Uyuni. Modified from Davis et al. (1982).

ulexite-rich tufa (Muessig, 1966; Alonso and Viramonte, 1985). Catalano (1964b) cites three analyses of thermal-spring waters from one area (Antuco) that show 14.9 to 16.4% NaCl, about 1% SO_4, and traces of CaO and MgO. B was found, but the amount was not determined.

The distribution of Li in brines of Salar de Uyuni (Figure 10) indicates that Rio Grande de Lipiz is the major source of Li. The brines show similar concentration patterns for K and Mg. The rocks in the Rio Grande de Lipiz drainage system are largely quartz-latite ash-flow tuffs and dacite to andesite flows of late Miocene and Pliocene age. In the Rio Grande de Lipiz delta (Figure 10), the stream waters, which have been slightly concentrated by evaporation, average about 3 mg/L Li and 10 mg/L B (Figure 6). The brines formed from these waters are superimposed on the residual brines in the salar crust, which formed during final desiccation of Lago Tauca (Figure 2) in early Holocene time. The present distribution of Li in Salar de Uyuni (Figure 10) indicates that the residual Lago Tauca brines contained less than 250 mg/L Li.

Leaching tests of pulverized samples of rhyolitic ash-flow tuffs and obsidian from northern Chile have shown that these rocks contain water-soluble saline material (Ericksen, 1963). Cold-water leachates were found to contain from about 500 mg/L to 3500 mg/L dissolved solids, consisting chiefly of Cl^-, SO_4^{2-}, Na^+, and Ca^{2+}. We believe that this saline material includes both contaminants, perhaps from wind-blown saline dust, and primary water-soluble constituents of the magma that were trapped at the time of eruption.

PRODUCTION AND RESOURCES

In terms of world consumption, Andean salars contain large resources of rock salt, lithium, and boron, and moderate to small resources of gypsum, sodium sulfate, potassium, and magnesium. With the exception of gypsum and magnesium, all these commodities have been produced. The greatest production, by far, has been of rocksalt and borate minerals, both of which have been exploited since the middle to late 19th century. In addition, rock salt probably has been exploited for local use since pre-Columbian times. Data bout rock-salt production are scanty, but existing production records indicate that annual production during the 1980s was about 800,000 t. Most has been high-purity rock salt quarried from Salar Grande, in the Coastal Range of northern Chile, and smaller amounts have been recovered as newly recrystallized salt from brines in Salar de Lagunas, in the Coastal Range east of Salar Grande, in Salar de Uyuni, Bolivia, and in Salar de Pocitos, Argentina. Additional large amounts of salt are obtained from brines in saline lakes and salars in Argentina that are outside the area discussed in this chapter. Production of borate minerals in 1984 was 166,889 t, which included 142,880 t of borax from the Tincalayu deposit and 13,000 t of ulexite from an unspecified salar, both in northwestern Argentina (Milka Brodtkorb, written communication, 1986), 10,000 t of ulexite from Laguna Salinas, Perú, and 1000 t of ulexite from Salar de Ascotán, Chile (U.S. Bureau of Mines, 1986; Ericksen, unpublished data, 1986). Production of lithium carbonate from Li-rich brines at Salar de Atacama, Chile, began in 1984, in an operation designed for an annual production of 7500 t of lithium carbonate (Ihor Kunasz, oral communication, 1986).

Production of the other saline commodities has been small. Since the 1950s, sodium sulfate has been recovered sporadically from deposits of mirabilite and thenardite in the northeastern part of Salar de Pintados, Chile. During the early 1940s, several thousand tons of potassium chloride and potassium nitrate were produced from the K-rich salt crust in the southern part of Salar de Bellavista, Chile. Efflorescences of natron and thermonatrite reportedly are harvested periodically at Laguna Collpa, Bolivia, by local inhabitants (Ballivian and Risacher, 1981). During the latter part of the 19th century and early part of the 20th century, two salars in northern Chile, Salar del Carmen and Salar de Pampa Blanca, were exploited for nitrate ore.

Although potentially exploitable materials are found in many salars, the largest resources are in relatively few. Li-rich brines, for example, are found in many salars, but by far the largest quantities and highest grades of Li-rich brines are in Salar de Atacama, Salar de Uyuni, and Salar de Hombre Muerto. The brines of these salars also contain large amounts of potentially recoverable K and B. The central halite body of Salar de Atacama has the greatest amount of Li- and K-rich brines, which average about 1.800 mg/L Li and nearly 28,000 mg/L K. Ihor Kunatz (oral communication, 1985) reported that this salt crust extends over an area of about 1400 km^2 and to a depth of as much as 426 m. He also reported that the brines are concentrated in the uppermost highly porous and permeable 10 m of this salt body, whereas the underlying more dense and less permeable salt contains relatively little recoverable brine. In contrast, Li-rich brines (500 mg/L Li and 10,000 mg/L K) are found in an area of about 2000 km^2 in the central part of Salar de Uyuni (Figure 10). We estimate the average thickness of the porous salt crust in this zone, which contains most of the recoverable brine, to be at least 5 m. A recent drill hole in Salar de Uyuni, by the Office de la Recherche Scientifique et Technique Outre-Mar, penetrated alternating layers of porous rock salt and clastic lacustrine sediments to a depth of 43 m (F. Risacher, oral communication, 1986). Brines in Salar de Hombre Muerto average 700-800 mg/L Li and 7000-8000 mg/L K over an area of 100-150 km^2 (Nicolli et al., 1982b). The salt body of this salar is known to extend to a depth of at least 15 m, but neither the average nor maximum thickness is known.

Potentially exploitable ulexite deposits also are known in many salars, and the largest deposits, many of which have been exploited, are in salars Surire, San Martín, Ascotán, Pedernales, and Pintados in Chile; Uyuni, Laguani, Pastos Grandes, and Capiña in Bolivia; and Salinas Grandes, Cauchari, Pastos Grandes, Rincon, and Pocitos in Argentina.

Salar Grande, in the coastal region of northern Chile, is by far the most valuable deposit of rock salt in the central Andes. Not only does it contain huge amounts of high-purity salt (averaging 99% NaCl), but it also lacks interstitial brine, so that the salt can be mined easily and cheaply. The salar is near the deep-water port of Iquique. As a result, the salt can be shipped at a competitive price on the world market. Since the 1960s, large amounts have been exported, chiefly to Japan.

The other potentially recoverable saline constituents of Andean salars—gypsum and anhydrite in salar crusts, Na_2SO_4 and Na_2CO_3 as minerals in salar crusts and dissolved in brine, and Mg and B in brine—may be exploited for internal markets, but for the foreseeable future probably will not be competitive in the world market, either because of small potential production or because of high costs of beneficiation and transport.

CONCLUSIONS AND RECOMMENDATIONS

Future production from Andean salars probably will include recovery of rock salt, lithium, borate minerals, and, perhaps, potassium for the manufacture of fertilizer. Production of rock salt and lithium may increase slowly in the near future, but major increases probably will come about only in response to major increases in world demand and prices. Of the borate minerals, only borax from the Tincalayu deposit is now highly competitive in the world market, and this deposit should continue to be a dominant source of borate minerals in the central Andes. Ulexite, which is virtually the only borate mineral in present-day salars, is relatively expensive to beneficiate. It is now produced largely for internal consumption. Because world resources of borate minerals such as borax, kernite, and colemanite, which can be processed relatively cheaply, are very large, ulexite from Andean salars is not likely to become a significant export commodity within the foreseeable future. Potassium and magnesium are potential byproducts of Li-rich brines. Recovery of potassium is planned for Salar de Atacama.

Investigations during the past two decades have shown the nature, distribution, and resource potential of the saline constituents of Andean salars. The principal near-future problem is one of development of these resources. Future exploration for saline materials in the central Andes, other than exploration required for development of known resources, should emphasize the search for buried deposits of borate minerals in late Tertiary and Quaternary playa sediments. Such deposits—of which Tincalayu, Argentina, is an example—may contain borax or other borate minerals that are more competitive on the world market than the ulexite of the present-day salars.

REFERENCES CITED

Alonso, R. N., and J. G. Viramonte, 1985, Geyseres boratíferos de la Puna Argentina: IV Congreso Geológico Chileno, Actas, v. 2, Area 3, p. 23-44.

Ballivian, O., and F. Risacher, 1981, Los salares del Altiplano boliviano: Paris, Office de la Recherche Scientifique et Technique Outre-Mer, 246 p.

Bárnabe, J. F., 1915, Los yacimientos minerales de la Puna de Atacama: Buenos Aires, Anales del Ministerio de Agricultura, Sección de Geología, Mineralogía, y Minería, v. 10, no. 5, 63 p.

Brüggen, J., 1918, El Salar de Pintados y sus yacimientos de potasa: Sociedad Nacional de Minería, Boletín Minero, v. 30, p. 3-20.

Brüggen, J., 1950, Fundamentos de la geologia de Chile: Instituto Geográficio Militar [Chile], 374 p.

Catalano, L. R., 1926, Geología económica de los yacimientos de boratos y materiales de las cuencas—Salar de Cauchari—Puna de Atacama: Buenos Aires, Ministerio de Agricultura, Dirección General de Minas, Geología e Hidrología, Publicación No. 23, 110 p.

Catalano, L. R., 1927, Geología química de los boratos: Buenos Aires, Ministerio de Agricultura, Dirección General de Minas, Geología e Hidrología, Publicación No. 28, 27 p.

Catalano, L. R., 1930, Boratera de Coyahuaima: Buenos Aires, Ministerio de Agricultura, Dirección General de Minas, Geología e Hidrología, Publicación No. 89, 57 p.

Catalano, L. R., 1964a, Estudio Geológico-Económico del Salar de Hombre Mueto (Puna de Atacama): Buenos Aires, Ministerio de Economía, Subsecretaría de Minería, Estudios de Geología y Minería Económica Serie Argentina No. 4, 131 p.

Catalano, L. R., 1964b, Boratera de Antuco, datos generales; Cuenca de Diabillos: Buenos Aires, Ministerio de Economía, Subsecretaría de Minería, Estudios de Geología y Minería Económica Serie Argentina No. 2, p. 1-32 and p. 1-70.

Chong, D., G., 1984, Die Salare in Nordchile—Geologie, Struktur und Geochemie: Geotektonische Forschungen No. 67, 146 p.

Cusicanqui, H., W. A. J. Majon, and A. J. Ellis, 1975, The geochemistry of the El Tatio geothermal field, northern Chile: San Francisco, Second U.N. Symposium on the Development and Use of Geothermal Resources, Proceedings, v. 1, p. 703-711.

Davis, J. R., K. A. Howard, S. L. Retting, R. L. Smith, G. E. Ericksen, F. Risacher, H. Alarcon, and R. Morales, 1982, Progress reports on lithium-related investigations in Bolivia: U.S. Geological Survey Open-File Report 82-782, 14 p.

Ericksen, G. E., 1963, Geology of the salt deposits and the salt industry of northern Chile: U.S. Geological Survey Open-File Report, 164 p.

Ericksen, G. E., M. E. Mrose, and J. J. Fahey, 1970, Ice-clear mirabilite from Salar de Pintados, northern Chile: Mineralogical Record, v. 1, p. 12-15 and 24-28.

Ericksen, G. E., G. Chong D., and T. Vila G., 1976, Lithium resources of salars in the central Andes: U.S. Geological Survey Professional Paper 1005, p. 66-74.

Ericksen, G. E., J. D. Vine, and R. Ballon A., 1977, Chemical composition and distribution of lithium-rich brines in Salar de Uyuni and nearby salars in southwestern Bolivia: Energy, v. 3, p. 355-363.

Gale, H. S., 1918, Potash in the Pintados Salar, Tarapacá, Chile: Engineering and Mining Journal, v. 105, p. 674-677.

Jochamowitz, A., 1917, Los boratos de América y en especial los boratos del Perú: Lima, Perú, Congreso Nacional de la Industria Minera, Sec. 4, 89 p.

Leiding, B., 1942, Inventario de reservas de minerales no metálicos en Chile: Santiago de Chile, Primer Congreso Panamericano de

Ingeniería de Minas y Geología, Anales, v. 4, p. 1651-1678.

Montes de Oca, I., 1982, Geografíco y recursos naturales de Bolivia: La Paz, Instituto Geográfico Militar, 628 p.

Morago B., A., G. Chong D., M. A. Fortt Z., and H. Henriquez A., 1974, Estudio Geológico del Salar de Atacama, Provincia de Antofagasta: Instituto de Investigaciones Geológicas (Chile), Bol. 29, 56 p.

Mordójovich K., C., 1965, Resena sobre las exploraciones petrolíferous de ENAP en la zona norte, anos 1956-1962: Minerales, v. 20, p. 32-61.

Mortimer, C., 1973, The Cenozoic history of the southern Atacama Desert, Chile: Journal of the Geological Society of London, v. 129, p. 505-526.

Mortimer, C., 1975, Cenozoic studies in northernmost Chile: Geologische Rundschau, v. 64, no. 2, p. 395-420.

Mortimer, C., and N. Saric R., 1972, Landform evolution in the coastal region of Tarapacá province, Chile: Revue de Geomorphologie Dynamique, Annee 21, no. 4, p. 162-170.

Muessig, S., 1958, Turilari, a borax crystal playa deposit in Argentina (abs.): Geological Society of America Bulletin, v. 69, p. 1696.

Muessig, S., 1966, Recent South American borate deposits, *in* J. L. Rau, ed., Second symposium on salt: Cleveland, Ohio, Northern Ohio Geological Society, v. 1, p. 151-159.

Muessig, S., and R. D. Allen, 1957, Ezcurrite ($2Na_2O.5B_2O_3.7H_2O$) a new sodium borate from Argentina: occurrence, mineralogy, and associated minerals: Economic Geology, v. 52, p. 426-437.

Nicolli, H. B., J. M. Suriano, J. F. Kimsa, and A. Brodtkorb, 1982a, Geochemical characteristics of brines in evaporitic basins, Argentinian Puna: 26th International Geological Congress, Paris, 1980, section S10 (geochemistry): Academía Nacional de Ciencias, Córdoba, Miscelanea No. 64, Reprint by the Comisión Nacional de Investigaciones Espaciales, Argentina, 19 p.

Nicolli, H. B., J. M. Suriano, V. Mendez, and M. A. Gomez P., 1982b, Salmueras ricas en metales alcalinos del Salar del Hombre Muerto, Provincia de Catamarca, Republica Argentina: V Congreso Latinoamericano de Geología, Buenos Aires, Actas: Reprint by the Comision Nacional de Investigaciones Espaciales, 20 p.

Pratt, W. P., 1961, Local evidence of Pleistocene to Recent orogeny in the Argentine Andes: Geological Society of America Bulletin, v. 72, p. 1539-1550.

Reichert, F., 1907, Los yacimientos de boratos y otros productos minerales explotables del Territorio de los Andes (Puna de Atacama): Buenos Aires, Anales del Ministerio de Agricultura, Sección de Geología, Mineralogía y Minería, v. 2, 104 p.

Rettig, S. L., B. F. Jones, and F. Risacher, 1980, Geochemical evolution of brines in the Salar of Uyuni, Bolivia: Chemical Geology, v. 30, p. 57-79.

Risacher, F., 1978, Le cadre géochimique des bassins à évaporites des Andes Boliviennes: Cahiers L'Office de la Recherche Scientifique et Technique Outre-Mer, Serie Géologie, v. 10, no. 1, p. 37-48.

Servant, M., and J. C. Fontes, 1978, Les lacs quaternaires des hauts plateaux des Andes Boliviennes—Premières interprétations paléoclimatiques: Cahiers L'Office de la Recherche Scientifique et Technique Outre-Mer, Serie Geologie, v. 10, no. 1, p. 9-23.

Stoertz, G. E., and G. E. Ericksen, 1974, Geology of salars in northern Chile: U.S. Geological Survey Professional Paper 811, 65 p.

Vila G., T., 1974, Geología y geoquímica de los salares andinos, Provincia de Antofagasta (Chile): Santiago, Univeridad de Chile, Facultad de Ciencias Físicas y Matemáticas, Departamento de Geología, thesis for degree in geology, 201 p.

Vila G., T., 1975, Geología de los depositos salinos andinos, Provincia de Antofagasta, Chile: Revista Geológica de Chile, No. 2, p. 41-55.

Vila G., T., 1976, Hidrogeología y distribución zonal de las costras salinas de Bellavista-Pintados, Norte Grande de Chile: Primer Congreso Geológico Chileno Actas, v. 2, p. E33-E52.

U.S. Bureau of Mines, 1986, Minerals yearbook, 1984: U.S. Government Printing Office, v. III, 1079 p.

Ericksen, G. E., Cañas Pinochet, M. T., and Reinemund, J. A., editors, 1989, Geology of the Andes and its relation to hydrocarbon and mineral resources: Houston, Texas, Circum-Pacific Council for Energy and Mineral Resources Earth Science Series, v. 11.

CHAPTER 11

ORIGIN OF LITHIUM IN SALAR DE ATACAMA, NORTHERN CHILE

Fernando Ide Y.
Comité de Sales Mixtas
Corporación de Fomento de la Producción
Santiago, Chile

Ihor A. Kunasz
Cyprus Specialty Metals Company

ABSTRACT

The Salar de Atacama of northern Chile is within a tectonic basin of internal drainage which originated in the late Tertiary and which has been the site of intensive volcanism from the Miocene to the present. The salar, the largest in Chile and one of the largest in the world, contains the world's richest lithium-brine deposit. Saline minerals and brines in the salar, including the lithium, were derived the following sources: (1) geothermal waters associated with active volcanism that produced huge amounts of rhyolitic ash-flow tuffs; (2) leaching of water-soluble salts from the volcanic rocks; (3) leaching of lithium-rich clays; and (4) saline waters of lakes and salars in the high Andes to the east, which moved into the Salar de Atacama basin through permeable fault zones of Pliocene(?) and Quaternary age. Long-term accumulation and evaporation of these waters, under desert conditions that have prevailed at least since the Miocene, gave rise to the large amounts of saline materials in this basin.

INTRODUCTION

Salar de Atacama, the largest salar or salt pan in Chile, is in a basin of internal drainage between the Cordillera de Domeyko, one of the front ranges of the Andes, and the Andean Highlands. The salar, at an altitude of 2300 m, is an elongate basin (Figure 1) filled with lacustrine clastic sediments and evaporites. It is 85 km long, has a maximum width of 50 km, and covers an area of about 3000 km^2. It is accessible by road from Calama in the north and from the Antofagasta-Socompa road in the south. The basin in which the salar occurs, referred to as the Atacama basin in this chapter, has its own microclimate that is intermediate between the climate of the Andean Highlands to the east and the Atacama Desert to the west. The annual rainfall is only 10 to 30 mm, and most of the precipitation occurs from December to March. The climate is characterized by westerly winds, predominantly in the afternoon, which at times attain velocities of as much as 100 km/hr. Relative humidity is low—commonly less than 10%—and temperatures fluctuate between a minimum of $-5\,^\circ$C in winter to about 35° in summer. Skies are generally cloudless, and solar radiation is about 630 langleys/day (6,300,000 cal/m^2/day). These climate conditions result in a high rate of evaporation of 10 L/m^2/day (metallic pan measurement), which favors concentration of lithium-rich brines by solar evaporation.

Lithium-rich brines are recovered and processed by a joint company owned by the Corporación de Fomento de la Producción (CORFO) (45%) and the Foote Mineral Company (55%), now the Cyprus Specialty Metals Division of Cyprus Minerals Company. Annual production capacity (1988) is about 8200 t of lithium carbonate.

REGIONAL GEOLOGY

Salar de Atacama is in a tectonic basin that originated in the Miocene. It has a basement of Paleozoic and Mesozoic marine rocks overlain by Tertiary-Quaternary continental sedimentary rocks, in part with abundant saline beds, rhyolitic ash-flow tuffs, and andesitic lavas and breccias. Unconsolidated clastic sediments have accumulated as erosion products of the bedrock, and huge amounts of saline minerals, predominantly of volcanic origin, have accumulated in the salar itself during late Tertiary and Quaternary. Saline minerals continue to accumulate today.

Tertiary sedimentary rocks, which are the sources of part of the saline materials, have been subdivided into six formations (from oldest to youngest): San Pedro, Tambores, El Campa-

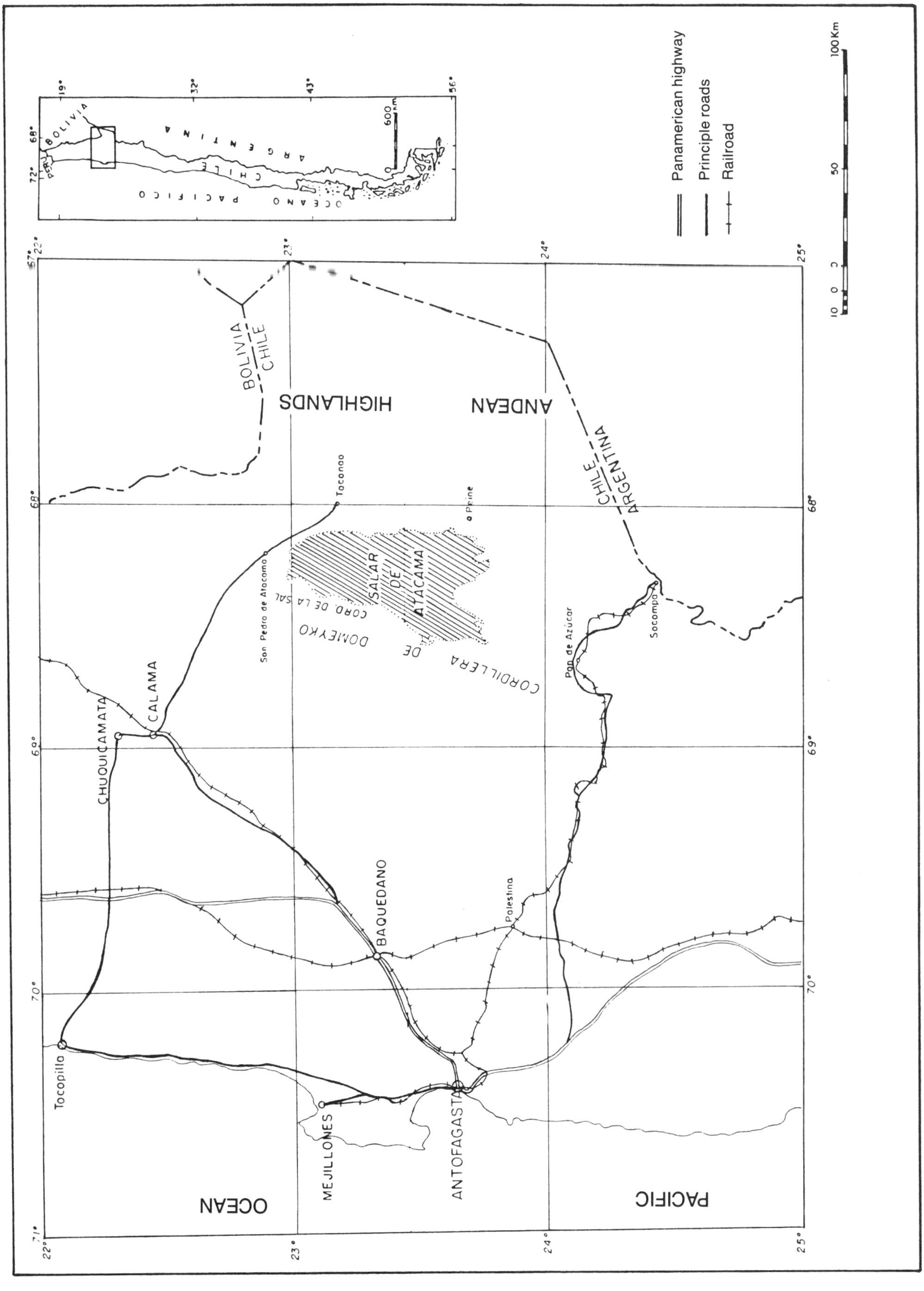

Figure 1. Sketch map showing location of the Salar de Atacama.

mento, Quepe, El Tambo, and Vilama. The San Pedro Formation is a sequence of reddish claystones and sandstones exposed in the Cordillera de la Sal, just west of the Salar de Atacama (Figure 2). It was deposited under conditions similar to the present, and includes erosion products from the Purilactis Formation (Triassic), which contains evaporite units. The overlying Tambores Formation, of Oligocene to late Miocene age, consists of conglomerates and interbedded sandstones, and locally abundant veinlets of gypsum (Ramirez and Gardeweg, 1982). This formation crops out along the western side of the Cordillera de Domeyko and west of Tilomonte (Figure 2). The El Campamento Formation is a sequence of interbedded halite and volcanic rock units that crops out in the Cordillera de la Sal. Ramirez and Gardeweg (1982) assigned an age of late Miocene to Quaternary to this formation. At the north end of the Cordon Lila, the El Campamento Formation consists of about 20 m of interbedded volcanic ash, sandstone, and siltstone firmly cemented by halite, gypsum, and calcium carbonate. The Quepe Formation, which is exposed at the western side of the Andean Highlands, along the eastern side of the Atacama basin, consists of sandstones, conglomerates, and siltstones of probable Tertiary age (Ramirez and Gardeweg, 1982). The El Tambo and Vilama formations are of late Miocene to Pleistocene age. El Tambo consists of brown calcareous rocks that crop out west of Tilomonte and near the village of Peine (Figure 2). The Vilama Formation consists of interbedded crystal-rich tuffs, volcanic ash, diatomite, and gravels. It crops out in the Cordillera de la Sal, the eastern side of the Cordillera de Domeyko, and the western side of the Andes. The formation unconformably overlies the San Pedro Formation.

The above formations are covered by extensive late Miocene to Quaternary alluvial deposits, chiefly piedmont sediments and mudflows, in the Llano de Paciencia, just west of the Salar de Atacama, and in the vicinity of Peine. The Andean Highlands, Atacama basin, and Cordillera de Domeyko are extensively covered with late Tertiary and Quaternary volcanic rocks of two dominant types—rhyolitic to dacitic ash-flow tuffs and andesitic lavas—the latter being associated with stratovolcanoes. Along the Andean front, the tuffs form extensive sheets that are gently inclined westward into the Atacama basin. They extend beneath the alluvial cover of the basin, and have been encountered in drill holes beneath as much as 250 m of alluvium. The ash-flow tuffs fill ancient valleys and depressions, and cover topographic highs on older rocks. At least 11 tuff sheets, as shown in Table 1, have been recognized. They range from late Miocene to Pleistocene in age, extend over areas of as much as 600 km^2, and are as thick as 250 m (Table 1). The tuffs are characteristically high in lithium, small amounts of which are readily soluble in water (Table 1).

Tectonic movements during the Pliocene and Quaternary caused extensive faulting (Figure 2), uplift of the Domeyko and Andean cordilleras, and development of several morphostructural units within the basin. Among the latter are the Cordillera de la Sal, Cordon Lila, Sierra Alameida, and the Tilocalar graben (Figure 2). The Cordillera de Domeyko is an uplifted block, bounded by north-trending reverse faults, in which folded Paleozoic and Mesozoic sedimentary and volcanic rocks are predominant. The Cordillera de la Sal, which consists chiefly of the San Pedro Formation disconformably

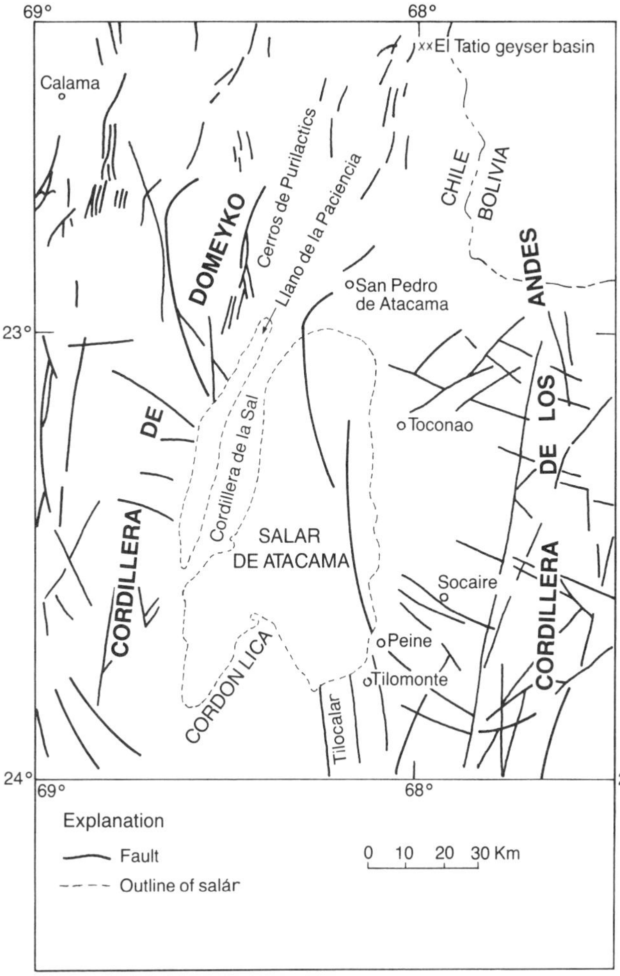

Figure 2. Sketch map showing faults and morphostructural units in the Salar de Atacama region.

overlain by the Vilama Formation and Miocene ash-flow tuffs, is bounded by north-trending normal faults. Northeast-trending anticlines and synclines are present in this cordillera, as are salt domes. The Cordon Lila, to the south of the Cordillera de la Sal, divides the Salar de Atacama into two subbasins. It is an uplifted block that probably also is bounded by faults that are buried by alluvium. The Tilocalar graben is bounded by north-trending, steeply dipping normal faults. Other faults within the graben cut both the Tambores Formation and the overlying Tucúcaro ash-flow tuff. Andesitic lavas from two small volcanoes in the central-western part of the graben cover both the ash flow tuffs and alluvium. In the Andean precordillera a number of west-northwest lineaments extend to the edge of the Salar de Atacama (Figure 2). The Sierra de Almeida, a north-south mountain range, extends southward from within the large embayment at the southern end of the Salar de Atacama (Figure 2) for nearly 100 km.

Hydrology of the Atacama Basin

The Salar de Atacama is constantly being recharged by streams and ground water from the Andes, where annual rainfall probably exceeds 50 mm, and to a lesser extent by

Table 1. Ash-flow tuff sheets in the Salar de Atacama Basin.

Tuff Sheet	Age	Area (km^2)	Thickness (m)	Total Li (ppm)[1] Soluble	Total
Sifón	Late Miocene (about 9 Ma)		69-120	1	28
Yerba Buena	Late Miocene		21	3	154
Pelón	Late Miocene		11	2	30
Toconao	Late Miocene	20		4	147
Puripicar	Late Pliocene	600	20-100	2	50
Atana	Pliocene	340	50		
Chaxas	Late Pliocene-Pleistocene		60	1	89
Tucúcaro	Pliocene	600		4	69
Patao	Late Pliocene-Pleistocene	490		15	83
Tatio	Pleistocene (about 1 Ma)		10-60	?	93
Cajón	Pleistocene	530	100-250		

[1]Analyses by the Servicio Nacional de Geología y Minería, Santiago, Chile.

rainfall in the center of the basin and in the Cordillera de Domeyko. The principal streams draining into the salar are the Río San Pedro and Río Vilama (Figure 3). These streams discharge water into the salar throughout the year but infiltrate before reaching the salar nucleus. Smaller streams, only part of which carry water the entire year, are along the western and southern sides of the salar. These smaller streams discharge water directly into the salar only during periods of relatively heavy rainfall in the Andes; at other times, waters disappear into the gravels east of the salar, and move into the salar as ground water. Annual flow rates of some of the streams are as follows (in L/sec):

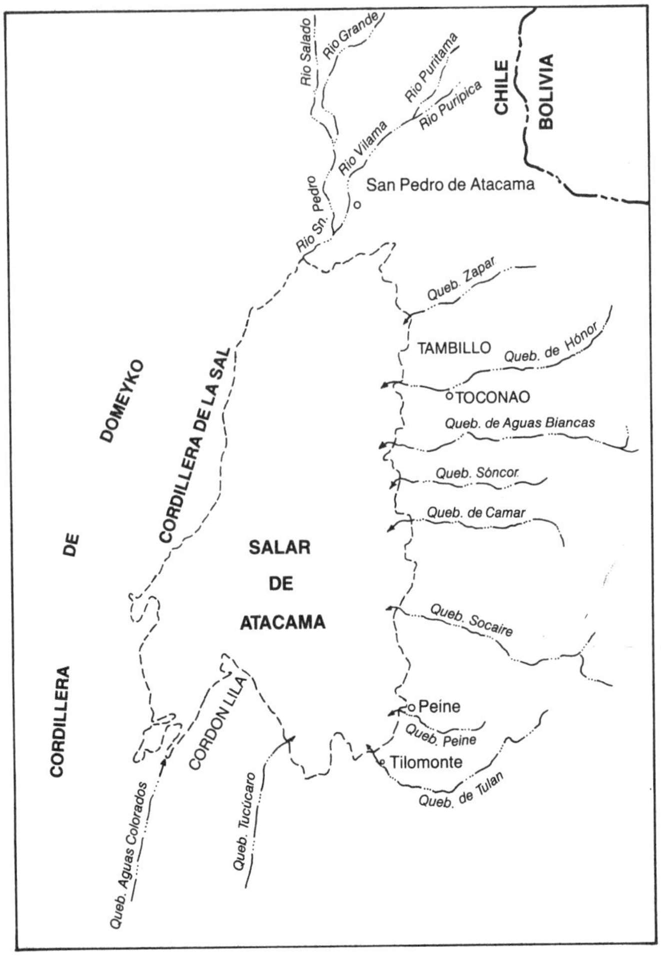

Figure 3. Map showing principal streams draining into Salar de Atacama.

San Pedro 1000
Vilama. 218
Hónar 50
Aguas Blancas. 134
Camar 3
Peine. 17
Tulan 59

The total average rate of discharge of these streams is 1677 L/sec.

Drilling by the Corporación de Fomento de la Producción has identified three aquifers in the central part of the Salar de Atacama basin. The uppermost is an open aquifer that is recharged by streams draining into the salar. The other two are locally artesian confined aquifers in and near the northern part of the salar. The waters in these confined aquifers have an average temperature of 26° C, which is higher than the average annual temperature of the area. The mechanism of recharge of confined aquifers is not known, but presumably is by infiltration of meteoric waters moving downward through permeable sediments, fractured tuffs, and bedrock faults along the Andean front to the east of Salar de Atacama. In addition, major east-west faults along the Andean front probably serve as conduits for waters from the Andean Highlands, including lithium-bearing saline waters from lakes and salars.

Both warm- and cold-water springs are found at many places in the Salar de Atacama basin, some in the salar itself. Some of these springs contain travertine deposits. The most unusual springs are within the halite nucleus of the salar where water from a deep, confined aquifer has dissolved the salt and formed open pools that Stoertz and Ericksen (1974) named "brine pools."

Most of the stream and ground water in the Atacama basin is slightly saline and tends to be relatively high in Li (Table 2). The Río San Pedro and Río Vilama contain 0.4 and 3 ppm Li, respectively, which reflects the composition of their sources. The Ríos Grande and Salado (Figure 3), which join to form the

Table 2. Chemical analyses of natural waters in the Salar de Atacama Basin.[1]

Muestra	Li (ppm)	K (ppm)
Río Salado (spring)	<1	<1
Río Salado	<1	3
Río Grande	<1	13
Río San Pedro	<1	13
Río Puripica	3	27
Río Puritama	3	29
Río Vilama	3	27
Río Blanco	<1	17
Río Incahuasi	<1	9
Tatio geyser	26	181
Quebrada Aguas Blancas	1.6	25.7
Quebrada Camar	1.5	35
Quebrada Peine	0.7	11
Quebrada Socaire	0.3	10
Quebrada Zápar	0	0
Quebrada Hónar	0	4.8
Socaire (spring)	0.9	10.3
Quelana (artesian well)	91	980
Tilomonte (well)	1.7	42
Tilopozo (well)	6.4	130
Tucúcaro (pit in alluvium)	6.3	140
San Pedro (artesian well)	3.8	50
Agua Colorada (spring)	0	10
Salar de Atacama (nucleus)	1,500	18,000

[1]Analyses by Foote Mineral Company.

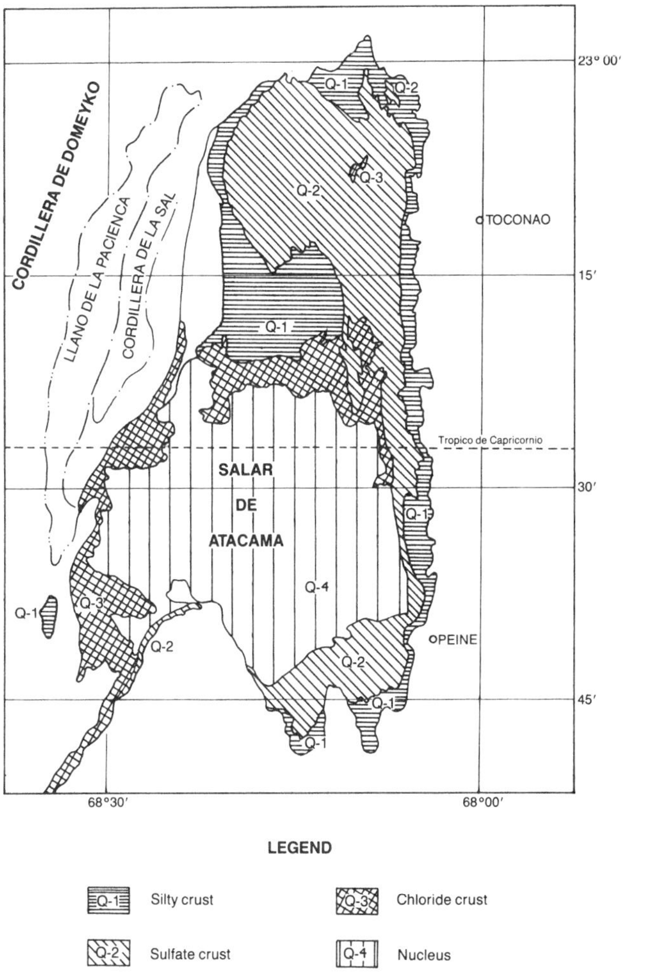

Figure 4. Map showing types of saline crusts of Salar de Atacama.

Río San Pedro, contain less than 1 ppm Li, whereas the Ríos Puripica and Puritama, which join to form Río Vilama, contain an average of 3 ppm Li. Thermal waters of the El Tatio geyser field, north of the Atacama basin (Figure 2), have unusual concentrations of as much as 47 ppm Li (Cusicanqui, Majon, and Ellis, 1975). The streams along the eastern side of the salar contain small amounts of dissolved salts but have Li contents of less than 1 ppm, except for Aguas Blancas and Camar, which contain about 1.5 ppm Li (Table 2).

The ground waters of the Salar de Atacama basin tend to have more dissolved salts than the surface waters. Lithium concentrations are generally greater than 3 ppm, and an exceptional concentration of 91 ppm Li was found in the artesian waters of the Quelana well north of Peine. This exceptional Li concentration is believed to be due to enrichment by dissolution of Li-rich evaporites at depth.

Geology of Salar de Atacama

Salar de Atacama is a tectonic basin filled with clastic sedimentary material and evaporites. It consists of a thick core or nucleus (Figure 4) of brine-saturated, porous to dense halite surrounded by zones where the saline crust tends to be thin and contains abundant sulfate minerals. The surface of the nucleus is rarely, if ever, flooded. The phreatic brine level is generally at a depth of 50 to 70 cm, whereas the marginal zone has many ponds, particularly during the rainy season, and the phreatic level is essentially at the surface in most unflooded areas.

As a consequence of the frequent flooding, the marginal zones of the salar undergo constant change due to dissolution and reprecipitation of the saline crust. Also, during periods of flooding, the saline waters in the marginal zones move toward the center of the salar to recharge the brines of the nucleus that are constantly lost by capillary evaporation in the porous upper halite crust. Such movement of brine is most pronounced in the northern and eastern sides of the salar, where the greatest inflow of fresh water takes place. The relatively constant brine level and lack of flooding in the salar nucleus indicates equilibrium between recharge and evaporation in the salar.

Four major types of saline crust are recognized in Salar de Atacama (Figure 4):

(1) A silty crust that consists principally of silt and sand firmly cemented by halite and underlain by moist to brine-saturated fine-grained silts and sands. This type of crust formed by capillary evaporation of brine within the clastic sediments to deposit the salt cement (Stoertz and Ericksen, 1974). Such crust is characteristic of relatively dry marginal areas of the salar.

(2) A sulfate crust consisting of soft saline efflorescences in marginal zones of the salar, where the phreatic level is essentially at the surface or where periodic flooding and desiccation take place. Sulphate minerals such as gypsum and thenardite tend to be abundant in such crust, as is halite. Minor amounts of other sulfate minerals and the borate mineral ulexite may also be present.

(3) A chloride crust that consists predominantly of halite and which is intermediate between the marginal zones of thin saline crust and the thick salar nucleus. The halite is porous and the phreatic brine level is generally at depths of 50 to 70 cm. The brine is saturated with NaCl.

(4) The salar nucleus, which consists of relatively pure halite and in which the brine generally has a density of more than 1.2 g/L. Drill holes in the nucleus have locally penetrated as much as 390 m of relatively massive halite without reaching the underlying clastic sediments or bedrock. Only the upper 20 to 30 m of the halite mass of the nucleus is sufficiently porous to yield the Li-rich brines now being exploited. In 1975, a helicopter-supported brine-sampling program that was carried out in the nucleus showed the distribution of high Li brine (Figure 5). In this program, brine samples were collected on a 5-km grid from small-diameter holes drilled to a depth of 3 m. In addition, brine samples were collected at 0.5-km intervals along all existing roads in the salar from pits dug to a depth of 1 to 1.5 m. In some places, such samples were collected as far as 7 km from the nearest road. Figure 5 shows that the highest average lithium concentrations are near the Chepica Peninsula in the southernmost part of the salar.

ORIGIN OF LITHIUM AT THE SALAR DE ATACAMA

The origin of lithium at the Salar de Atacama is clearly volcanic. However, analytical data obtained from a variety of samples and waters clearly indicate that no single mechanism or volcanic source is responsible for the presence and the introduction of lithium into the salar system; several mechanisms and sources must be considered. In an effort to discover these various potential mechanisms and sources, sampling of various rock types and waters within the hydrological basin of the salar was undertaken. Available data obtained by prior researchers have also been incorporated into the interpretation.

Samples of various ash-flow tuffs north of San Pedro de Atacama showed concentrations of 29 to 89 ppm total Li. These samples were also washed with distilled water to evaluate lithium solubility. In the short residence time of 24 hours, during which the ground sample was in contact with water at room temperature, 1 to 2 ppm Li was dissolved. Marinovic and Lahsen (1984) show that these tuffs extend over an area of 1700 km^2 in the Calama quadrangle, chiefly within the Salar de Atacama basin. The potential available lithium per meter of thickness of tuffs in this area is calculated to be 12,800 tonnes for an assumed minimum release of 2 ppm Li. The Sifón tuff has recorded thicknesses of as much as 120 m; the Río Salado tuff is even thicker. The tuff sheets of the Toconao quadrangle cover an area of 2100 km^2, also predominantly within the Salar de Atacama basin, and show thicknesses ranging from 20 m to more than 220 m (Ramirez and Gardeweg, 1982). Such volumes of ignimbrite could easily account for all the lithium in the Salar basin. The resources in the Salar nucleus have been estimated at 4.6 million tonnes Li. The total amount contained in dilute saline waters within the entire basin may be twice that amount.

The pattern of distribution of lithium has been well documented at the Salar de Uyuni (Risacher and Ballivan, 1981). In a study of the distribution of lithium over the salar, the highest anomaly was recorded at the mouth of the Río Grande de Lipez. There the concentration of lithium as well as boron is clearly in excess of the solute concentration trend prevalent in the northern Bolivian Altiplano. Risacher concluded that the major source of lithium must be the rhyolitic ash-flow tuffs that are prevalent in the drainage of the Río Grande. Ericksen and Salas (this volume) conclude that lithium and other saline materials in all salars of the central Andes are chiefly of volcanic origin, and were leached from the predominant rhyolitic ash-flow tuffs by meteoric waters as well as accumulating in thermal-spring waters.

The process of lithium release by weathering of voluminous ash-flow tuffs must be viewed, however, as a part of a much broader picture. The volcanic process does not provide only solids, but is also accompanied by significant volume of geothermal fluids of both meteoric and magmatic origin, which can and do contain enormous quantities of lithium. Lithium resources contained in the geothermal reservoirs of the United States have been estimated at 4.5 million tonnes Li. Clearly thermal-spring waters are major sources of lithium. The presently active Tatio geothermal field, north of Salar de Atacama, shows concentrations in the thermal-spring waters of 28 to 40 ppm (Cusicanqui, Majon, and Ellis, 1975).The Tatio system is an important source of lithium and, along with other now-extinct geothermal systems in the Salar de Atacama basin, has provided large quantities of Li-rich waters to the Salar de Atacama. Evidence for these older systems is found along the eastern side of the salar, where former thermal springs deposited large amounts of calcium-carbonate-rich sediments and probably released significant amounts of lithium into the salar. These carbonate sediments contain as much as 322 ppm Li, chiefly in the acid-soluble fraction. Present-day spring waters in this area contain only about 1 ppm Li.

Although the lithium is related to volcanic activity, some clearly predates the ash-flow tuffs. Argillaceous and saline-bearing sediments underlying the tuffs in the El Campamento Formation, Chepica Peninsula, and the San Pedro Formation, west of San Pedro de Atacama, show high lithium concentrations, especially in the saline and the fine-grained sediment fractions. The El Campamento Formation contains 25 to 484 ppm Li, of which 2 to 136 ppm is in the water-soluble fraction. Similarly, concentrations of 43 to 493 ppm Li are found in the carbonate sediments of this formation. This clearly indicates that significant lithium was available during deposition of the Campamento Formation, which points to a pre-existing lithium anomaly for the Salar de Atacama basin. The exact source of this lithium is unclear. Similar lithium anomalies have been documented for the Clayton Valley, Nevada basin (Kunasz, 1974), where Tertiary lacustrine sediments predating the Quaternary salar were found to contain anomalously high lithium as well as potassium concentrations. Therefore, the leaching of lacustrine sediments exposed around the Salar de Atacama must also be considered as an important mechanism for the introduction of lithium into the salar. The high-lithium source in the Chepica Peninsula is reflected in the presence of the highest average lithium concentration in brines of the entire Salar de Atacama basin (Figure 5).

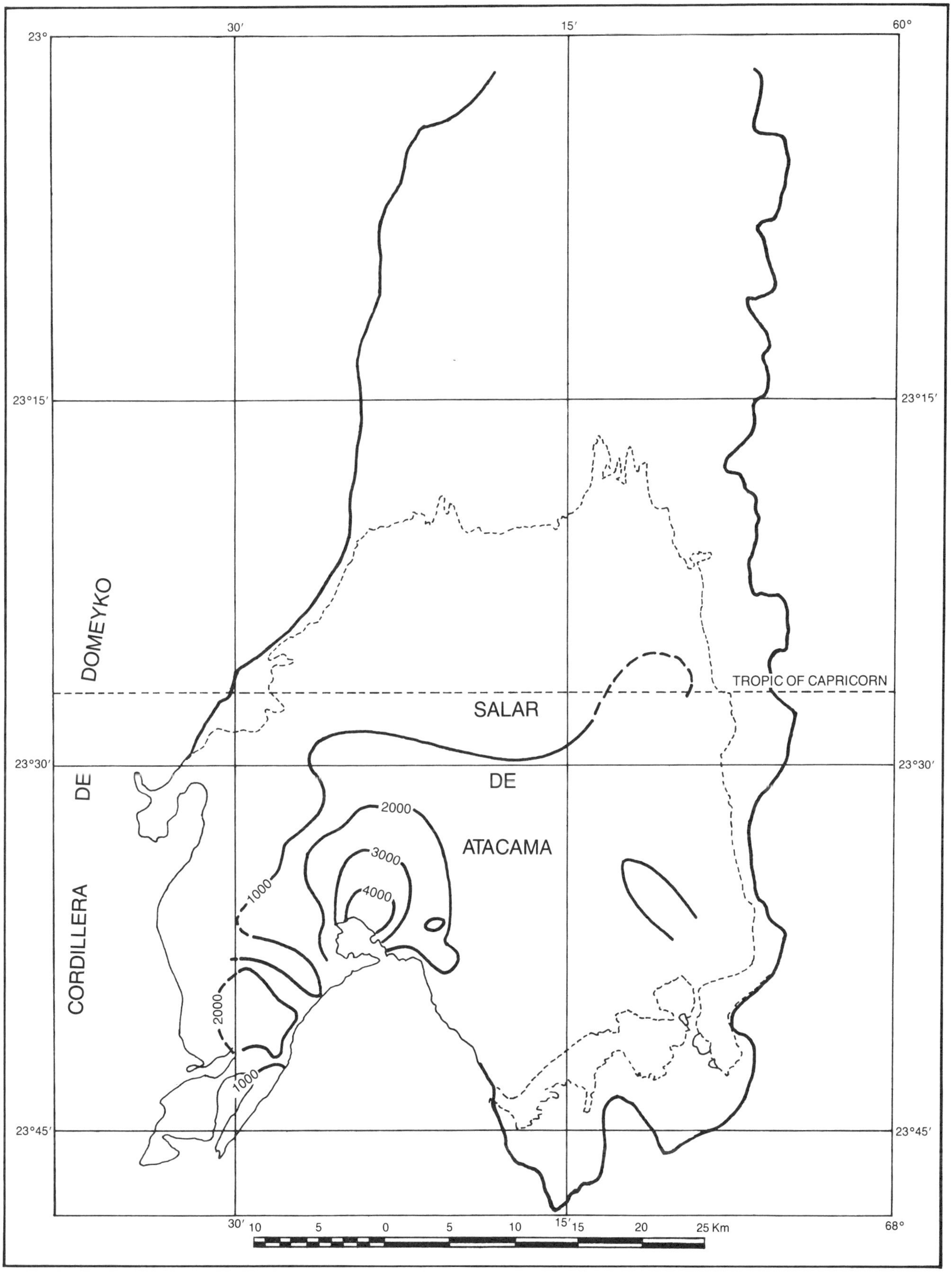

Figure 5. Map showing lithium concentration (ppm) isopleths in the southern part of Salar de Atacama.

Lithium-rich waters from lakes and salars in the Andes east of Salar de Atacama may have moved into the Atacama basin along permeable fault zones. Such faults in the basin have been documented by Frutos (1972). Vila (1975) has shown that brines in these lakes and salars commonly contain more than 100 ppm Li.

In conclusion, the origin of the lithium in the Salar de Atacama must be viewed from several perspectives. The leaching of volcanic rock was the principal source of this lithium, which was discharged into the salar by stream- and ground-water flow, High-lithium clays along the margins of Salar de Atacama probably resulted from weathering of the Tertiary volcanic rocks. Although mineralogic studies have not been made of the clays, Li-bearing hectorite probably is present. A second source of lithium in Salar de Atacama was weathering and leaching of lacustrine saline sediments predating the salar. Geothermal fluids were a third source of the lithium. Lithium-bearing brines from saline lakes and salars to the east of Salar de Atacama, which moved into the Atacama basin along permeable fault zones, supplied additional lithium. Long-term evaporation of all these waters has resulted in accumulation of the richest lithium brine deposit in the world.

REFERENCES CITED

Cusicanqui, H., W.A.S. Majon, and A.J. Ellis, 1975, The geochemistry of the El Tatio geothermal field, northern Chile: San Francisco, Second U.N. Symposium on the Development and Use of Geothermal Resources Proceedings, v. 1, p. 703-711.

Ericksen, G.E., and R. Salas O., 1989, Geology and resources of Andean salars, *in* G.E. Ericksen, M.T. Cañas P., and J.A. Reinemund, eds., Geology of the Andes and its relation to hydrocarbons and mineral resources, this volume.

Frutos, J., 1972, Ciclos tectónicos sucesivos y direcciones estructurales sobreimpuestas en Los Andes del Norte Grande de Chile: Simposio sobre los resultados de investigaciones del manto superior continental en América Latina, v. 2, p. 473-489.

Kunasz, I.A., 1974, Lithium occurrence in the brines of Clayton Valley, Esmeralda County, Nevada: Northern Ohio Geological Society International Salt Symposium, v. 1, p. 57-66.

Marinovic S., N., and A. Lahsen A., 1984, Geología de la Hoja Calama, Región de Antofagasta: Carta Geológica de Chile, v. 58.

Ramirez R., C., and M. Gardeweg P., 1982, Geología de la Hoja Toconao, Región de Antofagasta: Carta Geológica de Chile, v. 54.

Risacher, F., and O. Ballivian, 1981, Los salares de Altiplano Boliviano: Paris, Office de la Recherche Scientifique et Technique Outre-Mer, 246 p.

Stoertz, G.E., and G.E. Ericksen, 1974, Geology of salars in northern Chile: U.S. Geological Survey Professional Paper 811, 65 p.

Vila, G., T., 1975, Geología de los depósitos salinos andinos, Provincia de Antofagasta, Chile: Revista Geológica de Chile, v. 2, p. 41-55.

Ericksen, G. E., Cañas Pinochet, M. T., and Reinemund, J. A., editors, 1989, Geology of the Andes and its relation to hydrocarbon and mineral resources: Houston, Texas, Circum-Pacific Council for Energy and Mineral Resources Earth Science Series, v. 11.

CHAPTER 12

AREAS OF POTENTIAL NEW GOLD DEPOSITS IN COLOMBIA

Hernando Lozano Q. and Joaquin Buenaventura A.
Instituto Nacional de Investigaciones Geológico-Mineras
Bogotá, Colombia

ABSTRACT

Gold production in Colombia has been continuous since before Colonial times. In the 19th century, Colombia was the world's largest gold producer, but during the 1980s, in spite of showing an increase, its production has decreased in relation to that of other countries. Gold production in Colombia in 1984, 1985, and 1986 was 750,000, 1,142,386, and 1,279,274 oz, respectively, which ranks 14th in the world and 5th in Latin America. Most of the Colombian gold (80%) is produced from small- and medium-size placer operations, chiefly in the western part of the country. Studies by the Instituto Nacional de Investigaciones Geológico-Mineras (INGEOMINAS) have shown that recent faults have been important in determining the distribution of some placer deposits. These studies also show that paleosutures, such as the Romeral fault, are attractive exploration targets for disseminated and vein-type deposits of precious metal. The potential for discovery of new gold deposits in Colombia is believed to be excellent in both the Andes and the Guayana Shield (Colombian Orinoco-Amazon region).

INTRODUCTION

In recent years, the Instituto Nacional de Investigaciones Geológico-Mineras (INGEOMINAS) has compiled and interpreted available geological, tectonic, and mine information of Colombia for the purpose of defining the geological environments of known precious-metal deposits. As a result, deposit models have been prepared and potential areas for new vein-type, disseminated, and alluvial gold deposits have been selected. The compilation was restricted to lode and placer deposits in which the main element is gold. Other types of deposits in which base metals are prevalent—such as skarn, massive sulfide, and porphyry copper deposits—which are also important sources of gold, are not discussed. The studies indicate that the region west of the Magdalena River, in the Central and Western Cordilleras, is the most important gold area of Colombia. The Vetas-California district, in the Department of Santander, has the only economically significant gold deposits in the Eastern Cordillera. Important gold discoveries have been made in recent years in Guainia and Amazonas Comisarias near the border with Brazil.

Information in this report is based chiefly on studies by West (1972), geological and mining maps published by INGEOMINAS, and by Hall et al. (1970, 1971), Lopez (1971), Guarin (1971), and Mosquera and Buitrago (1971), Buitrago and Buenaventura (1975), Buitrago, Buenaventura, and Perez (1976), Buenaventura (1976a, 1976b), and Rodriguez and Pernett (1980).

PRODUCTION

Gold production in Colombia has been continuous since Colonial times. In the 19th century, Colombia was the world's largest producer; however, it is currently the 14th largest producer worldwide and is 5th in Latin America. Although gold production in Colombia has recently increased by 40%, its overall production has not increased as rapidly as in other countries where intense exploration and exploitation programs have been recently implemented. Gold production in Colombia from 1964 to 1986 (Figure 1) has followed international gold prices. Most of the Colombian gold is produced by medium and small mining operations, and more than 80% comes from alluvial deposits in the Cauca, Nechi, Atrato, San Juan, Saldaña, Patia, Porce, and Iscuande rivers located in the Central and Western Cordilleras. In 1983, 78% of the national production came from the Department of Antioquia in the northwestern part of the country (Figure 2). Important gold areas of this department, such as El Bagre, Zaragoza, Caucasia, and Nechi, have shown an increase in medium to small mining operations during recent years (Rodriguez et al., 1985).

AGE AND ORIGIN OF GOLD DEPOSITS

Most of the gold veins in the Central and Western Cordilleras are related to Tertiary diorite and quartz diorite intrusions, such as the Mande batholith and the Manizales and Hatillo

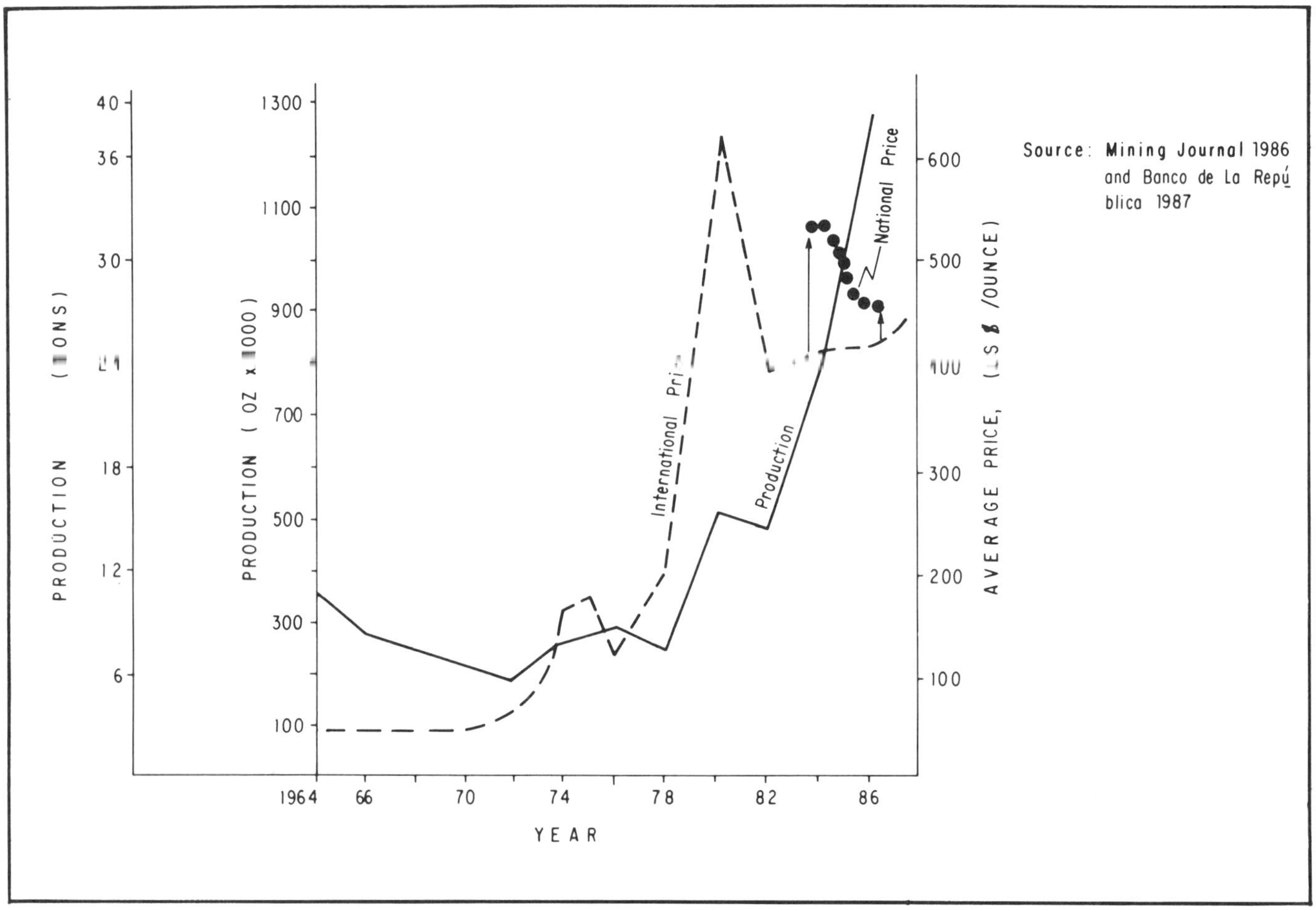

Figure 1. Production in gold in Colombia and gold prices (1964-1986).

stocks in the northern part of the country, and to dacite and andesite porphyries, such as in Cerro Bolivar, Lerma, Cerro Gordo, La Campana, and Paso de Bobo, in the departments of Cauca and Nariño in the southwestern part of the country. Although no radiometric-age studies have been done, these deposits are believed to be related chiefly to Jurassic and post-Jurassic magmatic events. As proposed by Utter (1984), the gold, possibly present originally in mafic marine volcanic rocks of various ages, has been remobilized and concentrated by metamorphic processes, felsitic intrusive activity, and local faults. The gold was deposited as veins, stockworks, and disseminations in different types of rocks. The Segovia, Remedios, and Sonson deposits, Department of Antioquia, are near or at margins of quartz diorite or granodiorite intrusions of Jurassic age. The deposits in the central part of the Department of Antioquia are apparently related to the Antioquia batholith of Cretaceous age.

The gold deposits of Colombia were emplaced chiefly during two major metallogenic epochs. The oldest deposits are related to the middle-upper Mesozoic plutons, largely in the Central Cordillera and the northern part of the Eastern Cordillera. Gold districts related to the Antioquia and Ibague batholiths, with some stocks intruding calcareous rocks, are found in the Central Cordillera and in the Vetas-California district in the northern part of the Eastern Cordillera. The second main metallogenic epoch, during the Tertiary, is characterized by gold deposits genetically related to intrusions in the Western Cordillera. These deposits tend to be polymetallic, rich in gold, and mostly associated with diorite porphyries of Tertiary age. Although the geology is not well known, this area contains important districts, such as Mande in the north and Marmato and Nariño-Cauca in the south.

Gold-quartz veins of Colombia commonly contain associated sulfide minerals, and the ores generally yield more than 5 g/t gold and relatively large amounts of silver. Such veins are generally epithermal and are irregular in thickness and length. The amount of ore in individual veins is relatively small.

In addition to the areas described above, the contact zones between oceanic crust and continental crust, such as the Romeral fault system (Figures 3, 4), are potential loci for lode-gold deposits.

AREAS OF POTENTIAL NEW DEPOSITS

Gold Veins

As a means of selecting areas having a potential for discovery of new lode-gold deposits, mineral deposit localities compiled by INGEOMINAS were plotted on geological maps. In the northeast Medellin zone (Figure 5), corresponding to the

Figure 2. Index map of Colombia showing departments and major rivers.

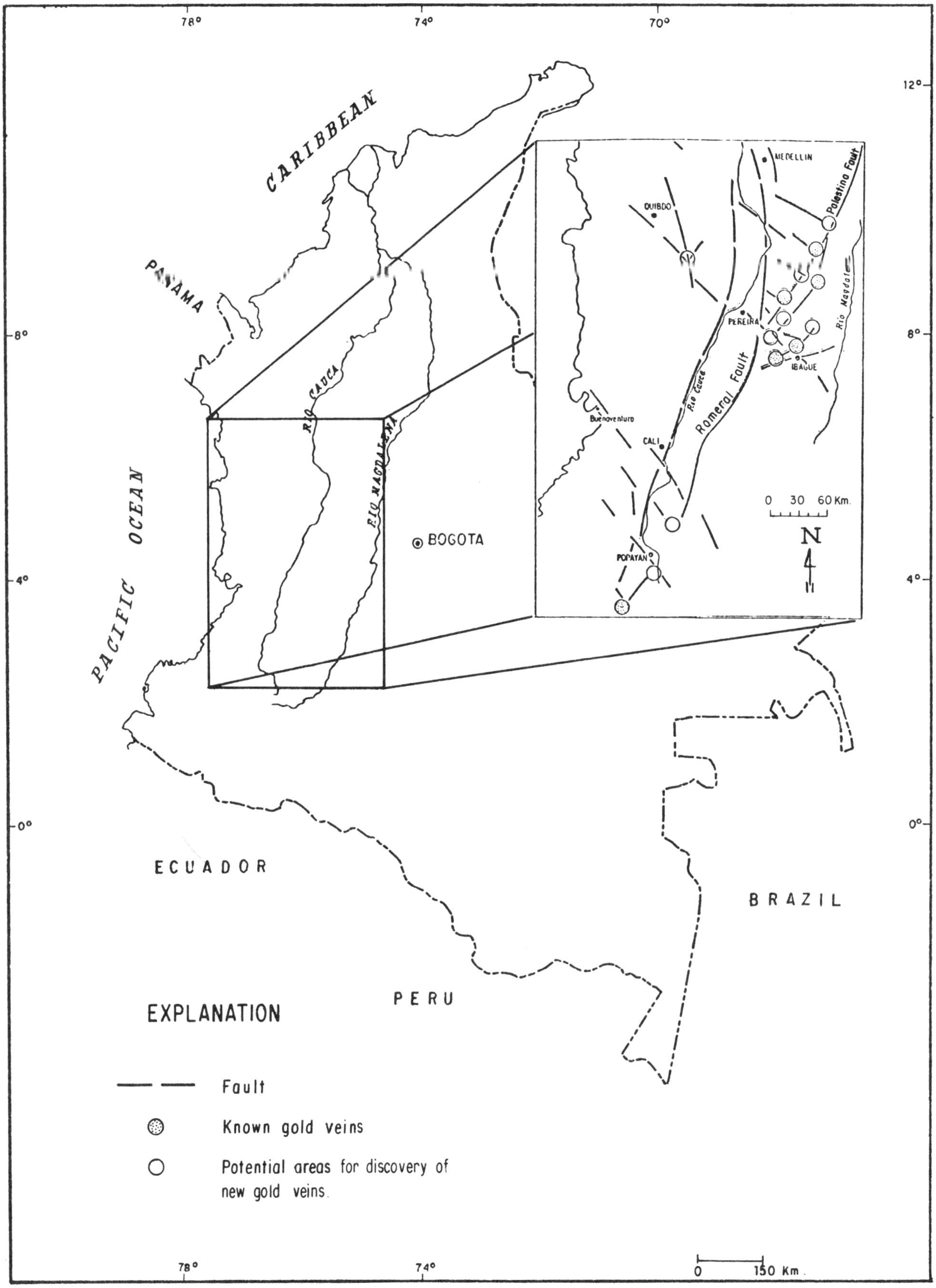

Figure 3. Areas having the greatest potential for discovery of new vein-type gold deposits.

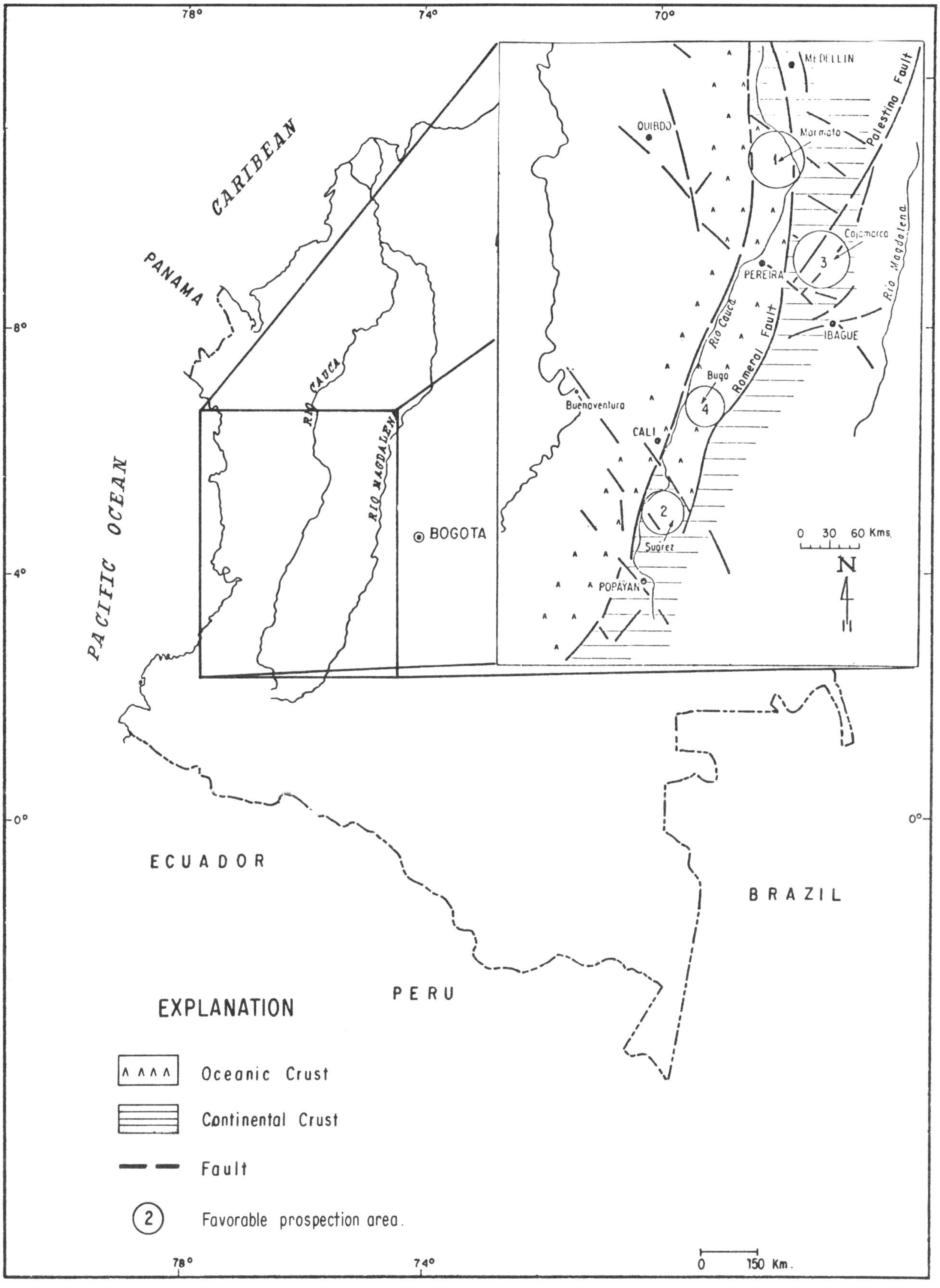

Figure 4. Areas having the greatest potential for discovery of disseminated gold deposits.

Frontino and Cañas Gordas areas, and in the central part of the Department of Antioquia, the gold veins are randomly distributed and it has not been possible to define any regional geological or tectonic control. These veins apparently are chiefly of Cretaceous age. South of the Montebello fault (Figure 5), between the cities of Ibague and Sonson, the veins are associated with stocks of Tertiary age, and tend to be clustered at intersections of major faults along and to the south of the Palestina fault system (Figure 5). In contrast, the Romeral fault apparently does not have associated vein-type gold deposits.

Based on the distribution of known deposits related to fault intersection zones, such as along the Palestina fault, the areas shown in Figure 3 are considered to be the most favorable exploration targets for gold veins in the Central Cordillera and eastern part of the Western Cordillera. In addition, host rocks have been important factors in determining the distribution of vein-type gold deposits, and serve as exploration guides. For example, the principal vein-type gold deposits in Antioquia are in the contact zone of a quartz diorite batholith of Cretaceous age that intrudes metamorphic rocks of Paleozoic age. Other deposits are in the upper part of this batholith. Elsewhere in the Colombian Andes, areas where the upper parts of intrusions have been less deeply eroded, such as those of Segovia in Antioquia, are more likely to contain vein-type deposits than the deeply eroded plutons. Favorable zones of this type are in the Sierra Nevada de Santa Marta, northernmost Colombia, and between the towns of Pitalito and Mocoa in the southern part of the Central Cordillera.

Disseminated Gold Deposits

In general, disseminated gold deposits are characterized by low gold values and large tonnages in comparison with vein-type deposits. Because they are not easily recognized, deposits of this type have not been found in Colombia in spite of the great potential for such deposits in several gold districts such as the Cajamarca district (Figure 4, area 3) (Pulido, 1985) and areas in the western and central parts of the Central Cordillera, where oceanic crust is faulted against continental crust along the Romeral fault system. Intensely fractured areas along the Romeral fault system produced by interaction with the Salento fault system and by the intrusion of stocks, dikes, and volcanic necks of intermediate composition are considered to be favorable for disseminated gold deposits. The Marmato area (Figure 4, area 1) is one of the zones where disseminated gold deposits appear to be associated with vein-type deposits. Another potentially important exploration target for disseminated gold deposits is the Buga batholith and surrounding area (Figure 4, area 4). This pluton, of Cretaceous age and intermediate composition, apparently was emplaced at the intersection of the Romeral and Palestina fault systems. It has been deeply eroded but possibly contains disseminated gold deposits. Other areas to the south that may have disseminated gold deposits are volcanic areas such as Ruiz, Tolima, Purace, and Galeras.

Alluvial Gold Deposits

In addition to being the major present source of gold in Colombia, placer deposits also are the major source of platinum. Exploitation, mainly by small to medium, hand-labor placer operations, is carried out in various parts of the country. The most important alluvial gold deposits now being exploited are along the Nechi River (Department of Antioquia), San Juan River (Department of Choco), and Saldaña River (Department of Tolima). Geological information available on the Nechi and Saldaña zones is more complete than that available on the San Juan zone. For this reason, the areas of El Bagre in the Nechi River zone, and Ataco zone, Saldaña River (Figure 6) are considered to contain types of gold placers that are most likely to be found elsewhere in Colombia. These deposits occur at recently reactivated faults where the river currents decrease in velocity and deposit sand and heavy minerals to form economically attractive gold placers.

The recently reactivated Zaragoza fault in the Nechi River zone (Figure 6) caused uplift of the southern block and depression of the northern block. Erosion of the many gold veins and small alluvial deposits in the southern block by the Porce and Nechi rivers yielded gold-bearing sediments that were carried northward and deposited along the Zaragoza fault zone to form the large Pato River and Nechi River placers. These placers have been exploited by small- and medium-size mining operations during the present century. In recent years, most of the gold production of Colombia has come from this area.

Conditions in the Saldaña River basin, Department of Tolima, are similar to those of the Nechi area. The Ataco fault (Figure 6) in this zone also has been affected by recent movement, which resulted in uplift of the southern block and depression of the northern block. Geologic units that crop out widely in the southern block are more restricted in the northern block where continental sediments of Tertiary age predominate. Sand, gravel, and gold transported by the Saldaña River have been deposited at the Ataco fault to form the rich gold placers.

Placer Gold Deposits

On the basis of the geological characteristics of fault-related alluvial deposits, the following nine large zones are considered favorable exploration targets for placer gold. These are located in the Andean and Guayana Shield areas.

Mutata-Barranquillita

The Mutata-Barranquillita area (Figure 7, zone 1) is in the northern part of the Atrato River basin, northwest of the municipality of Mutata, in the Western Cordillera. One of the last movements of the Mutata fault caused uplift of the southern block and depression of the northern block. As a consequence, geological units such as Cretaceous volcanic rocks and the Tertiary Mande batholith, which crop out in the southern block, do not crop out in the northern block. Movement along the Mutata fault caused widening of the Atrato River basin and changed the course of the Sucio River, which apparently formerly emptied directly into the Uraba Gulf through the Leon River basin. The Sucio River currently flows into the Atrato River, following the alignment of the Mutata fault.

The Sucio River basin consists of the Urama and Amparado River subbasins, where many vein-type deposits are known. Gold transported by the streams in these basins has been deposited in the northern part of the old Sucio River basin. For this reason, we consider that the polygonal area (BCDE) of the

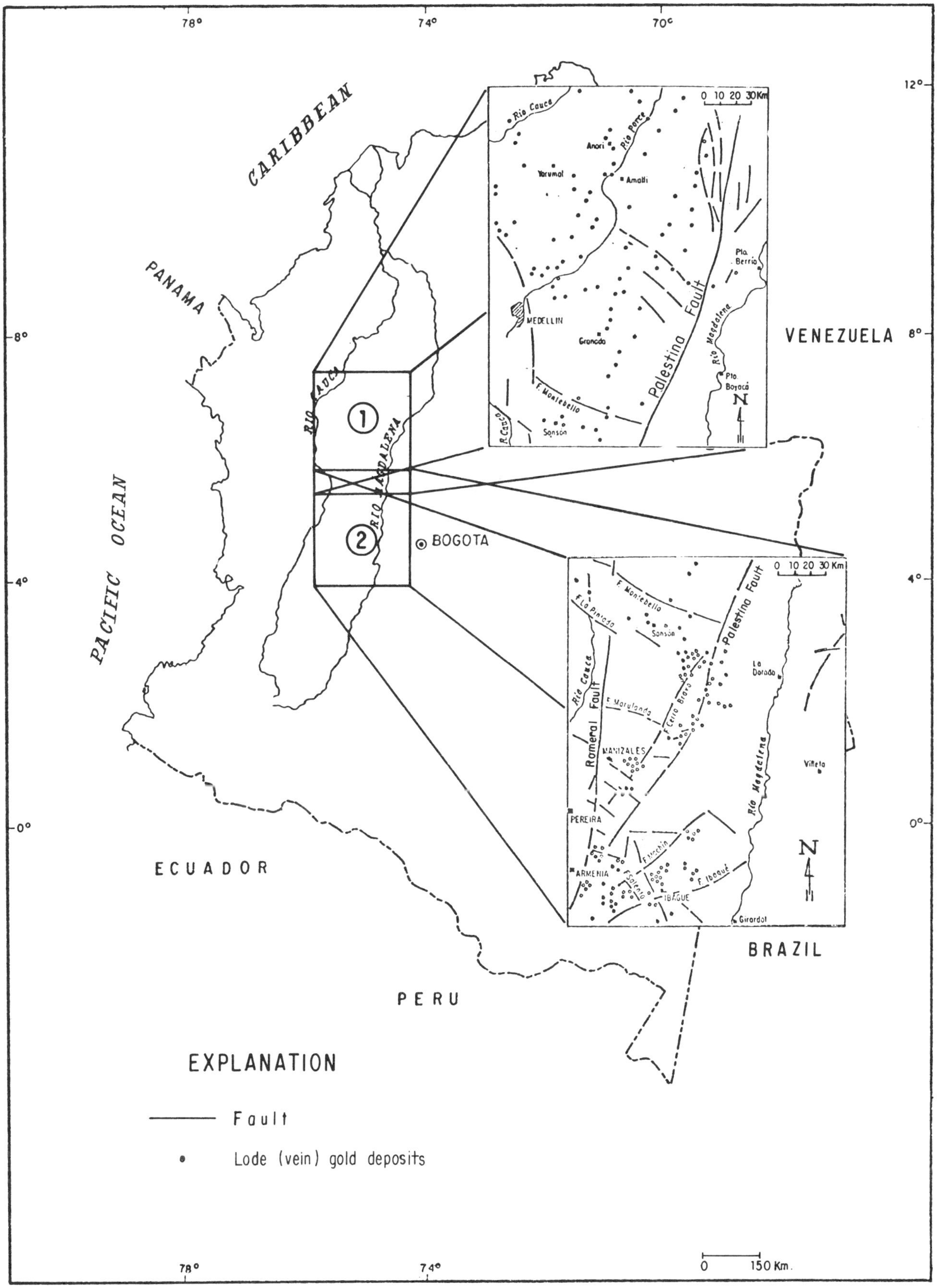

Figure 5. Lode-gold (vein) deposits in Colombia: (1) northeast Medellin zone and (2) Ibague-Sonson zone.

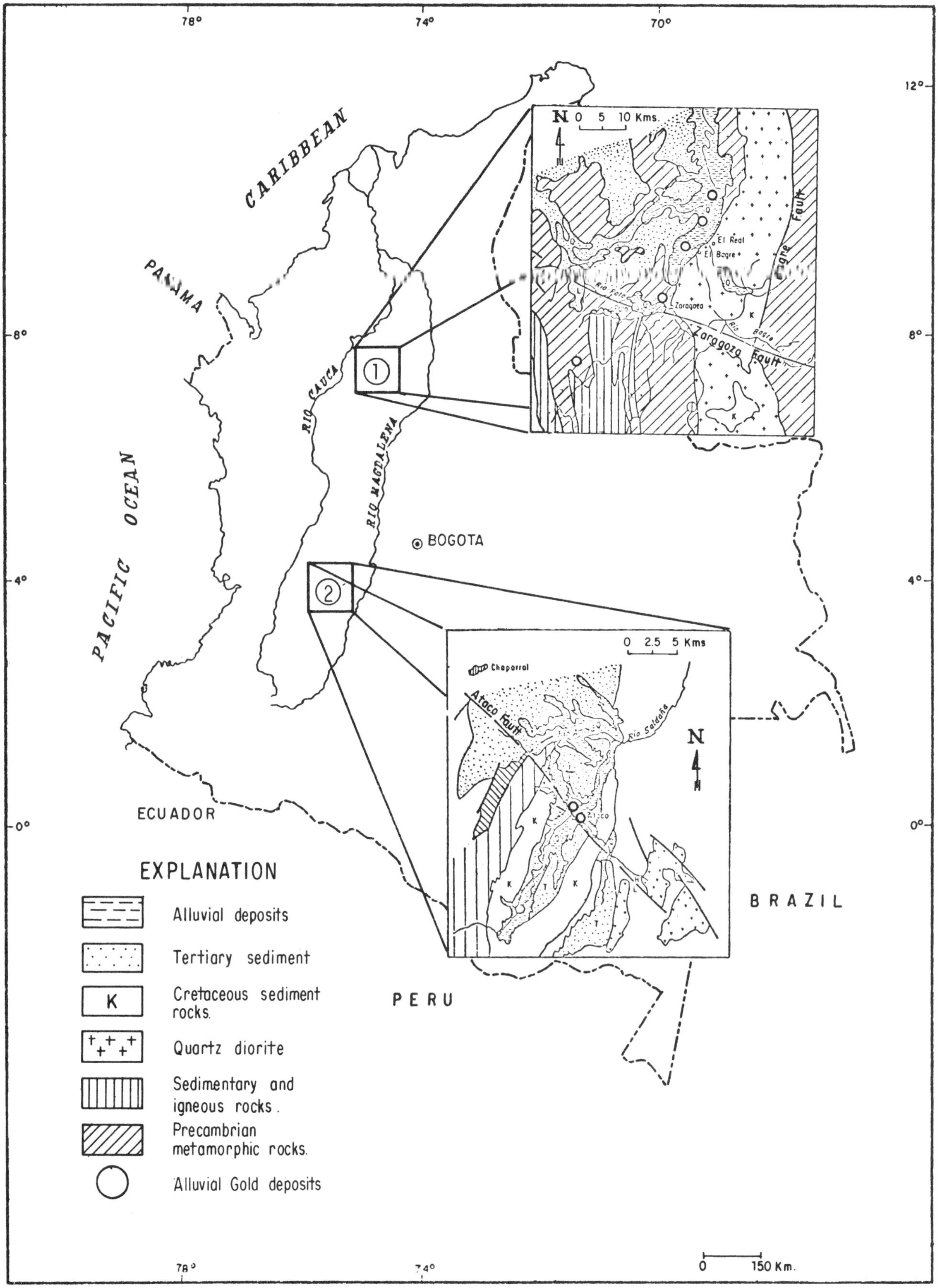

Figure 6. Placer gold deposits in Colombia: (1) Nechi zone and (2) Ataco zone.

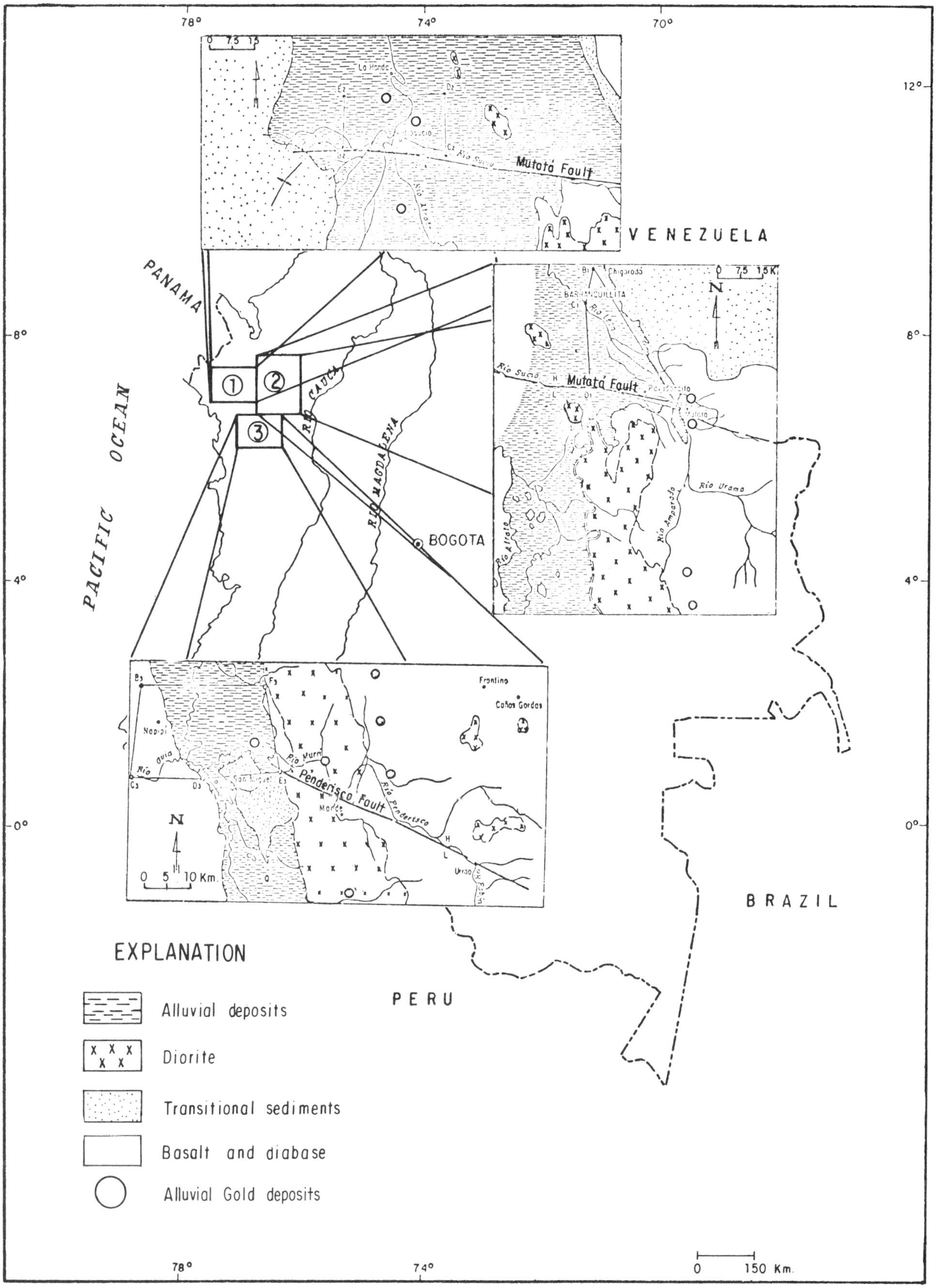

Figure 7. Areas having potential for discovery of new placer gold deposits in (1) Mutata-Barranquillita zone, (2) Rio Sucio zone, and (3) Murri-Penderisco zone.

Mutata-Barranquillita zone (Figure 7), which covers about 550 km^2, has the greatest potential for discovery of new placer deposits.

Rio Sucio

The Rio Sucio area (Figure 7, zone 2) is in the northern part of the Atrato River basin at its intersection with the Sucio River. As shown in Figure 7, this is the zone where the alluvial fill of the Atrato River is widest along the Mutata fault. In our opinion, a large part of the gold transported by the Atrato River may have been deposited in this zone because of loss of energy and transportation capability of the river at the Mutata fault. The most important zone, which covers an area of 500 km^2, is included within the BCDE polygon.

Murri-Penderisco

The Murri-Penderisco area (Figure 7, zone 3) is in the central part of the Atrato River basin, just north of the town of San Miguel. The Penderisco fault has experienced recent movement similar to that of the Mutata fault, the southern block being uplifted relative to the northern block. The Penderisco River, which originates in the Western Cordillera, drains a region having many gold deposits from which gold was eroded, transported, and deposited in the Atrato River basin. The most favorable zone, covering an area of 1000 km^2, is included within the BCEF polygon shown in Figure 7.

Quibdo

The Quibdo area (Figure 8, zone 4) is in the southern part of the Atrato River basin, near the city of Quibdo, capital of the Department of Choco. The Bagado fault, which is part of the large Salento fault system, has apparently been affected by several recent movements. The first caused the depression of the northern block, which favored the deposition of gold in the Atrato River, particularly southwest of Quibdo. Another movement caused the relative depression of the southern block, changing the course of the San Juan River—which in the past was apparently a tributary of the Atrato River—causing it to drain directly into the Atlantic Ocean. The alluvial deposits south of the town of Paimado (Figure 8) probably were deposited by the San Juan River before its course was changed. The most interesting zone, covering an area of 1060 km^2, is within the BCDE polygon shown on Figure 8.

Nariño

The Nariño zone (Figure 8, zone 5), Department of Nariño, contains a large amount of alluvial material consisting of metamorphic and volcanic rocks. This sector has the most widespread Quaternary alluvium of any area along the Pacific coast of Colombia. The alluvium apparently accumulated as the result of recent movement along the Policarpa fault, which caused uplift and inclination of the southern block toward the Pacific coast. As a consequence, several rivers—such as the Patia, Telembi, Iscuande, and Mira—transported abundant clastic material, produced by the erosion in the Andes, and deposited it in the foothills just west of the town of Barbacoas (Figure 8). Several dioritic stocks of Tertiary age which crop out in this area, together with gold from volcanic rocks and veins, contributed large amounts of gold to the rivers. The area of about 7200 km^2 has excellent possibilities for new discoveries of alluvial gold.

Caucasia

Although the Caucasia zone (Figure 9, zone 6), between the municipalities of Caucasia, Nechi, Bagre, and Zaragoza (Figure 9), is relatively well-known geologically and is the largest gold producing area of Colombia, it still has significant potential for discovery of new gold deposits. The north-flowing San Jorge and Cauca rivers show a dramatic change in direction at the Caucasia fault (Figure 9). The most interesting area is in the central block between a northwest-trending fault at Zaragoza, called the Zaragoza fault, and the Caucasia fault. The block between these two faults was dropped downward relative to blocks to the north and to the south. This downdropped block became an excellent geological trap for accumulation of alluvial gold. Placers currently under exploitation are chiefly in Quaternary alluvium and colluvium along the Zaragoza fault. A few other placers are found in upper Tertiary basal conglomerates, where some gold has been concentrated in segments of tree logs that have been converted into lignite. The zone of greatest interest (ABCD, Figure 9) has an area of 2200 km^2.

Tolima-Huila

The large (25,000 km^2) Tolima-Huila zone (Figure 9, zone 7), which corresponds to the Alto Magdalena sector of the Magdalena Valley between the Central and Eastern Cordilleras, has a significant potential for discovery of new placer gold deposits. The Gualanday Formation of Sheibe (1934, in de Porta, 1974) in this sector consists of Eocene sedimentary rocks of continental origin, called the Lower, Middle, and Upper Gualanday by Raasveldt (1956, in de Porta, 1974). White, medium- to coarse-grained sandstones predominate in the Lower; yellow, red, and blue claystones are abundant in the Middle; and chiefly sandstones and conglomerates are present in the Upper Gualanday Formation. Tellez and Navas (1962, in de Porta, 1974) estimated that the total thickness of the formation is 700-1500 m.

The Gualanday Formation consists of clastic material transported by rivers draining a large area in the Central Cordillera, which was uplifted during the Eocene by tectonic movements related to the Salento fault in the north and the Buenaventura-La Plata fault in the south (Figure 9). This zone, called the "Uplifted Block of Tolima Grande," apparently was tilted toward the east in the direction of the Upper Magdalena Valley, which favored accumulation of material eroded from the eastern flank of the Central Cordillera. The rocks that crop out in the Tolima Grande block are chiefly Paleozoic schists and Jurassic and Cretaceous igneous rocks of the Ibague batholith and the Buga stock, respectively. Although in the northern and southern parts of the Central Cordillera these lithological units have a large number of lode gold deposits, such deposits are not known in the Tolima Grande block. Therefore, if such deposits once existed, they were eroded from the Tolima Grande block when it was uplifted, and gold-bearing clastic material was transported and deposited to form the Gualanday Formation. Such gold may have accumulated in former stream channels in this formation.

The basin of the Saldaña River (Figure 9), which in recent

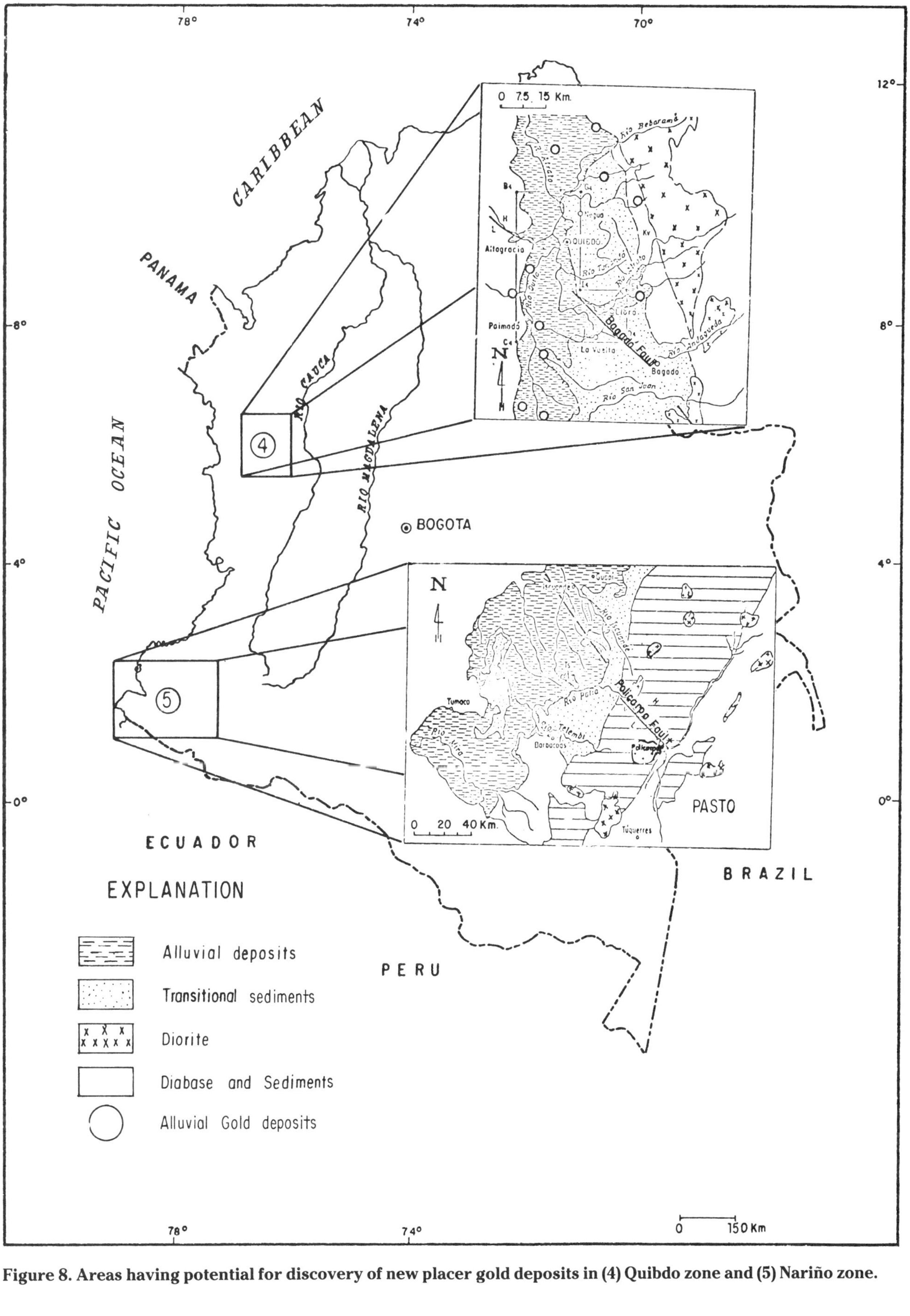

Figure 8. Areas having potential for discovery of new placer gold deposits in (4) Quibdo zone and (5) Nariño zone.

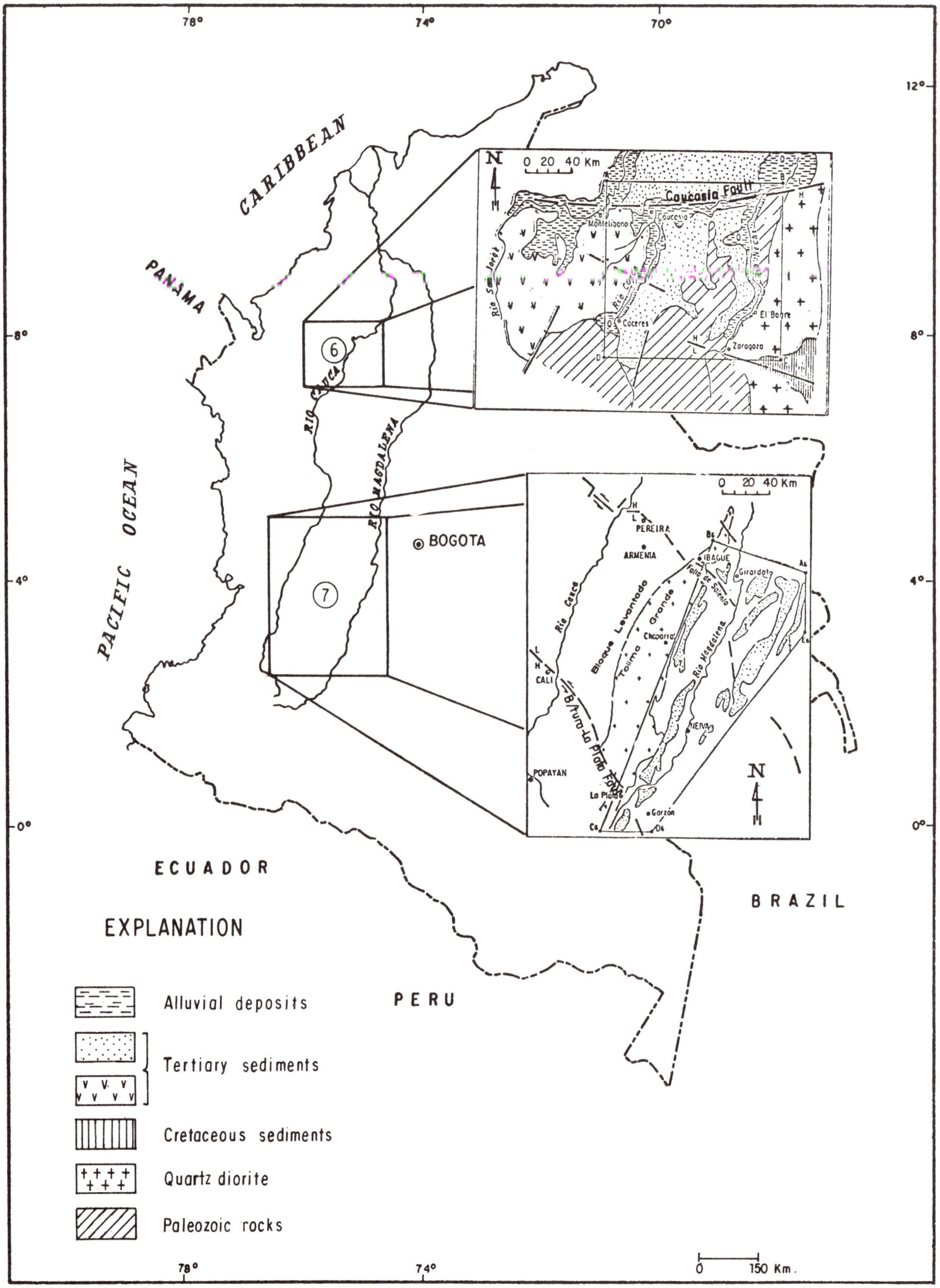

Figure 9. Areas having potential for discovery of new placer gold deposits in (6) Caucasia zone and (7) Tolima-Huila zone.

years has become the largest gold-producing zone in the Department of Tolima, illustrates accumulation in paleochannels. The alluvial gold deposits of Ataco, Santa Rosa, and others of this basin are related to such paleochannels in the Gualanday Formation. The basin is relatively closed in the southern part, and all the streams flowing into the Saldaña River drain the Gualanday Formation, which is the only possible source of the gold. Neither alluvial nor vein gold has been found to the west of this formation or in higher parts of the Central Cordillera, which is the headwater region of the most important rivers of the basin.

We suggest that a systematic study of the Gualanday Formation in the Tolima and Huila departments will discover new placer gold deposits.

Guainia

The southern Guainia region (Figure 10, zone 8) of Colombian Amazonas has a significant potential for discovery of new gold deposits, particularly in the drainage systems of the Inirida, Guainia, and Cuiari rivers. The zone of greatest potential, about 30,000 km^2, in certain geologic aspects, resembles the South African Witwatersrand, which now produces about 60% of the world's gold, and other large deposits such as Blind River in Canada and Cerro Jacobina in Brazil. These similarities are discussed by Utter (1984) and Lozano (1984), who consider the Guainia area to have an important gold potential.

The rocks in the Guainian zone (Figure 10) consist of the Mitu magmatic complex of Precambrian age (2000 Ma), which is overlain by a sedimentary sequence (also of Precambrian age) that is correlated with the Roraima Formation. The Mitu complex consists of feldspar gneisses, amphibolites, granitic plutons, and diabase dikes. The sedimentary sequence consists of a lower unit of conglomerates and quartzose sandstones and an upper unit of shales, both deposited in a fluvial-deltaic to marine environment, having a total thickness of 1000-1800 m.

Detrital gold and local associated pyrite have been observed in samples of folded and deformed, laminated sandstone beds of the sedimentary sequence. Gold is currently being recovered by small-scale placer operations in the Cuiari River, southern part of the Serrania de Caranacoa, and in Caño Tigre and San Nicolas region, Serrania de Naquen (Figure 10).

Traira

The Traira zone (Figure 10, zone 9) of southeasternmost Colombia has geological characteristics similar to those of the Guainia region and therefore has comparable gold potential. The area includes the Serrania del Traira, which is drained by tributaries of the Apaporis River at the border with Brazil. Placer gold has recently been recovered in this zone, chiefly by reworking of old placers. The rocks consist of the Mitu magmatic complex unconformably overlain by a sequence of regionally metamorphosed sedimentary rocks known as the Pedrera Formation (Figure 10), which correlate with the lower Proterozoic Tunui Formation of Brazil. The geology is little known, so detailed study is needed to determine the distribution and potential of gold-bearing bedrock and placers.

CONCLUSIONS

Although large amounts of gold have been recovered in Colombia, mostly in the Andean sector, starting before the discovery of the Americas, the country still has excellent potential for the discovery of new gold deposits. Three types of deposits are recognized—veins, disseminations, and placers—all of which have potential for new discoveries.

Areas considered to have the greatest potential for discovery of new gold veins in the Central Cordillera of Colombia are at intersections of faults of the Palestina fault system, striking N30 °E, with those of the Salento fault system, striking N55 °W. These zones of intersection show intense fracturing and were the sites of igneous intrusion and associated hydrothermal gold deposition. The principal known gold veins and, consequently, favorable prospect areas, are in the zone between the cities of Ibague and Sonson in the northern part of the Central Cordillera and between the cities of Popayan and Pasto in the south (Figure 2).

The best prospect areas for disseminated gold deposits are in the middle part and western flank of the Central Cordillera of Colombia (Figure 4). This region is characterized by the presence of both oceanic and continental crust separated by the Romeral fault zone in which are several plutons. The most favorable prospect areas are the sectors where the plutons are in major fracture areas of intersection of the Romeral, Salento, and Palestina fault systems. Such areas are the Marmato, Department of Caldis; Suarez, Department of Cauca; Norte de Cajamarca, Department of Tolima; and Buga, Department of Valle (Figure 4).

Placer deposits offer the best prospects for immediate discovery of potentially large gold resources. Recent fault movements have played an important role in the accumulation of gold placer deposits in the Nechi River, Department of Antioquia, and Saldaña River, Department of Tolima (Figure 6). These movements formed traps in which gold accumulated. Other favorable prospect areas are in the Mutata Barranquillita, Riosucio, Murri-Penderisco, Quibdo, Nariño, Caucasia, and Tolima-Huila zones (Figures 7-9).

The most favorable prospect areas for new placer deposits are in the Amazonas region of southeastern Colombia (Figure 10), where both alluvial and colluvial gold placers occur. Apparently, the sources of these deposits were Precambrian rocks similar to those from which large amounts of gold are currently being recovered in South Africa, Canada, and Brazil.

REFERENCES CITED

Buenaventura A. J., 1976a, Ocurrencias minerales en la región norte del Departamento del Tolima: INGEOMINAS, Inf. 1688, 397 p.

Buenaventura A. J., 1976b, Ocurrencias minerales en la región sur y oriental del Departamento del Tolima: INGEOMINAS, Inf. 1694, 392 p.

Buitrago, R. C., and A. J. Buenaventura, 1975, Ocurrencias minerales en la región central el Departamento del Tolima: INGEOMINAS, Inf. 1672, 837 p.

Buitrago, R. C., A. J. Buenaventura, and S. H. Perez, 1976, Ocurrencias minerales el Departamento del Huila: INGEOMINAS, Inf. 1698, 503 p.

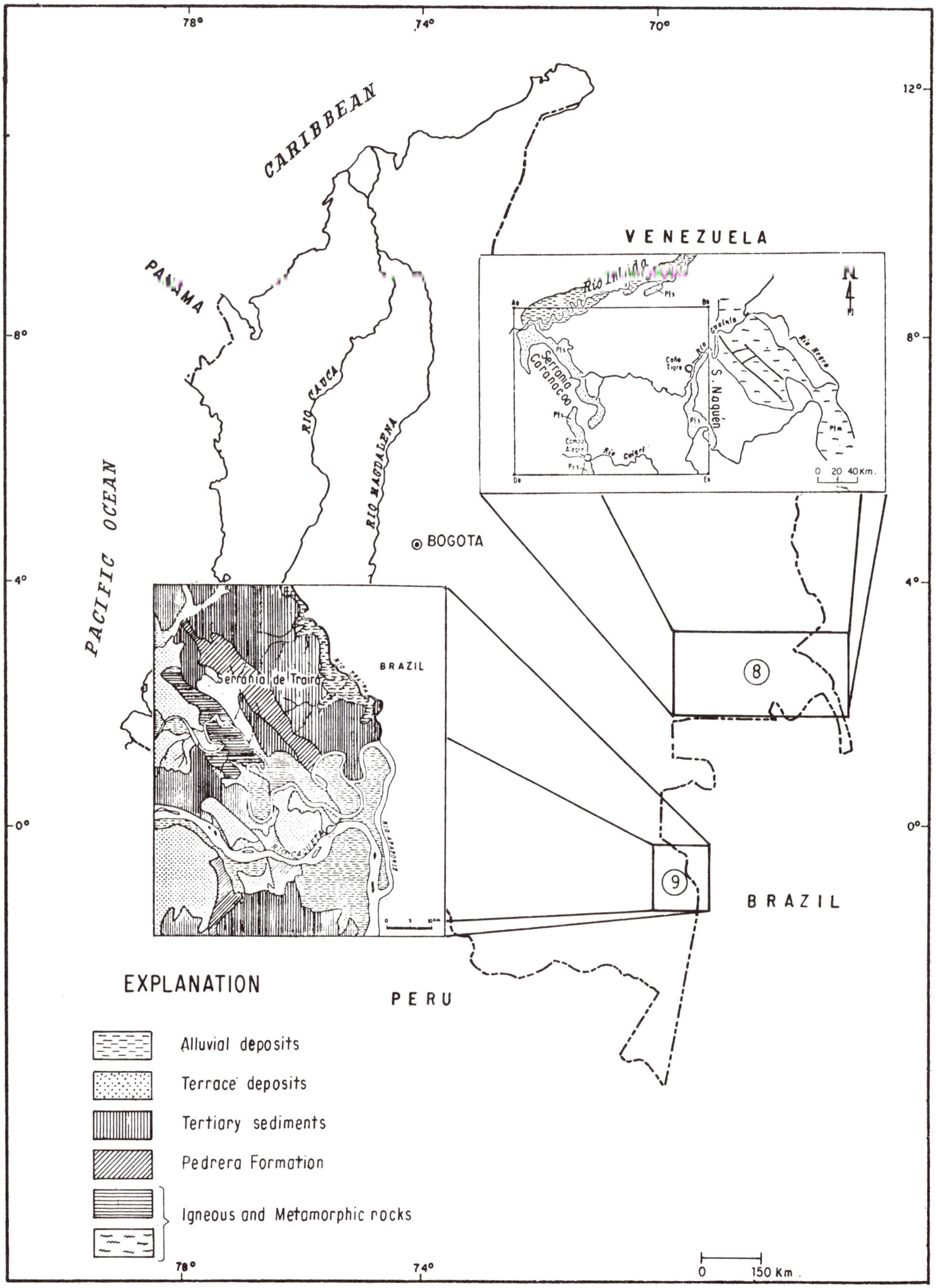

Figure 10. Areas having potential for discovery of new placer gold deposits in (8) Guainia zone and (9) Traira zone.

Guarin, G. G., 1971, Ocurrencias minerales del Departamento de Risaralda: INGEOMINAS, Inf. 1636, 66 p.
Hall, R. B., et al., 1971, Recursos minerales de parte de los Departamentos de Antioquia y Caldas: Inventario Minero Nacional, Tomo II, 90 p.
Hall, R. B., T. Feininger, D. Barrero, H. Rico, and J. Alvarez, 1970, Recursos minerales de parte do los Departamentos de Antioquia y Caldas: Bol. Geol., v. XVIII, no. 2, 90 p.
Lopez, J. H., 1971, Ocurrencias minerales del Departamento de Caldas: INGEOMINAS, Inf. 1602, 123 p.
Lozano Q., H., 1984, Notas sobre explotaciones de oro en la Comisaría del Guainía: INGEOMINEAS, Informe interno, 25 p.
Mosquera, T. D., and R. C. Buitrago, 1971, Ocurrencias minerales del Departamento del Quindío: INGEOMINAS, Inf. 1599, 88 p.
Porta, J. de, 1974, Lexique stratigraphique international: Paris, Centre National de la Recherche Scientifique, v. V, 634 p.
Pulido, U. O., 1985a, Reconocimiento regional para mineralizaciones de oro diseminado en tres zonas de los Departamentos de Caldas, Quindío y Tolima, Colombia: INGEOMINAS, Informe interno, 60 p.
Pulido, U. O., 1985b, Geología del área de San Antonio Cajamarca, Tolima: INGEOMINAS, Informe interno, 84 p.
Rodriguez, C. J., and A. Pernett, 1980, Mapa de recursos minerales de Antioquia: INGEOMINAS, 14 p.
Rodriguez, P. A., G. H. Gil, et al., 1985, Análisis de los efectos ocasionados por las explotaciones auríferas en la zona de Zaragoza-El Bagre-Cuturu, Departamento de Antioquia: INGEOMINAS, Informe interno, 54 p.
Utter, T., 1984, Gold deposits in Colombia—their metallogenesis and some fundamental facts about prospecting for precious metals: Tubingen, West Germany, Institute for Scientific Cooperation, Mineral Resources and Development, v. 20, p. 92-108.
West, R., 1972, La minería de aluvión en Colombia durante el período colonial, traducción de J. O. Melo: Bogotá, Imprenta Nacional, 157 p.

Ericksen, G. E., Cañas Pinochet, M. T., and Reinemund, J. A., editors, 1989, Geology of the Andes and its relation to hydrocarbon and mineral resources: Houston, Texas, Circum-Pacific Council for Energy and Mineral Resources Earth Science Series, v. 11.

CHAPTER 13

CENOZOIC TECTONIC EVENTS, MAGMATIC PULSES, AND BASE- AND PRECIOUS-METAL MINERALIZATION IN THE CENTRAL ANDES

Edwin H. McKee
U.S. Geological Survey
Menlo Park, California, U.S.A.

Donald C. Noble
Mackey School of Mines, University of Nevada-Reno
Reno, Nevada, U.S.A.

ABSTRACT

In the Andes of Perú, two periods of Paleogene deformation—Incaic I (Paleocene) and Incaic II (Eocene about 43 Ma)—and three periods of Neogene deformation—Quechua I (about 19 Ma), Quechua II (about 9.5 Ma), and Quechua III (about 6 Ma)—are recognized. These tectonic pulses, which appear to be present in other parts of the central Andes, are associated with pulses of igneous activity that peaked in the Paleocene, late Eocene, early Miocene, middle to late Miocene, and late Pliocene to Holocene. Hydrothermal base- and precious-metal mineralization is clearly related to the igneous activity and defines the same pulses, but is 1 to 5 m.y. younger.

In southern Perú, where the Andean chain widens markedly, the loci of magmatic and hydrothermal activity defines belts parallel to the structural trends of the mountain chain, but at different positions relative to its axis. Late Cretaceous and Paleocene magmatism of the Coastal batholith was superseded during the late Paleocene and Eocene by magmatic activity in the northeastern part of the Andean chain. This in turn shifted southwestward to the central part of the chain in the early Miocene and thence farther southwestward to the site of the active volcanic arc in the late Pliocene. These relationships, and the episodic nature of tectonism and magmatism, correspond to major changes in the plate-tectonic regime and suggest that the changes in the rate and direction of convergence of the Nazca and South American plates and the angle of subduction together have controlled the timing and location of tectonic and magmatic events.

INTRODUCTION

The purpose of this chapter is to summarize stratigraphic, structural, and geochronologic information for Cenozoic igneous rocks and associated mineral deposits in the central part of the Andes. Evaluation of these data indicates distinct patterns in time and space of tectonism, magmatic activity, and mineralization.

Field work was started in 1971 and has included mapping, regional stratigraphic and structural studies, and reconnaissance and detailed local studies of many precious- and base-metal districts. More than 200 samples from the Peruvian and Bolivian Andes have been dated by the K-Ar or $^{39}Ar/^{40}Ar$ methods. These data augment approximately 100 published dates. Many individual studies have been collaborative efforts with colleagues from Perú, France, Canada, and the United States, and we wish to specifically acknowledge their contributions to this review.

RADIOMETRIC AGE BASE

Radiometric age determinations from a variety of local and regional studies were evaluated and integrated for this study. These include both published data (e.g., Noble, 1977; Noble,

McKee, and Mégard, 1979; Tosdal, Farrar, and Clark, 1981; Mégard et al., 1984) and a large number of unpublished age determinations from our studies. Ages for volcanic and plutonic rocks were limited to those from Perú, whereas those for the mineral-district compilation included published ages from Bolivia and northern Chile (e.g., Grant et al., 1979; Sillitoe, 1981). A total of about 300 ages was available, about 250 of which were not duplicate age determinations on the same rock unit. As might be expected, most of the age determinations are for rocks of moderately young age (greater than about 1 Ma and less than about 30 Ma) because of sampling bias. Rocks less than 1 Ma in age commonly have not been dated because they typically form parts of active volcanic centers and have been inferred to be of Quaternary age. Rocks of this age also present analytical problems caused by large relative amounts of atmospheric argon. Rocks older than about 30 Ma are either widely eroded or are buried by younger rocks and consequently have more restricted exposure than the younger rocks. Even where exposed they commonly are altered, making radiometric dating difficult or impossible. Therefore, the general decrease in the number of age determinations with time does not imply a progressive decrease in the intensity or amount of magmatic activity through the middle and early Cenozoic. Because of these limitations, and because our studies have in large part not involved rocks of early Eocene and older age, the time frame for this review is from about 45 Ma to the present. Nevertheless, it is clear from geologic and geochronologic studies of the Coastal batholith of Perú and its associated mineral deposits (Beckinsale et al., 1985; Mukasa, 1986; Vidal, 1985) that the Late Cretaceous and Paleocene was a time of intense igneous activity during which a number of important mineral deposits were formed.

TECTONIC PULSES

Dating tectonic events requires carefully selected samples from geologic units whose relationships are clearly understood. In most cases, the strategy is to date units that predate and others that postdate a regional structural or erosional feature, generally an unconformity. The dated rocks above and below the unconformity bracket the age of the tectonic event. In some places, dates of volcanic units within clastic wedges formed by erosion of uplifted or otherwise tectonically disturbed terranes provide a minimum age. In other cases, dates of the youngest units in a deformed sequence provide a maximum age for deformation. Forty-eight ages are available that bracket, or in some way date, four major regional tectonic pulses in Perú (Figure 1). Each pulse seems to have spanned a relatively short period of time, commonly on the order of 1 to 2 m.y.

Steinmann (1929) recognized two tectonic pulses of Cenozoic age—the Incaic and Quechua—defined mainly by unconformities observed in the Rimac Valley and Ayacucho basin of central Perú. Although new stratigraphic, structural, and radiometric age data obtained over the past several decades indicate a larger number of discrete events, the data continue to support Steinmann's concept of discrete tectonic pulses separated by periods of relative tectonic quiescence. Two events of probable Paleogene age are recognized, an Incaic I event of Paleocene age (Bussell, 1983) and an Incaic II event of middle Eocene age (Noble et al., 1985).

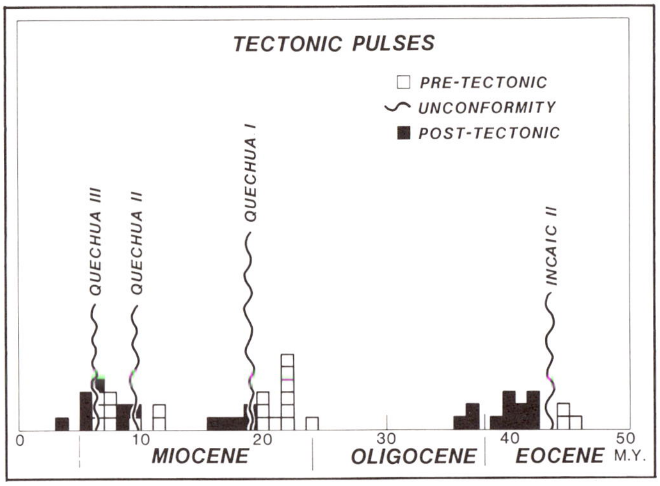

Figure 1. Ages of Tertiary pulses of deformation in the central Andes, based on ages of pre- and post-deformation rock units (each box is the age of an igneous unit) that bracket the age. Incaic II is the oldest pulse, and Quechua I, II and III are the youngest.

Three late Tertiary events are recognized, largely from ages obtained for rocks in the Ayacucho-Huancavelica-Castrovirreyna region of south-central Perú. The Quechua I pulse, which was the most intense, is of early Miocene age (McKee and Noble, 1982; Mégard et al., 1985), whereas the Quechua II and III pulses which, at least in Perú, appear to be less intense and of more local nature, are of late Miocene age (Mégard et al., 1984, 1985).

We believe that the Incaic II and Quechua I and III pulses can also be recognized in the Andes of Bolivia, northern Chile, and northwestern Argentina.

Incaic II

The age of the Incaic II deformation is based on igneous rock dates in northern and central Perú. Because of its relative antiquity, geologic relationships are obscure, and at only a few localities was it possible to obtain samples of pre- and post-deformation rocks that could be dated. Nevertheless, 15 dates are available that bear on the age of the Incaic II deformation. Of these, 11 span a short enough period to bracket the age of deformation. In northern Perú, rocks beneath the unconformity caused by Incaic II compression are about 45 m.y. old. In central Perú, rocks above the unconformity are as old as about 41 m.y. These dates place the Incaic II deformation at about 43 m.y. (Figure 1).

Quechua I

Fifteen dated rock units are used to date the Quechua I pulse of compressive deformation (Figure 1). Ten of the units predate and five postdate the unconformity formed after deformation. The age of deformation is bracketed at about 18 Ma. (Figure 1; McKee and Noble, 1982; Mégard et al., 1985).

Quechua II and III

Quechua II and III pulses of deformation are recognized in south-central Perú. In the Ayacucho area, two distinct unconformities are recognized (Mégard et al., 1984, 1985), and

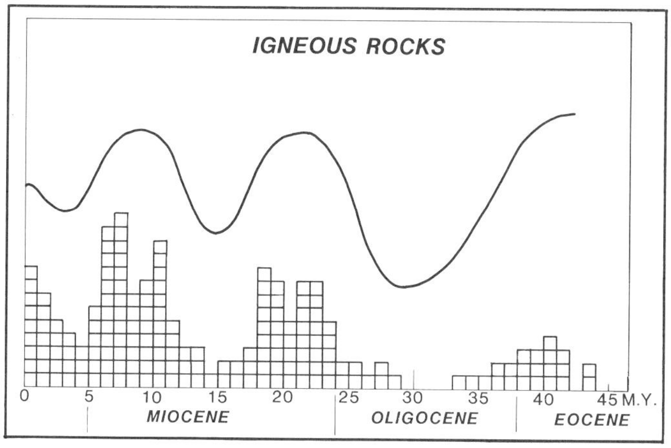

Figure 2. Histogram of radiometric ages of middle and late Cenozoic igneous rocks in the central Andes. Curve represents periods of high and low intensity of igneous activity. Each box is the age of an igneous unit—either a single age determination or the average of several radiometric ages.

radiometric ages for 15 rock units (Figure 1) show the Quechua II and III pulses to be bracketed at about 9.5 and 6 Ma, respectively. Elsewhere, these deformational pulses are less certain and as the region becomes better known geologically and as more radiometric ages become available, these pulses may prove to be local pulses of single regional late Miocene deformation.

IGNEOUS PULSES

A histogram for 181 single or average radiometric ages of middle and late Cenozoic igneous rocks clearly shows four maxima separated by three minima (Figure 2), a pattern first recognized by Noble et al. (1974). The oldest maxima spans a period from about 44 to 36 Ma (the older limit is poorly constrained) and tapers off to a time of near quiescence by about 34 Ma. Little igneous activity is indicated until about 24 Ma, when widespread igneous activity is recorded by many dated lavas, ash-flow tuffs, and plutonic rocks. This period of intense igneous activity lasted for about 7 m.y., ending about 17 Ma. The middle part of the Miocene from about 17 to 13 Ma was a time of decreased volcanism followed by renewed widespread igneous activity in late Miocene that lasted from about 13 to 5 Ma. The radiometric ages indicate a relative quiescence from about 5 to 2 Ma before resumption of the widespread volcanism that has continued to the present. Each of the pulses of intense igneous activity, therefore, lasted about 7 m.y. A curve drawn from the histogram which averages irregularities caused by nonuniform sampling or other problems introduced by a small sample population is symmetrically bell shaped (Figure 2), indicating similar rates in waxing and waning during each pulse.

Eocene and Early Oligocene Pulse

Major widespread igneous activity associated with and following the Incaic II period of compressive deformation (Figure 1) is recorded by 20 radiometrically dated units of voluminous ash-flow tuff, widespread intermediate lavas, and high-level silicic to intermediate plutons (Figure 2). Rocks from this pulse are especially prevalent in the central and eastern Andes.

Late Oligocene and Early Miocene Pulse

Many large silicic ash-flow sheets, silicic to intermediate lavas, and silicic domes and plutons were emplaced prior to, during, and after the Quechua I period of compressive deformation. Rocks of this age have been found throughout the length of the Peruvian Andes, and 51 rock units have been dated that are between 24 and 17 Ma (Figure 2). The decline of igneous activity was less pronounced and of shorter duration than that following the late Eocene-early Oligocene pulse.

Late Miocene Pulse

A renewal of mostly silicic volcanism marked by the eruption of voluminous silicic ash-flow tuffs and silicic to intermediate lavas and domes is documented by 69 radiometrically dated units ranging from 13 to 5 Ma. The most intense activity was during the 6-m.y. interval between 11 and 5 Ma (Figure 2). This pulse of igneous activity spans the Quechua II and Quechua III pulses of compressive deformation. Volcanic rocks of this age are distributed widely in the central part of the Peruvian Andes. The decline in igneous activity that marks the end of this pulse is less pronounced than that after the earlier pulses (Figure 2).

Late Pliocene and Quaternary Pulse

Major calc-alkalic, volcanic-arc igneous activity began about 3.5 Ma at a number of places in southern and central Perú. The presently active volcanoes, such as Coropuna and Sara-Sara in southern Perú and Carhuaraso east of the volcanic front in south-central Perú, are part of this igneous pulse. A total of 25 radiometrically dated rock units define the pulse (Figure 2).

Distribution of Igneous Rocks

Cenozoic igneous rocks of various ages in the central Andes define belts that trend north-northwest (Figure 3). In northern and central Perú, where the Andean chain is relatively narrow, the locus of magmatic activity has migrated eastward only a short distance since the beginning of the Cenozoic (Beckinsale et al., 1985). In southern Perú, however, where the Andes are appreciably wider, there have been major shifts with time in the locus of magmatic activity. During the early Cenozoic, magmatic activity shifted northward from near the coast (the Coastal batholith of Cretaceous and earliest Cenozoic age) to the eastern part of the Andes (Noble et al., 1984). Following middle Oligocene magmatic and tectonic quiescence, igneous activity was reinitiated in the late Oligocene and early Miocene in the region near the eastern margin of the Coastal batholith. In the middle Miocene, this igneous activity shifted toward the northeast and finally, in the Pliocene, shifted southwest to its present position marked by the active volcanic chain. These changes in the locus of volcanism apparently were relatively abrupt, and as magmatic activity ceased in one belt it broke out at the same time or a short time later along another belt.

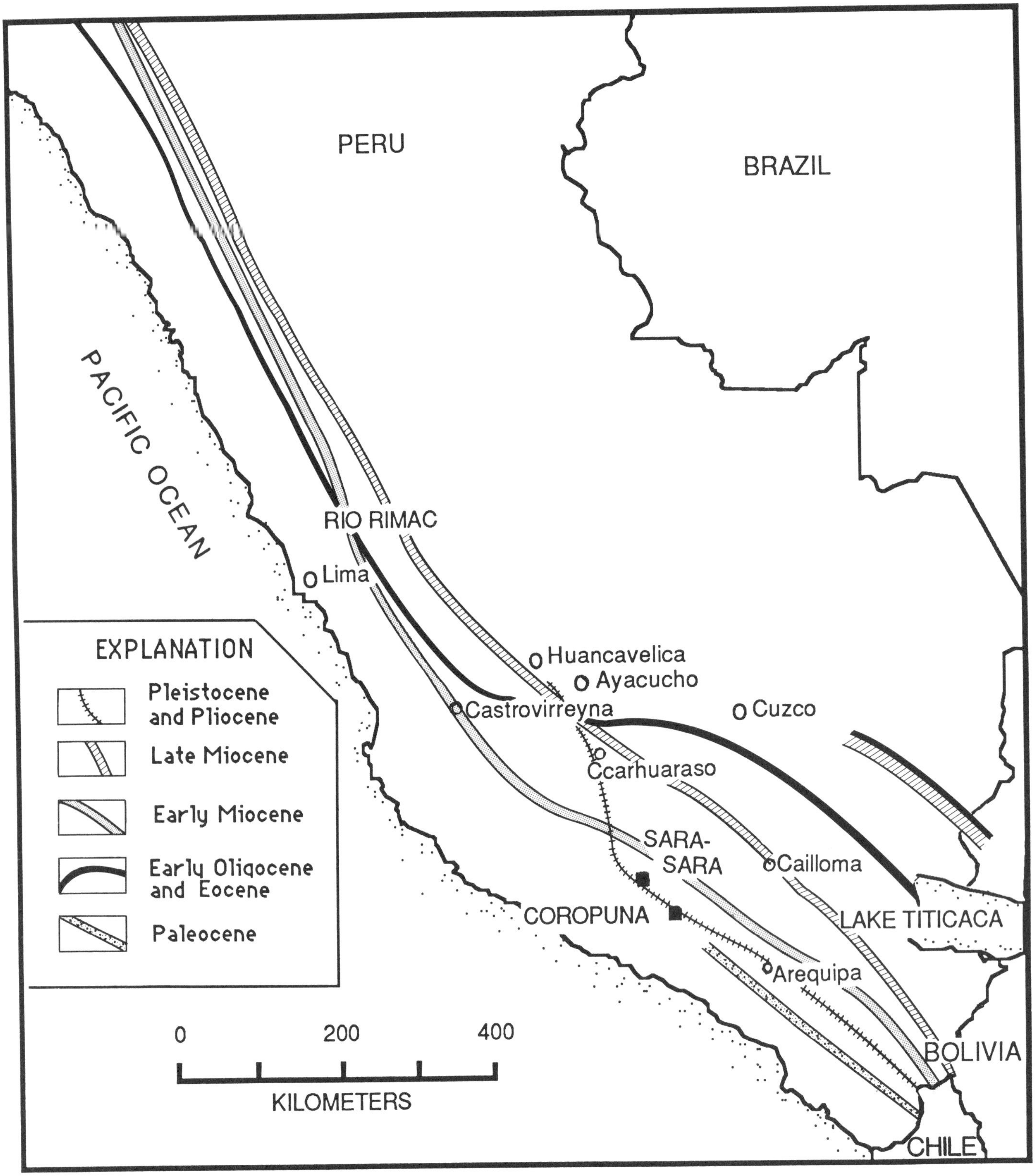

Figure 3. Schematic map showing axis of belts of Cenozoic igneous activity and base- and precious-mineralization in the central Andes.

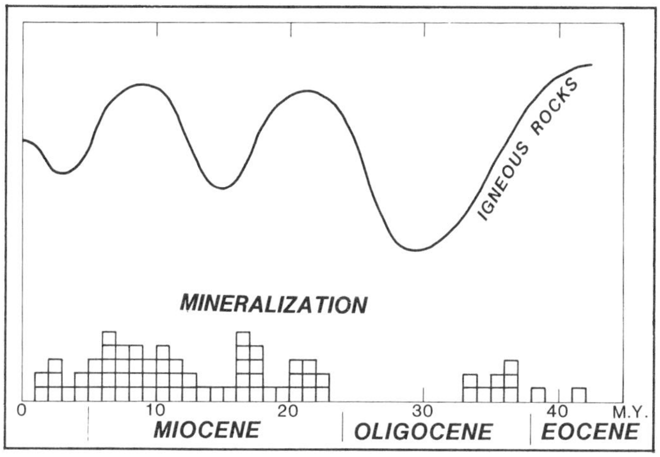

Figure 4. Histogram of radiometric ages of mineral deposits in the central Andes. Curve from Figure 2 shows variations in igneous activity. Each box is the age of a mining district, as determined from one or more age determination on rocks from the district.

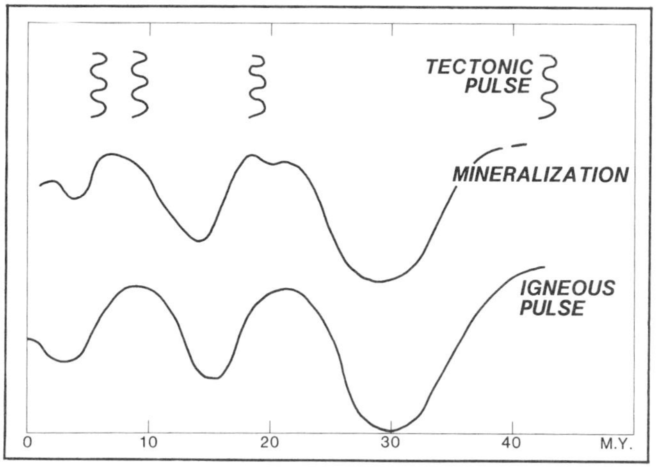

Figure 5. Comparison of ages of pulses of igneous activity, mineralization, and tectonism in the central Andes.

MINERALIZATION

Radiometric ages of vein and alteration minerals are used to evaluate the major times of mineralization (Figure 4). Most of the dates are from mines or mining districts in central and southern Perú representing samples of epithermal origin in volcanic and subvolcanic rocks, but some are from districts in northern Perú, northern Bolivia, and northern Chile. The ages define three periods of hydrothermal alteration and mineralization in the late Tertiary. These periods, although less well defined than the pulses of igneous activity because of fewer dates, nevertheless correspond to the pulses of igneous activity. The peaks for mineralization appear to be about 1 to 5 m.y. younger than corresponding peaks for magmatic pulses (Figure 5). This is consistent with observations showing that within individual districts the mineral deposits commonly formed during the latter part or shortly after magmatic activity.

The oldest episode of mineralization—between about 40 and 34 Ma—is associated with the final stages of igneous activity of the Eocene-early Oligocene pulse. Major mineral districts of this period occur along a 600-km-long belt trending north-northwestward from Lake Titicaca (Noble et al., 1984). Hydrothermal mineralization and associated alteration ceased in this region after this episode, as there are no deposits dated between about 23 to 33 Ma. This hiatus is also seen in the igneous rock record.

Hydrothermally altered or mineralized rocks appeared again at about 23 Ma and continued to form until the end of the Miocene (5 Ma). Dating of samples from 19 mining districts indicates a subperiod of mineralization that reached its peak at 17 to 16 Ma (Figure 4). Mineral deposits of this subperiod are widely spaced in the western Andes of Perú and in a belt in the east-central Andes extending from southeast of Cuzco southward into northernmost Bolivia. Few mineral deposits of middle Miocene age (13 to 15 Ma) are known in this part of the central Andes. A younger subperiod of mineralization, which began about 12 Ma and ended about 5 Ma, reached its peak at 6 to 8 Ma (Figure 4). Deposits of this subperiod are more numerous (28 were dated) and more widespread in the Peruvian Andes than deposits of any of the earlier periods of mineralization. Consequently, the period from 12 to 5 Ma was the major time of base- and precious-metal mineralization in the central Andes. Hydrothermal mineralization and alteration waned 3 to 4 Ma and then was followed by renewed mineralization.

CONCLUSIONS

Radiometric dating applied to studies of tectonic, magmatic, and hydrothermal processes has confirmed the episodic nature of compressive tectonism and magmatic activity in the central Andes. Tectonic events appear to have occurred during periods of more intense magmatic activity, and hydrothermal mineral deposits are closely related in space and time to pulses of intense volcanism and plutonism, particularly to the later phases of such pulses. These relationships, together with apparent synchroneity of events in much or all of the central Andes, suggest that observed magmatic and tectonic events reflect major changes in the rate and direction of convergence of the Nazca and South American plates. The major shifts in the locus of magmatic activity in southern Perú presumably reflect, at least in part, differences in the angle of subduction.

ACKNOWLEDGMENTS

The work of Noble was partly supported by grants from the National Science Foundation. We thank many companies in Perú, and particularly Cia. de Minas Buenaventura, S.A., and Mauricio Hochschild y Cia. Ltda., S.A., for assistance and support. We would especially like to thank Ing. Alberto Benavides, Ing. V. Raúl Eyzaguirra, and Dr. Francois Mégard for help during all stages of this work.

REFERENCES CITED

Beckinsale, R. D., A. W. Sanchez-Fernandez, M. Brook, E. J. Cobbing, W. P. Taylor, and N. D. Moore, 1985, Rb-Sr whole-rock isochron and K-Ar age determinations for the Coastal batholith of

Perú, *in* W. S. Pitcher, M. P. Atherton, E. J. Cobbing, and R. D. Beckinsale, eds., Magmatism at a plate edge; Glasgow, Blackie and Sons, p. 177-202.

Bussell, M. A., 1983, Timing of tectonic and magmatic events in the central Andes of Perú: Journal of the Geological Society, London, v. 140, p. 279-286.

Grant, J. N., C. Halls, W. Avila Salinas, and N. J. Snelling, 1979, K-Ar ages of igneous rocks and mineralization in parts of the Bolivian tin belt: Economic Geology, v. 74, p. 838-851.

McKee, E. H., and D. C. Noble, 1982, Miocene volcanism and deformation in the Western Cordillera and high plateaus of south-central Perú: Geological Society of America Bulletin, v. 93, p. 657-662.

Mégard, F., D. C. Noble, E. H. McKee, and H. Bellon, 1984, Multiple pulses of Neogene compressive deformation in the Ayacucho intermontane basin, Andes of central Perú: Geological Society of America Bulletin, v. 95, p. 1108-1117.

Mégard, F., D. C. Noble, E. H. McKee, and Y. Cuenod, 1985, Silicic dikes contemporaneous with latest Miocene Quechua 3 tectonism in the Rimac Valley, Western Cordillera of central Perú: Journal of Geology, v. 93.

Mukasa, S. B., 1986, Zircon U-Pb ages of super-units in the Coastal batholith, Perú: implications for magmatic and tectonic processes: Geological Society of America Bulletin, v. 97, p., 241-254.

Noble, D. C., 1977, A summary of radiometric age determinations on rocks from Perú: Isochron/West, v. 5, p. 576-578.

Noble, D. C., E. H. McKee, E. Farrar, and U. Petersen, 1974, Episodic Cenozoic volcanism and tectonism in the Andes of Perú: Earth and Planetary Science Letters, v. 21, p. 213-221.

Noble, D. C., E. H. McKee, and F. Mégard, 1979, Early Tertiary "Incaic" tectonism, uplift, and volcanic activity, Andes of central Perú: Geological Society of America Bulletin, pt. 1, v. 90, p. 903-907.

Noble, D. C., E. H. McKee, V. R. Eyzaguirre, and R. Marocco, 1984, Age and regional tectonic and metallogenetic implications of igneous activity and mineralization in the Andahuaylas-Yauri belt of southern Perú: Economic Geology, v. 79, p. 172-176.

Noble, D. C., M. Sebrier, F. Mégard, and E. H. McKee, 1985, Demonstration of two pulses of Paleogene deformation in the Andes of Perú: Earth and Planetary Science Letters, v. 73, p. 345-349.

Sillitoe, R. H., 1981, Regional aspects of the Andean porphyry copper belt in Chile and Argentina: Transactions of the Institution of Mining and Metallurgy, v. 90, sec. B, p. B15-B36.

Steinmann, G., 1929, Geologie von Perú: Heidelberg, Springer-Verlag, 448 p.

Tosdal, R. M., E. Farrar, and A. M. Clark, 1981, K-Ar geochronology of the late Cenozoic volcanic rocks of the Cordillera Occidental, southernmost Perú: Journal of Volcanol. Geothermal Research, v. 10, p. 157-174.

Vidal, C. E., 1985, Metallogenesis associated with the Coastal batholith of Perú: a review, *in* W. S. Pitcher, M. P. Atherton, E. J. Cobbing, and R. D. Beckinsale, eds., Magmatism at a plate edge: Glasgow, Blackie and Sons, p. 243-249.

Ericksen, G. E., Cañas Pinochet, M. T., and Reinemund, J. A., editors, 1989, Geology of the Andes and its relation to hydrocarbon and mineral resources: Houston, Texas, Circum-Pacific Council for Energy and Mineral Resources Earth Science Series, v. 11.

CHAPTER 14

GEOLOGY AND MINERAL DEPOSITS OF THE CENTRAL ANDES, REPUBLIC OF ARGENTINA

Vicente Mendez and Eduardo Zappettini[1]
Direccion General de Fabricaciones Militares
Buenos Aires, Argentina

ABSTRACT

The central Andean region of Argentina, an area of 130,000 km^2 between lat. 32° and 40°S, is characterized by two major orogenic domains: (1) the Paleoidic orogen, which consists of Cambrian to Triassic terranes deformed during continent-continent collision; and (2) the Neoidic orogen, which is characterized by widespread plutonism and volcanism in a passive margin intracratonic basin during the Jurassic-Quaternary. Several pulses of extrusive and intrusive magmatic activity are associated with these orogens. Deformation related to the Caledonian, Variscan, Cimerian, and Andean orogenies caused folding, overthrusting, and block-faulting, resulting in the typical north-south structural grain of the Andes.

Several types of mineral deposits are present in the central Andes. Deposits having the greatest economic potential are: (1) hydrothermal lead-silver-zinc, gold, and iron veins, which are the most numerous type, (2) porphyry copper deposits, which are the largest; and (3) copper-uranium-vanadium and barium-strontium stratabound deposits. These and other types of deposits tend to be clustered in specific belts and zones, each characterized by a single type of deposit according to the geotectonic environment in which it formed.

INTRODUCTION

This report provides a geologic-metallogenic synthesis of the central Andes in Argentina, between lat. 32° and 40°S (Figure 1). It is based on a study of aerial photographs at a scale of 1:50,000 and space imagery at a scale of 1:250,000, and on field investigations of geology and mineral deposits. A simplified geologic map of the region is shown in Figures 2a and 2b.

Geologic and Tectonic Environment

Two large geotectonic units have been recognized in the central Andes of Argentina, the first representing the Cambrian-Triassic interval and the second the Jurassic-Quaternary interval. These intervals correspond to the Paleoidic and Neoidic geotectonic epochs, respectively, as defined by Stille (1955) and recognized in Argentina by Borrello (1969). Features related to the Paleoidic orogen are best displayed in the Cordillera Frontal, Precordillera, San Rafael block, and Cordillera Neuquina (Figure 2b). The Neoidic orogen is displayed in the Cordillera Principal (Figure 2b) where it is superimposed on the Paleoidic orogen, which is exposed in the cores of large anticlinal structures involving Jurassic-Cretaceous sediments. These structures are well exposed in the Cordon del Infiernillo, Sierra Azul, Sierra de Cara Cura, Sierra de Reyes, Cordillera del Viento, and Sierra de Chachil (Figure 1).

The Paleoidic orogen developed in a tectonic environment (Haller and Ramos, 1984) involving collision of a passive continental margin with an active continental margin. During an early stage of evolution, two sedimentary sequences were deposited in a remnant basin, one of Cambrian-Ordovician age and the other of Late Ordovician-Early Devonian age. Ophiolitic complexes of Middle-Late Ordovician (Caradocian) age are associated with these sequences. The continent-continent collision occurred during the Middle Devonian, forming a suture zone and generating magmatic activity that ended in the Early Carboniferous. Thereafter, marine to continental sediments were deposited during the late Paleozoic, and a backarc magmatic belt developed during the Permian-Triassic. Cratonic compressive basins were formed and received Permian sediments: subsequently, intramountain basins were formed, in which Triassic molassic sediments were deposited.

The first stage in the development of the Neoidic orogen was formation of a backarc cratonic basin in the Early to Middle Jurassic, accompanied by incipient ophiolite assemblages in the region of north Neuquen. This backarc basin evolved

[1]Translated by S. Pierri.

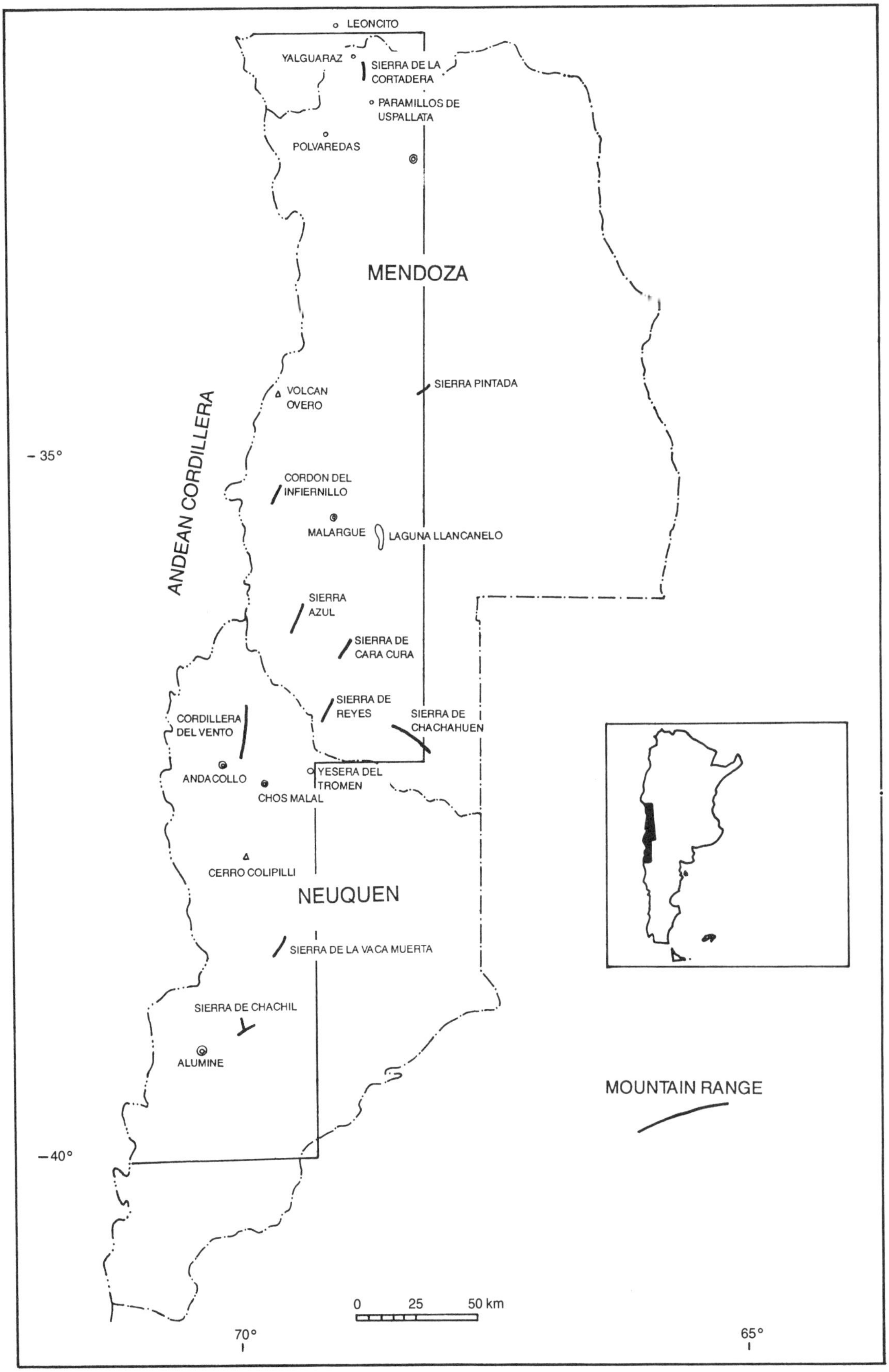

Figure 1. Index map of the central Andean region, provinces of Mendoza and Neuquen, Argentina.

into an intracratonic subsident basin in which two marine sedimentation cycles are recognized, one of Jurassic age and the other of Late Jurassic-Early Cretaceous (Andean) age. Thereafter, continental clastic sediments were deposited, followed by an Atlantic marine transgression during the Late Cretaceous-Early Tertiary (Riograndic) cycle. Continental sediments were deposited from late Eocene to Quaternary in backarc compressive basins of the piedmont region of the central Argentine Andes. A continental-margin magmatic arc developed during the Neoidic orogen, giving rise to widespread plutonism and volcanism during Middle to Late Jurassic, Late Cretaceous to Early Tertiary (middle Eocene), and Early Tertiary (late Eocene) to Quaternary.

MINERAL DEPOSITS

Paleoidic Orogen

Deposits Related to the Suture Zone

Syngenetic mineral deposits that are associated with ophiolites in the suture zone (Figure 3) include chromite, magnetite, massive sulfides of Cu-Zn-Fe, talc, and asbestos; the latter two types resulted from postophiolite metamorphism. These deposits were emplaced at various depths and are exposed at various erosion levels. The shallowest deposits are exposed in the northern part of the suture zone and the deepest in the southern part. For example, the northern (shallow) environment is characterized by deposits of massive sulfides of the Cyprus type; southward, these deposits give way to disseminated sulfide deposits and then to Fe and Cr deposits formed by deep-level magmatic segregation. From north to south the following types of mineral deposits are present (Figure 3): (1) remobilized massive sulfides that formed veins in mafic and ultramafic rocks (examples are San Antonio, San Francisco, Sierra de Cortadera, northern Mendoza); peridotite bodies, also on northern Mendoza, that contain podiform masses and granular chromite, both products of magmatic segregation; (2) lenticular bodies of massive pyrrhotite, chalcopyrite, and sphalerite, together with lesser amounts of pentlandite, some of which is cobaltiferous, and gold (the Salamanca deposit is typical); and (3) magmatic segregations of magnetite (Don Rocha deposit) in serpentinized peridotites.

Deposits Related to the Overthrust Belt

The mineral deposits in the overthrust belt (Figure 3) include deposits emplaced prior to, during, and after overthrusting (Mitchell and Garson, 1981). Thus, prethrust deposits such as those associated with the ophiolitic rocks are exposed as a result of tectonic imbrication and subsequent erosion. Devonian-Carboniferous syncollisional and postcollisional plutonic bodies are present in the overthrust belt, together with Permian-Triassic volcanic and plutonic rocks related to the subduction of the Pacific Plate, typical of a backarc magmatic environment. This belt contains concentrations of tungsten, molybdenum, and subordinate tin as well as polymetallic deposits of lead, zinc, silver, gold, copper, antimony, and arsenic. Characteristic deposits are as follows (see Figure 3):

Tungsten—Tungsten is present only in the Josefina Group, where it occurs as masses of wolframite in quartz veins associated with a granitic stock. Small amounts of scheelite, molybdenite, pyrite, chalcopyrite, and pyrrhotite also are present.

Molybdenum—Veinlets of molybdenite are found in granitic rocks at El Portillo and Arroyo Cuevas, and in an intrusive breccia body at Polvaredas. These occurrences have not been exploited.

Tin—The tin mineral cassiterite occurs in metamorphosed carbonaceous sedimentary rock at Arroyo Cuevas.

Polymetallic veins—Polymetallic vein deposits are found in the region of the Josefina Group (Figure 3) and Sierra de la Cortadera, within the San Rafael block (Figure 2b). The primary minerals are sulfides and sulfosalts of iron, lead, zinc, copper, silver, antimony, and arsenic; trace amounts of gold and cobalt also are present.

Deposits Related to Internal-Arc Compressive Basins

Stratabound uranium deposits occur as impregnations of sandstone beds in the sedimentary sequence exposed at Sierra Pintada (Figure 3). These deposits consist of mantos, 10 m or more in thickness, averaging as much as 0.1% U_3O_8 as uranophane and uraninite. Small amounts of chalcopyrite, coffinite, and brannerite also are present.

Deposits Related to Intermontane Basins

Mineral deposits that accumulated in intermontane basins include phosphatic beds such as those at El Risco (Figure 3), which are associated with bituminous shales.

Neoidic Orogen

Deposits Related to Passive-Margin Intracratonic Basins

Syngenetic, diagenetic, and epigenetic mineral deposits are found in sedimentary sequence of passive-margin intracratonic basins (Figure 3). They include stratiform red-bed deposits of copper, uranium, and vanadium (some with silver); bitumens and asphaltites; petroleum; evaporite sequences containing deposits of celestite, barite, and gypsum; sodium and potassium salts; sulfur in gypsum layers; and phosphatic beds. An iron-titanium occurrence, probably a paleoplacer, also is present. Evaporite deposits of barite, celestite, gypsum, and sodium and potassium salts accumulated in a sabkha-type environment (Brodtkorb, Ramos, and Ametrano, 1985). The deposits occur in the Aquilco and Huitrin formations (Figure 4) and show the following characteristics:

Barite-Celestite.—Several types of deposits occur within the evaporite facies (Figure 3), including syngenetic mantos of barite at San Eduardo; celestite in the El Minarete group; veins of barite at 4 de Noviembre, which consist of remobilized bedded barite; barite-celestite at Barda Klein; and epigenetic veins of barite at San Jose, which formed from descending solutions. Remobilized minerals of the evaporite facies also formed in situ deposits such as stalactites and other fillings of cavities in or near stratabound deposits as well as vein deposits, which originated by solutions moving away from the original deposit. Such migration of solutions caused separation of the more soluble $SrSO_4$ from less soluble $BaSO_4$, resulting in formation of relatively pure deposits of celestite and barite, respectively. Both the strontium and the barium in all these deposits had a marine origin as indicated by the Sr^{87}/Sr^{86}

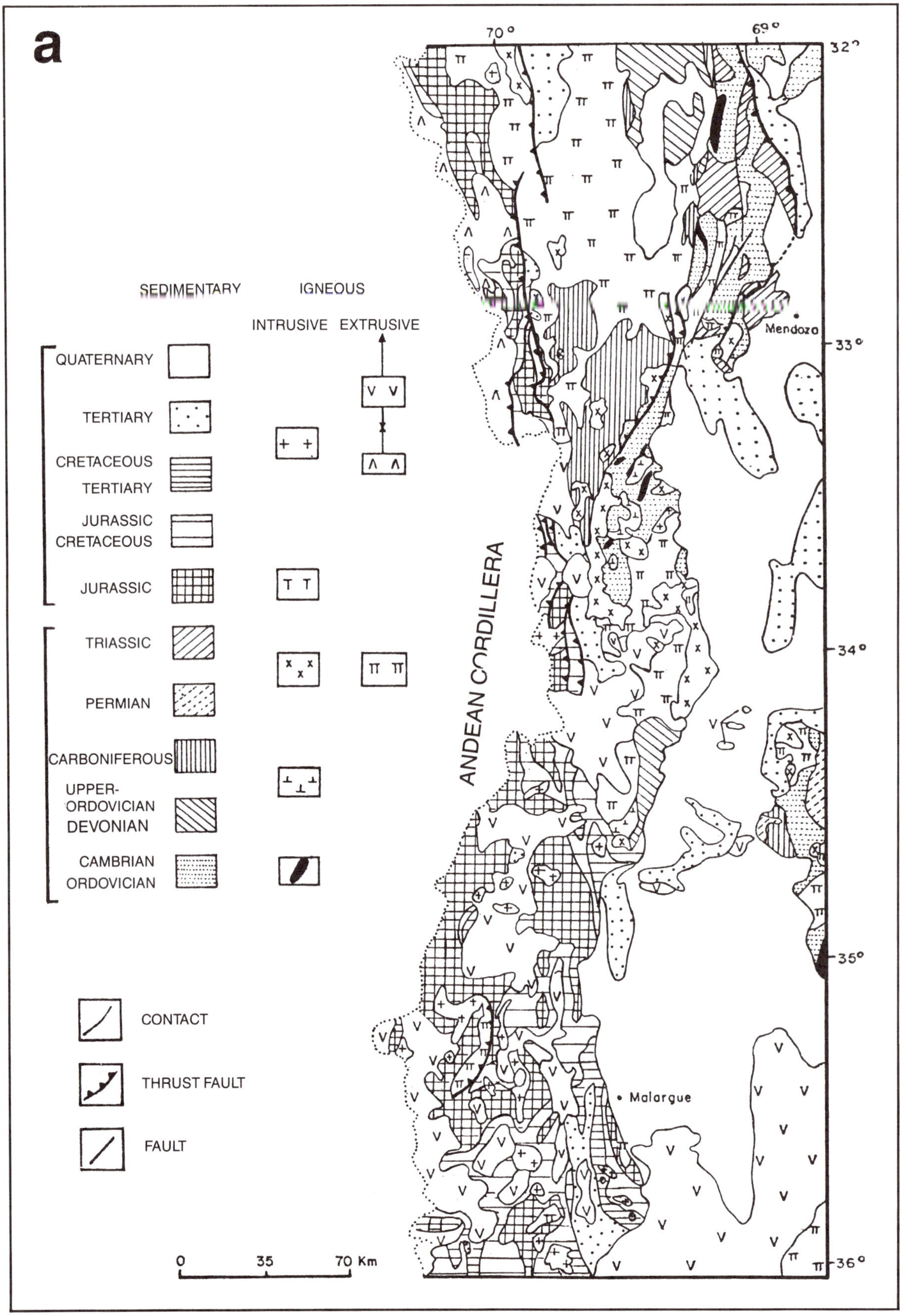

Figure 2. Age distribution of rocks in the central region of Argentina: (a) northern sector, (b) southern sector.

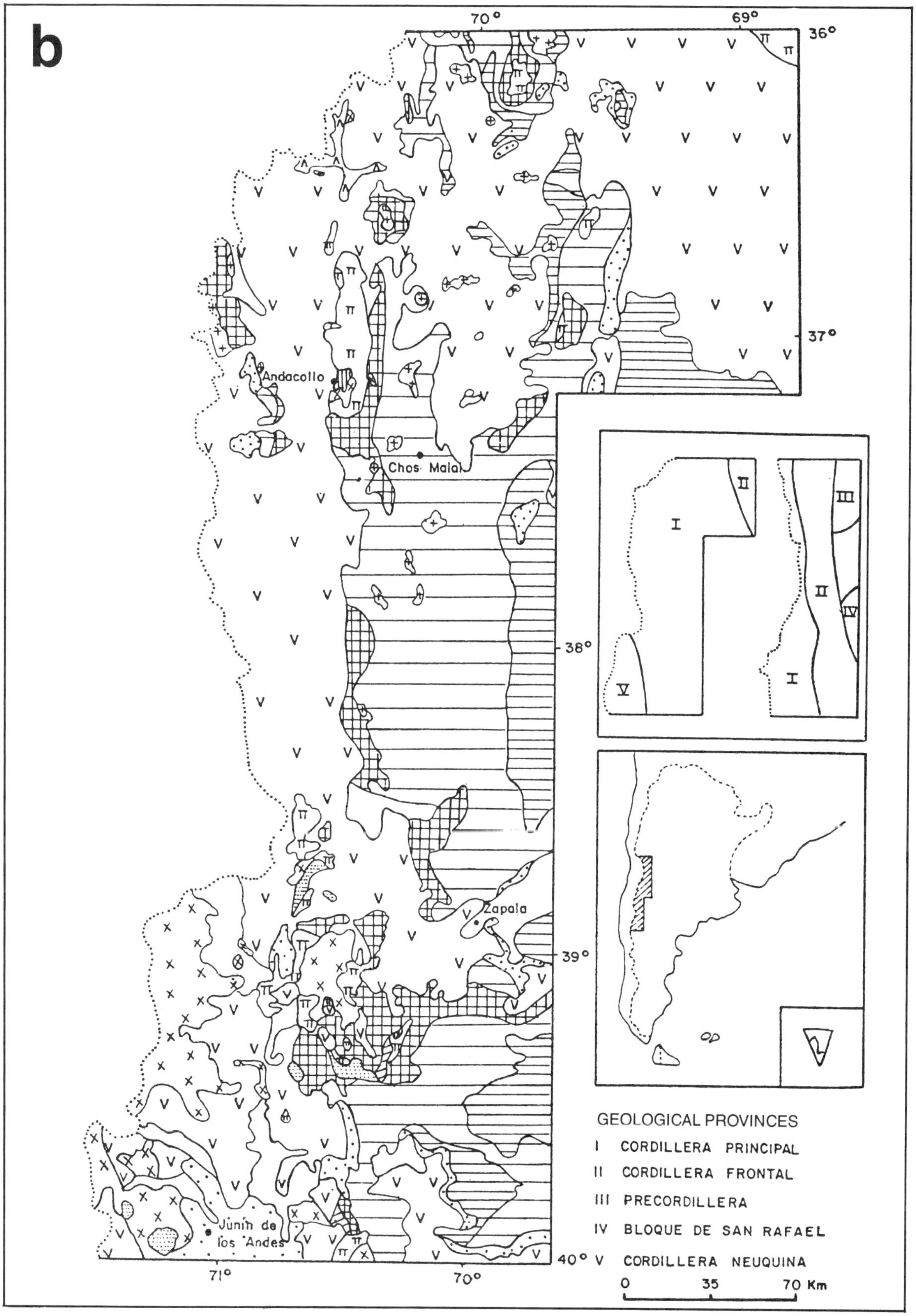

Figure 2 continued.

	TECTONIC DOMAIN	MINERAL DEPOSITS				GEOGRAPHIC COORDINATES		SIZE *
		METAL	ORIGIN	TYPE	EXAMPLES	S. LAT.	W. LONG.	
PALEOIDIC OROGENE	SUTURE ZONE	Cu	VOLCANOGENIC	VEIN (REMOBILIZED)	San Antonio	32° 17' 30"	69° 10' 00"	1
					San Francisco	32° 18' 00"	69° 07' 00"	1
		Cu-Zn-Fe		VEIN	La Salamanca	33° 21' 30"	69° 28' 00"	1
		Fe	MAGMATIC SEGREGATION	PODIFORM AND DISSEMINATED	Don Rocha	33° 28' 30"	69° 27' 30"	1
		Cr			La Cortadera	32° 17' 30"	69° 10' 00"	0
	OVERTHRUST BELT	W	HYDROTHERMAL	VEIN	Grupo Josefina	33° 28' 00"	69° 26' 00"	2
		Mo (Cu)		STOCKWORK	Arroyo Cuevas	33° 05' 00"	69° 28' 30"	0
		Mo			Alborada	33° 53' 00"	69° 19' 00"	0
				DISSEMINATED	El Portillo	33° 36' 00"	69° 32' 00"	0
		As-Ag-Au		VEIN	La Cortadera	32° 19' 00"	69° 09' 30"	1
		Pb-Zn-Cu			Grupo Las Picazas	34° 34' 00"	68° 50' 00"	2
		Cu	VOLCANOGENIC	MANTO	Mantos de Cobre	32° 29' 00"	69° 05' 00"	1
	INTERNAL ARC COMPRESSIVE BASIN	U	DIAGENETIC-EPIGENETIC	STRATA-BOUND	Sierra Pintada	34° 37' 00"	69° 30' 00"	4
	INTRA MONTANE BASIN	P	BIOCHEMICAL		El Risco	33° 02' 00"	69° 10' 00"	0
NEOIDIC OROGENE	PASSIVE-MARGIN INTRACRATONIC BASIN	Ba	SYNGENETIC — CHEMICAL	VEIN	Grupo San José	37° 52' 00"	69° 58' 00"	2
				STRATA-BOUND	Achalay	38° 32' 30"	70° 06' 00"	3
					4 de Noviembre	38° 20' 30"	70° 03' 30"	3
					San Eduardo	37° 44' 30"	70° 16' 30"	3
		Ba-Sr			Barda Klein	37° 38' 30"	70° 00' 30"	3
		Sr			Grupo El Minarete	37° 55' 00"	69° 38' 30"	2
					Grupo La Cecilia	37° 50' 00"	69° 54' 00"	0
		Na-K			Luncay	36° 35' 30"	69° 47' 00"	3
					Salinas del Huitrín	37° 39' 00"	69° 47' 30"	3
		P	SYNGENETIC — BIOCHEMICAL		Bajada del Agrio	38° 37' 00"	70° 09' 00"	0
		Fe-Ti	SYNGENETIC — PALEOPLACER		Aquada del Overo	39° 18' 30"	70° 09' 00"	0
		S	BIOCHEMICAL		Andalucía	35° 24' 00"	69° 44' 00"	0
					Cerro Partido	38° 36' 00"	69° 57' 20"	0
		Cu-U			Dona Juana	39° 13' 30"	69° 51' 00"	2

*Size: Five categories (0-4) according to economic importance. 0 = deposit not well known or information about reserves is lacking. The other categories are in accord with international categories. (Modified from Mapa Metalogenico de America Sur, 1983.)

Figure 3. Tectonic domains and associated mineral deposits in the central Andean region of Argentina.

ratios (Barbieri et al., 1981). Some of the barium, however, may be of exhalative volcanic origin, for example that in the Achalay deposit.

Gypsum.—Important gypsum deposits are found in the Auquilco Formation (Yesera del Tromen, Sierra de la Vaca Muerta) and in the Huitrin Formation (Malargue) (Figure 4). Unusually thick gypsum in the Auquilo Formation resulted from a reaction of volcanic H_2S with bicarbonate in sea water to form gypsum instead of limestone. Gypsum formed in this way augmented the gypsum formed by evaporation of normal sea water. That such an enrichment of gypsum has taken place is indicated by a gypsum/limestone ratio in the Auguilco Formation that cannot be attributed to simple evaporation of seawater.

Halite and Sylvite Deposits.—Deposits of halite and sylvite are found in the Huitrin Formation (Figures 3, 4). The most important deposit (Luncay) is in the province of Mendoza. In addition, important concentrations of sylvite were encoun-

	TECTONIC DOMAIN	MINERAL DEPOSITS: METAL	ORIGIN	TYPE	EXAMPLES	GEOGRAPHIC COORDINATES: S. LAT.	W. LONG.	SIZE *
NEOIDIC OREGENE	MAGMATIC ARC	Cu-Mo	HYDROTHERMAL	DISSEMINATED	Paramillos Sur	32° 29' 30"	69° 06' 00"	4
					Río de las Vacas	32° 34' 00"	69° 50' 00"	0
		Cu			Campana Mahuída	38° 13' 30"	70° 32' 30"	3
				VEIN	Valquaraz I y II	32° 08' 30"	69° 26' 00"	1
		Pb-Ag-Zn			Paramillos de Uspallata	32° 28' 00"	69° 08' 00"	2
					El Cajón	35° 24' 00"	69° 57' 00"	2
					Campana Mahuída	38° 13' 00"	70° 35' 30"	1
					Huayelón	38° 14' 00"	70° 28' 30"	2
					Carreri	38° 56' 00"	70° 34' 00"	2
					1° de Mayo	39° 24' 30"	70° 38' 00"	1
		F-Mn			Agua Escondida	36° 00' 00"	68° 50' 00"	3
		Au			Oro del Norte	32° 25' 30"	69° 06' 00"	0
					Bosques	32° 31' 00"	69° 05' 30"	0
		Au-(Hg)			Distrito Andacollo	37° 12' 00"	70° 32' 00"	2
				DISSEMINATED	Los Maitenes	37° 13' 30"	70° 41' 00"	0
		Cu	METASOMATIC	REPLACEMENT BODIES	Las Choicas	34° 56' 30"	70° 13' 00"	1
		Fe		SKARN	Hierro Indio	34° 59' 00"	69° 47' 00"	2
		S	VOLCANOGENIC	EXHALATIVE	Volcán Overo	34° 30' 30"	70° 01' 30"	3
		Cu		HYDROTHERMAL	El Burrero	34° 56' 30"	70° 13' 00"	1
	BACKARC COMPRESSIVE BASIN	Au-Hg	SEDIMENTARY – SYNGENETIC	PLACER	Distrito Andacollo	37° 12' 00"	70° 38' 00"	1
		Na		SALINE	Diamante	34° 56' 00"	68° 52' 00"	2
		Mn	SEDIMENTARY – EPIGENETIC	IMPREGNATION	Real de Piedra Ahorcada	33° 56' 30"	69° 40' 00"	0
		U-(Cu)		VEIN	Soberanía-Papagayos	32° 53' 00"	69° 00' 00"	1
		Cu	DIAGENETIC-EPIGENETIC – CHEMICAL-SEDIMENTARY	STRATA-BOUND	San Romeleo	36° 35' 00"	69° 27' 30"	1
		Cu-V			El Alambrado-El Manzano	36° 08' 00"	69° 46' 00"	0
					Calmucó-Río Barrancas	36° 35' 00"	70° 00' 00"	0
		Cu-U-Ag			Estancia Charahuilla	39° 26' 30"	70° 24' 30"	0
		U-Cu			Huemul-Agua Potada	35° 47' 00"	69° 40' 00"	3
		Mn			Sierra de Chachahuén	37° 01' 00"	68° 51' 00"	0

*Size: Five categories (0-4) according to economic importance. 0 = deposit not well known or information about reserves is lacking. The other categories are in accord with international categories. (Modified from Mapa Metalogenico de America Sur, 1983.)

Figure 3 continued.

tered during drilling at the top of the evaporite facies of the Huitrin in the region along the border between Mendoza and Neuquen provinces.

Phosphorite.—Three phases of phosphorite deposition are recognized in central western Argentina. The earliest consists of glauconite- and collophane-rich beds associated with limestones of the La Manga Formation (Figure 4) in the northern part of Neuguen province. The next youngest consists of accumulations of collophane-rich nodules, clastic fragments, and fossil fragments in limestones and sandstones of the Vaca Muerta Formation (Figure 4) (Mastandrea et al., 1982). Six phosphatic units, which are present in this formation at Bajada del Agrio (Figure 3), range from 0.2 to 4 m in thickness and contain 1.5 to 10.1% P_2O_5. These phosphorite units are well exposed on the eastern flank of the Vaca Muerta anticline. The youngest phosphatic material is at the top of the Mulichinco Formation and at the bottom of the overlying Agrio Formation (Figure 4).

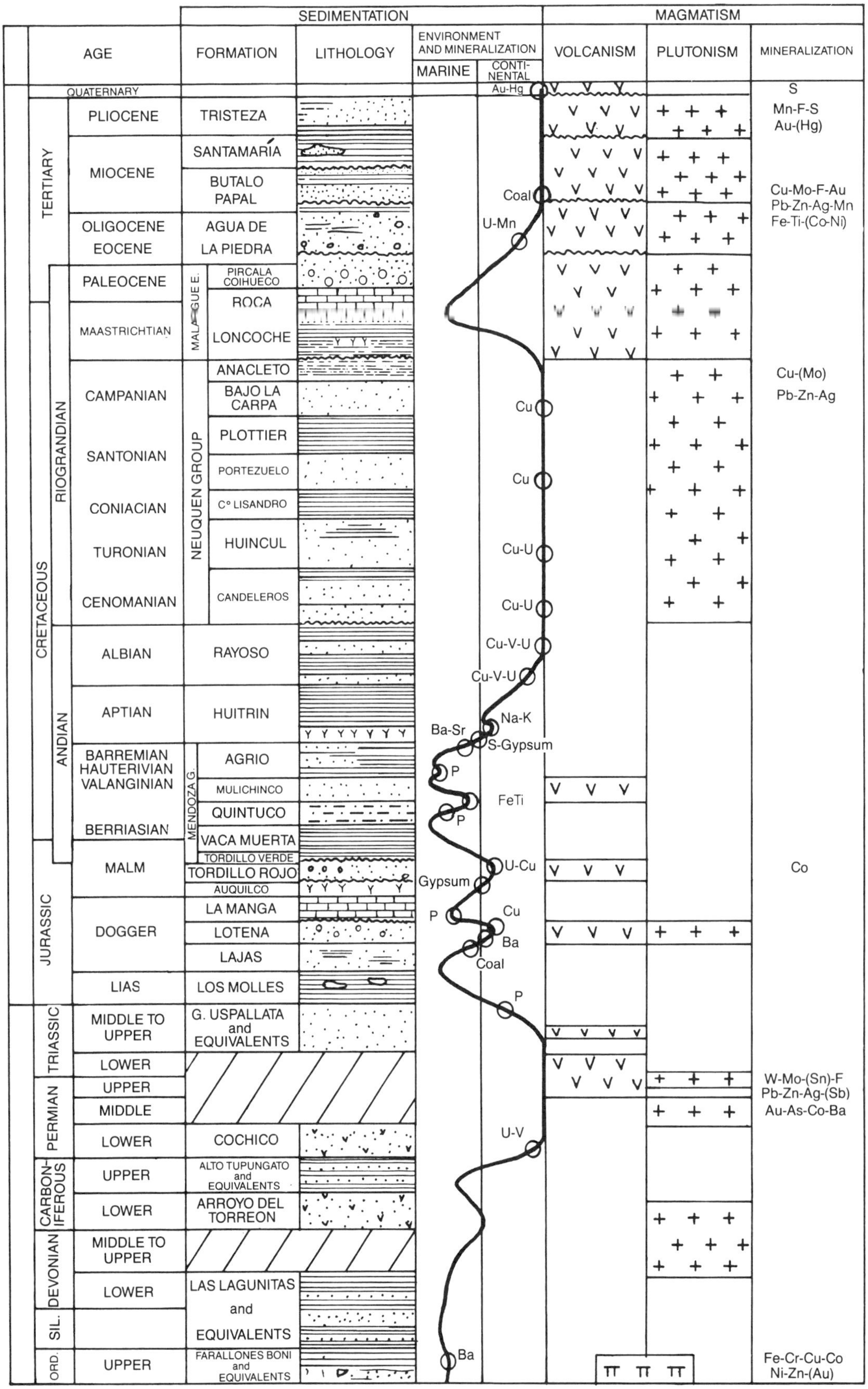

Figure 4. Sedimentary and magmatic cycles of the Paleoidic and Neoidic orogens in the central Andean region of Argentina.

Iron-Titanium.—Concentrations of magnetite and illmenite are found in the Cretaceous Aguada del Overo Formation (Figure 3), where they originally accumulated in a beach environment. They occur as magnetite- and illmenite-rich layers 50 to 70 cm in thickness interbedded with sandstones. These layers contain 29 to 37% Fe and, locally, as much as 10% Ti.

Sulfur.—Sulfur deposits formed as the result of reduction of gypsum by anerobic bacteria, and oxidation of hydrocarbon accumulations beneath the evaporate sequence of the Huitrin Formation (Figure 4). Such transformations required free circulation of meteoric waters. The sulfur deposits occur at Andalucia in the province of Mendoza, and at Cerro Partido in the province of Neuquen (Figure 3). At both of these areas, the sulfur is found in gypsum layers of the Huitrin Formation as irregular impregnations and grains of native sulfur over lateral distances of several hundred meters.

Copper-Uranium.—Diagenetic-epigenetic occurrences of copper and uranium are found in continental sedimentary sequences of the Lotena, Tordillo, and Rayoso formations (Figure 4). The deposits evidently formed by deposition of copper and uranium minerals (from metal-rich meteoric waters) in sedimentary sequences rich in organic matter. The host rocks include sandstones and shales, and commonly the metals accumulated in paleochannels such as those that are well exposed in the Tordillo and Rayoso formations.

Deposits Related to the Magmatic Arch

The metalliferous deposits with the greatest economic potential are those in the magmatic arc (Figure 3). Because of their large size, the copper-molybdenum porphyries are especially important among the various types of deposits. Lead-zinc-silver and gold veins occur around the peripheries of the porphyry copper deposits. Volcanic centers in the magmatic arc contain many exhaltive sulfur deposits and sparse stratabound copper deposits. Contact metamorphic deposits containing copper and iron are found near plutons in the magmatic arc but are of little economic importance. Some of the iron-rich skarn deposits also contain small amounts of manganese- and fluorine-bearing minerals.

Porphyry Copper.—Many porphyry copper deposits of Late Cretaceous and Tertiary age are related to igneous intrusive rocks of the magmatic arc. Mendez and Zappettini (1984) suggested that these deposits formed at shallow depths in areas undergoing tectonic compression. The deposits show alteration zoning around a central potassium-bearing argillic core. Supergene enrichment is poorly developed and, consequently, ore grades tend to be lower than in the porphyry copper deposits of Chile, many of which contain abundant supergene minerals. The host rocks of the porphyry coppers are of various ages and lithologies. The most important deposits appear to be related to regional shear zones or megafracture zones. The deposits of Paramillos Norte, Paramillos Sur, and Campana Mahuida exhibit such structural control. The porphyry copper deposits also show regional differences in geochemistry, notably an increase in gold and molybdenum toward the east.

Lead-Zinc-Silver Veins.—Several porphyry copper deposits are characterized by peripheral veins containing lead-, zinc-, and silver-bearing minerals. Among these veins are those of the Paramillos de Uspallata district, which is marginal to the Paramillos Norte and Paramillos Sur porphyry coppers, and the Compana Mahuida-Huayelon district, which is near the Campana Mahuida porphyry copper deposit. Typically, the lead-zinc-silver veins contain argentiferous galena, sphalerite, and chalcopyrite in a gangue of quartz and carbonate minerals, chiefly calcite and siderite.

Stratabound Copper.—Some copper occurrences in the magmatic arc consist of copper minerals as vesicle fillings in lavas and as crosscutting veins in interbedded lavas and red beds of the Tordillo Formation. They occur in two areas near the Argentina-Chile border, one at San Jose and the other at Burrero (Figure 3), both in the province of Mendoza. The deposits are similar to the manto-type copper deposits of central Chile, which may be syngenetic in origin (Ruiz, 1964). Zappettini (unpublished report, 1984) suggested that the Argentine deposits are of syngenetic-hydrothermal origin, the vesicular ore being volcanogenic and the veins consisting of hydrothermally mobilized minerals from the vesicular ore. The deposits contain native copper, chalcocite, and bornite associated with prehnite and datolite. Also present are the secondary minerals malachite, azurite, and chrysocolla.

Gold.—Gold deposits are found in dacitic to andesitic volcanic rocks of the magmatic arc, occurring in isolated districts, such as the Andacollo district (Figure 3) or in districts marginal to copper-molybdenum porphyry deposits, such as the Oro del Norte and Bosques districts (Figure 3), which are associated with the Paramillos deposits. Four types of gold deposits are present in the Andacollo district: (1) auriferous quartz veins (Rosario); (2) galena-sphalerite-auriferous pyrite veins (Sofia); (3) auriferous pyrite veins (Erika); and (4) disseminated-type gold deposits such as Los Maitenes; which was a major source of gold in nearby placer deposits.

Native Sulfur.—Exhalative native sulfur deposits are found on dacitic to andesitic stratovolcanoes of Late Tertiary and Quaternary age. The principal deposit in central western Argentina is on Volcan Overo, province of Mendoza, which has been second in sulfur production in Argentina.

Copper- and Iron-Bearing Skarns.—Two types of skarn deposits are recognized in central western Argentina, one in which copper is the principal economic metal and the other in which iron is the principal metal. The Las Choicas deposit (Figure 3), province of Mendoza, is an example of a copper-bearing skarn containing disseminated bornite and chalcopyrite, which formed in limestone at the contact with a diorite pluton. Also present are veins containing chalcopyrite, pyrite, tetrahedrite, barite, and siderite. Iron-bearing skarns are widespread in the southern part of the province of Mendoza. The most abundant iron-bearing minerals are magnetite, hematite, and pyrite; chalcopyrite also may be present. The principal gangue minerals are garnet, epidote, and calcite. Typical iron-bearing skarns are found in the Colipilli and Hierro Indio districts, province of Mendoza.

Manganese Oxide-Fluorite Veins.—The Agua Escondida district, in the southern part of the province of Mendoza, contains veins of manganese oxides and fluorite in a sequence of volcanic rocks of Permian-Triassic age. The veins appear to be genetically related to a Pliocene eruptive center. The principal manganese minerals are psilomelane, pyrolusite, and cryptomelane, and the gangue minerals are chalcedony, opal, quartz, and calcite. Fluorite occurs only in the deeper parts of

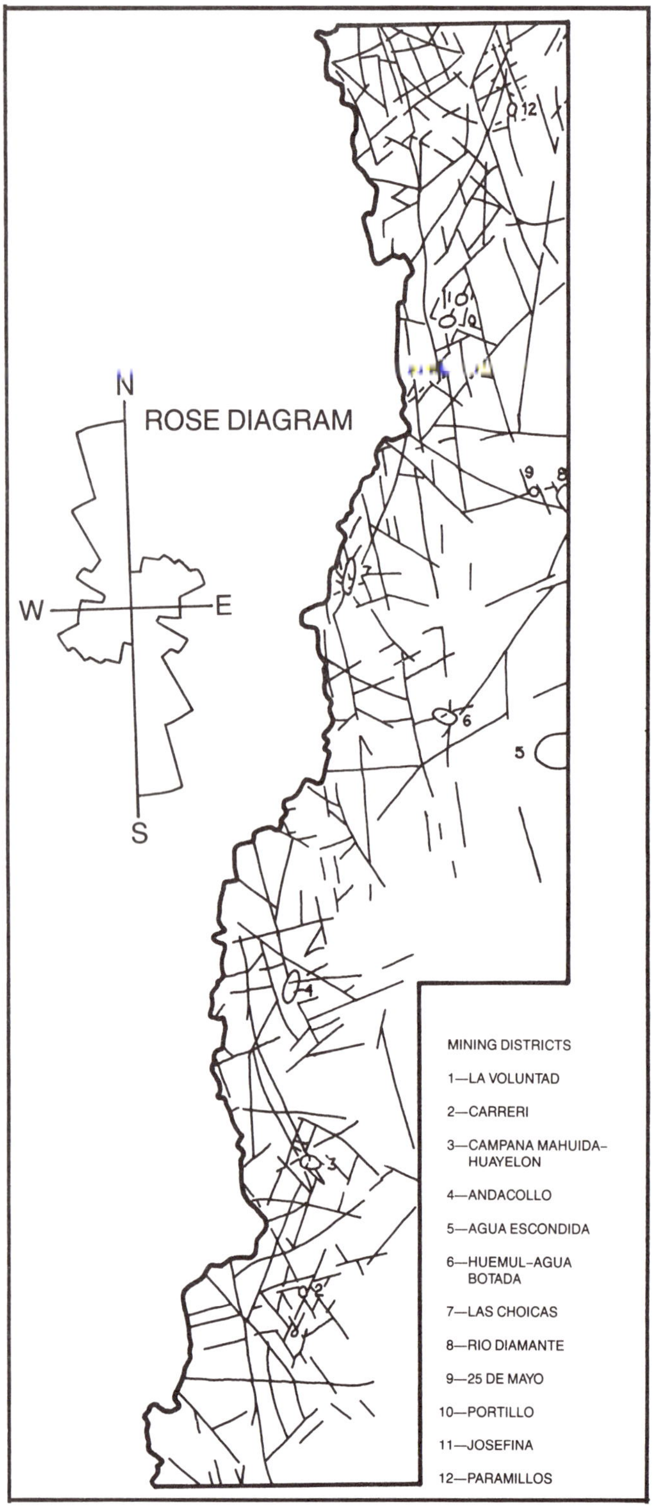

Figure 5. Map showing major lineaments and selected mining districts in the central Andean region of Argentina.

the veins, where it is associated with both manganese and iron oxides.

Deposits Related to Compressive Backarc Basins

The clastic sedimentary sequences of compressive backarc basins contain several types of metalliferous deposits—gold-mercury deposits of the Andacollo district; Real de Piedra Ahorcada manganese and Soberania uranium-copper deposits, both of which are epigenetic deposits formed by descending meteoric waters; and stratabound, diagenetic-epigenetic copper-vanadium-uranium, locally argentiferous, deposits in the Neuquen Group sedimentary sequence. Among these, the Huemul-Agua Botada uranium-copper deposit and the San Rameleo copper deposit have the greatest economic potential.

REGIONAL FRACTURE PATTERNS

Figure 5 shows the distribution of lineaments (> 10 km in length), which are believed to reflect major faults and fracture and shear zones. Some of the principal mining districts are located on these lineaments commonly at or near intersecting lineaments. As can be seen on the rose diagram on Figure 5, the lineaments follow two general trends, one north-northwest (N-N30 °W) and the other east-northeast (chiefly N60 °-70 °E). This fracture system evidently developed in a stress field generated by ocean-continent convergence during the Late Cretaceous-Tertiary Andean orogeny. The orientation suggests that the direction of convergence was along an ENE-WSW direction; lineaments perpendicular to this direction are compressive structures, whereas those parallel to this direction are tension structures. Other fractures shown on Figure 5 have diverse orientations, part of which appear to be oriented 45 ° to the principal stress directions. The mining districts shown on Figure 5 tend to be located at or near intersections of lineaments that trend north-northwest and east-northeast to west-northwest.

MINERAL BELTS

The different types of mineral deposits in the central Andean region of Argentina tend to cluster in well-defined belts or zones, as shown on Figure 6. Some of these belts follow the longitudinal structural grains of the Andes, as do the mineral belts in Chile (Ruiz and Ericksen, 1962); others, or parts of others, appear to be unrelated to either longitudinal or transverse structures. Among the features that determine the distribution of deposits are megastructures such as former magmatic arcs, sedimentary basins, and major fault systems; depths of formation and levels of erosion; and inhomogeneities of crustal and subcrustal rocks that were sources of magmas with which mineralizing hydrothermal fluids were associated.

Lead-Zinc-Silver Belt.—A discontinuous lead-zinc-silver belt parallels the Andes throughout the area, a longitudinal distance of about 900 km (Figure 6). The southernmost segment of this belt, in the province of Neuquen, is defined by many deposits through a longitudinal distance of about 380 km. The belt is relatively narrow, generally < 15 km wide. The ores are found in veins that tend to be clustered in mining districts located at points where major transverse structures cross the belt. The principal mining districts in this sediment are Campana Mahuida, Huayelon, Carreri, and 1 ° de Mayo. To the north the belt is discontinuous, with gaps where deposits are not known to occur.

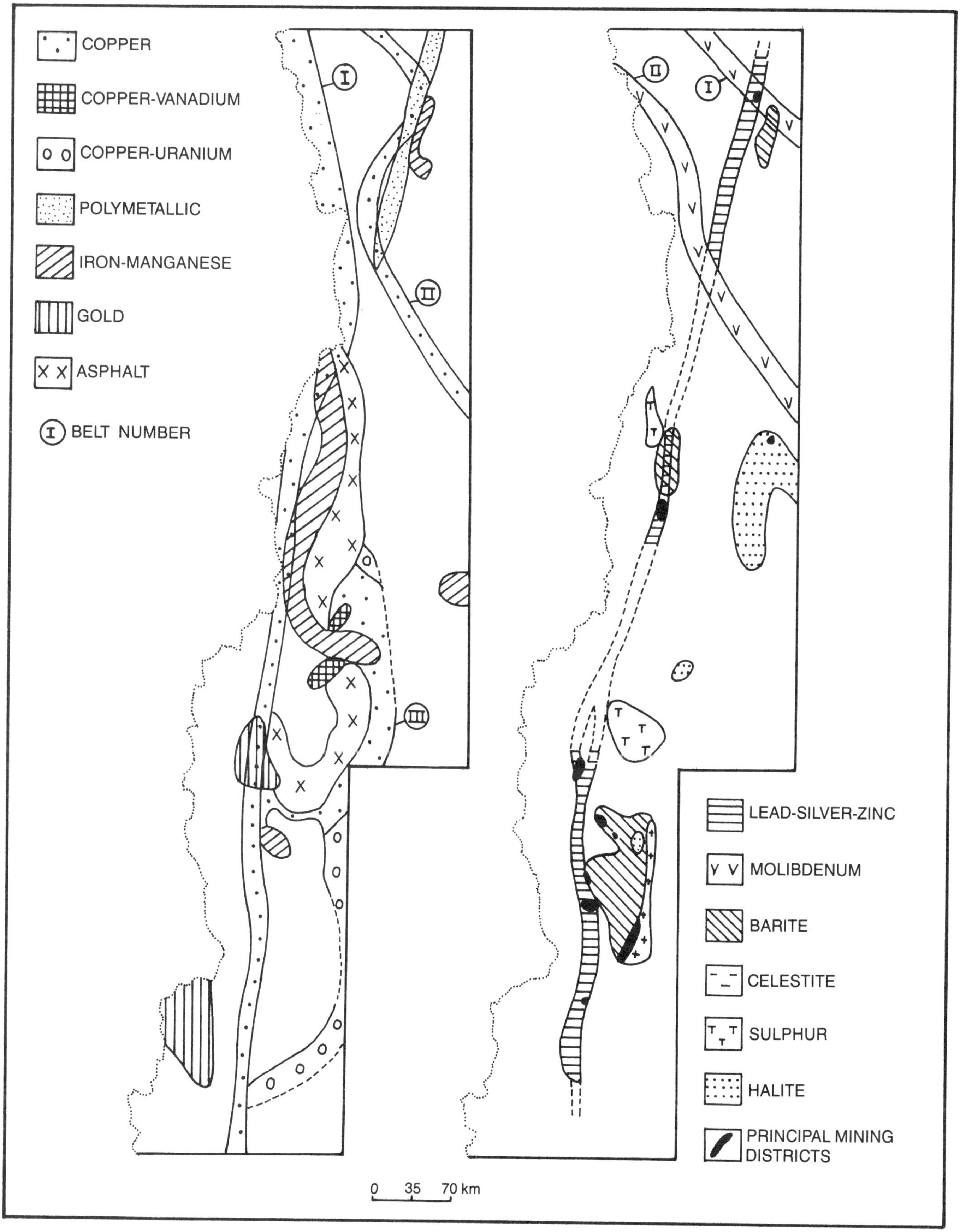

Figure 6. Mineral belts in the central Andean region of Argentina.

Molybdenum Belts.—Molybdenum deposits are found in two parallel belts—designated I and II on Figure 6—in the northern part of the area. The deposits formed during two epochs of mineralization, one during the Permian-Triassic and the other during the Tertiary. These deposits include molybdenite as disseminations (Santa Clara), in intrusive breccias (Polvaredas), and in stockworks (Alborada). The South Paramillos copper-molybdenum porphyry-type deposit, at the intersection of molybdenum belt I and the copper belt II, is by far the largest molybdenum deposit. The El Portillo mining district is located at the intersection of molybdenum belt II and the polymetallic belt. Fluorite deposits (Piedra de Afilar district) are found in the southern part of molybdenum belt II.

Copper Belts.—Copper deposits are found in three belts, (Figure 6). Belts I and II, which are defined by porphyry copper deposits, differ geochemically in that belt II has a greater Mo/Cu ratio than belt I. Belt I includes, among others, porphyry copper deposits of Cretaceous and Tertiary age (Bayo North, Campana Mahuida, La Voluntad), volcanogenic copper deposits of Jurassic age (San Jose, El Burrero), and copper skarn deposits (Las Choicas).

Polymetallic Belt.—The polymetallic belt is superimposed on the lead-silver-zinc belt and crosses the copper II and molybdenum belts (Figure 6). Deposits in this belt contain tungsten, cobalt, chromium, gold, copper, lead, silver, and zinc minerals. Copper-molybdenum vein deposits and copper-vanadium deposits also are present. The deposits in the polymetallic belt range in age from Ordovician to Tertiary.

Iron-Manganese Belts and Zones.—Iron and manganese deposits are found in a sinuous longitudinal belt in the central part of the region and in several isolated zones (Figure 6). The principal belt is characterized by iron-bearing skarn deposits related to two pulses of igneous intrusive activity, one in the Paleogene and the other in the Neogene. The principal primary iron-bearing minerals are magnetite, hematite, pyrite, and pyrrhotite. Iron- and manganese-bearing veins also are present. The principal isolated iron-manganese zones include skarn deposits at Colipilli, manganese associated with fluorite at Agua Escondida, and hematite-pyrolusite veins west of the city of Mendoza.

Gold.—Gold deposits occur chiefly in three areas (Figure 6): (1) the polymetallic belt in the northern part of the region, (2) the region of Andacollo, in the south-central part of the area, and (3) at Alumine, in the southwest. The Andacollo area, which is by far the most important economically, contains disseminated deposits, vein deposits, and placer deposits, all of which have been extensively exploited. Gold deposits also have been exploited in the polymetallic belt. The deposits of the Alumine area, chiefly gold-quartz veins and placers, are of little economic importance.

Barite-Celestite.—Three barite-celestite zones are recognized (Figure 6) of which only the deposits in the province of Neuquen are of economic importance. These contain syngenetic and epigenetic remobilized sedimentary deposits of Jurassic, Cretaceous, and Tertiary age. The celestite deposits are concentrated along the eastern side of the area, whereas the barite deposits are to the west.

Sulfur.—Volcanogenic exhalative sulfur deposits are found in two areas, one in the province of Neuquen and the other in the province of Mendoza (Figure 6).

Sodium-Potassium Salts.—Rock-salt beds crop out in the Huitrin Formation of Cretaceous age in two areas (Figure 6), one in southern Mendoza (Luncay) and the other in northern Neuquen (Salinas de Huitrin). A thick (as much as 180 m) salt unit in the region between Sierra de Reyes and Chachahuen includes beds of sylvite-rich salt as much as 3 m thick which contain 16 to 30% K_2O. These beds were encountered in drill holes at depths ranging from 700 to 1000 m. Sodium-chloride-rich brines occur in present-day salt pans at Laguna Llancanelo and Salinas del Diamante.

CONCLUSIONS

Several types of mineral deposits are found in the central Andean region of Argentina. Porphyry copper, lead-zinc-silver, barite, and uranium deposits have the greatest economic potential. The different types of deposits are genetically related to major tectonic units such as former magmatic arcs and sedimentary basins within two orogens, a Paleoidic orogen of Cambrian to Triassic age, and a Neoidic orogen of Jurassic to Quaternary age.

Mineral exploration in this region is in an early stage, and it is believed that many deposits remain the be found.

ACKNOWLEDGMENTS

The writers wish to thank Fabricaciones Militares for permission to publish this report and Mrs. Norma Alicia Coppollillo who assisted in preparation of the report.

REFERENCES CITED

Barbieri, M., M.K. Brodtkorb, S. Ametrano, and V. Ramos, 1981, Datos isotopicos de Sr^{87} and Sr^{86}, relacionados a los yacimientos de celestina y baritina de la formation Huitrin, Provincia de Neuquen: VIII Congreso Geologico Argentino, Actas II, p. 787-850.

Borrello, A.V., 1969, Los geosinclinales de la Argentina: Direccion Nacional de Geologia y Mineria, Anales XIV, p. 188.

Brodtkorb, M.K., V. Ramos, and S. Ametrano, 1975, Los yacimientos estratoligados de celestina-baritina de la formacion Huitrin y su origen evaporitico, Provincia del Neuquen, Argentina: II Congreso Iberoamericano de Geologia Economica, v. 2, p. 143-168.

Haller,, M. and V. Ramos, 1984, Las ofiolitas famatinianas (eopaleozoico) de las Provincias de San Juan y Mendoza: IX Congreso Geologico Argentino, Actas II, p. 66-83.

Mastandrea, O., H. Leanza, C. Hugo, and C. Oblitas, 1982, Prospeccion de fosfatos sedimentarios en la Republica Argentina: V Congreso Latinoamericano de Geologia, v. III: 159-176.

Mendez, V., and E. Zappettini, 1984, Los porfidos de cobre de la Republica Argentina: IX Congreso Geologico Argentino, Actas V, p. 589-610.

Mitchell, A.H., and M.S. Garson, 1981, Mineral deposits and global tectonic settings: London, Academic Press.

Mombru, C.A., and M.A. Uliana, 1978, Esquema tectosedimentario de la Cuencas Mesozoica de Mendoza y Neuquen: VII Congreso Geologico Argentino, Actas II. p. 239-256.

Routhier, P., 1980, Ou sont les metaux por l'avenir?: Memoire Bureau de Recherches Geologiques et Minieres No. 105, Orleans, p. 409.

Stille, H., 1955, Recent deformations of the earth's in the light of those earlier epochs: Geological Society of America Special Paper 62, 171 p.

Ericksen, G. E., Cañas Pinochet, M. T., and Reinemund, J. A., editors, 1989, Geology of the Andes and its relation to hydrocarbon and mineral resources: Houston, Texas, Circum-Pacific Council for Energy and Mineral Resources Earth Science Series, v. 11.

CHAPTER 15

PRECIOUS-METAL MINERALIZATION OF CENOZOIC AGE IN THE ANDES OF PERÚ

Donald C. Noble
Mackay School of Mines
University of Nevada-Reno
Reno, Nevada, U.S.A.

Victor Raúl Eyzaguirre
Empresa Mineral del Centro del Perú
Lima, Perú

Edwin H. McKee
U.S. Geological Survey
Menlo Park, California, U.S.A.

ABSTRACT

Although deposits of silver and gold are present throughout the central Andes, the major production—and particularly that of silver—has been from Perú. Important exceptions include Potosí, Bolivia, and Chamarcillo and El Indio, Chile. Deposits in Perú and Bolivia can be divided into four principal groups based on mineralogic characteristics and volcano-tectonic setting: (1) polymetallic vein and skarn Pb-Zn-Cu ± W-Ag deposits (for example, Casapalca, Antamina); (2) high-sulfur enargite-tetrahedrite (Ag-Cu ± Bi ± Au) deposits (for example, Julcani); (3) classic low-S epithermal Ag-Au deposits (for example, Arcata); and (4) Sn-Ag deposits (for example, Potosí). Almost all mineralization was closely associated with high-level magmatic activity; and the time when mineralization occurred was within less than 0.2 to about 1 m.y. of the major igneous event(s) of a given district. Deposits of groups 1 and 2 (above) typically are found where the Andean chain and the Cenozoic magmatic belt are narrow and are associated regionally with porphyry Cu and Cu-Mo systems. Group 2 deposits are associated with penocryst-rich dacite/quartz monzonite which, where studied in detail, is characterized by high fO_2 and fS_2 and distinctive trace-element patterns. Where classic epithermal Ag-Au ores are present, they typically occur as volumetrically minor paragenetic stages in base-metal-dominated systems (for example, mina Colqui). Deposits of group 3 are commonly associated with rhyolites—many of which contain abundant sanidine and possess evolved trace-element patterns—and with large-volume ash-flow sheets and collapse calderas. Such deposits are characteristic of southern Perú, where the Andes and the Cenozoic volcanic belt are wide. Group 4 deposits are restricted to the southern part of the Bolivian tin belt, which is located in an "inner cordilleran arc" characterized by S-type volcanic and plutonic rocks.

INTRODUCTION

Perú is one of the world's three major producers of silver and is becoming a significant producer of gold. The other countries of the central Andes have important present and/or past production of precious metals as well as definite potential for major economic discoveries. As our experience is primarily in Perú, in this chapter we will concentrate on mineral deposits and associated igneous rocks located there.

The principal precious-metal deposits in the central Andes are of Neogene age. These Miocene and younger precious-metal deposits define a belt centered on the Western Cordillera of northern and central Perú, and on the central part of the high-plateau region of southern Perú (Figure 1). These deposits formed on thick, preexisting, but by no means tectonically stable, continental lithosphere.

REGIONAL MAGMATIC VARIATIONS

Major changes in the nature of Neogene volcanic activity and mineralization are related to the widening of the Andean Cordillera from central to southern Perú. These include:

1. A general change from strongly calc-alkalic and dominantly intermediate volcanic-arc magmatism in the

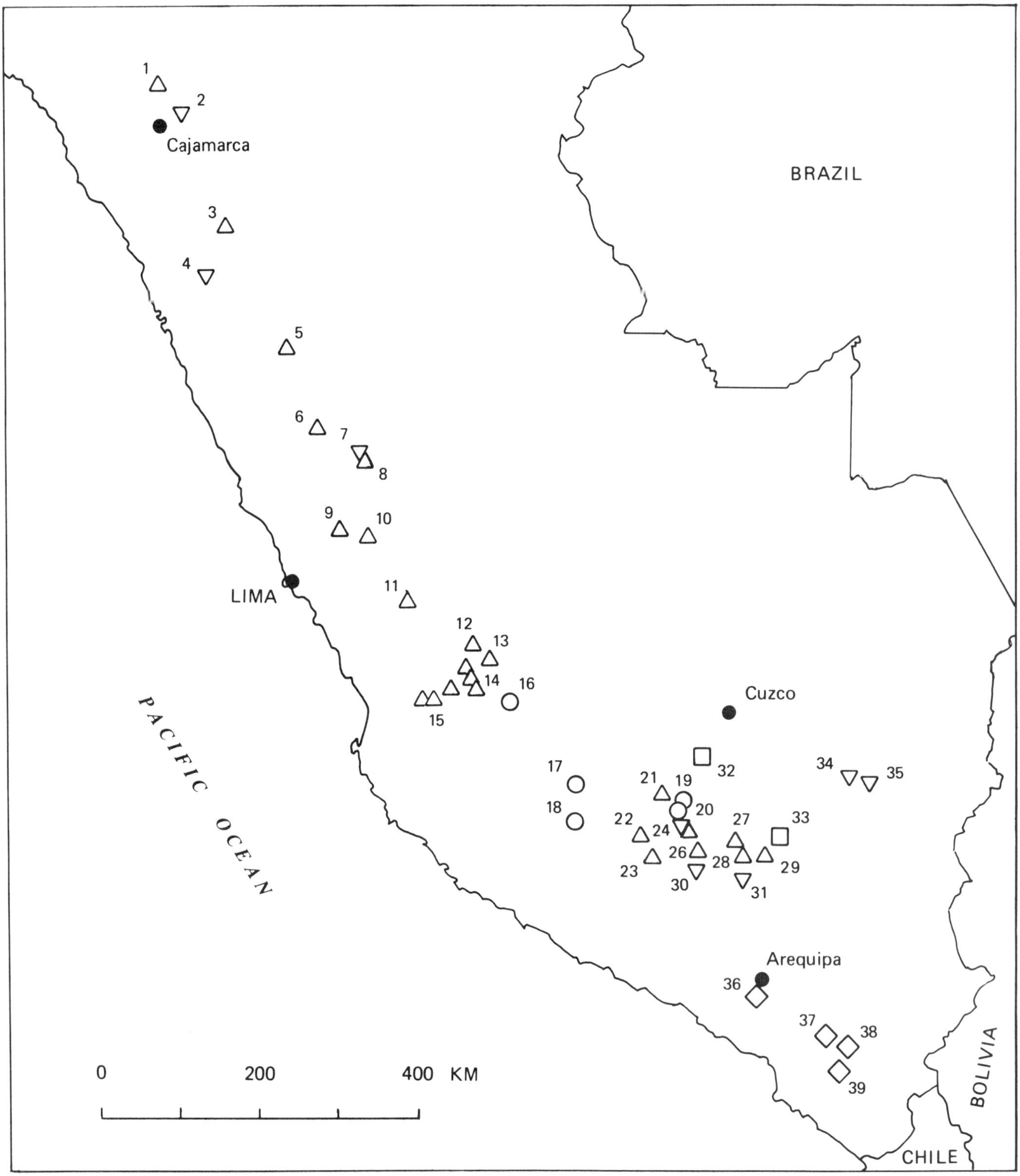

Figure 1. Map showing location and ages of mineral districts for which radiometric age control is available. Diamonds = Paleocene; squares = early Oligocene; inverted triangles = early Miocene; upright triangles = middle and late Miocene; circles = late Pliocene and Pleistocene. Mineral districts and prospects: 1, Hualgayoc; 2, Michiquillay; 3, Pasto Bueno; 4, Pashpap; 5, Antamina; 6, Raura; 7, Cerro de Pasco; 8, Colquijirca; 9, Colqui; 10, Morococha; 11, Yauricocha; 12, Huancavelica; 13, Julcani; 14, Huachocolpa-Tangana; 15, Castrovirreyna (San Genaro, Caudalosa, Astohuaraca, etc.); 16, Atunsulla (Nvdo. Portugueza); 17, Ccarhuaraso; 18, Palla-Palla; 19, Tetón; 20, Cerro Millo; 21, Sto. Domingo; 22, Esquillay; 23, Pararapa; 24, San Martín; 25, Farallón; 26, Arcata; 27, San Miguel; 28, Coramina, 29, Sucuitambo; 30, Orcopampa; 31, Cailloma; 32, Chalcobamba; 33, Tintaya; 34, Sto. Domingo; 35, San Rafael; 36, Cerro Verde; 37, Cuajone; 38, Quellaveco; 39, Toquepala.

north to areally more diffuse large-volume silicic volcanism in the south; and

2. A concomitant change from dominantly polymetallic base- and precious-metal and enargite-tetrahedrite-dominated high-sulfur mineral systems in the north to classic epithermal silver-gold mineralization in the south.

Neogene magmatic activity in the north is reflected by a relatively narrow belt of igneous rocks. In central Perú, rocks of this belt consist largely of dacitic and andesitic lavas that form stratovolcanoes, fields of volcanic domes, and compound volcanic structures. Ash-flow tuffs are relatively uncommon and, where present, are of rather small volume. Few collapse calderas have been recognized, and those known are small structures. At least one—Nevado Portugueza (Noble and McKee, 1982)—is centered on a large composite volcano in a manner reminiscent of Mt. Mazama in the Cascade arc of the northwestern United States (Bacon, 1983). In the northern half of central Perú, erosion is, in general, somewhat greater than to the south. In this area a number of stocks are exposed, for example at mina Raura (Petersen, 1965), and in northern Perú major faulting has exposed the 9-12 m.y.-old batholith of the Cordillera Blanca (Pitcher and Cobbing, 1985).

The belt of Miocene igneous activity becomes wider and more diffuse in southern Perú. Moreover, in southern Perú, large-volume ash-flow sheets of both low-silica rhyolite and chemically moderately to highly evolved high-silica rhyolite (D.C. Noble, unpublished data) are common. Where studied, these large ash-flow sheets are genetically associated with large collapse calderas (Figure 2) of the type present, for example, in various parts of the western United States. Large volumes of rhyolitic lava, mainly in the form of endogenous domes, also are present, as well as appreciable volumes of dacite and andesite.

REGIONAL VARIATIONS IN MINERAL DEPOSITS

The Neogene mineralization of the two areas exhibits significant differences. In central and northern Perú, precious-metal values are mainly found in the following types of deposits: (1) polymetallic Pb-Zn-Cu ± W-Ag deposits such as those of the well-known districts of Casapalca and Morococha; (2) Zn-Pb-Cu-Ag skarn deposits such as Antamina; and (3) high-sulfur enargite-tetrahedrite Ag-Cu ± Bi ± Au deposits such as Julcani, Quiruvilca, and the pyrite-enargite veins of Cerro de Pasco (Petersen, 1979). In these deposits silver is contained largely in tetrahedrite.

The high-sulfur enargite-tetrahedrite systems are quite remarkable geochemically and mineralogically. Such deposits are characterized by abundant coarse-grained alunite that formed from relatively high-temperature and moderately saline sulfate-rich solutions (Bruha and Noble, 1983; Noble and Silberman, 1984) in altered rocks that border vein structures. Moreover, the dacitic magmas possessed extremely high oxygen and sulfur fugacities, features that are reflected, for example, by the presence of phenocrysts of pyrrhotite and anhydrite preserved in glassy rocks (Drexler, 1982; Drexler and Noble, 1983).

In southern Perú, silver and gold are found largely in classic low-sulfur epithermal vein systems. Examples include such important districts as Cailloma, Arcata, and Orocopampa, as well as a large number of geologically similar minor districts and important prospects. In such deposits, silver sulfosalts and electrum are the more important economic minerals, although tetrahedrite is commonly present and in certain districts (for example, Orcopampa) is the dominant silver-bearing mineral. The veins exhibit classic epithermal banding and adularia, carbonates, and manganese silicates are characteristic gangue minerals. Potassium metasomatism of surrounding wall rocks also is typical.

Although classic epithermal Ag-Au deposits are known in central and northern Perú, they are not common, and are perhaps better developed in the southern part of central Perú (for example, in the Castrovirreyna district; Lewis, 1964) than farther north. Indeed, perhaps the best known epithermal occurrence in this region—that of mina Colqui in central Perú (Kamilli and Ohmoto, 1977)—occurs as the second of seven paragenetic stages in a vein system dominated by sphalerite, galena, and chalcopyrite.

Deposits in all parts of Perú show spatial relationship to igneous activity. The polymetallic and enargite-tetrahedrite occurrences in central and northern Perú are closely associated with dacitic domes, which typically occur in fields (Petersen et al., 1977) or parts of central volcanic structures. The eruption of these lavas is commonly preceded and accompanied by extremely explosive pyroclastic activity. Dacites associated with mineralization are petrographically distinctive and appear to have a characteristic geochemical signature.

The epithermal deposits of southern Perú, in contrast, show a definite association with rhyolitic rocks and with large collapse calderas. For example, at mina Arcata, epithermal Ag-Au mineralization, although hosted by dacitic lavas, is temporally most closely related to rhyolitic domes exposed along the southern margin of the district. At the Esquillay caldera, mineralization is present both in fractures of a well-developed central resurgent dome and along structures outside and radial to the caldera margin. Erosion is somewhat deeper in the middle Miocene Chonta caldera complex, and the intracaldera tuff prism contains many caldera-margin slide blocks of Mesozoic limestone and quartzite (Peterson et al., 1983).

Although classic low-sulfur epithermal mineralization is typical of southern Perú, several "high-sulfur" enargite-tetrahedrite systems are known. However, these are not temporally associated with the epithermal silver-gold occurrences, but rather are of late Pliocene and Quaternary age, and are associated in part with central volcanoes of the now-extinct northern extension of the presently active frontal arc. The best example is Ccarhuaraso, where gold and lesser amounts of silver are associated with dacitic lavas and intense premineral quartz-alunite alteration within the central part of a large central volcano of overall pyroxene andesite composition. Mineralization is dated at 1.2 m.y., only several tenths of a m.y. younger than dacite lavas that were erupted late in the history of the volcano. Other altered and mineralized young central volcanoes include Palla-Palla.

This very young mineralization would appear to be associated with the southwestward shift of magmatic activity during the middle to late Pliocene and the formation of a well-defined frontal arc, which includes such major active volcanoes as

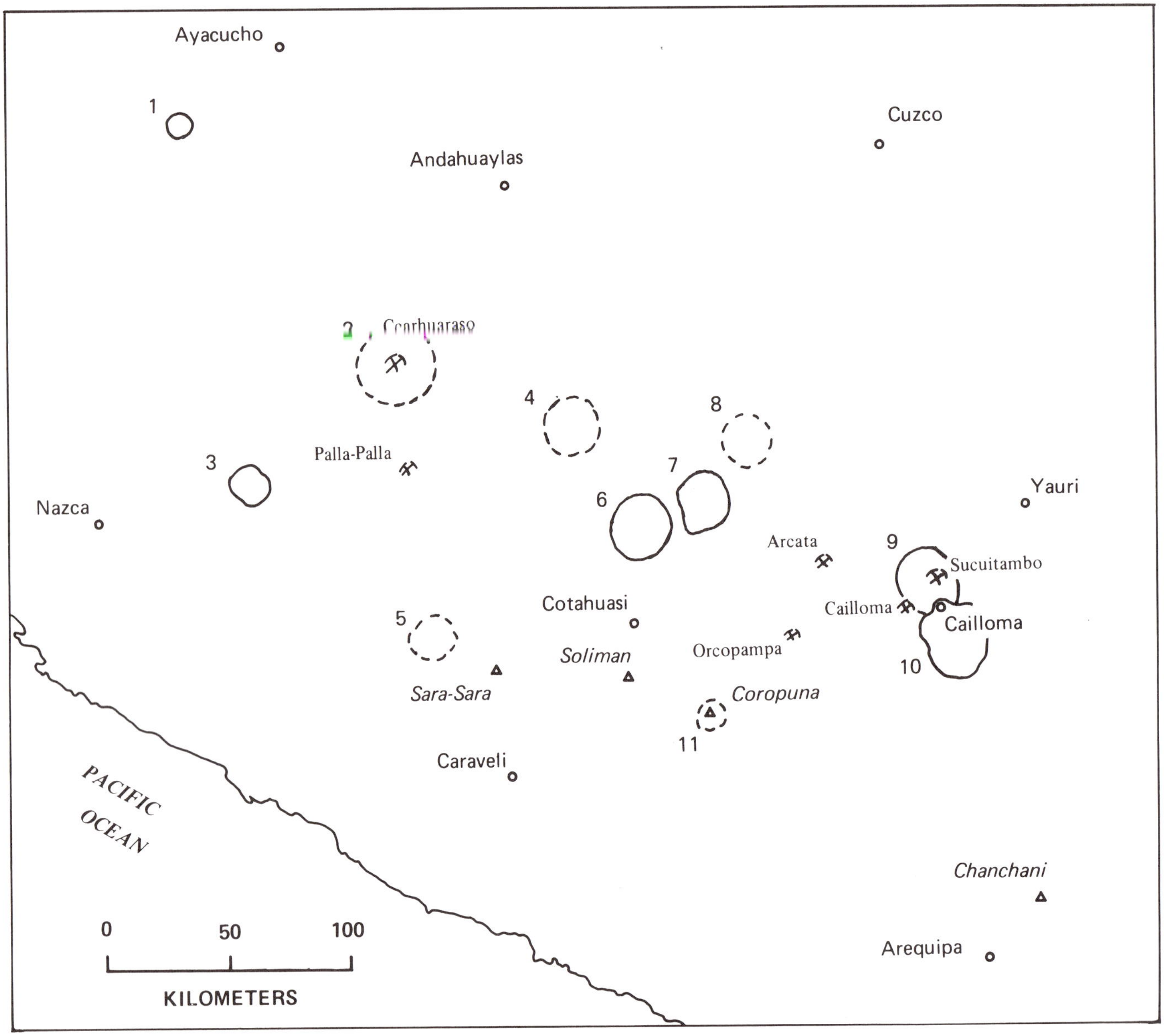

Figure 2. Map showing location of known and inferred (dashed lines) collapse calderas in south-central and southern Perú. 1, Nevado Portugueza; 2, Ccarhuaraso; 3, Pampa Galeras; 4, Tumiri; 5, Parinacocha; 6, Esquillay; 7, San Martin; 8, Tetón; 9, Chonta; 10, Cailloma; 11, Coropuna.

Coropuna and Soliman. Behind this arc a rather distinctive suite of andesite, low-silica andesite, and basalt were erupted, as well as locally significant volumes of dacite and rhyolite. In addition, there are a number of large central volcanoes of late Pliocene and Quaternary age behind the frontal arc, some of which (for example, Cerro Millo and Tetón) are characterized by extensive, strongly developed high-level quartz-alunite-native sulfur alteration.

Irrespective of the type of mineralization—whether polymetallic, enargite-tetrahedrite, or classic epithermal—silver is the most valuable precious metal. Although many districts contain modest gold values, and although major gold occurrences are known—a prime example is the Calera vein at mina Orcopampa (Silberman, McKee, and Noble, 1985)—gold is clearly subordinate to silver in the Peruvian Andes. The dominance of silver production over that of gold is reflected by the mineralogical and geochemical characteristics of many deposits. For example, Julcani is the silver analogue of such "high-sulfur" gold systems as Goldfield and Summitville in the United States, and in certain epithermal Ag-Au vein systems silver to gold ratios may average greater than 1000.

The association of inner-arc S-type magmas and tin, tungsten and, locally, silver mineralization provides another distinctive association of an igneous rock type with a specific type of mineral deposit. The Bolivian tin belt is located in the eastern part of the Bolivian Andes and extends into Argentina and southeasternmost Perú. In southeastern Perú, tin and copper mineralization at mina San Rafael is genetically associated with S-type granites of latest Oligocene age (Clark et al., 1983). During the Neogene large-volume ash-flow sheets of strongly

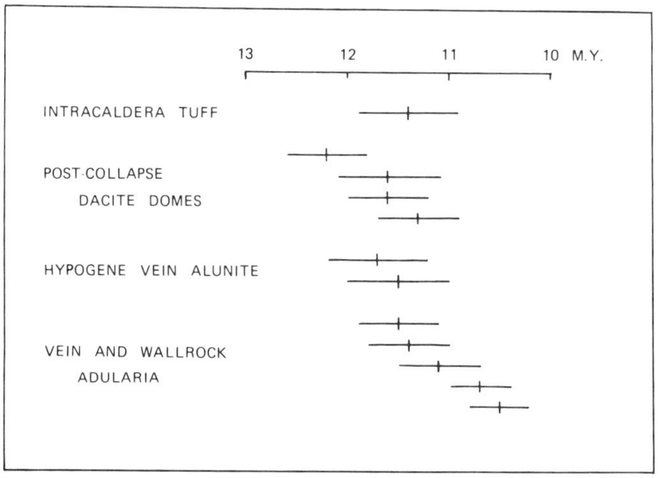

Figure 3. Diagram summarizing radiometric ages obtained on volcanic rocks and mineral deposits and prospects related to the Chonta caldera complex, southern Perú. Length of bar indicates estimated analytical uncertainty at the one-sigma level.

peraluminous S-type rhyolite containing phenocrysts of muscovite, andalusite, and sillimanite, and possessing high concentrations of Sn, Li, Rb, U, Cs, P, and other lithophile elements were erupted in the same region (Noble et al., 1984). Petrographically similar rocks of about the same age appear to be present to the south in the Bolivian part of the tin belt (Sillitoe, 1981; G. Ericksen, oral communication, 1985).

TEMPORAL ASSOCIATION OF MINERALIZATION WITH IGNEOUS ACTIVITY

There is a close temporal association between mineralization and volcanic activity. For example, at Julcani in central Perú, more than 30 dacitic domes and associated pyroclastic rocks were erupted 10.1 m.y. ago over an area of 25 km^2 (Petersen et al., 1977; Noble and Silberman, 1984). The economic vein systems were formed several tenths of a million years later in response to the upward movement of a new batch of chemically distinct rhyodacitic magma. Quartz-alunite-pyrite alteration at the margins of the veins has been dated at about 9.8 m.y., and dikes of the rhyodacite magma were emplaced very shortly thereafter. Hydrothermal activity then waned, but nevertheless continued to about 9.4 m.y. ago.

Similar relationships are apparent at Cerro de Pasco in central Perú, where the tetrahedrite-bearing pyrite-enargite veins formed at virtually the same times as the host pyroclastic rocks and lavas that fill the Cerro diatreme vent structure (Silberman and Noble, 1977). At Nevado Portugueza in central Perú, the time of mineralization along fractures of the central resurgent dome of the Nevado Portugueza caldera followed the last igneous activity—the emplacement of a rhyodacite dome along the ring-fracture system— by 0.1 m.y. or less (Noble and McKee, 1982).

A number of mineral systems are present within the Chonta caldera complex in southern Perú (Peterson et al., 1983). Age determinations on the hydrothermal systems range from indistinguishable (within the limits of analytical uncertainty) from those obtained on the intracaldera tuff and postcollapse ring domes to ages as much as 1 m.y. younger than caldera collapse (Figure 3). In a manner similar to that for other studied systems in the western United States and Perú, the systems with strong quartz-alunite-pyrite alteration are virtually the same age as igneous activity, whereas systems characterized by vein and wall-rock adularia are significantly younger than volcanic activity.

ACKNOWLEDGMENTS

The work of the first author has been supported in part by grants from the National Science Foundation. We thank the many companies who have assisted or facilitated our work, and in particular Mauricio Hochschild y Compañia, Ltda., S.A. and Compañia de Minas Buenaventura, S.A.

REFERENCES CITED

Bacon, C.R., 1983, Eruptive history of Mount Mazama and Crater Lake caldera, Cascade Range, U.S.A.: Journal of Volcanology and Geothermal Research, v. 18, p. 57-115.

Bruha, D.J., and D.C. Noble, 1983, Hypogene quartz-alunite ± pyrite alteration formed by moderately saline, ascendant hydrothermal solutions (abs.): Geological Society of America Abstracts with Programs, v. 15, p. 325.

Clark, A.H., V.V. Palma, D.A. Archibald, Edward Farra, M.J. Arenas F., and R.C.R. Robertson, 1983, Occurrence and age of tin mineralization in the Cordillera Oriental, southern Perú: Economic Geology, v. 78, p. 514-520.

Drexler, J.W., 1982, Mineralogy and geochemistry of Miocene volcanic rocks genetically associated with the Julcani silver-bismuth-lead-copper-gold-tungsten deposit, Perú: physio-chemical conditions of a productive magma body: Houghton, Michigan, Michigan Technological University, unpublished P h.D. dissertation, 275 p.

Drexler, J.W., and D.C. Noble, 1983, Highly oxidized sulfur-rich dacite-rhyolite magmas of the Julcani silver district, Perú: insight into a "productive" magmatic system (abs.): Geological Society of America Abstracts with Programs, v. 15, p. 325.

Kamilli, R. J., and H. Ohmoto, 1977, Paragenesis, zoning, fluid inclusion, and isotopic studies of the Finlandia vein, Colqui district, central Perú: Economic Geology, v. 72, p. 950-982.

Lewis, R.W., Jr., 1964, The geology, mineralogy, and paragenesis of the Castrovirreyna lead-zinc-silver deposits, Perú: Palo Alto, California, Stanford University, unpublished Ph.D. dissertation, 374 p.

Noble, D.C., and E.H. McKee, 1982, Nevado Portugueza volcanic center, central Perú: a Pliocene central volcano-collapse caldera complex with associated silver mineralization: Economic Geology, v. 77, p. 1893-1900.

Noble, D.C., and M.L. Silberman, 1984, Evolución volcánica y hidrotermal y cronología de K/Ar del Distrito Minero de Julcani: Sociedad Geológicas del Perú, Volumen Jubilar LX Aniversario, Fascículo 5, 35 p.

Noble, D.C., T.A. Vogel, P.S. Peterson, G.P. Landis, N.K. Grant, P.A. Jezek, and E.H. McKee, 1984, Rare-element enriched, S-type ash-flow tuffs containing phenocrysts of muscovite, andalusite, and sillimanite, southeastern Perú: Geology, v. 12, p. 35-39.

Petersen, U., 1965, Regional geology and major ore deposits of central Perú: Economic Geology, v. 60, p. 407-476.

Petersen, U., 1979, Metallogenesis in South America: progress and problems: Episodes, v. 1979, no. 4, p. 11-33.

Petersen, U., D.C. Noble, M.J. Arenas, and P.C. Goodell, 1977, Geology of the Julcani mining district, Perú: Economic Geology, v. 72, p. 931-949.

Peterson, P.S., D.C. Noble, E.H. McKee, and V.R. Eyzaguirre, 1983, A resurgent, mineralized caldera in southern Perú: preliminary report: EOS (American Geophysical Union Transactions), v. 64, p. 884.

Pitcher, W.S., and E.J. Cobbing, 1985, Phanerozoic plutonism in the Peruvian Andes, *in* W.S. Pitcher, M.P. Atherton, E.J. Cobbing, and R.D. Beckinsale, eds., Magmatism at a plate edge—the Peruvian Andes: Glasgow, Blackie and Sons, Ltd., 328 p.

Silberman, M.L., and D.C. Noble, 1977, Age of igneous activity and mineralization, Cerro de Pasco, central Perú: Economic Geology, v. 72, p. 925-930.

Silberman, M.L., E.H. McKee, and D.C. Noble, 1985, Age of mineralization at the Cailloma and Orcopampa silver districts, southern Perú: Isochron/West, no. 43, p. 17-18.

Sillitoe, R.H., 1981, Ore deposits in cordilleran and island-arc settings, *in* W.R. Dickinson, and W.D. Payne, eds., Relations of tectonics to ore deposits in the Southern Cordillera: Arizona Geological Society Digest, v. 14, p. 49-61.

Ericksen, G. E., Cañas Pinochet, M. T., and
Reinemund, J. A., editors, 1989, Geology of the Andes
and its relation to hydrocarbon and mineral resources:
Houston, Texas, Circum-Pacific Council for Energy
and Mineral Resources Earth Science Series, v. 11.

CHAPTER 16

GEOLOGICAL FRAMEWORK OF ANDEAN MINERAL RESOURCES

Ulrich Petersen
Harvard University
Cambridge, Massachusetts, U.S.A.

ABSTRACT

The long and complex geological history of the Andes has had a decisive role in generating a diverse set of mineral deposits in a variety of geological settings. Some of these deposits are well exposed and were readily discovered and developed, whereas other were either destroyed by erosion or are deeply buried and remain to be discovered.

In most places the Precambrian basement is deeply buried under younger rocks. It contains sedimentary manganese and banded iron deposits, as well as pegmatites.

The relatively extensive and partly metamorphosed lower Paleozoic clastic sedimentary rocks contain some important syngenetic iron deposits, lead-zinc and tungsten deposits, and small stratabound nickel-cobalt and copper-lead-zinc deposits. Exploration for similar deposits has been hampered by widespread soil and vegetation cover in their major outcrop areas along the eastern slopes of the Andes. During the late Paleozoic, arid conditions led to the formation of local salt deposits and small copper deposits in Permian red beds. Two important base metal deposits, Bailadores in Venezuela and Cobriza in Perú, are thought by some geologists to have formed syngenetically at this time. A number of epigenetic deposits of gold, chromite, nickel, cobalt, uranium, copper, lead, zinc, tin, tungsten, bismuth, molybdenum, and silver appear to have formed during various Paleozoic orogenic and magmatic episodes.

During the Mesozoic, volcanogenic copper deposits formed in the eugeosynclinal sequences, whereas coal and syngenetic polymetallic deposits formed in the adjoining miogeosyncline. Evaporites were formed mostly on the cratonic margin to the east of the geosyncline. The evaporites may contain as yet undiscovered potash deposits.

Magmatism during the Late Cretaceous and Tertiary generated a great variety of mineral deposits, including iron-bearing skarns and magmatic segregations, gold and silver veins, copper-molybdenum porphyries, copper-zinc-lead-gold-bearing skarns, copper-lead-zinc-silver veins and replacement deposits, enargite and precious metal veins, mineralized pipes and disseminated orebodies, tin-tungsten veins and porphyries, and mercury deposits. The original distribution of these various types of deposits was governed by regional geochemical and tectonic features, including the depth of cover at the time of formation. Their present surface distribution is further determined by the extent to which they were buried by later sedimentary and volcanic rocks and by the degree to which subsequent erosion has exposed the rocks in which they occur.

During the late Tertiary and Quaternary, phosphate and diatomite were deposited in marine basins along the Pacific coast, sulfur accumulated on volcanic cones along the continental divide, gold placers formed along the eastern slopes of the Andes from Colombia to Bolivia, gold and platinum placers formed in central Colombia, tin placers formed in Bolivia, nickel laterites formed by weathering of ultramafic rocks in the northern Andes, and continental borate, nitrate, and lithium-bearing evaporites formed in the deserts of southern Perú, western Bolivia, northern Chile, and northwestern Argentina.

INTRODUCTION

This chapter provides a synthesis of Andean mineral-deposit types and the geological environments in which they formed. More detailed discussions of Andean ore deposits are provided by Ahlfeld (1954, 1967), Ahlfeld and Schneider-Scherbina (1964), Angelelli (1950, 1984), Angelelli et al. (1970), Angulo (1978), Bellido, De Montreuil, and Girard (1969), Ericksen (1976, 1980), Fernandez Lima and Caminos (1980), Goossens (1972a, 1972b), Jaramillo and Escobar (1980), Petersen (1970, 1972, 1979), Ponzoni (1980), Ponzoni, Postigo, and Birkbeck (1969), Putzer (1971, 1976), Ruiz and Ericksen (1962), Ruiz et al. (1965), Samamé Boggio (1979-84), Soler, Grandin, and Fornari (1986), Terrones (1984), and Zeil (1964).

The Andes have been emergent for most of the last 70 million years, shedding their erosion products both to the east and to the west. To the east, terrigenous sediments formed thick accumulations along the Andean front from Venezuela to southern Argentina, and farther east they spread out in the Amazon trough. As a consequence, the pre-Tertiary basement is deeply buried in these regions, and can be observed only where it crops out as islands surrounded by the Tertiary and Quaternary debris (for example, in Argentina) or where it has been penetrated by deep drill holes. To the west, some

Andean detritus accumulated in alluvial fans and within longitudinal grabens in southern Perú and in northern Chile, but most accumulated as near-shore deposits that formed the continental shelf and slope. Some of these marine deposits are now exposed in small uplifted basins. In addition, intermontane valleys and basins in the Andes were filled with late Tertiary and Quaternary sediments and volcanic rocks.

The detrital and volcanic cover in the Andean region obscures many mineral deposits and complicates our understanding of the geological settings and distribution of the ore deposits. The search for these buried deposits is a formidable challenge to exploration geologists.

PRECAMBRIAN

Outcrops of Precambrian rocks are rare in the Andean realm (Cobbing, Ozard, and Snelling, 1977; Ponzoni, 1980; Davidson, 1984). It is, therefore, not surprising that Precambrian ore deposits represent a minority of the mineral deposits exposed at the surface. Clearly, the chances are small for finding sizable new deposits in these terranes. Nevertheless, some Precambrian ore deposits are known and others may be located in the future. Many important Precambrian deposits such as chromite- and platinum-bearing mafic and ultramafic intrusives, pegmatites, carbonatites, and diamond pipes do not contain abundant sulfide minerals and, therefore, do not have conspicuous weathering zones. Also, magnetite and ilmenite deposits may not be easily recognized, and prospectors, geologists, and miners accustomed to typical Andean ores may have overlooked or underestimated such deposits.

Some sedimentary iron and manganese formations are definitely of Precambrian age (Figure 1), including the Santa Bárbara manganese deposit in Precambrian quartzites in Colombia (Angulo, 1978) and ferriferous itabirites at Tarpuy, Perú (Fernández Concha, 1956; Fernández Concha and Amstutz, 1956).

Throughout the Andes there are a number of nonsedimentary ore deposits in metamorphic terranes where both the age of the sediments and of the metamorphism are debatable. These deposits are commonly grouped into a Precambrian-early Paleozoic metallogenic epoch. However, where the mineralization is related to metamorphism (as is likely for pegmatites and gold-quartz veins), where it is associated with intrusives (as in magmatic segregations or in spatially associated veins), and where it is clearly epigenetic (as in crosscutting veins), the ores could well have formed during the Hercynian or Andean orogenic cycles.

PALEOZOIC

Along the Andes, sedimentary sequences of gray and black shales, phyllites, schists, sandstones, and quartzites have been identified as Ordovician, Silurian, and Devonian. These rocks have undergone various degrees of metamorphism.

In northern Chile and northwestern Argentina a number of igneous rocks and thermal events have been dated between 398 and 485 Ma (Davidson, 1984). Early Paleozoic orogenic activity has been documented in Perú on the basis of a 440 Ma gneiss complex in the Arequipa massif and a 392 to 440 Ma batholith at San Nicholás (Pitcher, 1984).

The large Aguilar lead-zinc deposit, in northwestern Argentina (Figure 2), has been interpreted as a stratabound sedimentary-diagenetic deposit of Early Silurian age (Sureda and Amstutz, 1981). Silurian sedimentary iron formations exposed at Sierras de Zapala, Puesto Viejo, and Santa Bárbara occur in this same general region (Figure 1). In the south, Silurian-Devonian iron formations occur at Sierra Grande-Rio Negro (Angelelli et al., 1970). Stratiform scheelite deposits in limestone, considered to be syngenetic (Brodtkorb and Brodtkorb, 1979; Brodtkorb et al., 1985), occur within the metamorphosed Paleozoic schists of San Luís province, central Argentina.

The sedimentary iron deposits at Chilicuaquen, Relún, and Nueva Imperial, central and southern Chile (Figure 1), were formerly thought to be Precambrian (Ruiz et al., 1965), but are now assigned to the lower Paleozoic (C. R. Petersen, oral communication). On the other hand, the stratabound tin deposits of Bolivia, formerly thought to be Silurian (Schneider and Lehmann, 1977), are now considered to be epigenetic and Tertiary in age (Lehmann, 1979, 1985).

Stratabound nickel-cobalt-arsenic-tin and iron-copper-zinc-lead deposits are found within lower Paleozoic rocks in the Yauli dome of central Perú (Kobe, 1982b; Rivera and Kobe, 1983a, 1983b). In addition, it has recently been proposed that the lead-zinc ores of Cerro de Pasco are part of a Paleozoic massive sulfide deposit (Cheney, 1985), contrary to the traditional view that these ores are related to a Tertiary volcanic vent (Einaudi, 1977; Silberman and Noble, 1977).

Given that the lower Paleozoic sediments correspond to a time interval of about 150 m.y. and formed in a marine environment, it is puzzling that no other important syngenetic copper, zinc, or lead deposits have been reported from this setting. One can ask:

1. Was there insufficient magmatism to produce abundant hydrothermal fluids?
2. Were the conditions such that the hydrothermal fluids issuing from the seabed were generally dispersed so that economic metal concentrations could not form?
3. Have such deposits not been found yet because the major outcrop areas of the lower Paleozoic sediments are in regions of widespread soil and vegetation cover (i.e., in the northern Andes and along the eastern slopes of the Andes in Ecuador, Perú, and Bolivia) where a gossan can easily be confused with lateritic soil?

Our interpretation of metallogeny during the early and middle Paleozoic may be profoundly influenced by the outcome of the current controversy on the existence and extent of exotic or tectonostratigraphic terranes along the western margins of North and South America. It also will be influenced by new information from geophysical and geochemical prospecting, as well as from the use of remote-sensing techniques in exploring for blind ore deposits.

The gold-quartz veins in the lower Paleozoic metamorphic belt along the eastern slopes of the Andes may have formed during any one of the Paleozoic orogenies (Ponzoni, 1980; Pitcher, 1984). These veins were the source of much of the Cenozoic placer gold of this region. Some, such as the veins of the former Santo Domingo mine in southeastern Perú (Samamé Boggio, 1981), have been exploited.

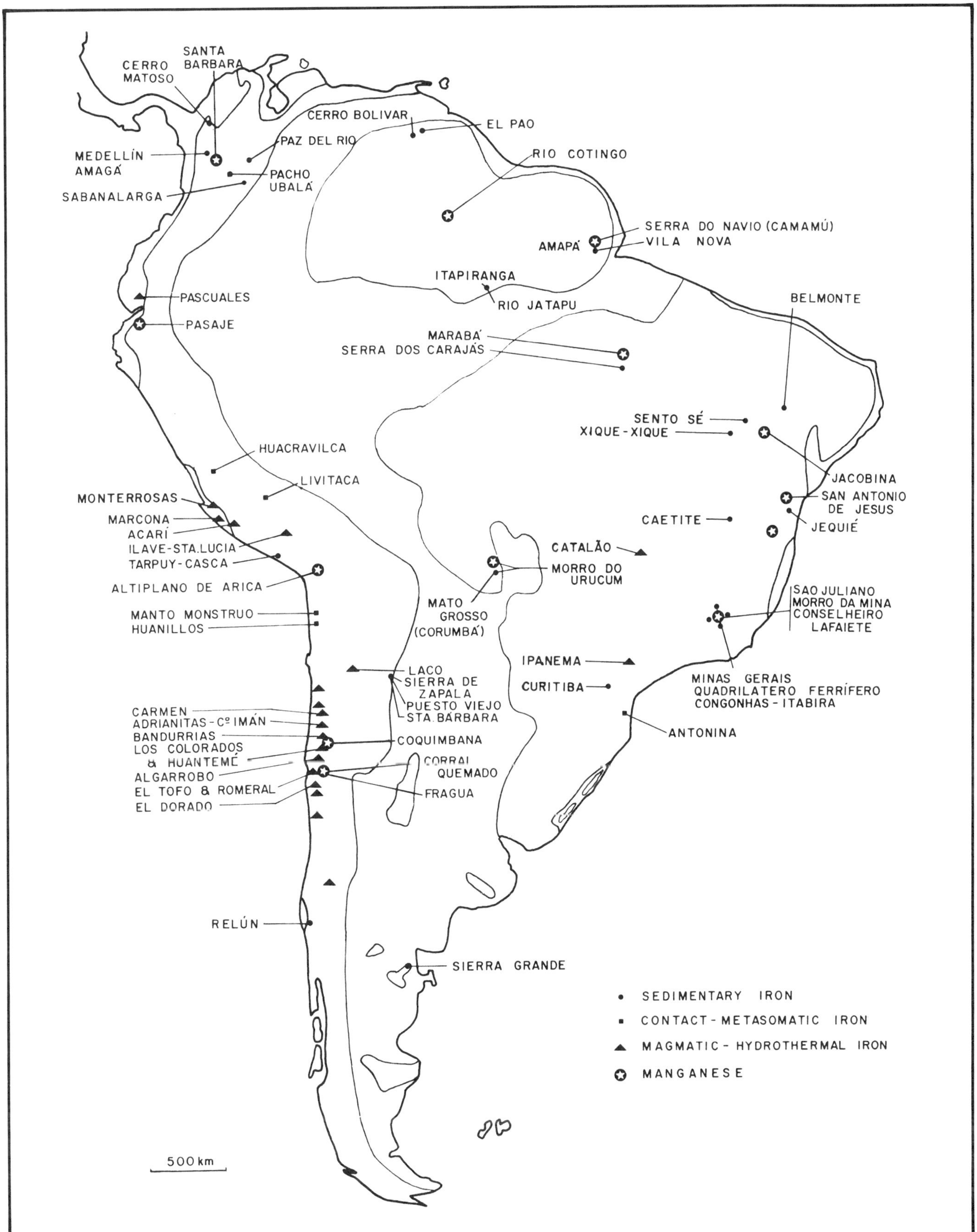

Figure 1—Iron and manganese deposits in South America.

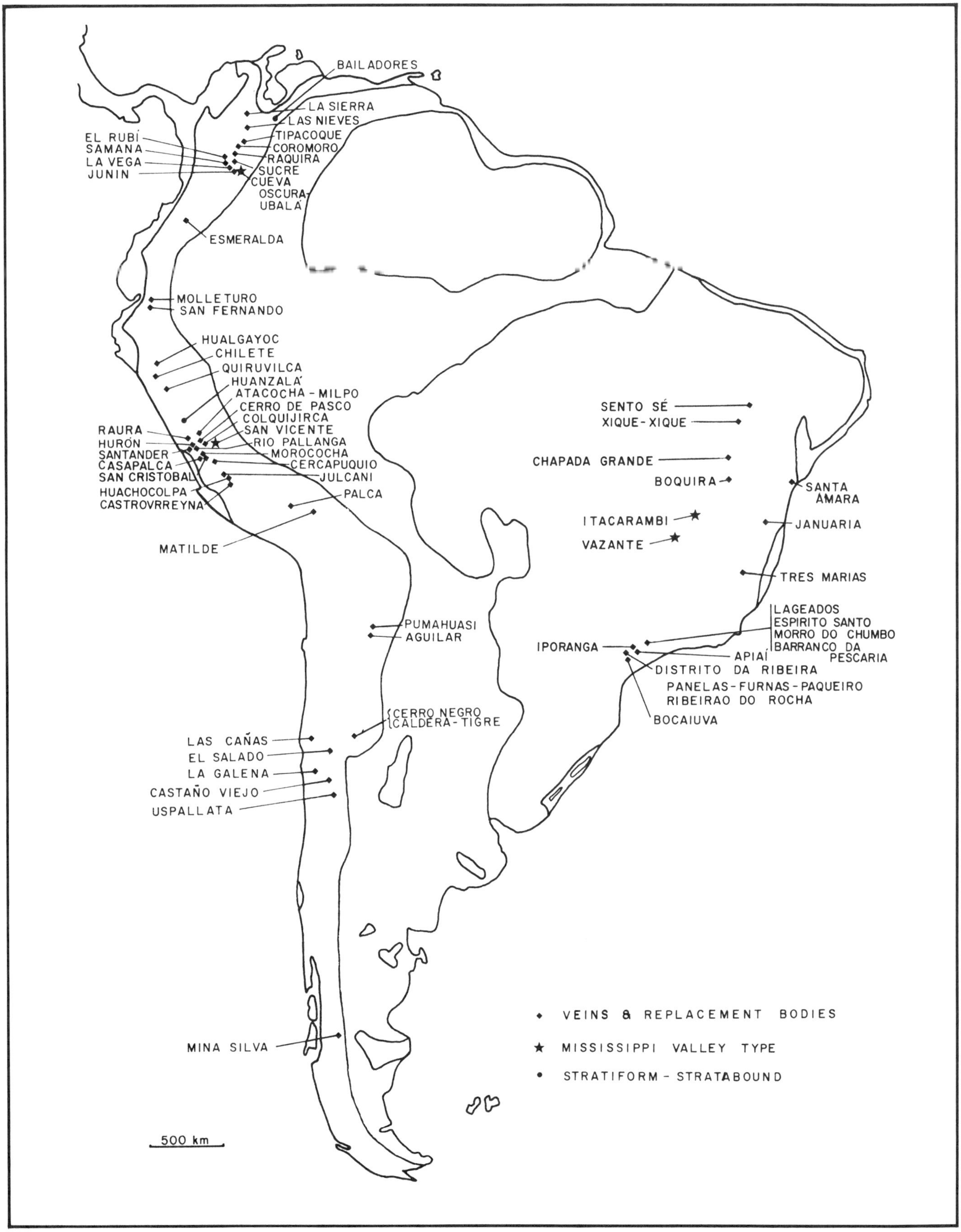

Figure 2—Lead-zinc deposits in South America.

During the late Paleozoic the area of the present Andes became largely emergent. As the land rose under desertic conditions, continental red beds were deposited in the central Andes (Jenks, 1951, 1956; Lancelot et al., 1980; Pitcher, 1984; Davidson, 1984) and became the hosts of small copper deposits in Perú (Figure 3), such as those at Negra Huanusha and Cerro Landa (Ponzoni, 1980). Salt deposits formed nearby (Benavides, 1968). A large part of southern South America was glaciated during the late Paleozoic, while magmatic intrusion and volcanism were widespread in the central Andes and marine conditions continued in the northern Andean region.

Several small tungsten, iron, copper, lead, silver, and zinc deposits in Argentina are thought to have formed in association with igneous stocks intruded during the Paleozoic-Mesozoic transition (Angelelli et al., 1970). Of special interest are four porphyry copper deposits recently dated as Permo-Carboniferous (Sillitoe, 1977).

Some mineral occurrences in southeastern Perú and northwestern Bolivia, particularly nickel, cobalt, chromite, and uranium (Figure 4), may be late Paleozoic (Ponzoni, 1980), although they are largely confined to lower Paleozoic rocks. In southeastern Perú, some copper-tin-tungsten deposits containing minor bismuth and molybdenum are related to upper Paleozoic intrusions (Ponzoni, 1980; Candiotti and Guerrero, 1983). However, the nearby Quenamari-San Rafael tin-copper veins are Tertiary (Palma, 1981), as are analogous polymetallic tin deposits in central and southern Bolivia (Evernden, Kriz, and Cherroni, 1977; Grant et al., 1979; Lehmann, 1979).

The Bailadores lead-zinc-copper massive sulfide deposit in Venezuela (Figure 2) is variously interpreted as occurring in and being contemporaneous with upper Paleozoic (Carlson, 1977) or Lower Cretaceous (Rodríguez, 1974) sediments. Similarly, the copper mineralization in upper Paleozoic limestone at Cobriza, Perú, has been interpreted by some geologists to be syngenetic (Husman et al., 1987) and not an epigenetic skarn, as generally believed (Petersen, 1965; Valdez, 1983).

MESOZOIC

During most of the Mesozoic, marine conditions prevailed over the Andean realm. During the Triassic and Jurassic, volcanic rocks were more abundant than sediments along the coast, conforming to the classic concept of a eugeosyncline or to the modern concept of a marginal volcanic arc related to a subduction zone. Granodiorite intrusions were emplaced in southeastern Perú and northern Bolivia during this time (Evernden, Kriz, and Cheronni, 1977). In Perú, the stratigraphic column to the east of the volcanic arc is dominated by limestone, conforming to the classic miogeosyncline concept. To the north, in Ecuador, and to the south, in Bolivia, limestone is scarce.

The Mesozoic eugeosynclinal environment was particularly favorable for the formation of volcanogenic massive sulfide deposits, mainly of copper (Figure 5), but also containing some lead, zinc, silver, and barite. Notable examples are La Plata, Sigchos, and Macuchi in Ecuador (Putzer, 1971) and Tambogrande (Ponzoni, 1980; Aye, 1985) and Leonila-Graciela (Vidal, 1980) in Perú.

Many stratabound deposits in the Triassic to Lower Cretaceous volcanic and sedimentary rocks of the coastal region (Figure 5) show evidence of metasomatism and vein mineralization that suggest that they represent a type that is different from the conventional volcanogenic massive sulfide or Kuroko-type deposits (Ponzoni and Vidal, 1982; Injoque et al., 1983). Deposits of this type include Raúl (Ripley and Ohmoto, 1977; Wauschkuhn and Ohnsmann, 1984), Condestable (Wauschkuhn and Thum, 1982), and Eliana-Rio Seco (Cardozo and Vidal, 1981; Agar, 1981, 1982; Ponzoni and Vidal, 1982) in Perú, as well as several copper "mantos" in Chile (Ruiz et al., 1965). Atkin, Injoque-Espinoza, and Harvey (1985) consider that the copper-iron deposit of Monterrosas and the iron deposits of Acarí and Marcona (Figure 1) in Peru belong to the same ore deposit type as Raúl and Condestable. This deposit type is characterized by copper-iron-amphibole (mainly actinolite) ores formed by replacement and local vein infilling under hydrothermal conditions. If so, they may be related to the important Mesozoic metasomatic iron deposits of northern Chile (Figure 1), exemplified by El Romeral (Bookstrom, 1977), where magnetite is microscopically intergrown with actinolite.

In recent years several geologists have proposed that the miogeosynclinal sedimentary sequence in Perú locally contains a significant volcanic component that may have been responsible for stratiform ("manto") lead-zinc mineralization at Hualgayoc, El Extraño, Pachapaqui, Pacllón-Llamac, Huanzalá, Machcán, San Cristóbal, Cercapuquio, and San Vicente (Lyons, 1968; Levin and Amstutz, 1973, 1976; Paredes, 1980; Fontboté, Amstutz, and Samaniego, 1981; Samaniego and Amstutz, 1981; Samaniego, 1982; Kobe, 1982a; Carrascal et al., 1983; Vera, 1983; Rivera and Kobe, 1982a, 1982b; Gunnesch et al., 1984; Soler, Carrascal, and Saez, 1986). These syngenetic deposits required considerable water depths to prevent the hot hydrothermal fluids from boiling, yet nearby—and close in time—significant coal (Dunin, 1981, 1983) and evaporite (Benavides, 1968) deposits were forming in shallow waters. This raises the question whether water depths were highly variable or whether these metalliferous deposits are epigenetic rather than syngenetic. If the zinc-copper deposit at Bailadores, Venezuela, is proved to be of Early Cretaceous age (Rodríguez, 1974), it might have had an origin similar to the above stratiform deposits in Perú, which it resembles.

In the northern Andes, porphyry copper deposits (Figure 6) formed in western Colombia during the middle Jurassic and earliest Cretaceous (Sillitoe et al., 1982). Tin-tungsten deposits in northern Bolivia are genetically related to the Huayna Potosí, Taquesí-Murata, and Sorata batholiths of Late Triassic to Early Jurassic age (Ahlfeld, 1954; Turneaure, 1960; Ahlfeld and Schneider-Scherbina, 1964; Evernden, Kriz, and Cherroni, 1977; Grant et al., 1979).

Although during the Mesozoic some evaporites formed along the Andes, the bulk were deposited between the Andes and the Brazilian shield, and subsequently formed numerous salt domes (Benavides, 1968; Ponzoni, Postigo, and Birkbeck, 1969). The geological setting of these evaporites is similar to that of the potash deposits of Saskatchewan, Canada, and it is tempting to speculate that evaporite formation may have progressed to the stage of potash deposition somewhere in the Amazon basin of Perú either during the Mesozoic or the Cenozoic. At present, potash-bearing fertilizers are seldom used in Perú, Ecuador, Bolivia, and adjoining parts of Brazil because potash ore deposits are locally unknown and it is too expen-

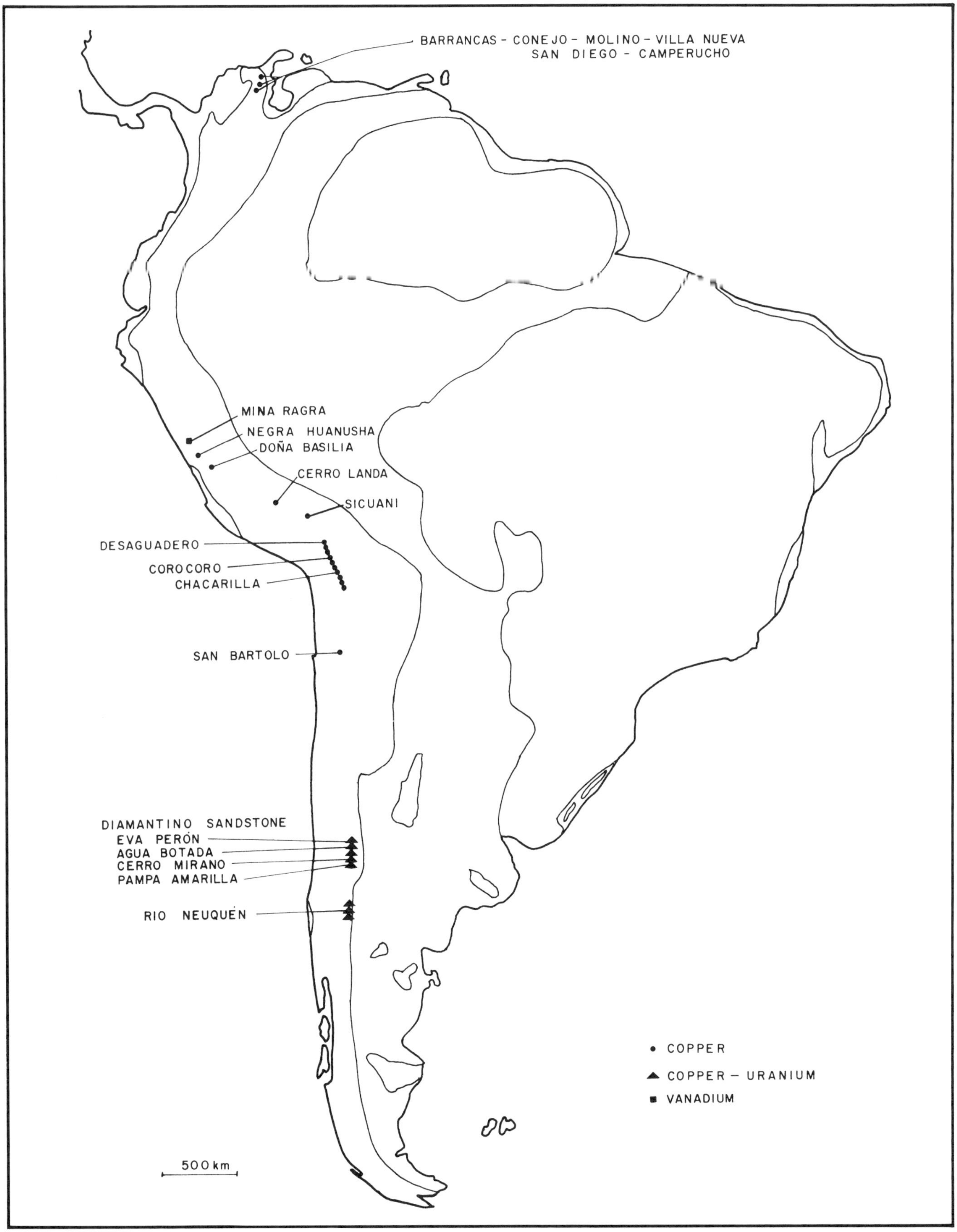

Figure 3—Red-bed copper and uranium deposits in the Andean region.

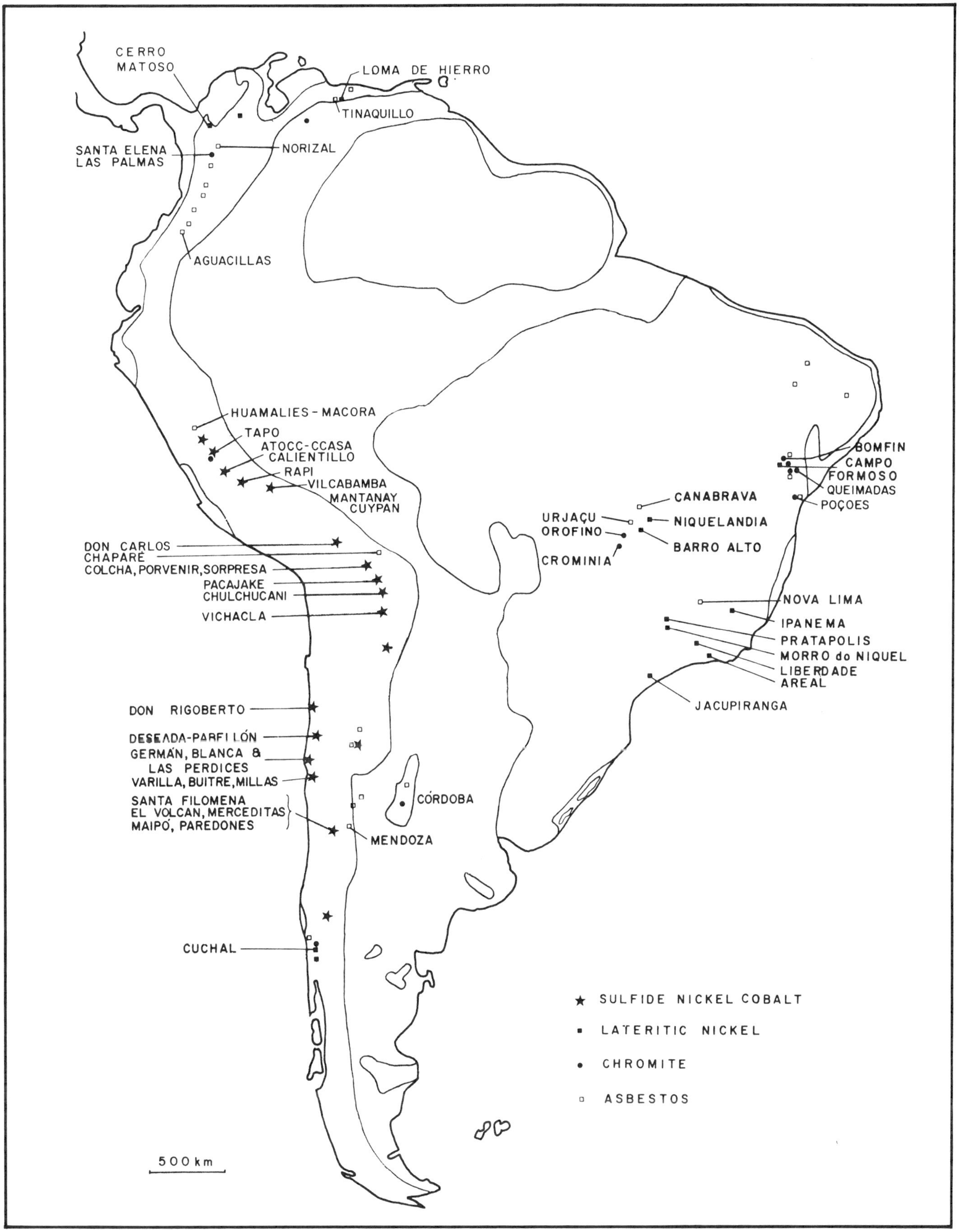

Figure 4—Nickel, chromite, and asbestos deposits in South America.

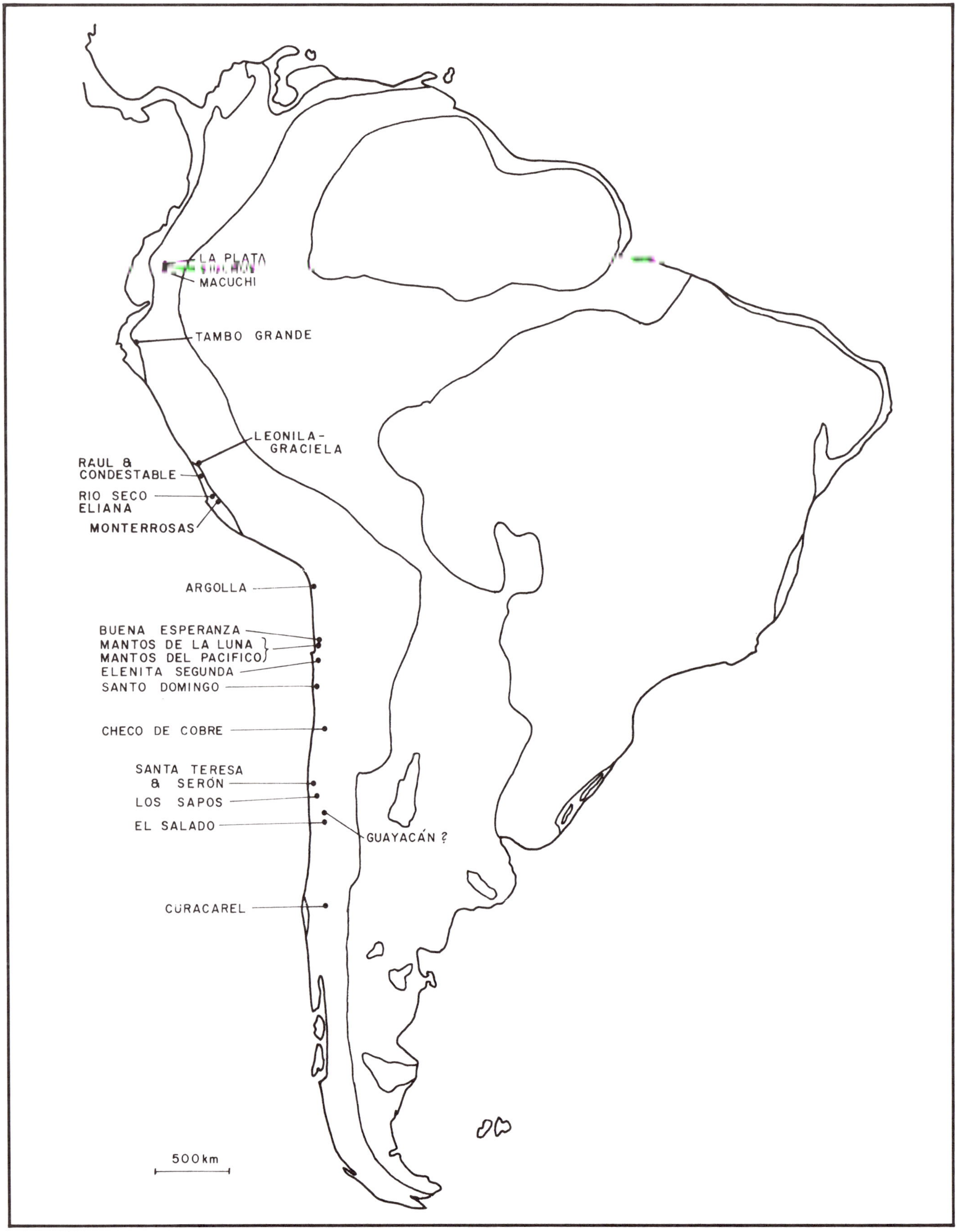

Figure 5—Stratiform copper deposits in andesitic volcanic rocks in the Andean region.

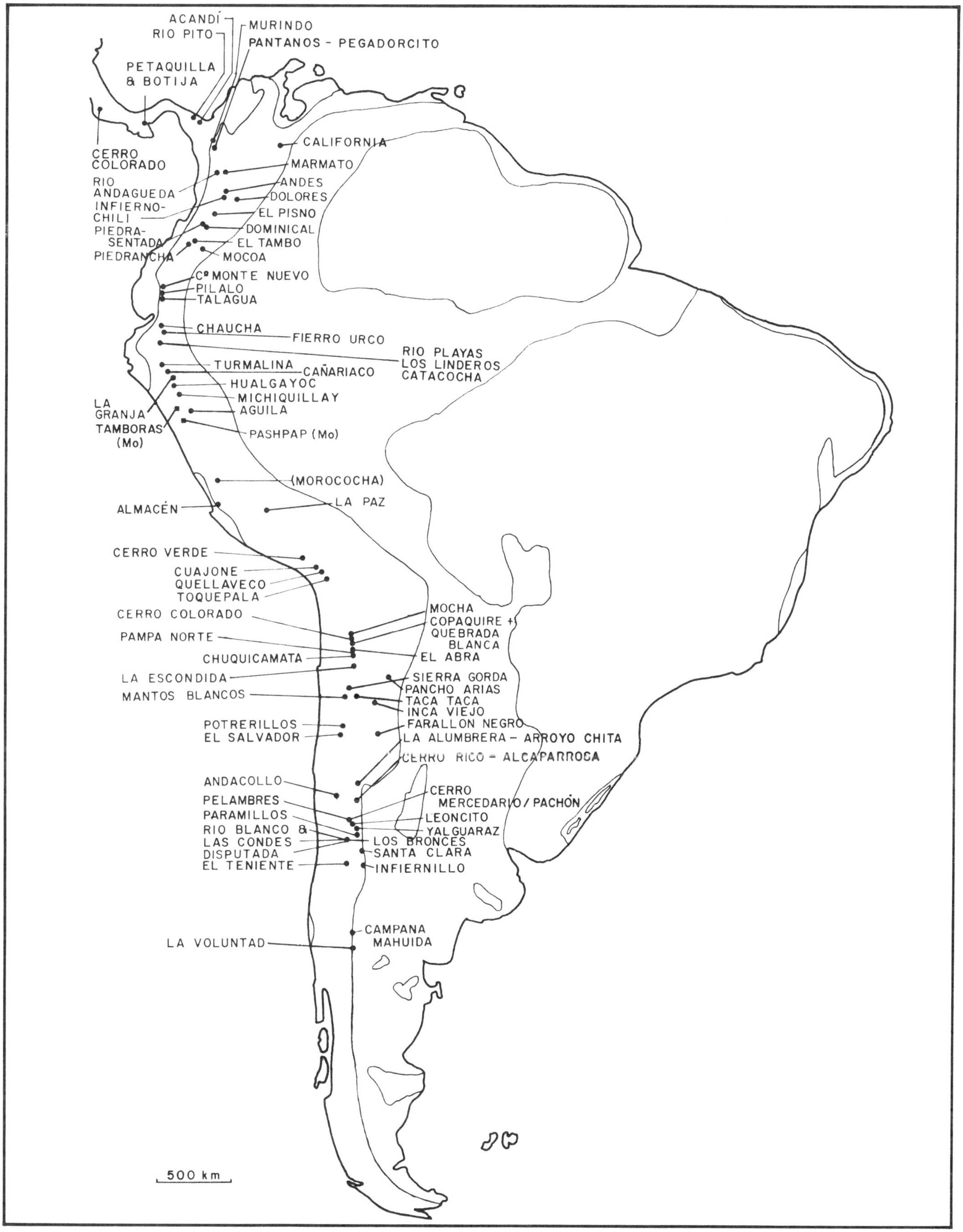

Figure 6—Porphyry copper deposits in the Andean region.

sive to import potash from North America or Europe. A local source of potash may well spur the use of potash fertilizers in this region, significantly increasing agricultural productivity. Hence, an exploration program directed at locating these potential potash deposits seems to be well justified.

CENOZOIC

The Andean orogeny started during the Cretaceous with multiple intrusions that eventually coalesced to form the Andean batholith (Davidson, 1984), referred to as the Coastal batholith in Perú (Cobbing and Pitcher, 1972; Pitcher, 1974, 1978, 1984; Pitcher et al., 1985; Atherton, 1984). The gradual uplift that led to the emergence of the Andes also began at this time. Access of the Pacific Ocean to the Amazon basin became restricted to an embayment in northern Perú that disappeared by Oligocene time (Weeks, 1947; Harrington, 1962). Shallow-marine sediments and red beds were deposited on the newly uplifted land and became widespread during the Cretaceous in Chile (Davidson, 1984) and during the early Tertiary in Perú (Jenks, 1956). However, these sediments were soon buried by a thick sequence of andesitic and rhyolitic volcanic rocks that were erupted from numerous volcanic centers along the eastern side of the Andean (Coastal) batholith. The Miocene Cordillera Blanca batholith (Harmon and Berreiro, 1984) could well be the uplifted root zone of some of the volcanic centers east of the Andean batholith.

The structural grain of the pre-Cenozoic sediments changes significantly along the Andes as the result of late Mesozoic-early Cenozoic deformation. Major structural deflections are at the Venezuela-Colombia border, at the Ecuador-Perú border, in northern Perú ("Huancabamba deflection"), in south-central Perú ("Abancay deflection"), at the Perú-Chile border, and in southernmost Chile-Argentina. Some of these deflections may be attributed to deformation at the South American plate as it converged with the subducting oceanic plates (Heki, Hamano, and Kono, 1983). Fracturing at the inflection points would be expected to stimulate magmatic and hydrothermal activity, but no convincing case has been made so far for ore deposits being preferentially located at these structural deflections.

Major Cenozoic strike-slip faults have been identified in the northern Andes, for example, the Boconó fault in Venezuela and the Romeral fault zone in Colombia. Major faults in other parts of the Andes were chiefly dip-slip faults, both reverse and normal. Many ore deposits appear to be related to such regional faults, and recent studies in southern Perú and northern Chile indicate a relationship between porphyry copper deposits and regional longitudinal faults.

The loci of magmatism shifted either eastward or westward with time in different parts of the Andes (Farrar et al., 1970; Francis et al., 1974; Kussmaul et al., 1977; Frutos, 1982; Pitcher, 1984). Volcanism was not necessarily continuous in distribution from north to south (Jordan et al., 1983; Thorpe, 1984), and major grabens formed adjacent to volcanic chains (Brüggen, 1950). Tectonism and magmatic activity intensified and waned episodically (Noble et al., 1984; Stewart, Evernden, and Snelling, 1974; Baker and Francis, 1978; Grant et al., 1979; Petersen and Vidal, 1983; McKee and Noble, 1985). The vast majority of ore deposits scattered along the Andes formed from magmatic emanations and from hydrothermal cells adjoining igneous centers, taking advantage of the permeability resulting from local fracturing. Hence, the location and abundance of many ore deposit types in the Andes is closely related to the location and intensity of the igneous activity during the Cenozoic (Trurnit, Fesefeldt, and Stephan, 1982; Francis, Halls, and Baker, 1983).

Certain groups of ore deposits are genetically related to the Late Cretaceous to early Tertiary magmatism that formed the Andean batholith. In Perú, such groups include the gold-quartz veins of the Nazca region (Samamé Boggio, 1981; Vidal, 1985) and some barite veins (Valera et al., 1983). However, most of the important hydrothermal deposits formed later, in association with small stocks and batholithic complexes east of the Andean (Coastal) batholith.

Porphyry copper deposits (Figure 6) are found along most of the Andes, but are especially abundant and high grade in northern and central Chile (Ruiz et al., 1965; Sillitoe, 1973; Lowell, 1974; Gustafson and Hunt, 1975; Alvarez and Aracena, 1982; Hunt, Bratt, and Marquardt, 1983; and in southern Perú (Samamé Boggio, 1980; Satchwell, 1983). Generally lower grade deposits, which have not been developed, are found in western Argentina (Sillitoe, 1977; Zantop, 1981), Ecuador (Goosens and Hollister, 1973), northern Perú (Samamé Boggio, 1980), and Colombia (Sillitoe et al., 1982). Porphyry copper deposits are relatively scarce in central Perú. The Almacén and La Paz deposits are rather insignificant, Pashpap is really a molybdenum deposit, and the truly porphyry-copper-type mineralization at Morococha is relatively limited.

The oldest porphyry copper deposit at this orogeny seems to be Campana Mahuída, in western Argentina, dated at 74 Ma (Sillitoe, 1977). Next in age are the porphyry copper deposits in southern Perú—Cerro Verde, Toquepala, Cuajone, and Quellaveco—which are on the order of 50-60 m.y. old. These deposits are variously interpreted as related either to the youngest intrusive phases of the Andean batholith or to the early phases of the next pulse of Tertiary magmatism (Satchwell, 1983; Zimmermann and Kihien, 1983; Le Bel, 1985; Vidal, 1985). The western porphyry copper deposits in Colombia are 43–55 m.y. old (Sillitoe et al., 1982). The porphyry copper deposits of Chuquicamata, Potrerillos, El Salvador, and El Abra, northern Chile, are 30 to 40 m.y. old (Ambrus, 1977). Most of the other porphyry copper deposits in central Colombia, Ecuador, northern Perú, Argentina, and the high Andes of Chile are less than 20 m.y. old. A molybdenum-rich porphyry copper deposit has been found in Colombia at Mocoa (Sillitoe, Jaramillo, and Castro, 1984), and two low-grade porphyry molybdenum deposits—Pashpap (Samamé Boggio, 1980) and Tamboras (Heintze, 1985)—have been explored in northern Perú.

An important group of enargite deposits (Figure 7) stretches from Pilzhum in southern Ecuador (Putzer, 1971) to Laurani in northern Bolivia (Ahlfeld and Schneider-Scherbina, 1964). The largest and richest of these deposits are in central Peru (Petersen, 1965, 1970, 1979). The Capillitas and Famatina deposits in Argentina (Angelelli, 1950, 1984) are also representatives of this group, as are El Indio (Araneda, 1982, 1984) and Las Hediondas (Ruiz et al., 1965) in Chile. Several porphyry copper deposits in northern and central Chile contain enargite, most abundantly at Chuquicamata (Ruiz et al., 1965),

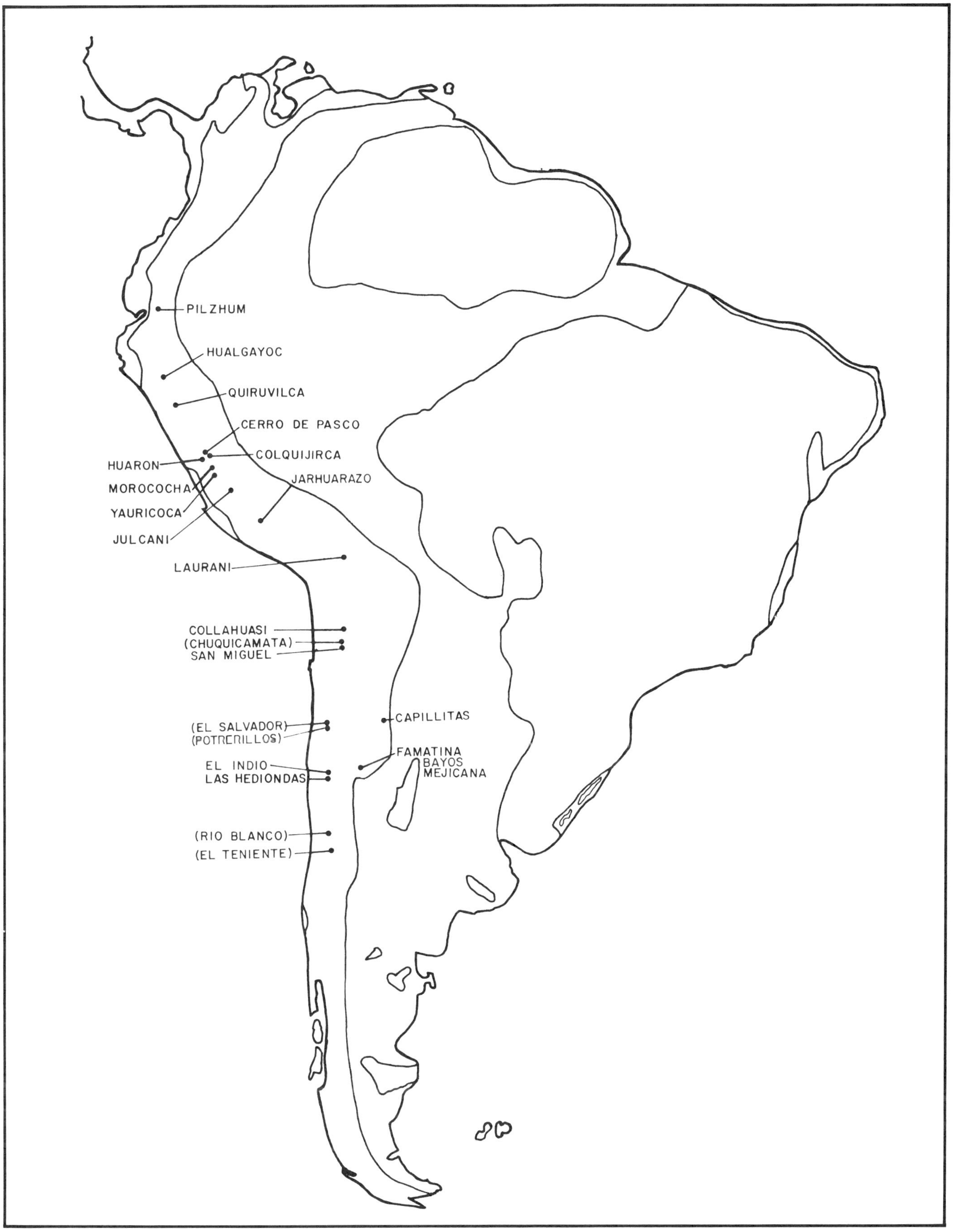

Figure 7—Enargite deposits in the Andean region.

and enargite ores coexist with low-grade porphyry copper ores at Morococha (Samamé Boggio, 1980). All the enargite ores are either in veins or disseminated, and are clearly epigenetic. However, the enargite "mantos" at Colquijírca, Perú, are thought to have a dominant syngenetic component by some (Lehne and Amstutz, 1982) and are believed to be epigenetic by others (Vidal et al., 1984).

Most of the Andean enargite deposits contain significant amounts of precious metals. Thus, at Cerro de Pasco, Morococha, and Yauricocha (Petersen, 1965) the sulfide ores were typically classed for mining purposes as copper-silver and lead-silver. Julcani (Petersen et al., 1977; Benavides, 1983; Noble and Silberman, 1984) is generally considered to be a silver deposit, having produced 56 million ounces of silver from 4.5 million tonnes of ore (A. Benavides, oral communication, 1985). The El Indio deposit (Anonymous, 1982; Araneda, 1982, 1984) is considered to be a gold deposit, although it also produces copper and silver.

The regions where the stratigraphic sequence contains abundant Mesozoic limestone—for example, in Perú and, to a lesser extent, northern and central Chile and western Argentina—were favorable to the formation of major lead-zinc deposits and a variety of skarn deposits genetically related to Cenozoic magmatism (Figures 2, 8). The lead-zinc deposits are mostly vein and replacement deposits in Mesozoic limestones, but some consist of veins in Tertiary red beds and volcanic rocks (Petersen, 1965). Silver is an important byproduct of these deposits, which account for about 59% of Peru's silver production (Benavides, 1984). Most of this silver is in galena and tetrahedrite, for example, at Casapalca (Wu and Petersen, 1977). Most of the lead-zinc deposits are clearly epigenetic, but some may be syngenetic. Among the latter, which continue to be controversial, are Huanzalá (Carrascal, Saez, and Soler, 1983; Tsuchiya et al., 1983; Imai et al., 1985; Soler, Carrascal, and Saez, 1986; Imai, 1986) and Hualgáyoc (Ericksen, Iberico, and Petersen, 1956; Paredes, 1980; Vidal and Cabos, 1983). Some of the skarn deposits, such as those at Magistral, Antamina, Chungar, Katanga, Tintaya, and Ferrobamba, are rich in copper, whereas others, such as those in the Adahuaylas-Yauri belt (Noble et al., 1984), contain iron. A few skarn deposits are gold-bearing, such as those at Katanga, in southeastern Perú. At Uchucchacua, in central Perú, silver-manganese mineralization is associated with a skarn (Alpers, 1980), and the Santander skarn contains zinc and lead.

The vast majority of silver-gold deposits are found in Tertiary volcanics, which are relatively widespread in Perú. About 36% of Peru's silver production comes from such deposits (Benavides, 1984). Millotingo, Colqui, Arcata, Orcopampa, Cailloma, San Genaro, and Caudalosa are among the principal districts of this type. Some silver deposits are found in Mesozoic sediments immediately below the Tertiary volcanic blanket. Uchucchacua, in central Perú, is an example of such a deposit. In the central and southern part of the tin belt in Bolivia, the silver tends to be associated with tin deposits, which in turn are associated with subvolcanic intrusives. A good example is the famous Potosí district (Ahlfeld 1954; Ahlfeld and Schneider-Scherbina, 1964; Turneaure, 1971; Sugaki et al., 1983). Gold predominates over silver in the Salpo-Millhuachaqui, Machacala, and Colquipucro deposits of northern Perú, and at Portovelo in Ecuador (Putzer, 1971). Association with caldera ring fractures has been proposed for the Cailloma silver deposit in Perú (Dávila, 1981) and for Potosí in Bolivia (Francis, Baker, and Halls, 1981). A common characteristic of the silver deposits is their formation in a near-surface environment, as indicated by the geological settings in which they occur and by evidence of widespread boiling of the hydrothermal fluids, which is well shown by the Colqui deposit, Perú (Kamilli and Ohmoto, 1977).

The tin deposits in the central and southern part of the Bolivian tin belt (Turneaure, 1960; Evernden, Kriz and Cherroni, 1977; Grant et al., 1979; Lehmann, 1979) probably formed at relatively shallow depth, as evidenced by the boiling of hydrothermal fluids (Kelly and Turneaure, 1970), and by the upward mushrooming of some of the intrusives with which they are associated (Ahlfeld, 1954; Ahlfeld and Schneider-Scherbina, 1964). Various combinations are observed of vein, stratiform ("manto"), and disseminated ("porphyry type") mineralization in Tertiary volcanics and stocks, as well as in the surrounding Paleozoic clastic sedimentary rocks (Sillitoe, Halls, and Grant, 1975; Grant et al., 1980; Lehmann, 1985). In contrast, the Mesozoic tin and tungsten deposits associated with batholithic masses in northern Bolivia appear to have been formed at somewhat greater depths; however, evidence of boiling in fluid inclusions indicates that their depth of formation was still less than 3 km (Kelly and Turneaure, 1970).

In general, the igneous rocks of the Bolivian tin belt are strongly enriched in tin, lithium, beryllium, rubidium, and phosphorus, and have high initial strontium 87/86 ratios, indicating significant crustal contamination (Lehmann and Pichler, 1980; Francis, Halls, and Baker, 1983). In addition, their lead isotopic composition suggests a mixture of a mantle component with upper or intermediate continental crust (Tilton et al., 1981). However, the extensive lower Paleozoic clastic sediments that host the intrusives are not particularly enriched in tin and lithium, but are rich in boron (Lehmann, 1979; Lehmann et al., 1987). This suggests that these magmas were not contaminated by the lower Paleozoic clastic sediments but rather were either contaminated by underlying crustal rocks, or that enrichment in tin and other elements was due to magmatic fractionation and differentiation during their ascent within the crust. In this context, it is interesting to note that the tin-tungsten deposits of southwest England and southeast Asia are associated with crustally derived granitic rocks with late volatile-enriched (fluorine, lithium, boron, phosphorous) differentiates (Manning, 1986).

The tin belt is largely confined to Bolivia, ending to the south at the Pirquitas deposit in northernmost Argentina. To the north, a Cenozoic segment of the tin belt reappears in southeastern Perú at the San Rafael tin-copper deposit (Arenas, 1980). Tin-rich ash-flow tuffs are found in the Macusani area, just north of San Rafael and in the Morococala plateau in the central part of the Bolivian tin belt (G. E. Ericksen, written communication, 1986). The Macusani tuffs probably mark the northernmost end of the Bolivian tin belt.

Mercury deposits are scattered through the Andes (Figure 9), but the Santa Bárbara deposit in Perú has been the only major mercury producer, particularly during Spanish colonial times (Yates, Kent, and Fernandez Concha, 1951). The relative scarcity of mercury in the Andes is contrary to what might be expected from the widespread volcanism during the Cenozoic. Also, associated with the Cenozoic volcanism were the

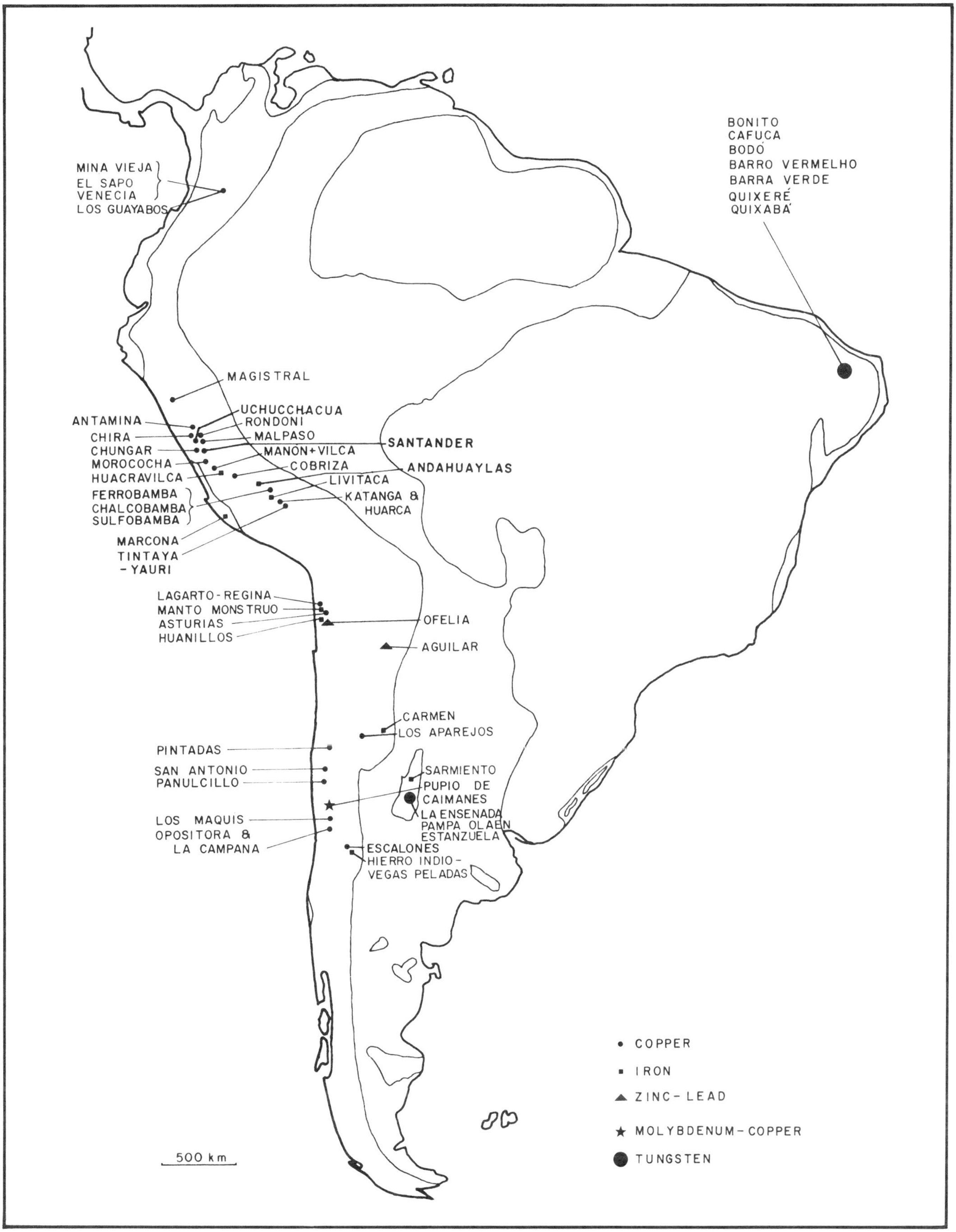

Figure 8—Skarn deposits in the Andean region.

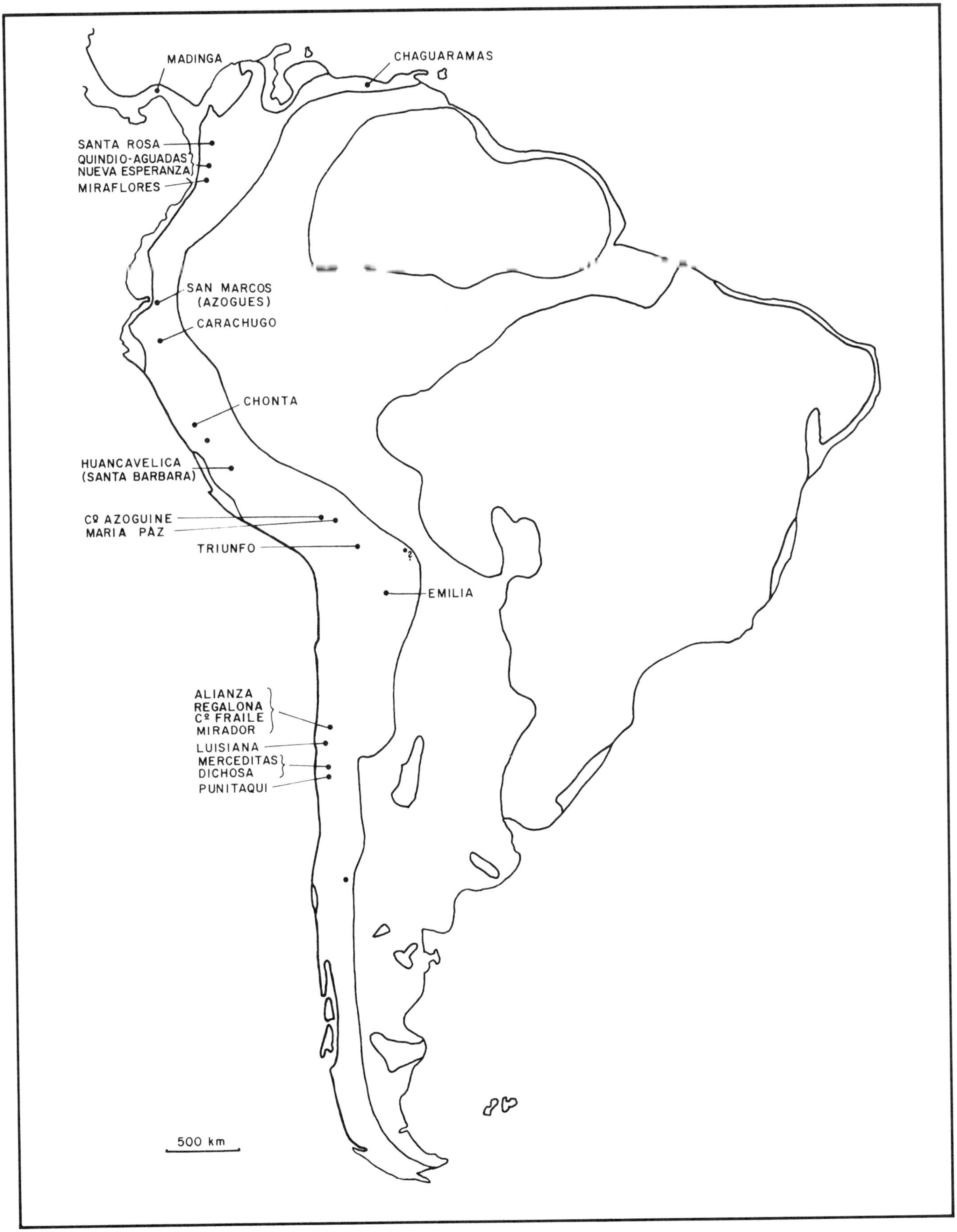

Figure 9—Mercury deposits in the Andean region.

unusual magnetite flows at El Laco, northern Chile (Park, 1961; Ruiz et al., 1965) and the exhalative native sulfur deposits on volcanic cones in the Andean regions of Colombia (Angulo, 1978), southern Peru (Samamé Boggio, 1981), northern Chile, western Bolivia, and northwestern Argentina (Ahlfeld, 1954; Ahlfeld and Schneider-Scherbina, 1964; Ruiz et al., 1965).

A large number of other important types of ore deposits formed along the Andes during the Cenozoic. In Colombia, oolitic iron deposits (Figure 1), such as the Paz de Rio deposit (Angulo, 1978), and coal beds (Durán and Arboleda, 1985) formed along the continental margin, and lateritic nickel deposits, such as the Cerro Matoso (Robertson and Patiño, 1985), became enriched during lateritic weathering (Figure 4). In Perú, phosphate and diatomite deposits formed in restricted marine basins along the Pacific coast. The Sechura basin in northern Perú is an example (Samamé Boggio, 1981). In the Andes, red-bed deposits of copper, uranium, and vanadium, such as those at Corocoro, Diamantino sandstone, Desaguadero, Sicuani, and Minas Ragra, were formed during the Tertiary (Figure 3) (Ahlfeld, 1954; Ahlfeld and Schneider-Scherbina, 1964; Ponzoni, 1980). In southernmost Perú, western Bolivia, northern Chile, and northwestern Argentina economic concentrations of nitrates, borates, and lithium formed in widespread and numerous salars (Ahlfeld, 1954; Ahlfeld and Schneider-Scherbina, 1964; Samamé Boggio, 1981; Ericksen, 1983; Ericksen and Salas, this volume). At places, gold placers such as those at San Antonio de Poto formed under glacial conditions. However, most of the economic gold placers formed under fluvial conditions in central Colombia (gold and platinum), along the eastern slopes of the Andes between Colombia and Argentina, and in northern and southern Chile (Ahlfeld, 1954; Ahlfeld and Schneider-Scherbina, 1964; Ponzoni, 1980). In southernmost Perú, western Bolivia, northern Chile, and northwestern Argentina economic concentrations of nitrates, borates, and lithium formed in widespread and numerous salars (Ahlfeld, 1954; Ahlfeld and Schneider-Scherbina, 1964; Samamé Boggio, 1981; Ericksen, 1983; Ericksen and Salas, this volume). At places, gold placers such as those at San Antonio de Poto formed under glacial conditions. However, most of the economic gold placers formed under fluvial conditions in central Colombia (gold and platinum), along the eastern slopes of the Andes between Colombia and Argentina, and in northern and southern Chile (Ahlfeld, 1954; Ahlfeld and Schneider-Scherbina, 1964; Ruiz et al., 1965; Angulo, 1978; Zegarra and Galloso, 1983).

CONCLUSIONS

The geological setting of Andean mineral deposits is complex, because many geological processes have taken place within the Andean realm during its long history. In addition, this setting is not well understood because in many areas the pertinent rock record has either been destroyed by erosion or is deeply buried under younger sediments. For these reasons, our knowledge of the Precambrian metallogeny is limited to a few sedimentary iron and manganese deposits. The Paleozoic metallogenic conditions are better known but are still rather poorly understood. In the lower Paleozoic, iron, lead-zinc, tungsten, and subordinate copper, nickel, and cobalt deposits were formed in the marine environment. Also in the Paleozoic, gold-quartz veins formed during orogenic metamorphism, and iron, copper, lead, zinc, silver, cobalt, nickel, uranium, tin, tungsten, bismuth, molybdenum, and chromite deposits were formed during magmatism. Small red-bed copper deposits formed under continental conditions during the late Paleozoic. In contrast, the geological environments in which ore deposits formed during the Mesozoic and the Cenozoic are much better known, although important problems requiring extensive research still remain.

During the first 100 million years of subduction of the oceanic plates under South America, there was a submarine volcanic arc west of the present continental divide. Volcanogenic massive sulfide deposits of the Kuroko type were formed along the arc, as were a number of copper-iron veins and amphibole-rich stratabound deposits that probably represent a different deposit type. At the same time, geosynclinal sedimentation took place to the east of this volcanic arc in basins of different size, water depth, and subsidence rate. Hydrothermal solutions of moderate temperatures invaded the deeper basins, giving rise to stratiform lead-zinc deposits, whereas coal and some evaporite deposits formed in the shallow and emerging basins. Magmas ascended to form batholiths with associated tin mineralization in northern Bolivia and porphyry copper stocks in western Colombia. Farther east, large volumes of evaporites formed on the cratonic shelf. During the late Mesozoic, the axis of granitic magmatism shifted to the west, forming the Andean (Coastal) batholith at the transition zone between the former volcanic arc and the adjoining geosyncline. Gold-copper and barite veins are associated with this batholith.

The vast majority of Andean mineral deposits formed during the Cenozoic, when the axis of magmatic activity was east of the Andean (Coastal) batholith. Many igneous stocks were intruded into the sediments of the geosynclinal prism, and some batholithic masses, such as the Cordillera Blanca batholith in Perú, were emplaced. At present, the magmatic axis generally lies in the Western Cordillera, just east of the Andean batholith but west of the belt of Cenozoic magmatism.

The great variety of hydrothermal deposits that formed during the Cenozoic can be attributed to the diversity of origins of the hydrothermal fluids, as well as to the variety of their source and host rocks. Thus, different hydrothermal fluids probably resulted from the expulsion of connate brines from the geosynclinal sediments, from metamorphism of the keel of the geosynclinal prism, from magmatic emanations, and from infiltrating meteoric waters. Metals could be scavenged by the magmas from the subducting oceanic plates and their sedimentary veneer, from the asthenospheric mantle, from the Precambrian and Paleozoic basement, and from the Mesozoic and Tertiary country rocks. The upper crustal rocks could also have contributed their metals directly to circulating hydrothermal fluids, and the different host rocks probably reacted with the hydrothermal fluids in various ways.

In order to sort out the many possible combinations of factors responsible for the variety in ore deposit types in the Andes, it is necessary to carry out many integrated geochemical and isotopic studies. For example, Lehmann and Pichler (1980) related the extent of the Bolivian tin belt to the tin content of mid-Andean volcanic rocks, and Audebaud and

Amosse (1981) correlated the distribution of copper, nickel, cobalt, lithium, rubidium, lead, and zinc in igneous rocks across the Andes of southern Perú with the general composition of ore deposits in the same transect. Lead isotopes have been used extensively to evaluate crustal contamination of igneous rocks and to infer the source of ore leads (McNutt, Clark, and Zentilli, 1979; Tilton et al., 1981), and more recent isotopic studies of strontium, neodymium, and oxygen have given additional information about sources (Harmon and Barreiro, 1984; Pitcher et al., 1985). The presence of lead-zinc and skarn deposits in regions with abundant limestone is well known, and the possible involvement of evaporite deposits in determining the presence of both tungsten and base-metal mineralization at San Cristóbal, Perú, has been indicated by sulfur isotopic studies (Campbell, 1983). The role of meteoric water as a hydrothermal fluid at San Cristóbal, Peru, was évaluated with hydrogen and oxygen isotopic studies (Campbell, Rye, and Petersen, 1984). Once sufficient information of this nature has been obtained, it should be possible to evaluate the importance of the "geostill" concept (Field and Dasch, 1981) for the generation of hydrothermal ore deposits in the Andes.

While hydrothermal deposits were forming within the Andean edifice during the Cenozoic, a number of sedimentary ore deposits formed in the adjoining seas and coastal environments (for example, oolitic iron deposits, phosphate and diatomite deposits, and coal seams) and on the emerging land (for example, lateritic nickel deposits; red-bed copper, uranium, and vanadium deposits; salars with nitrates, borates, and lithium; and gold, tin, and platinum placers).

REFERENCES

Agar, R. A., 1981, Copper mineralization and magmatic hydrothermal brines in the Rio Pisco section of the Peruvian coastal batholith: Economic Geology, v. 76, p. 677-693.

Agar, R. A., 1982, Copper mineralization and magmatic hydrothermal brines in the Rio Pisco section of the Peruvian coastal batholith—a reply: Economic Geology, v. 77, p. 1954-1955.

Ahlfeld, F., 1954, Los yacimientos minerals de Bolivia: Bilbao, Spain, Imprenta Industrial S. A., 277 p.

Ahlfeld, F., 1967, Metallogenic epochs and provinces of Bolivia: Mineralium Deposita, v. 2, p. 291-311.

Ahlfeld, F., and A. Schneider-Scherbina, 1964, Los yacimientos minerales y de hidrocarburos de Bolivia: La Paz, Bolivia, Ministerio de Minas y Petróleo, Departamento Nacional de Geologia, Boletín 5 (especial), 388 p.

Alpers, C. N., 1980, Mineralogy, paragenesis, and zoning of the Luz vein, Uchucchacua, Perú: Cambridge, Massachusetts, Harvard University, unpublished A.B. thesis, 137 p.

Alvarez, C., O., and I. Aracena P., 1982, El Abra; uno de los grandes yacimientos de cobre aún no explotados: Minerales, v. 37, 160, p. 25-35.

Ambrus, J., 1977, Geology of the El Abra porphyry copper deposit, Chile: Economic Geology, v. 72, p. 1062-1085.

Angelelli, V., 1950, Recursos minerales de la República Argentina I, Yacimientos metalíferos: Revista del Instituto Nacional de Investigaciones de las Ciencias Naturales "Bernardino Rivadavia," Ciencias Geologicas, Tomo II, 543 p.

Angelelli, V., 1984, Yacimientos metalíferos de la República Argentina: La Plata, Comision de Investigaciones Científicas, 704 p.

Angelelli, V., F. Fernandez Lima, A. Herrera, and L. Aristarain, 1970, Descripcion del mapa metalogenético de la República Argentina: Buenos Aires, Direccion Nacional Geología y Minería, Anales 25, 172 p.

Angulo C., R., ed., 1978, Recursos minerales de Colombia: Ministerio de Minas y Energía, INGEOMINAS, 544 p.

Anonymous, 1982, Chile's El Indio begins output of full product line: Engineering and Mining Journal, January, p. 39-43.

Araneda G., R., 1982, El Indio, yacimiento de oro, plata y cobre, Coquimbo, Chile: Minerales, v. 37, 160, p. 5-13.

Araneda, G., R., 1984, Nuevos aportes al conocimiento de la geologí de El Indio, yacimiento de oro, plata y cobre, Coquimbo, Chile: Minerales, v. 39, 168, p. 27-39.

Arenas, M. J., 1980, El distrito minero San Rafael, Puno: estaño en el Perú: Sociedad Geológica del Perú, Boletín 66, p. 1-11.

Atherton, M. P., 1984, The coastal batholith of Perú, *in* R. S. Harmon and R. A. Barreiro, eds., Andean magmatism—chemical and isotopic constraints: American Geophysical Union, Special Symposium Proceedings, Shiva Geology Series, p. 168-179.

Atkin, B. P., J. L. Injoque-Espinoza, and P. K. Harvey, 1985, Cu-Fe-amphibole mineralization in the Arequipa segment, *in* W. S. Pitcher et al., Magmatism at a plate edge—the Peruvian Andes: New York, Blackie Halsted Press, John Wiley & Sons, p. 261-270.

Audebaud, E. J. M., and J. Amosse, 1981, Distribution of trace elements in igneous rocks of Peruvian Andes—metallogenic implications, *in* M. T. Halbouty, ed., Energy resources of the Pacific region: American Association of Petroleum Geologists Studies in Geology 12, p. 55-61.

Aye, F., 1985, Tambo Grande: a volcanogenic massive sulphide deposit in the Cretaceous of the Peruvian coast (abs.): Symposium on Geology of the Andes and Its Relation to Hydrocarbon and Mineral Resources, Santiago, Chile, p. 25.

Baker, M. C. W., and P. W. Francis, 1978, Upper Cenozoic volcanism in the Central Andes—ages and volumes: Earth and Planetary Science Letters, v. 41, p. 175-187.

Bellido, E., L. De Montreuil, and D. Girard, 1969, Aspectos generales de la metalogenia del Perú: Servicio de Geología y Minería, XI Convencion de Ingenieros de Minas, 96 p.

Benavides, A., J., 1983, Alteración y mineralización en un sector del distrito minero de Julcani: Sociedad Geológica del Perú, Boletín 72, p. 99-110.

Benavides Q., A., 1984, Ocurrencia de la plata en el Perú: Minería, 180, p. 5-10.

Benavides, V., 1968, Saline deposits of South America, *in* R. B. Mattox, ed., Saline deposits: Geological Society of America Special Paper 88, p. 249-290.

Bookstrom, A. A., 1977, The magnetite deposits of El Romeral, Chile: Economic Geology, v. 72, p. 1101-1130.

Brodtkorb, M. K., and A. Brodtkorb, 1979, Consideraciones genéticas de la scheelita presente en las metamorfitas de la provincia de San Luis: Asociacion Geologica Argentina, Rev., XXXIV (2), p. 131-140.

Brodtkorb, M. K., N. Pezzutti, A. Brodtkorb, and M. Schidlowski, 1985, Tourmaline schists and their relationship to Precambrian scheelite deposits from the San Luis province, Argentina: Berlin-Stuttgart, Gebrueder Borntraeger, Monograph Series on Mineral Deposits, 25, p. 151-160.

Brüggen, J., 1950, Fundamentos de la geología de Chile: Instituto Geografico Militar, 374 p.

Campbell, A. R., 1983, Genesis of the tungsten-base metal ores at San Cristobal, Perú: Cambridge, Massachusetts, Harvard University, unpublished Ph.D. dissertation, unpublished 174 p.

Campbell, A., D. Rye, and U. Petersen, 1984, A hydrogen and oxygen isotope study of the San Cristobal mine, Perú: implications of the role of water to rock ratio for the genesis of wolframite deposits: Economic Geology, v. 79, p. 1818-1832.

Candiotti, H., and T. Guerrero, 1983, Ocurrencia de un yacimiento tipo albita-greisen de Cu, W, Sn y Mo - Carabaya, Puno: Sociedad Geológica del Perú, Boletín 71, p. 69-78.

Cardozo, M., and C. Vidal, 1981, Metalogénesis y magmatismo en el

flanco occidental de los Andes centrales: revisión: Sociedad Geológica del Perú, Boletín 68, p. 13-24.
Carlson, G. G., 1977, Geology of the Bailadores, Venezuela, massive sulfide deposit: Economic Geology, v. 72, p. 1131-1141.
Carrascal, R., J. Saez, and P. Soler, 1983, El yacimiento estratoligado (Pb-Zn-Cu-Ag) e Huanzalá (Huánuco, Perú Central): discusión genética: Sociedad Geológica del Perú, Boletín 71, p. 1-15.
Cheney, E. S., 1985, Cerro de Pasco and other massive sulfide deposits of Central Perú (abs.): Geological Society of America 98th Annual Meeting, p. 543.
Cobbing, E. J., and W. S. Pitcher, 1972, The coastal batholith of central Peru: Journal of the Geological Society of London, v. 128, p. 421-460.
Cobbing, E. J., J. M. Ozard, and N. J. Snelling, 1977, Reconnaissance geochronology of the crystalline basement rocks of the Coastal Cordillera of southern Peru: Geological Society of America Bulletin, v. 88, p. 241-246.
Davidson, J., 1984, Introducción a la geología de Chile: Santiago, Servicio Nacional de Geología y Minería, Seminario, Actualización de la Geología de Chile, 6-10 Ago., p. B1-B24.
Dávila M., D., 1981, Cailloma: caldera tipo valle: Sociedad Geológica del Perú, Boletín 68, p. 25-31.
Dunin, E., 1981, Los carbones en el Perú: posibilidades de su explotación: Sociedad Geológica del Perú, Boletín 68, p. 33-51.
Dunn, E., 1983, Ensayo preliminar sobre el carbón del grupo Goyllarisquizga (facies meridionales y orientales): Sociedad Geológica del Perú, Boletín 70, p. 13-24.
Durán R., R. A., and C. A. Arboleda O., 1985, Potencial carbonífero de Colombia: Santiago, Chile, Symposium on Geology of the Andes and Its Relation to Hydrocarbon and Mineral Resources, 35 p.
Einaudi, M. T., 1977, Environment of ore deposition at Cerro de Pasco, Perú: Economic Geology, v. 72, p. 893-924.
Ericksen, G. E., 1976, Metallogenic provinces of the southeastern Pacific region: American Association of Petroleum Geologists Memoir 25, p. 527-538.
Ericksen, G. E., 1980, Metallogenesis in the Andes, *in* Metalogenesis en Latinoamerica: International Union of Geological Sciences Publication 5, p. 289-312.
Ericksen, G. E., 1983, The Chilean nitrate deposits: American Scientist, v. 71, p. 366-374.
Ericksen, G. E., M. Iberico, and U. Petersen, 1956, Geología del distrito de Hualgáyoc, departamento de Cajamarca: Instituto Nacional de Investigación y Fomento Mineros, Boletin 16, 99 p.
Ericksen, G. E., and R. Salas O., 1985, Geology and resources of salars in the central Andes (abs.): Santiago, Chile, Symposium on Geology of the Andes and Its Relation to Hydrocarbon and Mineral Resources, p. 52-53.
Ericksen, G. E., and R. Salas O., Salars in the Central Andes, *in* Geology of the Andes and its relation to hydrocarbon and mineral resources: this volume.
Evernden, J. F., S. J. Kriz, and C. Cherroni, 1977, Potassium-argon ages of some Bolivian rocks: Economic Geology, v. 72, p. 1042-1061.
Farrar, E., A. H. Clark, S. J. Haynes, G. S. Quirt, H. Conn, and M. Zentilli, 1970, K-Ar evidence for the post-Paleozoic migration of granitic intrusion foci in the Andes of northern Chile: Earth and Planetary Science Letters, v. 10, p. 60-66.
Fernández Concha, J., 1956, El yacimiento de fierro del cerro Casca o Tarpuy, Arequipa: Sociedad Geológica del Perú, Boletín 30, p. 167-175.
Fernández Concha, J., and G. C. Amstutz, 1956, El yacimiento de hierro sedimentario del Cerro Casca o Tarpuy: III Convención Ingenieros de Minas del Perú, Anales, T. II, p. 8-15.
Fernandez Lima, J. C., and R. Caminos, 1980, Reseña de la metalogénesis de Argentina, *in* Metalogenesis en Latinoamerica: International Union of Geological Sciences Publication 5, p. 175-196.
Field, C. W., and E. J. Dasch, 1981, Epilogue: Geostill reconsidered: Geological Society of America Memoir 154, p. 817-824.
Fontboté, L., G. C. Amstutz, and A. Samaniego, 1981, Zur faziellen Stellung und zum diagenetischen Kristallisationsprozess von Erzmineralien in schichtgebundenen Zn-Pb-Lagerstaetten (am Beispiel von San Vicente im zentralen Ostperu): Stuttgart, Zentral Blatt fur Geologia und Palaontologie, Tiel I. H. 3/4, p. 365-477.
Francis, P. W., M. C. W. Baker, and C. Halls, 1981, The Kari Kari caldera, Bolivia, and the Cerro Rico stock: Journal of Volcanology and Geothermal Research, v. 10, p. 113-124.
Francis, P. W., C. Halls, and M. C. W. Baker, 1983, Relations between mineralization and silicic volcanism in the Central Andes: Journal of Volcanology and Geothermal Research, v. 18, p. 165-190.
Francis, P. W., M. J. Roobol, G. P. L. Walker, P. R. Cobbold, and M. P. Coward, 1974, The San Pedro and San Pablo volcanoes of North Chile and their hot avalanche deposits: Geologische Rundschau, 63, p. 357-388.
Frutos, J., 1982, Andean metallogeny related to the tectonic and petrologic evolution of the Cordillera—some remarkable points, *in* G. C. Amstutz et al., eds., Ore genesis—the state of the art: New York, Springer-Verlag, p. 493-507.
Goossens, P., 1972a, Los yacimientos e indicios de los minerales metalicos y no-metálicos de la República del Ecuador: Guayaquil, Dept. Facul. Ciencias Naturales, 123 p.
Goossens, P., 1972b, Metallogeny in Ecuadorean Andes: Economic Geology, v. 67, p. 458-468.
Goossens, P., and V. F. Hollister, 1973, Structural control and hydrothermal alteration pattern of Chaucha porphyry copper, Ecuador: Mineralium Deposita, v. 8, p. 321-331.
Grant, J. N., C. Halls, W. A. Salinas, and N. J. Snelling, 1979, K-Ar ages of igneous rocks and mineralization in part of the Bolivian tin belt: Economic Geology, v. 74, p. 838-851.
Grant, J. N., C. Halls, M. F. Sheppard, and W. Avila, 1980, Evolution of the porphyry tin deposits of Bolivia: Mining Geology, Special Issue, no. 8, p. 151-173.
Gunnesch, K. A., M. Gunnesch, A. Baumann, and H. Delgado, 1984, Investigaciones mineralógicas y metalogenéticas en las areas mineras de Milpo, Atacocha y Machcán (Perú Central): Sociedad Geológica del Perú, Vol. Jubilar LX Aniversario Homenaje Dr. G. Petersen, 4, p. 1-14.
Gustafson, L. B., and J. P. Hunt, 1975, The porphyry copper deposit at El Salvador, Chile: Economic Geology, v. 70, p. 857-912.
Harmon, R. S., and B. A. Barreiro (eds.), 1984, Andean magmatism—chemical and isotopic constraints: American Geophysical Union Special Symposium Proceedings, Shiva Geology Series, 250 p.
Harrington, H. J., 1962, Paleogeographic development of South America: American Association of Petroleum Geologists Bulletin, v. 46, p. 1773-1814.
Heintze, L., 1985, Geologie und geochemie der porphyrischen Stockwerk-Molybdaen-Lagerstaette Tamboras, Zone La Negra (Perú): Hannover, Bundesanstalt für Geowissenschaften und Rohstoffe und der Landesaemtern der Bundesrepublik Deutschland, Geologisches Jahrbuch, R. D., H. 72, p. 3-94.
Heki, K., Y. Hamano, and M. Kono, 1983, Paleomagnetism of the central Andes: a preliminary report, *in* Andes Science 2: Tokyo, Geophysical Studies of the Central Andes, p. 67-76.
Huaman G., M. A., A. Rivera Z., E. Antunez De Mayola R., and H. W. Kobe, 1987, Manto Cobriza: mineralizacion estratoligada y estratiforme en los sedimentos del grupo Tarma (abs. YM-11): Lima, VI Congreso Peruano de Geologia, 19-24 Julio, p. 117.
Hunt, J. P., J. A. Bratt, and J. C. Marquardt L., 1983, Quebrada Blanca, Chile: an enriched porphyry copper deposit: Mining Engineering, June, p. 636-644.
Imai, H., 1986, Mineralization and paragenesis of the Huanzalá mine—a reply: Economic Geology, v. 81, p. 196-199.

Imai, H., M. Kawasaki, M. Yamaguchi, and M. Takahashi, 1985, Mineralization and paragenesis of the Huanzalá mine, central Perú: Economic Geology, v. 80, p. 461-478.

Injoque E., J., J. Mendoza D., A. Aranda V., L. Ramirez Y., and R. Andrade T., 1983, Petrología y mineralogía del yacimiento Raúl—Condestable y reinterpretación de su génesis: Sociedad Geológica del Perú, Boletín 72, p. 233-244.

Jaramillo, L., and R. Escobar, 1980, Cinturones de porfidos en las cordilleras Colombianas, *in* Metallogenesis en Latinoamerica: International Union of Geological Sciences Publication 5, p. 75-100.

Jenks, W. F., 1951, Triassic to Tertiary stratigraphy near Cerro de Pasco, Perú: Geological Society of America Bulletin, v. 62, p. 203-220.

Jenks, W. F., 1956, Handbook of South American geology: Geological Society of America Memoir 65, 378 p.

Jordan, T. E., B. L. Isacks, R. W. Allmendinger, J. A. Brewer, V. A. Ramos, and C. J. Ando, 1983, Andean tectonics related to geometry of subducted Nazca plate: Geological Society of America Bulletin, v. 94, p. 341-361.

Kamilli, R. J., and H. Ohmoto, 1977, Paragenesis, zoning, fluid inclusion, and isotopic studies of the Finlandia vein, Colqui district, central Perú: Economic Geology, v. 72, p. 950-982.

Kelly, W. C., and F. S. Turneaure, 1970, Mineralogy, paragenesis and geothermometry of the tin and tungsten deposits of the eastern Andes, Bolivia: Economic Geology, v. 65, p. 609-680.

Kobe, H. W., 1982a, El ambiente de la mineralización estratoligada de Zn-Pb-Ag-Ba-Mn-Fe-Cu en los sedimentos de la cuenca occidental del Pucará, Perú Central (Tentativa de una integración): Sociedad Geológica del Perú, Boletín 69, p. 41-69.

Kobe, H. W., 1982b, A strata-bound Ni-Co arsenide/sulphide mineralization in the Paleozoic of the Yauli dome, Central Perú, *in* G. C. Amstutz et al., Ore genesis—the state of the art: New York, Springer-Verlag, p. 150-160.

Kussmaul, S., P. K. Hormann, E. Ploskonka, and T. Subieta, 1977, Volcanism and structure of southwestern Bolivia: Journal of Volcanology and Geothermal Research, v. 2, p. 73-111.

Lancelot, J. R., G. Laubacher, R. Marocco, and U. Renaud, 1980, Radiocronología U/Pb de dos plutones graníticos de la Cordillera Oriental (Perú)—Extension de actividad magmática permiana y consecuencias: Sociedad Geológica del Perú, Boletín 66, p. 43-50.

Le Bel, L. M., 1985, Mineralization in the Arequipa segment: the porphyry-Cu deposit of Cerro Verde/Santa Rosa, *in* Pitcher et al., Magmatism at a plate edge—the Peruvian Andes: New York, Blackie Halsted Press, John Wiley and Sons, p. 250-260.

Lehmann, B., 1979, Schichtgebundene Sn-Lagerstaetten in der Cordillera Real/Bolivien: Berliner Geowissenschaftliche Abhandlungen, R. A., B. 14, 136 p.

Lehmann, B., 1985, Formation of the strata-bound Kellhuani tin deposits, Bolivia: Mineralium Deposita, v. 20, p. 169-176.

Lehmann, B., and H. Pichler, 1980, Tin distribution in mid-Andean volcanic rocks: Mineralium Deposita, v. 15, p. 35-39.

Lehmann, B., U. Petersen, R. Santivanez, and L. Winkelmann, 1987, Tin and boron distribution in the lower Paleozoic sequence of the Cordillera Real, Bolivia: Manuscript in preparation.

Lehne, R. W., and G. C. Amstutz, 1982, Sedimentary and diagenetic fabrics in the Cu-Pb-Zn-Ag deposit of Colquijirca, Central Perú, *in* G. C. Amstutz et al., eds., Ore genesis—the state of the art: New York, Springer-Verlag, p. 161-166.

Levin, P., and G. C. Amstutz, 1973, Neue Untersuchungen ueber schichtgebundene Lagerstaetten im zentralen Ost-Perú: Münstersche Forschungen zur Geologie und Palaontologie, H. 31/32, p. 233-259.

Levin, P., and G. C. Amstutz, 1976, Kristallisation und Bewegung in Erzrhythmiten am Beispiel triassisch-jurassischer Lagerstaetten in Ostperú: Munstersche Forschungen zur Geologie und Palaontologie, H. 38/39, p. 111-128.

Lowell, J. D., 1974, Three new porphyry copper mines for Chile?: Mining Engineering, November, p. 22-28.

Lyons, W. A., 1968, The geology of the Carahuacra mine, Perú: Economic Geology, v. 63, p. 247-256.

Manning, D. A. C., 1986, Contrasting styles of Sn-W mineralisation in peninsular Thailand and SW England: Mineralium Deposita, v. 21, p. 44-52.

McKee, E. H., and D. C. Noble, 1985, Tectonic events, magmatic pulses, and base and precious metal mineralization in the central Andes (abs.): Symposium on Geology of the Andes and Its Relation to Hydrocarbon and Mineral Resources, Santiago, Chile, p. 26.

McNutt, R. H., A. H. Clark, and M. Zentilli, 1979, Lead isotopic compositions of Andean igneous rocks, latitudes 26° to 30°S: petrologic and metallogenic implications: Economic Geology, v. 74, no. 4, p. 827-837.

Noble, D. C., and M. L. Silberman, 1984, Evolución volcánica e hidrotermal y cronología de K-Ar del distrito minero de Julcani, Perú: Sociedad Geológica del Perú, Vol. Jubilar LX Aniversario Homenaje Dr. G. Petersen, 5, p. 1-35.

Noble, D. C., E. H. McKee, V. R. Eyzaguirre, and R. Marocco, 1984, Age and regional tectonic and metallogenetic implications of igneous activity and mineralization in the Andahualylas-Yauri belt of southern Perú: Economic Geology, v. 79, p. 172-176.

Noble, D. C., E. H. McKee, E. Farrar, and U. Petersen, 1974, Episodic Cenozoic volcanism and tectonism in the Andes of Perú: Earth and Planetary Science Letters, v. 21, p. 213-220.

Palma, V. V., 1981, The San Rafael tin-copper deposit, Puno, S. E. Perú: Kingston, Ontario, Queen's University, unpublished M.S. thesis, 226 p.

Paredes, J., 1980, Informe definitivo del estudio geo-económico del distrito minero de Hualgayoc-Cajamarca: Private report, 117 p.

Park, C. F., Jr., 1961, A magnetite "flow" in northern Chile: Economic Geology, v. 56, p. 431-436.

Petersen, G., and C. Vidal, 1983, Trés épocas metalogenéticas evidenciadas en el Cenozoico del Perú: Sociedad Geológica del Perú, Boletín 71, p. 107-116.

Petersen, U., 1965, Regional geology and major ore deposits of central Perú: Economic Geology, v. 60, p. 407-476.

Petersen, U., 1970, Metallogenic provinces in South America: Geologische Rundschau, B. 59, p. 834-897.

Petersen, U., 1972, Geochemical and tectonic implications of South American metallogenic provinces: New York Academy of Science Annals, v. 196, p. 1-38.

Petersen, U., 1979, Metallogenesis in South America: progress and problems: Episodes, v. 1979, no. 4, p. 3-11 (also in Metalogenesis en Latinoamerica: International Union of Geological Sciences Publication 5, p. 249-274, 1980).

Petersen, U., D. C. Noble, M. J. Arenas, and P. C. Goodell, 1977, Geology of the Julcani mining district, Perú: Economic Geology, v. 77, p. 931-949.

Pitcher, W. S., 1974, The Mesozoic and Cenozoic batholiths of Perú: Pacific Geology, v. 8, p. 51-63.

Pitcher, W. S., 1978, The anatomy of a batholith: Journal of the Geological Society of London, v. 135, p. 157-182 (reprinted in Spanish: Sociedad Geológica del Perú, Boletín 60, p. 13-60, 1979).

Pitcher, W. S., 1984, Phanerozoic plutonism in the Peruvian Andes, *in* R. S. Harmon and B. A. Barreiro, Andean magmatism—chemical and isotopic constraints: American Geophysical Union Special Symposium Proceedings, Shiva Geology Series, p. 152-167.

Pitcher, W. S., M. P. Atherton, E. J. Cobbing, and R. D. Beckinsale, 1985, *in* Pitcher et al., Magmatism at a plate edge—the Peruvian Andes: New York, Blackie Halsted Press, John Wiley and Sons, 328 p.

Ponzoni, E., 1980, Metalogenia del Perú, *in* Metalogenesis en Latinoamerica: International Union of Geological Sciences Publica-

tion 5, p. 101-140.
Ponzoni, E., and C. Vidal, 1982, Copper mineralization and magmatic hydrothermal brines in the Rio Pisco section of the Peruvian coastal batholith—a discussion: Economic Geology, v. 77, p. 1951-1955.
Ponzoni, E., A. Postigo, and J. F. Birkbeck, 1969, Guía para el mapa metalogenético del Perú: Lima, Sociedad Nacional Mineria y Petróleo, 128 p.
Putzer, H., 1971, Mineralische Rohstoffe in Ecuador, *in* W. Sauer and H. Putzer, eds., Geologie von Ecuador, Hdb. Region: Geologische der Erde, v. 11, p. 205-257.
Putzer, H., 1976, Metallogenetische Provinzen in Sudamerika: Stuttgart, E. Schweizerbart'sche Verlagsbuchhandlung, 316 p.
Ripley, E. M., and H. Ohmoto, 1977, Mineralogic, sulfur isotope, and fluid inclusion studies of the stratabound copper deposits at the Raul mine, Perú: Economic Geology, v. 72, p. 1017-1041.
Rivera G., N., and H. W. Kobe, 1983a, Evolución geológica del domo de Yauli: Sociedad Geológica del Perú, Boletín 72, p. 159-175.
Rivera G., N., and H. W. Kobe, 1983b, Metalogenia del domo de Yauli: Sociedad Geológica del Perú, Boletín 72, p. 177-195.
Robertson, R. R., and I. Patiño V., 1985, Colombia's Cerro Matoso: Engineering and Mining Journal, May, p. 18-22.
Rodríguez, S. E., 1974, Sulfur isotopes in sulfides from the Bailadores ore deposits, Venezuelan Andes, and the Aroa mining district, western central range, Venezuela: Society of Mining Engineers of AIME Annual Meeting, Dallas, Texas, Preprint 74-S-15, 16 p.
Ruiz, F., C., and G. E. Ericksen, 1962, Metallogenetic provinces of Chile, S. A.: Economic Geology, v. 57, p 91-106.
Ruiz, F., C., L. Aguirre, T. Corvalan, C. Klohn, E. Klohn, and B. Levi, 1965, Geologia y yacimientos metalíferos de Chile: Santiago, Instituto Investigaciones Geológicas, 305 p.
Samamé Boggio, M., 1979-84, El Perú minero: Lima, Editora Perú, v. I (1979) 311 p.; v. II (1979) 618 p.; v. III (1980) 347 p.; v. IV (1980, 1981) 1726 p.; v. V (1981) 969 p.; v. X (1983) 584 p.; v. XII (1982) 379 p.; v. XIII (1984) 410 p.
Samaniego, A., 1982, Correlation of strata-bound mineral deposits in the early Cretaceous Santa metallotect of North and Central Perú, *in* G. C. Amstutz et al., eds., Ore genesis the state of the art: New York, Springer-Verlag, p. 508-527.
Samaniego, J. A., and G. C. Amstutz, 1981, Metalogenese des Santa-Formation-Metallotekts: Stuttgart, Zentral Blatt fur Geologia und Palaontologie, Teil I, H. 3/4, p. 456-464.
Satchwell, P. C., 1983, Geología de la mina Cuajone: Sociedad Geológica del Perú, Boletín 72, p. 127-146.
Schneider, H. J., and B. Lehmann, 1977, Contribution to a new genetical concept on the Bolivian tin province, *in* D. D. Klemm and H. J. Schneider, eds., Time- and strata-bound ore deposits: New York, Springer-Verlag, p. 153-168.
Silbermann, M. L., and D. C. Noble, 1977, Age of igneous activity and mineralization, Cerro de Pasco, Central Perú: Economic Geology, v. 72, p. 925-930.
Sillitoe, R. H., 1973, Geology of the Los Pelambres porphyry copper deposit, Chile: Economic Geology, v. 68, p. 1-10.
Sillitoe, R. H., 1977, Permo-Carboniferous, Upper Cretaceous, and Miocene porphyry copper-type mineralization in the Argentinian Andes: Economic Geology, v. 72, p. 99-109.
Sillitoe, R. H., C. Halls, and J. N. Grant, 1975, Porphyry tin deposits in Bolivia: Economic Geology, v. 80, p. 913-927.
Sillitoe, R. H., L. Jaramillo, and H. Castro, 1984, Geologic exploration of a molybdenum-rich porphyry copper deposit at Mocoa, Colombia: Economic Geology, v. 79, p. 106-123.
Sillitoe, R. H., L. Jaramillo, P. E. Damon, M. Shafiqullah, and R. Escovar, 1982, Setting, characteristics, and age of the Andean porphyry copper belt in Colombia: Economic Geology, v. 77, p. 1837-1850.
Soler, P., R. Carrascal, and J. Saez, 1986, Mineralization and paragenesis of the Huanzalá mine, central Perú—a discussion: Economic Geology, v. 81, p. 195-196.
Soler, P., G. Grandin, and M. Fornari, 1986, Essai de synthèse sur la métallogénie du Péru: Geodynamique 1 (1), p. 33-68.
Stewart, J. W., J. F. Evernden, and N. J. Snelling, 1974, Age determinations from Andean Perú: a reconnaissance survey: Geological Society of America Bulletin, v. 85, p. 1107-1116.
Sugaki, A., H. Ueno, N. Shimada, I. Kusachi, A. Kitakaze, K. Hayashi, S. Kojima, and O. Sanjines V., 1983, Geological study on the polymetallic ore deposits in the Potosí district, Bolivia: Science Reports of the Tohoku University, S. III, v. XV, no. 3, p. 409-460.
Sureda, R. J., and G. C. Amstutz, 1981, New investigations on the stratabound Pb-Zn deposits in the Sierra de Aguilar, province of Jujuy, Argentina: Zentral Blatt fur Geologia und Palaontologie, Teil I, H. 3/4, p. 494-504.
Terrones, A., 1984, Overview of the mineral resource potential of Latin America in relation to global tectonic and metallogenic controls: Global Tectonics and Metallogeny, v. 2, nos. 3 and 4, p. 213-256.
Thorpe, R. S., 1984, The tectonic setting of active Andean volcanism, *in* R. S. Harmon and B. A. Barreiro, eds., Andean magmatism—chemical and isotopic constraints: American Geophysical Union Special Symposium Proceedings, Shiva Geology Series, p. 4-8.
Tilton, G. R., R. J. Pollak, A. H. Clark, and R. C. R. Robertson, 1981, Isotopic composition of Pb in central Andean ore deposits: Geological Society of America Memoir 154, p. 791-816.
Trurnit, P., K. Fesefeldt, and S. Stephan, 1982, A caldera of Neogene age and associated hydrothermal ore formation, Ticapampa-Aija mining district, Cordillera Negra, Department of Ancash, Perú, *in* G. C. Amstutz et al., eds., Ore genesis—the state of the art: New York, Springer-Verlag, p. 528-552.
Tsuchiya, Y., M. Toledo T., D. Mendoza Z., and R. Soto M., 1983, Geologia del yacimiento minero de Huanzalá: Sociedad Geológica del Perú, Boletín 71, p. 17-40.
Turneaure, F. S., 1960, A comparative study of major ore deposits of central Bolivia: Economic Geology, v. 55, p. 217-254, 574-606.
Turneaure, F. S., 1971, The Bolivian tin-silver province: Economic Geology, v. 66, p. 215-225.
Valdez C., M., 1983, Alteracion y mineralizacion hidrotermal del manto Cobriza—Mina Cobriza: Sociedad Geológica del Perú, Boletín 72, p. 111-126.
Valera, J., J. Injoque, C. Vidal, and J. Zegarra, 1983, Manifestaciones de baritina hidrotermal en el batolito de la costa: Sociedad Geologica del Perú, Boletín 70, p. 47-50.
Vera R., 1983, Reinterpretación genética de la mineralización en el domo de Yauli: Sociedad Geológica del Perú, Boletín 71, p. 41-52.
Vidal, C., 1980, Mineral deposits associated with the Peruvian coastal batholith: Liverpool, England, University of Liverpool, unpublished Ph.D. thesis.
Vidal, C., 1985, Metallogenesis associated with the Coastal batholith of Perú: a review, *in* Pitcher et al., Magmatism at a plate edge—the Peruvian Andes: New York, Blackie Halsted Press, John Wiley and Sons, p. 243-249.
Vidal, C., and R. Cabos, 1983, Zoneamiento regional de las alteraciones y menas hidrotermales en Hualgáyoc, Cajamarca: Sociedad Geológica del Perú, Boletín 71, p. 117-120.
Vidal, C., O. Mayta, D. C. Noble, and E. H. McKee, 1984, Sobre la evolución de soluciones hidrotermales desde el centro volcánico Marcapunta en Colquijirca-Pasco: Sociedad Geológica del Perú, Vol. Jubilar LX Aniversario Homenaje Dr. G. Petersen, 10, p. 1-14.
Wauschkuhn, A., and M. Ohnsmann, 1984, Petrology and geochemistry of Mesozoic volcanic rocks and related ore deposits in central Perú and their plate tectonic significance: Sociedad Geológica del Perú, V. Jubilar LX Aniversario Homenaje Dr. G.

Petersen, 11, p. 1-36.
Wauschkuhn, A., and R. Thum, 1982, The Condestable mine, a volcano-exhalative deposit in the Mesozoic coastal belt of central Perú, *in* G. C. Amstutz et al., eds., Ore genesis—the state of the art: New York, Springer-Verlag, p. 174-190.
Weeks, L. G., 1947, Paleogeography of South America: American Association of Petroleum Geologists Bulletin, v. 31, p. 1194-1241.
Wu, I., and U. Petersen, 1977, Geochemistry of tetrahedrite and mineral zoning at Casapalca: Economic Geology, v. 72, p. 993-1016.
Yates, R. G., D. F. Kent, and J. Fernandez Concha, 1951, Geology of the Huancavelica quicksilver district, Perú: U.S. Geological Survey Bulletin 975-A, p. 1-45.
Zantop, H., 1981, Argentina's porphyry copper potential: Mining Engineering, February, p. 137-142.
Zegarra N., J., and A. Galloso C., 1983, Yacimientos auriíferos de la zona sur oriental del Perú: Sociedad Geológica del Perú, Boletín 71, p. 125-133.
Zeil, W., 1964, Geologie von Chile: Berlin, Gebr. Borntraeger, 233 p.
Zimmermann, J.-L., and A. Kihien C., 1983, Détermination par la méthode K/Ar de l'âge des intrusions et des minéralisations associées dans le porphyre cuprifère de Quellaveco (sud ouest du Pérou): Mineralium Deposita, v. 18, p. 207-213.

Ericksen, G. E., Cañas Pinochet, M. T., and Reinemund, J. A., editors, 1989, Geology of the Andes and its relation to hydrocarbon and mineral resources: Houston, Texas, Circum-Pacific Council for Energy and Mineral Resources Earth Science Series, v. 11.

CHAPTER 17

PROSPECTS FOR MINERAL EXPLORATION AND DEVELOPMENT IN THE ANDES

Juan A. Proaño
Inter-American Development Bank
Washington, D.C., U.S.A.

HISTORY AND EVOLUTION OF MINING IN THE ANDEAN REGION

When Christopher Columbus stepped ashore in the New World, he never imagined the mineral riches that would be discovered in these lands in the next five centuries. He returned from his first voyage bringing some gold nuggets and golden handicrafts as presents for the Catholic monarchs. However, he was aware of the potential riches that could issue from additional gold discoveries, and he brought with him, on his second voyage in 1493, tools and 1500 men, among whom were what Friar Bartolomé de las Casas described as "a large quantity of workmen to take the gold from the mines." The following year the monarchs were petitioned to send to the New World gold miners and panners from Almaden in order to "gather (gold) from the sands and scrape it from the land." Events moved quickly, and by 1535 the cities of Mexico and Lima had been founded and the conquistadors had begun the first successful and effective campaign of mineral exploration. During the next several decades many of the major mining districts in the Andean region were discovered.

Prominent among the mining districts of the 16th and 17th centuries were Cerro Rico de Potosí in Bolivia, Cerro de Pasco in Perú (Figure 1), and Guanajuato in Mexico, all of which are still in production. The Cerro Rico de Potosí mine, discovered in the mid-16th century, has so far produced at least one billion ounces of silver and, according to some authors, production may have been closer to two billion ounces. Interestingly, Potosí in the 17th century had a population exceeded only by that of Paris. The Cerro de Pasco mine has so far yielded 870 million ounces of silver, in addition to substantial quantities of copper, lead, and zinc. For geographic, historical, strategic, and economic reasons, Latin America has played an important role in world affairs since the 16th century. Latin America now is a major supplier of essential raw materials and an important market for goods and services, a market which in 1984 was represented by a population of 378 million. Estimates show that this population may exceed 550 million by the year 2000. For capital goods, the current market for equipment and machinery in Latin America is about US$30 billion per year.

RECENT DEVELOPMENTS IN MINING AND THEIR SIGNIFICANCE

According to the Inter-American Development Bank (1985), the gross domestic product of the mining sector of Latin America in 1984 was $19,702 million dollars. This represented 3.2% of the gross domestic product of the region as a whole, which was equivalent to US$611,087 million (in 1982 dollars). According to information gathered mainly from the Mining Journal (June 1985), mineral output (not including hydrocarbons) of the Andean countries of South America in 1984 amounted to about US$4671 million equivalent, as follows:

Chile	US$2314 million
Perú	US$1763 million
Bolivia	US$343 million
Colombia	US$155 million
Argentina	US$96 million

The 19th and 20th centuries have seen a substantial shift in the priorities accorded mining of metals. From the time the Europeans arrived in South America through the first half of the 19th century, both prospecting and mining operations emphasized the search for and production of gold and silver. During the 19th century, systematic exploitation of the guano and nitrate deposits began in response to developing technology in agriculture and industry. Nitrate became increasingly important for the manufacture of munitions for 19th and 20th century conflicts, up to and including World War I. Systematic development of copper and tin deposits began in the 19th century, and the discovery of the method of concentration by flotation led to exploitation of base-metal sulfide deposits in the late 19th century. In response to industrial development, new mining technology and metallurgical methods were applied in the 20th century to bring about large-scale exploitation of copper, tin, lead, zinc, and other metalliferous deposits.

This intensification of mining activity has resulted in discovery and initiation of production of a number of new mineral deposits in Latin America. Among these successes have been (1) the start of operations of the Real de los Angeles deposit, Mexico, which in 1984 yielded 9 million ounces of silver, and (2) the beginning of phosphate mining operations in Baja California Sur, which could bring Mexico out of a position of nearly total dependence on imported phosphates in 1980 to virtual self-sufficiency in phosphates. In 1985, mining operations began in the El Cerrejón coal basin, Colombia (Figure 1), and production from this basin is expected to make Colombia the fifth largest coal producer in the world by the year 2000. According to estimates, exports from El Cerrejón in 1986 will amount to 6 million tonnes of coal, and coal production is expected to reach 15 million tonnes in 1990. This coal, because of its low sulfur and ash content and its high calorific value, is of excellent quality and enjoys widespread acceptance.

Significant new gold, silver, and zinc mines were opened or had expanded productions in Perú during the 1975-1986 period. Among these are the Uchucchacua mine (Figure 1), which in 1984 produced 2.7 million ounces of silver and which, with the recent expansion of plant capacity to 1000 tonnes per day, has further increased output. The Orcopampa mine (Figure 1) produced 2.3 million ounces of silver and 28,261 ounces of gold in 1984, and increased its treatment capacity to 700 tonnes per day in 1985. Zinc production from the San Vicente mine (Figure 1) has been greatly expanded since 1979; in 1984 the mine produced 67,200 tonnes of zinc in concentrates, making it the second largest zinc producer in Perú.

Of particular interest in Chile was the inauguration in 1982 of the El Indio mining complex (Figure 1), which in that year produced 380,000 ounces of gold, 516,000 ounces of silver, and 8093 tonnes of copper. In 1984, production from the El Indio mine was about 280,000 ounces of gold, one million ounces of silver, and 19,000 tonnes of copper. Also in 1984, the production of carbonate from lithium-rich brines of Salar de Atacama (Figure 1) began at a plant having a capacity of producing 6400 tonnes of Li_2CO_3 per year. This output is equivalent to 14% of world consumption of Li_2CO_3. Another significant event in Chilean mining was the inauguration in 1984 of the El Toqui mine (Figure 1), which brought about an increase in this country's production of zinc from 3000 tonnes in 1981 to 36,000 tonnes in 1983. All these deposits have served to diversify the Chilean mining industry, which previously had been concerned largely with production of copper and nitrates. Additionally, the discovery of the huge and high-grade Escondida porphyry copper deposit (Figure 1) is the most significant development in the copper sector. This deposit has estimated reserves of 1.7 billion tonnes of ore having an average grade of 2.6% copper in addition to substantial amounts of molybdenum.

Among new developments in Argentina are the discovery and initial feasibility studies for exploiting potassium-rich salt deposits in the region south of Mendoza (Figure 1). The government of Argentina has issued an international invitation for bids to develop the Huemules gold and silver deposit (Figure 1), and hopes to attract foreign capital for this mine and for other mines.

A gold rush in southeastern Ecuador is expected to greatly increase gold placer production. The Ecuadorian Mining Code, which has recently been amended, is expected to encourage mining activity.

The increase in gold production in Latin America deserves special attention. In 1974 the region produced only 40 tonnes of gold, whereas in 1985 gold production had increased nearly fourfold to about 150 tonnes. The increase was greatest in Brazil, where production rose from 6 tonnes in 1970 to 55 tonnes in 1984. In Chile, gold production rose from 1.56 tonnes in 1970 to nearly 17 tonnes in 1984, largely as a result of production from the new El Indio mine and stimulation of mining under the National Gold Plan to develop small placer mining. Furthermore, placer gold production has been greatly expanded in the rain forest region east of the Andes in Bolivia and Perú as well as in Ecuador.

APPLICATION OF MODERN EXPLORATION, MINING, AND METALLURGICAL TECHNIQUES

New mineral discoveries and production would have been impossible without the development and employment of new concepts in geology, mining, metallurgy, and management as well as changes in national policy, and improved training and greater use of young national technicians engaged in the mining sector. The involvement of these technicians is well demonstrated by their excellent presentations about tectonics, stratigraphy, metallogeny, regional geology, and geochemistry at this symposium.

Employment of geologic models should be encouraged in exploration efforts, provided the model is adaptable to the geologic realities of the given area of investigation. The successful exploration work leading to the development of the El Indio deposit has created great international interest in the search for epithermal deposits of precious metals in the volcanic and subvolcanic rocks that occur widely in the central Andean region of southern Perú, western Bolivia, and northern Chile. Deposits of precious metals, formed by thermal-spring systems, such as Carlin-type deposits in sedimentary and/or volcanic rocks, constitute another model that ought to be used more widely in the Andean region. In the Dominican Republic, the Pueblo Viejo gold-silver deposit produces about 1000 ounces of gold and 4000 ounces of silver daily from a Carlin-type deposit. Models for volcanogenic Kuroko-type deposits and other volcanic sedimentary deposits should prove useful for exploration in Ecuador, Perú, and northern Chile, and perhaps the southernmost regions of Chile and Argentina.

To a large degree the success of a mineral exploration program is a measure of the experience and competence of the explorationists and their ability to apply techniques—geologic mapping, geochemistry, geophysics, photointerpretation, stratigraphy, paleontology, satellite imagery, diamond drilling, sampling, calculation of reserves, and economic evaluation of deposits—to ore discovery and evaluation. The geologist must know which tools to use and how to interpret and correlate the findings. At the present time, when the mining industry is facing a metals price crisis and the mining sector is being radically restructured, it is essential to utilize the most experienced mining and exploration geologists. Additionally, innovative methods of mining and metallurgy as well as management are required to establish and maintain profitable operations.

In the age of specialization in which we live, and given the close relationship between geologic, mining, metallurgical, and economic problems in the mining industry, I think it is essential that mining and exploration geologists be familiar with basic concepts of mining, metallurgy, and economics. These geologists need to have a clear perception that the purpose of their work is not only to detect a geochemical or geophysical anomaly, a zone of alteration, or to determine geologic reserves, but also to ascertain the volume of economically exploitable ore reserves. Such ore reserves must be suitable for mining and for recovery of their metallic content at a cost sufficiently lower than the market value of the metals to give a reasonable profit on investment. Only in such cases can it be said that the geologic work is actually leading to mine development, and that the purpose and desired goals of exploration are being achieved.

The advent of new and improved methods of mining and metallurgy are making it possible to reduce production costs and to make a mineral concentrate or metal acceptable to the buyer. The ongoing mechanization and modernization of mining equipment enables, for example, such expansion of production as is typical of the medium-scale mining sector of Perú. Likewise, at the El Teniente mine in Chile changes being made in the conventional mining method of block caving should bring about a reduction of the mining costs in this large copper mine.

A noteworthy development in metallurgy has been the introduction of leaching processes that will be used to recover metals from former mine waste and low-grade ore. Leaching of waste dumps, which have accumulated for more than 50 years at the Chuquicamata mine in Chile, with the sulfuric acid recovered from the smelter gases is expected to recover about 50,000 tonnes of copper a year. At the Uchucchacua mine in Perú, the leaching of manganese from silver concentrates has reduced the risk of spontaneous combustion and yields a more acceptable silver concentrate that is safe for ocean transportation. At the San Vicente mine, leaching of zinc concentrates reduces their magnesium content and has eliminated smelter penalties that were imposed because of high magnesium contents. The method of in-situ leaching applied to accumulated tailings having low gold and silver content is being employed successfully at Otuzco, Perú.

ROLE OF NATIONAL GOVERNMENTS IN MINERAL EXPLORATION AND DEVELOPMENT

In order to maintain a balance of effort between the public and private sectors so as to encourage investment in exploration and exploitation of mineral deposits, governments must enact and maintain mining, tax, and labor legislation that will protect the interests of the state as well as the interests of investors and workers, and also provide for effective protection of the environment. The enactment of laws governing foreign investment, maintenance of labor peace, repatriation of profit, and reinvestment in mining operations, as well as the opening of former national reserves to exploration, constitute incentives for private investment in the mining sector. In Chile, the development of the El Indio precious-metal deposit and the Salar de Atacama Li-rich brines is due in large part to laws establishing an environment favorable to private investment.

In August 1985, the government of Ecuador enacted a new mining law designed to encourage foreign and national investment in the mining industry. Under this new law, royalties are established at 1% of the gross value of the production during the first 4 years of mining operations, and then increased to 3% beginning in the 5th year. The law further encourages investment in the mining sector by national investors by providing for the waiver of as much as 50% of taxes on income that is invested in mining. It also extends the life of a mining concession from 10 to 30 years. In an effort to make the government more responsive to the needs of the mining industry, the Instituto Nacional de Minería (IMEN) was established to replace the former Dirección General de Geología y Minas.

In Argentina, the government is engaged in the reorganization of the mining industry in order to stimulate exploration of the national territory which, in spite of its great mineral potential, produces a very small quantity of minerals. The reorganization is aimed at increasing mine productivity and mineral exports, promoting new mine investment, implementing preinvestment programs, and carrying out an intensive mineral exploration program.

ROLE OF THE INTER-AMERICAN DEVELOPMENT BANK (IDB)

The Inter-American Development Bank completed 25 years of service in Latin America in 1985, and has played an important role in mobilizing domestic and foreign resources for the economic development of the region. Up to and including 1984, the IDB had made loans and grants of nearly US$28 billion of its own resources in the financing of projects having a total investment of about US$100 billion. The Bank has had great interest in the development of Latin America's mining industry. So far, it has made 32 loans in the mining sector, totaling US$849.7 million, of which 6 loans were for studies and regional exploration, 13 were for specific mining projects, and 14 were for support of small- and medium-scale mining. In addition, at the end of 1986, one loan for a mining project and another for general support of the mining sector were pending approval by the Board of Executive Directors of the Bank.

An agreement for establishing the Inter-American Investment Corporation was completed in draft form in November 1984 and will be enacted during 1986. As an affiliate of the Bank, this corporation will promote economic development of the developing regional member countries by encouraging the establishment, expansion, and modernization of private companies, preferably small- and medium-scale enterprises. Companies that are partly owned by the government, or other public institutions whose activities are designed to enhance the private sectors of the economy, are also eligible to borrow from this newly created corporation.

STRATEGY OF THE INTER-AMERICAN DEVELOPMENT BANK IN THE MINING SECTOR

Latin America is one of the richest sources of minerals in the world, and mining is especially important because most of the mineral output is exported to earn foreign exchange and strengthen the balance of payments. The Bank supports countries with established mining traditions by granting credit for projects to maintain and improve the levels of efficiency and productivity of mining operations. This results in a reduction of production costs so as to maintain the competitive standing of these industries on the world market. In countries which do not traditionally engage in mining, but whose territories have such geologic conditions that indicate the possible presence of mineral deposits, the Bank encourages surveys to identify the areas of greatest potential, finances those that involve preinvestment projects, and eventually contributes to the financing of specific projects, particularly those having the potential for establishing new work centers or for creating new development centers. In addition, the Bank lends support to private sector development of medium- and small-scale mining activity through global loans, which are disbursed to the small and medium mining sector through intermediate lending institutions such as mining banks.

CIRCUM-PACIFIC MAP PROJECT

The Circum-Pacific Map Project is an excellent example of how cooperation can be achieved between national geological institutions, private mining and oil companies, and mining sector scientists from many organizations. The papers delivered at this symposium and the cooperation between geologists from North, Central, and South America, whose efforts have made possible the preparation of many Circum-Pacific Province Maps, demonstrate the rewards of the work carried out by the Circum-Pacific Council in promoting international cooperation among geoscientists. These maps are incentives for the geologic services of the Andean countries to continue to bring up-to-date the store of geologic information on their respective countries and to attract the interest of mining investors in areas where the geology favors the presence of certain types of mineral deposits.

REFERENCES

Andean Airmail and Peruvian Times, 1985, Carta Minera: Andean Airmail and Peruvian Times weekly publication, v. 4, no. 1-51.

Barton, P. B., 1983, Unconventional mineral deposits: a challenge to geochemistry, *in* W.C. Shanks, ed., Cameron volume on unconventional mineral deposits: American Institute of Mining and Metallurgical Engineers, p. 1-4.

Bernstein, M., D. Thompson, and P. Ly, 1984, Exploration in the Andes for gold and silver deposits: Huaraz, Perú, 27th Mining Engineering Convention, unpublished report, 4 p.

Griffith, W.A., 1985, Proper management during prosperous times will lessen impact of rough times: Mining Engineering, v. 38, p. 535-536.

Inter-American Development Bank, 1985, Economic and social progress in Latin America—external debt, crisis and adjustment: Inter-American Development Bank, 471 p.

Mining Journal, 1985, Mining annual review: Mining Journal, 544 p.

Terrones, A. J., 1984, Overview of the mineral resource potential of Latin America in relation to global tectonics and metallogenetic controls: Global Tectonics and Metallogeny, v. 2, p. 213-256.

Ericksen, G. E., Cañas Pinochet, M. T., and
Reinemund, J. A., editors, 1989, Geology of the Andes
and its relation to hydrocarbon and mineral resources:
Houston, Texas, Circum-Pacific Council for Energy
and Mineral Resources Earth Science Series, v. 11.

CHAPTER 18

MINERAL DEPOSITS IN THE VENEZUELAN ANDES

Simon E. Rodriguez
The Geological Survey of Venezuela
Ministry of Energy and Mines
Caracas, Venezuela

ABSTRACT

The Venezuelan Andes have a variety of metalliferous, nonmetalliferous, and coal deposits, some of which have had a significant production. Metalliferous deposits, which are sparse and small compared with such deposits elsewhere in the Andes, include volcanogenic massive sulfide deposits; sedimentary iron and stratabound copper and uranium-copper deposits; and hydrothermal, locally argentiferous, base-metal veins. The principal nonmetalliferous deposits are phosphate rock, gypsum, silica sand, flint clay, asphalt, and feldspar. Deposits of subbituminous coal, some of which is coking coal, are present in early Tertiary sedimentary rocks.

METALLIFEROUS DEPOSITS

Copper

Distinctly different copper deposits have been found in three areas of the Venezuelan Andes, as follows: (a) volcanogenic massive sulfide deposits near the border between Estado Táchira and Estado Mérida; (b) sedimentary deposits in the western part of Estado Táchira; and (c) hydrothermal deposits in the northern part of Estado Mérida (Figure 1). Of these deposits, the massive sulfides have by far the greatest economic potential. They consist of masses of copper-lead-zinc sulfides in a thick sequence of metavolcanic and metasedimentary rocks of Paleozoic age (Figure 2). The deposits show metal zonation. One of the deposits, Bailadores, was studied by the Venezuela Geological Survey between 1970 and 1975, and was determined to have reserves of more than 4 million tonnes of ore. The chalcopyrite zone of this deposit has values ranging from 1.00% to 4.28% copper, in addition to lead and zinc (Nogueira et al., 1972).

The sedimentary copper deposits of Estado Táchira consist chiefly of chalcocite, chalcopyrite, and pyrite impregnations in coarse-grained gray sandstone and conglomerate of Triassic-Jurassic age. The secondary copper minerals malachite, azurite, and chrysocolla are present in near-surface parts of the orebodies. Several deposits have been studied and drilled. Among these, it was determined that the Cerro Mono deposit has more than 500,000 tonnes of ore containing 2.5% copper, and the Seboruco deposit has more than 12 million tonnes of ore with 1.5% copper. Apparently, the ores consist of remobilized copper minerals originally deposited in closed basins.

The hydrothermal copper deposits in northern Mérida (Figure 1) consist of disseminated copper minerals, chiefly bornite, associated with Triassic granites that intrude schists and gneisses of Precambrian age (Woznessensky, 1972). The rocks of this region are intensely fractured, propylitized, and silicified. Deposits of this area were recently studied by the Geological Survey of Venezuela, but no estimates of ore reserves were made. None of the deposits have been exploited.

Iron

Sedimentary iron deposits occur in the southern part of Estado Lara, near the northern margin of the Venezuelan Andes (Figure 1). They are within the Quebrada Arriba Formation of Eocene age, which crops out extensively in both Lara and Trujillo, and consist of layers of massive oolitic hematite and goethite interbedded with chamositic, hematitic, and glauconitic sandstone (Figure 3). The ores contain 45 to 65% iron, and regional studies indicate resources of more than 50 million tonnes of ore available for open-pit mining (Rodriguez et al., 1977).

Small hydrothermal hematite deposits occur in the area of Los Almogres, southern Mérida (Figure 1). The deposits consist of thick veins of gray hematite rich in quartz, and are associated with Precambrian granites. More than 3 million tonnes of ore have been outlined.

Lead, Zinc, and Silver

As previously mentioned, lead and zinc are associated with copper in the marine sulfide deposits of the Bailadores region, southern Mérida (Figure 1). In addition, hydrothermal veins of argentiferous galena are associated with late Precambrian intrusives of this same region. The veins contain late fracture fillings of sphalerite. Such veins were exploited, chiefly for

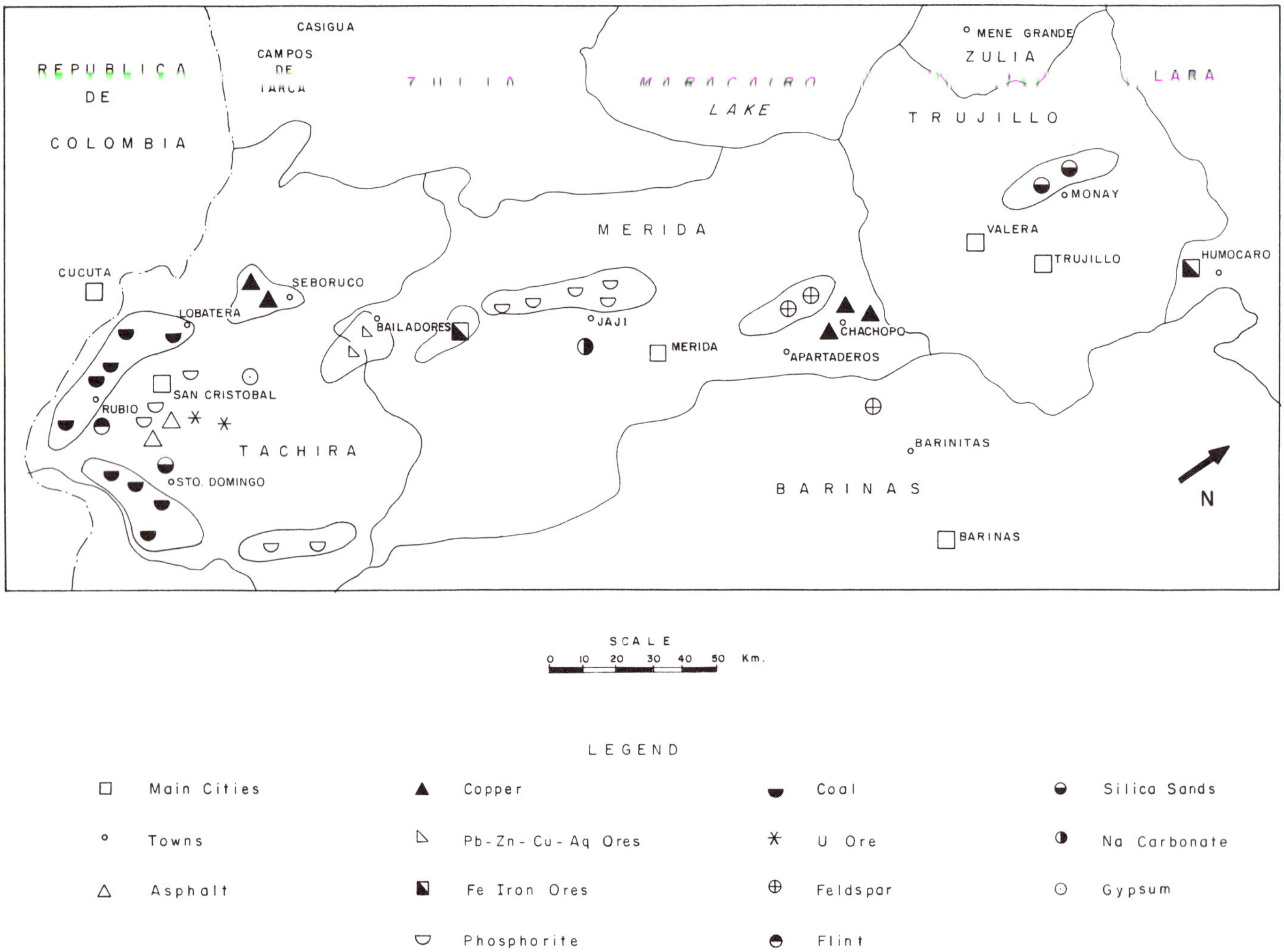

Figure 1. Mineral deposits of the Venezuelan Andes.

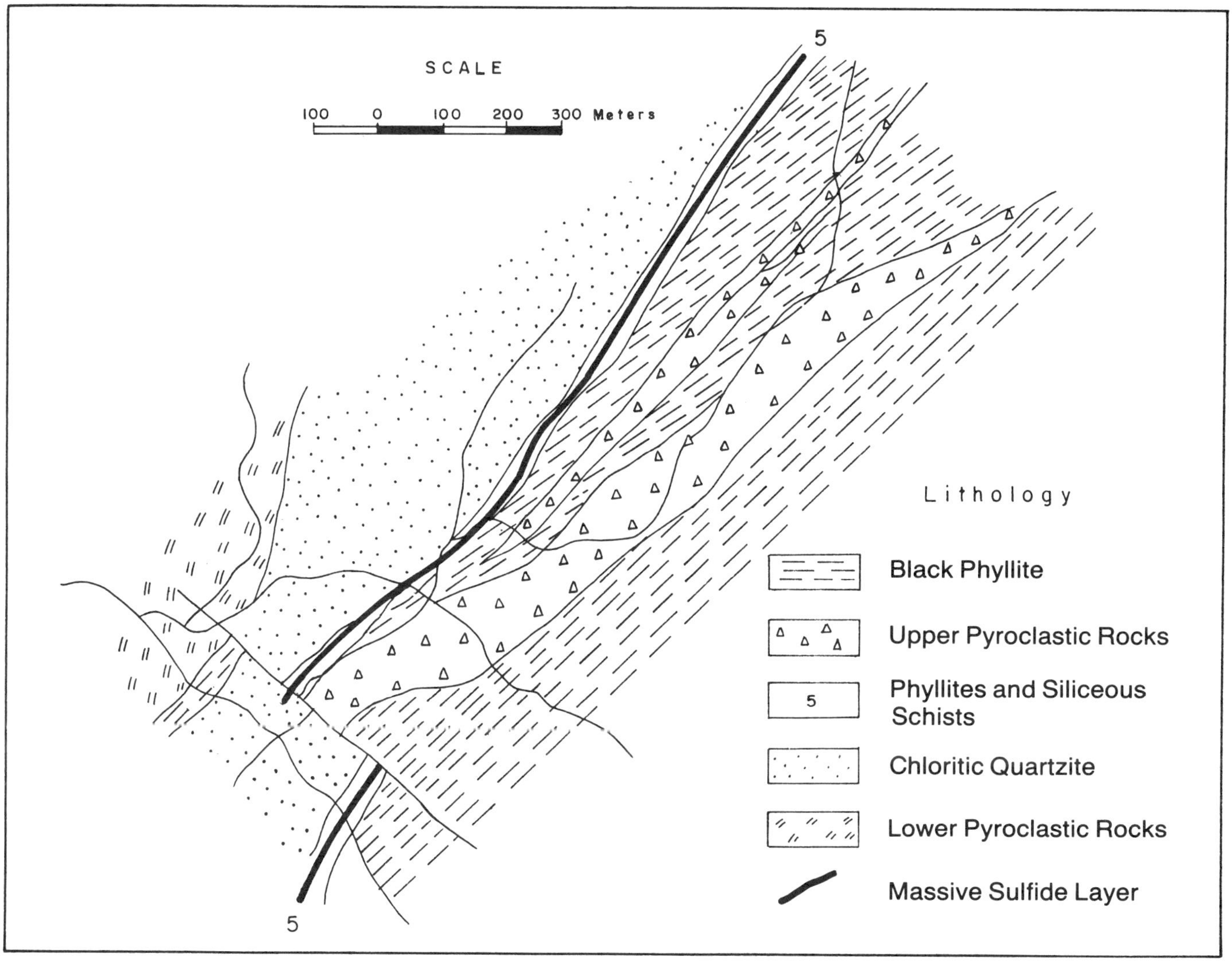

Figure 2. Geologic sketch map showing the Bailadores sulphide deposits, Estado Mérida. (After Carlson, 1977.)

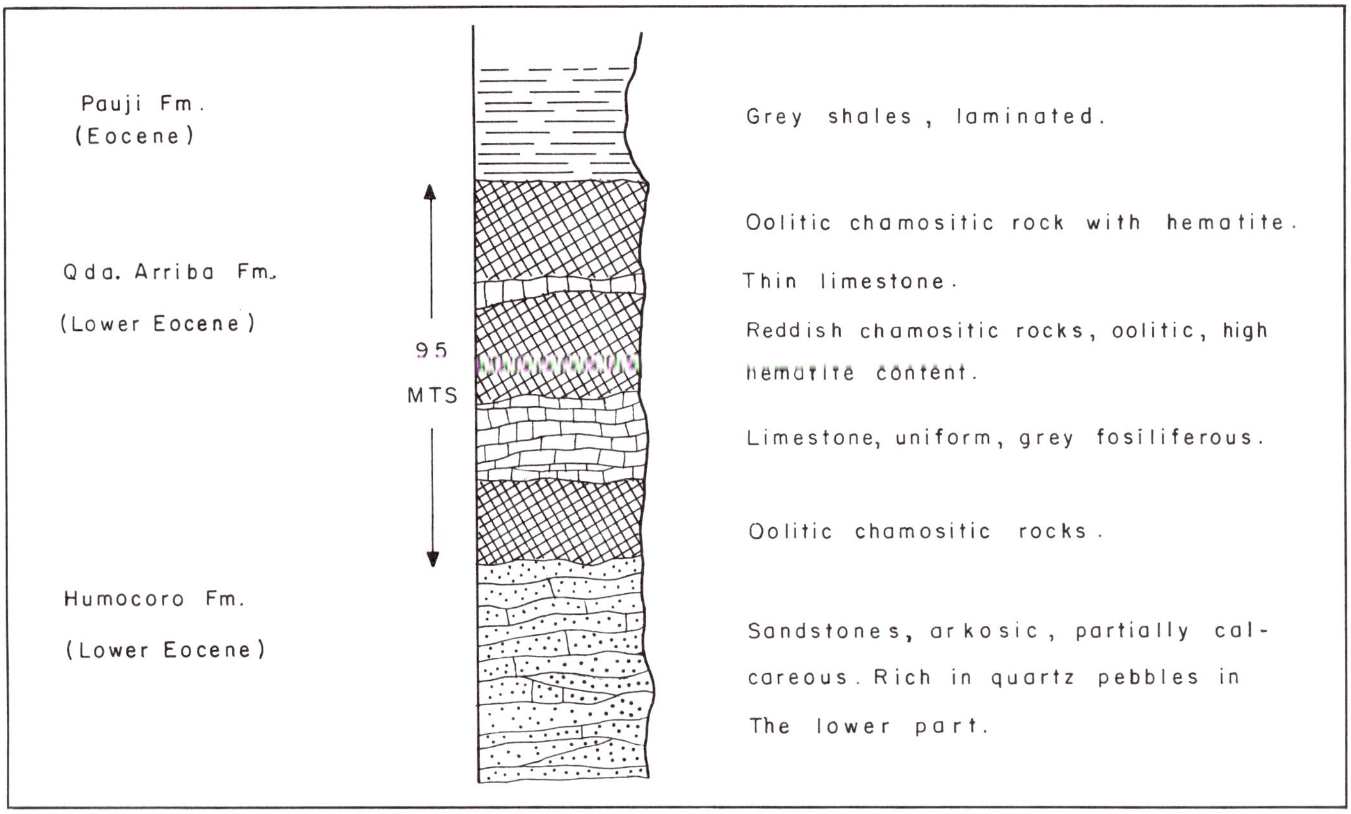

Figure 3. Columnar section showing the distribution of sedimentary iron units in Eocene formations in the southern part of Estado Lara, Venezuela. (After Muñoz, 1966.)

lead ore, until 1966. Reserves have been estimated at 300,000 tonnes of ore averaging 7% lead and 90 grams of silver per tonne.

Uranium

Uranium deposits, near the city of San Cristobal, Táchira (Figure 1), are in the Triassic-Jurassic La Quinta Formation, which consists of a thick sequence of red clastic rocks and carbonaceous shales. The richest deposits, near the towns of Los Caños and El Viso (Figure 4), contain as much as 2% uranium, 1% copper, and 100 ppm silver per tonne. Uraninite, the principal uranium mineral, is generally in veinlets.

Uranium is also present in the La Blanca area and the Indio-Mesa de Chaucha area, both near San Cristobal (Figure 1). The La Quinta Formation in the La Blanca area contains four uraniferous horizons, whereas uraniferous material at Indio-Mesa occurs in a 1.5-m unit of gray carbonaceous mudstone, conglomerate, and asphaltite. Chemical analyses of 26 samples taken along this unit show an average uranium content of 0.338%. The uranium mineral is pitchblende (Cárdenas, 1980).

NONMETALLIFEROUS DEPOSITS

Phosphorite

Phosphorite deposits are known in three areas in the Venezuelan Andes (Figure 1): (a) central Mérida; (b) eastern Táchira; and (c) southern Táchira. The principal deposits, near the town of Jají, central Mérida (Figure 1), are in the upper Tres Esquinas member of the middle Cretaceous La Luna Formation. The deposits have a relatively high content of potassium and sulfur. Regional studies indicate resources of more than 150 million tonnes of phosphate rock having a P_2O_5, content of 18 to 28% (Useche, 1985). In the region of San Joaquín de Navay (Figure 5), eastern Táchira, phosphorite is found in the Navay Formation, which correlates with the La Luna Formation of central Mérida. The phosphorite layers, which consist of fluorapatite oolites in a fine-grained cherty matrix, are interbedded with white chert and siliceous shale. An estimated 80 million tonnes of phosphorite, containing 13 to 25% P_2O_5, is available to open-pit mining (Cárdenas, 1985). In western Táchira, small, high-grade phosphorite deposits are found in the middle part of the La Luna Formation. The deposits consist of tectonically deformed lenses of purple oolitic phosphorite, each containing 100,000 to 2,000,000 tonnes of phosphatic rock. Total phosphorite resources are not known, but may be large. During the 1980s, several of these deposits have been exploited in open-pit mines.

Gypsum

High-quality gypsum occurs in the Lower Cretaceous Río Negro Formation exposed in the Andean region of central Táchira (Figure 1). This formation consists chiefly of a sequence of clastic sediments. Gypsum forms layers 2 to 10 m thick that were deposited locally when "sabhka" environments prevailed. As the result of regional deformation the

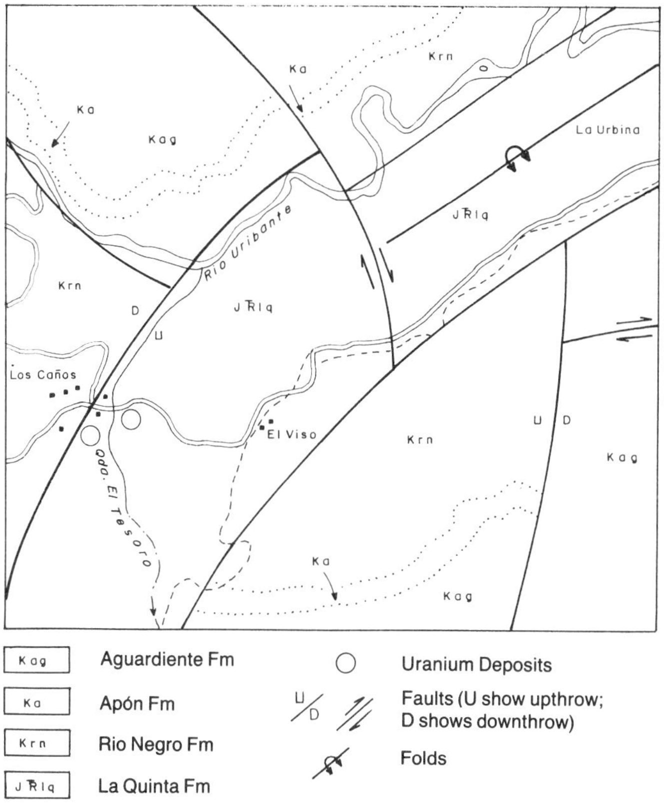

Figure 4. Uranium deposits, Los Caños-El Viso areas, Estado Táchira (Cárdenas, 1980).

gypsum layers are folded and faulted, making mining difficult. Typically, layers consist of 90% or more gypsum. An estimated 20 million tonnes or more of gypsum is available in this area.

Silica Sand

Middle Tertiary sedimentary rocks in the central part of Trujillo (Figure 1) contain large resources of high-quality silica sand. In the 1980s, the deposits of this area were widely exploited, in open-pit mines, for silica sand used by the cement, glass, ceramic, and foundry industries. Silica sand also is present in Tertiary sedimentary rocks in the southern part of Estado Táchira.

Flint Clay

The flint clay deposits of the Venezuelan Andes are found in lower Tertiary sedimentary rocks in western Táchira, especially in coal mining areas near the city of Rubio (Figure 1). They are 2 to 7-m-thick underclays of subbituminous coal beds. The clays contain 30 to 37% Al_2O_3 and are suitable for the manufacture of refractory brick. It is estimated that more than 10 million tonnes of flint clay in this region is available for open-pit mining.

Asphalt

Large deposits of natural asphalt are present in beds of Tertiary sandstone in the southern Venezuelan Andes of southwestern Táchira (Figure 1). These deposits formed where reactivated deep-seated faults brought the Tertiary sandstone into contact with Cretaceous petroleum reservoir rocks, after which the petroleum seeped into the sandstone to form the asphalt deposits. Regional studies indicate that more than 100 million tonnes of natural asphalt material is present in this area. Several deposits are currently (1985) being mined in open pits.

Feldspar

The Venezuelan Andes have many commercial deposits of feldspar, and a large part of the Venezuelan feldspar production comes from these deposits. Two types of deposits are present: (1) large Precambrian pegmatite bodies in eastern Mérida, and (2) extensive, feldspar-rich pegmatite dikes in the northern Estado Barinas (Figure 1). The Pena Altas pegmatites, north of Apartaderos in central Mérida, are large bodies consisting chiefly of feldspar, quartz, and mica. Estimated reserves are more than 5 million tonnes having an average chemical composition of 75.09% SiO_2, 16.61% Al_2O_3, 3.15% NaO_2, and 4.50% K_2O. The feldspar-rich dikes of Estado Barinas intrude metamorphic rocks of Precambrian age. Resources are large, and deposits have been exploited as sources of feldspar for the ceramic industry of central Venezuela (Schubert, 1962).

Coal

The Venezuelan Andes contain significant resources of coal, including coking coal, in the Carbonera Formation of early Tertiary age in the southern part of Táchira (Figure 1). Coal beds are as much as 4.7 m thick (Figure 6), and some of the coal can be mined by open-pit methods. The western area, shown in Figure 1, contains an estimated 115 million tonnes or more of coal (Useche, 1985), whereas the eastern area contains 220 million tonnes or more. The coals are subbituminous and are suitable for thermic use (Tadeusz and Renz, 1985).

REFERENCES CITED

Cárdenas, H., 1980, Resultado de la exploración para depósitos de uranio en el Estado Táchira: Caracas, Venezuela, Informe Interno 000483, División de Geofísica, Dirección de Geología, Ministerio de Energía y Minas, 112 p.

Cárdenas, H., 1985, Los depósitos de fosfatos arenosos de la Formación Navay, Táchira Sur-Oriental: Memoria VI Congreso Geológico Venezolano, Caracas, 1985. Sociedad Venezolana de Geologos, Tomo VI, p. 3854-3892.

Carlson, G.G., 1977, Geology of the Bailadores, Venezuela, massive sulfide deposit: Economic Geology, v. 72, no. 6, p. 1131-1142.

Muñoz, N., 1966, Descripción de las Formaciones Humocaro y Quebrada Arriba en sus localidades tipo de los Andes y Lara: Boletín Informativo, Asociación Venezolana de Geología, Minería y Petróleo, v. 9, no. 5, p. 134-146.

Nogueira, A., A. Santana, E. Lavie, and E. Garcia, 1972, El yacimiento de plomo, zinc y cobre de Bailadores, Mérida, Venezuela: Caracas, Venezuela, Boletín de Geología, Ministerio de Energía y Minas, Publicación Especial 5, p. 2728-2748.

Rodriguez, S., 1978, Depositos sedimentarios de cobre asociados con la Formación La Quinta, Estado Táchira, Venezuela: Memoria del II Congreso Colombiano de Geología, Bogota, Colombia, 1978, Instituto Nacional de Investigaciones Geológico-Mineras, 12 p.

Rodriguez, S., D. Rodriguez, C. Velasco, and N. Torrealba, 1977,

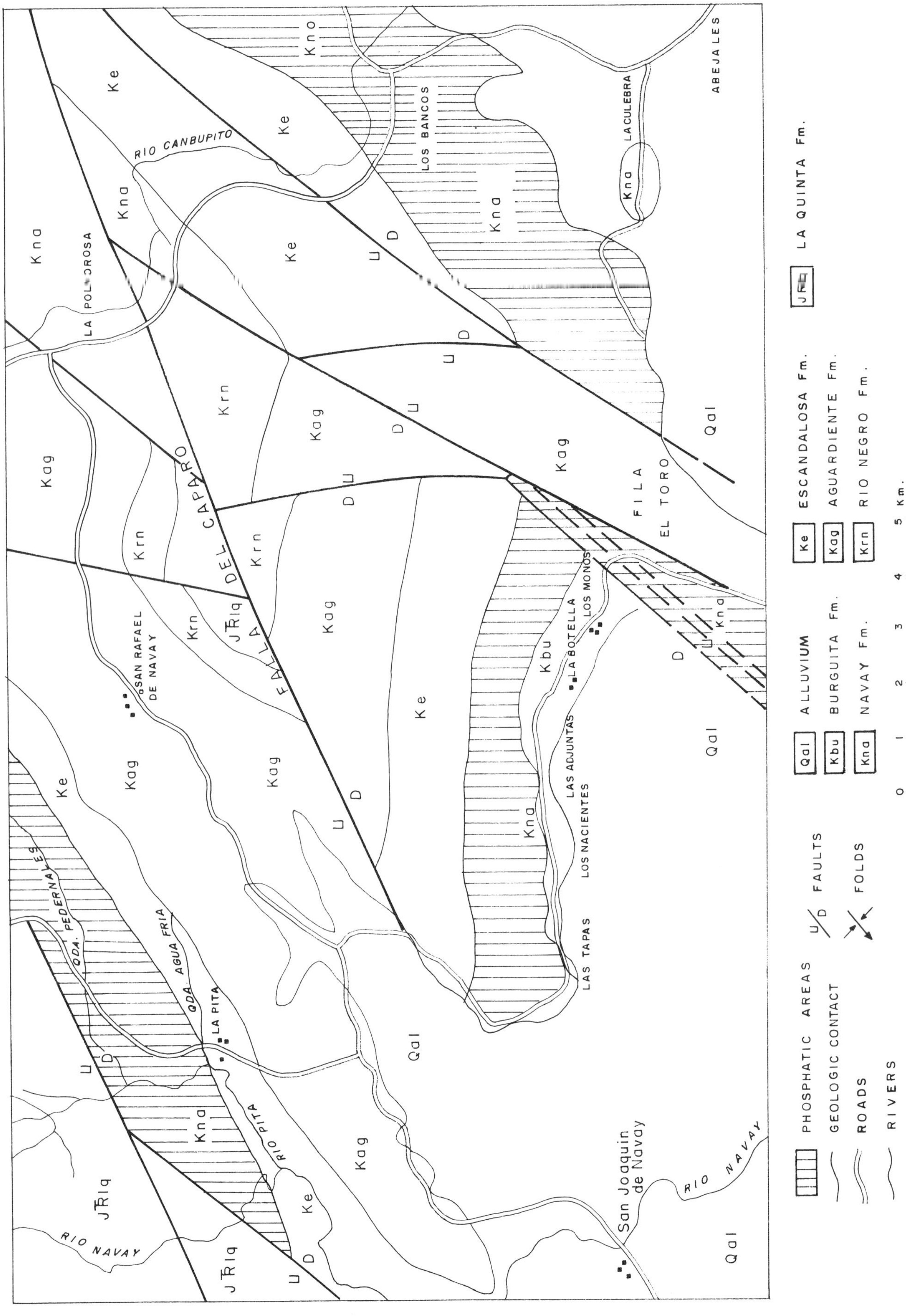

Figure 5. Geology of the phosphorite belt of San Joaquin de Navay, Estado Táchira (Cárdenas, 1985).

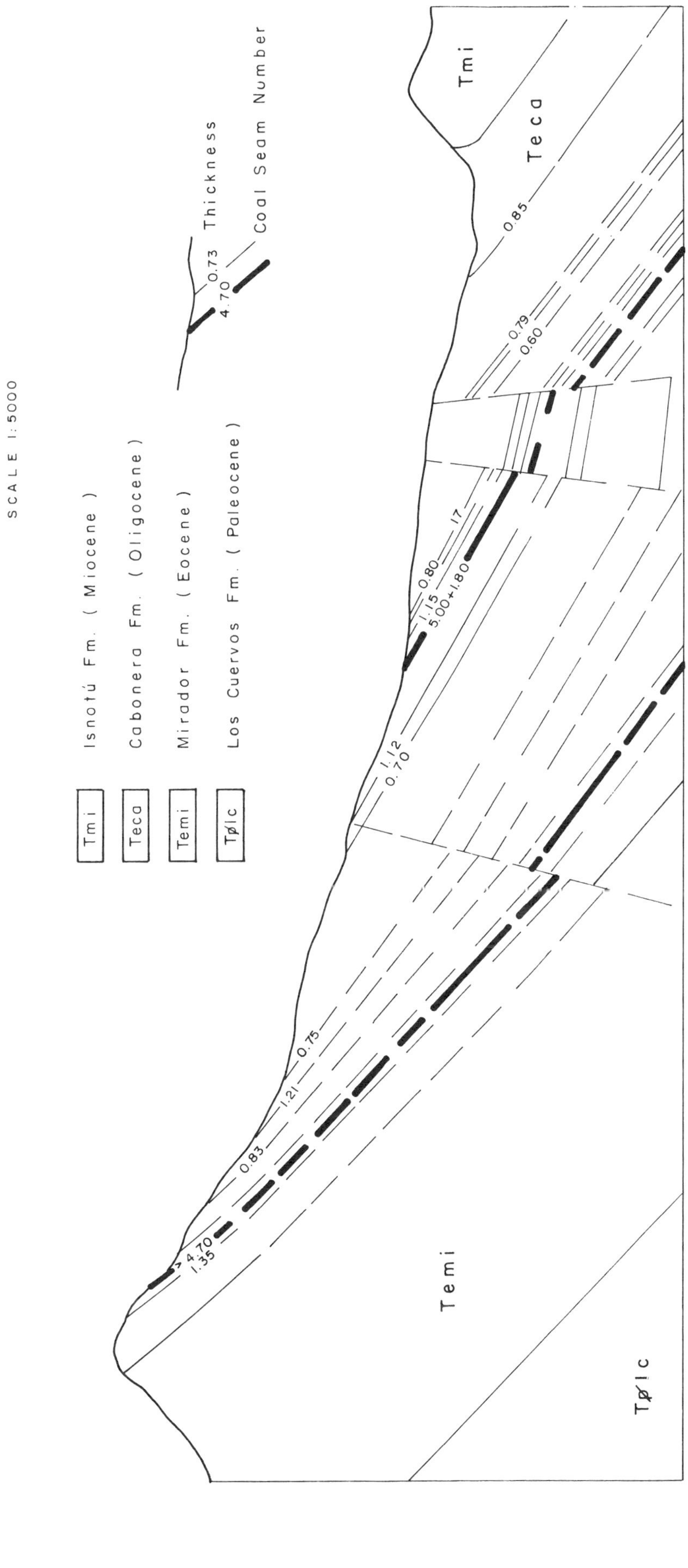

Figure 6. Geologic section showing coal deposits at Quebrada Pajuila, Santo Domingo area, Estado Táchira.

Mineralogía, génesis e importancia económica de los depósitos de hierro sedimentario de Lara Sur-Occidental, Venezuela, *in* A. Espejo, ed., Memoria V Congreso Geológico Venezolano: Caracas, Venezuela, Tomo III, p. 1101-1119.

Schubert, C., 1962, Geología de la región de Barinitas-Santo Domingo, Andes Venezolanos surorientales: Caracas, Boletín Geológico, v. 9, no. 19, p. 182-261.

Tadeusz, B., and R. Renz, 1985, Yacimiento de carbón en Santo Domingo, Táchira, Venezuela, *in* Memoria VI Congreso Geológico Venezolano, Caracas, 1985: Sociedad Venezolana de Geológos, Tomo VI, p. 3799-3832.

Useche, A., 1985, Estudio Geológico preliminar, Depósito de roca fosfática, Estado Mérida, *in Memoria* VI Congreso Geologico Venezolano, Caracas, 1985: Sociedad Venezolana de Geológos, Tomo II, p. 4395-4431.

Woznessensky, B., 1972, Investigaciones metalogénicas en los Andes Venezolanos: Caracas, Boletín Geológico, Publicatión Especial, v. 5, no. 5, p. 2827-2844.

Ericksen, G. E., Cañas Pinochet, M. T., and Reinemund, J. A., editors, 1989, Geology of the Andes and its relation to hydrocarbon and mineral resources: Houston, Texas, Circum-Pacific Council for Energy and Mineral Resources Earth Science Series, v. 11.

CHAPTER 19

DISTRIBUTION AND CHARACTERISTICS OF CHILEAN COPPER DEPOSITS

Carlos Ruiz Fuller
Department of Geology
University of Chile
Santiago, Chile

ABSTRACT

Almost all Chilean copper resources are found in three types of deposits: porphyry copper deposits, stratabound deposits, and vein deposits. The largest copper resources of Chile are those in the porphyry copper deposits, which account for 83% of the national copper production. These deposits, which include recoverable amounts of molybdenum, are distributed through a 2,300 km longitudinal belt from Perú to lat. 38°45′ S. They consist of mineralized stockworks or intrusive breccia bodies in hydrothermally altered tonalite to monzonite quartz porphyry stocks whose age ranges from Eocene to Pliocene. Pyrite, chalcopyrite, bornite, and molybdenite are the principal primary ore minerals. In the majority of cases, secondary enrichment processes have been of prime importance in formation of porphyry copper ores, giving rise to large amounts of antlerite, atacamite, brochantite, and other secondary minerals in the oxide zone, and chalcocite and covellite in the sulfide enrichment zone. The hydrothermal alteration associated with the mineralization was intense and widespread, with the quartz-sericitic, argillic potassic, and propylitic phases being most important.

The principal stratabound deposits, representing the second largest source of copper in Chile, are found in andesitic to rhyolitic volcanic rocks and associated sedimentary rocks. The principal copper minerals are bornite and chalcopyrite. The two most important deposits, Mantos Blancos, near the city of Antofagasta, and El Soldado, to the northwest of Santiago, have reserves of tenths of millions of tonnes.

Vein-type deposits are numerous and widespread, and although their current production is relatively small, they were the principal type of deposits exploited during the 19th century when Chile was the world's largest copper producer. Of the largest vein deposits, Carrizal Alto originally had an estimated 3 million tonnes of 5% copper ore, and Tamaya 2 million tonnes of estimated 12% copper ore. The ores also contain small amounts of recoverable gold.

INTRODUCTION

The first copper production known in Chile dates back to 1601, but production did not gain any real importance until 1844 when Lambert introduced the metallurgy applicable to the fusion of copper sulfide minerals, thereby allowing an increase in production to 10,000 tonnes of fine copper per year. Production then progressively increased and reached a maximum of approximately 50,000 tonnes per year between 1860 and 1876, when Chile became the world's great copper producer. A large part of this production came from the large vein deposits at Tamaya and Carrizal Alto. In 1876, the production began to decline slowly because of depletion of the high-grade vein deposits, chiefly in the Coastal Cordillera, and by 1910 production had declined to 20,000 tonnes per year.

The era of exploitation of porphyry copper deposits began in 1912-1915 when the Braden Company and the Chile Exploration Company started producing copper from the huge El Teniente (1912) and Chuquicamata (1915) deposits. A record production of 102,000 tonnes of fine copper was reached in 1917. Thereafter, production continued to increase to a million tonnes of fine copper per year by 1982. This increase was due chiefly to successive enlargement of operations at El Teniente and Chuquicamata, and opening of new porphyry copper mines of Potrerillos, El Salvador, and Andina. Exploitation of the stratabound deposits at Mantos Blancos and El Soldado contributed significantly to production in recent years. In 1985, copper production reached 1,360,000 tonnes of fine copper, making Chile the world's largest producer of primary copper.

In 1974, the Copper Corporation of Chile (CODELCO) made an estimate of the identified copper resources of deposits under its control as well as porphyry copper deposits controlled by other companies, which were in various stages of

Table 1. Copper resources of Chile (Ruiz, 1977).

IDENTIFIED ECONOMIC RESOURCES				
Measured (tonnes of ore)	Inferred (tonnes of ore)	Total (tonnes of ore)	Grade (% Cu)	Fine Copper (tonnes)
6,012,000,000	3,253,000,000	9,265,000,000	1.02	94,500,000

IDENTIFIED SUBECONOMIC RESOURCES (based on 1977 economic conditions)							
Paramarginal (tonnes of ore)	Grade (% Cu)	Fine Copper (tonnes)	Submarginal (tonnes of ore)	Grade (% Cu)	Fine Copper (tonnes)	Total (tonnes of ore)	Total Fine Copper (tonnes)
1,970,000,000	0.51	10,030,000	4,625,000,000	0.24	10,870,000	6,595,000,000	20,904,000

exploration. The results of this estimate are summarized in Table 1, which shows a total of 94.5 million tonnes of fine copper in identified economic resources and 20.9 million tonnes of fine copper in identified subeconomic resources. These constitute the world's largest copper resources.

The copper resources shown in Table 1 may be compared with the total amount of copper produced in Chile through the first half of 1985, which is approximately 26 million tonnes of fine copper. Although the porphyry copper deposits of Chile have yielded increasingly large amounts of copper during the last 10 years, which is estimated to be about 8 million tonnes, exploration in and near these deposits, as well as other deposits, has revealed new resources that probably more than compensate for the production of copper during this period. In addition, the porphyry copper metallogenic subprovince (described hereinafter) is far from being totally explored, and undoubtedly many deposits remain to be found.

The stratabound deposits represent the second largest source of Chilean copper, whereas the numerous vein deposits now yield only a small part of Chile's copper production, in contrast with the 19th century when nearly all the copper came from veins.

CHILEAN PORPHYRY COPPER DEPOSITS

Distribution

The main porphyry copper deposits in Chile are distributed through a longitudinal, slightly sinuous north-trending belt, extending from the Chile-Perú boundary in the north to lat. 38°45′S, a longitudinal distance of 2,300 km (Figure 1). As has been pointed out (Ruiz et al., 1965), the position of the porphyry copper belt shows a clear relationship to the Andean geosynclinal basin at its maximum development during the early Jurassic. To the south of Chuquicamata the belt lies just to the east of the basin axis, whereas to the north of Chuquicamata it is near the eastern margin of the basin. Furthermore, the porphyry copper belt lies at or near the transition between the Jurassic eugeosynclinal facies in the west and the miogeosynclinal facies in the east.

Of the Chilean porphyry coppers, Chuquicamata and El Teniente are the largest copper deposits in the world, each containing several billion tonnes of ore. The other deposits currently being exploited are El Salvador, Andina, and Los Bronces. The Potrerillos deposit was exploited until 1959, when ore that could be mined economically became exhausted and the mine was closed. Other porphyry copper deposits of Chile in various stages of exploration are (listed from north to south): Cerro Colorado, Quebrada Blanca, Collahuasi, El Abra, La Escondida, and Los Pelambres (Figure 1).

Geologic Characteristics

Although all the porphyry coppers have been included in the same group, they show remarkable differences with respect to age, composition of the porphyry intrusions with which they are associated, and characteristics of mineralization and hydrothermal alteration. Because of these differences, the Chilean porphyry copper deposits do not fit a specific ore-deposit model. The differences among the four largest deposits of Chuquicamata, El Teniente, Andina-Los Bronces, and El Salvador are shown by the following descriptions.

Chuquicamata

Chuquicamata, located 240 km northeast of Antofagasta (Figure 1), is the largest copper mine in Chile. It produced 549,052 tonnes of copper and 11,867 tonnes of molybdenum in 1985, which includes a relatively small production from the nearby Sur or Exotic mine. It was originally developed by the Anaconda Company, but in 1972 it was nationalized and ownership was transferred to CODELCO.

Chuquicamata consists of a complex of three elongate, north-trending porphyry intrusions that are almost entirely the mineralized host rocks (Figure 2). The mineralized intrusive complex is younger than the Fortuna granodiorite of Tertiary age (38.0 ± 0.3 to 36.2 ± 1.4 Ma), which is exposed to the west of the West Fissure, a major postmineral structure. To the east the complex intrudes the Elena granodiorite, presumably of Jurassic age, and undifferentiated porphyries (Figure 2). The mineralized complex is almost 3,500 m long, more than 1,100 m wide, and extends to a depth of at least 1,500 m below the original surface. The three porphyry bodies are similar in composition, consisting of adamellite or quartz monzonite, but are of different ages. Potassium-argon ages made on biotite concentrates range from 31.8 to 33.8 (± 0.9 to 1.3) Ma. In addition, a K-Ar age made on a sericite concentrate from sericite quartz rock gave 28 Ma (Alvarez, Miranda, and Guzman, 1980). The age of mineralization may span the interval indicated by the ages of the biotite and sericite. The porphyry complex is highly altered and contains both primary and secondary sulfide minerals.

The entire porphyry complex was affected by late-stage magmatic alteration and mineralization. It contains quartz veins with alteration halos of alkaline feldspar in which are

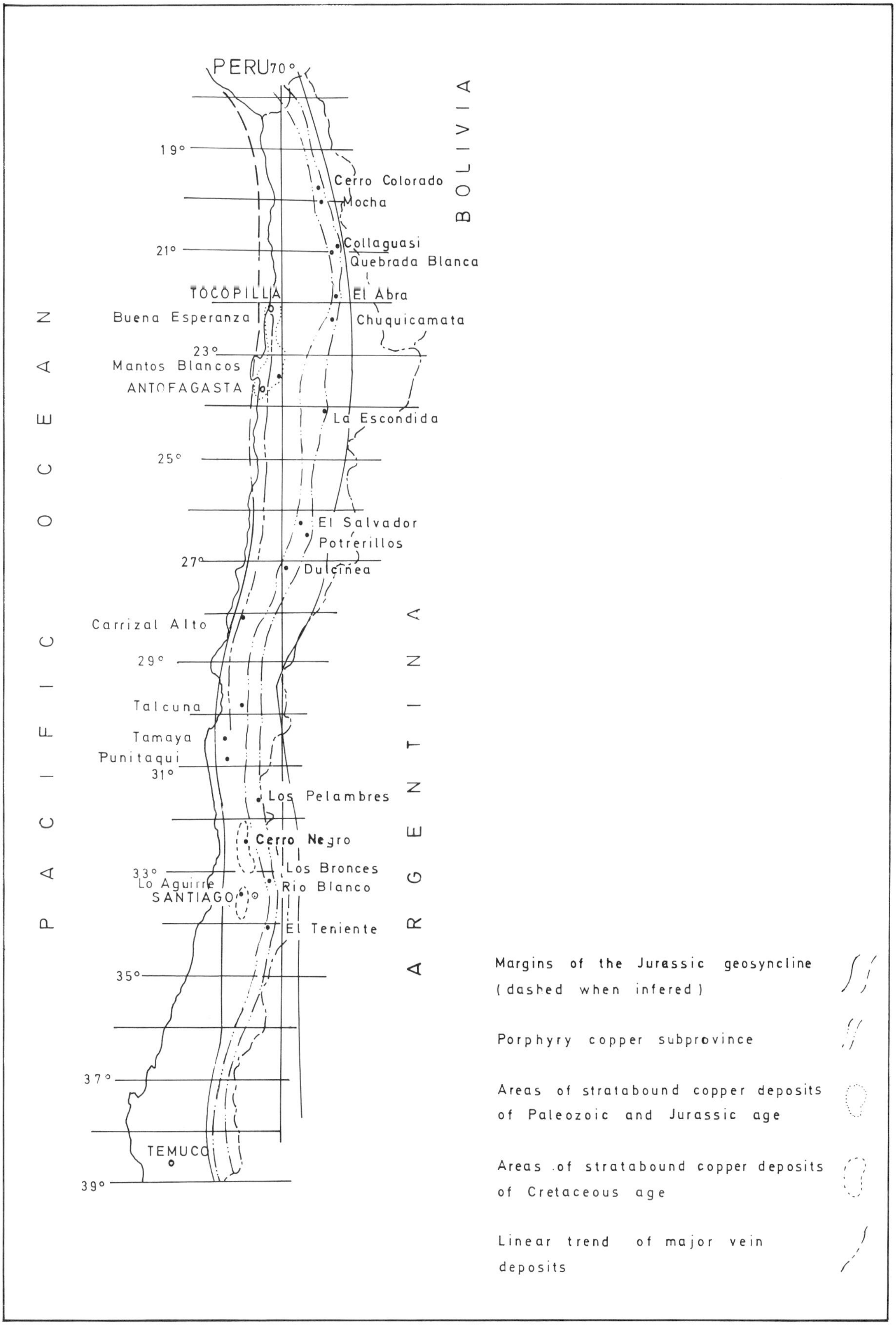

Figure 1. Distribution of the principal copper deposits of Chile.

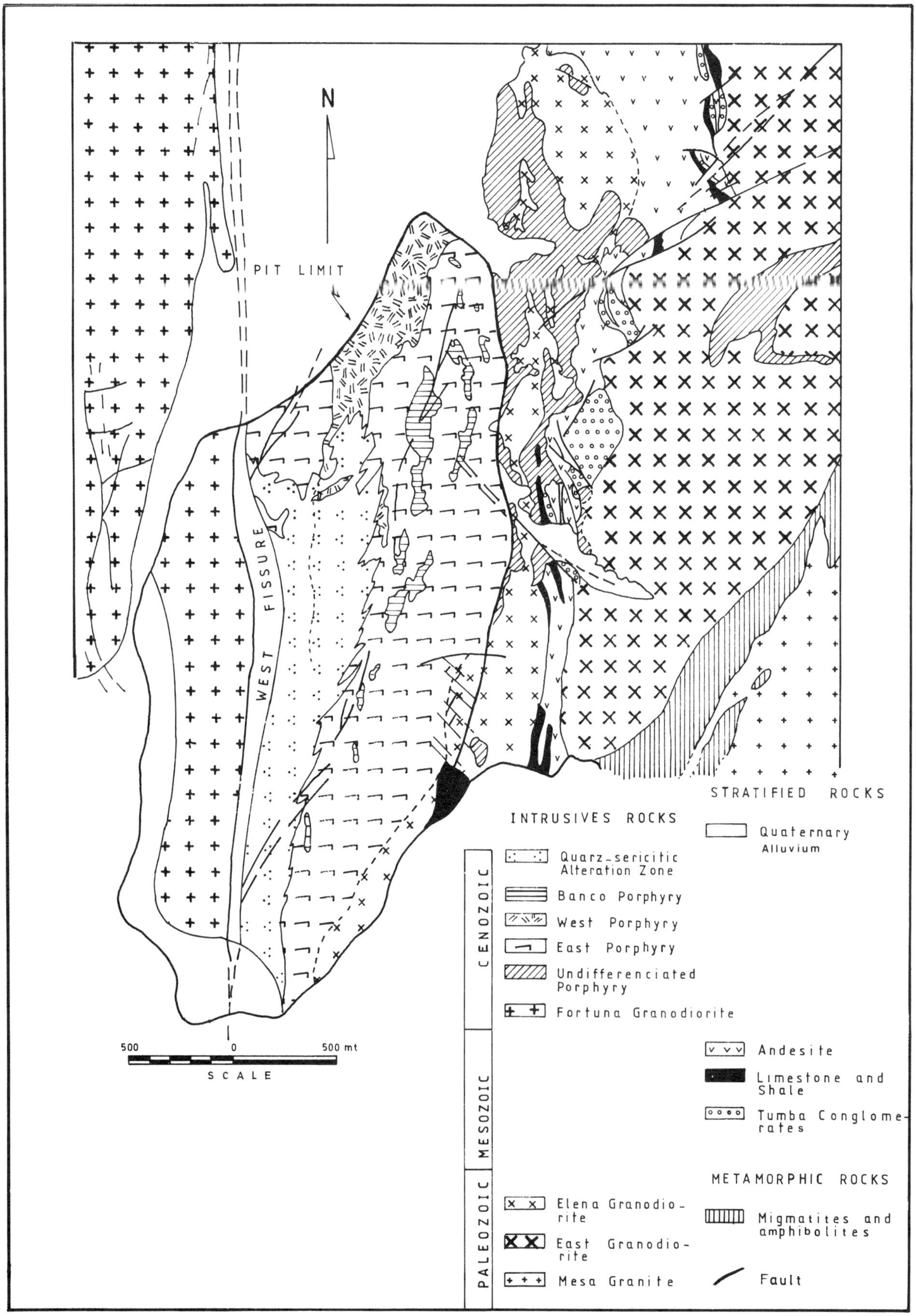

Figure 2. Geologic sketch map of the Chuquicamata mine area.

abundant sulfide minerals such as chalcopyrite, bornite, digenite, and pyrite. The eastern part of the complex shows propylitic alteration.

The sericite-quartz alteration zone is important in that it has the major concentration of secondary copper sulfide minerals, which constitute the richest copper ore. This type of alteration probably followed the late-stage magmatic and propylitic alterations. The zone contains veins and closely spaced veinlets having abundant pyrite, chalcopyrite, molybdenite, and enargite, and smaller amounts of galena, sphalerite, and tennantite. It consists of a large volume of altered and mineralized rock in a longitudinal belt along the West Fissure (Figure 2), which is about 2.5 km long and 100-150 m wide. The eastern margin of the sericite-quartz zone migrates toward the West Fissure at depth, and is still recognizable at a depth of 1,400 m below the surface. The copper minerals have been strongly leached in the upper part of this zone, leaving a variety of iron oxides such as goethite, jarosite, copiapite, and chenevixite. The sulfide enrichment zone, which contains the major portion of the high-grade ore of the deposit, lies in the depth. The secondary sulfides in this zone consist of a chalcocite-dominant upper unit and a covellite-dominant lower unit. The secondary enrichment zone extends to a depth of as much as 750 m, over a longitudinal distance of approximately 3,500 m; the average width is 500 m. Leaching was less intense in the upper parts of the potassic and propylitic alteration zones in the eastern part of the deposit. An enriched zone of copper oxides, which has been almost entirely mined out, occurred in this part of the deposit. The principal copper minerals in this oxide zone are antlerite, chalcanthite, brochantite, atacamite, and turquoise. The oxide ore overlies a sulfide enrichment zone similar to, though smaller than, the sulfide enrichment zone to the west.

Undoubtedly the postmineral West Fissure cuts off part of the Chuquicamata orebody. The West Fissure is a vertical to steeply dipping belt of gray, fine- to medium-grained, laminated mylonite consisting chiefly of sericite, quartz, and sulfide minerals. The mylonitic zone is 1-2 m thick. It separates the mineralized quartz-sericitic zone to the east from the Fortuna granodiorite to the west. The Fortuna granodiorite is fractured but totally sterile. Premineral or synmineral faults occur in the mineralized body to the east of the West Fissure. These faults trend northeastward, some curving toward the east. They have been mineralized, and in the central part of the orebody lose their identity. They seem to have provided the structural control for the formation of mineralized stockworks, and some were the loci of high-grade mineralized bodies (Alvarez, Miranda, and Guzman, 1980). It may be concluded that the faults to the east of the West Fissure were the major controls of mineralization, particularly those related to sericitic alteration. Regional structural control, which might explain the alignment between Chuquicamata, Chuquicamata Norte, and El Abra, could have been a premineral fault that was reactivated as the postmineral West Fissure.

El Teniente

El Teniente porphyry copper deposit (Howell and Malloy, 1960; Enrione et al., unpublished report, 1983), located 75 km south-southeast of Santiago (Figure 1), contains copper resources comparable in size with those of Chuquicamata. However, production from El Teniente is smaller, being only 319,302 tonnes of fine copper and 4,435 tonnes of molybdenum in 1985. El Teniente is currently the world's largest underground copper mine. It originally was developed by the Braden Copper Company; in 1972 it was nationalized and ownership was transferred to CODELCO.

The mineralized body (Figure 3) consists of veins and stockworks in a host consisting chiefly of andesite lava and lesser amounts of tonalite and dacite intrusive rocks. The orebody surrounds a pipelike mass of clastic volcanic material, the so-called Braden chimney or Braden Formation, of which a marginal zone consists of tourmalinized breccia (Figure 3). This chimney is cone-shaped, and the clastic fill probably resulted from an explosive-type eruption in a volcanic vent. The shape of the ore deposit in plan, as shown in the Teniente 5 level (Figure 3), one of the main mine levels, is similar to the longitudinal section of a pear elongated in a northwest direction. On this level, the body is 2,900 m long and its maximum width, in the south, is 1,700 m. The Braden chimney on this level is circular, having a diameter of 900 m. The margins of the 0.5% Cu body are vertical and extend to depths of 1,000 m or more below the surface. Mineralization decreases gradually in intensity with depth, and exploitable mineralized rock terminates at approximately 1,500 m below the surface.

The host rocks consist of: (1) a sequence of andesitic rocks of probable early Miocene age, the Farellones Formation, which has been subdivided into three members (Howell and Malloy, 1960) and which is cut by tonalite and dacite porphyry bodies and several minor dikes; and (2) a tourmaline breccia pipe that was partly destroyed by the explosive eruption that formed the Braden chimney. Only a marginal ring remains of the original tourmalinized breccia pipe (Figure 3).

The dacite porphyry occurs as a vertical tabular body to the north of the Braden chimney. It is about 1,500 m long and has an average width of 200 m. The southern tip is cut off by the marginal ring of the tourmaline breccia pipe (Figure 3).

The remnant of the original tourmaline breccia pipe consists of a marginal ring 5 to 60 m wide of dense tourmalinized and silicified breccia containing chalcopyrite. Also present are anhydrite, molybdenite, and copper sulfosalts. The presence of mineralized clasts indicates that the tourmalinized breccia is younger than the main mineralization, whereas the presence of fragments of this breccia in the Braden Formation indicates that it is older than this formation.

The Braden Formation is a pipelike breccia body located somewhat eccentrically with respect to the El Teniente orebody (Figure 3). In plan, it is almost circular, having a diameter ranging from 1,200 m at the surface to 650 m at depth. Its vertical extent, as shown by drilling, is 1,800 m. It extends to still greater depth as a latite neck or plug. The fragmental material in the Braden Formation consists of andesite, dacite, and latite, which are the host rocks of the El Teniente orebody. Some of the latite occurs as large blocks of brecciated material. The matrix consists chiefly of pulverized rock. Concentrically arranged pebble dikes are found around the chimney, especially in the area to the north and west. These dikes formed during the youngest hydrothermal phase.

Much of the El Teniente orebody is a mineralized stockwork, which at places consists of more than 30 veinlets per meter. The stockwork fracturing is most intense in the andesite and less intense in the tonalite and dacite porphyries. The

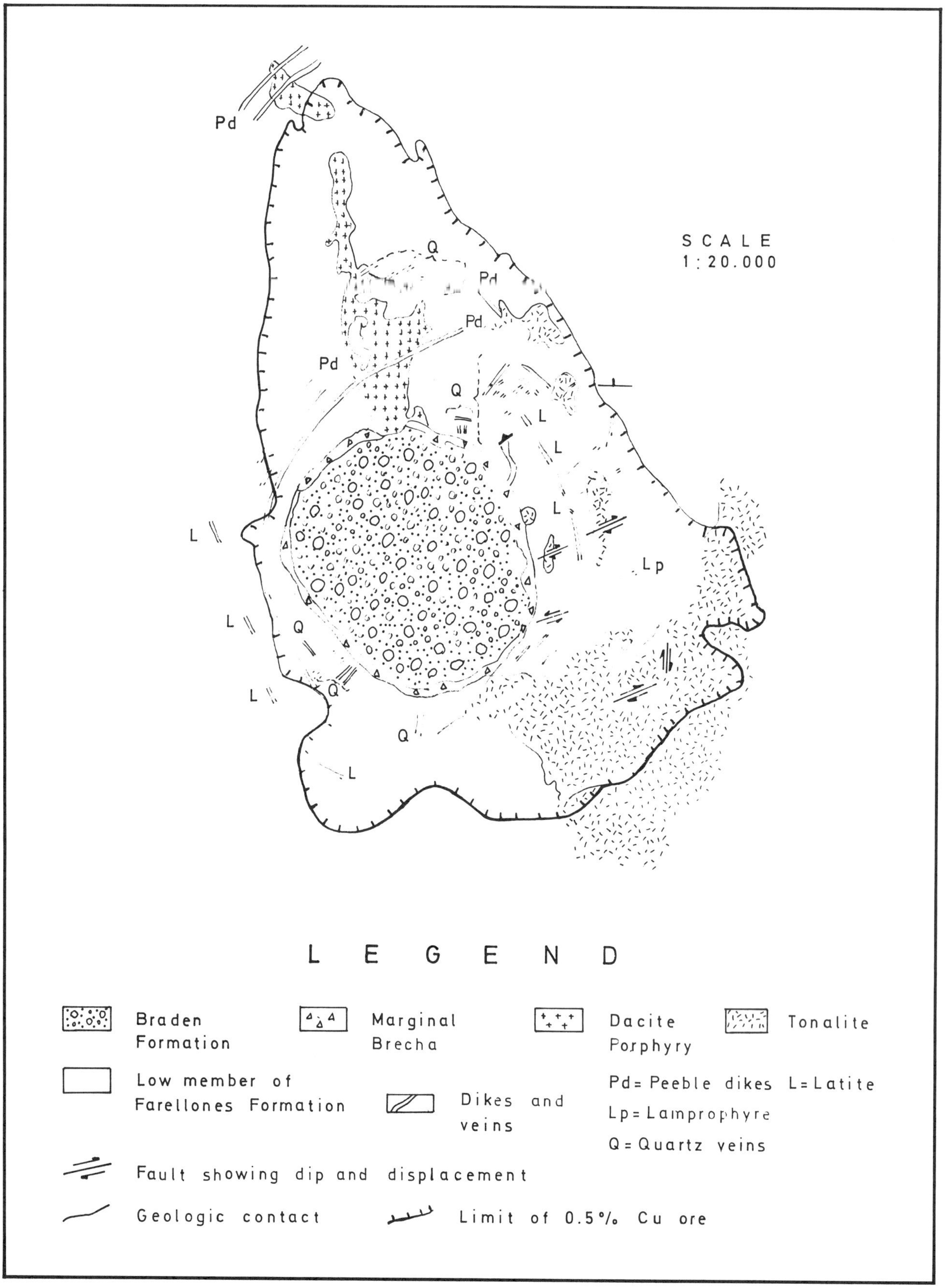

Figure 3. Geologic sketch map of the El Teniente orebody (> 0.5% cu) on level T5, S.E. (From unpublished report by Enrione et al., 1983.)

stockwork fracturing has not affected the Braden chimney.

Late-stage magmatic alteration and mineralization may have caused formation of stockwork veinlets, potassic alteration of the intrusive rocks and andesite, and introduction of bornite, chalcopyrite, molybdenite, and pyrite. During this late magmatic phase irregular veins of quartz, potassic feldspar, and anhydrite formed in the dacite porphyry and tonalite intrusions. Secondary biotite occurs in the andesite unit. Propylitic alteration, represented by concentrations of epidote, tourmaline, magnetite, and specularite, extends to distances of as much as 1,500 m outward from the orebody.

The main hydrothermal stage is characterized by chlorite and quartz-sericite alteration, which is superimposed on the late-stage magmatic alteration, and deposition of chalcopyrite, pyrite, and molybdenite. A still later stage of hydrothermal activity was probably associated with the intrusion of the latite porphyry neck and formation of tourmalinized breccia pipe. Mineralization during this stage consisted of deposition of bornite, chalcopyrite, molybdenite, sphalerite, galena, and base-metal sulfosalts, together with abundant tourmaline, anhydrite, quartz, and calcite. The final hydrothermal stage caused formation of the Braden chimney. Large cavities in this chimney commonly contain crystals of tetrahedrite-tennantite. One 300 m^3 megacavity is lined with large crystals of sphalerite, galena, pyrite, barite, quartz, and apatite.

An oxidized and leached zone of no economic value caps the El Teniente orebody, extending to depths ranging from 40 to 350 m. Beneath this zone is a sulfide enrichment zone, 100 to 500 m thick, in which chalcocite is the principal secondary copper mineral. Below this is a zone, from which anhydrite has not been leached, constituting the primary mineral zone.

Los Bronces-Rio Blanco

The Los Bronces-Rio Blanco, or Disputada-Andina copper district (Figure 1), is located at the headwaters of Rio Blanco and Rio San Francisco at altitudes ranging from 3,000 to 4,000 m, and lies about 50 km northeast of Santiago. Los Bronces is owned and operated by Exxon Minerals and Andina by CODELCO. In 1985, these two mines produced 154,700 tonnes of fine copper. The following description of this area is taken from Stambuck (1982), and Warnaars (1980), and Warnaars, Holgrem, and Barassi (1985).

The oldest rocks in the area are part of a stratified volcanic sequence corresponding to the lower Miocene Farellones Formation. A large intrusive body, the Rio Blanco-San Francisco stock, which consists chiefly of granodiorite and quartz monzonite, cuts the Farellones Formation. The Farellones Formation is exposed chiefly near the eastern and southern margins of the district, and also as two roof pendants in the stock. Recent radiometric age dates of the stock (Warnaars, Holgrem, and Barassi, 1985) indicate an age of 20.1 $\pm$ 2 Ma.

Extensive breccia bodies, containing tourmaline, specularite, anhydrite, pyrite, and chalcopyrite, occur in both the pluton and the Farellones andesite. The clasts show quartz-sericitic and at places propylitic alteration. Dikelike dacite porphyry subvolcanic intrusions were emplaced after the breccia bodies. These dikes trend N10W. Both the breccia bodies and the dacite porphyry dikes become narrower with depth. A volcanic neck, which cuts the mineralized body, and a dacite sill are at the northern end of the Andina deposit. A final stage of hydrothermal activity caused formation of an extensive breccia body at the western side of the mineralized complex. This young breccia body has a matrix of pulverized rock and locally has been tourmalinized. Primary copper mineralization of the Los Bronces-Rio Blanco deposit took place within the Rio Blanco-San Francisco stock and nearby Farellones andesite. The host rock of the deposit shows quartz-sericitic to propylitic alteration near the surface and potassic alteration at depth. The quartz-sericite alteration zones contain veins of quartz, pyrite, chalcopyrite, and molybdenite. Intruding the porphyry copper body are several breccias containing abundant tourmaline, specularite, and sulfide minerals. These constitute the largest orebodies currently being exploited. The largest breccia complex is that of the Los Bronces area in the northwestern part of the district. It is elliptical, being 2 km long in a north-northwestern direction and a maximum of 700 m wide. The complex consists of seven individual breccia bodies, six of which are in quartz monzonite and the seventh in Farellones andesite. In general, the breccia clasts are variable in size and are in a matrix of pulverized rock containing tourmaline, quartz, specularite, anhydrite, sericite, chlorite, pyrite, chalcopyrite, and molybdenite. The clasts show hydrothermal alteration, chiefly quartz-sericitic and weak prebreccia mineralization. The primary copper ore is relatively low grade, whereas the enriched ore contains an average of about 0.9% Cu to depths of as much as 500 m.

The breccia body currently being exploited at Andina has the shape of a reversed cone. In the access level it is rectangular, having a maximum length of 300 m and a width of 150 m. Breccia fragments are chiefly of andesite in the upper and middle part, and granodiorite in the lower part, which reflect the contact between andesite and the underlying granodiorite. The matrix consists of pulverized rock containing tourmaline, specularite, pyrite, chalcopyrite, quartz, and chlorite.

A new, strongly mineralized tourmaline breccia body, called Sur-Sur, was recently discovered at Andina beneath a glacier 1.6 km to the south-southeast of the Andina mine. This orebody contains several million tonnes of ore averaging 2% Cu, and in 1985 it was being prepared for open-pit exploitation. The breccia clasts in the orebody consist of quartz monzonite, and the minerals, which are in the matrix, are specularite, tourmaline, chalcopyrite, and pyrite. Secondary chalcocite and chalcanthite are found near the surface. The minerals only partly filled open cavities between angular breccia fragments, an indication that the deposit formed near the surface.

El Salvador

El Salvador porphyry copper deposit, in the north-central part of Chile (Figure 1), was described by Gustafson and Hunt (1975). It is in a terrane characterized by Upper Cretaceous sedimentary and andesitic volcanic rocks and lower Tertiary rhyolitic and andesitic volcanic rocks. Rhyolite domes dated at 50 Ma and more or less contemporary voluminous eruptions of rhyolite and andesite were the earliest manifestations of Tertiary magmatism. A granodioritic porphyry complex in the center of the district, which is closely associated with the main mineralization and alteration, has been dated at 41 Ma. The oldest granodioritic porphyry (X porphyry) of this complex consists of a fine-grained, weakly porphyritic rock.

Thereafter, several feldspar porphyry bodies, characterized by visible phenocrysts of plagioclase and ferromagnesium minerals, were emplaced. The early intrusions are designated as K porphyries, whereas the later intrusions are the L porphyries. The X and K porphyries are strongly altered and mineralized, whereas the L porphyry is relatively fresh and unmineralized. Nearby andesites contain secondary biotite and also have been mineralized. In the lower mine levels this andesite is the major mineralized host rock. The youngest igneous rocks of the district consist of latite sills and pebble dikes that cut all the older rocks.

The earliest alteration and mineralization occurred before the intrusion of the L porphyry, and contributed three fourths of the 5,000,000 tonnes of copper that the deposit is estimated to contain. At this time there were emplaced veins and impregnations of quartz, feldspar, biotite, anhydrite, chalcopyrite, and bornite or chalcopyrite-pyrite. This phase of mineralization was followed by formation of molybdenite-bearing quartz veins as fillings of planar fractures. Still later veins, which cut the earlier veins, contain pyrite and lesser amounts of bornite, chalcopyrite, enargite, sphalerite, and galena.

The El Salvador orebody consists of about 300 million tonnes of secondarily enriched ore averaging about 1.6% Cu. This blanketlike body of enriched ore ranges from few meters to about 300 m in thickness. The upper limit of ore is well defined, but the lower limit is gradational with the underlying protore. In enriched ore, chalcocite, and covellite show pervasive replacement of bornite and chalcopyrite and weak replacement of pyrite.

STRATABOUND COPPER DEPOSITS

The many stratabound copper deposits of Chile have a relatively simple mineralogy consisting chiefly of chalcopyrite, bornite, and pyrite, and lesser amounts of primary chalcocite, covellite, and specularite. Many of these deposits are stratiform; others, including the most important, have irregular shapes, although they are everywhere limited to a specific stratigraphic unit of either volcanic or sedimentary rock. Copper sulfide minerals occur chiefly as disseminated grains, sulfide veinlets, and as cementing material in volcanic or tectonic breccias. The stratiform deposits range from few meters to several dozen meters in thickness. If the stratabound body has an irregular shape as, for example, the El Soldado deposit, all three dimensions are highly variable.

Many stratabound copper deposits are in volcanic, locally hypabyssal rocks, chiefly andesite, dacite, and trachyte lavas and breccias and rhyolitic ignimbrites. In many of the deposits occurring in lavas, the copper minerals are in the upper amygdaloidal part of the lava flow as fillings of vesicles and small fractures as well as disseminated grains in the more dense lavas. Buena Esperanza and Guayacan (Figures 1, 4) are examples of this type of deposit. Some of the largest stratabound copper deposits such as the Mantos Blancos, El Soldado, and Lo Aguirre deposits (Figures 1, 4) consist of disseminations of sulfide minerals and veins in irregularly shaped orebodies.

Other stratiform deposits are in sequences of volcaniclastic sedimentary rocks. The Cerro Negro and Talcuna deposits (Figure 1) are examples. The host rocks consist of fine- to coarse-grained volcanic sandstone or sedimentary volcanic breccia. The clasts consist chiefly of porphyritic andesite lava and the matrix, if any, is calcitic. If matrix material is absent, as is the case of the Cerro Negro deposit (Figure 1), the rounded clasts appeared to be welded. The ores consist of finely disseminated bornite and chalcopyrite either in the volcanic clasts or in the calcitic matrix, as at the Talcuna deposit. Several copper deposits of this type also are found in skarns formed from marine limestones. The copper in these deposits was probably present in the limestones before they were metamorphosed to skarn. Other stratiform deposits consist of oxidized copper minerals in continental rocks without volcanic affiliation. The most important are the [illegible] copper deposits, where the copper was leached from a primary deposit and redeposited as a cement in gravels and sandstones and as fracture fillings in underlying bedrock. The most important deposit of this type is the Exotic or Mina Sur deposit downslope from the Chuquicamata deposit.

Mantos Blancos

The Mantos Blancos deposit (Figure 1) (Chavez, 1985; Department of Geology, Mantos Blancos Mining Corporation) is located 45 km northeast of the city of Antofagasta. From 1961, when open-pit exploitation of oxidized copper ore started, until 1985, more than 45 million tonnes of oxidized ore averaging 1.65% Cu have been extracted. As the result of discoveries of large amounts of sulfide ore at depth, it was decided in 1974 to open an underground mine to exploit this ore. In 1985, the sulfide ore being exploited had an average grade of 1.9% Cu and 22 gr/t of silver[1]. It is estimated that the total reserves of the district exceed 1,600,000 tonnes of fine copper, contained in more than 100 million tonnes of ore. The current production (1985) is 70,000 tonnes of fine copper per year.

The Mantos Blancos host rock is a sequence of andesitic to rhyolitic volcanic rocks having a minimum thickness of 500 m. The volcanic sequences show sodium metasomatism, which caused formation of albitic ($An_{0\text{-}2}$) feldspar. The principal ores occur as lenticular bodies in the eastern part of the mine area, which contain copper and copper-iron sulfides as disseminated grains and, at places, small veins. The minerals are distributed through a 300-m thick sequence of volcanic rocks. The main units of sequence, from bottom to top, are: (1) quartz dacite, a homogeneous-to-layered rock containing abundant quartz eyes; (2) porphyritic dacite lavas having few to abundant phenocrysts of albitic plagioclase and microcrystals of quartz; and (3) an upper purple andesite, which is locally vesicular and fragmented. The copper minerals occur only in units 1 and 2.

The intense alteration of the volcanic sequence has inhibited direct radiometric age determinations. However, a K-Ar age determination of amphibole from a postmineral and postalteration andesite sill indicates a minimum age for the main phase of mineralization to be Late Jurassic. The volcanic host rocks have been correlated with the Jurassic La Negra Formation, but in lithology and texture the rocks more closely

[1]The silver (Chavez, 1985) is contained as a solid solution in the copper minerals, especially covellite, bornite, and chalcocite.

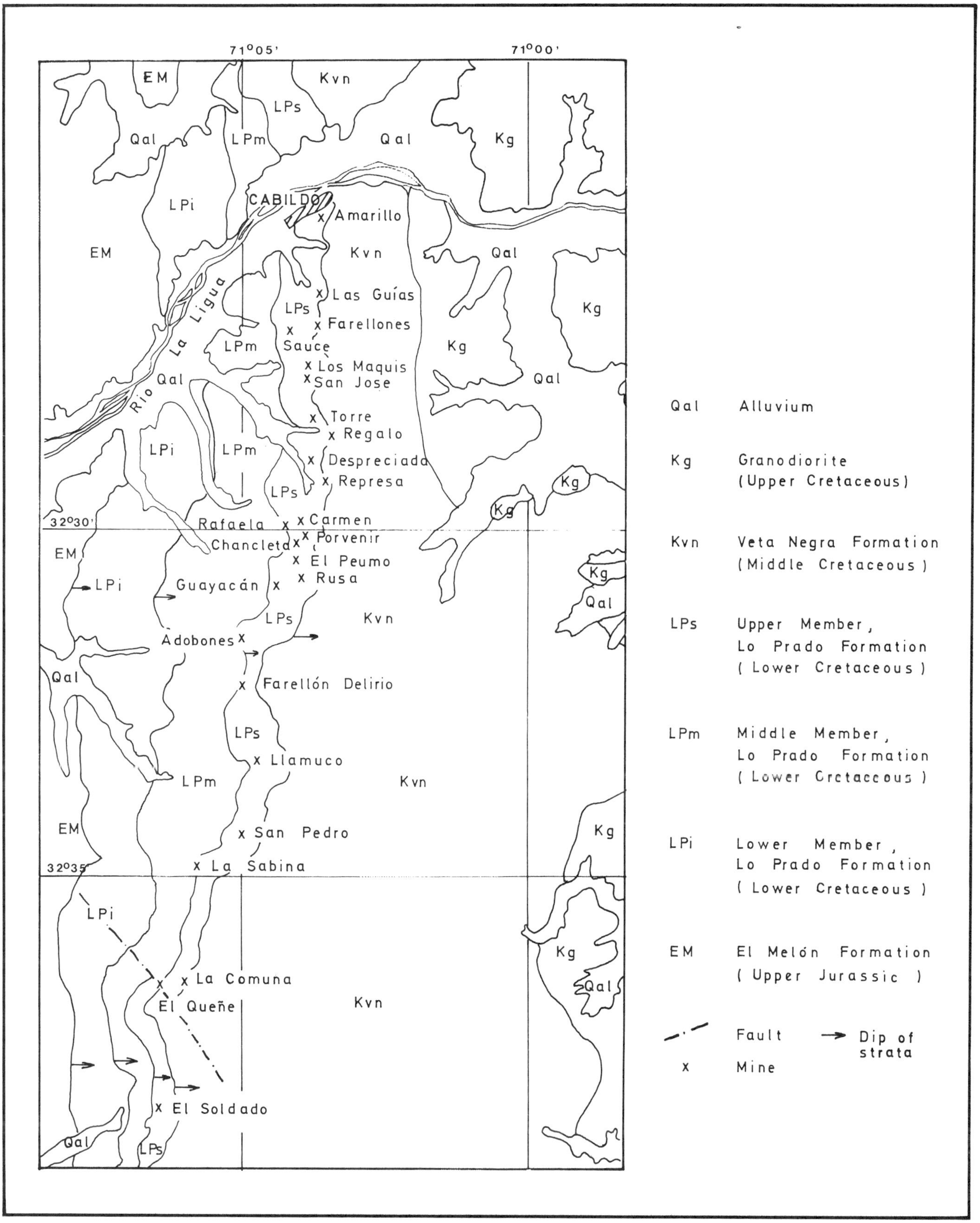

Figure 4. Stratabound copper deposits in the upper member of the Lo Prado Formation. (Compiled from unpublished geologic quadrangle maps by W. D. Carter et al., 1961.)

resemble Triassic or upper Paleozoic volcanic rocks of this region. The volcanic rocks of the Mantos Blancos district are cut by many postmineral faults having trends ranging from N40 °E to N25 °W; most dip to the east. Most of the faults show normal displacements, and blocks between faults are inclined gently toward the west.

Four main orebodies are present at Mantos Blancos, the two principal bodies being Marina and Nora. The Marina body is the largest of the district and consists almost entirely of mineralized quartz dacite or porphyritic dacite, suggesting lithologic control of the mineralization. The Nora body is in the same stratigraphic unit, but probably separated from the Marina body by a fault.

The Marina body shows a vertical zonation of sulfide mineralization, which from top to bottom is as follows: (1) chalcocite, digenite, and hematite; (2) digenite, covellite, and hematite; (4) bornite and chalcopyrite; and (5) chalcopyrite, galena, and pyrite. Paragenetic studies show that the first copper minerals to be formed were bornite and chalcopyrite, which were converted to digenite, chalcocite, and covellite. Hematite was exsolved during hypogene alteration of the primary sulfide minerals.

Other Jurassic Deposits

Several stratabound copper deposits occur in andesites of the Jurassic La Negra Formation in the coastal region between Antofagasta and Tocopilla (Figure 1). This formation consists of several thousand meters of andesitic lava flows ranging from several meters to about 50 m in thickness. Sulfide minerals are concentrated in the upper amygdaloidal and fragmental part of each lava flow, forming successive stratiform bodies in which the main primary minerals are chalcocite and bornite. The dense underlying parts of lava units are weakly mineralized. During subsequent deformation, possibly because of emplacement of volcanic necks, the mineralized layers were partially fragmented and then recemented by remobilized copper minerals.

In a region about 100 km northwest of Santiago, many stratabound copper deposits occur along a longitudinal belt about 30 km long (Figure 4). Most of these deposits are associated with a sequence of andesitic and trachytic lavas and interbedded marine limestone, which correspond to the upper member of the Lower Cretaceous Lo Prado Formation. Many of the deposits consist of mineralized amygdaloidal upper parts of andesite lava flows. A typical deposit is that at the Guayacan mine, which is part of a group of deposits along the same volcanic horizon for a distance of about 5 km. The mineralized unit at Guayacan shows a maximum thickness of 3 m. The deposit consists chiefly of bornite and lesser amounts of chalcopyrite and chalcocite occurring as disseminated fine grains in dense lava and as fillings of vesicles and fractures. The upper parts of some lavas are cut by stream channels that are filled with sparsely mineralized breccia and sandstone overlain by bioclastic marine limestone. Apparently, the upper part of the lavas was mineralized before excavation of the stream channels. Other deposits occurring in the belt consist of irregular mineralized bodies in trachytic and andesitic lavas, which are at the same stratigraphic horizon. El Soldado and La Comuna deposits (Figure 4) are examples.

El Soldado

The following description of the El Soldado deposit has been taken from reports by Holmgren (1985) and Ahumada (1985). The discussion of the origin of this deposit is based on studies by the author of this chapter.

El Soldado (Figure 4) is located 100 km north of Santiago. From 1920 to 1960 it was exploited by the French company M'Zaita, which developed the mine and built a concentrator and smelter. In 1971, the mine was nationalized and later turned over to Exxon Minerals, which carried out an exploration program that resulted in establishing the reserves of several million tonnes of 1.8% Cu ore, and a construction program that enlarged the concentration plant from 3,500 to 5,500 tonnes/day. A new enlargement plan is currently (1985) under way to increase mine and plant capacity to 11,500 tonnes/day. Reserves are estimated to be 70 million tonnes of ore averaging about 1.6% Cu, extractable by underground mining methods. The current production (1985) is 33,000 tonnes of fine copper per year, which may increase to more than 60,000 tonnes in the enlarged operation.

The host rocks of the deposit, as mentioned, correspond to the upper member of the Lo Prado Formation, which ranges from 300 to 1,000 m in thickness. These rocks strike northward and have an average dip of 30 °E. They consist of four trachytic porphyry units, each having a thickness ranging from 25 to 200 m, interbedded with andesites, breccias, tuffs, and thin discontinuous layers of sedimentary rocks. The trachyte is porphyritic, consisting of albite phenocrysts in a groundmass of microcrystalline potassic feldspar, albite, and quartz. Four trachytic feeder dikes have been identified.

The main structural features are three large faults striking north, northwest, and northeast. These faults have been recognized over vertical ranges of several hundred meters, and can be traced for strike distance of 2,000 m or more. Zones of breccia and mylonite along the faults are as much as 3.5 m thick. Some contain copper minerals.

The mineralized body at El Soldado is 1,800 m long, 700 m wide, and extends to depths of as much as 350 m. The distribution of ore is irregular, constituting individual bodies whose width, length, and depth fluctuate, respectively, from 30 to 150 m, 100 to 450 m, and 80 to 250 m. The hypogene sulfide minerals consist of about 60% chalcopyrite, 25% bornite, 10% pyrite, and 5% chalcocite. Hematite is abundant. More than half the copper minerals occur as disseminated grains less than 0.9 mm in diameter, and in lesser amounts as massive aggregates or veinlets.

Some orebodies are stratiform, but in general they show crosscutting relationships with the host rock. They are, however, confined to the upper member of the Lo Prado Formation, and preferably are in the trachytic porphyry. There is a close space relationship between the large mineralized bodies and the trachytic feeders.

It has been suggested that El Soldado deposit probably is of volcanic origin and that the mineralization was epigenetic and controlled by the faults, especially those of the north-south system. Considering the limited copper mineralization observed in the faults, the author suggests, as an alternative, that the original copper sulfide contribution was syngenetic. In this case, the metallic sulfides would have occurred as segregated droplets in the trachytic and andesite magmas. This would explain distribution of disseminated grains of sulfide minerals in the dense volcanic rocks. Subsequent leaching by heated meteoric waters may have mobilized the sulfide minerals and deposited them in the vicinity of faults. A syngenetic origin is further supported by the close association of these deposits and the upper member of the Lo Prado Formation (Figure 4).

VEIN-TYPE COPPER DEPOSITS

Copper-bearing veins were the principal sources of copper produced in Chile up to the early part of this century. They were of maximum importance during the second half of the 19th century when Chile became the world's largest copper producer. This production came from many relatively small smelters that treated high-grade ore averaging 10-15% Cu. The most important production at that time was at the Tamaya and Carrizal Alto districts and the Dulcinea mine (Figure 1).

Other relatively important vein deposits, which were mined during the 20th century, are deposits in the vicinity of Tocopilla, the Delirio de Punitaqui deposit, and La Africana deposit (near Lo Aguirre) (Figure 1), all of which had closed prior to 1985. Several other small vein deposits were being worked in 1985. Total production was small.

The veins consist of minerals that were deposited in open fractures or fissures and which replaced the adjoining fractured wallrock. Two types of veins can be distinguished. The first type is characterized by the presence of abundant chalcopyrite, actinolite, and magnetite and associated small amounts of cobaltite, cobaltiferous arsenopyrite, uraninite, and molybdenite. Deposits of this type are aligned along an axis, parallel to the coast, which extends for 900 km from near La Serena in the south to Tocopilla in the north (Figure 1). The veins in these deposits characteristically strike between N50 °E and N60 °W, and thus trend obliquely to both the axis along which the deposits occur and the dominant parallel, north-trending structural grain of Chile. The veins are as much as 3 km long, and commercial mineralization extends to depths of as much as 700 m. The veins tend to occur in, but near the margins of, dioritic to granodioritic intrusions of Jurassic and Cretaceous age. A few veins are found in metamorphized rocks adjacent to these intrusions. The principal deposits are in the vicinity of Tocopilla and at the Gatico, Carrizal Alto, and La Higuera districts to the south.

The Carrizal Alto district had an annual production, during its most active period in the latter part of the 19th century, of as much as 60,000 tonnes of selected ore averaging between 10 and 15% copper. It has been estimated that a total of 3,000,000 tonnes of ore, averaging 5% copper, have been extracted from veins in this district. The veins occur in a dioritic intrusive body bounded on the west by quartzites, phyllites, and micaceous schists. The veins strike east-northeast and dip 35°-80°N. The most important vein, La Principal, had several ore shoots, which commonly were 4-5 m and exceptionally as much as 13 m in thickness. This vein is 2,800 m long and has been exploited to a depth of as much as 500 m. Vein minerals are actinolite, chlorite, quartz, calcite, pyrite, cobaltiferous arsenopyrite, pyrrhotite, chalcopyrite, cobaltite, and uraninite. The veins show oxidation and leaching to depths ranging from 10 to 40 m. Secondary enrichment was slight, and the primary ore zone was generally encountered at depths ranging from 20 to 60 m. Chalcopyrite is by far the most abundant copper mineral, although at depth the amount of chalcopyrite decreases as pyrrhotite increases.

The second type of vein deposits, which are the most numerous, are characterized by a simple metallic mineral association consisting almost exclusively of specularite (in part replaced by magnetite), pyrite, and chalcopyrite. Many of the deposits contain gold, which was introduced during a late-stage mineralization characterized by deposition of chalcopyrite and small amounts of sphalerite, galena, and tetrahedrite. Almost all of the veins occupy faults that trend northward. Some of these faults are major regional structures such as the Atacama fault along which are many copper and gold deposits over a longitudinal distance of 120 km. Large faults also are present in the Tamaya and Punitaqui districts (Figure 1), and at La Africana mine near Santiago. The Tamaya district was Chile's major copper mine during the second half of the 19th century. The Dulcinea mine also produced large amounts of copper during this period. Dulcinea is exceptional in that ore extends to depths of as much as 1,000 m, and in having two superimposed sulfide enrichment zones of significant size.

The veins at Tamaya are hosted by both volcanic and sedimentary rocks. The main vein, which strikes north and dips 60 °W, is 4,000 m long and extends to depths of 500 m or more. The vein mineralogy is relatively complex, but the only abundant minerals are calcite, hematite, pyrite, bornite, and chalcopyrite. Ore shoots average 8-10% copper over widths of 1-2 m, but which exceed 6 m at places. Oxide and enrichment zones are poorly developed to a maximum depth of 30 m. The vein shows a vertical zonation wherein bornite is the dominant copper mineral in the upper part of the vein to depths of as much as 400 m below the outcrop; chalcopyrite is dominant in the underlying 200 m, and sterile pyrite is dominant in the deepest parts of the vein. During the most productive period, from 1850 to 1890, the Tamaya district produced an estimated 900,000 tonnes of ore averaging 20% copper. It is estimated that a total of more than 2 million tonnes of ore averaging about 10% copper has been extracted from the district.

REFERENCES CITED

Ahumada, R., 1985, Eventos intrusivos en el yacimiento cuprifero El Soldado: Antofagasta, Anales IV Congreso Geologico Chileno, v. 4, p. 752-773.

Alvarez, O., J. Miranda, and P. Guzman, 1980, Geologia del complejo Chuquicamata: Santiago, Anales del Congreso Cincuentenario, Instituto Ingenieros de Minas, v. 2, p. 689.

Chavez, M., 1985, Geologic setting and the nature and distribution of disseminated copper mineralization of the Mantos Blancos district, Antofagasta, Chile: Berkeley, University of California, unpublished Ph.D. dissertation.

Gustafson, L., and J. Hunt, 1975, The porphyry copper deposits at El Salvador, Chile: Economic Geology, v. 70, p. 857-912.

Holmgren, C., 1985, Antecedentes para un modelo genetico del yacimiento El Soldado: Antofagasta, Anales del IV Congreso Geologico Chileno, v. 4, p. 626-650.

Howell, F. H., and J. S. Malloy, 1960, Geology of the Braden orebody, Chile, South America: Economic Geology, v. 55, p. 863-905.

Ruiz, C., 1977, Copper reserves and potential copper resources in

the porphyry copper metallogenic province of Chile: Paris, Proceedings of the World Conference "Towards a Plan of Actions for Mankind," v. 1, p. 1-20.
Ruiz, C., L. Aguirre, J. Corvalan, C. Klohn, E. Klohn, and B. Levi, 1965, Geologia y yacimientos metaliferos de Chile: Instituto de Investigaciones Geologicas, p. 305.
Stambuk, V., 1982, Geologia del yacimiento Rio Blanco: Concepcion, Anales de III Congreso Geologico Chileno, v. 2, p. E419-439.
Warnaars, F., 1980, Brechas de cobre y turmalina en Los Bronces, Chile: Santiago, Congreso Cincuentenario del Instituto de Ingenieros de Minas, Anales, III, p. 175-200.
Warnaars, F., C. Holgrem, and F. Barassi, 1985, Porphyry copper and tourmaline breccias at Los Bronces-Rio Blanco: Economic Geology, v. 80, p. 1544-1563.

Ericksen, G. E., Cañas Pinochet, M. T., and Reinemund, J. A., editors, 1989, Geology of the Andes and its relation to hydrocarbon and mineral resources: Houston, Texas, Circum-Pacific Council for Energy and Mineral Resources Earth Science Series, v. 11.

CHAPTER 20

GEOLOGIC OCCURRENCE, PALEOGEOGRAPHY, AND RESOURCES OF PHOSPHATE ROCK IN THE ANDEAN REGION

Richard P. Sheldon
Geologic Consultant
Washington, D.C., USA

William C. Burnett
Department of Oceanography
Florida State University
Tallahassee, Florida, U.S.A.

Victor Ricalde R.
Asociacion de Geoscientificos para el Desarrollo Internacional, AGID
Cochabamba, Bolivia

ABSTRACT

The Andean region of western South America was the site of four major episodes of marine phosphate sedimentation as the result of the low paleolatitudinal position and upwelling ocean currents along the margin of the South America craton. Three of these episodes, occurring in the Quaternary, Miocene, and Early Jurassic, were due to trade-wind belt, eastern-boundary current upwelling during times of polar glaciation. The fourth, in the Late Cretaceous, was due to equatorial-belt upwelling along the southern margin of the Tethys Sea on the continental shelves of Venezuela, Colombia, and Ecuador.

Small igneous or contact-metamorphic apatite deposits of Jurassic age are known in the Chilean Andean orogen. The phosphate resources of the region are large but mostly undeveloped, and the potential exists for discovering additional deposits.

INTRODUCTION

No phosphate deposits, other than phosphate-rich guano deposits, were known in the Andean region before the middle part of this century. This contrasts with other regions, where major phosphate rock deposits were discovered in the late 19th and early 20th centuries (Krauss, Saam, and Schmidt, 1984). The rapid post-World War II increase in demand for phosphate stimulated phosphate exploration worldwide, and during this period the major Andean phosphate deposits were found. Petroleum geologists discovered Upper Cretaceous phosphate deposits in Venezuela in the 1950s, and during the next two decades phosphate geologists found additional deposits in other parts of the country as well as in Colombia and Ecuador. In the late 1950s, the Sechura phosphate deposit of Perú was discovered by application of Kazakov's (1937) upwelling hypothesis (McKelvey, 1963). In 1960, the Jurassic deposit of Perú was discovered by using the previously established association of vanadiferous black shale and phosphate rock (L.T. Grose, personal communication, 1968).

Production from the deposits of the Andean region has been small. The combined production in 1981 from Colombia,

Perú, and the Andean part of Venezuela amounted to about 17,000 metric tons of P_2O_5 (in 50,000 t of concentrate). This was approximately 5% of the phosphate consumption of the total region, which includes Venezuela, Colombia, Ecuador, Perú, Bolivia, and Argentina, so that 95% was imported. The application rates of phosphate to arable land in countries of this region ranged from 600 to 19,300 grams P_2O_5 per hectare in 1979, which as 2 to 56% of the average application rate of the developed countries of the world.

Successful phosphate exploration continues in the Andean region. For example, a new Upper Cretaceous deposit was recently found in Venezuela and is now being delineated (S.E. Rodriguez, personal communication, 1984). The Miocene Mejillones phosphate deposit was found in the 1970s and may prove to be a potentially valuable deposit (Valdebenito, in press).

Is this stage of exploration almost finished, or are there more undiscovered phosphate deposits to be found? Is it possible that some of the existing small deposits are just the tips of larger "icebergs?" Will improved extraction, benefication, and processing technology make some of the larger, low-grade known deposits economical to exploit? Will increased economic demand stimulate such development? None of these questions can be answered with certainty at the present time. It seems possible that 30 years from now the phosphate supply situation in the Andean region may be as different from the present situation as the present situation is from that of 30 years ago. However, to get to that position 30 years from now, an assessment of the present stage of knowledge is essential for the direction of the geologic work required now. This review hopefully contributes to that goal.

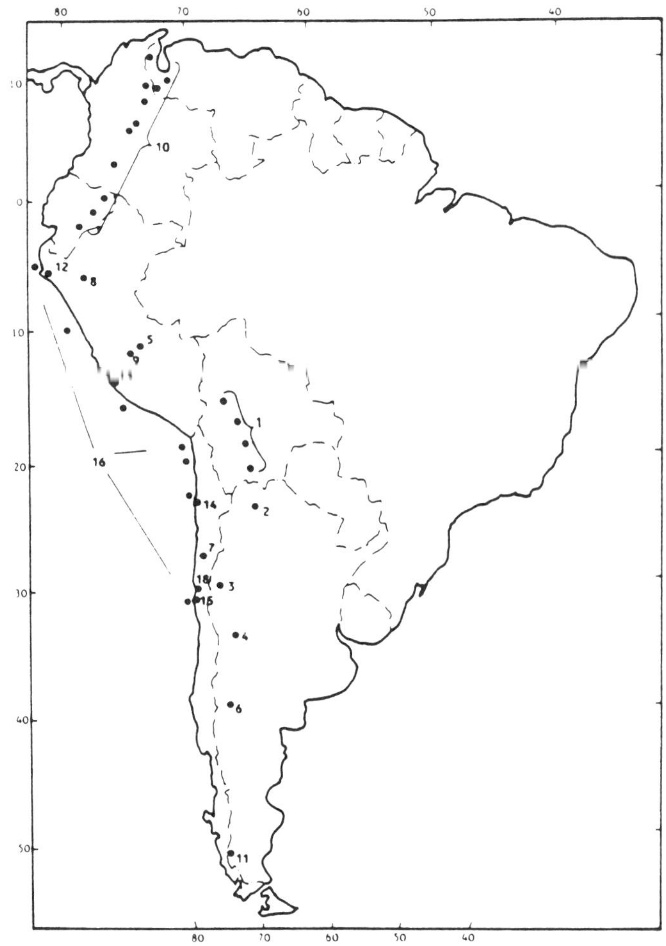

Figure 1. Locations of phosphate deposits and occurrences in the Andean region. See Table 1 for index of localities.

PHOSPHATE RESOURCES OF THE ANDEAN REGION

Phosphate rock occurs in every country of the Andean region, which for the purposes of this paper includes the submerged continental shelf and slope on the west. The deposits of the Andean region are listed in Table 1, which gives the general location, deposit and formation name, resource (if any), and the source of information. The locations of the deposits are plotted in Figure 1. The phosphate deposits are limited to (a) sedimentary phosphorites deposited on the continental shelf (or paleoshelf) in either open marine or semirestricted basins, and (b) small intrusives or contact-metamorphic deposits in the Andean orogen. The marine phosphorites range in age from Ordovician to Holocene, and in general match the major global episodes of phosphogenesis in Phanerozoic time (Figure 2) (Cook and McElhinny, 1979; Sheldon, in press-a). Only the Proterozoic-Cambrian and Permo-Carboniferous episodes are missing from the Andean stratigraphic sequence. The igneous apatite deposits occur in Chile as separate lens-shaped magnetite- and amphibole-rich apatite deposits (Harrington, Word, and McKelvey, 1966) associated with the Jurassic plutons.

Mesozoic and Cenozoic phosphorites of the Andean region have been deposited on the continental margin of South America at its active plate boundary with the oceanic Pacific Plate. Sedimentation, sea-level variations, volcanism, tectonism, and climate variations have complexly interacted, and the history of sedimentation was the result of this interaction. The northern part of South America has been at low latitudes for the last 300 million years (Ziegler et al., 1982), so it is expected that the Mesozoic and Cenozoic phosphogenic episodes would be represented. The Permian phosphogenic episode also should be represented, but it is not. The westernmost facies of Permian rocks are generally terrestrial clastic beds, suggesting that marine continental shelf sediments were either unimportant or have been removed by plate-tectonic activity.

It has been recognized that certain geologic times were more favorable than others for marine phosphorite sedimentation (Figure 2) (Gimmel'farb, 1958; Strakhov, 1960; Bushinski, 1966; Cook and McElhinny, 1979; Sheldon, 1981), and in order to describe the global context of each Andean deposit, relevant global marine phosphogenetic episodes also are discussed. Although the underlying causes of the empirical episodicity of phosphorite sedimentation are not completely understood, it appears to have been controlled by variations in a number of interrelated factors including paleoceanography and paleogeography, relative sea level, paleocurrents, paleoclimate, and temporal changes in oceanic chemistry. The underlying cause of all this would appear to be plate-tectonic processes that reconfigure oceans and continents and change ocean-basin geometry and land/sea

Table 1. Sedimentary and igneous phosphate rock deposits, occurrences, and resources in the Andean region.

Geologic Age and Location	Deposit Name and Formation (Deposit numbers refer to Figure 1)	Resources[a] Identified[b]	Hypothetical	References
ORDOVICIAN				
BOLIVIA				
Tarija Province	1. Izcayachi	Occurrence		Rtz, 1986, London
Potosi Province	Betanzon, Anzaldo Formation	Occurrence		Geobol, 1979
Cochambamba Province	Capinota	4 (6-25%)	?	
La Paz Province	Caranavi	Occurrence		
ARGENTINA				
Jujuy and Salta Provinces	2. Labrado and Centinela Formations	Occurrence		Masteandrea and Leanza, 1975; Fernandez, 1984; Leanza et al., in press
SILURIAN AND DEVONIAN				
ARGENTINA				
San Juan Province	3. Los Espeyos and Talacasto Formations	Occurrence		Leanza et al., 1979; Leanza et al., in press
TRIASSIC				
ARGENTINA				
Mendoza Province	4. Cacheuta Formation	Occurrence		Leanza et al., in press
LOWER JURASSIC				
PERU				
Junin Province	5. Mantaro Deposit, Aramachay Formation	Occurrence[c]		Szekeley and Grose, 1972; Loughman and Hallam, 1982; Loughman, 1984; Grose, in press
LOWER CRETACEOUS				
ARGENTINA				
Neuquen Province	6. Vaca Muerta Formation	Occurrence		Mastandrea et al., 1975
CHILE				
Atacama Province	7. Fabellon Formation	Occurrence		Sheldon and Perez, 1963
PERU				
Northern	8. Muerto Formation	Occurrence		Fischer, 1956; Cathcart, 1977;
Northern	9. Pariatambo Formation[d]			Harrison, 1940; Wilson, 1963; Grose, in press
UPPER CRETACEOUS				
VENEZUELA				
Zulia St.	10. La Luna Formation	18 (3-16%)	?	Rodriguez, in press
Merida St.	La Luna Formation	110 (18%)	?	Rodriguez, in press
Tachira St.	La Luna Formation	1 (22%)	?	Rodriguez, in press
Tachira St.	Navay Formation	22 (16%)	?	Rodriguez, in press

Table 1. Sedimentary and igneous phosphate rock deposits, occurrences, and resources in the Andean region.

Geologic Age and Location	Deposit Name and Formation (Deposit numbers refer to Figure 1)	Resources[a] Identified[b]	Hypothetical	References
COLOMBIA				
Cordillera Oriental	10. La Luna and Guadalupe Formations	674 (5-37%)[e]	500-1000 +	Mojica, 1979; Cathcart, 1977
		100 (25 + %)		Maughan et al., 1979
		57 (20-25%)		Maughan et al., 1979
		158 (15-20%)		Maughan et al., 1979
				Cathcart and Zambrano, 1967, [illegible]
ECUADOR				
Napo Province	10. Napo Formation	Occurrence		Wilkinson, 1982; Torrez, 1984
ARGENTINA				
Santa Cruz province	11. Las Hayas Formation	Occurrence		Leanza et al., in press
MIOCENE				
PERU				
Piura Dept.	12. Sechura Deposit	250 (25-32%)	38,000[f]	Cheney, 1961; Cheney, McClellan, and Montgomery, 1979; McClellan, in press
Ica Dept.	13. Pisco Formation	Occurrence		Abele, 1957
CHILE				
Antofagasta Province	14. Mejillones Deposit Caleta Herradura Formation	56 (7%)	?	Valdebenito, in press
Coquimbo Province	15. Tongoi area	Occurrence		Sheldon and Perez, 1963
QUATERNARY				
PERU AND CHILE				
Continental shelf	16. Sea-floor sediments		?	Burnett, 1977; Baturin, 1982
IGNEOUS APATITE				
CHILE				
Coquimbo and Atacama	18. Los Charos, Domeyko, and Los Baros	1 (22-26%)	?	Harrington, Ward, and McKelvey, 1966

[a]In million metric tons.
[b]Percent figure is percent P_2O_5.
[c]Question mark indicates a significant but unestimated hypothetical resource.
[d]Phosphorite has not been identified, but is probably present, in this fish tooth and bone containing black shale and phosphorite-bearing sequence.
[e]Different resource figures are for different grade categories.
[f]Tons of concentrate.

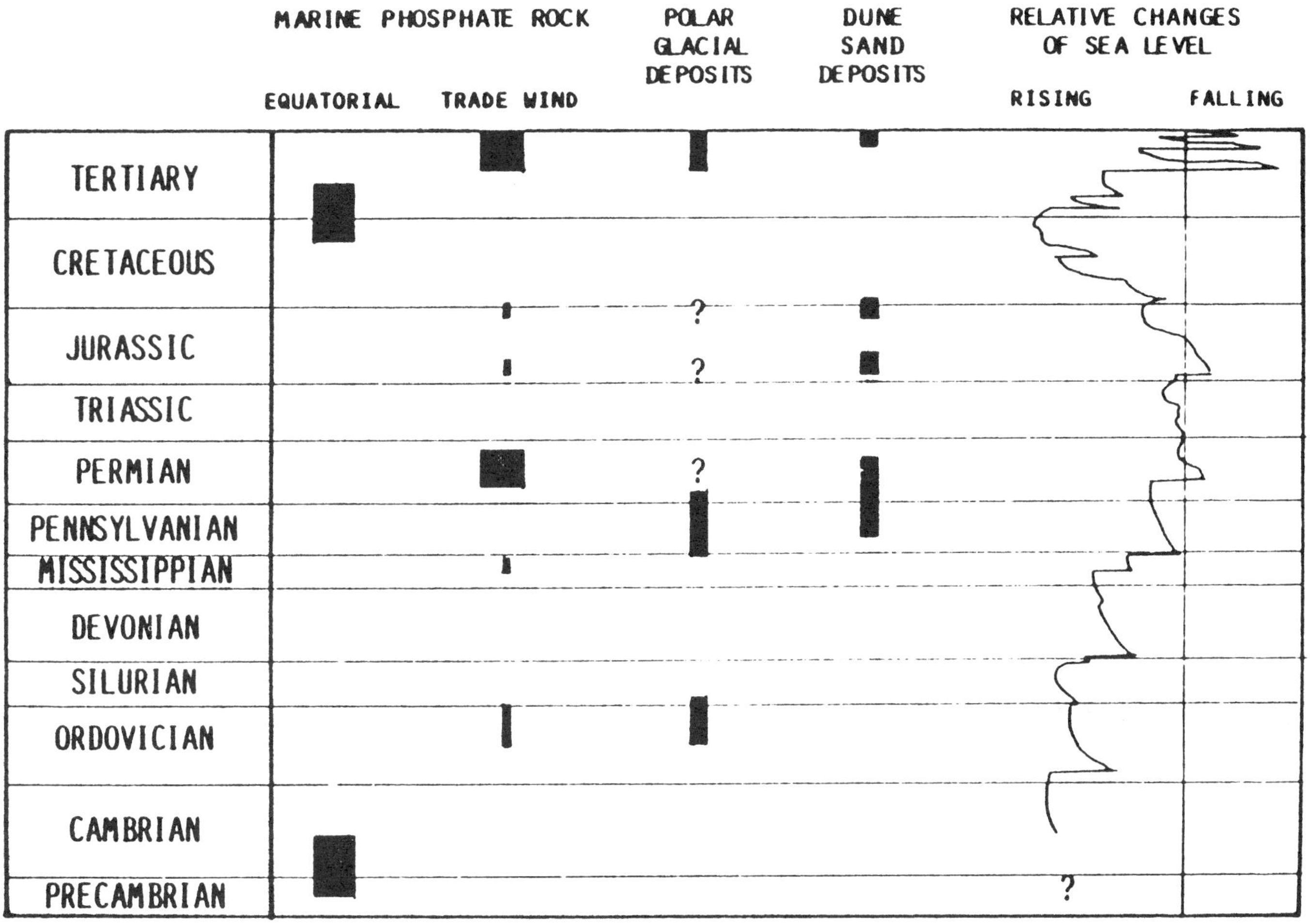

Figure 2. Correlation of ancient phosphorite deposits with ancient polar glacial deposits, interior desert dune deposits, and sea-level changes. Distribution of glacial deposits from Hambrey and Harland (1981); dune sand deposits from McKee (1980). Sea-level curve from Vail, Hardenbol, and Todd (1977). Major phosphogenic events indicated by wider bars.

surface-area ratios. Some large phosphate deposits—such as those of the Triassic Shublik Formation of the North Slope of Alaska and the Lower Ordovician phosphorites of Estonia, Poland, and Sweden—do not fit the scheme of phosphogenic episodes. Nevertheless, episodic phosphogenesis is a useful concept for initial understanding of the temporal phosphate distribution in a large region. Deposits of each phosphogenic episode of the Andean region, from youngest to oldest, are discussed below.

QUATERNARY PHOSPHORITES

Phosphatic nodules, slabs, and pellets commonly occur along the outer continental shelf and upper slope of Perú and Chile in the area of coastal upwelling (Burnett, 1977; Burnett, Veeh, and Soutar, 1980). The phosphorites extend over a wide latitudinal range from about lat. 5°-32°S, and in water depths ranging from 100-450 m. The nodules are more common on unconformities, although there is also evidence for nodules "floating" on soft muds. The distribution of phosphorites on the present sea floor seems to mimic the intersection of the oxygen-minimum zone with the continental margin in this area. In addition, the youngest phosphorites tend to be located near the upper and lower boundaries of the zone of minimum dissolved oxygen. It is not clear at this time if this is primarily a chemical effect or a secondary effect (such as enhanced bioturbation in well-preserved organic-rich sediment). Whatever the cause, it is now clear that this observation applies to the surface sediments only—cores have penetrated phosphatic-rich sequences throughout the oxygen-minimum layer.

Recent studies of sediment from the so-called "upper slope mud lens," which underlies the areas of persistent coastal upwelling off Perú and Chile, have shown that phosphatic pellets, commonly oolitic, are fairly common constituents. This is significant because most large marine deposits now on land are primarily pelletal rather than nodular. The advantages of studying modern pellets are clear. Because these pellets appear to be currently forming, an analysis of their environment should provide answers to the conditions, mechanisms, and rates under which the geologic deposits formed—an "actualistic" approach.

Geochemical and radiochemical studies have shown that these phosphate pellets form quickly in the uppermost part of the sediment column (probably the top centimeter or so) during early diagenesis (Froelich and Vathakanon, in press;

Burnett and Baker, in press). The source of the phosphorus is unclear, but it specifically does not appear to be related to release from normal marine organic matter during anoxic diagenesis. Instead, there is a very fast release of phosphorus in relation to nitrogen and carbon. The source of this phosphorus may be from benthic productivity in the form of sulfur-oxidizing bacterial mats (Thioploca), which occur commonly on the sea floor in this area (Reimers, 1985).

Phosphorus fluxes into pellets have been estimated by scaling the burial rates of pellets to the concentration of phosphorus contained in the pellets. This assumes in situ formation of the pellets, which is supported by recent geochemical and radiochemical results (Burnett and Baker, in press). Four cores from the Perú shelf show that the highest phosphorus fluxes occur in the cores with the highest net sediment accumulation rate (Table 2). Although these sediments have a lower concentration of pellets, their higher sedimentation rate buries a larger amount of authigenic phosphorus per unit time.

Using a medium phosphorus uptake rate from these data, we can calculate an approximate mineralization rate for phosphate pellets on the entire Perú shelf:

$$(8\ \mu\text{mol-P cm}^{-2}\text{yr}^{-1})\ (31 \times 10^{-6}\text{g}/\mu\text{mol-P})(1.5 \times 10^{14}\text{cm}^2) = 3.7 \times 10^{10}\text{gPyr}^{-1}$$

P uptake rate	gP/μmol	Perú shelf upper slope mud lens	Aprox. mineral. of P on Perú shelf per year

How does this rate compare with those of the large geologic deposits? It is difficult to say because those rates cannot be measured directly in the ancient deposits. Sheldon (1981), however, has estimated that the richest section of the Phosphoria Formation in southeast Idaho contains approximately 1.5×10^{17}g of phosphorus in the area covering 2.15×10^{14} cm^2. This material is all contained within the Meade Peak Phosphatic Shale Member, which was deposited over an interval estimated to have lasted from 2-6 million years. The range in mineralization rates for this area of the Phosphoria is thus $2.4\text{-}7.3 \times 10^{10}$ gPyr^{-1}, similar in magnitude to that estimated for the modern environment.

Another important aspect of the work on modern pellets is the conclusion that rich (although thin) beds of phosphatic pellets may accumulate without reworking. Pellets within one bed may be of uniform size but are unsorted relative to the enclosing sediment. Thus, the "sorting," i.e., uniform grain size of the phosphate pellets, is some effect inherent in the mechanism of precipitation of the pellets rather than an effect of mechanical processes.

Table 2. Estimates of sediment accumulation rates, S, and burial flux of phosphorus in pellets, F_p in four cores from the Peru shelf. An illustration of the calculation is shown for the case of Core BX-2.

Core Number	S (cm/10^3 yrs)	F_p (μmol-P cm^{-2} yr^{-1})
GS-6	1.6	4.2
BX-2	1.9	5.9
BX-3	160	7.2
BX-6	230	15.0

Burial Flux of P in Pellets Core BX-2

Sediment bulk density, ρ	=	1.2 g/cm^3
Sediment accumulation rate, S	=	1.9 cm/10^3 yrs
Average pellet concentration	=	76 wt. %
Phosphorus in pellets	=	10.5% P

$$\text{Flux of phosphorus, } F_p = \frac{(1.2\ \text{g/cm}^3)\ (0.76)\ (1.9\ \text{cm}/10^3\ \text{yrs})\ (0.105)}{31 \times 10^{-6}\ \text{g}/\mu\text{mol-P}} = 5.9\ \mu\text{mol-P cm}^{-2}\ \text{yr}^{-1}$$

MIOCENE PHOSPHORITES

Four semirestricted coastal basins in Perú and Chile contain Miocene phosphatic sediments. From north to south, they are (1) Sechura basin, Perú, (2) Pisco basin, Perú, (3) a basin in Chile associated with the Mejillones Peninsula, and (4) a basin near Tongoi, Chile.[1] These basins lie between lat. 6° and 30°S and were formed by foreland islands composed of Jurassic intrusives and country rocks. Only the Sechura basin is known to contain large phosphate resources, but the Mejillones basin, though apparently a small deposit, has not been adequately explored and could contain significant resources. The other localities contain only small amounts of phosphatic sediments, and it is doubtful that exploration will reveal any significant resources of phosphatic material. Each is described below.

Sechura Phosphorite Deposit.—The Sechura phosphorites were laid down in the Sechura coastal basin, which covers 22,000 km^2 and is located about 800 km north of Lima (Cheney, 1961; Cheney, McLellan, and Montgomery, 1979; McClellan, in press). The basin contains about 1500 m of Eocene to Miocene marine sediments, Pliocene nonmarine sediments, and Quaternary eolian sands and alluvium. This sequence overlies Cretaceous sedimentary and intrusive igneous rocks in the eastern part of the basin. On the west, the basin is terminated by the Illescas Mountains, which were formed by Paleozoic rocks intruded by Jurassic-Cretaceous granites and granodiorites. The center of the modern basin is a desert depression 22.5 m below sea level. The basin sediments are nearly horizontal, but have been slightly deformed by tectonic movements including gentle folding and minor faulting, some of which occurred during sedimentation of the phosphorite.

The phosphatic sequence is middle Miocene in age and consists of four members having a total thickness of about 170 m. It contains 13 phosphorite beds that exhibit marked lateral continuity in both petrographic character and thickness throughout their extent in the basin. These phosphorites are interbedded primarily with diatomites, some of which are slightly phosphatic, and a few crystal-rich tuff beds. Minor uplift and erosion occurred in the middle of the phosphorite-

[1]The writers have just been advised that a fifth basin containing phosphate of probable Miocene age has been located near Bahía Inglesia 876 km north of Santiago, Chile (Pavalovic and Osses, personal communication, 1985).

diatomite deposition, resulting in deposition of the Clam Bore Sandstone, a fine to medium-grained sandstone with clam and other fossil molds, unconformably overlying the Lower Diatomite and Phosphorite Member. The Clam Bore Sandstone is conformably overlain by the Upper Diatomite and Phosphorite Member and the uppermost Barren Diatomite Member. Post-Miocene erosion truncated this sequence prior to deposition of Pliocene continental sediments.

Three ore zones have been identified—the Diana ore zone in the Lower Diatomite and Phosphorite Member, and the Zero and Minerva ore zones, separated by the Inca Diatomite of the Upper Diatomite and Phosphorite Member. The Diana, the richest and thickest of the three ore zones, is 35-40 m thick and contains 7-8% P_2O_5; individual phosphorite beds contain as much as 18 percent P_2O_5. The Zero ore zone averages 6.3 m in thickness and contains 9% P_2O_5. The Minerva ore zone averages 6.7 m in thickness and contains 10.3% P_2O_5 in the west, and is 25 m thick and contains 5-6% P_2O_5 in the east, reflecting an increase of diatomite content form west to east.

The phosphorites are composed of fluorhydroxyapatite pellets that, while showing differences in sorting, are generally well sorted. The exogangue is composed of diatom fragments with minor amounts of volcanic ash, very fine quartz and feldspar grains, sponge spicules, and organic matter. In the subsurface, both the diatomite and phosphorite are dark colored because of their organic content.

From the paleogeographic setting, petrographic character, and stratigraphic "layer-cake" continuity of the beds of the Sechura phosphatic sequence, it is apparent that sedimentation occurred in a sheltered embayment of shallow depth that was open to marine surface currents to supply silica, phosphate, and nitrate nutrients. Cheney, McLellan, and Montgomery (1979) postulated:

> The phosphate pellets in the phosphorite beds were subjected to the movement of currents after their formation; this removed smaller grained material from the area of deposition. The action of currents is demonstrated by (a) the differences in sorting of the phosphorite pellets in the individual beds; a higher degree of sorting is accompanied by a higher P_2O_5 content, i.e., more pellets and less fine-grained interpellet dilutants, (b) tangential contacts of pellets, (c) erosional basal contacts of phosphorites with diatomites, and (d) local cross-bedding.

However, current action must have been minor to allow the deposition of widespread beds of similar grain-size distribution, and could best be described as winnowing. Most of the phosphorite beds are structureless, without signs of current activity, and the fragile diatomite exogangue argues against even extensive winnowing. Very fine quartz and feldspar sand grains could have been windblown to the site of deposition. As discussed above, Quaternary phosphorites now being deposited in a few hundred meters of water on the Perú-Chile continental shelf under an algal mat are composed of well "sorted" apatite pellets in a poorly sorted sediment and have not been or are not being reworked or winnowed (Burnett and Baker, in press; Froelich and Vathakanon, in press). The process of formation to give well "sorted" pellets appears to be dominantly, if not completely, early diagenetic.

The strong cyclicity characteristic of the diatomite and phosphorite sedimentation, which gave rise to 13 identifiable phosphorite beds, is comparable to the phosphorite cyclicity identified by Riggs (1984) in the middle Miocene phosphatic sequence of the offshore Pungo River Formation in North Carolina. Following Riggs' analysis, one can hypothesize that the diatomite-phosphorite cyclicity of the Sechura deposit is caused by differences in upwelling intensity and sea level, rather than differences in wave-generated current activity.

A major sedimentologic problem is the cause of alternating diatomite and phosphorite sedimentation. This is a general unsolved problem for Cretaceous and younger phosphorites throughout the world (Sheldon, in press-a), which may be related to dinoflagellate and diatom ecology and sedimentation, with dinoflagellate sedimentation resulting in phosphorite formation (Fauconnier and Slansky, 1980) and diatom sedimentation resulting in diatomite. Alternately, if diatoms are the source of both silica and phosphate, their separation may be due to alkaline diagenetic conditions fluxing silica out of diatomaceous sediments and acid diagenetic conditions fluxing phosphate out of diatomaceous sediments; the different conditions are perhaps determined by whether the sediments were deposited within or outside the oxygen-minimum zone of ocean water (Govean and Garrison, 1981; Burnett, Veeh, and Soutar, 1980).

McClellan (in press) estimated the phosphate resources of the well-studied southern part of the Sechura deposit as at least 250 million tons of apatite concentrates. This part is only a small fraction of the total area of the deposit, for which Cheney (1961) estimated 38 billion tons of concentrate for the combined blocked-out and indicated reserves, geologically probable reserves, and geologically possible reserves. Most of this larger resource would occur below normal strip-mining depths. However, the successful development of bore-hole mining technology (McKelvey, in press) might allow mining of this deeper resource.

Miocene Phosphatic Rocks in the Pisco Basin.—The Pisco Formation of Miocene age crops out along the south-flowing Ica River in the Ocucaje region in the Department of Ica, Perú. The formation was deposited in a coastal basin similar to the Sechura basin. On the west, it abuts against the Coastal Range headland, which consists of Jurassic-Cretaceous granitic rocks.

The Pisco Formation is composed of sandy, argillaceous shales with thin beds of sandy limestone. The formation contains three beds, 0.5-0.8 m thick, which contain abundant apatite nodules 1-6 cm in diameter (Abele, 1957). The nodules are round or ovoid and contain abundant fossil remains. The phosphate content of the nodules is low—5-15% P_2O_5, with an average of 8% P_2O_5.

This occurrence of phosphatic rocks is apparently not economic, although the lower parts of the Pisco Formation in the deeper parts of the basin remain to be explored. It is significant, however, that this basin, which is similar paleogeographically to the Sechura basin, also contains Miocene phosphatic rocks.

Phosphatic Rocks in the Miocene Basin of Mejillones Peninsula, Chile.—The headland of the Mejillones Peninsula is formed by folded Paleozoic and Jurassic sedimentary rocks intruded by Permo-Triassic and Jurassic igneous rocks. The Upper Tertiary basin to the east contains the Miocene Caleta Herradura Formation, formed of bedded marine sediments

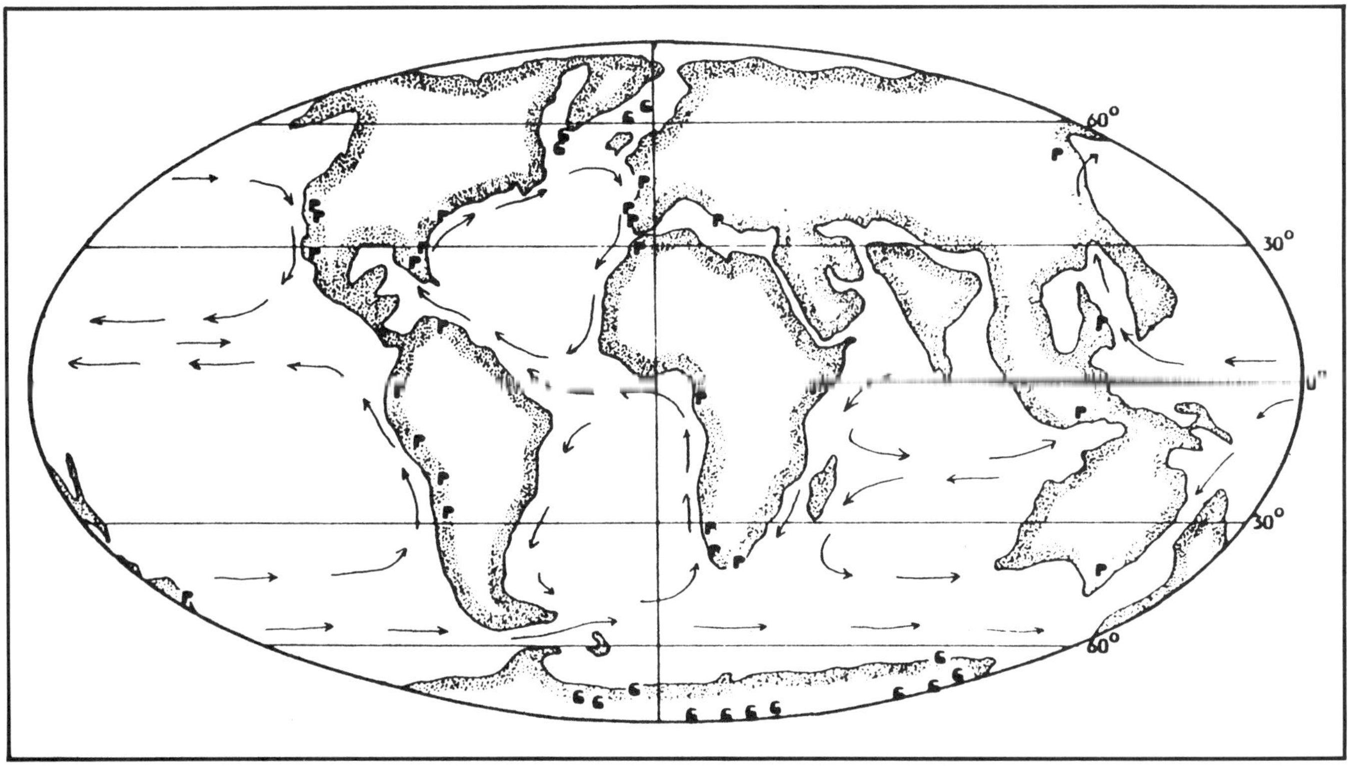

Figure 3. Global distribution of Miocene marine phosphorites and diamictites of glacial origin. P = phosphorite, G = glacial diamictites. Hypothetical surficial ocean currents shown by arrows. Data on diamictites from Hambrey and Harland, 1981. Data on phosphorites from Sheldon, in press-d. Reconstructed Middle Miocene continents (10 my bp) by paleogeographic base map from Ziegler, Scotese, and Barrett, 1982.

with a thickness of 422 m, consisting of sandstone, limestone, and shale with important diatomite and phosphatic sandstone layers (Valdebenito, in press). The formation covers an area of 405 km^2 and underlies thin Quaternary continental sediments in places. The Miocene rocks were affected by later tectonic block faulting.

A mudstone unit near the top of the formation includes diatomite and phosphatic sandstone, which contains foraminifera, siliceous sponge spicules, and bryozoans. The phosphate content of the sandstone is 6-7% P_2O_5, occurring as apatite pellets. The bed is unconsolidated and is 3-5 m thick. Apatite concentrates of the phosphatic sandstone contain 20-25% P_2O_5. An area of about 12 km^2 was mapped and drilled and found to contain 56 million tons of phosphatic sandstone averaging about 7% P_2O_5. The resource of the entire 405-km^2 area underlain by the formation has not been prospected, but the total resource probably is considerably larger.

Tongoi area, Chile.—A small Upper Tertiary basin near Tongoi, Chile, was found to contain a bed of sandstone 1.4 m thick with 2% P_2O_5 in the form of apatite pellets and bone fragments (Sheldon and Perez, 1963). Diatomite also occurs in the sequence. This occurrence is of interest because it is paleogeographically similar to the phosphatic sediments in the Miocene basins to the north, and it marks the southernmost known occurrence of Upper Tertiary phosphatic rocks. It apparently does not contain exploitable phosphatic material.

Global Setting of Miocene Phosphorites.—Antarctica was glaciated during the middle Miocene, and perhaps as early as late Oligocene (Mercer, 1983). This initiated a strong temperature gradient from equator to pole, which increased the intensity of zonal winds and resulted in the deposition of eastern-boundary-current phosphorites in the trade-wind belt (Sheldon, in press-d). These conclusions regarding paleoclimate and paleoceanography are more firm for the Miocene than for the Ordovician or Jurassic, as discussed below, because these processes are operating today in the same general area as the Miocene phosphorites and have been observed and studied. The Quaternary phosphorites of the Chile-Perú continental shelf, which are discussed above, are related to the Humboldt current and serve as a contemporary analog for interpreting the Miocene phosphorites. Thus, the general paleoceanographic and paleoclimatic conditions giving rise to the Miocene phosphorites can confidently be hypothesized.

The global distribution of Miocene phosphorites and glacial deposits is shown in Figure 3. The Miocene phosphorites of the Andean region are clearly eastern-boundary-current, trade-wind-belt phosphorites. Their distribution from 6° to 30°S paleolatitude (approximately the same as their present latitudes) matches closely the distribution of Quaternary continental-shelf phosphorites, as discussed in the last section.

The onshore Miocene marine phosphorites of Chile and Perú are shallow, inner continental-shelf, protected-basin deposits. It is possible that Miocene phosphorites were also deposited in the deeper water of the continental shelf, but at present, Quaternary sediments cover Upper Tertiary continental-shelf sediments (Riggs, in press). This possibility can be verified only by drilling, but if Miocene phosphorite

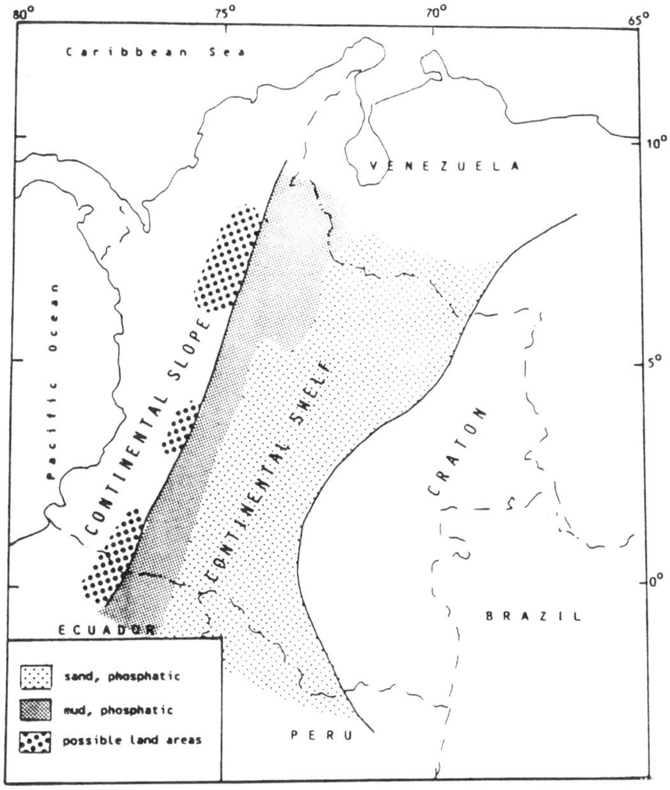

Figure 4. Late Cretaceous paleogeography of Colombia. After Maughan et al. (1979) and Wilkinson (1982).

does indeed occur and if offshore bore-hole mining technology is perfected (McKelvey, in press), these speculative resources might become significant.

UPPER CRETACEOUS PHOSPHORITES

Upper Cretaceous phosphorite deposits are known in Venezuela, Colombia, and Ecuador, where they constitute a single phosphate province. These phosphorites were deposited in the northeast-trending foreland basin between the South American craton and the Cretaceous Andean volcanic arc (Figure 4) (Cathcart, 1968; Maughan et al., 1979; Wilkinson, 1982), and now crop out in the folded and uplifted belt of the Cordillera Oriental and part of the Cordillera Central. West of the Cretaceous volcanic arc, a thick sequence of Upper Cretaceous slope and trench volcanogenic sediments and volcanics was deposited and is now exposed in Cordillera Central and Cordillera Occidental.

The phosphorites occur in two facies. In the northwestern foreland basin, a sequence of black organic-rich shale, chert, and black phosphorite forms the La Luna Formation in both Venezuela (Rodriguez, in press) and Colombia (Cathcart and Zambrano, 1967, 1969; Cathcart, 1968; Maughan et al., 1979). In the southeastern foreland basin, a less organic-rich, sandy facies containing phosphorite occurs in the Navay Formation in Venezuela (Rodriguez, in press), in the Guadelupe Formation in Colombia (Cathcart and Zambrano, 1969; Maughan et al., 1979), and in the Napo Formation in Ecuador (Wilkinson, 1982), all equivalent to the La Luna Formation. These formations interfinger with a nonphosphatic sandstone facies in the southeastern part of the foreland basin near the craton.

Phosphorite occurs in multiple horizons in the Upper Cretaceous sequence of the province. In Venezuela, as many as 18 horizons of phosphorite occur in the Santonian Tachira Member of the La Luna Formation, and phosphorite occurs at several horizons in the Navay Formation in Santonian to Campanian and possible Maestrichtian rocks (Rodriguez, in press). In Colombia, phosphorite occurs at many horizons from Coniacian to Campanian age, in the La Luna and Guadelupe Formations, with the major phosphorite beds occurring in rocks of Santonian age (Maughan et al., 1979, F. Etayo S., personal communication, 1983). A thick glauconitic phosphorite in the northern part of the province occurs at the base of the Umir Shale, which overlies the La Luna Formation and is probably Maestrichtian in age. In Ecuador, the Napo Formation has been dated as mid-Albian to mid-Maestrichtian. Seven thin beds and one major bed of phosphorite occur in the upper Napo Formation shales, which are considered as extensions of the Guadelupe Formation of Colombia (Wilkinson, 1982). Thus, the phosphorites of the Napo Formation are probably also of Coniacian to Campanian age.

In Cordillera Oriental of Colombia, Upper Cretaceous rocks form a nearly continuous sequence that exhibits cyclic sedimentation (Sheldon, 1983). A typical cycle consists of a basal sandstone overlain by a condensed chemical sedimentary unit of chert-phosphorite-dark limestone, and at the top, a black shale. In Boyaca, an excellent stratigraphic section along the Iza-Cuitiva road ranges in age from Turonian to Santonian (F. Etayo S., personal communication, 1983). Six cycles occur in this sequence, the lower five of which contain phosphorite, and all of which contain chert of siliceous mudstone. Within each cycle, there is evidence from bed sequences that the sedimentation was probably transgressive and regressive on a small scale. On a larger scale, determined from regional correlations along the Cordillera Oriental, the cycles are episodic; that is, they are broadly synchronous. Farther to the north, in Santander and Norte de Santander, a Maestrichtian to possibly Paleocene cycle contains glauconitic phosphorite and, in the Pesca area, an Eocene hematite oolitic ironstone contains as much as 5% P_2O_5. Thus, in the Cordillera Oriental as many as 8 episodes occurred, the upper two of which were iron rich. These episodes ranged from Turonian to Eocene in age and match the episodes of phosphogenesis in North and West Africa and the Middle East.

The Venezuelan-Colombia-Ecuador Upper Cretaceous phosphate province is a part of the Late Cretaceous-early Tertiary, circumglobal, low-latitude phosphogenic episode (Sheldon, 1980) (Figure 5). Phosphorite deposits of Late Cretaceous age occur in the Middle East (Turkey, Syria, Iran, Iraq, Jordan, Israel, and Saudi Arabia), in North Africa (Egypt, Morocco, and Western Sahara), Gabon, eastern Brazil, and on numerous seamounts in the Pacific. The deposits of Jordan, Israel, Saudi Arabia, and Egypt, in particular, are economically important. Minor deposits are known in India and eastern USSR (not shown on Figure 5). Lower Tertiary phosphorites have a generally similar distribution.

The paleogeographic distribution of the Upper Cretaceous phosphate province of northwestern South America is different from that of the Miocene phosphate province of western South America. At times of polar glaciation, strong trade

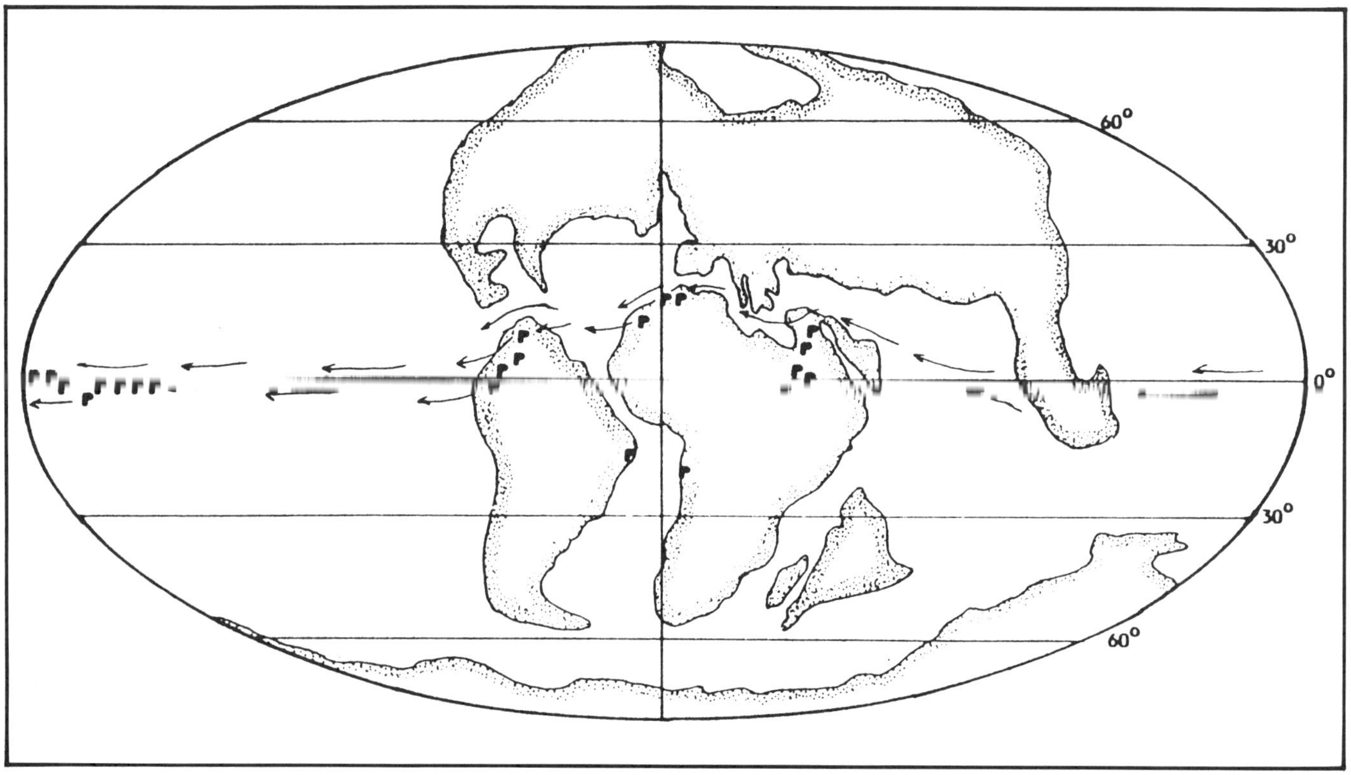

Figure 5. Global distribution of Upper Cretaceous major marine phosphorites (P). Hypothetical surficial ocean currents shown by arrows. Reconstructed Late Cretaceous (80 my bp) continents and continental shelves from Ziegler, Scotese, and Barrett (1982).

winds developed as the result of the large equator-pole temperature gradients, causing the formation of trade-wind desert eolian sand deposits and eastern-boundary-current, trade-wind-belt phosphorites. Trade-wind-belt phosphogenesis is characterized also by generally low sea levels. This was apparently true for the Late-Middle Ordovician, Early Carboniferous, Late Permian, Early and Late Jurassic, and Oligocene to Quaternary (Figure 2). During the Late Cretaceous, polar glaciation and extensive trade-wind desert eolian sands are not known, and sea level was variable but generally high. The paleogeographic distribution of Upper Cretaceous-lower Tertiary phosphorites is also quite different from the trade-wind-belt phosphorites in that they occur in a circumglobal, low-latitude belt and are unimportant at the localities where trade-wind-belt, eastern-boundary-current coastal upwelling would be expected to occur. It seems apparent that paleocean currents were different, caused by different paleoclimatic conditions. For one thing, the polar areas and the deep ocean waters were probably warmer, and world epeiric seas were more extensive than they were during glacial episodes. The phosphorite distribution points to major circumglobal equatorial upwelling that was much stronger than the present-day weak equatorial upwelling that does not produce significant phosphate deposits. One hypothesis to account for this is the existence of a circumglobal, east-west equatorial seaway (Sheldon, 1980, 1981). Another is the hypothetical existence of a ring system orbiting Earth, causing a low-latitude shadow that allowed equatorial upwelling (Sheldon, in press b, c). But regardless of the causal factors, the empirical distribution of Upper Cretaceous-Lower Tertiary phosphorites is different from that of the trade-wind-belt phosphorites, and this has important implications for phosphate rock assessment and exploration.

The Upper Cretaceous phosphorites of northwest South America have a paleogeographic position analogous to the Upper Cretaceous-Lower Tertiary phosphorites of the Middle East province and the Western Sahara-Morocco province (Figure 5). The deposits are located on the western sides of major promentories that extended northward into the path of the probable westward-flowing paleocurrent of the Tethys Sea.

The present phosphate rock resources of the Upper Cretaceous province of Venezuela and Colombia are significant, with total identified phosphate resources of about 300 million metric tons of phosphate rock in beds 0.15-12 m thick and P_2O_5 contents averaging about 20% (Rodriguez, in press; Cathcart, 1977; Maughan et al., 1979; Mojica, 1979). Hypothetical undiscovered resources have been estimated for Colombia at 500-1000+ million tons (Cathcart, 1977), but have not been estimated for Venezuela. The phosphate resources (either identified or hypothetical) of the Nabo Formation of Ecuador have not been estimated, but they appear to be small (Wilkinson, 1982).

A phosphate occurrence of Late Cretaceous age, but unrelated to the northwestern South America Upper Cretaceous phosphate province, has been reported in the Austral basin of southwestern Santa Cruz province in western Argentina at a latitude of about 50 °S (Leanza et al., in press). Thin beds of dark brownish phosphatic limestone occur in anoxic marine black shales of the Las Hayas Formation, and they contain a few tenths of a percent to 15% P_2O_5. The deposition of phosphate at such high paleolatitudes, which were about the same at the present latitudes, is difficult to explain with the present

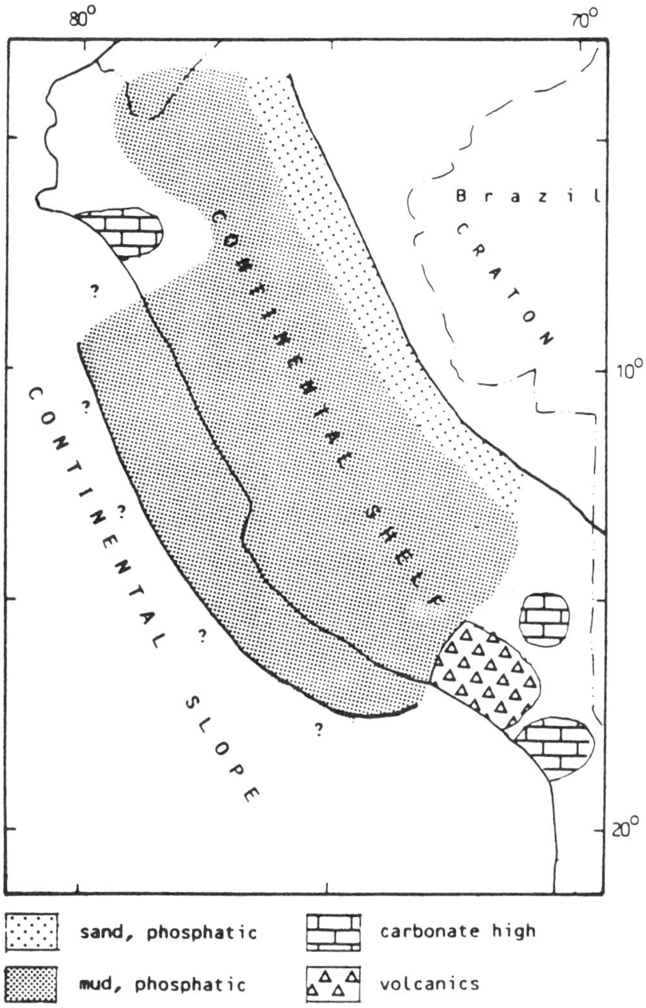

Figure 6. Early Jurassic (Sinemurian) paleogeography of Perú. After Loughman and Hallam (1982).

paleogeographic sedimentation models developed for major phosphorites. The deposit is small and probably not significant as a resource.

LOWER CRETACEOUS PHOSPHATIC ROCKS

Several occurrences of Lower Cretaceous phosphatic rocks have been reported in the Andean region. In northwestern Perú, the Middle Albian Muerto Formation contains a phosphatic limestone tens of meters thick with 5% P_2O_5 (Cathcart, 1977; Fischer, 1956). The correlative Middle Albian Pariatambo Formation, which is widespread in the central Andes of Perú (Harrison, 1940; Wilson, 1963), is an organic-rich cherty shale and limestone, containing fish scales and teeth, that seems likely to contain phosphatic beds (Grose, in press). At Quebrada Melendez south of Copiapo, northern Chile, the Lower Cretaceous Pabellon Formation contains phosphatic limestone associated with black shale and chert (Sheldon and Perez, 1963). Several hundred meters of the lower part of the Pabellon Formation are slightly phosphatic. At Quebrada Tofoval near Vallenar to the south, thin beds of impure black phosphorite occur in the Pabellon Formation, one bed measuring 35 cm thick. In Neuquen province of western Argentina, the Vaca Muerta Formation of the Upper Jurassic-Lower Cretaceous Mendoza Group contains a zone of lenticular beds of phosphatic limestone and phosphatic sandstone of Berriasian age (Leanza et al., in press). Thicknesses range from a few centimeters to a few meters, and P_2O_5 contents range from a few percent to 10%. A second zone of lesser importance occurs in the upper Valanginian to lower Hauterivian sequence.

Lower Cretaceous rocks in many parts of the world contain beds of phosphatic sandstone or limestone, but none of these occurrences is of more than local economic value. Of the Lower Cretaceous phosphatic rocks in the Andes, only the Muerto Formation of northwestern Perú appears to have possible economic potential. The global distribution of these occurrences has not been analyzed, but in the Andean region, the Lower Cretaceous phosphorites are distributed from lat. 60 ° to 38 °S, which was approximately their original paleolatitude. This suggests that there were eastern-boundary-current, trade-wind-belt phosphorites.

LOWER JURASSIC

The phosphatic Aramachay Formation of Early Jurassic (Sinemurian) age is the middle of three formations of the Pucara Group, which crops out in the Altiplano and intermontane valleys of the Department of Junin, Perú (Szekely and Grose, 1972; Grose, in press; Loughman and Hallam, 1982). The Aramachay is made up of two members—the lower Ichpachi Member (300 m thick), a bituminous shale, and the upper Alata Member (100 m thick), which is phosphatic sandstone, limestone, and chert (Loughman and Hallam, 1982). The Alata Member contains two persistent phosphatic zones, a lower zone 10-25 m thick and an upper zone 30 m thick, both containing more than 3% P_2O_5. The upper zone contains a single phosphatic bed 17 m thick averaging 11% P_2O_5, in which the phosphate occurs as apatite pellets in a medium-grained feldspar-apatite-quartz sandstone (Grose, in press; Loughman, 1984). It displays greater differences in lithology and bedding than the lower zone and locally contains cross-bedded phosphatic sandstone with lithic volcanic clasts.

Loughman and Hallam (1982) present evidence that the Aramachay was deposited in a shallowing-upward sequence on an extensive continental shelf open to the Pacific Ocean (Figure 6). In the Middle and Late Jurassic, a volcanic arc developed at the continental margin, modifying the shelf. Many authors (Sheldon, 1974; Szekely and Grose, 1972; Ziegler et al., 1982; Loughman and Hallam, 1982) have suggested that the phosphatic sediments were deposited by trade-wind-belt upwelling of the ancestral Humboldt current flowing from south to north along the coast.

From a global perspective, the supercontinent Pangaea, formed of all the present continents, stretched from the north pole to the south pole, dividing the Earth into a single ocean (Panthalassia) and a single continent (Figure 7). If the Aramachay phosphatic rocks formed by trade-wind-belt upwelling, then by analogy with the present Earth climate-ocean conditions, eastern-boundary-current upwelling in the trade-wind belt should have been accompanied by polar glaciation and trade-wind-desert eolian sands. Furthermore, similar phosphorites of the same age should have formed in the northern

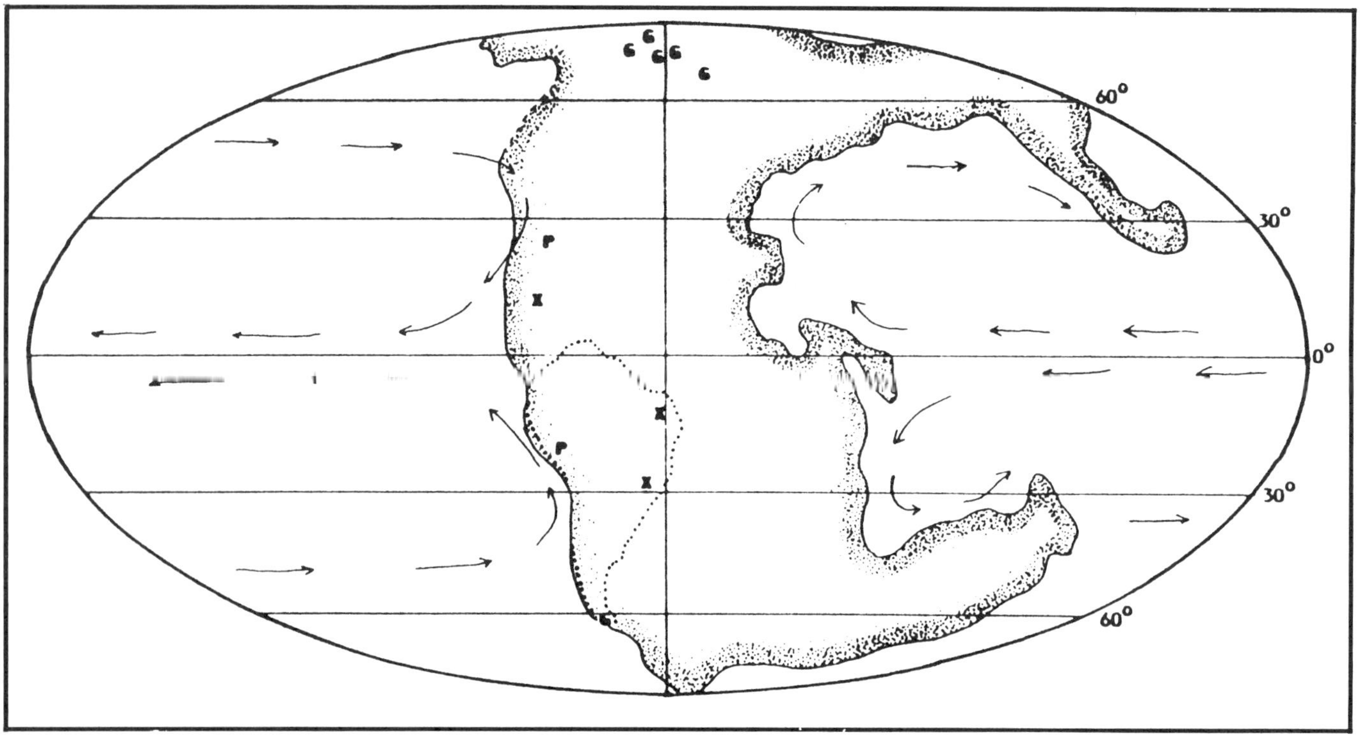

Figure 7. Global distribution of Lower Jurassic marine phosphorites, eolian sandstones, diamictites of possible glacial origin. P = phosphorite, X = eolian sandstone, G = diamictite, possibly of glacial origin. Hypothetical surficial ocean currents shown by arrows. Data on eolian sandstones from McKee (1980) and Bigarella and Salamuni (1961). Diamictite data from Cecioni (1981) and Chumakov (1981). Reconstructed Early Jurassic (190 my bp) continents and continental shelves from Ziegler, Scotese, and Barett (1982). Dotted line shows outline of present South American continent.

hemisphere in a paleoceanographic position comparable to the Aramachay Formation. Such seems to be the case.

Glaciation during the Jurassic is not known with certainty, but Chumakov (1981) proposed that pebbly shales at numerous localities in North Siberia (Figure 7) of Middle and Late Jurassic age are of ice-marine origin, although other authors disagree. In Tierra del Fuego, Chile, the Upper(?) Jurassic Flamenco and Seno Rodrigues Formations contain diamictites of possible glacial origin (Cecioni, 1981). Indirect evidence of glaciation comes from global sea-level curves. Vail, Hardenbol, and Todd (1984) show a global eustatic sea-level low during Sinemurian time as the lowest sea level of the 16 sea-level cycles of the Jurassic supercycle. They do not postulate a glacial origin for the cyclicity, but an increasing number of geologists have done so.

Strong Jurassic trade winds in the northern hemisphere are indicated by thick eolian sandstone of the Navajo Sandstone of the southwestern U.S. (McKee, 1980), and in the southern hemisphere by the thick eolian sandstones of the Botocatu Sandstone of Brazil and Uruguay (Bigarella and Salamuni, 1961) (Figure 7).

In British Columbia, Canada, Sinemurian phosphorites of the Fernie Group were deposited on the western continental shelf of North America (Christie, in press) (Figure 7).

From the perspective of global paleogeography, paleoclimates, and paleoceanography, the Lower Jurassic rocks of low-latitude western continental shelves of North and South America are excellent terranes to prospect for phosphorites. Upper Jurassic rocks also offer a potential in South America. In Mexico, the Upper Jurassic La Caja and La Casita Formations contain phosphate in sequence similar to the Aramachay Formation, indicating that trade-wind-belt phosphogenesis occurred on the western continental shelf of North America in the Late Jurassic (Rogers et al., 1961; Sheldon, 1964), and possibly also in the southern hemisphere at a comparable position and time.

TRIASSIC PHOSPHORITES

Phosphate concretions containing as much as 36% P_2O_5 occur in bituminous, lacustrine shales of the Triassic Cacheuta Formation in the Mendoza Province of Argentina (Leanza et al., in press). Lacustrine phosphate associated with black shales is not uncommon in rocks of many ages in the world, but large economic deposits of this type are not known.

SILURIAN-DEVONIAN PHOSPHORITES

A phosphatic bed, 0.1-0.65 m thick and containing as much as 14% P_2O_5, occurs at the base of the Silurian Los Espejos Formation, which consists chiefly of olive-green sandstones, siltstones, and shales, in the San Juan Precordillera of Argentina (Leanza et al., in press). The phosphatic rock consists of glauconite and collophane pellets in a siliceous sandstone cemented by apatite. Three additional but less important phosphatic beds occur in the overlying Lower Devonian Talacasto Formation, which consists of green sandstones and red-

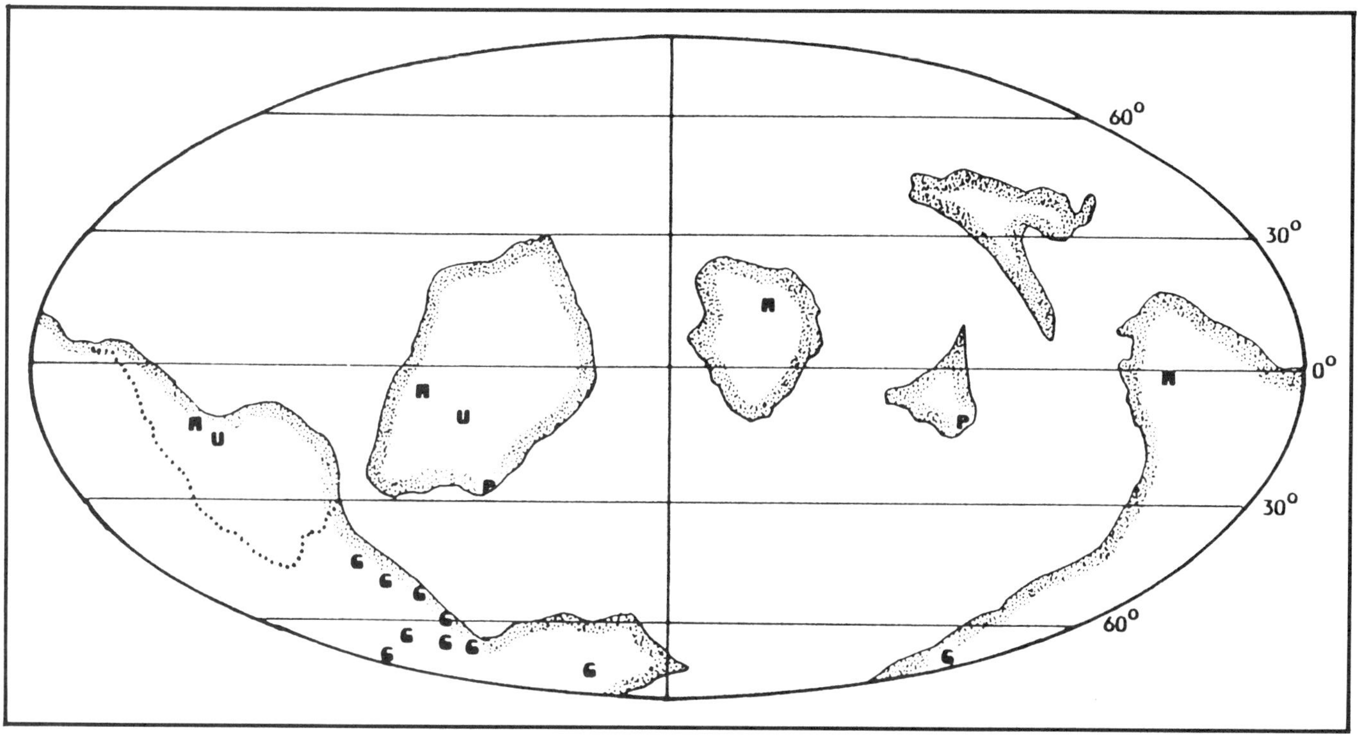

Figure 8. Global distribution of Middle and Upper Ordovician marine phosphorites and diamictites of glacial origin. P = phosphorites of undifferentiated Ordovician age, M = Middle Ordovician phosphorites, U = Upper Ordovician phosphorites, G = diamictite of glacial origin. Reconstructed Middle Ordovician continents and continental shelves from Scotese (1979). Dotted line shows outline of present South American continent.

dish shales. In the Rio Negro province of eastern Argentina, the Silurian-Devonian Sierra Grande Formation contains iron-rich beds containing as much as 2-3% P_2O_5 (Leanza et al., in press).

These phosphate and iron-rich beds are similar in some respects to the Silurian Red Mountain Formation of Alabama (U.S.), which contains the Clinton hematitic-chamoisitic iron ores that are also slightly phosphatic (Sheldon, 1970). One thin bed of phosphorite occurs in the Red Mountain Formation north of the Birmingham mining district. However, the Silurian and Early Devonian were not periods of significant global phosphogenesis, and no phosphate rock of these ages has been mined.

ORDOVICIAN PHOSPHORITES

Some Ordovician phosphatic rocks in Bolivia contain as much as 20% P_2O_5 (Harrington, Ward, and McKelvey, 1966; Gonzales, 1984), and in Argentina contains as much as 14% P_2O_5 (Leanza et al., in press; Masteandrea and Leanza, 1975; Fernandez, 1984). These rocks constitute a single phosphogenic province. The phosphatic material is predominantly apatitic brachiopod shell fragments occurring in sandstone. The phosphatic strata are of Middle and Late Ordovician age in Argentina and Late Ordovician age in Bolivia.

Similar phosphatic rock, composed predominantly of brachiopod shell fragments, occurs in the Middle Ordovician Swan Peak Quartzite of Utah and Idaho (western U.S.). A number of other Middle and Upper Ordovician sedimentary phosphatic rocks of different textures and compositions occur in other parts of the World, including Australia, Kazakhstan and Siberia of USSR, and Kentucky, Tennessee, and Iowa of the U.S. These deposits are shown according to their epoch on a Middle Ordovician paleogeographic base map (Figure 8). The Middle and Upper Ordovician phosphate deposits were deposited at less than 30° paleolatitude. Most deposits were formed on continental shelves, but Australian, Siberian, and Iowan deposits appear to be intracratonic. The Middle and Upper Ordovician phosphorites were deposited at the time of high-latitude Gondwana glaciation (Hambrey and Harland, 1981; Caputo and Crowell, 1985), which occurred in the Middle Ordovician to Early Silurian (Figure 8). Thus, they may have been formed by trade-wind-belt upwelling during a period of high temperature gradient from equator to pole, favoring strong trade winds.

IGNEOUS APATITE

Many igneous apatite occurrences have been found in the Andean region. In Chile, Harrington, Ward, and McKelvey (1966) report:

> A belt of granodiorite along the coast range and longitudinal valley in Coquimbo and Atacama Provinces contains about 20 known deposits of apatite which are associated with local intrusives or dikes. The deposits are in the area bounded by lat. 28°40 'S and lat. 30°00 'S.
>
> Los Choros, on the railroad about 80 km north of La Serena, is the shipping center for the area. About 20 lens-shaped apatite deposits are associated with amphibole and magnetite.

Composite samples assayed as follows: 26.3 percent P_2O_5, 36.5 percent CaCO, 31.1 percent Fe_2O_3, 1.7 percent Al_2O_3, and 4.8 percent Cl. Reserves are estimated to be 420,000 tons. Similar deposits are found in the north near Vallenar.

In Atacama Province, intrusives containing apatite have been worked at Los Barron and Domeyko. Reserves at Domeyko are estimated by the British Sulfur Corporation to be 540,000 tons of material. The ore contains from 22.5 to 32% P_2O_5.

An intriguing hypothesis was advanced by Ziegler et al. (1982). They point out that western South America must have had a shelf zone facing the deep ocean in all of Phanerozoic time, and it could very well have been the site of upwelling and phosphorite accumulation at times other than the known continental-shelf phosphorite deposits, primarily the Paleozoic, and particularly the Permian. They state:

> It is our belief that the absence of this shelf zone is due to the phenomenon of "tectonic erosion" operating off western South America. By this process, the margin of the continent is continually rasped or shaved by the descending oceanic plate. Up to 250 km of the width of South America may have been removed and carried down the subduction zone during the past 200 million years. This figure is based on the observations that the calc-alkaline volcanoes have migrated 250 km east during this interval. If phosphorites had been deposited on the outer shelf margin in earlier times, they would have been carried down the subduction zone as well. If significant amounts had been carried down, one might expect that they would reappear in the volcanic edifices of the Andean Mountains. Two deposits of igneous apatite have been discovered in northern Chile, and these could have been formed by this potential recycling process. More deposits of igneous origin may await discovery.

SUMMARY OF RESOURCES AND POTENTIAL

The identified phosphatic deposits of the Andean region that are either presently being mined or have a significant potential resource were deposited during times of global phosphogenesis. They include deposits of Ordovician, Early Jurassic, Late Cretaceous, Miocene, and Quaternary age. Except for the Late Cretaceous phosphorites, they belong to the trade-wind-belt family of phosphorites, which were deposited on continental shelves by eastern-boundary-current upwelling currents.

The Miocene Sechura phosphorites of Perú and the Upper Cretaceous La Luna phosphorites of Colombia and Venezuela are presently being mined. The identified reserves of both provinces are determined to be hundreds of millions of tons, and both have hypothetical resources that are estimated in the billions of tons. Good prospects exist for developing significant amounts of phosphate resources in the Lower Jurassic phosphorites of Perú, the Miocene phosphatic sandstones of Chile, and perhaps the offshore Perú-Chile Quaternary deposits.

Fair prospects exist for developing significant resources of Ordovician phosphatic sandstone deposits in Bolivia and Argentina. There appears to be a fair chance of locating significant additional igneous apatite resources in the Andean igneous terrane. The Upper Cretaceous rocks of Ecuador have a fair chance of containing undiscovered phosphate deposits better than those so far identified.

A number of phosphorite occurrences of other ages have been identified in the Andean region, including Silurian-Devonian, Triassic, and Lower Cretaceous, that at present appear to have a small chance of containing larger resources than so far identified.

Speculative phosphate resources, i.e., phosphorite provinces not yet identified in the Andean region, perhaps include the following. There appears to be a fair chance that significant deposits of Miocene phosphorite exist on the Perú-Chile continental shelf offshore, which could in some places be shallow enough and near enough to the sea floor for dredging or bore hole mining. Upper Jurassic phosphorites were deposited in Mexico in the northern hemisphere eastern-boundary current, trade-wind belt, and a southern hemisphere counterpart deposit would not be unlikely. The same is true of Permian rocks, which in the U.S. include the giant Phosphoria Formation phosphorites. No southern hemisphere counterpart has yet been identified.

ACKNOWLEDGMENTS

This review article on phosphorites of the Andes depended mainly on previous work, which has been acknowledged in the text. In addition, critical reviews of the manuscript were made by George Ericksen, David Davidson, Arthur Notholt Hector Leanza, and John Reinemund, for which the writers are highly grateful. The work on modern phosphorites of the Perú shelf was supported by a grant to William C. Burnett from the National Science Foundation (OCE8317181). This paper is a contribution to the International Geological Correlation Program (IGCP) Project 156—Phosphorites.

REFERENCES

Abele, C. G., 1957, Reconocimento por fosfatos en el Perú: Soc. Geol. Perú, B.T., v. 32, p. 5-19.

Baturin, G. N., 1982, Phosphorites on the sea floor: New York, Elsevier, 343 p.

Bigarella, J. J., and R. Salamuni, 1961, Early Mesozoic wind pattern as suggested by dune bedding in the Botucatu sandstone of Brazil and Uruguay: Geological Society of America Bulletin, v. 72, p. 1089-1106.

Burnett, W. C., 1977, Geochemistry and origin of phosphorite deposits from off Perú and Chile: Geological Society of America Bulletin, v. 88, p. 813-823.

Burnett, W. C., and K. B. Baker, in press, Rates of formation of modern phosphate nodules and pellets, *in* W. C. Burnett and S. R. Riggs, eds., Neogene to modern phosphorites: Cambridge, England, Cambridge University Press.

Burnett, W. C., H. H. Veeh, and A. Soutar, 1980, U-series, oceanographic and sedimentary evidence in support of recent formation of phosphate nodules off Perú, *in* Y. K. Bantor, ed., Marine phosphorites, geochemistry, occurrence, genesis: Society of Economic Paleontologists and Mineralogists Special Publication, v. 29, p. 61-72.

Bushinski, G. I., 1966, Old phosphorites of Asia and their genesis: Israel Prog. Science Trans. Jerusalem, Israel, 1969, 266 p. (Transl.

of Trudy Geol. Inst., 1966, vol. 149, 142 p.)

Caputo, M. V., and J. C. Crowell, 1985, Migration of glacial centers across Gondwana during Paleozoic Era: Geological Society of America Bulletin, v. 96, p. 1020-1036.

Cathcart, J. B., 1968, Phosphate exploration in Colombia—a case history, *in* Proceedings of the seminar of sources of mineral raw material for the fertilizer industry in Asia and the Far East: United Nations Mineral Resources Development Series, v. 32, p. 248-254.

Cathcart, J. B., 1977, Distribution and resources of phosphate in the eastern United States and in Latin America, *in* Proceedings of the geology and food conference, with related U.S. Geological Survey projects and a bibliography: United States Geological Survey Circular 768, p. 3-7.

Cathcart, J. B., and O. F. Zambrano, 1967, Roca fosfática en Colombia: Servicio Geol. Nac. Bol. Geol., v. 15, p. 65-162.

Cathcart, J. B., and O. F. Zambrano, 1969, Phosphate rock in Colombia: a preliminary report: United States Geological Survey Bulletin 1271-A, p. 3-7.

Cecioni, G., 1981, Jurassic Flamenco and Seno Rodrigues Formations of southern Chile, *in* M. J. Hambrey and W. B. Harland, eds., Earth's pre-Pleistocene glacial record: Cambridge, England, Cambridge University Press, p. 835-836.

Cheney, T. M., 1961, Geology of phosphate deposits in the western Sechura Desert: Unpublished company report.

Cheney, T. M., G. H. McClellan, and E. S. Montgomery, 1979, Sechura phosphate deposits, their stratigraphy, origin, and composition: Economic Geology, v. 74, p. 232-259.

Christie, R. L., in press, Jurassic phosphorite of the Fernie synclinorium, southeastern British Colombia, *in* A. J. G. Notholt, R. P. Sheldon, and D. F. Davidson, eds., World phosphate resources: Cambridge, England, Cambridge University Press.

Chumakov, N. M., 1981, Scattered stones in Mesozoic deposits of North Siberia, U.S.S.R., *in* M. H. Hambrey and W. B. Harland, eds., Earth's pre-Pleistocene glacial record: Cambridge, England, Cambridge University Press, p. 264.

Cook, P. J., and M. W. McElhinny, 1979, A reevaluation of the spatial and temporal distribution of sedimentary phosphate deposits in the light of plate tectonics: Economic Geology, v. 74, p. 315-330.

Fauconnier, D., and M. Slansky, 1980, Relations entre le developpement des dinoflagellas et la sedimentation phosphatee du basin de Gafsa (Tunisie), *in* Geologie comparee des gisements de phosphates et de petrole: Bureau de Recherches Geologiques et Mineres, Document 24, p. 184-204.

Fernandez, R., 1984, Bioestratigrafía y mineralogía de Neuvos Descubrimientos de Roca Fosforica en la Provincia de Jujuy, Argentina, *in* La Roca Fosforica: Cochabamba, Bolivia, GLIRF, v. 1, p. 61-70.

Fischer, A. G., 1956, Desarollo Geologico del noreste Peruano durante el Mesozoico: Bol. Soc. Geol. Perú, v. 30, p. 177-190.

Froelich, P., and C. Vathakanon, in press, Diagenesis of phosphorus in sediments from upwelling areas: the magic mechanism of phosphorite precipitation, *in* W. C. Burnett and S. R. Riggs, eds., Neogene to modern phosphorites: Cambridge, England, Cambridge University Press.

Geobol, 1979, Prospeccron de los yacimientos fostaticos del anticlinal de Capinota: Unpublished report.

Gimmel'farb, B. M., 1958, Regularity of the tectonic distribution of phosphorite deposits in the USSR: Zakonomern. Razmeshcheniya Polezn. Iskop., v. 1, p. 487-516 (in Russian): Int. Geol. Rev., 1964, vol. 6, p. 317-343.

Gonzales, A., 1984, Descripción de los Yacimientos de Roca Fosfórica en Bolivia, *in* La Roca Fosfórica: Cochabamba, Bolivia, GLIRF, v. 1, p. 1-12.

Govean, T. M., and R. E. Garrison, 1981, Significance of laminated and massive diatomites in the upper part of the Monterey Formation, California, *in* R. E. Garrison and R. G. Douglas, eds., The Monterey Formation and related siliceous rocks of California: Pacific Section, Society of Economic Paleontologists and Mineralogists Special Publication 0, p. 181-198.

Grose, L. T., in press, Geologic summary of the occurrence of phosphatic rocks in the Mantaro phosphate field, central Andes, *in* A. J. G. Notholt, R. P. Sheldon, and D. F. Davidson, eds., World phosphate resources: Cambridge, England, Cambridge University Press.

Hambrey, M. J., and W. B. Harland, 1981, Earth's pre-Pleistocene glacial record: Cambridge, England, Cambridge University Press, p. 1004.

Harrington, J. F., D. E. Ward, and V. E. McKelvey, 1966, Sources of fertilizer minerals in South America—a preliminary study: United States Geological Survey Bulletin 1240, p. 66.

Harrison, J. V., 1940, A preliminary note on the geology, Central Andes of Perú: Soc. Geol. del Perú, v. 10, p. 31-53.

Kazakov, A. V., 1937, The phosphorite facies and the genesis of phosphorites: Trans. Sci. Inst. Fertilizers and Insecto-Fungicides, v. 142, p. 95-113 (published for the 17th Internat. Geol. Cong., Leningrad).

Krauss, U. H., H. G. Saam, and H. W. Schmidt, 1984, International strategic minerals inventory summary report—phosphate: United States Geological Survey circular 930-C, p. 41.

Leanza, A. F., H. A. Leanza, C. A. Hugo, and A. T. Spiegelman, 1979, Hallazgo de rocas fosfáticas marinas en el Precordillera San Juanina: Sexto Congresso Geológico Argentino, Bahía Blanca (21-27 septiembre, 1975), Actas II, p. 317-330.

Leanza, H. A., A. T. Spiegelman, C. A. Hugo, O. O. Mastandrea, and C. J. Oblitas, in press, Sedimentary phosphatic rocks of Argentina, *in* A. J. G. Notholt, R. P. Sheldon, and D. F. Davidson, eds., World phosphate resources: Cambridge, England, Cambridge University Press.

Loughman, D. L., 1984, Phosphate authigenesis in the Aramachay Formation (Lower Jurassic) of Perú: Journal of Sedimentary Petrology, v. 54, p. 1147-1156.

Loughman, D. L., and A. Hallam, 1982, A facies analysis of the Pucara Group (Norian to Toarcian carbonates, organic-rich shale and phosphate) of central and northern Perú: Sediment. Geol., v. 32, p. 161-194.

Mastandrea, O., H. A. Leanza, C. A. Hugo, and A. T. Spiegelman, 1975, Manifestaciones fosfáticas en la Sierra de Vaca Muerta y adyacencias Prov. del Neuquen, Rep. Argentina: II Cong. Iberoam. Geol. Econ. Buenos Aires, Actas II, p. 461-483.

Maughan, E. K., O. F. Zambrano, G. P. Mojica, M. J. Abozaglo, P. F. Pachon, and R. R. Duran, 1979, Paleontologic and stratigraphic relations of phosphate beds in upper Cretaceous rocks of the Cordillera Orienta, Colombia: U.S. Geological Survey Project Report Series, p. 97.

McClellan, G. H., in press, Geology of the phosphate deposits at Sechura, Perú, *in* A. J. G. Notholt, R. P. Sheldon, and D. F. Davidson, eds., World phosphate resources: Cambridge, England, Cambridge University Press.

McKee, E. D., 1980, A study of global sand seas: U.S. Geological Survey Professional Paper 1052, p. 29.

McKelvey, V. E., 1963, Successful new techniques in prospecting for phosphate deposits, *in* Natural resources: U.S. Government Printing Office, United States papers prepared for the United Nations conference on the application of science and technology for the benefit of the less developed areas, v. 2, p. 163-172; reproduced in 1967, Colombian Boletin Geologico, v. XV, p. 45-64 (in Spanish).

McKelvey, V. E., in press, Present and future role of SE phosphates in World economy, *in* W. C. Burnett and S. R. Riggs, eds., Neogene to modern phosphorites: Cambridge, England, Cambridge University Press.

Mercer, J. H., 1983, Cenozoic glaciation in the southern hemisphere: Annual Review of Earth and Planetary Science, v. 11, p.

99-132.

Mojica, G. P. E., 1979, Fosfatos, Resursos minerales de Colombia: Inst. Nacional de Investigaciones Geológica-Mineras, Publ. Geol. Esp., v. 1, p. 237-268.

Notholt, A. J. G., 1980, Igneous apatite deposits, mode of occurrence, economic development and world resources, *in* R. P. Sheldon and W. C. Burnett, eds., Fertilizer mineral potential in Asia and the Pacific: Honolulu, Hawaii, East-West Center, p. 263-268.

Reimers, C., 1985, Relationship of bacterial mats to phosphate mineralization, *in* W. C. Burnett and S. R. Riggs, eds., Neogene to modern phosphorites: Cambridge, England, Cambridge University Press.

Riggs, S. R., 1984, Paleoceanographic model of Neogene phosphorite deposition, U.S. Atlantic continental margin: Science, v. 223, p. 123-131.

Riggs, S. R., in press, Model of Tertiary phosphorites on the world's continental margins. NATO Advanced Research Institute, *in* P. G. Taleki, ed., Marine minerals: resource assessment strategies: Boston, Reidel Publishing Company.

Rodriguez, S. E., in press, Phosphorite deposits of Venezuela, *in* A. J. G. Notholt, R. P. Sheldon, and D. F. Davidson, eds., World phosphate resources: Cambridge, England, Cambridge University Press.

Rogers, C. L., Z. DeCserna, E. Tavera, R. Van Vloten, and J. Ojeda, 1961, Reconocimento geológico y depositos de fosfatos del norte de Zacetac y areas adyacentes en Coahuila, Nuevo Leon y San Luis Potosi: Consejo de Recursos Naturales No Renovables Boletin, v. 56, p. 322.

Scotese, C. R., R. K. Bamback, C. Barton, R. Van Der Voo, and A. M. Ziegler, 1979, Paleozoic base maps: Journal of Geology, v. 87-3, p. 217-277.

Sheldon, R. P., and E. Perez, 1963, Phosphate potential of northern Chile: U.S. Geological Survey Project Report series, p. 7.

Sheldon, R. P., 1964, Paleolatitudinal and paleogeographic distribution of phosphorite: U.S. Geological Survey Professional Paper, v. 01-C, p. C105-C113.

Sheldon, R. P., 1970, Sedimentation of iron-rich rocks of Llandovery age (Lower Silurian) in the southern Appalachian Basin: Geological Society of America Special Paper, v. 102, p. 107-112.

Sheldon, R. P., 1980, Episodicity of phosphate deposition and deep ocean circulation—a hypothesis: Society of Economic Paleontologists and Mineralogists Special Publication, v. 29, p. 239-247.

Sheldon, R. P., 1981, Ancient marine phosphorites: Annual Review of Earth and Planetary Sciences, v. 9, p. 251-284.

Sheldon, R. P., 1983, Phosphate resource development in Colombia—a preliminary report: unpublished report to Instituto Nacional de Investigaciones Geologico-Mineras.

Sheldon, R. P., in press-a, The association of phosphatic and siliceous sedimentary deposits, *in* Sedimentary siliceous rock hosted ores and petroleum: Stroudsburg, Pennsylvania, Van Nostrand Reinhold Company, Inc.

Sheldon, R. P., in press-b, Equatorial upwelling origin of the Early Tertiary equatorial-belt family of phosphorites—were they caused by the shadow of a tektite ring system orbiting the Earth?, *in* Proceedings of International Geological Correlations Project 156: Annual international field workshop/seminar, Morocco-Senegal, October 1983.

Sheldon, R. P., in press-d, The temporal and geographic distribution of post-Eocene marine phosphorites and their paleoceanographic and paleoclimatic problems, *in* W. C. Burnett and S. R. Riggs, eds., Neogene to modern phosphorites: Cambridge, England, Cambridge University Press.

Strakhov, N. M., 1960, Principles of lithogenesis: Izd. AN SSSR, Moskva, v. 3.

Szekely, T. S., and L. T. Grose, 1972, Stratigraphy of the carbonate, black shale, and phosphate of the Puccara Group (Upper Triassic-Lower Jurassic), central Andes, Perú: Geological Society of America Bulletin, v. 83, p. 407-428.

Torez, L., 1984, Exploracion de fosfato en el Ecuador, *in* La Roca Fosfórica: Cochabamba, Bolivia, GLIRF, v. 1, p. 53-60.

Vail, P. R., R. M. Mitchum, Jr., and S. Thompson, III, 1977, Global cycles of relative change of sea level, *in* C. E. Payton, ed., Seismic stratigraphy—applications to hydrocarbon exploration: American Association of Petroleum Geologists Memoir 26, p. 83-89.

Vail, P. R., J. Hardenbol, and R. G. Todd, 1984, Jurassic unconformities, chronostratigraphy, and sea-level changes from seismic stratigraphy and biostratigraphy, *in* J. S. Schlee, ed., Interregional unconformities and hydrocarbon accumulation: American Association of Petroleum Geologists Memoir 36, p. 129-144.

Valdebenito, M. E., in press, Phosphate deposits of the Caleta Herradura Formation, Mejillones Peninsula, Chile, *in* A. J. G. Notholt, R. P. Sheldon, and D. F. Davidson, eds., World phosphate resources: Cambridge, England, Cambridge University Press.

Wilkinson, A. F., 1982, Exploration for phosphate in Ecuador: Trans. Inst. Mining and Metallurgy, v. 91, sec. B, p. 130-145.

Wilson, J. J., 1963, Cretaceous stratigraphy of central Andes of Perú: American Association of Petroleum Geologists, v. 47-1, p. 1-34.

Ziegler, A. M., S. F. Barrett, and C. S. Kazmer, 1982, Geologic history of the upwelling zones off western South America, *in* Geology of phosphorite deposits and problems of phosphorite genesis: Novosibirsk, U.S.S.R., Academy of Sciences of the USSR Siberian Branch, Institute of Geology and Geophysics.

Ziegler, A. M. C. R. Scotese, and S. F. Barrett, 1982, Mesozoic and Cenozoic paleogeographic maps, *in* P. Broche and J. J. Sunderman, eds., Tidal friction and the earth rotation II: Berlin, New York, Springer-Verlag, p. 240-252.

Ericksen, G. E., Cañas Pinochet, M. T., and
Reinemund, J. A., editors, 1989, Geology of the Andes
and its relation to hydrocarbon and mineral resources:
Houston, Texas, Circum-Pacific Council for Energy
and Mineral Resources Earth Science Series, v. 11.

CHAPTER 21

GOLD-SILVER OCCURRENCES OF THE EL INDIO BELT, CHILE

Gordon Siddeley
Toronto, Canada

Ramon Araneda
Compañía Minera San José
La Serena, Chile

ABSTRACT

Gold-silver prospects of the El Indio belt in the high Andes of the IV Region occur over a discontinuous strike length of 150 km, and have remarkable lithological, mineralogical, and tectonic similarities with the El Indio mine area. A comparison is made among some of these prospects and El Indio in the light of exploration experience over the past 5 years, with the conclusion that they are genetically similar. The areas described—El Indio, Tambo, Sancarrón, and Nevada—are upper Tertiary epithermal volcanic deposits, characterized by widespread hydrothermal alteration under strongly acidic and oxidizing conditions. Pyroclastic volcanics (dacite tuffs) and lavas are the dominant rock types, which are concentrated in well-defined eruptive centers elongated north-south.

Ores are found in veins, shoots, and breccias that have strong tectonic control. Gold, silver, and copper deposits are accompanied by either massive sulfides, silicification, or sulfate deposition. In common with some occurrences of the U.S.A. (Summitville, Colorado; Goldfield, Nevada), a predominant feature is the abundant sulfur occurring as sulfates, sulfides, sulfosalts, and native sulfur.

The potential economic importance of deposits of these types is clearly shown by El Indio, where combined production since 1979 and proven reserves (1988) are more than 100 t of fine gold.

INTRODUCTION

Prior to 1975, little attention was paid to the Miocene/Pliocene volcanics of the high Andes as potential hosts for precious-metal deposits. The first mention that a precious-metal "belt" might exist was made by Llaumett and Henriquez (1976) at the First Chilean Geological Congress, following the recent discovery of El Indio deposits (Figure 1). Further geological studies were carried out by Maksaev et al. (1984) for the Servicio Nacional de Geología y Minería (SERNAGEOMIN). As El Indio developed, the St. Joe Minerals staff published information on the detailed geology and mineralogy (Araneda, 1982; Jannas and Araneda, 1985; Walthier, Araneda, and Crawford, 1982).

The potential of geologically young volcanic centers of the Chilean Andes was recognized by St. Joe Minerals Company geologists, and that company took a lead in acquiring property for exploration and development. The spectacular increase in gold prices during the latter half of the 1970s stimulated the search for mineral deposits in the high Cordillera, away from the porphyry copper prospects of the open, more accessible lowlands of Chile. The best guides to identification of favorable prospect ores for precious metals were determined to be the well-defined color anomalies, visible from aircraft, and in air photos and satellite imagery. These color anomalies identify many volcanic centers that have undergone hydrothermal alternation in the Andes of easternmost Chile from south of Santiago to Perú. The centers are variously eroded, but most retain remnants of volcanic landforms. To date, gold prospects and deposits have been found to be more prevalent in the zone between lat. 26° and 30°S. To the north, in similar host rocks, silver appears to be the more important precious metal, whereas to the south, copper is of greatest economic importance.

EL INDIO

Geological Environment

The El Indio mine is on the northern side of the volcanic dome of Cerro Canto (Figure 2), on the inner wall of a well-defined crater now marked by the small stream, Rio Malo. Tambo mines are located 5 km away, on the southeastern flank of the same formation. Cerro Canto is the dominant topographical feature of the mine area, rising to 4550 m above sea level. Its strongly elliptical shape in plan view suggests a

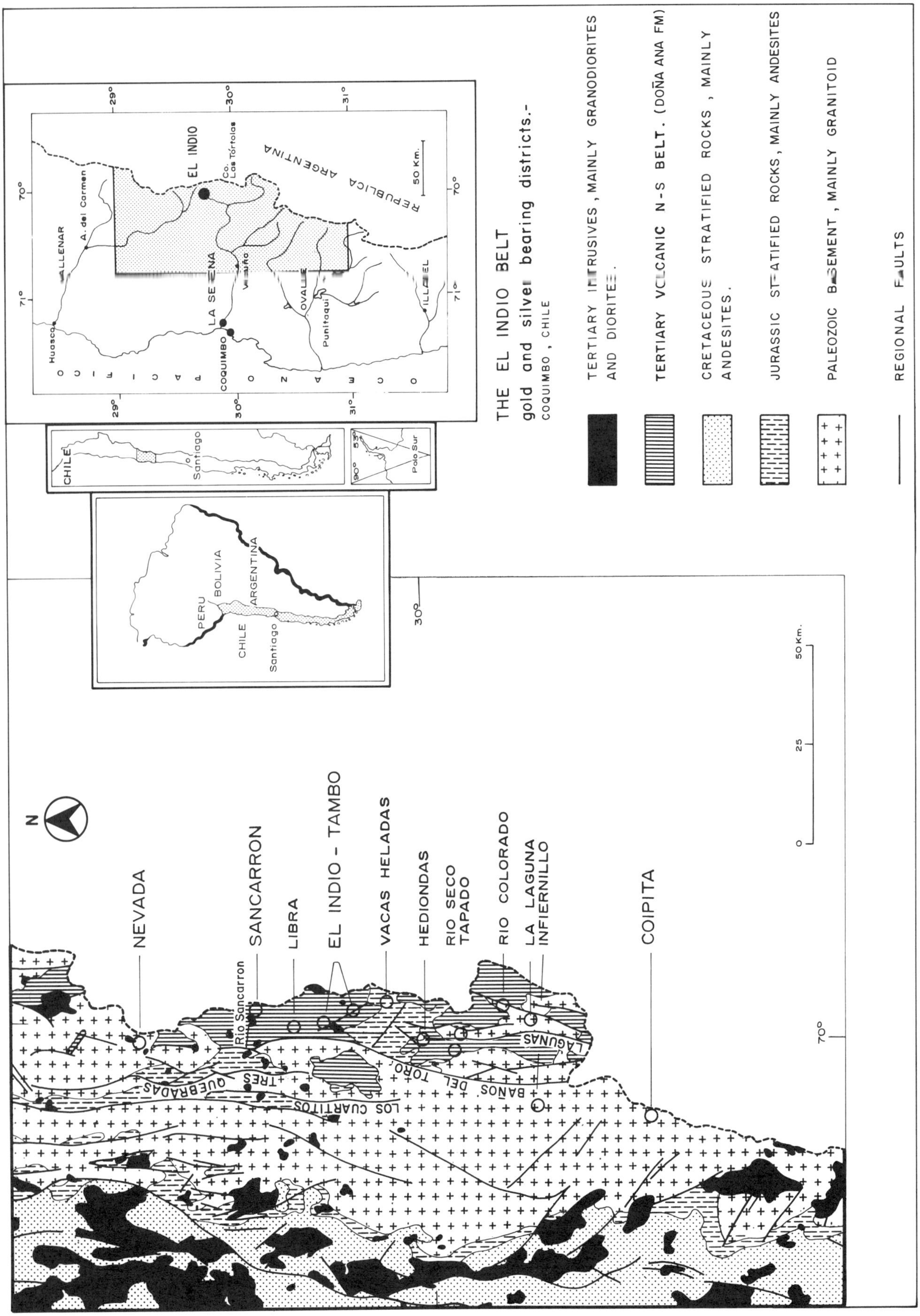

Figure 1. Generalized geological setting of the Cordillera de los Andes, Coquimbo region (after Maksaev et al., 1984). Main gold and silver occurrences of the El Indio belt are shown.

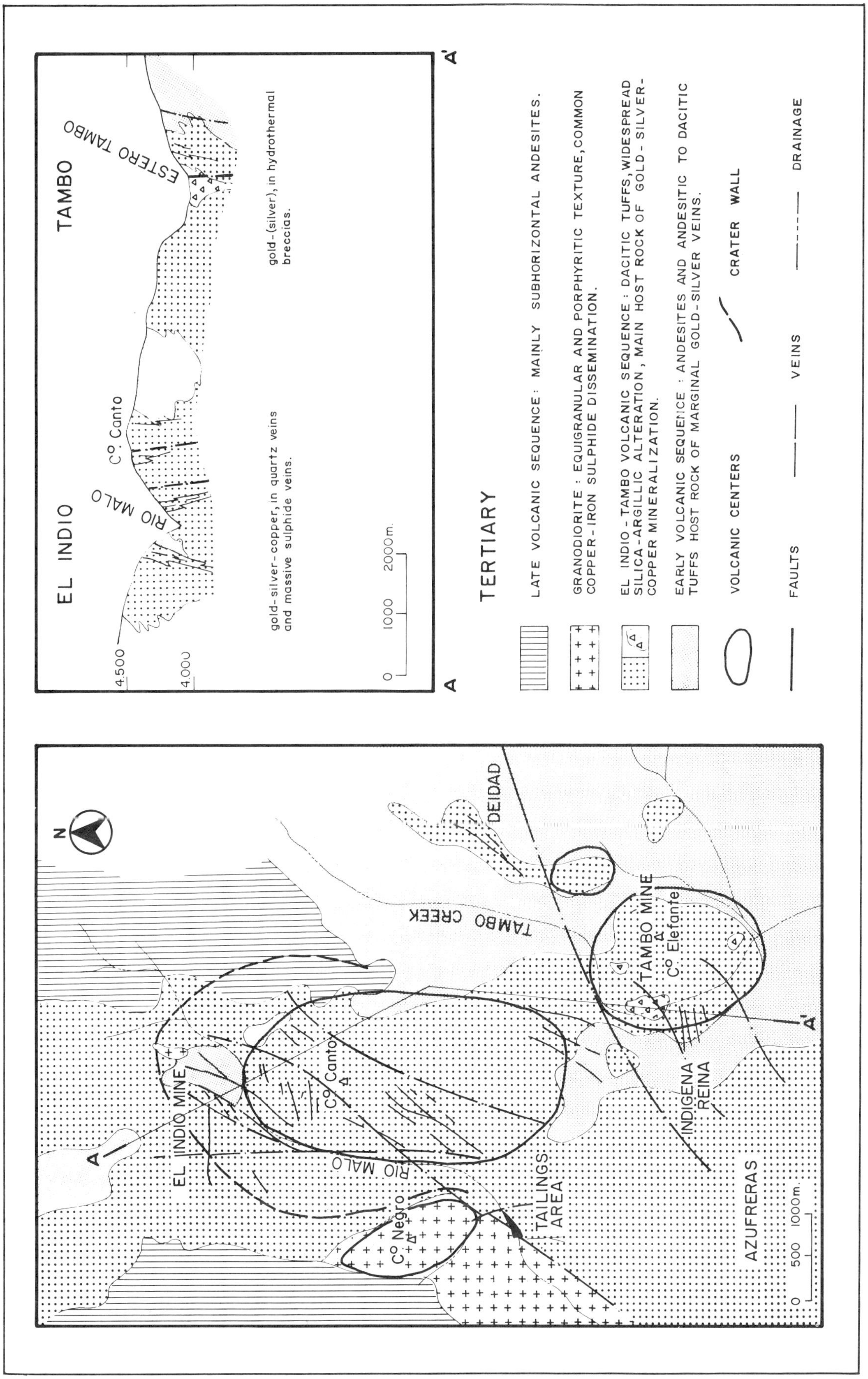

Figure 2. Generalized geological plan and section of El Indo-Tambo district, Coquimbo, Chile.

north-south fissure accumulation rather than a simple cone accumulation from a central vent. Smaller satellite or parasitic cones occur peripheral to or superimposed upon the earlier volcanic body. Cerro Negro, west of the approach road to El Indio, is the site of late intrusive activity, representing a subvolcanic neck of grandiorite. This neck presumably was emplaced beneath a now-eroded satellite volcanic cone.

The present-day height of the crater wall on the north side indicates that a block subsidence of some 350 m is likely to have affected the Cerro Canto dome, along the circular fault structure marked by the Rio Malo. However, movement must have occurred prior to mineralization, because veins of the Viento/Viejo system cross the Rio Malo between the crater wall and the northern part of the Cerro Canto dome without apparent dislocation. To the south, the crater wall is breached by the Rio Malo, which has been responsible for a considerable amount of erosion of the volcanic pile on its southwestern side. The Rio Malo cut down into the "basement" andesites, a layered series of lavas, agglomerates, and andesitic tuffs considered by Maksaev et al. (1984) to be between Late Jurassic and middle Tertiary in age.

The volcanic rocks of Cerro Canto are primarily pyroclastic types, dacitic and andesitic tuffs, agglomerates, and ashes, with minor rhyolite/dacite flows near the El Indio mine area. Pumiceous tuffs and almost flat-lying native sulfur/alunite thermal-spring deposits are the youngest manifestations of volcanism at Cerro Canto. These spring deposits, on the south side of Cerro Canto at Azufreras, have been worked locally for sulfur. Northwest of the mine the dacite tuffs wedge-out horizontally over the basement andesites, but to the east the contacts are unclear, and fine-grained black trachytes in that area may cover the hydrothermally altered Cerro Canto volcanic rocks.

Hydrothermal alteration affects the entire volcanic sequence at Cerro Canto, as well as a considerable zone of basement andesites along a north-south trend. This prominent belt of alteration extends some 30 km north-south by 5 km east-west (Figure 1), on both sides of Cerro Canto, from Rio Sancarrón to south of the Hediondas. Younger volcanic rocks of the Doña Ana Formation are little affected by the hydrothermal alteration, which suggests that hydrothermal activity was largely finished by the end of the Cerro Canto volcanism. The strong argillic alteration, affecting both the dacitic pyroclastic rocks of Cerro Canto and nearby andesites, destroyed original rock textures, which has caused problems in mapping. Less altered basement andesites are found in the Vacas Heladas area (Figure 1) and along the approach road south of El Indio. This alteration is propylitic and hematitic, in keeping with the regional alteration of these rocks. Blocks of considerably altered basement andesite are found in the overlying Cerro Canto Formation near El Indio mine.

Erosion over the summit of Cerro Canto is not likely to have been more than 200 to 300 m, as is suggested by the still-sharp topographic profile of the volcanic center and the undisturbed sinters at Azufreras. Preliminary studies of the thermometry of fluid inclusions in El Indio quartz veins also indicate very shallow deposition of these associated ores.

Radiometric (K/Ar) dating of the altered rocks in the area by Maksaev et al. (1984) and Araneda (1982) show two age groups—16 to 17 Ma (early Miocene) and 8.7 to 11 Ma (late Miocene). At El Indio, propylitic andesite, possibly basement andesite, gave a whole-rock K/Ar age of 10.7 + 0.7 Ma, and altered wall rock of the Indio Sur vein gave 8.6 + 0.4 Ma. These ages suggest two separate phases of hydrothermal alteration in the Miocene. They also indicate that the basement andesites of the Baños del Toro Formation may be younger than formerly thought, and might even be transitional into the Miocene volcanics of Cerro Canto.

Mineralization

Most of El Indio's ore comes from a 200 × 400 m area between the Rio Malo and the rim of the north crater wall. The "block" is defined by two northeast-trending faults known as the Inca Norte and Inca Sur faults, both of which are mineralized. Between these faults are somewhat arcuate secondary faults, trending generally east-northeast, which have also mineralized. Some of these veins show a "cymoid loop" configuration (Araneda, 1984). Beyond the Inca block, other steeply dipping veins such as Campana, Viento, Viejo, and Maria occupy northeast- to east-trending faults. The strong structural pattern is evident in air photos of the area. Faulting must have occurred not long after the period of active volcanism in order to accommodate the late-stage epithermal vein deposition. Geological surveys of mine workings show that limited dislocation of about 50 to 80 m occurred along these same structures after mineralization.

Sulfide Ores

Sulfide ores have accounted for the bulk of El Indio's production. Two veins—Mula Muerta and Viento—provide more than 80% of this type of ore. Vein widths range from centimeters to 30 meters at the widest stopes. The maximum vein length of mineable ore is 150 m for Mula Muerta and 300 m for Viento. Exploration suggests that Viento may be as much as 800 m long.

The dominant minerals in the sulfide ores are enargite (Cu_3AsS_4) and pyrite, which are intimately mixed in a coarsely crystalline aggregate; minor tetrahedrite, chalcopyrite, sphalerite, and hubnerite also are present. These minor minerals increase slightly with depth, but enargite and pyrite continue to dominate. The sulfide ores average 8% Cu, 3 g/t Au, and 150 g/t Ag. They are the source of El Indio's copper, and the precious-metal content is of lesser economic importance. The gold is associated with pyrite (analyses have shown that pyrite may contain as much as 20 g/t Au), or with small amounts of later-introduced quartz. The Viento vein locally contains as much as 15 g/t gold, in zones of massive sulfides where quartz occurs as a minor constituent.

Quartz Ores

Gold-bearing quartz veins contain the high-grade precious-metal ores of El Indio. The gold content of individual veins ranges widely, with a typical gold content of 12 to 20 g/t. The bonanza ores of the "3500" Inca Sur and Contacto veins have yielded direct smelting ores containing more than 150 g/t Au. Isolated values in excess of 7000 g/t Au have been encountered. The high-grade parts of the bonanza veins are everywhere enclosed by lower grade "plant" ores, and the current mining system is to selectively extract the high-grade material 1 or 2 m in advance of the plant ore. The "3500" Inca Sur vein (described in detail by Jannas and Araneda, 1985) has been the most productive of the bonanza veins, yielding more than 30 MT of fine gold. This represents 30% of the total (mine production + known reserves) precious-metal values of El Indio, all the more astonishing when one considers that the vein is only 130 m long, 3 m wide, and extends to a presently known depth of 270 m.

The quartz veins cut the massive sulfide veins at several locations and are therefore younger. They also dip more steeply (70° to 80°) as compared with the sulfide veins (50° to 70°). Both types of veins appear to steepen with depth.

Gold occurs mostly as free gold, in the size range of 5 to 50 microns. Visible gold is rare, but is seen in fine-grained aggregates or in hairlike clusters in quartz or around clasts. The silica host is generally the gray cryptocrystalline variety, but chalcedonic and drusy forms also occur. At least 7 pulses of silicification are identifiable in hand specimens, caused by repeated fracturing of the vein and recementation by silica. Pyrite and enargite are also present in minor amounts, generally evenly disseminated throughout the quartz vein. In the Indio Norte ore zone, enargite has been leached to a depth of more than 100 m below the surface. This may have been the result of hydrothermal acid-leaching from below, rather than supergene alteration. In other parts of the deposit, supergene alteration penetrates only about 10 m.

The wall rocks—dacitic or andesitic tuffs—are moderately to strongly silicified and sericitized, forming an ill-defined siliceous envelope around the gold-quartz orebodies. In contrast, wall rock alteration around the massive sulfide veins is strongly argillic and is not silicified.

Weathering

Both the sulfosalts and the sulfides disintegrate on weathering, and where abundant colluvium from the fragmented volcanics is mixed with near-surface weathered mineralized material, the orebody may be camouflaged. The primary ore minerals leach readily in the weathered zone, and their former presence may be indicated only by crystal-shaped voids in quartz and by local staining of secondary minerals such as the sulfates chalcanthite ($CuSO_4.5H_2O$) and osarizawite ($PbCuAl_2(SO_4)_2(OH)_6$). Near-surface metal values indicate various degrees of leaching or enrichment, as follows:

Indio Norte (Open pit, at 4200 m altitude)—58 g/t Au, 366 g/t Ag, 0.1% Cu
Indio Norte (Underground mine, at 4000 m altitude)—53 g/t Au, 44 g/t Ag, 3.7% Cu

Whereas the behavior of copper (leached) and silver (enriched) at the surface is known, that of gold is difficult to determine. Various 1978 assays for the El Indio, Campana, and Viento veins gave a 2:1 "surface enrichment factor" for gold over samples from the same veins 30 m below. However, because of the normally erratic distribution of gold in the primary ores, these few samples may not be statistically valid.

Geochemical surveys for gold, silver, and arsenic were useful for detecting orebodies. Arsenic is the most widely dispersed, silver is less dispersed, and gold is least dispersed. Anomalous values of Au (+1 ppm) and Ag (+15 ppm) were found over an area of 500 × 300 m near the discovery site, whereas anomalous As values (+1000 ppm) are over a 1500 × 750 m area that includes the area of anomalous Au and Ag values.

TAMBO

Geological Environment

Tambo is southeast of the Cerro Canto dome, near its southern margin (Figure 2), and occupies a geological habitat similar to that of El Indio. The crater limits are not well defined at Tambo, where they are obscured by haphazard pyroclastic accumulations and possible avalanche deposits originating from the Cerro Canto vent area. Ambrus (1983) believes that the lack of stratification over most of the Tambo area resulted from "intravolcanic" activity, whereby the pyroclastic material came from many local vents in the Tambo area. Such deposits were reworked during phases of multiple eruptions with nearby volcanic units. Under certain circumstances, such reworking would have obliterated evidence of earlier cone accumulations. However, there is evidence that two major satellite cones—the largest being Cerro Elefante and the other a neighboring unnamed peak about 500 m east—were responsible for the accumulation of at least 500 m of dacite pyroclastics and andesite lavas.

In the Tambo valley, an estimated 200 to 300 m of erosion has exposed the roots of several former parasitic vents on Cerro Elefante, now exposed as hydrothermal breccia pipes and small intrusive feldspar-porphyry necks. An extrusive equivalent of this feldspar porphyry has not been recognized. This leads to the conclusion that at the time of emplacement, the ascending magma had a high gas content so that it disintegrated explosively, upon reaching the surface, to form the dacite tuff typical of this area.

Along the valley 1 to 5 km south of Tambo are well-stratified ash-flow tuffs which contain some densely welded and silicified units. These are tuffs that originated from nearby vents but flowed southward away from the zone of intense volcanic activity where stratification could develop.

The predominant rock type is dacitic (lithic) tuff, with subordinate dark rhyolite and quartz latite interbedded with dacite in the area of Deidad. At Azufreras (Figure 2) reddish jasperoidal andesites and andesite porphyries underlie volcanic ash impregnated with native sulfur and alunite. Jasperoidal domes at Azufreras may represent surface deposits (silica caps?) covering potential ore-bearing breccia pipes. Ash layers with pumice are the younger volcanic materials in this area.

Mineralization

Two types of ores—veins and hydrothermal breccias—are present at Tambo. Both are structurally controlled, favoring sinuous, north to northeast fractures that show no evidence of strong displacement. The orebodies are discrete features, occurring between altitudes of 4000 and 4550 m (a vertical distance of at least 550 m). The veins occupy simple tension fractures (with or without postore movement) or fractures related to the breccia bodies.

The Reina vein is the best studied orebody at Tambo to date, having been explored to a depth of 210 m from the surface. Old prospect pits at this site prompted the first underground testing by Compañia Minera El Indio at Tambo. The vein crops out in the center of a siliceous breccia pipe which is otherwise barren; at depth the vein separates into three lens-shaped ore shoots (approximately 14 m long × 1.5 m wide), which are aligned east-west and dip steeply. At the deepest level, there is evidence that the shoots once more converge. Coarsely crystalline barite and alunite are the major gangue minerals, accompanied by silicified fragments of wall rock. Gold is strongly associated with the barite, occurring as visible grains along cleavage planes and also as disseminated grains throughout crystals. Assay values of more than 1 kg/t Au are common where barite is the major vein constituent. Fine-grained gray native tellurium accompanies the gold. Neither the alunite nor the silicified clasts contain more than traces of gold or silver.

Wall rock in the immediate vicinity of the veins is intensely silicified, but of diminishing intensity away from them, and breccia bodies are moderately silicified throughout. At the surface the zone of intense silicification may be 50 m wide, whereas at a depth of 150 m it narrows to 15 m. Intense silicification near veins obliterates rock textures, and the rock now consists of a cryptocrystalline aggregate of quartz and finely disseminated pyrite. Despite the proximity to ore, precious-metal values in this altered rock are rarely above assay detection level (100 ppb Au, 1 ppm Ag).

Two hundred meters north of the Reina vein the Indigena vein (Figure 2) follows a cymoid fault structure over a 110-m strike length (width 0.4 to 2 m), displaying the same barite/alunite mineralogy as Reina near the surface. However, on the 4100-m level, 40 m below, the surface barite and alunite are almost nonexistent, giving way to a spongy silicified vein with scattered enargite, the first to be detected at Tambo. The texture is similar to the leached Indio Norte vein at El Indio. The vertical change from barite/alunite to silica/enargite takes place over a few meters and implies sharp changes during ore deposition, from a reducing sulfide environment at depth to an oxidizing sulfate environment near the surface.

At Deidad (Figure 2), veins occupy five parallel, brecciated fault structures, in an en echelon pattern extending over 300 m horizontally. The most persistent individual vein is 175 m long; widths range from 0.2 to 2 m. Gold values range from 11 to 17 g/t, and are related to barite content of the barite/alunite/silica/hematite gangue. Locally, the veins are affected by postore faulting and brecciation.

Breccia ores occur in the hydrothermal breccia pipes and bodies of Tambo. At least a dozen have been identified and mapped, and probably tens more remain undetected beneath the shallow colluvial cover. A typical breccia body is roughly round or elliptical and about 100 m in diameter. The centers of the bodies contain rotated blocks and pebble breccias, indicating zones of strongest movement and attrition. Toward the edges of the bodies the host rock is less fractured and disturbed, with a consequent reduction in void space for the ore deposition that followed fracturing. The bulk of the breccia is formed of blocks, clasts, and fine debris of the host dacite tuff, variously silicified. Drill evidence indicates that these breccias narrow or funnel at depth, giving way to restricted shoot or vein deposits, such as Reina, in the narrow feeder zones.

After brecciation, the clasts and host rock were subjected to intense silicification. Individual clasts developed 3- to 4-cm "rinds" of cryptocrystalline silica. Away from the central zone of brecciation, silicification diminished as permeability decreased. Subsequently, crystalline barite was deposited in voids in the breccia, bringing with it gold that accumulated along cleavage planes and surfaces of barite crystals. Alunite, lesser jarosite, and rare rodalquilarite ($H_3Fe_2(TeO_3)_4Cl$) completed the paragenesis.

The gold grains range from micron size to 1/2 mm, but average about 50 microns. Bulk metallurgical testing indicates more than 80% free gold, including significant amounts of coarse gold; 40% is more than 200-mesh. Although strongly correlated with the barite, some gold grains are encapsulated by quartz and alunite, indicating that gold deposition started earlier and finished later than the barite stage. Free gold predominates among the gold-bearing minerals, but calaverite and gold tellurates also are present. Native tellurium and the rare silver minerals iodyrite and matildite have been recognized. Silver is contained also in other sulfosalt minerals of 50-micron size or less.

Both in the vein and breccia ores, silver has a general, but not close, affinity to gold, and ratios of Au/Ag range from 1000:1 to 1:1000. The Tambo ores contain only traces of copper, typically 100 ppm, except for of the lower Indigena vein, which contains as much as 0.5% Cu.

SANCARRÓN

Geological Environment

The Sancarrón volcanic center is 12 km north of El Indio (Figure 3), where the central peaks, Cerros Don Lucho and Tio Pepe, consist of a sequence of hydrothermally altered andesite tuffs, dacite tuffs, and crackle breccias more than 1000 m thick. These altered volcanic rocks extend north-south over a 4-km-long narrow belt. As at El Indio, the sequence cuts through a basement of mauve-colored andesite flows and agglomerates. At Sancarrón the rocks were mapped by Maksaev et al. (1984) as part of the Cerro Tórtolas Formation, considered to be middle to upper Miocene. With the overprint of hydrothermal alteration, no clear contacts are visible between the basement and the overlying volcanic sequence. Toward the base, however, andesitic tuffs and agglomerates predominate, perhaps indicating a transition from older andesites to the more felsic dacite tuffs found at the upper levels of Sancarrón. Limited stratigraphic evidence suggests that the Sancarrón volcanic pile has been tilted more than 40° from its original position. This movement probably caused the intense and widespread shattering of the sequence.

Stratification is largely obscured by the coarse textures of the extrusives, the effects of widespread alteration, shearing, and the disruptive late-stage intrusive breccias. The most common rock type—dacite tuff—is a moderately to densely welded, devitrified ignimbrite having various amounts of pumice. The tuffs have undergone considerable argillic alteration, and the shards and feldspar groundmass is almost completely converted to dickite, kaolin, and sericite. Superimposed on this argillic alteration is an intense solfataric alteration and introduction of predominantly sulfate minerals, chiefly alunite, but also anhydrite, gypsum, selenite, jarosite, and native sulfur. Alunite replaces the groundmass of the tuffs, but also occurs as pure porcellaneous veins as much as 1 m thick, and is present within the fine-grained matrix of stock-

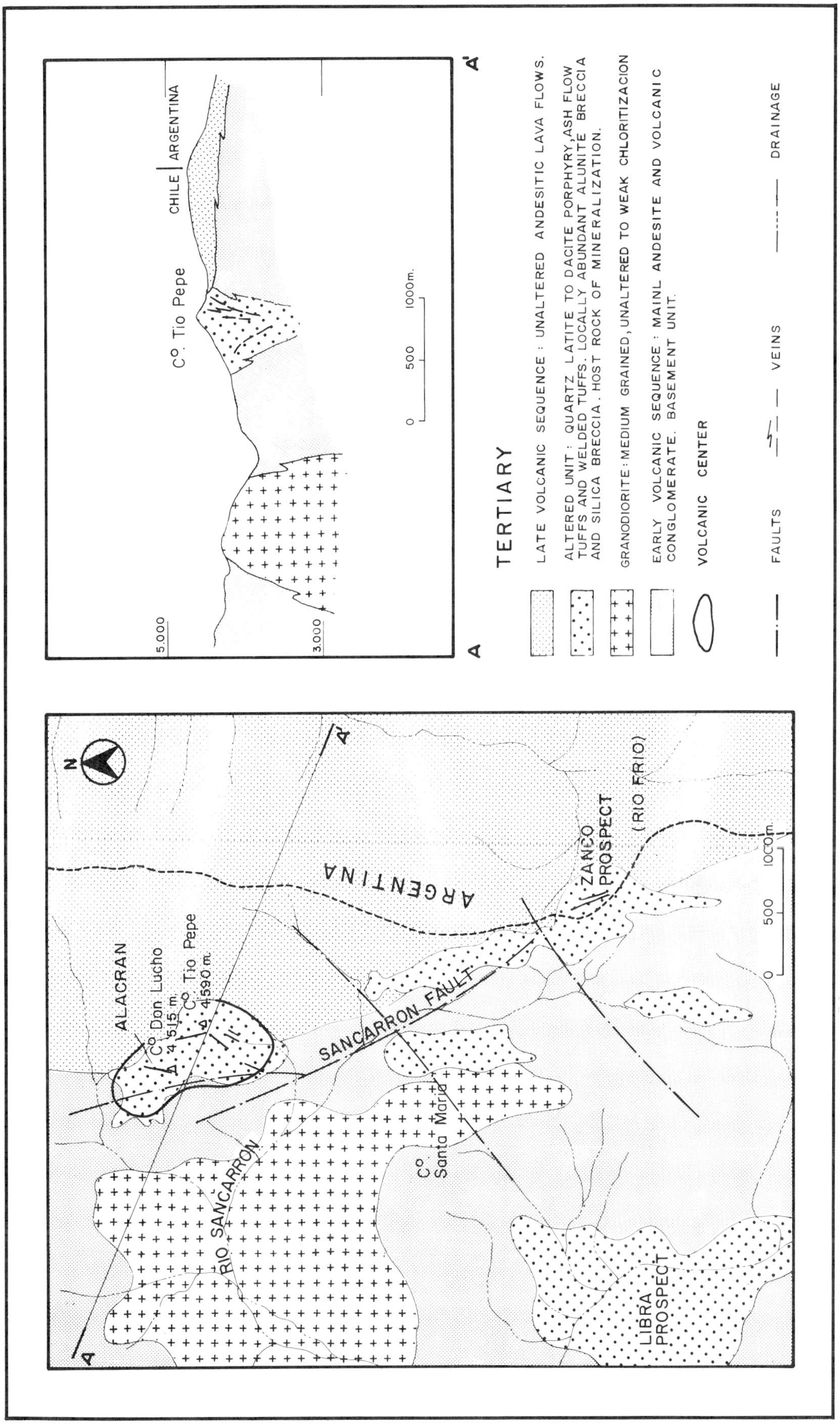

Figure 3. Generalized geological plan and section of Sancarrón project, Coquimbo, Chile.

works and breccias. Vugs in the tuffs contain coarsely crystalline alunite and native sulfur. Nearby, at the Libra prospect, sedimentary textures of alunite in tuffs indicate deposition as sulfate sinters. On the flanks of Cerro Tio Pepe, fragmentary evidence from drill cores indicates that the sulfates diminish at depth.

Silicification is much less pronounced, and less evident at Sancarrón than at any of the other areas described, tending to be concentrated toward the higher ground and associated with alunite veining. Generally, the silicification is weak to moderate, affecting shear zones along a few meters length in any direction. Small amounts of native sulfur and lenses and pods of coarsely crystalline barite have been found along one such shear zone high in the volcanic sequence.

Mineralization

Sancarrón contains several types of epithermal ores, and the extremely permeable host rock promoted diffusion of mineralization over wide areas. Gold is erratically distributed in the silicified shear zones and the silica-barite pods previously mentioned. The small number of silica-bearing veins, at the summit of Tio Pepe and across the Argentine border at Rio Frio, once again display the barite-gold relationship high in the volcanic sequence. Barite ores may assay more than 100 g/t Au, but quantities of such ore are small because veins appear to be restricted at most to a few tens of meters horizontally and vertically, and widths are less than 30 cm. Native gold is concentrated in grains along cleavage planes of barite and in small quartz-lined vugs in strongly silicified wall rock.

One occurrence of disseminated gold (3 to 10 g/t Au) is found in a banded silica-alunite sinter at the summit of Cerro Don Lucho. This sinter, within dacite tuffs dipping at 60° to 70°, was interpreted as a fissure vein until microscopic examination revealed sedimentary features such as graded bedding and slump structures. The unit consists of many thin (1 mm) layers of devitrified silica, colloidal and spongy in hand specimen, with much powdery sulfate minerals in the matrix. Visible gold is present, but metallurgical tests show that 70% is <50 micron size, and 40% of the total is <6 microns. The steep dip of this ore-bearing horizon implies substantial post-depositional tilting.

Low values of gold and silver accompany enargite-pyrite-bearing stockworks and veins which are less than 50 m in length. These veins, which dip from 20° to vertical, commonly "horse-tail" and pinch out.

Micron-size native gold is disseminated throughout the more silicified alunite breccias. In some of the alunite breccias, clasts of enargite-pyrite ore as large as 1 m have been caught up in the intrusive debris of pre-existing tuff. The breccias in turn are cut by unbroken stringers of enargite-pyrite. Therefore, sulfide mineralization both predates and postdates brecciation. At the surface, the enargite is almost entirely oxidized to scorodite ($Fe^{+3}ASO_4.2H_2O$). Most of the copper has been leached, but the silver content is enriched (as much as 2000 g/t has been reported from assays).

Essentially, gold values are linked to silicification, whether in the weakly altered shear zones or in the silicified matrix of alunite-bearing breccias. The most persistent gold-bearing structure is the sinter on Cerro Don Lucho, which has a 120-m strike length. Together with patchy gold values in adjacent tuffs, this body has the potential of a 1-million ton ore reserve of 2 to 3 g/t Au.

Sulfide minerals appear to be restricted to small veins at lower levels in the volcanic sequence.

NEVADA (SIERRA NEVADA, ATACAMA REGION)

Geological Environment

The Nevada prospect (Figure 4), 53 km north of El Indio, lies on the Chile-Argentine border at an altitude of 4800 m. Unlike the El Indio/Tambo and Sancarrón pyroclastic accumulations, Nevada is characterized by a thick (+800 m) sequence of Tertiary lavas which are predominantly dacites, rhyolites, and monzonites (quartz porphyries). These rocks, which cover an area of 4 × 2 km, may be best observed on the Argentine side of the border where considerable erosion has exposed the sequence. The severe erosional features and the lack of definition of volcanic doming indicate that the sequence probably was originally much more extensive. The basement rocks are coarsely crystalline granites (adamellites) and monzonites typical of the Chilean Paleozoic elsewhere. The volcanic trough of Tertiary andesites which hosted the El Indio and Sancarrón volcanics does not continue to Nevada, but appears to swing east into Argentina as a result of major flexuring of the boundary faults. Many felsic to mafic dikes of Tertiary age cut the basement but not the overlying volcanic sequence.

Hydrothermal alteration of the Tertiary volcanic sequence at Nevada is widespread, covering 15 km^2 on both sides of the international boundary. The argillic and sericitic alteration affects the basement rocks, and granites show considerable bleaching near contacts. Silicification is less widespread but much more intense, locally causing wholesale rock metasomatism. Most affected by silicification is a 1-km area of basement granites and overlying dacites and monzonites at Pedro's Creek. In places, the intensity of silicification is such that original rock textures have been completely obliterated and replaced by microcrystalline quartz. Widely spaced whole-rock samples have assayed as much as 97% SiO_2.

Hydrothermal breccias cut the volcanics in steeply dipping, ill-defined bodies with horizontal or gently dipping branches along pre-existing joints or bedding planes. Typically breccia bodies are on the order of 200 × 50 m in plan view. Seven varieties have been recognized, based on details of clast components and type of groundmass. The clasts originated both from basement intrusives and from Tertiary volcanics, and are cemented by quartz ("jasper" breccias) or alunite. The breccias display evidence of a high-energy emplacement, from areas of rounded pebble dikes through zones of rotated blocks to stockworks and veinlets with pure cementing material, generally alunite. Pedro's Creek, the area of intense silicification, also is the area most affected by such brecciation.

In terms of volcanic chronology, the phase of silicification predates emplacement of the alunite-bearing breccias, since the breccias contain intensely silicified clasts of the host rock without having a correspondingly silicified matrix. Alunite, as well as being a major hydrothermal constituent of the breccias, also occurs in the wall rocks for several meters away

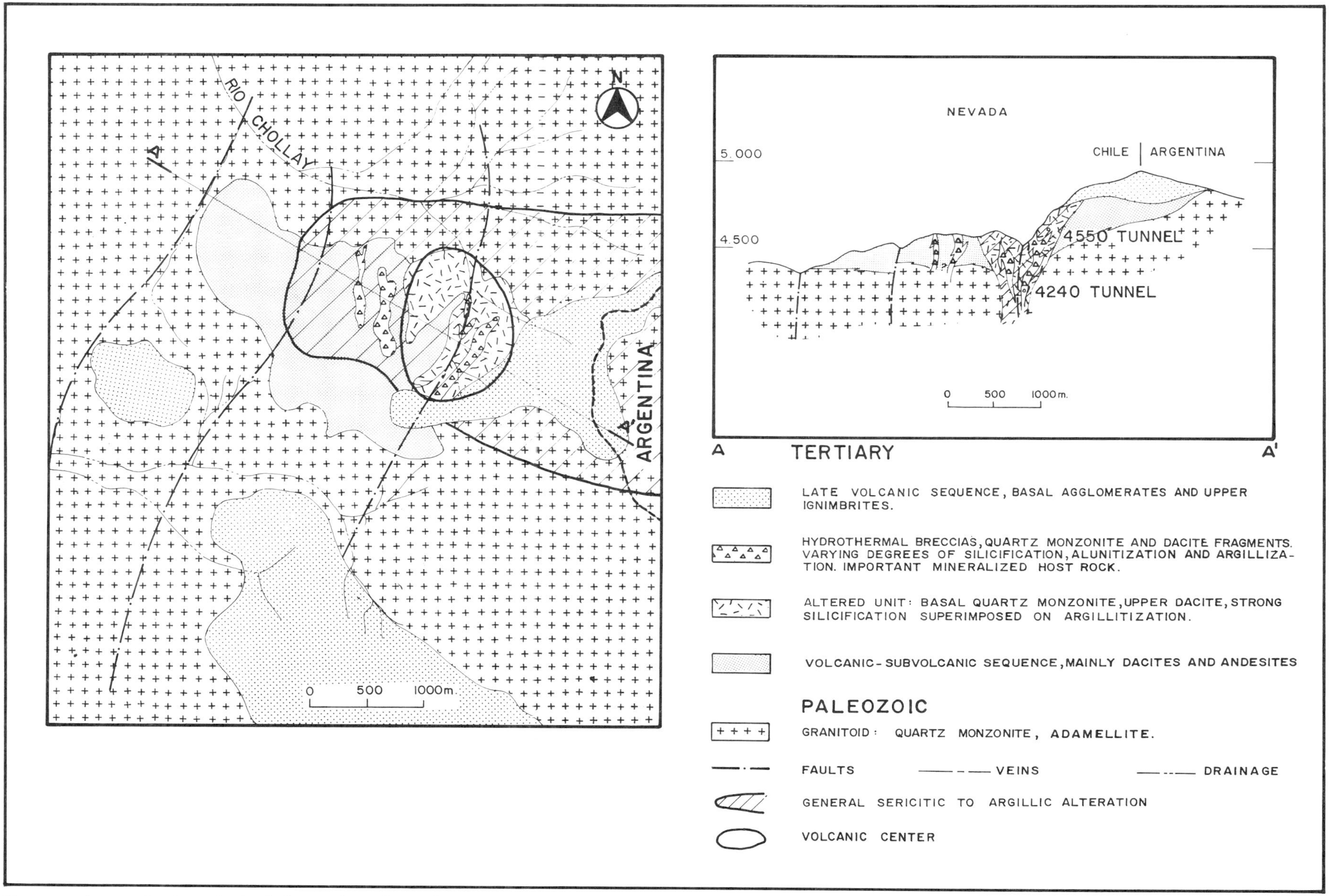

Figure 4. Generalized geological plan and section of Nevada project, Atacama, Chile.

from the breccia bodies, where it selectively replaced feldspars in the volcanic rocks.

Hydrothermal alteration appears to have proceeded in three stages: a wide aureole of argillation, inner cores of silicification, and, finally, alunitization. Proximity to ore is characterized by microcrystalline quartz and jasperoid, sericite, alunite, gypsum, jarosite, limonite, and finely disseminated pyrite.

Mineralization

Precious-metal ores were deposited in veins, stockworks, and breccia bodies that in part were formed during late-stage volcanic activity. Quartz veins and stringers, which host erratic values of gold and silver, pinch and swell, and are difficult to identify in the zones of intense wall-rock silicification. A typical vein is 50 m long, less than 1 cm to 50 cm wide, and contains 4 to 20 g/t Au. Some veins contain only trace amounts of gold but high silver (as much as 550 g/t Ag). Approximately 50 veins have been identified, some of which are clustered in groups or "swarms." They strike from northwest to northeast. Feathering is common, both vertically and horizontally. Small stockworks containing pyrite, melanterite (iron sulfate), and trace amounts of enargite, but only small amounts of Au and Ag, are associated with the quartz veins.

The breccia bodies were formerly considered to be postore, because many early assays showed that they contained only traces or were lacking in gold and silver. However, one large alunite-bearing body was found to contain disseminated fine gold at the 4240-m level, assaying as high as 110 g/t Au, but typically ranging from 1 to 4 g/t Au in the core. This gold is concentrated in the alunite-silica-rock-flour matrix, a powdery material occupying 20 to 30% of the breccia-body volume. The minus-200-mesh fraction of this matrix material has gold values as high as 10 times the overall grade of the breccia body. This gold is probably associated with the soft alunite of the matrix. The silicified clasts on the other hand, which occupy 70 to 80% of the breccia body, contain only traces of gold, if present, in amounts below detection limits.

The geochemical anomaly at Nevada covers a larger area than any of the other project areas, including El Indio. Surface samples of 80-mesh rock debris (soils are not present) have anomalous values of more than 1 ppm Au and 20 ppm Ag in an area of 800 × 400 m.

Exploration on both sides of the international borders has shown that precious-metal and minor enargite-pyrite mineralization extends at least 500 m vertically, between altitudes of 4220 and 4720 m.

CONCLUSIONS

A study of the geological, structural, and petrological features of the late Tertiary volcanic centers of El Indio (I), Tambo (T), Sancarrón (S), and Nevada (N) allows general conclusions to be made regarding the characteristics of ores and their genesis.

Along the high Cordillera of Chile, epithermal ores are hosted by dacitic pyroclastic rocks (I, T, S) and dacite and latite lavas (N) in well-defined volcanic centers showing relatively little erosion. Whereas the regional basement rocks are propylitically altered (I, T, S), both the pyroclasts and lavas of the volcanic hosts show extensively argillic alteration. Argillic alteration is a feature of the volcanism, inasmuch as the basement rocks are similarly affected in the vicinity of vents, whereas younger, overlapping volcanic rocks from other centers are unaltered (I). Alteration proceeded in several stages. Widespread argillic alteration was followed by local silicification that was most intense in the vicinity of faults and breccia zones. Finally, alunite was deposited throughout the volcanic pile, generally filling open spaces, but also replacing wall-rock feldspars in some areas (S, N). Wall-rock alteration during mineralization was not significant, perhaps because the host rocks have low porosity.

Ground preparation, which was vital to ore deposition, involved open-space jointing, tensional-fault fissuring, and formation of breccia bodies which contain abundant cavities. These openings were the major loci of ore deposition.

Mineralization occurred in various pulses, as shown by multistage silica in the El Indio quartz veins, as well as by the crosscutting relationship of these veins with the massive sulfide veins. At Sancarrón, sulfide veins cut mineralized breccias and earlier sulfide veins.

Field evidence shows that massive sulfides were followed by quartz veins, which in turn were followed by sulfates. Although all three types of veins reached the surface, limited observations suggest that the sulfides predominate at deeper levels and the sulfates predominate at or very close to the surface (T, S). Quartz veins cut both sulfide and sulfate veins, and extend over a vertical distance of at least 500 m.

Gold deposition spanned all three vein stages, with the most favorable hosts being quartz veins (I) or barite (S, T). The voluminous alunite deposits of all areas are essentially barren of precious metals, with some exceptions (N, S, T). Bonanza Au ores containing more than 100 g/t Au are found in quartz veins (I) and in barite-bearing veins and breccias (T, N).

Figure 5 attempts to place each of the areas discussed into one theoretical geological section. The position of each center is based upon an interpretation of field evidence, taking into account the level of erosion and depths of ore deposition. Fluid-inclusion studies presently under way may allow more accurate estimates to be made in the future.

ACKNOWLEDGMENTS

The information presented is the result of exploration by personnel of St. Joe Minerals and its subsidiaries over a period of several years. We wish to acknowledge the work of the many geologists who have contributed to these studies, and particularly those who were in charge of projects at various times, including P. Valenzuela, C. Gacitúa, D. Carmichael, P. Paleczeck, D. Williams, M. Luján, and E. Viteri.

REFERENCES CITED

Ambrus, J., 1983, Geology of the Deidad-El Tambo-Cerro Canto area: Internal report of Compañia Minera San José.

Araneda, R., 1982, El Indio, yacimiento de oro, plata y cobre, Coquimbo, Chile: Minerales, v. 37, p. 5-13.

Araneda, R., 1984, Nuevos aportes al conocimiento de la geología de El Indio, yacimiento de oro, plata y cobre: Minerales, v. 39, p. 27-39.

Jannas, R., and R. Araneda, 1985, Geología de la veta Indio Sur

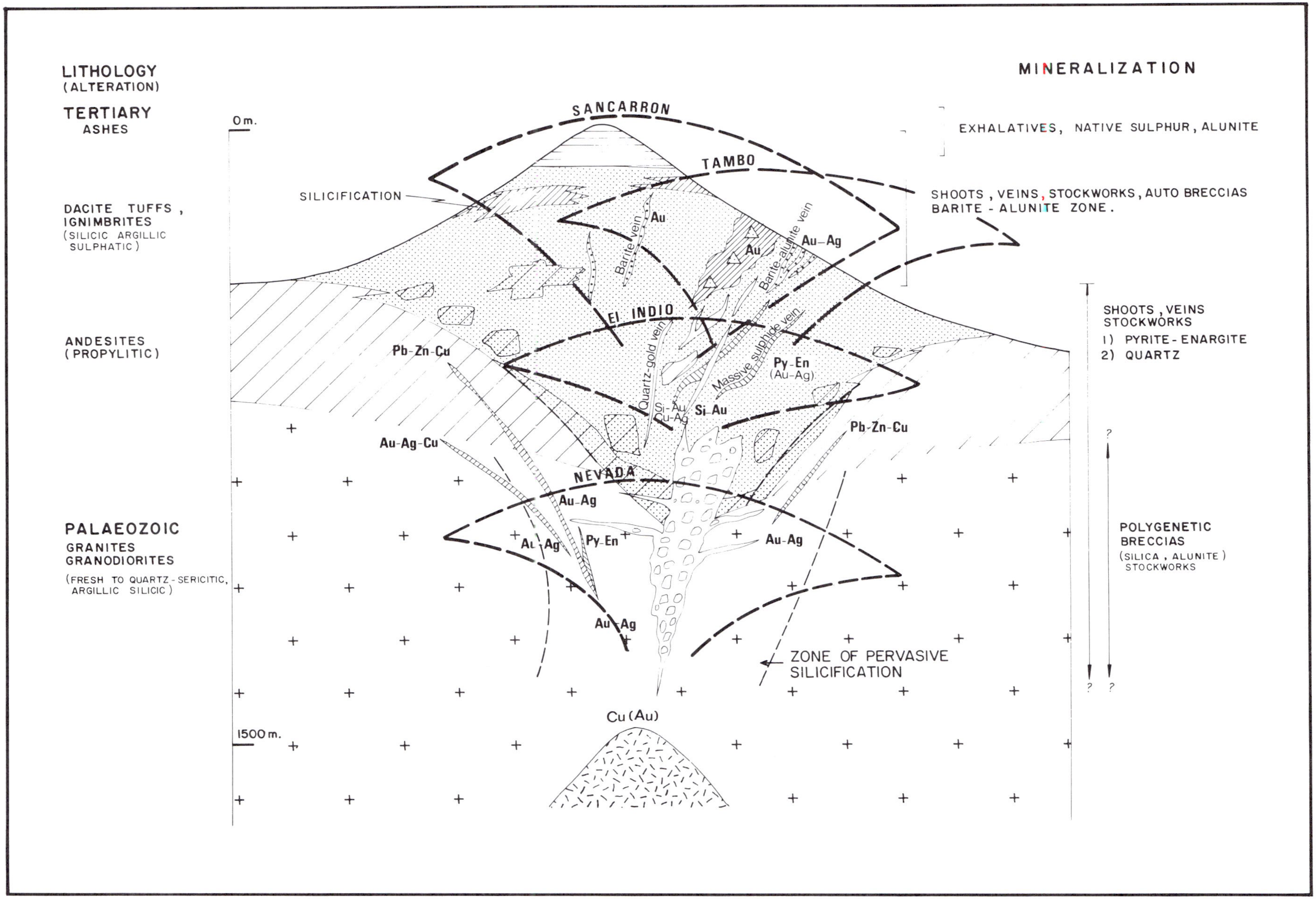

Figure 5. Vertical zonation and relative position of ore deposits. (El Indio, Tambo, Sancarrón, and Nevada).

3500; una estructura tipo bonanza del yacimiento El Indio: Revista Geológica de Chile, no. 24, p. 49-62.

Llaumett, C., and F. Henriquez, 1976, Faja de mineralización de oro-plata-cobre-arsénico en la Cordillera Andina de las III y IV Región de Chile: Santiago, I Congresso Geológico Chileno, Tomo II, p. E175-E191.

Maksaev, V., R. Moscoso, C. Mpodozis, and C. Nasi, 1984, Las unidades volcánicas y plutónicas del Cenozoico superior en la Alta Cordillera del Norte Chico (29°-31° S); Geología, alteración hidrotermal y mineralización: Revista Geológica de Chile, no. 21, p. 11-51.

Walthier, T., R. Araneda, and J. Crawford, 1982, The El Indio gold, silver and copper deposit, region of Coquimbo, Chile: Honolulu, Circum-Pacific Congress.

Ericksen, G. E., Cañas Pinochet, M. T., and Reinemund, J. A., editors, 1989, Geology of the Andes and its relation to hydrocarbon and mineral resources: Houston, Texas, Circum-Pacific Council for Energy and Mineral Resources Earth Science Series, v. 11.

CHAPTER 22

COPPER DEPOSITS AND ANDEAN EVOLUTION

Richard H. Sillitoe
London, United Kingdom

ABSTRACT

Approximately 97% of Andean copper resources are present in deposits related directly to Mesozoic-Cenozoic intrusive activity. This category is dominated throughout the Andes by porphyry copper deposits but also includes skarn deposits and enargite-bearing replacement deposits, especially in southern and central Perú. The remaining Andean copper resources are in volcanic and/or sedimentary sequences of Late Carboniferous through early Miocene age and lack any direct relationship to intrusive activity. This second category includes manto-type deposits, mainly in Chile, as well as major red-bed and Kuroko-type copper deposits in western Bolivia and northern Perú, respectively.

Deposits of the first category were generated from copper-bearing magmatic-hydrothermal fluids liberated during the consolidation of shallow-level stocks of intermediate composition. Restriction of most major intrusion-related deposits in the Andes to Cenozoic arcs is believed to reflect erosion level. Deposits of the second category were probably emplaced from fluids generated during burial metamorphism, dissolution of evaporites, or convective circulation of seawater. The ages of these deposits reflect the intervals during which tectonic and paleogeographic conditions for the generation of these three types of ore fluid were optimized.

Many intrusion-related deposits underwent supergene enrichment, which acted most efficiently in northern Chile-southern Perú as a consequence of Andean uplift and an intensification of aridity during the Oligocene-early Miocene.

INTRODUCTION

The western side of the central Andes in northern and central Chile and southern Perú has long been considered as a metallogenic province dominated by copper (e.g., Stoll, 1965). During the last two decades, however, copper mineralization also has been discovered elsewhere in the Andes, including western Argentina, northern Perú, Ecuador, Colombia, and Venezuela.

Chile and Perú, which have some of the largest and highest grade copper deposits in the world, make a major contribution to world copper production and reserves. Chile is currently the western world's principal producer, and Perú is the sixth largest (Table 1). It may be estimated that the production to date plus reserves of copper in Chile exceed 240 million metric tons (Tables 2 and 3). As a whole, the Andes constitutes the world's premier copper province, and accounts for some 40% of world copper resources.

The Andes is characterized by many styles of copper mineralization, a fact that is particularly evident in Chile and Perú (Ruiz and Ericksen, 1962; Ruiz et al., 1965; Sillitoe, 1976; Ponzoni, 1980; Cardozo and Vidal, 1981). In this overview, Andean copper deposits are assigned to two broad categories depending upon whether or not they possess a direct genetic connection with intrusive activity. In the case of both categories, attention is confined to the economically most important deposits, which are summarized in Tables 2 and 3. For each of eight deposit types, selected information is provided on geographic and geologic settings, deposit characteristics, and preferred genetic models; the less well-known deposit types are treated in somewhat greater detail. The review of deposit types is concluded with an assessment of the timing and effects of supergene copper enrichment. Finally, ore deposit models are related briefly to the geodynamic development of the Andean orogen.

COPPER DEPOSITS RELATED TO INTRUSIVE ACTIVITY

The Andes is characterized by three principal and several minor types of copper deposits that possess a direct connection to intrusive rocks. This category is dominated by deposits of the porphyry type, but in Perú it also includes important examples of skarn-type deposits and enargite-bearing replacement-type deposits (Table 2). Minor deposit types

Table 1. Copper production of the Andean nations for 1987.

Nation	Production (metric tons)	Percentage of Western World Production
Chile	1,418,000	21.4
Perú	371,000	5.6

Source: Mining Annual Review, 1988.

Table 2. Characteristics of principal porphyry-type, skarn-type, and replacement-type copper deposits of the Andes.

Deposit	Production + reserves[1] (million metric tons)	Cu, %	Mo, %	Other metals	Status	Related intrusion	Age m.y.[2]	Principal ore-bearing alteration	Style of mineralization	Hypogene Cu minerals(s)	Supergene enrichment	Data source
Moccoa	300	0.4	0.062		Prospect	Largely concealed dacite porphyry stock	166	Kspar-chlorite, sericitic	Stockwork, breccia	Chalcopyrite	Absent	Sillitoe, Jaramillo, and Castro (1984)
Chaucha	200	0.4	0.03		Prospect	Granodiorite + diorite pluton, quartz, manzonite, porphyry stock	11	Sericitic	Stockwork, minor breccia	Chalcopyrite	Present	Goossens and Hollister (1973)
La Granja	300	0.9	0.015		Prospect	Quartz-feldspar porphyry stock	Mid-Tertiary	Sericitic, advanced argillic	Breccia, stockwork minor skarn	Chalcopyrite, enargite	Major (25-250 m thick)	Schwartz (1982), C. Pointon (personal communication, 1985)
Cañariaco	380	0.6	Present		Prospect	Tonalite porphyry stock	Mid-Tertiary	Sericitic, K-silicate	Stockwork, minor tournaline breccia	Chalcopyrite	Minor	Flores, Zelaya, and Mamani, (1975)
Michiquillay	544	0.69	0.02		Prospect	Quartz monzonite porphyry stock	19(?)	Sericitic, K-silicate	Stockwork	Chalcopyrite	Present (as much as 90 m thick)	Hollister and Sirvas (1974), Minero Peru (1980)
Pashpap	62	0.8	0.052		Prospect	Quartz diorite porphyry stocks	15	K-silicate sericitic, calc-silicate	Stockwork, minor skarn	Chalcopyrite	Minor	E. Barrantes (written communication, 1985)
Antamina	105	1.42	0.04	1.1% Zn, 15 g/t Ag	Prospect	Quartz monzonite porphyry stock	10	Calc-silicate	Skarn	Chalcopyrite	Absent	Terrones (1958), Petersen (1965)
Cerro de Pasco	35[3]	2.47		124 g/t Ag	In production (Cu ore exploited)	Quartz latite porphyry domes	14	Advanced argillic	Veins, pipes; mainly limestone replacement	Enargite, luzonite, chalcopyrite, chalcocite	Important locally	Petersen (1965), Einaudi (1977)
Toro Mocho	200	0.85	0.013	0.02% WO_3, 8.6 g/t Ag	Prospect, pilot open-pit	Quartz monzonite porphyry stock, quartz porphyry	7	K-silicate, sericitic, calc-silicate	Stockwork, minor breccia and skarn	Chalcopyrite	Present	Eyzaguirre and Montoya (1973)
Morococha	37	1.2		4.0% Zn, 1.4% Pb, 180 g/t Ag	In production	Quartz diorite pluton, quartz monzonite porphyry stocks	7	Sericitic, advanced argillic, calc-silicate	Veins, pipes, and mantos; mainly limestone replacement	Chalcopyrite, tetrahedrite, enargite	Absent	Petersen (1965)

Table 2 continued.

Yauricocha	11.8	1.5		5.5%, Zn, 3.0% Pb, 135 g/t Ag	In production	Granodiorite to granodiorite porphyry stocks	8	Silicification, advanced argillic	Pipes and mantos; limestone replacement	Enargite, chalcopyrite, bornite	Important locally	Petersen (1965), E. Barrantes (written communication, 1985)
Cobriza	31	1.52		20 g/t Ag	In production	Unknown	Mid-late Miocene(?)	Calc-silicate	Skarn	Chalcopyrite	Absent	Petersen (1965), Centromin (1980), E. Barrantes (written communication, 1985)
Chalcobamba	28	2.0			Prospect	Quartz monzonite stock	33	Calc-silicate	Skarn	Chalcopyrite, bornite	Absent	Santa Cruz et al. (1979)
Tintaya	10 (oxides) 41 (sulfides)	2.2 2.0		0.6 g/t Au, 13 g/t Ag	In production	Quartz monzonite pluton	33	Calc-silicate	Skarn	Chalcopyrite, bornite	Minor	Terrones (1958), Santa Cruz et al. (1979)
Coroccohuayco	14	2.9			Prospect	Quartz monzonite stock	30	Calc-silicate	Skarn	Chalcopyrite, bornite, chalcocite	Absent	Santa Cruz et al. (1979), Aizawa and Tomizawa (1986)
Berenguela	16.5	1.36		140 g/t Ag	Small-scale production stopped	Unknown	Early Oligocene (?)	Silicification	Limestone replacement manto (entirely oxidized)	Trace chalcopyrite	Absent	E. Barrantes (written communication, 1985)
Cerro Verde-Santa Rosa	61 (oxides) 812 (sulfides)	1.01 0.66	 0.014		Oxides in production	Dacite porphyry stock	58	K-silicate, sericitic	Stockwork, minor breccia	Chalcopyrite	Major (as much as 140 m)	Kihien (1975), La Bel (1985)
Cuajone	22 (oxides) 426 (sulfides)	1.3 1.0	 0.013		In production	Quartz latite porphyry stock, andesite bodies	51	Sericitic, K-silicate	Stockwork	Chalcopyrite	Important (20 m thick)	Manrique and Plazolles (1975), Satchwell (1983)
Quellaveco	385	0.85	0.014		Prospect	Granodiorite and granodiorite porphyry stock	54	Sericitic, K-silicate	Stockwork, minor breccia	Chalcopyrite	Major (50-60 m thick)	Estrata (1975) Zweng and Clark (1984)
Toquepala	509	0.86	0.013		In production	Dacite porphyry stock	58	Sericitic	Stockwork, tourmaline breccia	Chalcopyrite	Major (as much as 150 m thick)	Richard and Courtright (1958),
Cerro Colorado	60	1.2	0.015		Prospect	Dacite porphyry stock	Paleocene	Sericitic	Stockwork	Chalcopyrite	Major (as much as 120 m thick)	G. Mato (personal communication, 1983)
Quebrada Blanca	400 (100)	0.82 (1.5)	0.02		Prospect	Quartz monzonite stock, granodiorite porphyry + quartz diorite porphyry dikes	38	K-silicate, sericitic	Stockwork, minor breccia	Chacopyrite, bornite	Major (as much as 200 m thick	Hunt, Bratt, and Marquardt (1983), Hunt (1985)

Table 2. Characteristics of principal porphyry-type, skarn-type, and replacement-type copper deposits of the Andes.

Deposit	Production + reserves[1] (million metric tons)	Cu, %	Mo, %	Other metals	Status	Related intrusion	Age m.y.[2]	Principal ore-bearing alteration	Style of mineralization	Hypogene Cu minerals(s)	Supergene enrichment	Data source
Collahuasi	200-250 +25-50 +3-5	1.5 1.5-2.0 10-15	0.04	5-9 g/t Ag 180 g/t Ag	Prospect	Concealed porphyry stock	Late Eocene-Early Oligocene	Sericitic, K-silicate, calc-silicate	Stockwork, veins, skarn	Chalcopyrite, bornite, chalcocite	Major (60 m thick)	Hunt et al. (1980), Hunt (1985)
El Abra	1,122	0.63	0.01		Prospect	Diorite stock, dacite porphyry bodies	34	K-silicate	Stockwork	Chalcopyrite, bornite	Absent	Ambrus (1977)
Chuquicamata	544 (oxides) 3,600 (sulfides)	1.55 1.0	 0.035		In production	Granodiorite -monzogranite porphyry (3 phases) stock	31	K-silicate, sericitic	Stockwork, minor veins	Chalcopyrite, bornite, enargite	Major (as much as 750 m)	Ambrus, (1978), Alvarez, Miranda, and Guzman (1980), Alvarez and Flores (1985), Flores (1985)
LaEscon-dida	545 (1,875)	2.16 (1.6)	0.015		Prospect	Quartz diorite porphyry (3 phases) and rhyolite porphyry bodies	34	Sericitic, advanced argillic	Stockwork, minor breccia	Chalcopyrite, bornite	Major (200-300 m thick)	Brimhall, Alpers, and Cunningham (1985), Ortiz et al. (1985), Ojeda (1986)
El Salvador	477	1.19	0.017		In production	Granodiorite porphyry (4 phases) stock	41	K-silicate, sericitic	Stockwork	Chalcopyrite, bornite	Major (as much as 300 m)	Gustafson and Hunt (1975)
Potrerillos	200	1.4			Largely worked out	Dacite porphyry stock	34	K-silicate, sericitic	Stockwork, minor skarn	Chalcopyrite, enargite	Major (as much as 100 m)	March (1935)
Andacollo	200	0.64	0.015	0.23 g/t Au	Prospect	Tonalite porphyry stock	112	K-silicate, sericitic	Stockwork	Chalcopyrite	Major (40 m thick)	Llaumett et al. (1975)
Los Pelam-bres	3,311	0.63	0.015		Minor production	Granodiorite stock, quartz diorite-quartz monzonite porphyry bodies	10	K-silicate	Stockwork, minor breccia	Chalcopyrite, bornite	Minor	Sillitoe (1973a), Atkinson, Souviron, and Vehrs (1984)
Río Blanco (Andina)	1,565	1.15	0.025		In production	Quartz monzonite pluton, quartz monzonite porphyry body, dacite porphyry dike	5	Sericitic, K-silicate	Tourmaline and other breccias, stockwork	Chalcopyrite, bornite	Absent	Stambuk, Blondel, and Serrano (1982), Stambuk, Serrano, and Vargas (1985)
Los Bronces (Disputada)	1,320	0.81	0.016		In production	Quartz monzonite pluton	5	Sericitic, propylitic	Tournaline and other breccias	Chalcopyrite, bornite	Important locally	Warnaars (1983), Warnaars, Holmgren, and Barassi (1985)

Table 2 continued.

El Teniente	2,855	1.31	0.018		In production	Quartz diorite, dacite porphyry bodies	5	K-silicate	Stockwork, minor tourmaline breccia	Chalcopyrite, bornite	Important locally	Howell and Molloy (1960), Camus (1975)
El Pachón	800	0.70	0.016		Prospect	Diorite porphyry stock, dacite porphyry bodies	10	K-silicate, sericitic	Stockwork	Chacopyrite	Important locally	Fernandez, Brown, and Lencinas (1974)
Bajo de La Allumbrera	340	0.49	0.007	0.62 g/t Au	Prospect	Dacite porphyry stock	7	K-silicate	Stockwork	Chalcopyrite, bornite	Absent	R.H. Sillitoe (unpublished report, 1973), González (1975)

[1]Proven + probable reserves.
[2]Taken from compilation by Sillitoe (1988).
[3]Cu-Ag orebodies only; excludes Pb-Zn-Ag orebodies.

Table 3. Characteristics of principal Kuroko-type, manto-type, and red-bed copper deposits of the Andes.

Deposit	Production + reserves[1] (million metric tons)	Cu, %	Other metals	Status	Host rocks	Age of host rocks	Style of mineralization	Principal hypogene Cu mineral(s)	Data Source
Bailadores	3.5	2.0	17% Zn, 6% Pb, 60 g/t Ag	Prospect	Felsic volcanics	Late Carboniferous	Lenses of pyritic massive sulfide	Chalcopyrite	Carlson (1977)
Tambo Grande	4.2	2.04	1.1% Zn, 37 g/t Ag, 3 g/t Au, 0.7% Pb	Prospect	Dacitic to rhyolitic volcanics	Late Jurassic-Early Cretaceous	Flat-lying lens of pyritic massive sulfide	Chalcopyrite	Injoque, Miranda, and Dunin-Borkowski (1979), Aye (1985)
Raúl-Condestable	18	2.0		In production	Andesitic pyroclastics + lavas, clastic sediments, limestone	Early Cretaceous	Disseminations, stockworks, massive replacements + breccia fillings in a series of mantos	Chalcopyrite	Ripley and Ohmoto (1977), Wauschkuhn (1979, Wauschkuhn & Thum (1982) E. Barrantes (written communication, 1985)
Buena Esperanza	7	2.9		In production	Andesitic flows + volcaniclastics	Middle-Late Jurassic	Disseminations in 30 mantos and cement to hydrothermal breccias abutting basic plug	Chalcocite, bornite	Definis and Palacios (1981)
Susana de Michilla	8	2.0	8-14 g/t Ag	In production	Andesitic flows, pyroclastics + volcaniclastics	Middle-Late Jurassic	Disseminations in 23 mantos and cement to hydrothermal breccia abutting basic plug	Chalcocite, chalcopyrite, bornite	Palacios and Definis (1981), Soto and Dreyer (1985)

Table 3 continued.

Deposit	Production + reserves[1] (million metric tons)	Cu, %	Other metals	Status	Host rocks	Age of host rocks	Style of mineralization	Principal hypogene Cu mineral(s)	Data Source
Mantos Blancos	54 (oxides) 40 (sulfides)	1.9 2.3	20 g/t Ag	In production	Dacitic flows, volcaniclastic sandstone, + minor andesitic flows + sill	Middle-Late Jurassic	Steeply plunging, cigar-shaped bodies with veinlets + breccia cement	Chalcocite, bornite, chalcopyrite	Tobar (1981) Chávez (1983)
Punta del Cobre	14.5	2.9		In production	Andesitic flows, conglomerates	Early Cretaceous	Cement to conglomerates + breccias as mantos; cross-cutting breccias, stockworks + veins	Chalcopyrite	Camus (1980)
Talcuna	15[2]	1.4[2]		Mines largely abandoned	Sedimentary breccia, sandstones	Early Cretaceous	Disseminations in mantos, cross-cutting veins	Bornite, chalcopyrite, chalcocite	Ruiz et al. (1971)
El Soldado	76	1.8		In production	Andesitic + trachytic	Early Cretaceous	Irregular disseminated bodies	Chalcopyrite, bornite,	Olcay and Alarcón (1975), Holmgren (1985),
					flows + tuffs calcareous sediments		controlled by lithology + structure	chalcocite	Ruge (1985)
Lo Aguirre	18	1.7		In production	Andesitic flows + tuffs, ignimbrites	Early Cretaceous	Stratabound bodies with disseminations, veinlets, and breccia fillings	Bornite, chalcocite, chalcopyrite	N. Saríc (personal communication, 1985)
Corocoro	10.5	4.1	30 g/t Ag	Abandoned	Red sandstones, conglomerates + claystones	Eocene-Oligocene, Early Miocene	Stratabound disseminated horizons abutting steep fault	Chalcocite, native copper	Entwistle and Gouin (1955), Ljunggren and Meyer (1964), Ahlfeld and Schneider-Scherbina (1964)

[1]Proven + probable reserves.
[2]Reserves only.

include mainly breccia pipes (unrelated to porphyry copper systems) and veins. All these ore deposit types are widely recognized and require no further definition.

Porphyry Copper Deposits

Andean porphyry copper deposits have much in common with representatives of this well-documented ore deposit class elsewhere. They are spatially, temporally, and genetically related to porphyry stocks, which range in composition from diorite to quartz monzonite, and are commonly composite in character (Table 2). All stocks are composed of typical I-type, magnetite-series rocks (Ishihara et al., 1984). Most of the mineralized stocks occur in long, linear subbelts roughly parallel to the Perú-Chile trench and the continental margin (Figures 1, 2).

As reviewed elsewhere (Sillitoe, 1988), each subbelt of porphyry copper deposits (and prospects) was generated during a discrete time interval. Most of the porphyry copper mineralization is Mesozoic-Cenozoic in age, although low-grade prospects of probable Early Carboniferous and Late Carboniferous-Early Permian ages have been documented in western Argentina and contiguous parts of Chile (Sillitoe, 1977). The largest and highest-grade porphyry copper deposits are all located in central and northern Chile and southern Perú (Table 2), and are parts of subbelts dated radiometrically as Paleocene-early Eocene (65-51 m.y.), late Eocene-early Oligocene (41-31 m.y.), and middle-late Miocene (16-5 m.y.) (Figure 1; Table 2). A composite Cretaceous (132-73 m.y.) subbelt also has been documented in Chile, but it includes only one significant prospect—Andacollo (Figure 1; Table 2). The middle-late Miocene subbelt also encompasses an appreciable part of western Argentina, where the important El Pachón and Bajo de La Alumbrera prospects plus a number of other occurrences are found (Figure 1; Table 2).

Important prospects in the same middle-late Miocene age range also are found in central and northern Perú, and in southern Ecuador (Figure 2; Table 2). In southern Perú, in addition to the major Paleocene-early Eocene deposits, porphyry-type occurrences of late Eocene-early Oligocene age also are known, although their restricted subbelt is better known for its skarn-type copper resources (see below).

In the northern Andes, a number of porphyry copper prospects are known in Colombia, where longitudinal subbelts of early-late Miocene (17-6 m.y.), early-middle Eocene (55-42 m.y.), and Middle Jurassic-earliest Cretaceous (166-131 m.y.) ages have been defined; the principal prospect, Mocoa, is Middle Jurassic (Figure 2; Table 2).

The porphyry copper subbelts coincide with belts of unmineralized plutons and/or stocks of essentially the same compositions as their mineralized counterparts. These intrusive belts mark the former sites of magmatic arcs, and commonly include volcanic rocks accumulated during the same time intervals. For example, in central Chile, the westernmost part of the middle-late Miocene porphyry copper subbelt was generated at the close of a volcanic episode which resulted in accumulation of the Farellones Formation between 22 and 7.4 m.y. (Vergara et al., 1985). Similarly, in southern Perú, Paleocene-early Eocene plutonism, volcanism, and porphyry copper emplacement were broadly coetaneous (Zimmermann and Kihien Collado, 1983), as were construction of an isolated stratovolcano and intrusion of porphyry copper-bearing stocks in the Farallón Negro district of Argentina (Sillitoe, 1973b). Elsewhere, however, as in the case of the late Eocene-early Oligocene porphyry copper subbelt of northern Chile, volcanic precursors have not been recognized. It is not known if this is due to the complete removal by erosion of a relatively restricted volume of volcanic products or to the absence of any volcanic counterparts, although the former alternative is favored by this writer.

As in porphyry deposits in general, mineralization in the Andean examples is present mainly in veinlet stockworks accompanied by subsidiary disseminations. Hydrothermal breccias carry major amounts of copper and molybdenum at Mocoa, La Granja, Toquepala, and Río Blanco-Los Bronces, and subsidiary amounts at Toro Mocho, Cerro Verde-Santa Rosa, Quellaveco, Quebrada Blanca, La Escondida, Los Pelambres, and El Teniente (Table 2). Skarn developed from carbonate wallrocks (and roof pendants) contributes relatively minor amounts of copper and molybdenum at La Granja, Pashpap, Toro Mocho, Collahuasi, and Potrerillos. Discrete, late-stage, ore-bearing veins are present at Chuquicamata and Collahuasi (Table 2).

Hypogene mineralization is an integral part of one or more of K-silicate, sericitic, or advanced argillic alteration. Calc-silicate assemblages constitute the skarn bodies. K-silicate alteration, generally characterized by biotite and alkali feldspar, is the earliest and deepest alteration type. It was overprinted and overlain by sericitic and, in turn, advanced argillic assemblages. K-silicate alteration is scarce in some deposits (Table 2), especially those (Toquepala, Río Blanco-Los Bronces) characterized by large, ore-bearing breccias, whereas sericitic and advanced argillic alteration are only minor in some deeply eroded deposits (El Abra).

Chalcopyrite is the principal hypogene copper mineral, commonly accompanied by subsidiary bornite (Table 2). Bornite, with minor hypogene chalcocite[1], is most abundant in K-silicate assemblages, as at El Abra, Chuquicamata, La Escondida, El Salvador, Los Pelambres, and El Teniente, although it also occurs with pyrite in feldspar-destructive alteration, as at El Salvador. Enargite is reported from advanced argillic alteration at La Granja and Cerro Verde-Santa Rosa, and accompanies the last stage of a widespread sericite-forming event at Chuquicamata. Molybdenum is the principal accompanying metal in Andean porphyry copper deposits and prospects; molybdenum contents generally average 0.01 to 0.035% but attain 0.06% at Mocoa (Table 2). The only gold-rich porphyry copper deposit known in the Andes is Bajo de La Alumbrera, where the copper mineralization carries an average of 0.62 ppm Au (Table 2). Locally, however, gold was concentrated in the propylitic halos of porphyry copper systems, as at Andacollo (Llaumett, 1980).

The advanced argillic alteration (with or without enargite mineralization) at La Granja, La Escondida, and elsewhere overprinted preexisting ore-bearing assemblages (Figure 3; Schwartz, 1982; Brimhall. Alpers, and Cunningham, 1985),

[1]In this chapter, chalcocite, whether hypogene or supergene in origin, is used as a generalized term for a family of copper sulfides including chalcocite (sensu stricto), digenite, djurleite, anilite, covellite, and blaubleibender covellite.

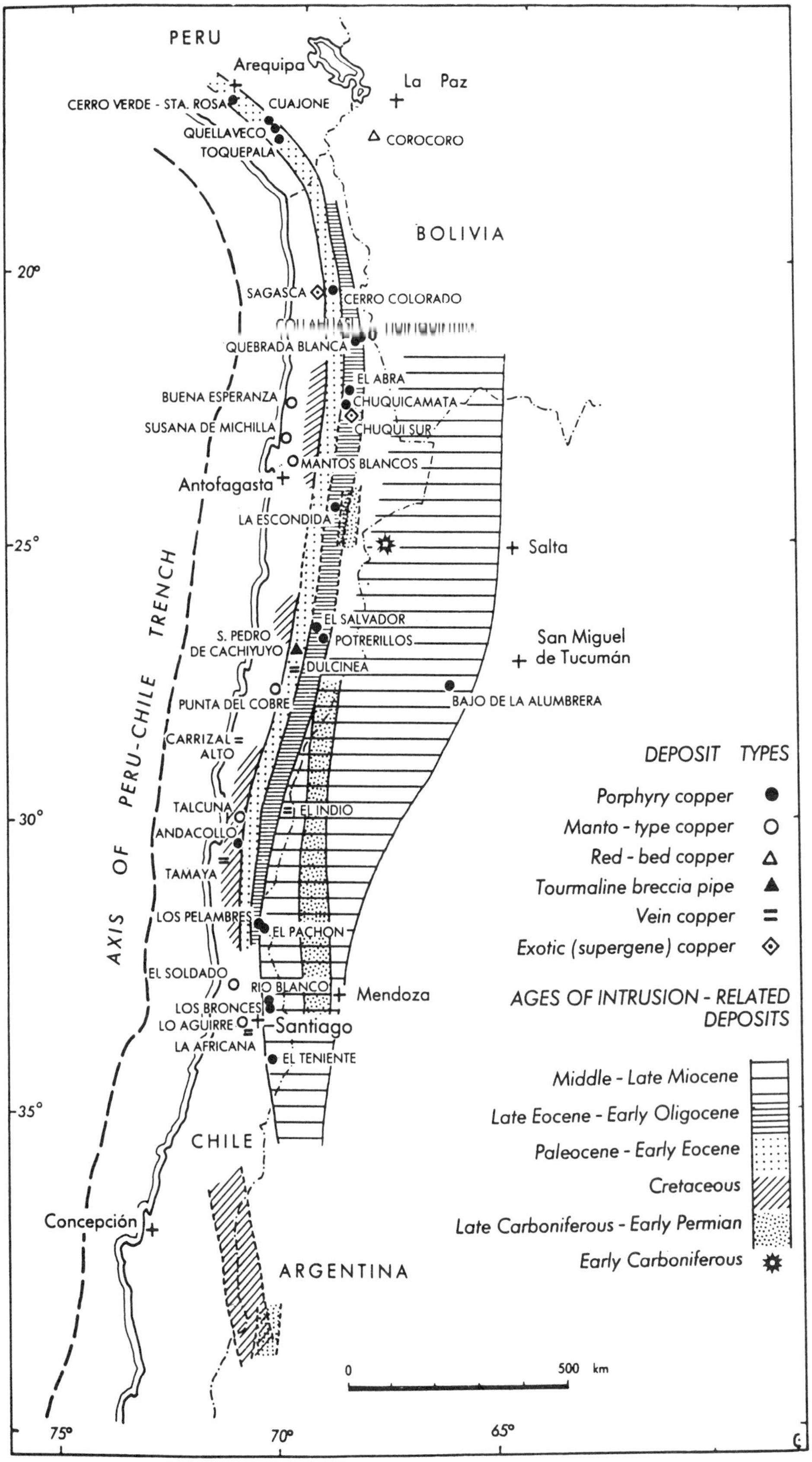

Figure 1. Locations of major copper deposits in the Andes of Chile, Bolivia, and Argentina. Ages of subbelts for intrusion-related deposits taken from Sillitoe (1988).

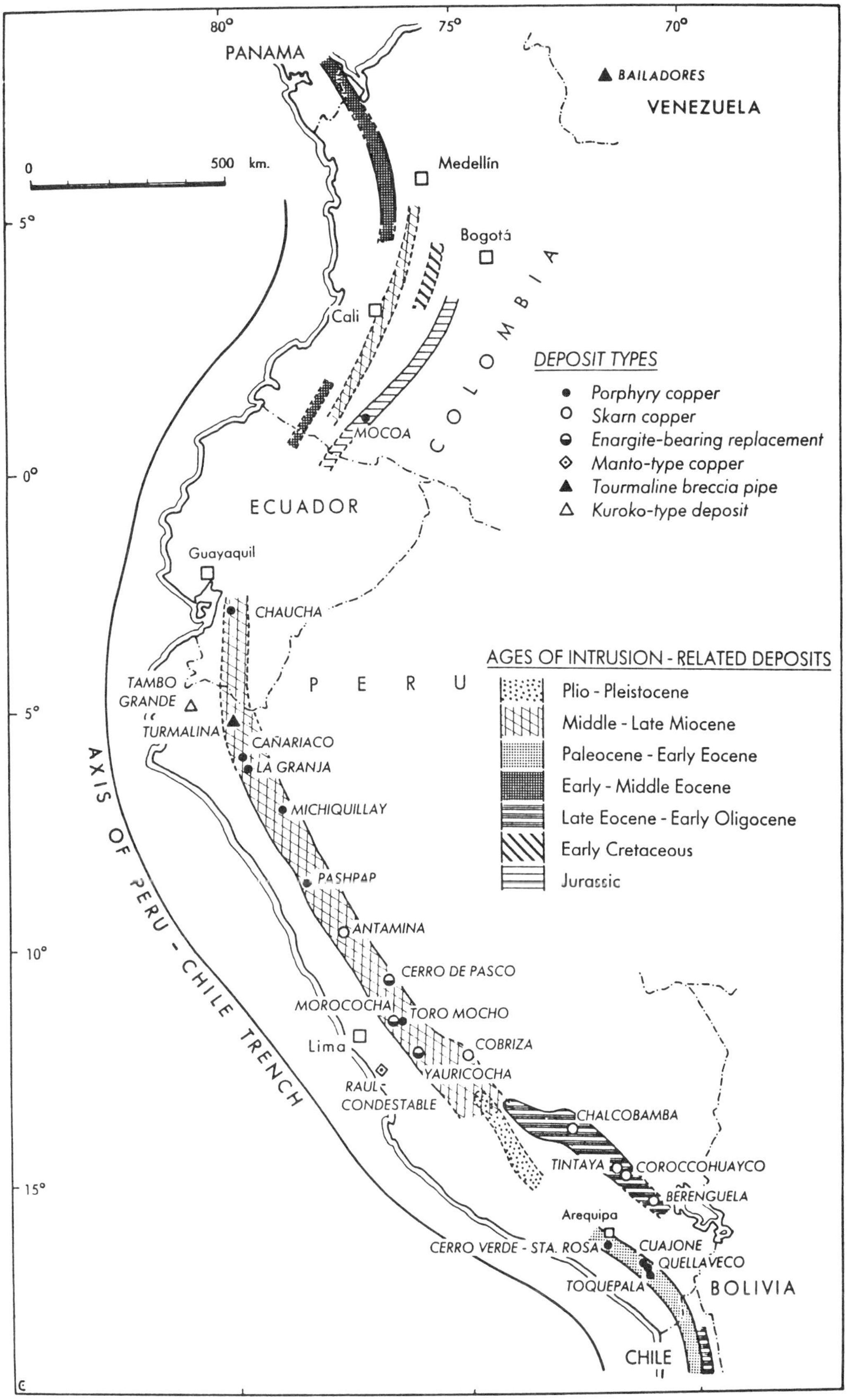

Figure 2. Locations of major copper deposits in the Andes of Colombia, Ecuador, and Perú. Ages of subbelts for intrusion-related deposits taken from Sillitoe (1988).

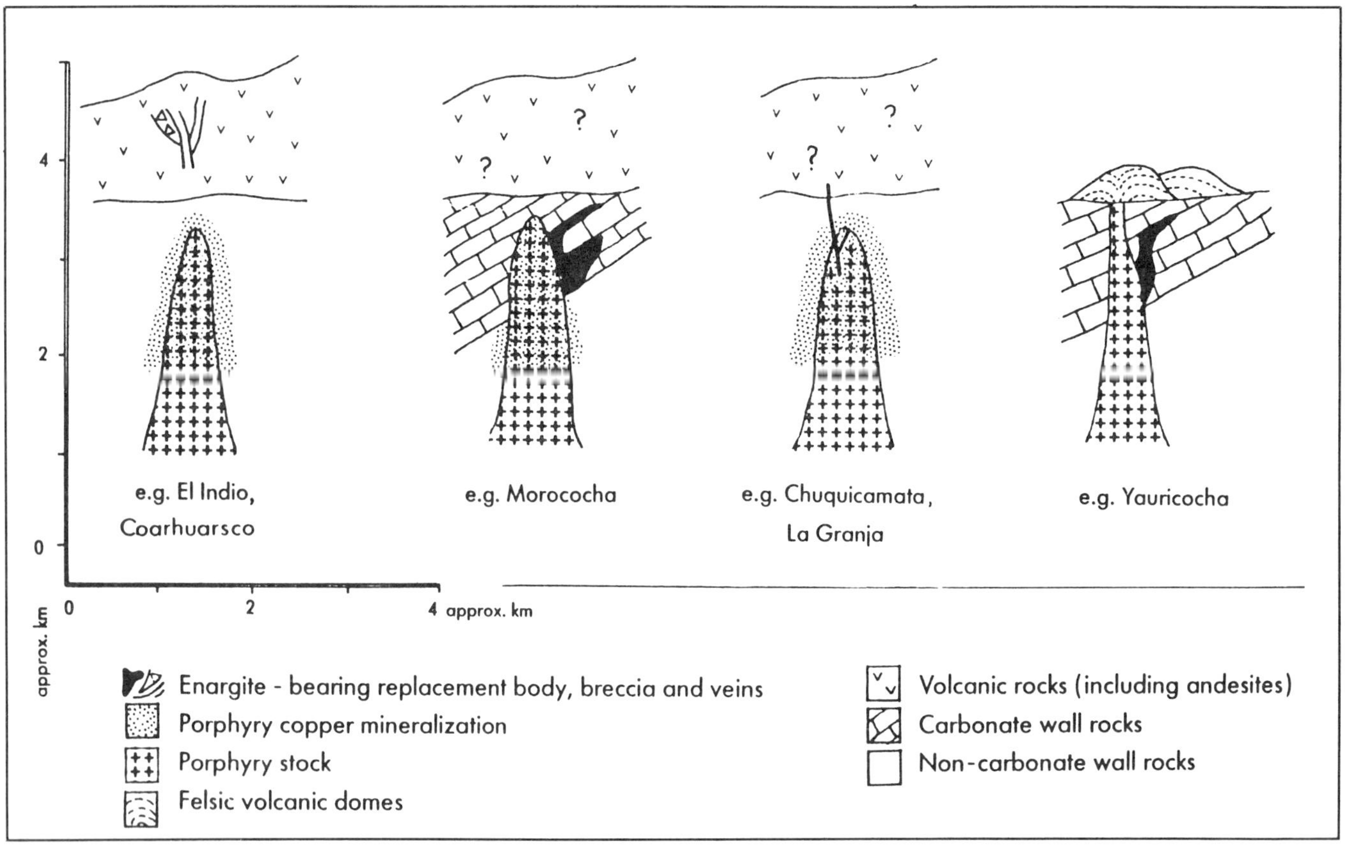

Figure 3. Scheme of various types of enargite-bearing deposits in the Andes: Volcanic-hosted veins and breccias above porphyry copper-bearing stocks; carbonate replacement bodies abutting porphyry copper-bearing stocks; veins overprinting porphyry copper-bearing stocks; and carbonate replacement bodies adjoining unmineralized stocks/domes.

whereas that at El Salvador mainly affected volcanic rocks topographically higher than the main hypogene copper-molybdenum zone (Gustafson and Hunt, 1975). By analogy with zones of advanced argillic alteration elsewhere, it is concluded that the advanced argillic alteration at El Salvador is an erosional remnant of an originally far more extensive zone. Such widespread zones of pyrite-rich advanced argillic alteration have been widely recognized recently in the andesite-dominated Cenozoic volcanic terranes of the Andes. Several of them, as at El Indio, Chile (Araneda, 1982; see below), Coarhuarsco, Perú (C. Vidal and D.C. Noble, unpublished talk, 1983), and El Queva, Argentina (Sillitoe, 1975), are known to carry epithermal, enargite-bearing precious-metal mineralization in bodies of hydrothermal breccia or lodes (Figure 3). It is concluded that such high-sulfur, advanced argillic-hosted gold and silver mineralization is commonly a near-surface manifestation of porphyry copper-bearing stocks at depth.

It is widely accepted that most of the copper and molybdenum in early K-silicate assemblages was transported and precipitated by magmatic-hydrothermal fluids liberated during consolidation of an associated stock. Geologic evidence in support of this concept was presented in detail for the El Salvador deposit by Gustafson and Hunt (1975), and further support is provided by the results of light stable isotope studies conducted at Cerro Verde-Santa Rosa (Le Bel, 1985), El Salvador (Sheppard and Gustafson, 1976), and El Teniente (Kusakabe et al., 1984). This is not to deny, however, that overprinted feldspar-destructive alteration induced by ingress of meteoric fluids to hydrothermal systems did not cause important redistribution of metals (e.g., Gustafson and Hunt, 1975).

More contentious, however, is the ultimate source of the copper and molybdenum—crustal, as a result of either magmatic assimilation or hydrothermal leaching; or subcrustal, from the subducted slab and/or the overlying subcontinental mantle. A subcrustal source for porphyry copper magmas and metals in the Andes is favored by the following lines of evidence:

1. Initial $^{87}Sr/^{86}Sr$ ratios for porphyry copper-bearing stocks at Toquepala (Beckinsale et al., 1985), Chuquicamata (Shibata, Ishihara, and Ulriksen, 1984), El Salvador (Gustafson and Hunt, 1975; Halpern, 1978; Shibata, Ishihara, and Ulriksen, 1984), Andacollo (Munizaga, Huete, and Hervé, 1985), and Río Blanco (Halpern, 1979) range from 0.7033 to 0.7044, thereby confirming the absence of any significant contamination of the magmas by sialic crust.

2. The rare-earth and other trace-element contents of porphyry copper-bearing stocks at El Abra, El Salvador, and Río Blanco-Los Bronces are interpreted to show that magma was derived from the mantle, to which small amounts of hydrous melt were contributed by low degrees of partial fusion of the subducted slab (Lopez-Escobar, 1982).

3. Ishihara et al. (1984) ascribed the oxidized (high Fe_2O_3/

FeO ratio) character of Chilean porphyry copper-bearing stocks to involvement of altered ocean-floor basalts in their genesis. The anomalously high copper contents of Chilean magnetite-series intrusive rocks (Ishihara et al., 1984) can also be explained in the same way (Sillitoe, 1972).

4. Hydrothermal sulfate-sulfide assemblages from the Toquepala, El Salvador, Chuquicamata, Ró Blanco, and El Teniente porphyry copper deposits are enriched in ^{34}S, probably as a result of the involvement of isotopically heavy marine sulfur (Field et al., 1983; Sasaki et al., 1984). Field et al. (1983) attributed the heavy sulfur in Andean porphyry-type mineralization to contamination by fluids derived from marine evaporites. However, similar heavy sulfur also characterizes the Atlas and Santo Niño porphyry copper deposits in the Philippines (Sasaki et al., 1984), where evaporites are absent. ^{34}S enrichment could be due therefore to a contribution of marine sulfur in melts produced by partial fusion of subducted oceanic crust.

5. The lead-isotopic signatures of pyrite from porphyry copper-type mineralization in the oceanic and continental domains of the Colombian Andes are interpreted to show that subducted pelagic sediments are the principal source of the trace lead, with only a relatively minor contribution from upper crustal rocks (Sillitoe and Hart, 1984).

6. There is no obvious relationship between the Cu/Mo ratios of principal porphyry copper deposits in the central Andes and the age, composition, or thickness of the continental crust in the immediate vicinities of the deposits (Sillitoe, 1986).

Skarn Copper Deposits

All major skarn copper deposits in the Andes are confined to southern and central Perú, although minor examples are also present in Chile, Ecuador, and Colombia. Skarn deposits are probably more abundant in Perú than elsewhere in the Andes because of the more widespread distribution there of limestone sequences (Petersen, 1970).

The largest concentration of calcic skarn copper-iron deposits is located in southern Perú (Figure 2; Table 2), in the so-called Andahuaylas-Yauri subbelt of late Eocene-early Oligocene age (radiometric dates of 34-32 m.y.; Noble et al., 1984). The skarn deposits, including Tintaya, Chalcobamba, and Coroccohuayco (Table 2), were localized by Mesozoic carbonate rocks at the contacts with small quartz monzonite stocks, which postdated extensive diorite and granodiorite plutons (Terrones, 1958; Santa Cruz et al., 1979; Aizawa and Tomizawa, 1986). Chalcopyrite, magnetite, and lesser pyrite, bornite, and chalcocite are present as disseminations, veinlets, and patches in garnet-diopside-actinolite-tremolite skarn, which constitutes stratabound mantos and smaller transgressive bodies (Terrones, 1958; Santa Cruz et al., 1979).

The Berenguela copper-silver deposit (Figure 2; Table 2) also might be considered as part of the Andahuaylas-Yauri subbelt, although supporting age data are not available. The deposit is interpreted as a mantiform replacement of Late Cretaceous limestone beyond the skarn front, although the apparent lack of an associated intrusion and the totally oxidized state of the manganese-rich ore combine to render the deposit rather enigmatic.

The other two major Andean skarn deposits, Antamina and Cobriza in central Perú, are located in the middle-late Miocene subbelt, although the age of mineralization at the latter remains in doubt. The deposits are somewhat different in character from the skarns in the Andahuaylas-Yauri subbelt (Table 2).

The Antamina deposit is located at the contact between a quartz monzonite porphyry stock and Cretaceous limestones (Terrones, 1958; Petersen, 1965). The calcic skarn, in which prograde garnet and retrograde chlorite are the most abundant silicates, contains a variety of metallic minerals, predominantly pyrite, chalcopyrite, and sphalerite (Terrones, 1958; Petersen, 1965). Molybdenite becomes more abundant at the expense of copper and zinc minerals toward the stock (McKee et al., 1979), and in its marginal part is present in stockwork veinlets.

The Cobriza deposit consists of well-banded, garnet-rich skarn carrying pyrrhotite, arsenopyrite, magnetite and chalcopyrite, and occupies parts of a contorted horizon of Early Permian limestone (Petersen, 1965). The presence of pyrrhotite and arsenopyrite demonstrates that ore formation at Cobriza took place under more reducing conditions than is customary for copper skarns. The Cobriza deposit is located 1 km from a Late Permian granite batholith but shows no evidence of having any genetic relationship with it (E.T. Barrantes G., 1985, written communication).

The major Peruvian skarn deposits summarized above were all considered by Einaudi, Meinert, and Newberry (1981) to be associated with barren rather than porphyry copper-bearing stocks. However, many of them are clearly transitional with porphyry copper-related skarn deposits, which include the relatively small Andean examples noted above at La Granja, Pashpap, Toro Mocho, Collahuasi, and Potrerillos (Table 2). The stocks associated with skarns in the Andahuaylas-Yauri subbelt and with the Antamina deposit exhibit K-silicate alteration and low-grade chalcopyrite-molybdenite mineralization (McKee et al., 1979; Noble et al., 1984).

A magmatic-hydrothermal origin for the copper and other metals in skarn deposits is favored by most investigators (Einaudi, Meinert, and Newberry, 1981), and is further supported by the association of several Peruvian examples with low-grade porphyry-type mineralization. However, given its distinctive mineralogic features and the absence of a known intrusive progenitor, Cobriza may prove not to be a true intrusion-related skarn of magmatic-hydrothermal parentage.

Enargite-Bearing Replacement Copper Deposits

Enargite-bearing replacement bodies are found in the central parts of polymetallic ore districts that Petersen (1970) termed "zoned copper-zinc-lead-silver deposits." All the major copper deposits of this type, including Cerro de Pasco, Morococha, and Yauricocha, are located in the middle-late Miocene subbelt of central Perú (Figure 2; Table 2). Similar enargite deposits, probably of the same general age but much smaller and commonly lacking the marked polymetallic character of the central Peruvian deposits, are known from Ecuador, Bolivia, Chile, and Argentina.

The large Peruvian enargite-bearing deposits occur mainly

in Triassic-Jurassic limestones in proximity to shallow granodiorite to quartz monzonite stocks, as at Morococha and Yauricocha (Figure 3; Petersen, 1965), or to their shallower volcanic equivalents, as at Cerro de Pasco, where quartz latitic domes and a diatreme are present (Silberman and Noble, 1977). Elsewhere in the Andes, however, enargite-bearing mineralization is commonly hosted by noncarbonate rocks, especially volcanics.

The deposits possess complex geometries that combine pipes, mantos, veins, and less regular forms. Hydrothermal breccias are commonly present. Pyrite and chalcedonic quartz are major components of the orebodies, which are enveloped in most cases by advanced argillic alteration. Enargite is associated with lesser quantities of tetrahedrite-tennantite, luzonite, bornite, and hypogene chalcocite. Outward from the centers of deposits, copper (with or without silver) gives way to zinc, lead, and silver and, where present, copper occurs mainly in tetrahedrite-tennantite and chalcopyrite rather than in enargite.

Enargite-bearing replacement deposits clearly are transitional with both skarn-type and porphyry-type copper deposits. This was pointed out by Petersen (1970) in the case of Morococha, where remnants of early calcic skarn are widespread and the Toro Mocho porphyry copper deposit (Table 2) is centered on one of the smallest stocks in the polymetallic mineral district; other stocks also carry weak porphyry-type mineralization (Petersen, 1970). In fact, Einaudi, Meinert, and Newberry (1981) classified Morococha as a porphyry-copper-related skarn deposit. Many, but not necessarily all, of the smaller, volcanic-hosted enargite deposits (Figure 3) in the Andes are thought to occupy structurally determined sites in the upper parts of porphyry copper systems, generally above the levels of any porphyry-copper-bearing stocks (see above).

Petersen (1979) suggested that the abundance of enargite-bearing deposits in the Miocene subbelt of Perú might be due to its coincidence with anomalous concentrations of arsenic in the crust. In contrast, this writer prefers to believe that enargite-bearing deposits are a widespread but, until recently, little-emphasized type of mineralization in high-level volcanic to subvolcanic settings, especially in the upper parts of porphyry copper systems (cf., Sillitoe, 1983). Where limestones are present as host rocks, as they commonly are in Perú (Petersen, 1970), the enargite-bearing bodies appear to attain larger sizes, although it should be recalled that large volcanic-hosted bodies are also known in other parts of the world (Sillitoe, 1983). Einaudi, Meinert, and Newberry (1981) also pointed out that some limestone-hosted, enargite-bearing replacement deposits are the shallow-level equivalents of cupriferous skarns.

As a result of their shallow levels of emplacement, enargite-bearing deposits stand a far better chance of escaping removal by erosion in middle to late Cenozoic magmatic arcs, as in the middle to late Miocene subbelt of the Andes, and in arc terranes of similar age elsewhere (e.g., Philippines, Sulawesi, Papua New Guinea). For this reason, enargite mineralization is scarce in the more deeply eroded early Tertiary and older porphyry copper provinces of southwestern North America and British Columbia. As emphasized by Einaudi (1977) in a discussion of the copper ores at Cerro de Pasco, pyrite-rich enargite mineralization accompanied by advanced argillic alteration was generated by hydrothermal fluids that attained a low pH and a high sulfur fugacity. Such fluid characteristics are likely where a plume of acid volatiles released from a subjacent stock impinged on the base of a zone of cool, oxygenated ground water (Sillitoe, 1983). The volatile elements, including sulfur and arsenic, might be expected to have partitioned preferentially into such magmatic plumes, and therefore they became concentrated at shallow levels above, or in the upper parts of, their parent stocks (Figure 3).

Other Types of Deposits

Veins and tourmaline breccia pipes are two widely recognized types of relatively small copper deposits in the Andes. Although small, many of these deposits contain(ed) high-grade ore: hand-cobbed material commonly carries more than 10% copper. However, it is difficult to determine the total copper contents of most of these deposits because of their long histories of generally unstructured exploitation, and the lack of any formal ore-reserve estimates. A few of the most important Andean examples shown in Figures 1 and 2 probably range in size from 2 to 5 million metric tons, or perhaps even more.

Copper-bearing tourmaline breccia pipes unrelated to porphyry copper systems are common in Chile and Perú (Sillitoe and Sawkins, 1971). Many of the breccia pipes in Chile and southern Perú are Paleogene in age, but older and younger examples also are known. The pipes occur singly, as at Turmalina in northern Perú, or as clusters, as at San Pedro de Cachiyuyo in northern Chile (Figures 1, 2). The pipes were emplaced near the roofs of phaneritic-textured quartz diorite to granodiorite plutons or, less commonly, are associated with small stocks. The breccias that filled the pipes are characterized generally by angular fragments cemented by tourmaline, quartz, chalcopyrite and pyrite. Molybdenite is locally important, as at Turmalina (Carlson and Sawkins, 1980), and scheelite is a common subsidiary mineral.

Copper-bearing quartz vein deposits separated from other deposit types are widespread throughout the Andes, especially in the coastal regions of central and northern Chile and southern Perú, where most deposits are clearly related to phaneritic-textured plutons of intermediate composition and of Jurassic to early Cenozoic age (e.g., Ruiz et al., 1965; Ponzoni, 1980). Chalcopyrite is the principal hypogene copper mineral, and is accompanied by bornite, pyrite, specularite, and magnetite. Many of these veins (e.g., Carrizal Alto, Tamaya, La Africana; Figure 1) carry only copper in economic quantities, whereas others have been or are worked for other commodities as well, such as gold at Dulcinea (Figure 1).

In addition to this pluton-related category of vein copper deposits, a much smaller number of veins in which copper is a major component is related to volcanism. Most of these volcanic-hosted veins carry enargite as the main copper mineral, and have been discussed in preceding sections. Part of the copper exploited at Cerro de Pasco was present in vein systems (Table 2; see above), as it is at the El Indio gold-silver-copper deposit (Figure 1), where a recently announced ore reserve of 4.7 million metric tons averages 5.1% Cu.

VOLCANIC- AND SEDIMENT-HOSTED COPPER DEPOSITS LACKING DIRECT CONNECTIONS TO INTRUSIVE ACTIVITY

Three well-defined styles of copper mineralization unrelated to intrusive activity may be recognized in the Andes—manto type, red-bed type, and Kuroko type. The characteristics of the principal deposits of these three types are summarized in Table 3, from which it is clear that the manto-type category is economically the most significant; it also provides many more small deposits and prospects than the other two types (see Ruiz et al., 1965).

The name "manto-type" copper deposit was formalized by Carter (1960) for tabular, stratabound deposits in Chile. Although it is realized that a number of examples, including one of the largest (El Soldado; Table 3), in part transgress stratigraphy, the name is still the best available and is retained here because of its widespread usage. The red-bed and Kuroko types of copper deposits are fully described in the literature, and need no further introduction. It should be mentioned, however, that Kuroko-type deposits may owe their origin to magmatic-hydrothermal fluids evolved from subjacent intrusions rather than simply to leaching effected by heated seawater (Sawkins, 1982); if so, these deposits might be added to the intrusion-related suite of copper deposits dealt with above.

Manto-Type Copper Deposits

The manto-type deposits of Chile occur in volcanic and volcano-sedimentary sequences, which accumulated under both submarine and subaerial conditions in magmatic arcs during the Middle-Late Jurassic, Early Cretaceous, Late Cretaceous, and early Tertiary (Table 3). The sequences are only gently folded and are characterized by homoclinal dips. They commonly attain 10 km or more in aggregate thickness (e.g., Levi, 1970), and contain an appreciable red-bed component. Deposits hosted by Jurassic rocks are most common in coastal northern Chile (lat. 22°-25°S), whereas those of Cretaceous and early Tertiary ages are most wide-spread in northern (lat. 27°-28° and 30° S) and central (lat. 32°30′-33°30′S) Chile (Figure 1).

Andesitic flows and pyroclastics are the most common host rocks of manto-type deposits; also acting as host rocks are the more felsic volcanics (including ignimbrites), volcaniclastic sandstones and siltstones, and limestones. At some deposits, several distinct lithologies are ore bearing. Mineralization is generally wholly or partly stratabound, at least in a gross sense, and is commonly present in permeable lithologic units, such as amygdaloidal and/or brecciated tops of andesite flows, unwelded ignimbrite carrying gas cavities, and volcaniclastic or hydrothermal breccias. Ore is commonly overlain by impermeable horizons, such as the chilled base of a flow or a mudstone. Stocks or larger plutons of intermediate composition are generally absent from the immediate vicinities of manto-type deposits; where they do occur, they are either barren or are characterized by a different type of mineralization (Ruiz et al., 1965). Minor intrusions are, however, common components of deposits, and include a gabbro plug at Buena Esperanza (Definis and Palacios, 1981); a gabbro plug, a diorite body, and andesite dikes at Susana de Michilla (Soto and Dreyer, 1985); andesite dikes at Mantos Blancos (Chávez, 1983) and Lo Aguirre; and andesite and trachyte dikes at El Soldado (Ruge, 1985). A number of large deposits are associated with intensely faulted rocks having structures of diverse orientations. Partial control of ore formation by faults is obvious at El Soldado, Buena Esperanza, and Susana de Michilla, probable at Lo Aguirre, but unproven at Mantos Blancos.

Small manto-type deposits commonly are confined to one or more thin (< 10 m) stratigraphic horizons, which may continue for several km along strike. Most large deposits occur either as a number of conformable ore shoots in thick (as much as 400 m) lithologic units (e.g., Mantos Blancos, Lo Aguirre), or as a combination of transgressive mineralization and a stacked series of thin stratabound ore horizons abutting minor intrusions (Figure 4; e.g., Buena Esperanza, Susana de Michilla) or faults (e.g., El Soldado). As many as 30 closely spaced stratigraphic horizons are mineralized at Buena Esperanza (Figure 4; Ruiz et al., 1971). These large deposits commonly also contain pipes, veins, or irregular bodies of hydrothermal breccia, which may contain monolithologic or—as at Mantos Blancos, Punta del Cobre, Susana de Michilla, and Buena Esperanza—heterolithologic fragment populations (R.H. Sillitoe, 1983, unpublished report). Manto-type deposits commonly range in average grade from 2 to 3% Cu, and some deposits contain appreciable amounts of silver (Table 3). Argentiferous manto-type copper deposits are transitional with copper-poor silver deposits, such as El Jardín, which carries 200 ppm Ag. Parts of some deposits are substantially higher in grade, especially where the ores occur in hydrothermal breccias or in close proximity to lithologic contacts or premineral dikes. At Punta del Cobre, for example, 3.5 million metric tons of 6% Cu ore have been exploited from transgressive breccias and veins (Camus, 1980). Dikes and plugs are generally impermeable and poorly mineralized, although exceptions occur (e.g., Mantos Blancos).

Much of the sulfide mineralization in manto-type deposits is present as finely disseminated grains or larger blebs in volcanic and sedimentary rocks, or as open-space fillings in volcanic, volcaniclastic, or hydrothermal breccias or in amygdules. Short veinlets and gash fractures are less commonly mineralized. The hypogene sulfide assemblages are iron and sulfur poor, and are typified by chalcocite, bornite, and chalcopyrite. At the scale of mineralized bodies or shoots, chalcocite-bornite tends to give way outward, beyond the ore zone, to chalcopyrite, pyrite-chalcopyrite, or pyrite (e.g., Mantos Blancos; Chavéz, 1983). Specularite is ubiquitous in ore, but at some localities it also occurs in halos around orebodies. For example, at Mantos Blancos, a specularite-rich andesitic unit overlies the ore-bearing dacitic or rhyodacitic unit (Chávez, 1983); a similar situation also characterizes Punta del Cobre (Camus, 1980). At the hand-specimen scale, bornite blebs or small nodules are rimmed by chalcocite, which in turn is enveloped by hematite; locally, the bornite contains remnant grains of chalcopyrite.

Ore, particularly in the high-grade parts of larger deposits, is accompanied by distinctive alteration assemblages dominated by quartz-albite, quartz-albite-chlorite, or sericite. Calcite is ubiquitous. Albite has been emphasized as a major alteration mineral (e.g., Olcay and Alarcón, 1975; Definis and

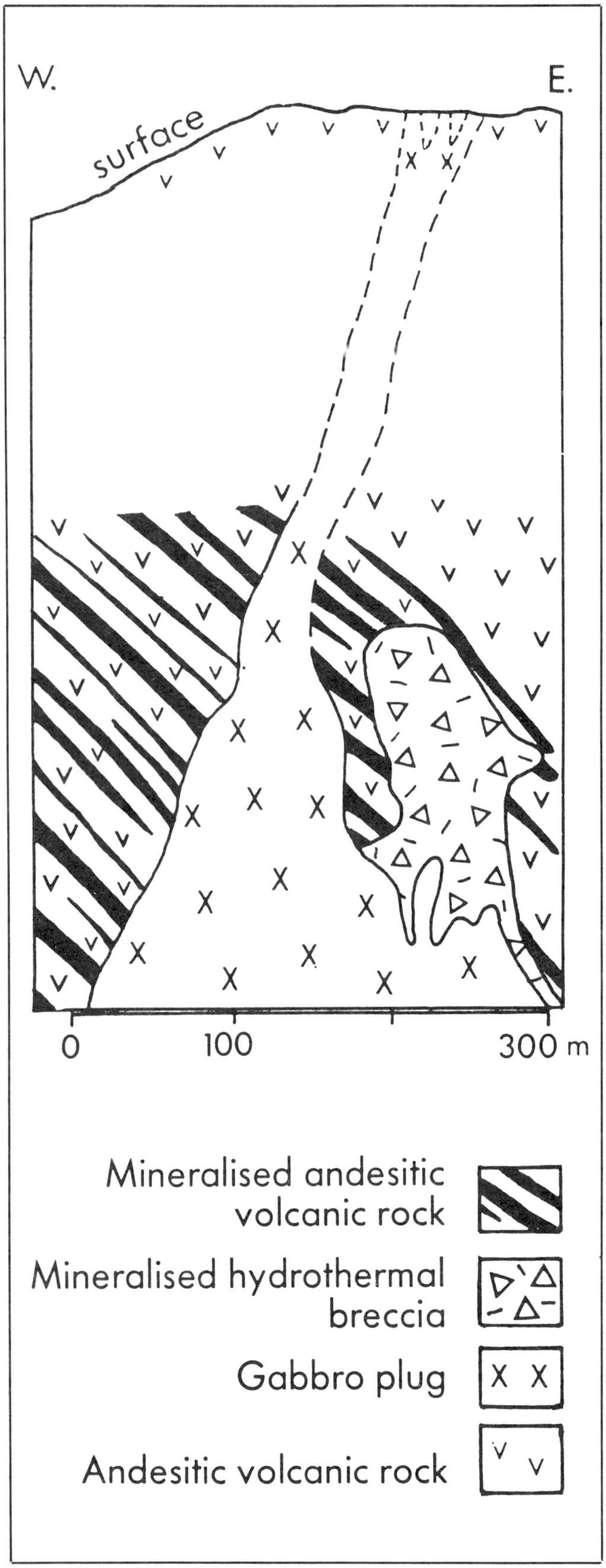

Figure 4. Cross section of Buena Esperanza manto-type copper deposit, northern Chile. Simplified from Definis and Palacios (1981).

Palacios, 1981; Chávez, 1983, 1984; Soto and Dreyer, 1985). Sericite appears to have overprinted earlier albite at Lo Aguirre, Buena Esperanza, and Mantos Blancos. The pyritic halos of deposits tend to exhibit chloritization. Although albite is enriched in many manto-type deposits, the alteration assemblages appear to be compatible with, and to grade into, the regionally developed burial-metamorphic assemblages. However, it is noteworthy that greenschist-facies assemblages characterized by epidote generally are encountered only beneath manto-type mineralization, which in many cases lacks epidote.

As summarized in Table 4, a variety of hypotheses has been advanced during the last 35 years for the origin of Chilean manto-type deposits. Arguments for a syngenetic origin are seriously weakened by the occurrence, even within restricted areas, of broadly similar deposits in host rocks of different ages, lithologies, and formational environments. All the evidence supports epigenetic emplacement, although the lack of any consistent association of manto-type deposits with intrusive rocks is thought to rule out magmatic-hydrothermal models. A metamorphogenic model, with generation of metalliferous fluids during regional burial metamorphism, best fits the data base (R.H. Sillitoe, unpublished report, 1983; Sato, 1984). Aspects of this model were first alluded to by Levi (1970) and Losert (1974).

Burial metamorphism is a characteristic feature of the Jurassic through early Tertiary volcano-sedimentary sequences of the western Andes (e.g., Levi, 1970; Aguirre, Levi, and Offler, 1978). Dewatering is an unavoidable facet of burial metamorphism, especially under the greenschist-facies conditions that prevailed in the deeper parts of the unconformity-bounded volcano-sedimentary sequences. It has been shown that both copper and sulfur were liberated from the volcanic rocks during epidotization (Losert, 1974). Copper (and silver) would have been scavenged and transported as chloride complexes; their presence in the ore fluid is supported by the high-salinity inclusions detected in quartz and calcite at El Soldado and Punta del Cobre (Holmgren, 1985), and by the occurrence of scapolite as an alteration mineral at Buena Esperanza (Losert, 1974). Metal-bearing fluids would have been expelled along permeable channelways—mainly up faults, gently dipping stratigraphic horizons, and contacts of minor intrusions—to sites at shallow levels conducive to ore deposition. Locally, the fluids became overpressured, probably as a result of self-sealing by quartz and albite precipitation, and generated hydrothermal breccias. The principal fluid conduits are invariably characterized by higher grade ore and more intense quartz-albite alteration.

The precise mechanism(s) of sulfide precipitation in manto-type deposits has not been clarified, although decreases of fluid temperature, perhaps involving mixing with cooler fluids, and fluid/wallrock reactions during ascent from epidote-bearing rocks to zones of prehnite-pumpellyite stability are a possibility. In a few deposits, sulfides were precipitated by localized concentrations of plant trash, but this is not a widespread mechanism. Evidence for preexisting disseminated pyrite has been obtained from a few deposits, such as Buena Esperanza (Losert, 1974), and pyrite may have acted as a more widespread precipitant of copper than is appreciated currently. Certainly the common in-situ replacement of chalcopyrite by bornite and bornite by chalcocite, with liberated

Table 4. Evolution of ideas on the origin of manto-type copper deposits, Chile.

Deposits	Timing of mineralization relative to host rocks	Ore fluid	Source of copper	Reference
Cabildo district (lat. 32°30′S)	Epigenetic	Magmatic (from intrusions)	Intrusive magma	Carter (1960), Ruiz and Ericksen (1962)
Throughout Chile	Syngenetic	Volcanogenic (fumarolic)	Volcanic host rocks	Ruiz et al. (1965, 1971), Stoll (1965)
Buena Esperanza (lat. 22°S)	Epigenetic	Metamorphogenic ± magmatic (from intrusions at depth)	Volcanic rocks during burial metamorphism	Losert (1974)
Amolanas & El Jardín (Lat. 28°S)	Syngenetic	Volcanogenic (fumarolic)	Volcanic host rocks	Lortie and Clark (1974)
Throughout Chile	Epigenetic	Magmatic (from intrusions at depth)	Unspecified	Olcay (1976)
Punta del Cobre (lat. 27°30′S)	Syngenetic	Seawater ± magmatic component (volcanic-exhalative)	Unspecified, but comparable to massive sulfide deposits	Camus (1980)
Northern Chile (lat. 22-24°S)	Epigenetic	Magmatic (from plugs and dikes of volcanic affiliation)	Intrusive plugs + volcanic host rocks	Definia and Palacios (1981)
Throughout Chile (especially lat. 22-26°S)	Epigenetic	Volcanogenic (subvolcanic)	Unspecified, but comparable to massive sulfide deposits	Espinoza and Palacios (1982), Espinoza (1982)
Mantos Blancos (lat. 23°30′S)	Epigenetic	Seawater (heated by dike or pluton emplacement)	Underlying volcanic pile	Chávez (1983)
Throughout Chile	Epigenetic	Metamorphogenic	Underlying volcanic pile during burial metamorphism	Sillitoe (unpublished report, 1983), Sato (1984)
El Soldado (lat. 32°30′S) and comparable deposits throughout Chile	Epigenetic	Magmatic (from subvolcanic alkaline intrusions)	Underlying volcanic pile	Holmgren (1985), Ruge (1985), Klohn, Holmgren, and Ruge (1986)

iron present as a hematite halo, demonstrates that there was progressive addition of copper to hypogene ore zones. The fact that $d^{34}S$ values for sulfides from Buena Esperanza, Mantos Blancos, Susana de Michilla, and El Soldado are similar to those for total sulfur from Chilean porphyry copper deposits (Sasaki et al., 1984; Holmgren, 1985) suggests that Chilean manto-type deposits contain sulfur of magmatic parentage. Much of the sulfur (and copper) could have been released from magmatic sulfide globules in volcanic rocks during their epidotization (Sato, 1984).

The contiguous Raúl and Condestable copper deposits in coastal central Perú (Figure 2; Table 3) are similar in many respects to the Chilean manto-type deposits. Disseminated, veinlet and breccia filling, and locally massive replacement chalcopyrite, magnetite, and pyrite are present as stacked series of stratabound lenses in andesitic volcanic and sedimentary rocks of Early Cretaceous age. Metamorphic minerals in the host rocks and the gangue minerals include amphibole, chlorite, clinopyroxene, epidote, albite, scapolite, quartz, and calcite, an assemblage that is indicative of somewhat higher-temperature conditions than those that prevailed in the immediate vicinities of the Chilean manto-type deposits. Although a syngenetic, volcanic-exhalative origin generally is favored for Raúl and Condestable (Ripley and Ohmoto, 1977; Wauschkuhn, 1979; Wauschkuhn and Thum, 1982), an epigenetic origin during burial metamorphism merits serious consideration (Atkin, Injoque-Espinoza, and Harvey, 1985).

Native copper-chalcocite mineralization in amygdaloidal volcanic rocks interbedded with Jurassic red beds of the Sierra de Perijá, Venezuela, were assigned a similar metamorphogenic origin by Viteri (1978).

Red-Bed Copper Deposits

The only red-bed copper deposit of any consequence in the Andes is at Corocoro in the Altiplano of Bolivia (Figure 1). The red-bed copper mineralization is hosted by a Tertiary molasse sequence, at least 10 km thick, that accumulated in an ensialic backarc basin. The mineralized units comprise fluviatile and lacustrine sandstones (some conglomeratic) and claystones ranging in age from Eocene-Oligocene to early Miocene (Evernden, Kriz, and Cherroni, 1977). Claystones are gypsiferous, and the mineralized area is pierced by gypsum diapirs, which are inferred to have been derived from thick evaporites at depth (Evernden, Kirz, and Cherroni, 1977). The host rocks of early Miocene age contain a significant tuffaceous component of andesitic composition (Ahlfeld, 1967), and mineralized rocks contain various amounts of plant trash (Ljunggren and Meyer, 1964). The Corocoro deposit, and smaller copper occurrences to both north and south, are confined to a complex, north-northwest-striking faulted antiform (Figure 5); the fault has been interpreted to follow a high-angle unconformity (Entwistle and Gouin, 1955). Structural complexity is at least partly attributable to diapirism.

At Corocoro, copper mineralization is confined to within 600 m of the fault over a strike length of 4000 m and a vertical interval of at least 550 m. Mineralization occurs in as many as 20 stratabound lenses, up to 5 m thick, that thin away from the fault (Figure 5). The geometry is reminiscent of that in the manto-type deposits (see Figure 4). Hypogene mineralization west of the fault is mainly chalcocite, whereas that to the east is exclusively native copper (Entwistle and Gouin, 1955). Chalcocite is present mainly as disseminated grains in bleached sandstone, whereas native copper occurs as a cement in hematitic sandstone and as veinlets along fractures and joints in sandstone and claystone, as well as locally in fault gouge (Ahlfeld, 1967). In a few places, galena, sphalerite, tetrahedrite, and bornite are associated with chalcocite. Minor aragonite, barite, celestite, quartz, and gypsum constitute the gangue.

It is clear that copper mineralization at Corocoro took place during or soon after the early Miocene, and before Miocene-Pliocene compression of the Altiplano gave rise to reverse movement on the Corocoro fault. Relatively low-temperature, oxidized, chloride-sulfate brines capable of generating the Corocoro deposit could have been expelled from evaporite-bearing red beds (Rose, 1976) and channeled upward along the fault or its precursor unconformity. Highly saline mine waters were encountered by Entwistle and Gouin (1955). The high chloride contents of the fluids would have facilitated leaching of copper from the red-bed sequence, in which the andesitic volcanic component could have been a principal source. Copper was precipitated in porous beds adjoining the fault/unconformity conduit. The reducing environment required to precipitate copper from solution was most likely provided by organic matter in the red beds. The discrete chalcocite and native copper domains (Figure 5), which must have been generated under different redox conditions, probably were juxtaposed by postmineral fault movements.

Kuroko-Type Deposits

The only major copper-bearing Kuroko deposit known to date in the Andes is at Tambo Grande in northern Perú (Figure 2; Table 3), although other examples an order of magnitude smaller have been recognized elsewhere in Perú as well as in Venezuela, Ecuador, and Colombia.

The Tambo Grande deposit is part of a submarine volcanic sequence of Late Jurassic-Early Cretaceous age, and seems to be closely associated with felsic volcanic rocks (Aye, 1985). The orebody is a massive lens of copper-, zinc-, and lead-bearing pyrite, the surface of which is completely oxidized. Banded rocks containing layers of chert, barite, and limonite (after sulfides) in the vicinity of the deposit were interpreted as exhalites by Injoque, Miranda, and Dunin-Borkowski (1979).

The Bailadores volcanogenic massive sulfide prospect in the Mérida Andes of Venezuela (Figure 2; Table 3) is essentially a zinc-lead-silver deposit, although it also contains 2% Cu as chalcopyrite. The prospect, which is hosted by felsic pyroclastic rocks of Late Carboniferous age, is in a zone of massive sulfides underlain by a chloritic stock (Carlson, 1977).

It is generally accepted that the bodies of massive sulfide that are the principal components of Kuroko-type deposits, like those at Tambo Grande and Bailadores, were emplaced syngenetically on, or immediately beneath, the seafloor from hydrothermal fluids dominated by seawater but perhaps containing a crucial magmatic component (Sawkins, 1982).

SUPERGENE COPPER ENRICHMENT

Parts of many intrusion-related copper deposits in Chile and Perú were significantly upgraded by supergene enrichment; enrichment factors of 2 to 3 are not uncommon. As summa-

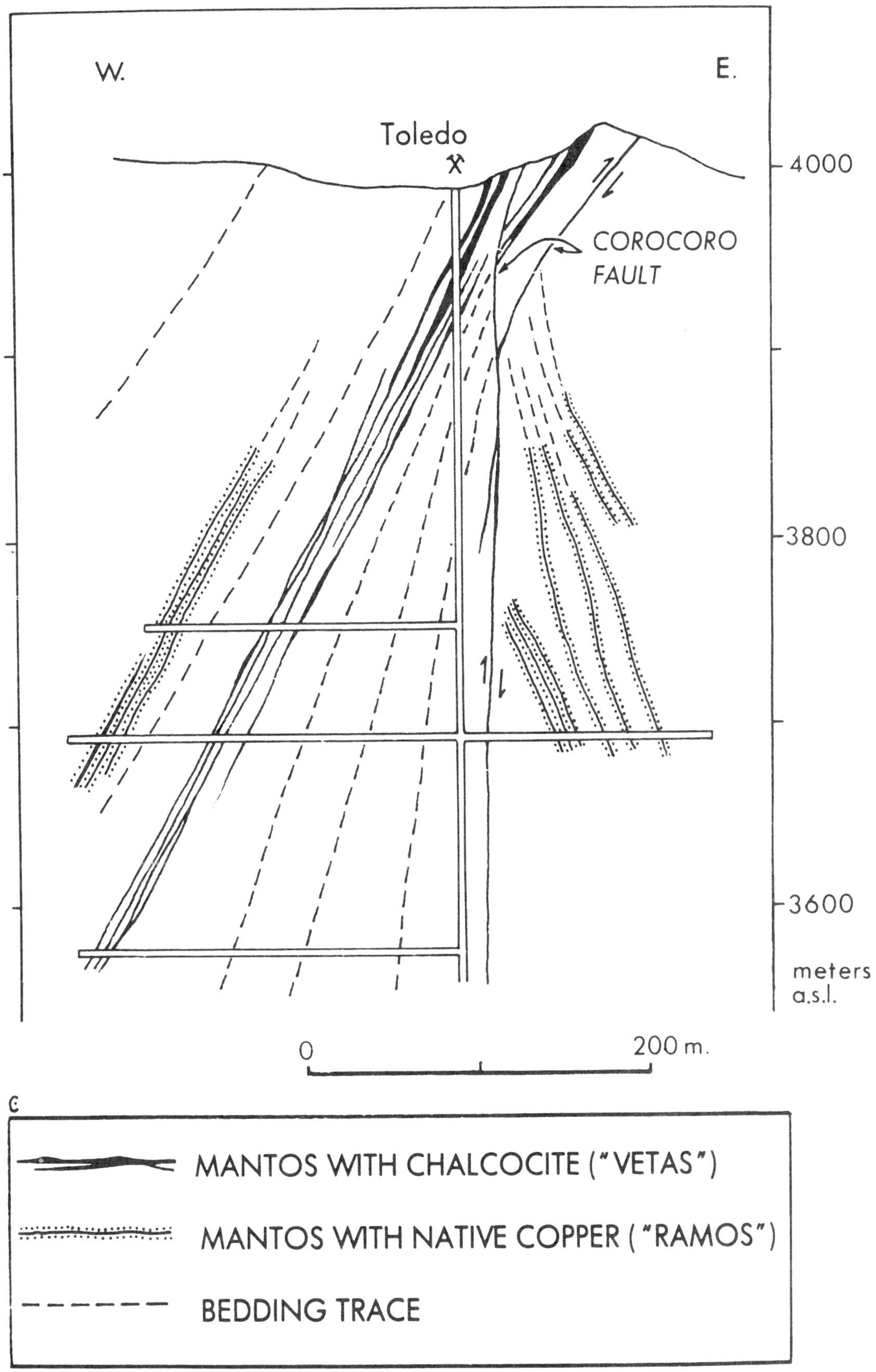

Figure 5. Cross section of Corocoro red-bed copper deposit, Bolivia. Taken from Entwistle and Gouin (1955).

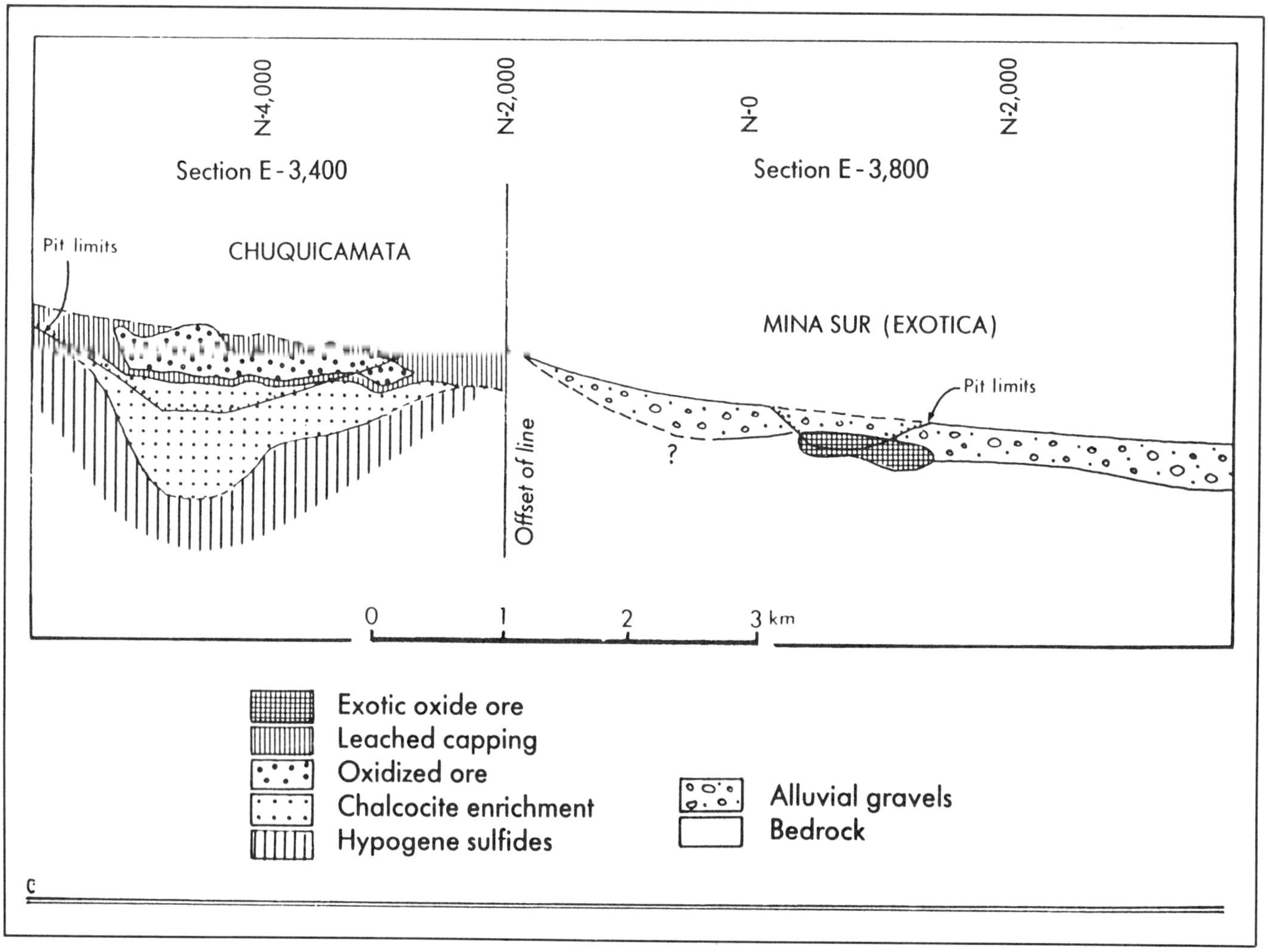

Figure 6. Cross section of supergene profile developed at Chuquicamata porphyry copper deposit, and its spatial relationship with Mina Sur exotic copper deposit. Adapted from Alvarez, Miranda, and Guzman (1980).

rized in Table 2, major chalcocite blankets range in thickness from 40 to 300 m, and at Chuquicamata attain 750 m. In these blankets, copper-bearing sulfides and (to a lesser extent) pyrite, are replaced, usually only partially, by chalcocite, which ranges from massive and steely to sooty in texture. The enrichment processes involved in development of major blankets were multicyclic, and required repeated enrichment and reoxidation of sulfides. Cerro Verde-Santa Rosa, Chuquicamata (Figure 6), and Potrerillos are examples of deposits which, prior to mining, were characterized by thick zones of oxidized copper minerals overlying the chalcocite ore; these economically important oxidized zones typically contain an abundance of the basic copper sulfates, antlerite and brochantite.

Mineralogic, structural, and geomorphologic factors all significantly influenced the supergene responses of Andean copper deposits. It is well known that the pyrite/copper-bearing sulfide ratio in copper deposits should exceed a critical value (say 2 or, in some cases, more) for the enrichment processes to operate efficiently. For this reason, porphyry copper deposits characterized by pyrite-rich sericitic and/or advanced argillic alteration generally underwent copper enrichment, whereas those dominated by pyrite-deficient K-silicate alteration developed little supergene chalcocite (Table 2). K-silicate zones, such as that at El Abra, underwent oxidation largely in situ, as did pyrite-poor skarn deposits (e.g., Tintaya). Sericitization, exhibiting a deep, high-grade enrichment blanket, and K-silicate alteration, characterized mainly by oxide mineralization, are both present in different parts of the Chuquicamata deposit (Figure 6; Flores, 1985).

The inhibition of copper leaching and subjacent chalcocite enrichment by the low pyrite contents of most K-silicate zones is further enhanced by the ability of alkali feldspar and biotite to neutralize the acid generated during oxidation of any pyrite that was present. In contrast, the neutralizing capacities of minerals such as sericite, clays, pyrophyllite, and alunite are minimal (cf., Brimhall, Alpers, and Cunningham, 1985).

The deficiency of pyrite (and of sericitic and advanced argillic assemblages) and the abundance of iron-poor sulfides such as chalcocite and bornite in most manto-type and red-bed copper deposits precluded the development of major enrichment blankets. As in the case of K-silicate assemblages, total or partial oxidation in situ is the normal situation, thereby resulting in major tonnages of oxidized copper (or copper-silver) ore (e.g., Mantos Blancos).

The closely spaced stockwork fractures and associated

breccias in porphyry copper deposits provide the high permeability required for thorough sulfide oxidation and effective enrichment. Moreover, the appreciable vertical extents (as much as 2 to 3 km) of many porphyry copper deposits further facilitated multicyclic enrichment, a parameter that also favored efficient enrichment of veins (e.g., Tamaya, Dulcinea) and breccia pipes (e.g., San Pedro de Cachiyuyo). In contrast, skarn and replacement-type copper deposits commonly had complex geometries and limited vertical extent (as in mantos), which tend to inhibit development of multicyclic blankets (Table 2). However, oxidation and enrichment are well developed in throughgoing pyritic veins and pipes intersected by the present surface at both Cerro de Pasco and Yauricocha (Petersen, 1965; Table 2).

The flow paths of the copper-bearing supergene solutions responsible for oxidation and enrichment are likely to be influenced by both structure and surface (water table) topography. At La Escondida, an appreciable (1 km) lateral component of fluid flow during the development of a major chalcocite blanket (Table 2) was documented recently by using innovative mass-balance calculations (Alpers, Brimhall, and Cunningham, 1984; Brimhall, Alpers, and Cunningham, 1985). Lateral drainage of fluids from porphyry copper deposits undergoing supergene alteration into permeable sequences of piedmont gravels may result in accumulation of large, lenslike oxide copper orebodies. Copper was precipitated, mainly as chrysocolla, wad, copper pitch, and atacamite, during the progressive neutralization of acid solutions flowing at, or immediately above, the gently inclined interfaces between the gravels and bedrock; the flow commonly was guided by buried paleochannels on the bedrock surfaces (Fam, 1979). Major exotic copper deposits of this type (Table 5) were developed for distances of 5 to 6 km from the Chuquicamata (Figure 6) and Collahuasi porphyry copper deposits, whereas the hypogene source of the copper in the Sagasca deposit has not yet been located.

The interplay between the timing of hypogene mineralization, climatic regime, and landform development exerted a strong influence on supergene alteration in the Andes (Figure 7). The age of hypogene mineralization, ranging from Mesozoic to late Miocene (Figures 1, 2; Table 2), is particularly important in the central Andes, where it is possible to decipher a geomorphologic chronology extending back to the Eocene (Mortimer, 1973; Noble, Farrar, and Cobbing, 1979). Climate, with its influence on erosion rates, is a fundamental control of supergene alteration. Where erosion rates were rapid, rock removal outpaced enrichment, and only low-grade, first-cycle chalcocite zones are encountered. As a consequence, the tropical, high-rainfall climate and rapid uplift during the late Cenozoic in Colombia precluded the generation (or preservation) of enrichment blankets at Mocoa and elsewhere (Sillitoe et al., 1982). Pliocene-Pleistocene glaciation in central Perú and central Chile, as at Antamina, Morococha and Río Blanco, had essentially the same effects. In northern Chile and southern Perú (between approximately lat. 15° and 28°S), and intensification of aridity during and since the Miocene seems to have effectively halted enrichment because of a lack of water for downward transport of copper. To both the north and south of this region, where precipitation is greater and existing supergene profiles tend to be less mature, enrichment is probably still active.

As summarized in Table 6, most if not all of the major chalcocite blankets in northern Chile and southern Perú (Table 2) were generated during the Oligocene-early Miocene, and were effectively fossilized by the beginning of the late Miocene. It is clear, therefore, that the chalcocite blankets in northern Chile and southern Perú were developed before the middle-late Miocene copper deposits in central Chile were even emplaced. In northernmost Chile and southern Perú, all or parts of the mature supergene profiles at Cuajone and Cerro Colorado were buried and preserved beneath ignimbrites and piedmont gravels of late Oligocene-early Miocene age (Figure 7; Table 6), and the exotic copper deposits at Mina Sur (Chuquicamata) and Sagasca (and probably also at Huinquintipa) accumulated in the basal piedmont gravels. Farther south in northern Chile (lat. 26°-32°S), development of the Atacama pediplain and its partial veneer of piedmont gravels and ignimbrite sheets was completed by the beginning of the late Miocene; it postdated all major supergene enrichment in the region, including that at El Salvador (Sillitoe, Mortimer, and Clark, 1968; Gustafson and Hunt, 1975).

Enrichment during the Oligocene-early Miocene is thought to have taken place at much lower elevations and under a more pluvial climate than those prevailing in northern Chile and southern Perú today (cf., Mortimer, Munchmeyer, and Urqueta, 1977; Hunt, Bratt, and Marquardt, 1983). Rainfall higher than at present and active uplift of the Andes were prerequisites for the widespread accumulation of piedmont gravels in northern Chile and southern Perú during the late Oligocene-early Miocene (Paskoff and Naranjo, 1983). Incremental Andean uplift, and a concomitant increase in aridity, would have combined to cause a progressive fall in the levels of regional water tables. A fall in water-table levels would have led to an increase in the thicknesses of copper-bearing rock subjected to oxidation and downward copper removal, and therefore enhanced the efficiency of the enrichment process (Brimhall, Alpers, and Cunningham, 1985). Pulses of more rapid water-table descent during active enrichment could be reflected by thin, copper-poor leached zones (Figures 6, 7) overlain by high-grade copper oxide ore and underlain by high-grade chalcocite ore at both Chuquicamata and Potrerillos (Jarrell, 1944; March, 1935). The leached zones are attributed to oxidation of hypogene sulfides isolated, along with overlying enrichment blankets, above markedly lower water tables.

South of lat. 30°S in Chile, and in central and northern Perú and southern Ecuador, supergene alteration during the late Miocene to Recent interval gave rise to thinner, more erratic, and less mature zones of chalcocite enrichment. Late Miocene to Recent chalcocite is mainly the sooty variety in contrast with that precipitated during the Oligocene-early Miocene, which is commonly more massive and steely. It is proposed that aging produces a transition from sooty to steely chalcocite. Chaucha, Michiquillay, La Granja, Cerro de Pasco, Yauricocha, Andacollo, Los Bronces, and El Teniente are examples of deposits that exhibit youthful, immature chalcocite zones. The chalcocite enrichment that underlies the pyrite-rich parts of the Los Bronces porphyry copper system was thought to be postglacial by Warnaars, Holmgren, and Barassi (1985). However, by analogy with the nearby El Teniente deposit (Figure 1), where chalcocite crops out on the lower slopes of a U-shaped glacial valley that was partly incised by the late Plio-

Table 5. Characteristics of principal exotic copper deposits of the Andes.

Deposit	Production + reserves, million metric tons	Cu, %	Status	Host rocks	Age of host-rock gravels	Principal supergene Cu minerals	Source deposit	Data source
Mina Sur (Exótica)	150	1.5	Production ceased	1/3 gravels, 2/3 Paleozoic bedrock	Early Miocene	Chrysocolla, copper pitch, atacamite, wad	Chuqui-camata	Mortimer, Munchmeyer, and Urqueta (1977) Alvarez, Miranda, and Guzman (1980)
Sagasca (Cascada)	30	1.8	In production	Gravels	Early Miocene	Chrysocolla, wad	Unknown	Fam (1979)
Huinquintipa	8	3.0	Prospect	Gravels	Early Miocene (?)	Chrysocolla, copper pitch	Collahuasi	Fam (1979)

Table 6. Constraints on age of supergene alteration of porphyry copper deposits, Peru and Chile.

Deposit	Age (m.y.) of hypogene mineralization[1]	Unit dated	Relationship to supergene profile	Age (m.y.)	Data source
Cuajone	51	Ignimbrite at base of Huaylillas Formation	Overlies erosion surface which truncated supergene profile	22.8 ± 0.7	Tosdal, Farrar, and Clark (1981)
Cerro Colorado	Paleocene	Ignimbrite in Altos de Pica Formation	Overlies erosion surface which truncated supergene profile	21.1 ± 0.3	Mortimer, Farrar, and Saric (1974)
Quebrada Blanca	38	Ignimbrite	Overlies gravels of probable early-middle Miocene age which, in turn, post-dated supergene alteration	4.3	Hunt, Bratt, and Marquardt (1983)
Collahuasi	Late Eocene-Early Oligocene	Ignimbrite	Overlies gravels of probable early-middle Miocene age which, in turn, host Huinquintipa exotic Cu deposit	6.0	M. Christensen, G. Pérez, F. Montecinos and G. Curtis (unpublished report, 1969)
Chuquicamata	31	Felsic ash bed 5-10 m below top of Fortuna gravels, Mina Sur	Gravels containing ash bed unconformably overlie gravels that host exotic Cu ore	8.4 ± 0.4	Mortimer, Munchmeyer, and Urqueta (1977), Alvarez, Miranda, and Guzman (1980)
La Escondida	34	Supergene alunite near top of enrichment blanket	Generated immediately prior to cessation of supergene enrichment	14.7 ± 0.5	Alpers, Brimhall, and Cunningham (1984)
El Salvador	41	Supergene alunite	Generated during supergene alteration	36.1 ± 0.6	Gustafson and Hunt (1975)
		Ignimbrite interbedded with gravels overlying Atacama pediplain	Ignimbrite-gravel sequence post-dated supergene alteration	12.6 ± 0.5	Clark et al. (1967)
Andacollo	112	Atacama pediplain	Truncated old supergene profile, but underlain by 40-m sooty chalcocite blanket	pre-12.6	Llaumett et al. (1975), R.H. Sillitoe, (unpublished data, 19[illegible])
El Teniente	5	Andesite flows	Occupy canyon that dissected supergene profile	2.3-1.8	Charrier and Munizaga (1979)

[1]Taken from compilation by Sillitoe (1988).

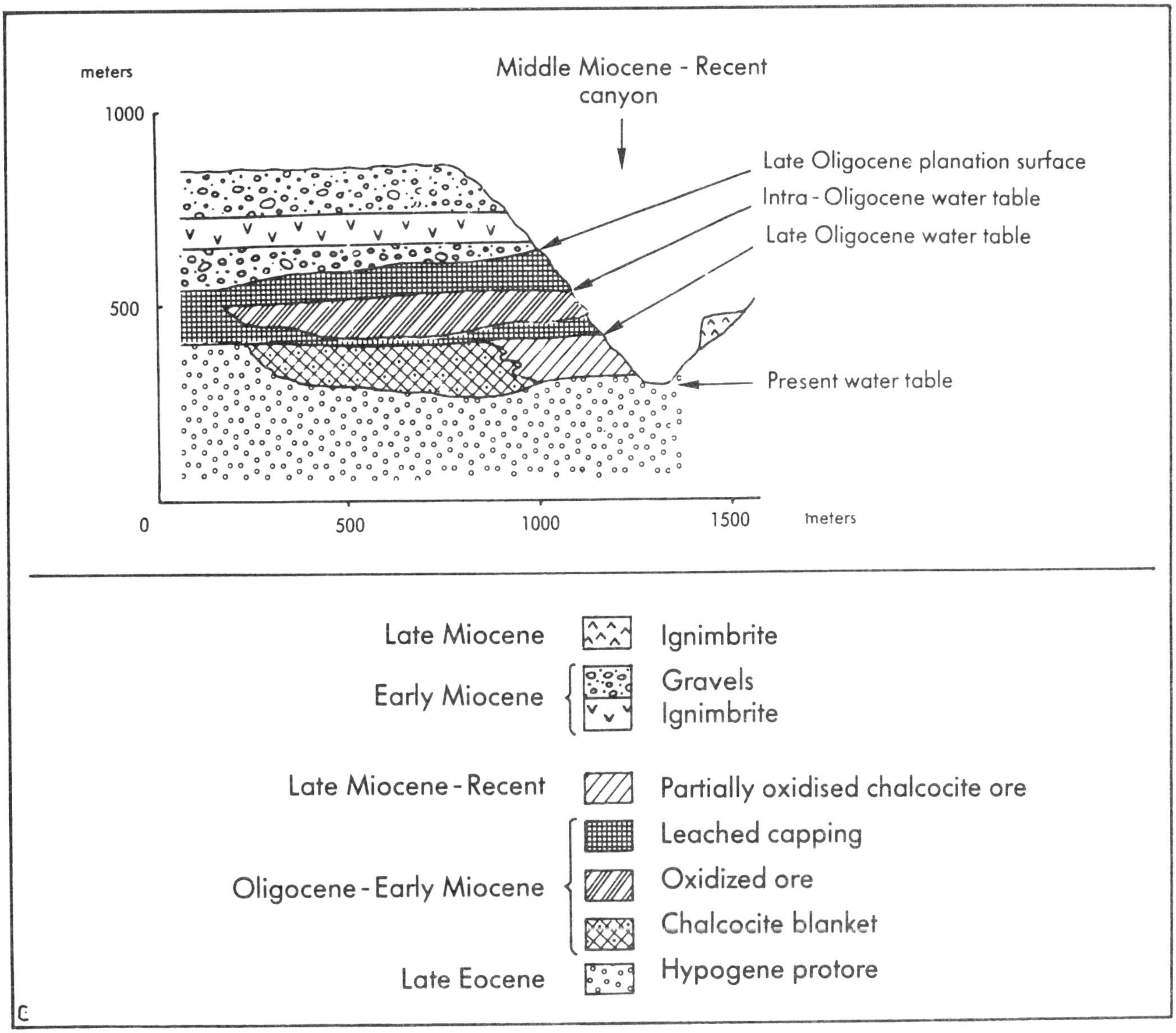

Figure 7. Composite cross section to illustrate supergene alteration of copper deposits in northern Chile and southern Perú.

cene (Table 6), enrichment is thought to have been largely preglacial or interglacial in age.

RELATIONSHIP TO ANDEAN EVOLUTION

Coira et al. (1982) and a number of subsequent workers emphasized that about 100 m.y. ago, during the inception of Atlantic spreading and westward motion of the South American Plate, a major tectonic transition affected the central Andes—the convergent margin changed from Mariana type to Chilean type in the sense of Uyeda and Kanamori (1979) (Figure 8). During the Jurassic and Early Cretaceous, the central Andean margin was extensional, and was characterized by partly submarine magmatic arcs paralleled along their eastern sides by ensialic backarc sedimentary basins. Moreover, at this time, in both central Chile (Åberg et al., 1984) and central Perú (Atherton, Pitcher, and Warden, 1983), the magmatic arcs themselves may have been constructed on thinned and rifted crust in ensialic marginal basins. During this period, the angle of subduction is inferred to have been steep. In contrast, from the Late Cretaceous onward a series of eastward-migrating magmatic arcs was subaerial, and was not coupled to backarc basins. Extensional features were less prevalent, and the oceanic slab is inferred to have underthrust the continental margin at a relatively low angle. The rate of convergence also increased markedly (Pitman, 1978). Chilean-type subduction still prevails (hence the name), with two segments of the Benioff zone at such a low angle that magmatism has been essentially inactive since the Pliocene (Barazangi and Isacks, 1976).

Most major porphyry copper deposits, as well as all the major skarn and replacement copper deposits, in the central Andes were emplaced during the Cenozoic, a time of Chilean-type subduction (Figure 8). In contrast, the Kuroko-type

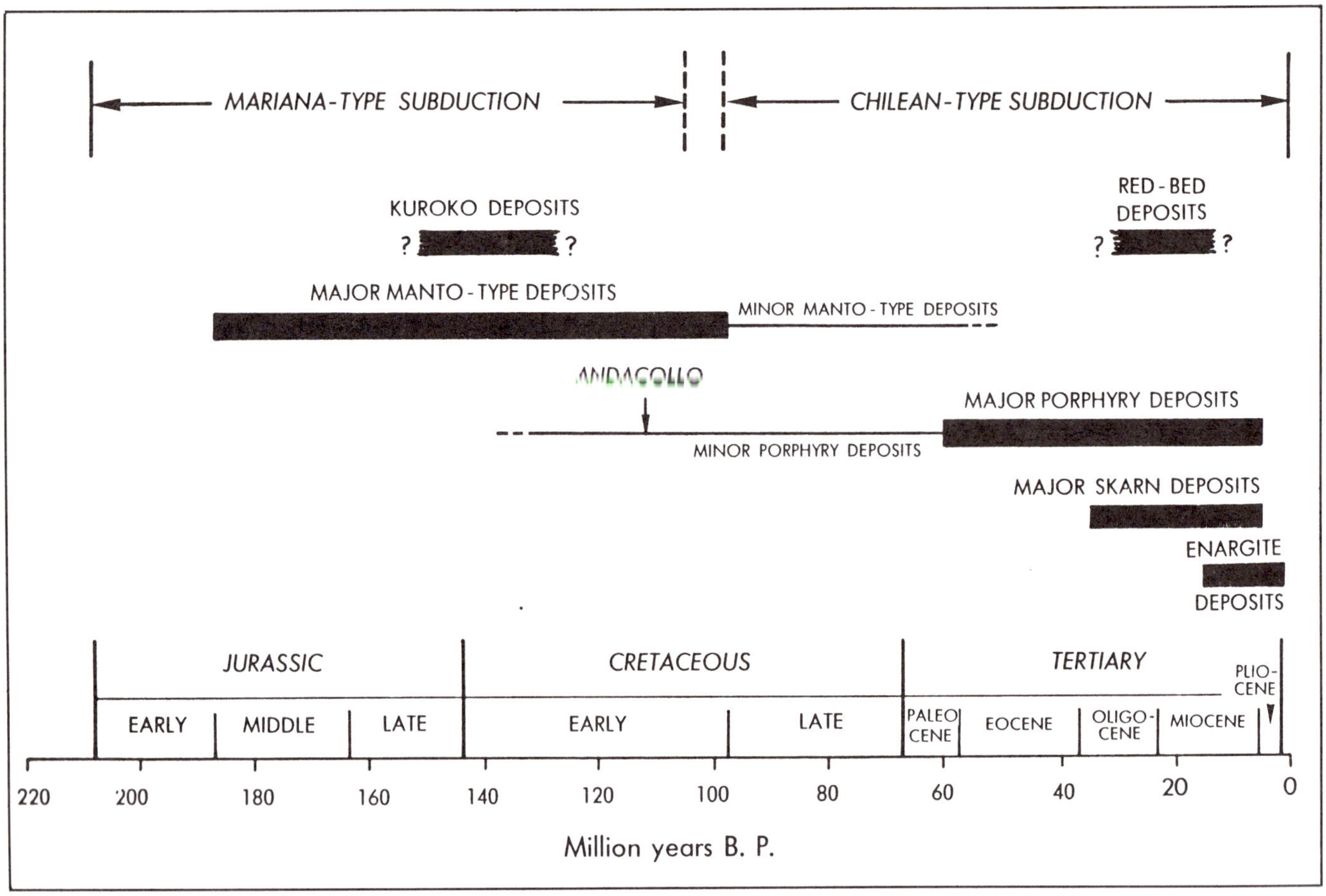

Figure 8. Timing of emplacement of principal types of copper deposits, and modes of subduction for the Mesozoic-Cenozoic central Andes. Time-scale after Geological Society of America (1983).

deposit at Tambo Grande was generated in the Late Jurassic-Early Cretaceous during Mariana-type subduction (Figure 8). These relationships comply with the proposal of Uyeda and Nishiwaki (1980) that porphyry copper deposits characterize Chilean-type margins, whereas Kuroko-type deposits characterize Mariana-type margins. In the case of the central Andes, however, the generally deeper erosion levels of the Jurassic-Early Cretaceous (and Late Cretaceous) magmatic arcs offer an alternative and preferred explanation for the deficiency of pre-Cenozoic porphyry copper deposits (Sillitoe, 1972). This alternative is thought to be supported by recently obtained K-Ar ages of 112 ± 10 m.y. for the Andacollo porphyry copper deposit (Figure 1) and 132 ± 8 m.y. for the Puntillas porphyry copper prospect in coastal northern Chile (Munizaga, Huete, and Hervé, 1985), which demonstrate that porphyry copper emplacement did take place in the central Andes during the Early Cretaceous. The precise nature of the porphyry copper-bearing Middle-Late Jurassic volcano-plutonic arc in Colombia (Figure 2) is poorly known, as are the Early Carboniferous and Late Carboniferous-Early Permian arcs that carry minor porphyry copper occurrences in Argentina and contiguous Chile (Figure 1). It is reasonable, however, to consider them all as the consequences of eastward subduction beneath western Gondwana (Sillitoe et al., 1982; Sillitoe, 1981). The extensional regime that characterized the Jurassic arc in Colombia (Etayo, 1985) may be taken to suggest that the subduction angle was steep and in the Mariana mode.

Kuroko-type deposits in the central Andes could have formed only in pre-Late Cretaceous times because of their necessary restriction to submarine volcanic environments. A control of their emplacement by the extensional stress regime at that time remains unproven. The timing of emplacement of manto-type deposits s even less amenable to correlation with regional stress conditions and mode of subduction (Sato, 1984). Manto-type deposits were emplaced in volcano-plutonic arcs both before and after the mid-Cretaceous change in subduction style, although the largest deposits (Table 3) are all pre-Late Cretaceous (Figure 8). It could be suggested that the high geothermal gradients during ensialic extension and emplacement of basaltic (gabbroic) magmas at a Mariana-type margin caused metamorphic dewatering of greater thickness of volcanic and volcano-sedimentary rocks. A greater volume of copper-bearing fluids might therefore have been available, and more readily released to form manto-type deposits.

The generation of red-bed copper deposits required the presence of subaerial, evaporite-bearing molasse sequences, which are most likely to be formed in the backarc regions of Chilean-type margins, as they were during the Tertiary on the Bolivian Altiplano (Figure 8). Subaerial conditions during the Tertiary were also a prerequisite for the unroofing of intrusion-related deposits, as well as for supergene enrich-

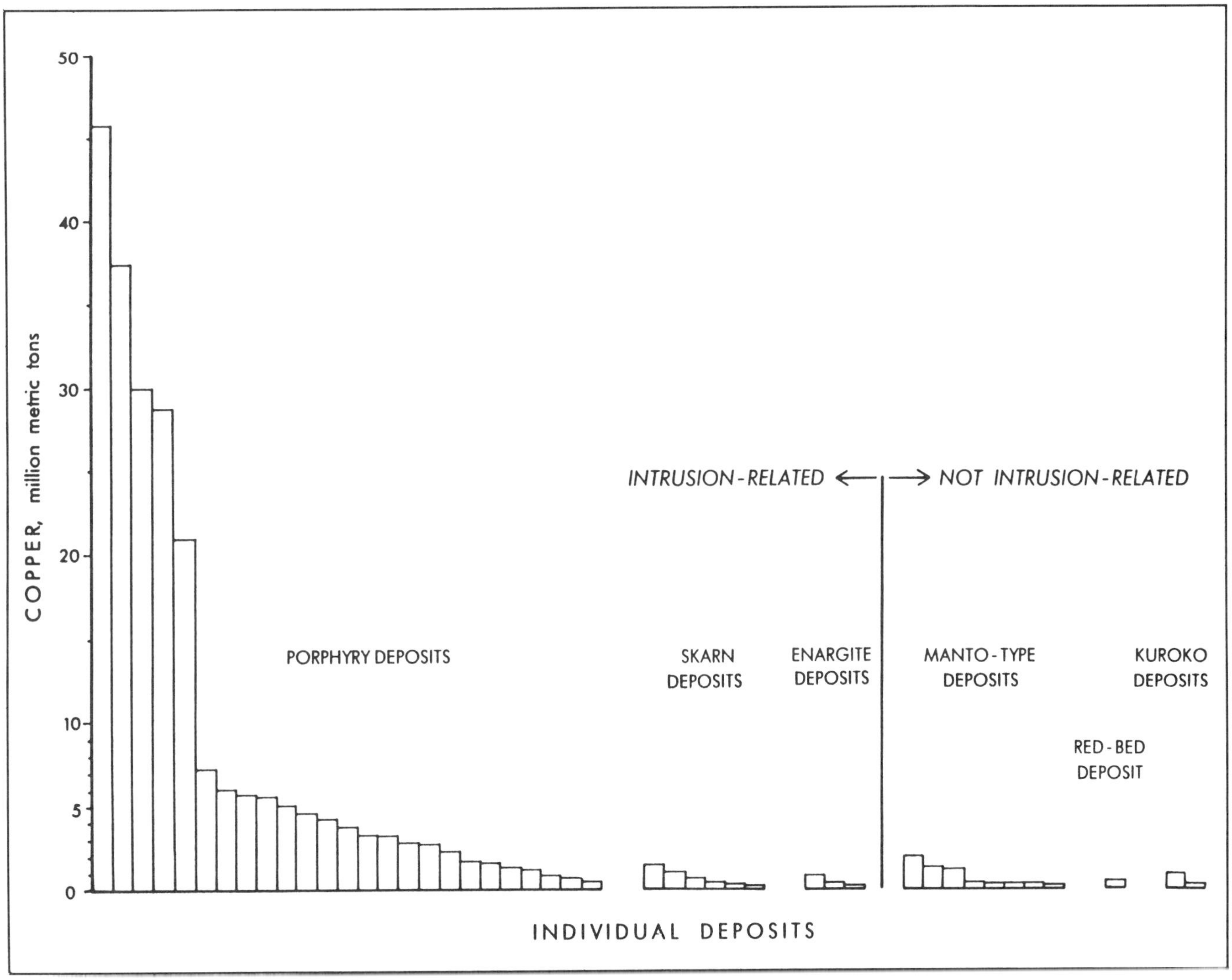

Figure 9. Comparisons of the tonnages of copper metal in principal Andean copper deposits. Note dominance of intrusion-related deposits over those generated in volcano-sedimentary formations, and of porphyry copper category by five deposits in Chile (Chuquicamata, El Teniente, La Escondida, Río-Blanco-Los Bronces, and Los Pelambres). Data taken from Tables 2 and 3, but with the contiguous Río Blanco and Los Bronces orebodies considered as a single deposit.

ment and generation of exotic copper deposits to take place in parts of Chile and Perú.

The fundamental reason for the dominance of post-Paleozoic Andean metallogeny by copper seems to be an overall copper enrichment of its subduction-related, I-type, magnetite-series magmas (Ishihara et al., 1984). Given this widespread availability of copper, ore formation became heavily dependent on adequate supplies of chloride-bearing fluids for copper mobilization and transport, and of sulfur for copper precipitation. It is clear that in the Andes (and elsewhere), magmatic-hydrothermal processes acting during the consolidation of high-level porphyry stocks were the most efficient means of providing upper-crustal concentrations of chloride and sulfur. It is within, alongside, and above such stocks that 97% of Andean copper resources occur as porphyry, skarn, and replacement deposits (Table 2; Figure 9).

In volcanic and volcano-sedimentary environments, however, sulfur is not so readily available, as evidenced by the low-sulfur sulfide assemblages in manto-type and red-bed copper deposits. Probably for this reason, combined with a less efficient focusing of the metal-bearing chloride brines, substantially less Andean copper—only about 3% (Figure 9; Table 3)—occurs in deposits unrelated to intrusive activity.

However, the copper in all the deposit types is considered to have been supplied ultimately by subcrustal, subduction-related magmatism. The copper in the large intrusion-related deposits was contributed directly from associated high-level cupolas as a component of chloride-bearing magmatic-hydrothermal fluids. In the case of manto-type deposits, a chloride-bearing fluid liberated during burial metamorphism is believed to have scavenged trace amounts of magmatic copper from andesitic (and basaltic) volcanic rocks. For red-bed deposits, a chloride brine, developed by dissolution of evaporites, leached magmatic copper from andesitic debris in a molasse sequence. In the specific case of Corocoro, this andesitic material was transported fluvially to the Altiplano backarc basin during the erosion of early to mid-Tertiary magmatic arcs farther west in Chile (Figure 1). Copper in Kuroko-type deposits was either scavenged from the underlying volcanic (or volcano-sedimentary) piles by convectively circulating seawater, or was contributed to such evolved seawater-dominated fluids by magmatic-hydrothermal brines released from subjacent stocks.

In contrast to the Mesozoic-Cenozoic Andes, the Paleozoic Andes are poorly endowed with copper mineralization although, as noted above, low-grade porphyry copper occurrences are known, especially in Argentina. The paucity of intrusion-related copper deposits may be attributed, in part, to a deeper erosion level than that characteristic of the Cenozoic Andes, and to the widespread concealment of Paleozoic magmatic arcs beneath Mesozoic-Cenozoic cover. However, a more fundamental explanation for the copper deficiency is suggested by the work of Ishihara et al. (1984) in central and northern Chile. They showed that Paleozoic intrusive rocks there were more reduced, and therefore less able to supply copper, than their Mesozoic-Cenozoic counterparts; in fact, a number of the Paleozoic plutons consist of ilmenite-series rather than magnetite-series rocks. The processes that controlled the oxidation states and copper contents of magmas at convergent margins remain to be elucidated, although it might be suggested that the more oxidized, copper-rich magmas required a contribution from the downgoing slab.

ACKNOWLEDGMENTS

Presentation of this invited paper at the symposium on "Geology of the Andes and its Relation to Hydrocarbon and Mineral Resources" in Santiago de Chile was facilitated by financial support provided by the Circum-Pacific Council for Energy and Mineral Resources. Valuable geologic and tonnage/grade information incorporated in the paper were furnished by Enrique Barrantes, José Cabello, Christopher Pointon, Nicolas Sarić, Enrique Tidy, and Fernando Urquidi. The manuscript benefited from reviews by José Cabello and Frederick J. Sawkins.

REFERENCES CITED

Åberg, G., L. Aguirre, B. Levi, and J. O. Nyström, 1984, Spreading-subsidence and generation of ensialic marginal basins: an example from the early Cretaceous of Central Chile, *in* B. P. Kokelaar and M. F. Howells, eds., Marginal basin geology: Geological Society (London) Special Publication 16, p. 185-193.

Aguirre, L., B. Levi, and R. Offler, 1978, Unconformities as mineralogical breaks in the burial metamorphism of the Andes: Contributions to Mineralogy and Petrology, v. 66, p. 361-366.

Ahlfeld, F., 1967, Metallogenetic epochs and provinces of Bolivia: Mineralium Deposita, v. 2, p. 291-311.

Ahlfeld, F., and A. Schneider-Scherbina, 1964, Los yacimientos minerales y de hidrocarburos de Bolivia: La Paz, Departamento Nacional de Geología Boletin No. 5 (Especial), 388 p.

Aizawa, N., and N. Tomizawa, 1986, Prospecting of Coroccohuayco copper deposit, Perú—with special reference to southern skarn zone: Mining Geology, v. 36, p. 83-100 (in Japanese with English abstract).

Alpers, C. N., G. H. Brimhall, and A. B. Cunningham, 1984, Mass balance and timing of supergene enrichment at La Escondida, Antofagasta province, Chile (abs.): Geological Society of America Abstracts with Programs, v. 16, p. 428.

Alvarez C., O., and R. Flores V., 1985, Alteración y mineralización hipógena en el yacimiento de Chuquicamata, Chile: Congreso Geológica Chileno, 4th, Antofagasta, Actas, v. 2, p. 78-100.

Alvarez C., O., J. Miranda M., and P. Guzman V., 1980, Geología del complejo Chuquicamata, *in* Minería de cobres porfídicos, v. 2: Santiago, Instituto de Ingenieros de Minas de Chile Congreso Cincuentenario 1930-1980, p. 314-363.

Ambrus, J., 1977, Geology of the El Abra porphyry copper deposit, Chile: Economic Geology, v. 72, p. 1062-1085.

Ambrus, J., 1978, Chuquicamata deposit, *in* A. Sutulov, ed., International molybdenum encyclopaedia 1778-1978: Santiago, Intermet Publications, v. 1, p. 87-93.

Araneda G., R., 1982, El Indio, yacimiento de oro, plata y cobre. Coquimbo, Chile: Santiago, Minerales, v. 37, no. 160, p. 5-13.

Atherton, M. P., W. S. Pitcher, and V. Warden, 1983, The Mesozoic marginal basin of central Perú: Nature, v. 305, p. 303-306.

Atkin, B. P., J. L. Injoque-Espinoza, and P. K. Harvey, 1985, Cu-Fe-amphibole mineralization in the Arequipa segment, *in* W. S. Pitcher, M. P. Atherton, E. J. Cobbing, and R. D. Beckinsale, eds., Magmatism at a plate edge: the Peruvian Andes: Glasgow & London, Blackie & Son Ltd., p. 261-270.

Atkinson, W. W., Jr., A. Souviron, and T. I. Vehrs, 1984, Geological results of recent exploration of the porphyry copper deposit at Los Pelambres, Chile (abs.): Geological Society of America Abstracts with Programs, v. 16, p. 434.

Aye, F., 1985, Tambo Grande: a volcanogenic massive sulphide deposit in the Cretaceous of the Peruvian coast, *in* Symposium on geology of the Andes and its relation to hydrocarbon and mineral resources (abs.): Santiago, Abstracts, p. 25.

Barazangi, M., and B. L. Isacks, 1986, Spatial distribution of earthquakes and subduction of the Nazca Plate beneath South America: Geology, v. 4, p. 686-692.

Beckinsale, R. D., A. W. Sanchez-Fernandez, M. Brook, E. J. Cobbing, W. P. Taylor, and N. D. Moore, 1985, Rb-Sr whole-rock isochron and K-Ar age determinations for the Coastal Batholith of Perú, *in* W. S. Pitcher, M. P. Atherton, E. J. Cobbing, and R. D. Beckinsale, eds., Magmatism at a plate edge: the Peruvian Andes: Glasgow & London, Blackie & Son Ltd., p. 177-202.

Brimhall, G. H., C. N. Alpers, and A. B. Cunningham, 1985, Analysis of supergene ore-forming processes and ground-water solute transport using mass balance principles: Economic Geology, v. 80, p. 1227-1256.

Camus, F., 1975, Geology of the El Teniente orebody with emphasis on wall-rock alteration: Economic Geology, v. 70, p. 1341-1372.

Camus, F., 1980, Posible modelo genético para los yacimientos de cobre del distrito minero Punta del Cobre: Revista Geológica de Chile, No. 11, p. 51-76.

Cardozo, M., and C. Vidal, 1981, Metalogénsis y magmatismo en el flanco occidental de los Andes centrales: Revisión: Boletin de la Sociedad Geológica del Perú, No. 68, p. 13-24.

Carlson, G. G., 1977, Geology of the Bailadores, Venezuela, massive sulfide deposit: Economic Geology, v. 72, p. 1131-1141.

Carlson, S. R., and F. J. Sawkins, 1980, Mineralogic and fluid inclusion studies of the Turmalina Cu-Mo-bearing breccia pipe, northern Perú: Economic Geology, v. 75, p. 1233-1238.

Carter, W. D., 1960, Origin of "manto-type" copper deposits of the Cabildo mining district, central Chile: International Geological Congress, 21st, Norden 1960, Report, Pt. 16, p. 17-28.

Centromin, 1980, Cobriza, *in* Samamé Boggio, ed., El Perú minero, v. 4: Yacimientos: Lima, Instituto Geológico Minero y Metalúrgico, p. 314-328.

Charrier, R., and F. Munizaga, 1979, Edades K-Ar de volcanitas cenozoicas del sector cordillerano del Río Cachapoal, Chile (34°15′ de lat. sur): Revista Geológica de Chile, No. 7, p. 41-51.

Chávez, W. X., Jr., 1983, The geologic setting of disseminated copper sulfide mineralization of the Mantos Blancos copper-silver district, Antofagasta province, Chile: Society of Mining Engineers of AIME Preprint 83-193, 20 p.

Chávez, W. X., Jr., 1984, Alteration mineralogy and chemistry of rhyolitic and andesitic volcanic rocks of the Mantos Blancos copper-silver district, Chile: Society of Mining Engineers of AIME Preprint 84-153, 5 p.

Clark, A. H., A. E. S. Mayer, C. Mortimer, R. H. Sillitoe, R. U. Cooke, and N. J. Snelling, 1967, Implications of the isotopic ages of ignimbrite flows, southern Atacama Desert, Chile: Nature, v. 215, p. 723-724.

Coira, B., J. Davidson, C. Mpodozis, and V. Ramos, 1982, Tectonic and magmatic evolution of the Andes of northern Argentina and Chile: Earth-Science Reviews, v. 18, p. 303-332.

Definis, M., A., and C. Palacios M., 1981, Geologia del yacimiento de cobre "Buena Esperanza," norte de Chile. Un ejemplo de mineralización asociada a un cuello volcánico: Madrid, Estudios Geológicos, v. 37, p. 177-185.

Einaudi, M. T., 1977, Environment of ore deposition at Cerro de Pasco, Perú: Economic Geology, v. 72, p. 893-924.

Einaudi, M. T., L. D. Meinert, and R. J. Newberry, 1981, Skarn deposits: Economic Geology 75th Anniversary Volume, p. 317-391.

Entwistle, L. P., and L. O. Gouin, 1955, The chalcocite-ore deposits at Corocoro, Bolivia: Economic Geology, v. 50, p. 555-570.

Espinoza R., S. T., 1982, Definición del tipo "diseminado infravolcánico de sulfuros de cobre": Congreso Geológico Chileno, 3rd. Concepción, Actas, v. 3, p. E201-E219.

Espinoza R., S. T., and C. Palacios M., 1982, Metalogénesis de los yacimientos de cobre en la Cordillera de la Costa entre Taltal y Tocopilla (22°-26°S), Chile: Congreso Latinoamericano de Geología, 5th, Argentina, Actas, v. 3, p. 51-63.

Estrada Ch., F., 1975, Geología de Quellaveco: Boletin de la Sociedad Geológica del Perú, v. 46, p. 65-86.

Etayo S., F., 1985, The stratigraphic-tectonic framework of the Colombian Andes, *in* Symposium on geology of the Andes and its relation to hydrocarbon and mineral resources: Santiago, Abstracts, p. 11-13.

Evernden, J. F., S. J. Kriz, and C. Cherroni M., 1977, Potassium-argon ages of some Bolivian rocks: Economic Geology, v. 72, p. 1042-1061.

Eyzaguirre P., V. R., and D. E. Montoya A., 1973, Geología del yacimiento de cobre diseminado de Toro Mocha, Morococha, Perú: Congreso Latinoamericano de Geología, 2nd, Venezuela, preprint.

Fam R., R., 1979, Mineralización de cobre del tipo "exótico" en el norte de Chile: Congreso Geológico Chileno, 2nd, Arica, Actas, v. 2, C235-C263.

Fernandez, R. R., R. F. Brown, and A. N. Lencinas, 1974, "Pachón". Un nuevo pórfiro cuprífero argentino, Dto. Calingasta, Prov. de San Juan, República Argentina: Congreso Geológico Argentino, 5th, Buenos Aires, Actas, v. 2, p. 77-89.

Field, C. W., R. O. Rye, J. R. Dymond, J. F. Whelan, and R. G. Senechal, 1983, Metalliferous sediments of the East Pacific, *in* W. C. Shanks, III, ed., Cameron volume on unconventional mineral deposits: New York, American Institute of Mining, Metallurgical & Petroleum Engineers, p. 133-156.

Flores, G., A. Zelaya, and F. Mamani, 1975, Geología de Cañariaco: Boletin de la Sociedad Geológica del Perú, v. 46, p. 87-102.

Flores, V., R., 1985, Control del enriquecimiento supérgeno en el yacimiento Chuquicamata, Chile: Congreso Geológico Chileno, 4th, Antofagasta, Actas, v. 2, p. 228-249.

Geological Society of America, 1983, Geologic time scale.

González, O. E., 1975, Geología y alteración en el cobre porfídico "Bajo La Alumbrera," República Argentina: Congreso Ibero-Americano de Geología Económica, 2nd, Buenos Aires, Actas, v. 2, p. 247-270.

Goossens, P. J., and V. F. Hollister, 1973, Structural control and hydrothermal alteration pattern of Chaucha porphyry copper, Ecuador: Mineralium Deposita, v. 8, p. 321-331.

Gustafson, L. B., and J. P. Hunt, 1975, The porphyry copper deposit at El Salvador, Chile: Economic Geology, v. 70, p. 857-912.

Halpern, M., 1978, Geological significance of Rb-Sr isotopic data of northern Chile crystalline rocks of the Andean orogen between latitudes 23° and 27° South: Geological Society of America Bulletin, v. 89, p. 522-532.

Halpern, M., 1979, Strontium isotope composition of rocks from the Disputada copper mine, Chile: Economic Geology, v. 74, p. 129-130.

Hollister, V. F., and E. Sirvas B., 1974, El pórfido de cobre Michiquillay: Boletin de la Sociedad Geológica del Perú, v. 44, p. 11-27.

Holmgren, D. C., 1985, Antecedentes para un modelo genético del yacimiento El Soldado, V Región, Chile: Congreso Geológico Chileno, 4th, Antofagasta, Actas, v. 4, p. 626-651.

Howell, F. H., and J. S. Molloy, 1960, Geology of the Braden orebody, Chile, South America: Economic Geology, v. 55, p. 863-905.

Hunt, J. P., 1985, Applied geology at Quebrada Blanca and Collahuasi, Chile, and in the future of U.S. metal mining: Economic Geology, v. 80, p. 794-800.

Hunt, J. P., J. A. Bratt, and J. C. Marquardt L., 1983, Quebrada Blanca, Chile: an enriched porphyry copper deposit: Mining Engineering, v. 35, p. 636-644.

Hunt, J. P., C. Munchmeyer F., J. C. Marquardt L., and H. Soto P., 1980, El proyecto Quebrada Blanca, *in* Minería de cobres porfídicos, v. 2: Santiago, Instituto de Ingenieros de Minas de Chile Congreso Cincuentenario 1930-1980, p. 263-289.

Injoque, J., C. Miranda, and E. Dunin-Borkowski, 1979, Estudio de la génesis del yacimiento de Tambogrande y sus implicancias: Boletin de la Sociedad Geológica del Perú, v. 64, p. 73-99.

Ishihara, S., C. E. Ulriksen, K. Sato, S. Terashima, T. Sato, and Y. Endo, 1984, Plutonic rocks of north-central Chile: Bulletin of the Geological Survey of Japan, v. 35, p. 503-536.

Jarrell, O. W., 1944, Oxidation at Chuquicamata, Chile: Economic Geology, v. 39, p. 251-286.

Kihien C., A., 1975, Alteración y su relación con la mineralización en el pórfido de cobre de Cerro Verde: Boletin de la Sociedad Geológica del Perú, v. 46, p. 103-126.

Klohn, E., C. Holmgren, and H. Ruge, 1986, El Soldado, a peculiar copper deposit associated with anomalous alkaline volcanism in the central Chilean Coastal Range: Society of Mining Engineers of AIME, 115th Annual Meeting, New Orleans, March 1986, Preprint, 10 p.

Kusakabe, M., S. Nakagawa, M. Hori, Y. Matsuhisa, J. M. Ojeda, and L. Serrano, 1984, Oxygen and sulfur isotopic compositions of quartz, anhydrite and sulfide minerals from the El Teniente and Río Blanco porphyry copper deposits, Chile: Bulletin of the Geological Survey of Japan, v. 35, p. 583-614.

Le Bel, L. M., 1985, Mineralization in the Arequipa segment: the porphyry-Cu deposit of Cerro Verde/Santa Rosa, *in* W. S. Pitcher, M. P. Atherton, E. J. Cobbing, and R. D. Beckinsale, eds., Magmatism at a plate edge: the Peruvian Andes: Glasgow & London, Blackie & Son Ltd., p. 250-260.

Levi, B., 1970, Burial metamorphic episodes in the Andean geosyncline, central Chile: Geologische Rundschau, v. 59, p. 994-1013.

Ljunggren, P., and H. C. Meyer, 1964, The copper mineralization in the Corocoro basin, Bolivia: Economic Geology, v. 59, p. 110-125.

Llaumett P., C., 1980, Antecedentes sobre producción, recursos y expectativas de la minería del oro en Chile: Santiago, Minerales, v. 35, no. 150, p. 9-20.

Llaumett, C., et al., 1975, El yacimiento cobre porfídico "Andacollo," Provincia de Coquimbo, Chile: Revista Geológica de Chile, no. 2, p. 56-66.

Lopez-Escobar, L., 1982, Características geoquímicas de rocas igneas asociadas con pórfidos cupríferos chilenas: Revista Geológica de Chile, No. 17, p. 3-19.

Lortie, R. B., and A. H. Clark, 1974, Stratabound fumarolic copper deposits in rhyolitic lavas and ash-flow tuffs, Copiapó district, Atacama, Chile: IAGOD Symposium, 4th, Varna, Proceedings, v. 1, p. 256-264.

Losert, J., 1974, Alteration and associated copper mineralizations in the Jurassic volcanic rocks of the Buena Esperanza mining area (Antofagasta province, northern Chile), *in* E. Klohn, ed., Coloquio sobre fenómenos de alteración y metamorfísmo en rocas volcánicas e intrusivas: Departamento de Geología, Universidad de Chile Publicación No. 41, p. 51-85.

Manrique C., J., and A. Plazolles V., 1975, Geología de Cuajone: Boletin de la Sociedad Geológica del Perú, v. 46, p. 137-150.

March, W. S., Jr., 1935, Ores at Potrerillos, Chile, *in* Copper resources of the world: International Geological Congress, 16th, Washington, v. 2, p. 485-500.

McKee, E. H., D. C. Noble, D. A. Scherkenbach, J. W. Drexler, J. Mendoza, and V. R. Eyzaguirre, 1979, Age of porphyry intrusion, potassic alteration, and related skarn mineralization, Antamina district, northern Perú: Economic Geology, v. 74, p. 928-930.

Minero Perú, 1980, Michiquillay, *in* M. Samamé Boggio, ed., El Perú minero, v. 4: Yacimientos: Lima, Instiuto Geológico Minero y Metalúrgico, p. 146-154.

Mortimer, C., 1973, The Cenozoic history of the southern Atacama Desert, Chile: Journal of the Geological Society (London), v. 129, p. 505-526.

Mortimer, C., E. Farrar, and N. Sarić, 1974, K-Ar ages from Tertiary lavas of the northernmost Chilean Andes: Geologische Rundschau, v. 63, p. 484-490.

Mortimer, C., C. Munchmeyer F., and I. Urqueta D., 1977, Emplacement of the Exótica orebody, Chile: Institution of Mining and Metallurgy Transactions, section B., v. 86, p. B121-B127.

Munizaga, F., C. Huete, and F. Hervé, 1985, Geocronología K-Ar y razones iniciales Sr^{87}/Sr^{86} de la "Faja Pacifica" de "Desarrollos Hidrotermales": Congreso Geológico Chileno, 4th, Antofagasta, Actas, v. 4.

Noble, D. C., E. Farrar, and E. J. Cobbing, 1979, The Nazca Group of south-central Perú: age, source, and regional volcanic and tectonic significance: Earth and Planetary Science Letters, v. 45, p. 80-86.

Noble, D. C., E. H. Mckee, V. R. Eyzaguirre, and R. Marocco, 1984, Age and regional tectonic and metallogenic implications of igneous activity and mineralization in the Andahuaylas-Yauri belt of southern Perú: Economic Geology, v. 79, p. 172-176.

Ojeda, F., J. M., 1986, Escondida porphyry copper deposit, II Región, Chile: exploration drilling and current geological interpretation, *in* Mining Latin America, Santiago, Chile, 1986: London, Institution of Mining and Metallurgy, p. 299-318.

Olcay, L., 1976, Relación entre alteración hidrotermal y mineralización cuprífera asociadas a rocas volcánicas andesíticas chilenas: Congreso Geológico Chileno, 1st, Santiago, Actas, v. 2, p. E53-E63.

Olcay V., L., and B. Alarcón F., 1975, Alteración hidrotermal sodica de mina El Soldado, Provincia de Valparaiso, Chile: Congreso Ibero-Americano de Geología Económica, 2nd, Buenos Aires, Actas, v. 2, p. 541-556.

Ortiz, F. J., J. D. Lowell, N. D. Rojas, and P. J. Burns, 1985, Reseña técnica del descubrimiento del yacimiento de pórfido cuprífero Escondida, II Región, Chile: Congreso Geológico Chileno, 4th, Antofagasta, Actas, v. 4.

Palacios M., C., and A. Definis M., 1981, Geología del yacimiento estratiforme "Susana," distrito Michilla, Antofagasta: Coloquio sobre volcanismo y metalogénesis, 1st, Universidad del Norte, Antofagasta, Actas, p. 80-91.

Paskoff, R., and J. A. Naranjo, 1983, Formation et évolution du piémont andin dans le désert du nord du Chili (18-21° latitude sur) pendant le Cénozoique supéieure: Compte Rendu de l'Académie des Sciences de Paris, ser. 2, v. 297, p. 743-748.

Petersen, U., 1965, Regional geology and major ore deposits of central Perú: Economic Geology, v. 60, p. 407-476.

Petersen, U., 1970, Metallogenic provinces in South America: Geologische Rundschau, v. 59, p. 834-897.

Petersen, U., 1979, Metallogenesis in South America: Progress and problems: Episodes, v. 1979, no. 4, p. 3-11.

Pitman, W. C., III, 1978, Relationship between eustacy and stratigraphic sequences of passive margins: Geological Society of America Bulletin, v. 89, p. 1389-1403.

Ponzoni S., E., 1980, Metalogenia del Perú, *in* Metalogénesis en Latinoamerica, Simposium internacional, México, D.F., 1980: International Union of Geological Sciences Publicación No. 5, p. 101-140.

Richard, K., and J. H. Courtright, 1958, Geology of Toquepala, Perú: Mining Engineering, v. 10, p. 262-266.

Ripley, E. M., and H. Ohmoto, 1977, Mineralogic, sulfur isotope, and fluid inclusion studies of the stratabound copper deposits at the Raúl mine, Perú: Economic Geology, v. 72, p. 1017-1041.

Rose, A. W., 1976, The effect of cuprous chloride complexes on the origin of red-bed copper and related deposits: Economic Geology, v. 71, p. 1036-1048.

Ruge R., H., 1985, Geología y mineralización del yacimiento de cobre El Soldado, V Región: Congreso Geológico Chileno, 4th, Antofagasta, Actas, v. 4, p. 854-872.

Ruiz F., C., and G. E. Ericksen, 1962, Metallogenic provinces of Chile, S.A.: Economic Geology, v. 57, p. 91-106.

Ruiz F., C., L. Aguirre, J. Corvalán, C. Klohn, E. Klohn, and B. Levi, 1965, Geología y yacimientos metalíferos de Chile: Santiago, Instituto de Investigaciones Geológicas, 305 p.

Ruiz, C., A. Aguilar, E. Egert, W. Espinosa, F. Peebles, R. Quezada, and M. Serrano, 1971, Strata-bound copper sulphide deposits of Chile, *in* Proceedings IMA-IAGOD Meetings '70, IAGOD Volume: Society of Mining Geologists of Japan Special Issue 3, p. 252-260.

Santa Cruz, S., T. Guerrero, F. Castilla, E. Caro, and H. Candiotti, 1979, Geología de yacimientos de cobre en "skarn" en la región sur oriental del Perú: Boletin de la Sociedad Geológica del Perú, v. 60, p. 153-174.

Sasaki, A., C. E. Ulriksen, K. Sato, and S. Ishihara, 1984, Sulfur isotope reconnaissance of porphyry copper and manto-type deposits in Chile and the Philippines: Bulletin of the Geological Survey of Japan, v. 35, p. 615-622.

Satchwell, P. C., 1983, Geología de la mina Cuajone: Boletin de la Sociedad Geológica del Perú, no. 72, p. 127-146.

Sato, T., 1984, Manto type copper deposits in Chile—a review: Bulletin of the Geological Survey of Japan, v. 35, p. 565-582.

Sawkins, F. J., 1982, The formation of Kuroko-type deposits viewed within the broader context of ore genesis theory: Mining Geology, v. 32, p. 25-33.

Schwartz, M. O., 1982, The porphyry copper deposits at La Granja, Perú: Economic Geology, v. 77, p. 482-488.

Sheppard, S. M. F., and L. B. Gustafson, 1976, Oxygen and hydrogen isotopes in the porphyry copper deposit at El Salvador, Chile: Economic Geology, v. 71, p. 1549-1559.

Shibata, K., S. Ishihara, and C. E. Ulriksen, 1984, Rb-Sr ages and initial $^{87}Sr/^{86}Sr$ ratios of late Paleozoic granitic rocks from northern Chile: Bulletin of the Geological Survey of Japan, v. 35, p. 537-545.

Silberman, M. L., and D. C. Noble, 1977, Age of igneous activity and mineralization, Cerro de Pasco, central Perú: Economic Geology, v. 72, p. 925-930.

Sillitoe, R. H., 1972, A plate tectonic model for the origin of porphyry copper deposits: Economic Geology, v. 57, p. 184-197.

Sillitoe, R. H., 1973a, Geology of the Los Pelambres porphyry copper deposit, Chile: Economic Geology, v. 68, p. 1-10.

Sillitoe, R. H., 1973b, The tops and bottoms of porphyry copper deposits: Economic Geology, v. 68, p. 799-815.

Sillitoe, R. H., 1975, Lead-silver, manganese, and native sulfur mineralization within a stratovolcano, El Queva, northwest Argentina: Economic Geology, v. 70, p. 1190-1201.

Sillitoe, R. H., 1976, Andean mineralization: a model for the metallogeny of convergent plate margins, *in* D. F. Strong, ed., Metallogeny and plate tectonics: Geological Association of Canada Special Paper 14, p. 59-100.

Sillitoe, R. H., 1977, Permo-Carboniferous, Late Cretaceous and Miocene porphyry copper-type mineralization in the Argentinian Andes: Economic Geology, v. 72, p. 99-103.

Sillitoe, R. H., 1981, Regional aspects of the Andean porphyry copper belt in Chile and Argentina: Institution of Mining and Metallurgy Transactions, sec. B, v. 90, p. B15-B36.

Sillitoe, R. H., 1983, Enargite-bearing massive sulfide deposits high in porphyry copper systems: Economic Geology, v. 78, p. 348-352.

Sillitoe, R. H., 1986, Space-time distribution, crustal setting and Cu/Mo ratios of central Andean porphyry copper deposits: metallogenic implications, *in* G. Friedrich, A. D. Genkin, A. J. Naldrett, J. D. Ridge, R. H. Sillitoe, and F. M. Vokes, eds., Geology and metallogeny of copper deposits: Society for Geology Applied to Mineral Deposits Special Publication 4, Berlin, Springer-Verlag, p. 235-250.

Sillitoe, R. H., 1988, Epochs of intrusion-related copper mineralization in the Andes: Journal of South American Earth Sciences, v. 1, p. 89-108.

Sillitoe, R. H., and S. R. Hart, 1984, Lead-isotopic signatures of porphyry copper deposits in oceanic and continental settings, Colombian Andes: Geochimica et Cosmochimica Acta, v. 48, p. 2135-2142.

Sillitoe, R. H., and F. J. Sawkins, 1971, Geologic, mineralogic, and fluid inclusion studies relating to the origin of copper-bearing tourmaline breccia pipes, Chile: Economic Geology, v. 66, p. 1028-1041.

Sillitoe, R. H., L. Jaramillo, and H. Castro, 1984, Geologic exploration of a molybdenum-rich porphyry copper deposit at Mocoa, Colombia: Economic Geology, v. 79, p. 106-123.

Sillitoe, R. H., C. Mortimer, and A. H. Clark, 1968, A chronology of landform evolution and supergene mineral alteration, southern Atacama Desert, Chile: Institution of Mining and Metallurgy Transactions, sec. B., v. 77, p. B166-B169.

Sillitoe, R. H., L. Jaramillo, P. E. Damon, M. Shafiqullah, and R. Escovar, 1982, Setting, characteristics, and age of the Andean porphyry copper belt in Colombia: Economic Geology, v., 77, p. 1837-1850.

Soto P., H., and H. Dreyer P., 1985, Geología de Mina Susana. Un yacimiento novedoso en Carolina de Michilla: Congreso Geológico Chileno, 4th, Antofagasta, Actas, v. 2, p. 354-382.

Stambuk, C., V., J. Blondel C., and L. Serrano S., 1982, Geología del yacimiento Río Blanco: Congreso Geológico Chileno, 3rd, Concepción, Actas, v. 2, p. E419-E442.

Stambuk C., V., L. Serrano S., and R. Vargas R., 1985, Geología del sector Sur-Sur, yacimiento Río Blanco: Congreso Geológico Chileno, 4th, Antofagasta, Actas, v. 2, p. 383-404.

Stoll, W. C., 1965, Metallogenic provinces of South America: Mining Magazine, v. 112, p. 22-33, 90-99.

Terrones L., A. J., 1958, Structural control of contact metasomatic deposits in the Peruvian Cordillera: American Institute of Mining and Metallurgical Engineers Transactions, v. 211, p. 365-372.

Tobar V., A., 1981, Geología de Mantos Blancos (resumen): Coloquio sobre Volcanismo y Matalogénesis, 1st, Universidad del Norte, Antofagasta, Actas, p. 68-69.

Tosdal, R. M., E. Farrar, and A. H. Clark, 1981, K-Ar geochronology of the late Cenozoic volcanic rocks of the Cordillera Occidental, southernmost Perú: Journal of Volcanology and Geothermal Research, v. 10, p. 157-173.

Uyeda, S., and H. Kanamori, 1979, Back-arc opening and the mode of subduction: Journal of Geophysical Research, v. 84, p. 1049-1061.

Uyeda, S., and C. Nishiwaki, 1980, Stress field, metallogenesis and mode of subduction, *in* D. W. Strangway, ed., The continental crust and its mineral deposits: Geological Association of Canada Special Paper 20, p. 323-340.

Vergara, M., R. Charrier, F. Munizaga, S. Rivano, P. Sepulveda, R. Thiele, and R. Drake, 1985, Miocene volcanism in the central Chilean Andes, Cordillera Principal (31°30′-34°30′ S), *in* F. Hervé and F. Munizaga, eds., Magmatic evolution of the Andes: Universidad de Chile, Santiago, Comunicaciones, no. 35, p. 249-251.

Viteri A., E., 1978, Genesis del cobre nativo asociado a rocas volcánicas de la Formación La Quinta en la Sierra de Perijá, Venezuela: Dirección General Sectorial de Minea y Geología, Caracas, Boletin de Geología, v. 13, no. 24, p. 47-82.

Warnaars, F. W., 1983, Copper tourmaline breccias at Los Bronces, Chile: Society of Mining Engineers of AIME Transactions, v. 272, p. 1902-1911.

Warnaars, F. W., C. Holmgren D., and S. Barassi F., 1985, Porphyry copper and tourmaline breccias at Los Bronces-Rio Blanco, Chile: Economic Geology, v. 80, p. 1544-1565.

Wauschkuhn, A., 1979, Geology, mineralogy and geochemistry of the stratabound copper deposits in the Mesozoic coastal belt of Perú: Boletin de la Sociedad Geológica del Perú, v. 62, p. 161-192.

Wauschkuhn, A, and R. Thum, 1982, The Condestable mine, a volcano-exhalative deposit in the Mesozoic coastal belt of central Perú, *in* G. Amstutz et al., eds., Ore genesis, the state of the art: Berlin, Springer-Verlag, p. 174-188.

Zimmerman, J.-L., and A. Kihien Collado, 1983, Détermination par la méthode K/Ar de l'âge des intrusions et des minéralisations associées dans le porphyre cuprifère de Quellaveco (sud oest du Pérou): Mineralium Deposita, v. 18, p. 207-213.

Zweng, P. L., and A. H. Clark, 1984, Impact of major tourmaline breccia formation in the evolution of the Toquepala Cu-(Mo) porphyry, Perú (abs.): Geological Society of America Abstracts with Programs, v. 16, p. 706.

Ericksen, G. E., Cañas Pinochet, M. T., and Reinemund, J. A., editors, 1989, Geology of the Andes and its relation to hydrocarbon and mineral resources: Houston, Texas, Circum-Pacific Council for Energy and Mineral Resources Earth Science Series, v. 11.

CHAPTER 23

TIN AND TUNGSTEN DEPOSITS OF THE BOLIVIAN TIN BELT

Fernando Urquidi-Barrau
Academia de Ciencias de Bolivia
La Paz, Bolivia

ABSTRACT

The Bolivian tin belt is characterized by a unique assemblage of hypogene mineral deposits that have yielded most of the metals produced in Bolivia since mining began in the 16th century. The belt extends along the Cordillera Oriental for about 1100 km, from southernmost Perú to northernmost Argentina, is generally less than 100 km wide, and contains hundreds of mineral deposits. Tin is the most widespread and economically important metal, but significant amounts of tungsten, silver, bismuth, antimony, lead, zinc, and gold also have been produced. The deposits are genetically related to intrusive bodies of Late Triassic to late Tertiary age in a terrane dominated by a thick sequence of intensively folded and faulted marine clastic sedimentary rocks of Ordovician to Devonian age.

The belt may be divided into two segments, each with distinctly different types of mineral deposits and associated intrusive rocks. The northern segment is along the Cordilleras Real and Quimsa Cruz where erosion, following Tertiary uplift, has exposed several large granitic plutons that were emplaced during two magmatic phases, Late Triassic-Early Jurassic and Oligocene-Miocene. Associated mineral deposits, chiefly veins in the Paleozoic rocks near plutonic contacts, tend to have relatively simple mineral suites dominated by tin and tungsten minerals. Essentially all of Bolivia's tungsten production has been from these deposits. The southern segment, from Oruro southward, has been less deeply eroded and is dominated by polymetallic tin deposits associated with subvolcanic porphyry intrusions of Miocene age. These deposits tend to have complex mineral suites, lack tungsten, and many show mineral zoning in which a tin-rich core gives way outward to lead-zinc or antimony zones. Many antimony and lead-zinc deposits, not related to tin deposits, also are present in this part of the tin belt.

Tin production has been declining in recent years, chiefly because of depletion of high-grade ores, and the possibilities of finding significantly large amounts of high-grade ore in known deposits are remote. However, many tin deposits remain to be found beneath the extensive Late Tertiary volcanic cover in the southern segment of the tin belt. A recent investigation indicates that intracaldera areas within these volcanic complexes also may contain tin deposits.

INTRODUCTION

During the 1970s and early 1980s, Bolivia accounted for roughly 14% of the world's primary tin production, 11% of the world's tungsten production, and about 27% of the world's antimony production, from the primary deposits in the so-called Bolivian tin belt. The tin belt, known also as the central Andes tin provinces, extends for 1100 km through the Eastern Cordillera from southern Perú through Bolivia to northwest Argentina (Figure 1). It is generally less than 100 km wide and contains hundreds of ore deposits. Besides ores exploited exclusively for tin, ores of silver-tin, tin-zinc, tin-tungsten, tungsten, bismuth, lead-zinc, and antimony also have been mined. In the northern segment of the belt, in the Cordilleras Real and Quimsa Cruz, ore deposits are genetically related to granitic intrusive bodies of Late Triassic to Early Jurassic and middle Tertiary (late Oligocene) age, respectively, whereas in the southern segment, which accounts for 80% of Bolivia's tin production, ore deposits are associated with subvolcanic quartz latite porphyry stocks of Miocene age.

The Eastern Cordillera is made up of a thick sequence of intensively folded and faulted Paleozoic marine clastic sedimentary rocks (Figure 2), which are slightly metamorphosed along pluton contacts. In the southern segment the Paleozoic rocks are locally overlain by Cretaceous and Tertiary red beds and Miocene-Pliocene volcanic rocks.

Systematic study of the tin belt and its ore deposits was started early in the 20th century and has continued to the present. Hundreds of reports describing the geology and mineral deposits have been published.

TECTONIC FRAMEWORK AND REGIONAL GEOLOGICAL SETTING OF TIN BELT

Geological and geophysical evidence indicate that the Precambrian rocks that form the Brazilian shield extend westward under the full width of the Andean orogen. This cratonic basement is overlain by a thick sequence of more than 4500 m of lower Paleozoic (Cambrian-Devonian) clastic marine sedimentary rocks that were deposited in a deep intracratonic basin (Figure 2). Only Ordovician to Devonian rocks are now exposed throughout the Bolivian Eastern Cordillera. Limited

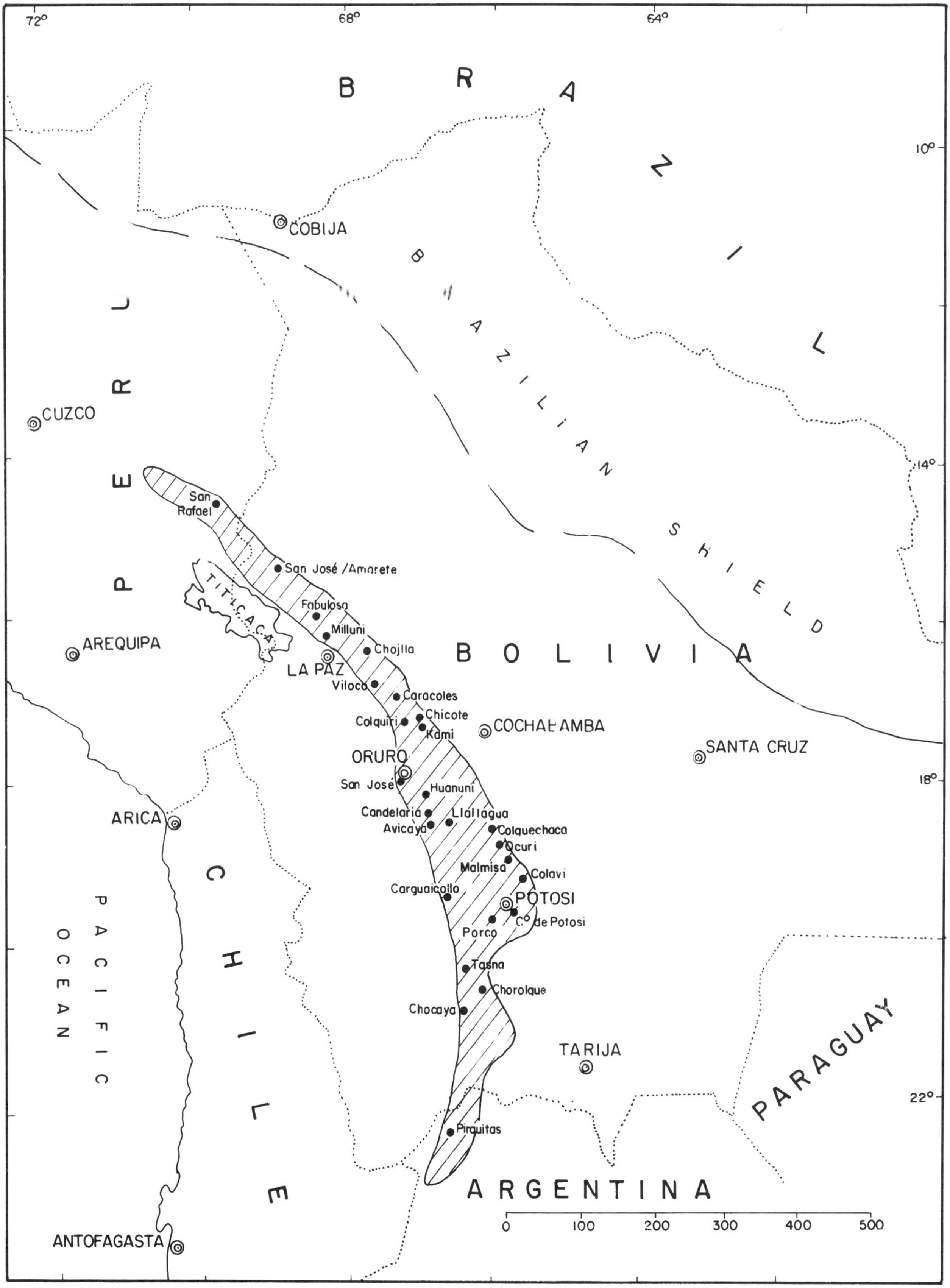

Figure 1. Map showing central Andes Bolivian tin belt (after Lehmann, 1979) and major tin-tungsten, tin, and tin-silver deposits.

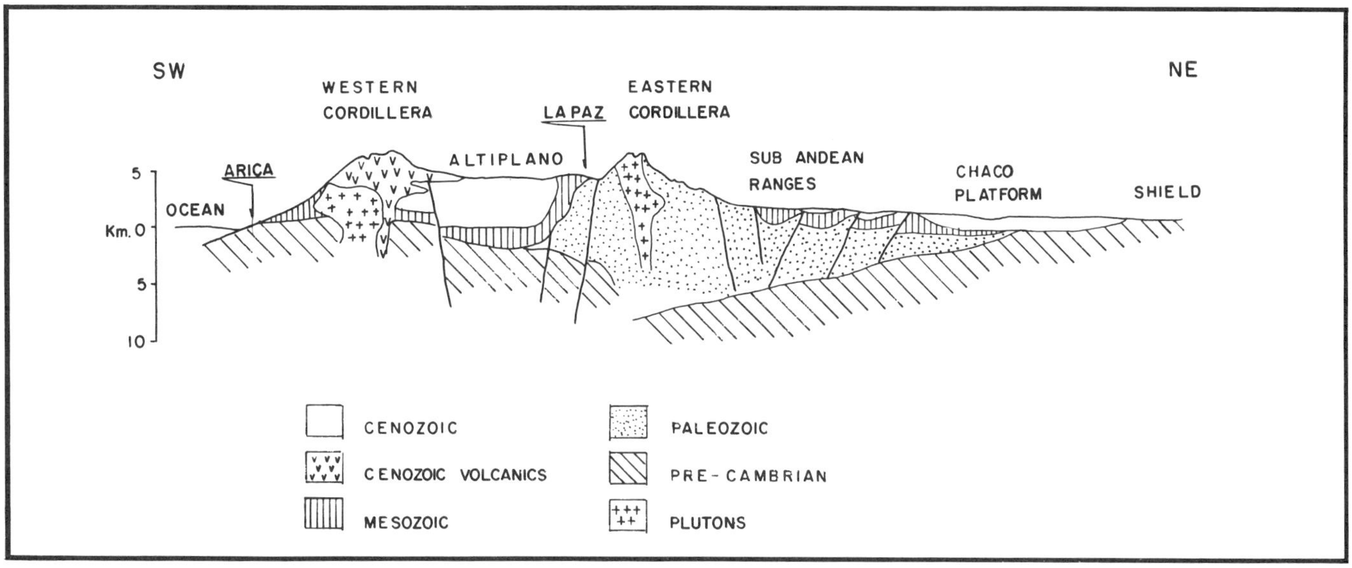

Figure 2. Schematic cross section through the central Andes (modified from James, 1971).

exposures of Cambrian rocks are found in the southernmost part of the Cordillera. The lower Paleozoic sequence is unconformably overlain by epicontinental sediments of late Paleozoic age. Several periods of deformation during the Paleozoic are reflected by a series of large east-west faults crossing the Paleozoic rock sequence in the Bolivian tin belt. Igneous rocks of Paleozoic age are not known in the tin belt, although granitic plutons of Paleozoic age are found in southern Perú and northern Argentina. In Bolivia, the oldest plutons are of Late Triassic age (Grant et al., 1980).

It is known that Andean orogeny began in the Triassic, in direct response to the beginning of eastward subduction of Pacific oceanic crust beneath the South American continental margin. This early-phase orogenesis was characterized by predominantly vertical block tectonics with very little crustal shortening, continental sedimentation in local basins, and intense igneous activity. Widespread faulting and volcanism continued into the late Cenozoic, and the Late Triassic-late Cenozoic orogeny constitutes the Andean cycle. The present-day tectonic framework was developed during this cycle.

The morphotectonic subdivisions of the central Andes (Figure 3) consist of (1) the coastal belt, consisting chiefly of sedimentary, volcanic, and plutonic rocks of Precambrian to Mesozoic age; (2) the Western Cordillera, consisting chiefly of Mesozoic rocks beneath an extensive cover of Tertiary-Quaternary volcanic rocks; (3) the Altiplano, a Cretaceous-Tertiary intermontane basin filled with molassic sediments; (4) the Eastern Cordillera, consisting of Paleozoic marine clastic rocks intruded by Late Triassic-Early Jurassic and Oligocene granitic plutons and Miocene porphyry stocks; and (5) the sub-Andean belt of Mesozoic to Tertiary sedimentary rocks that overlap the Brazilian Precambrian shield to the east. The Eastern Cordillera contains the Bolivian tin belt, which is the most easterly and most complex of the central Andean metallogenic belts (Figure 4).

The Bolivian tin belt is a well-defined metallogenic province, which extends for 1100 km throughout the length of the Eastern Cordillera in Bolivia and is generally less than 100 km wide (Figure 1). It extends from southern Perú, where tin ores are currently exploited at the San Rafael mine, to northern Argentina, where the principal tin production has been from the Pirquitas mine. In addition to today's output of tin, important amounts of tungsten, silver, zinc, lead, antimony, bismuth, and gold have been produced from deposits in the tin belt. Huge amounts of silver were produced in Spanish Colonial times, chiefly from polymetallic tin-silver deposits.

Most of the metalliferous deposits of the Bolivian tin belt are genetically related to plutonic rocks of Late Triassic-Early Jurassic and Oligocene age in the northern segment of the belt, and to subvolcanic porphyry stocks of Miocene age in the southern segment. Not only do the deposits in the northern and southern segments of the tin belt show different ages and are genetically related to different types of intrusive rocks, but they also have distinctly different metal and mineral suites and occur in terranes having distinctly different erosion levels and morphology (Everden, Kriz, and Cherroni, 1977; Clark and Farrar, 1973).

Northern Segment of Tin Belt

In the northern segment of the tin belt, the Paleozoic terrane is more elevated and more deeply eroded than it is in the south. The plutonic rocks are predominantly large granitic batholiths that form the cores of high mountain ranges (Figure 5). This segment can be divided into three mountainous units: (1) the northernmost Apolobamba and Muñecas Cordilleras; the former lacks igneous rocks and the latter contains a quartz-monzonite stock of unknown age; (2) the Cordillera Real, which, from north to south, contains the Late Triassic to Late Jurassic batholiths (Figure 6) of Yani (169 Ma), Illampu (214-180 Ma), Zongo (172-144 Ma), Huayna Potosí (213-191 Ma), and Mururata-Taquesi (206-199 Ma) (Lehmann, 1979). To the south of the Cordillera Real (Figure 6) are the Illimani, Quimsa Cruz, and Vera Cruz plutons, all of which are of late

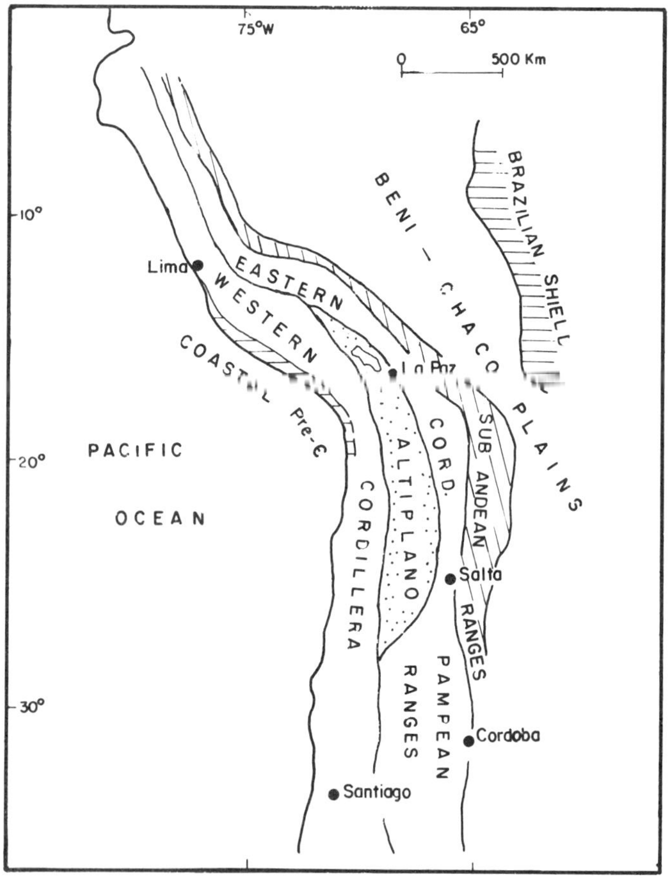

Figure 3. Physiographic provinces of the central Andes (after Grant et al., 1980).

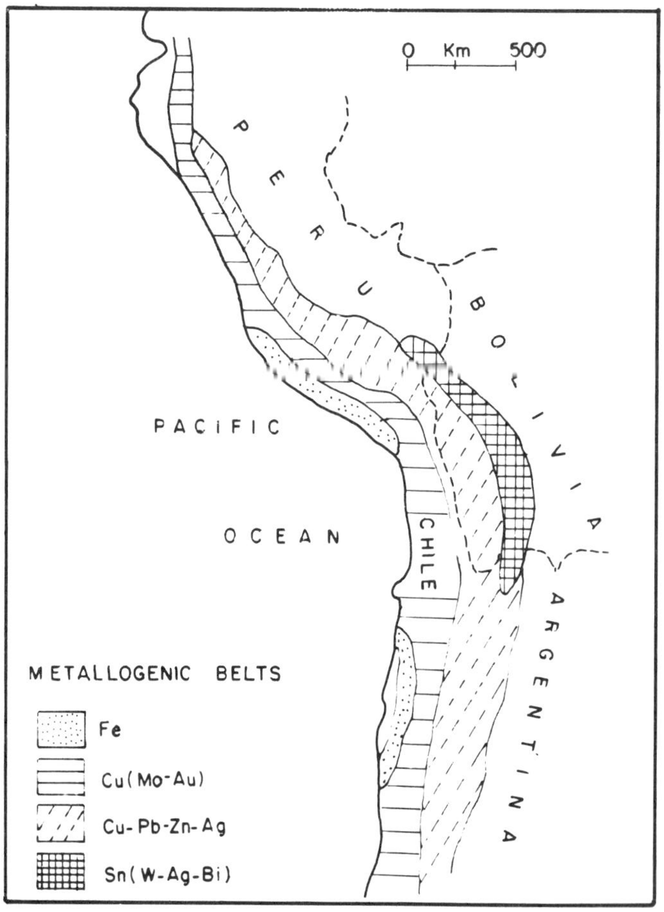

Figure 4. Metallogenic provinces of the central Andes (after Sillitoe, 1976).

Oligocene age (Lehmann, 1979). These plutons are of granodiorite to granite composition and are associated with numerous hypogene vein-type tin and tin-tungsten deposits. Although numerous, most of these deposits are small and have accounted for only a small part of Bolivia's tin production; however, they have been the source of almost all Bolivian tungsten production.

Southern Segment of Tin Belt

In the southern segment, to the south of the Santa Vera Cruz pluton (Figure 6), the Eastern Cordillera is more than 100 km wide—much wider than in the northern segment—and the Paleozoic sedimentary rocks have been uplifted to a lesser degree. The cordillera is characterized by a series of broad, plateaulike blocks rather than a single, prominent mountain chain as in the north. The igneous rocks form small isolated porphyry stocks and dikes and large eruptive complexes in which rhyolitic ash-flow tuffs predominate. Plutonic rocks are absent in the southern segment; the Kari Kari volcanic field, formerly thought to be a batholith, is now recognized as the ash-flow tuff fill of a resurgent caldera. Two extensive volcanic fields—the Morococala and Los Frailes Mesetas—cover about 35% of the northern, and most productive, part of the southern segment. The Morococala field, in the north, extends over an area of about 1500 km^2, whereas the larger Los Frailes Mesetas field to the south covers an area of 8500 km^2. The rocks of these two volcanic fields and a few satellitic fields consist of a sequence of silicic, nonwelded and welded ash-flow tuff sheets that unconformably overlie the Silurian-Devonian basement complex. They are related to the eruptive centers of major caldera in each field (Ericksen et al., 1985).

Radiometric (K-Ar) dating shows three apparent distinct periods of igneous activity in this segment of the Eastern Cordillera (Figure 7). In the central part, from Oruro south to Potosí, the mineralized subvolcanic eruptive complexes and the Kari Kari and associated volcanics are of early Miocene age (23-17 Ma). These overlap with the ages of the southernmost plutons of the northern segment of the tin belt. From Potosí southward, including the Cerro Rico stock at Potosí, the ages are all middle Miocene (17-12 Ma). This shows an apparent time break of 2 million years between these two age groupings (Grant et al., 1980). The late Miocene (8-6 Ma) Morococala and Los Frailes Mesetas volcanic fields belong to the third period of igneous activity.

MAJOR MINERAL DEPOSITS OF NORTHERN SEGMENT OF TIN BELT

The tin and tungsten deposits in the northern segment of the tin belt are largely fissure-filling veins along fault and fracture systems in sedimentary rocks of Paleozoic age peripheral

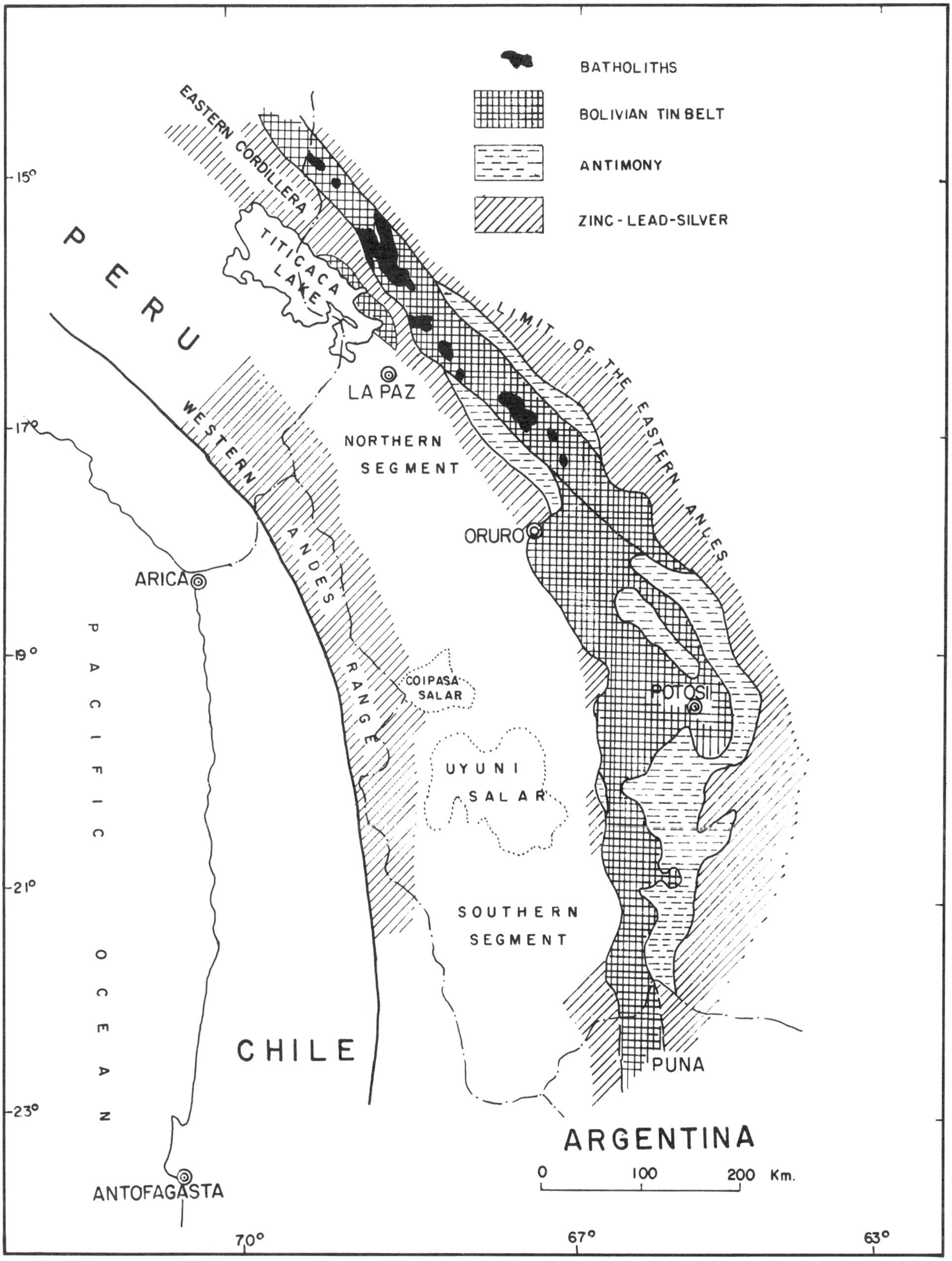

Figure 5. Areas of different types of mineral deposits in northern segment of tin belt, extending from southern Perú to near Oruro, and southern segment from Oruro to Argentina. The northern segment is characterized by tin-tungsten deposits, whereas the southern segment is characterized by silver-rich polymetallic tin deposits.

to the granitic plutons and stocks and within these igneous bodies. Mineralogy is relatively simple—cassiterite and wolframite are the dominant ore minerals, which generally are associated with abundant quartz and pyrite and various (though generally lesser) amounts of arsenopyrite, chalcopyrite, sphalerite, and pyrrhotite. Tourmaline, stannite, galena, and siderite also may be present, and some deposits contain small amounts of bismuthinite, jamesonite, and apatite.

San Rafael Tin Mine

The San Rafael tin mine, located 150 km north of Puno, Perú (Figure 1), is the largest tin producer in Perú and the northernmost tin deposit of the Bolivian tin belt. Its cassiterite-quartz lodes are genetically related to a highly tourmalinized granitic pluton of early Miocene age. Ore averaging about 2% Sn is concentrated in a 800 t/day plant, which has an annual output of about 2500 t of tin concentrates (J. McNamee, oral communication, 1985). Reserves are estimated to be sufficient for 10 years at the current (1985) production level.

Palaca XI Tungsten Mine

The Palaca XI tungsten deposit (not shown on Figure 1) is in the Department of Puno, Perú, south of the San Rafael mine. The mine was originally worked for copper, bismuth, and silver, and later for tungsten. The tungsten minerals are mainly ferberite and scheelite, which are associated with quartz, kaolin, and a variety of sulfide minerals. The host rock is predominantly shale of Ordovician age. Current exploitation is along the La Palca vein, extending for a distance of 500 m horizontally and 220 m vertically. Ore shoots have an average width of 7 m. Reserves are estimated at 1.5 million t of ore averaging 1.34% WO_3. Ore is treated in a 100 t/day plant. Annual production during the early 1980s was about 150 t WO_3.

San José de Ayata-Amarete Tin Mine

The San José de Ayata-Amarete is the largest tin deposit in the Cordillera de Las Muñecas (Figure 1), although in comparison with other Bolivian tin deposits, it is relatively small. The mine is owned and operated by the Empresa Minera Pabon. The deposit consists of six parallel, gently dipping quartz veins that contain fine-grained cassiterite together with pyrite, calcite, and chalcopyrite (Ahlfeld and Schneider-Scherbina, 1964). The host rock is Paleozoic shale; intrusive rocks are not known to occur in the vicinity of the deposit.

La Fabulosa Tin Mine

Several pegmatitic-type tin deposits have been found in the contact zone on the southern side of the Sarata batholith. The La Fabulosa deposit (Figure 1) is the largest of these deposits, although it is small in comparison with other tin deposits in Bolivia. It consists of cassiterite-bearing veins that grade laterally into white quartz which contains large pink cassiterite crystals.

Candelaria Tungsten Mine

The Candelaria deposit (not shown on Figure 1) is the largest tungsten deposit currently (1985) being exploited in Bolivia; production is 10 to 15 t of contained WO_3 per month. The ore consists of 15- to 60-cm-wide quartz veins in Paleozoic sedimentary rocks within the contact zone of the granitic pluton of the Cordillera Real. The scheelite-rich pods contain 3 to 6% WO_3.

Milluni Tin Mine

The Milluni mine, located 30 km north of the city of La Paz (Figure 1) at an elevation of 4400 m, is in the contact zone of the Huayna Potosí massif, one of the most spectacular snow-covered peaks of the Cordillera Real. The mine belongs to the Compañia Minera del Sur S.A. (COMSUR). The Milluni deposit is one of the richest in the Cordillera Real, and has been under exploitation since the early part of this century. The principal Rothschild vein has been exploited over a strike length of 1500 m. The ore from this vein averages 1.37% Sn. Current (1985) production comes both from the deepest part of the Rothschild vein and from the narrower Compañia and Salvadora veins. The veins consist chiefly of quartz, pyrite, and cassiterite; some sphalerite and minor arsenopyrite are present. Production in the early 1980s was about 1000 t of Sn contained in concentrates averaging about 15% Sn.

Kellguani Tin Mine

The Kellguani deposit (not shown on Figure 1) is one of several manto-type tin deposits in the Chacaltaya area along the southern edge of the Cordillera Real. Fine-grained cassiterite, together with quartz, was deposited by hydrothermal solutions along sheared quartzites interbedded with unmineralized shales. The quartzite units range from 4 to 35 m in thickness. The deposit is mined by both open-pit and underground methods. Production in the early 1980s was 300 t of ore per day.

Chojlla Tin-Tungsten Mine

The Chojlla mine (Figure 1), the largest tungsten producer in Bolivia, belongs to the International Mining Company, a subsidiary of the Empresa Minera Estalsa Ltd. The deposit is in the contact zone of the Taquesi-Mururata batholith, which intrudes sedimentary rocks of Ordovician and Silurian age. It consists of quartz veins containing irregular concentrations of cassiterite and wolframite. The tin to tungsten ratio is about 1:1, although tin decreases slightly in the upper mine levels. The deposit has three main veins cut by two large faults. Galena, chalcopyrite, sphalerite, arsenopyrite, pyrite, and fluorite are accessory minerals.

The mine has a 1200 t/day concentration plant that treats ores averaging 0.72% WO_3 and an equivalent amount of Sn. Mill recovery of both tungsten and tin is 74%. Average annual production (1985) is 446 t of WO_3 and about 350 t Sn. Reserves (1985) are estimated at 1.5 million tonnes with an average grade of 0.54% WO_3.

Bolsa Negra Tungsten Mine

The Bolsa Negra mine, one of the three tungsten mines belonging to the Corporación Minera de Bolivia (COMIBOL), is at the southern edge of the Taquesi-Mururata batholith (Figure 6). It consists of manto-type ferberite- and scheelite-bearing quartz veins parallel to foliation or layering in schists and hornfels. More than 20 mantos are found within a sequence about 60 m thick. Ore reserves are calculated at 1.3

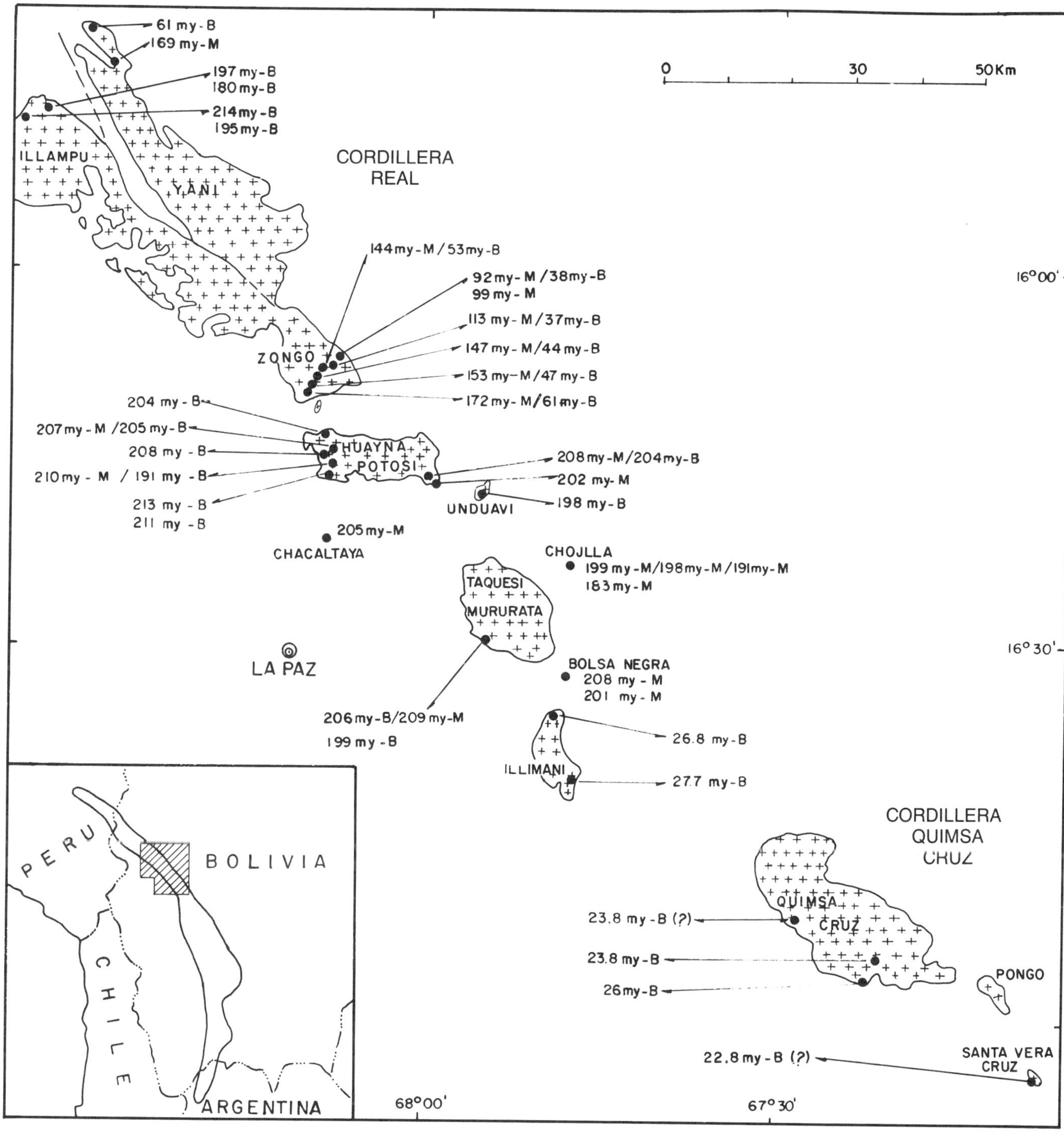

Figure 6. Map showing granite plutons in northern segment of Bolivian tin belt and radiometric ages (after Lehmann, 1979). Age determinations were made on biotite (B) and muscovite (M).

million tonnes averaging 0.78% WO_3. Average annual production in the early 1980s was 206 t of WO_3.

Viloco Tin Mine

The Viloco mine, also owned by COMIBOL, is 70 km south of La Paz, in the northwest flank of the Quimsa Cruz batholith (Figures 1, 6). The Viloco orebody was discovered in 1878 and has been worked since for both tin and tungsten. Viloco is the largest tin mine in the northern segment of the tin belt and is famous for its large and perfect cassiterite crystals, which are known as Araca crystals. The mineralized area covers 5 km^2 and shows well-defined zoning in which a core zone rich in molybdenite is surrounded by a narrow wolframite-rich zone near the contact zone with the Quimsa Cruz batholith, fol-

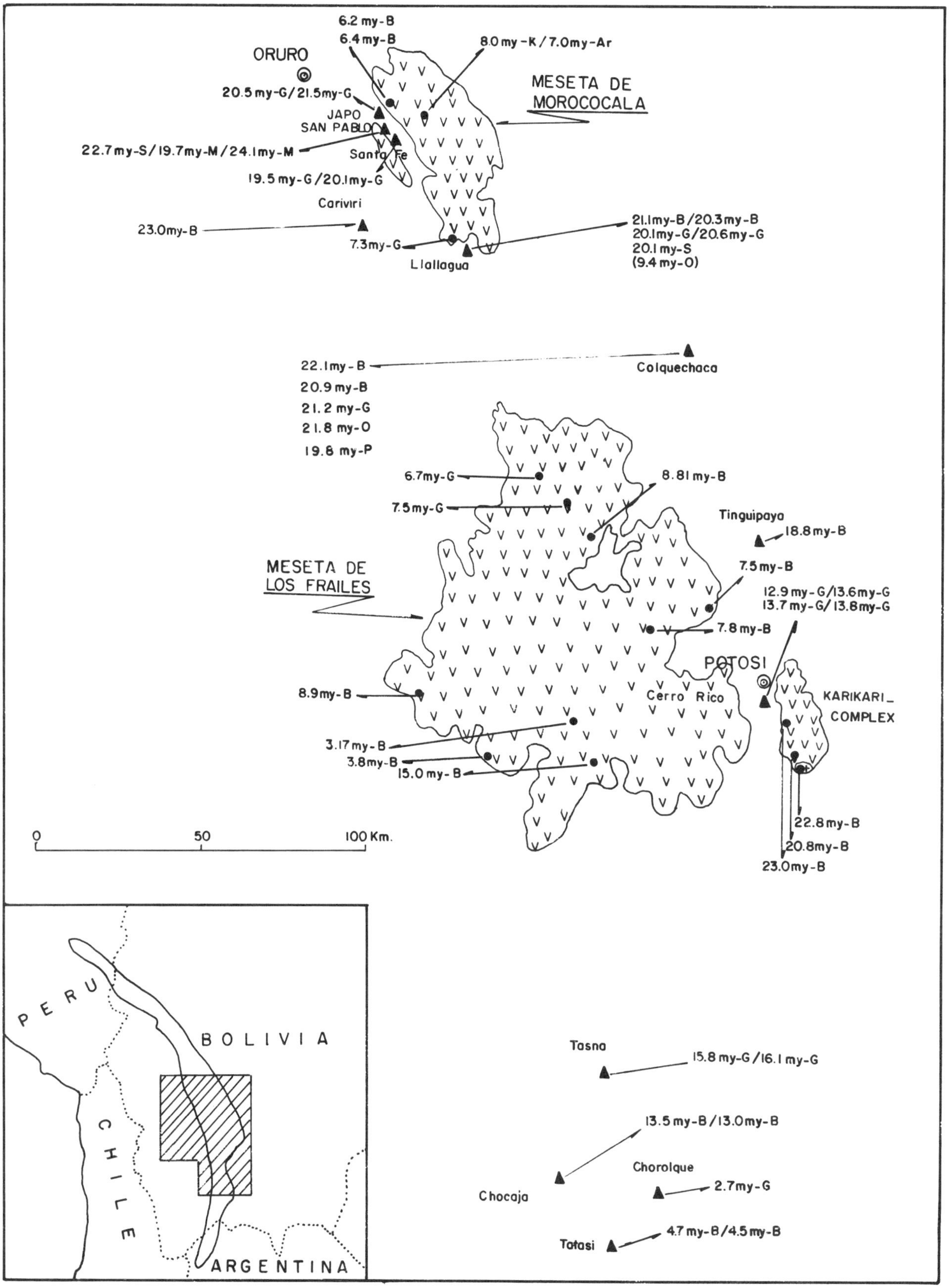

Figure 7. Radiometric age dates for mineral deposits and igneous rocks in southern segment of Bolivian tin belt (modified from Lehmann, 1979). Ages were determined for: Biotite (B), Muscovite (M), Orthoclase (O), Potassium (K), Plagioclase (P), Sanidine (S), Whole rock (G), and Argon (Ar).

lowed by a wider cassiterite-rich zone extending 500 to 2000 m from the batholith margin. Orebodies occur in Lower Devonian quartzites that are intensely faulted and sheared. Interbedded shales are barren (Ahlfeld and Schneider-Scherbina, 1964). Ore is treated in a 200 t/day concentration plant.

Caracoles Tin Mine

The Caracoles deposit, discovered at the end of the 19th century, is located at the southern edge of the Quimsa Cruz batholith (Figures 1, 6) at altitudes ranging from 4500 to 4950 m. In addition to tin, the deposit has yielded small amounts of bismuth and tungsten. Current production comes from three sectors—Pongo, Mina Argentina, and Pacuni—all of which supply ores to the 300 t/day Molinos concentration plant. A 400 t/day preconcentration plant, which was recently constructed, treats ore from 16 new veins found in the 2150-m-long Gran Recorte crosscut.

Caracoles consists of fissure-filling veins in batholiths, and contains chlorite, tourmaline, quartz, fine-grained cassiterite, pyrite, and some bismuthinite and wolframite (Ahlfeld and Schneider-Scherbina, 1964).

Colquiri Tin Mine

The Colquiri mine, at the southernmost tin deposit in the northern segment of the tin belt, is 55 km north of Oruro (Figure 1), southeast of the Quimsa Cruz and Santa Vera Cruz batholiths (Figure 6). Colquiri began producing tin in 1880 and was one of the richest tin mines in Bolivia early in the present century. Currently (1985) the tin production comes from the sectors of Incalacaya, San Juanillo, Estación Bomba, and Rea Rea. Ores occur in fissure-filling veins in folded and faulted sandy shales and sandstones of Silurian age. The main vein, Veta Principal, is along a normal fault having a horizontal extension of more than 3000 m. Minerals present in the veins are chlorite, cassiterite, fluorite, quartz, arsenopyrite, pyrite, chalcopyrite, sphalerite, and galena (Ahlfeld and Schneider-Scherbina, 1964). Ores are preconcentrated in a 2500 t/day sink and float plant that produces material which is then concentrated in the 200 t/day Hornuni plant.

Chicote Grande Tungsten Mine

The Chicote Grande tungsten deposit (Figure 1) was one of the most productive in Bolivia during the World Wars I and II, but since the 1940s, production has declined and recent exploration has not been successful in discovering significant quantities of new ore. Chicote is owned by the Anschutz Mining Company of Denver. The deposit consists of tungsten-bearing quartz veins which cut Ordovician, Silurian, and Devonian folded sedimentary rocks. In addition to the tungsten mineral, wolframite, and quartz, the veins contain pyrrhotite, sphalerite, and arsenopyrite. Wolframite is confined to a fault-bounded block of tourmaliniferous hornfels. Placer deposits of wolframite, locally called "llamperas," are associated with glacial moraines on nearby hillsides. Reserves of the llamperas are calculated at several million tonnes which contain about 200 g WO_3/m^3.

Kami Tungsten Mine

The Kami mine, owned by COMIBOL, is currently (1985) one of the largest tungsten producers in Bolivia. It is 7 km southwest of the Chicote Grande mine (Figure 1), and geologically is similar to it. The deposit consists of quartz veins which contain arsenopyrite, pyrrhotite, and wolframite.

MAJOR MINERAL DEPOSITS OF SOUTHERN SEGMENT OF TIN BELT

The polymetallic tin deposits in the southern segment of the tin belt occur within the Eastern Cordillera from Oruro to northernmost Argentina in the south (Figure 1). Most of these deposits have been described in detail by Turneaure (1935, 1960) and Ahlfeld and Schneider-Scherbina (1964). The ore deposits are associated with high-level porphyry stocks and eruptive complexes, which range from centers having scattered dikes to others having small porphyry stocks, presumably subvolcanic intrusives or volcanic necks associated with rhyolitic volcanic complexes.

Several of the largest tin deposits of Bolivia—Llallagua/Catavi, Huanuni, San José/Oruro, and Cerro Rico de Potosí—are at or near the margins of the two large volcanic fields of Morococala and Los Frailes. It is suspected that tin deposits, some of which may be comparable in size and grade with the largest known deposits, occur in rocks beneath the volcanic cover of these fields. In addition, the volcanic complexes of these two areas are not post-tin mineralization, and buried intracaldera intrusions in these complexes may contain associated tin deposits (Ericksen et al., 1985).

San José (Oruro) Tin-Silver Mine

The classic account of the San José orebody, known also as the Oruro stock, is that by Chase (1948). Subsequent reports describing this deposit are those by Turneaure (1960, 1971) and Ahlfeld and Schneider-Scherbina (1964). The Oruro stock is a composite stock capped by rhyolitic volcanic rocks of probable early Miocene age in a terrane of Devonian sedimentary rocks. The largest stock in Oruro is the San Pedro stock, but the major orebodies are associated with the smaller San José and Itos stocks. The stocks are considered to be hydrothermally altered quartz latite porphyry. The San José and Itos stocks, which contain abundant hydrothermal breccia, are intensely sericitized, silicified, and pyritized. As in most of the deposits in the southern segment of the belt, an early tin mineralization of Oruro was followed by a silver-rich phase. In Oruro the early vein stage consists of minor quartz with pyrite, cassiterite, and arsenopyrite; the later, silver-rich stage began with the deposition of sphalerite, chalcopyrite, pyrite, stannite, franckeite, and teallite followed by tetrahedrite and andorite, and finally by lead sulfosalts (Sillitoe, Halls, and Grant, 1975).

The San José tin-silver mine belongs to COMIBOL. Annual production during the early 1980s averaged 400 t of tin and 16 t of silver in concentrates.

Huanuni Tin Mine

Huanuni is Bolivia's largest tin producer and in the early 1980s was one of the few profitable mines of COMIBOL. It is located 50 km southeast of Oruro (Figure 1). Tin has been produced from this deposit since 1840. The ores occur in many relatively narrow veins in an asymmetrical, northwest-striking anticline having Silurian quartzites in its core and shales and lutites of the same age on its flanks. The 600-m-long Veta Grande, one of the major veins that is perpendicular to the anticlinal axis, consists chiefly of quartz, cassiterite, pyrite, some marmatite, chalcopyrite, and fluorite, and minor amounts of jamesonite, boulangerite, franckeite, and cylindrite. Many other, generally smaller, veins are present. A dike is the only associated igneous rock, but it is assumed that a porphyry stock lies at shallow depth beneath the deposit.

The Huanuni ores, which average 1.0% tin, are treated in the 1200 t/day Santa Elena concentration plant. Annual production in the early 1980s was about 3000 t of tin in concentrates.

Llallagua (Catavi-Siglo XX) Tin Mine

The Llallagua deposit is considered to be the largest underground tin deposit in the world, and since its discovery in 1872, it has produced in excess of 0.5 million tonnes of metallic tin. The classic description of Llallagua was made by Turneaure (1935), from which the following was taken. The Lallagua ores are associated with the intensely altered Salvador stock, a quartz latite porphyry stock intruded into the axis of a sharply overturned, northwest-striking anticline involving Silurian-Devonian graywackes, sandstones, and shales. Sillitoe, Halls, and Grant (1975) considered the Llallagua stock to be a tin porphyry formed by deep hydrothermal activity which produced a large volume of feldspar-destructive altered rocks containing widespread veins and veinlets of cassiterite. Llallagua consists of main and branching veins, narrow stringers, and sheeted zones, all with a predominantly northeasterly strike. The earliest mineralization in the main veins consisted of quartz, bismuthinite, and cassiterite. This early stage was followed by pyrrhotite, franckeite, stannite, sphalerite, and chalcopyrite. Late hypogene veinlets consist of sphalerite and siderite and, finally, hydrous phosphates (Sillitoe, Halls, and Grant, 1975).

Annual production in 1983-1984 by the Empresa Minera de Catavi, a COMIBOL subsidiary, was about 2500 tonnes of tin content, down from 5000 tonnes in the 1970s. The mine was closed in 1986.

Cerro Rico de Potosí Tin-Silver Mine

The conical Cerro Rico de Potosí (Figure 1) has been a center of mining activity since 1543, when it was first worked for silver by the Spaniards. Tin mining did not commence until the second half of the 19th century. The geology of the Cerro Rico and the Potosí area was described by several authors, among whom are Turneaure (1960) and Rivas and Carrasco (1968). Mineralization is genetically related to the Cerro Rico stock, a domelike intrusion in the ring-fracture zone of the Kari Kari caldera. This stock cuts through the axis of a northwest-striking anticline in shales and minor interbedded sandstones of Ordovician age and conglomeratic tuff of Tertiary age. The stock is pervasively altered, and alteration has also affected the sedimentary rocks (Sillitoe, Halls, and Grant, 1975). Veins that have yielded most of the tin and silver ores at Potosí generally trend northeasterly and postdate both intense alteration and stockwork tin mineralization. The early vein stage consists of quartz, pyrite, cassiterite, and arsenopyrite, with some wolframite and bismuthinite in deeper parts of veins. This stage was followed by a sulfide stage in which stannite, chalcopyrite, tetrahedrite, sphalerite, andorite, matildite, pyrargyrite, jamesonite, and boulangerite were deposited. The bonanza-type silver mineralization formed still later, chiefly in veins in the upper part of the stock.

In the 1980s, COMIBOL mined tin-silver ores in the lower part of the Cerro Rico deposit, and small miners and mining cooperatives mined tin-silver ores in the upper part. COMIBOL's Compañia Unificada del Cerro Rico operates a 1500 t/day preconcentration plant and four concentrators at Pailaviri, Ingenio Velarde, Colavi, and Canutillos. The company had an annual production in the early 1980s of about 1100 t of fine tin and 4 t of silver in concentrates.

Chorolque Tin Mine

The Chorolque deposit (Figure 1) is one of the southernmost of the important tin producers, although its production has been much smaller than that at Llallagua, Huanuni, and Potosí. The tin mineralization in the Chorolque deposit is within a volcanic breccia pipe, perhaps the vent of a former volcano, in a thick sequence of pyroclastic rhyolite that extends downward into underlying Ordovician sedimentary rocks. At present, mining is limited to this breccia pipe. Ore occurs in a system of veins within the pipe which strike easterly and dip steeply. These veins, which are in complex fractures that cut the altered breccia, contain quartz and cassiterite (Sillitoe, Halls, and Grant, 1975).

The Chorolque mine is the largest tin producer of COMIBOL's Empresa Minera Quechisla. In the early 1980s, annual production from Chorolque was 1000 t of tin in concentrates. The company also operates the Tasna mine, which produced about 200 t of fine tin per year in the early 1980s, and Santa Ana, which produced about 150 t of fine tin per year during this period.

MINING IN BOLIVIA AND ITS FUTURE

Bolivia's tin belt is essentially confined to rather inaccessible regions of the central Andes. Mines are generally located at high altitudes, and access at times poses formidable infrastructural and logistical problems. Hard rock tin deposits now yield relatively low-grade ores, and Bolivian mines show the highest costs of all the world's lode-tin mines. The complex ores in the southern segment of the tin belt have relatively high smelting and refining costs.

The future of tin mining in Bolivia depends on two separate but related sets of factors: (1) international market prospects for tin and (2) domestic performance and policies in the mining sector as a whole. The first set of factors, medium- and long-term international prospects for tin, is one over which Bolivia can exercise little control, except through its indirect influence in the International Tin Council (ITC), and the Association of Tin Producing Countries. By contrast, Bolivia has substantial control over domestic factors—namely, the government's objectives for the mining sector, plans for the rehabilitation of the nationalized sector operated by COMIBOL, tax policies, and credit availability

Short-term prospects for increased output and improved productivity in Bolivian tin mining are not bright. Past neglect of exploration and mine development, the chaotic state of COMIBOL (the state mining company the produces about

65% of Bolivia's tin), lack of credit, and social unrest are some of the factors that could prevent a recovery of tin production from its estimated 1985 low level of about 15,000 t. In short, even under the most optimistic scenario, prospects for tin mining in Bolivia appear dim.

REFERENCES

Ahlfeld, F., 1967, Metallogenetic epochs and provinces of Bolivia: Mineralium Deposita, v. 2, p. 291-311.

Ahlfeld, F., and A. Schneider-Scherbina, 1964, Los yacimientos minerales y de hidrocarburos de Bolivia: Bolivia, Departamento Nacional de Geología Boletín 5, 388 p.

Chase, F. M., 1948, Tin-silver veins of Oruro, Bolivia: Economic Geology, v. 43, p. 333-383, 435-470.

Clark, A. H., and E. Farrar, 1973, The Bolivian tin province: notes on the available geochronological data: Economic Geology, v. 68, p. 102-106.

Clark, A. H., E. Farrar, J. C. Caelles, S. J. Haynes, R. B. Lortie, S. L. McBride, G. S. Quirt, R. C. R. Robertson, and M. Zentilli, 1976, Longitudinal variations in the metallogenetic evolution of the central Andes: a progress report, *in* D. Strong, ed., Metallogeny and plate tectonics: Geological Association of Canada Special Paper 14, p. 22-58.

Ericksen, G. E., R. L. Smith, R. G. Leudke, M. Flores, A. Espinosa, F. Urquidi, and F. Saravia, 1985, Preliminary geochemical study of ash-flow tuffs in the Morococala and Los Frailes volcanic fields, central Andes Bolivian tin belt: U.S. Geological Survey Open-File Report 85-258, 8 p.

Everden, J. F., S. J. Kriz, and C. Cherroni, 1977, Potassium-argon ages of some Bolivian rocks: Economic Geology, v. 72, p. 1042-1061.

Grant, J. N., C. Halls, S. M. F. Sheppard, and W. Avila, 1980, Evolution of the porphyry tin deposits of Bolivia: Tokyo, Mining Geology Special Issue No. 8, p. 151-173.

James, D. E., 1971, Plate tectonic model for the evolution of the central Andes: GSA Bulletin, v. 82, p. 3325-3346.

Lehmann, B., 1979, Schichgebundene Sn-Lagerstätten in der Cordillera Real/Bolivien: Berliner Geowissenschaftliche Abhandlungen, Reih A/Band 14, 135 p.

Rivas, S., and R. Carrasco, 1968, Geología y yacimientos minerales de la regiön de Potosí, parte geologíca: Bolivia, Servicio Geológico de Bolivia Boletín 11, v. 1, 95 p.

Sillitoe, R. H., 1976, Andean mineralization: a model for the metallogeny of convergent plate margins, *in* Metallogeny and plate tectonics: Geological Association of Canada Special Paper 14, p. 59-100.

Sillitoe, R. H., C. Halls, and J. N. Grant, 1975, Porphyry tin deposits in Bolivia: Economic Geology, v. 70, p. 913-927.

Turneaure, F. S., 1935, The tin deposit of Llallagua, Bolivia: Economic Geology, v. 30, p. 14-60, 170-190.

Turneaure, F. S., 1960, A comparative study of the major ore deposits of central Bolivia: Economic Geology, v. 55, p. 217-254, 274-606.

Turneaure, F. S., 1971, The Bolivian tin-silver province: Economic Geology, v. 66, p. 215-225.

Ericksen, G. E., Cañas Pinochet, M. T., and Reinemund, J. A., editors, 1989, Geology of the Andes and its relation to hydrocarbon and mineral resources: Houston, Texas, Circum-Pacific Council for Energy and Mineral Resources Earth Science Series, v. 11.

CHAPTER 24

CHARACTERISTICS OF COLOMBIAN PHOSPHATE DEPOSITS

Francisco Zambrano-Ortiz and Pedro E. Mojica G.
Ministry of Mines and Energy
National Institute of Geological-Mining Investigations
Bogotá, Colombia

ABSTRACT

The phosphate rock deposits of Colombia are associated with Upper Cretaceous marine sedimentary rocks in the Eastern Cordillera and southernmost part of the Central Cordillera's eastern flank. They occur over an area approximately 600 km long that extends from the Department of Norte de Santander in the north, through Santander, Boyacá, Cundinamarca, and Tolima, to the Department of Huila in the south. Phosphate beds are present in (1) the upper part of the La Luna Formation in the Maracaibo basin; (2) the Galembo Member of the La Luna Formation in the middle part of the Magdalena River valley; and (3) the Plaeners Formation in the central part and the Monserrate Formation in the southernmost part of the phosphate zone. The phosphatic units also contain interbedded shale, sandstone, and chert, and range in thicknesses from a few centimeters to 5 m. The phosphatic material consists of nodules, pellets, and oolites of apatite. In general, the phosphate rock is of low to medium grade, containing as much as 25% P_2O_5. Weathered phosphate rock, from which calcium carbonate has been leached, generally contains the highest grade material.

INTRODUCTION

Prospecting and exploring for phosphatic materials in Colombia were first carried out in the 1940s and 1950s by the Servicio Geológico Nacional (SGN). These early studies involved a search for seabird guano in offshore islands and for sedimentary rock phosphate in various regions of the country. Unfortunately, they did not result in discovery of commercial phosphate deposits. Systematic investigations for phosphate rock, which began in the early 1960s, emphasized stratigraphic studies and mapping of sedimentary rocks in the central and northern part of Colombia. It was then determined that the most favorable hosts for phosphate rock were the Upper Cretaceous marine sediments (Bürgl, 1967). Additional studies in the Departments of Tolima and Huila, in the southern part of the country, also showed the presence of phosphate rock in the Cretaceous sequence (Slansky, 1963; Zambrano-Ortiz, 1966).

As a result of these investigations, a project for detailed study was planned and carried out by the SGN in cooperation with the U.S. Agency for International Development (AID) and with the technical and scientific assistance of the U.S. Geological Survey (USGS). This project began in 1964, and soon thereafter the most favorable areas for phosphate rock were identified in the Departments of Norte de Santander, Santander, Boyacá, Tolima, and Huila (Cathcart and Zambrano, 1967). In 1970 the national government requested the Colombian mine agency ECOMINAS to conduct studies to determine the industrial utilization of phosphate rock in the country. In addition, several regional companies were created to promote studies related to the use of phosphate rock for the production of fertilizers and for direct application to the agricultural soils.

Currently, the best known deposits (Figure 1) are those in the areas of Sardinata (Norte de Santander), Azufrada-Conchal-Vanegas (Santander), Iza-Cuitiva and Pesca (Boyacá), Aipe-Baraya-Yaguara (Huila), and Tesalia (Huila).

The Colombian Andes consist of three north-trending cordilleras—the Western, Central, and Eastern cordilleras—separated by the inter-Andean valleys of the Cauca and Magdalena rivers. The Eastern Cordillera and southern tip of the Central Cordillera, where the phosphate rocks occur, consist chiefly of Paleozoic, Mesozoic, and Cenozoic sedimentary rocks. Phosphorite beds are in marine sedimentary rocks of Late Cretaceous age in a belt about 600 km long, extending through the Eastern Cordillera from the Venezuelan border in the north to the southeastern flank of the Central Cordillera, Department of Huila, in the south.

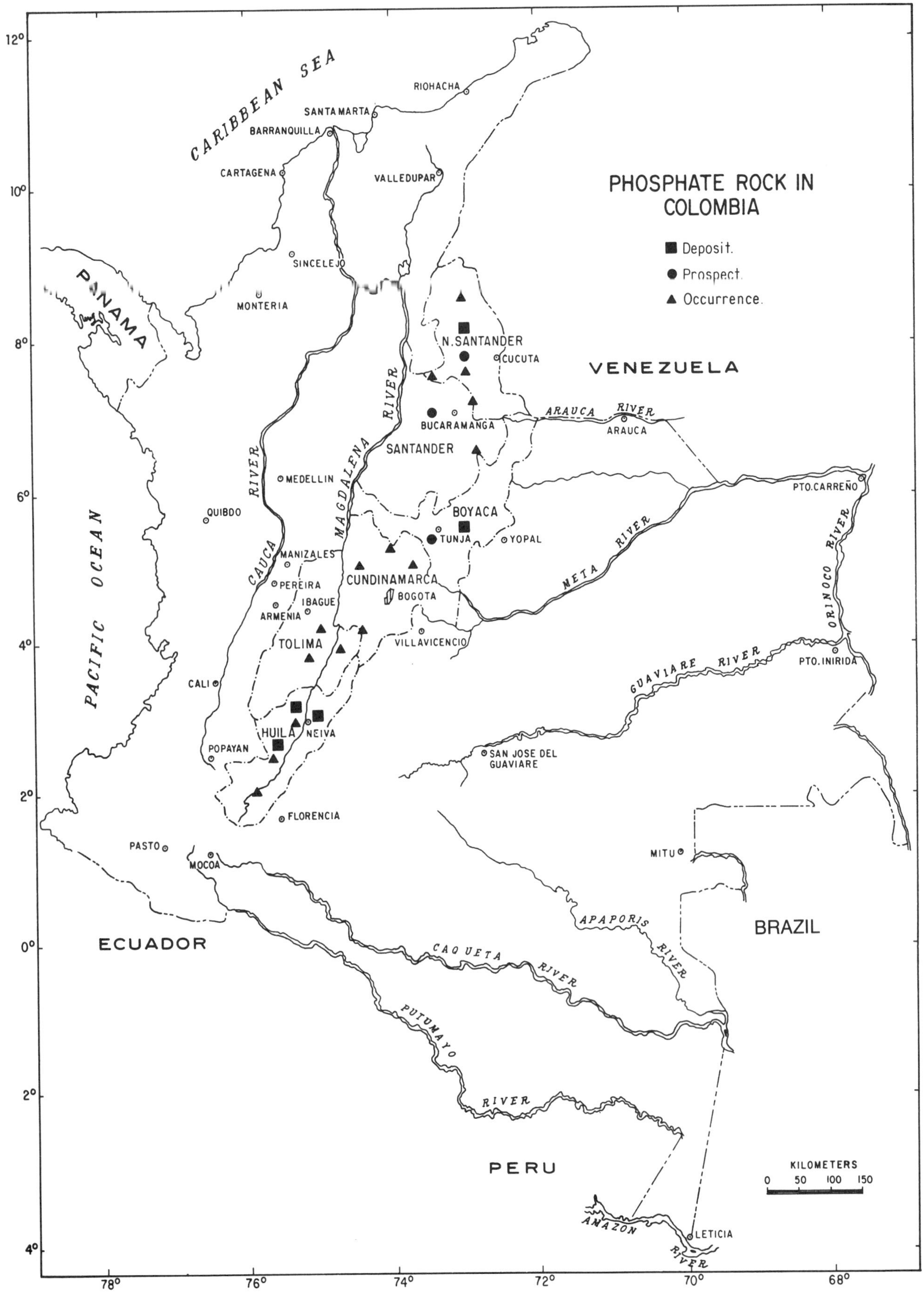

Figure 1. Map showing distribution of phosphate rock in Colombia.

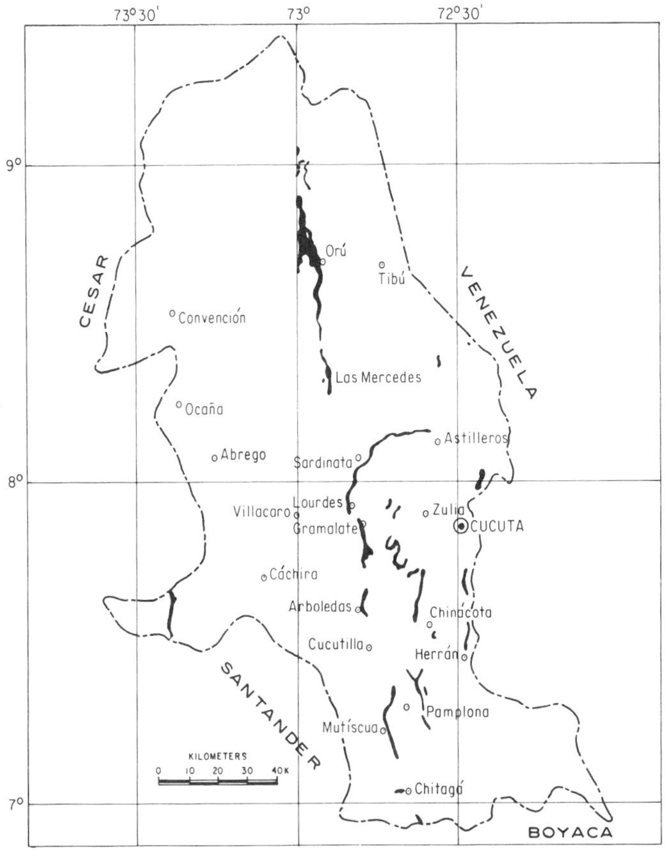

Figure 2. Phosphate rock in the Department of Norte de Santander. After Mojica (1978).

CHARACTERISTICS OF THE PHOSPHATE ROCKS

Although the lithology of the Cretaceous rocks changes considerably from north to south, the phosphate beds are associated with similar lithologic sequences of silt, chert, dark shale, and laminar claystone throughout the phosphatic belt. The presence of phosphate beds in the Upper Cretaceous formations of Colombia is explained by the hypothesis of Kazakov (1937), which was first proposed to explain the origin of the Permian Phosphoria Formation in Idaho, U.S.A. (McKelvey, Swanson, and Sheldon, 1953; Sheldon, 1963), as well as phosphorite deposits in other parts of the world. This hypothesis is based on the remarkable differences in phosphate solubility in deep oceanic bottom waters and shallow waters, the solubility being greatest in bottom waters of lower temperature and pH than in the warmer, higher pH surface waters. As a consequence, phosphate is precipitated from phosphate-rich bottom waters carried by currents into shallower, near-shore waters as the result of increase of temperature and pH and loss of CO_2. This precipitation is partly biologic because shallow, near-shore waters are in the zone most favorable for the existence of benthonic organisms. In addition to direct chemical precipitation of phosphatic material, the phosphate replaces shells and organic remains of sea life. Marine limestones, sandstones, carbonaceous shale, and glauconite also are deposited in this zone, and phosphorite units are interbedded with, and grade laterally into, such sediments. The phosphate beds generally form in zones of low sedimentation rates.

The phosphate rock in Colombia generally has a low to medium phosphorous content (P_2O_5), but where weathered, with resultant leaching of calcium carbonate, it may contain 22% or more P_2O_5. Phosphate beds range from 5 m to 1.5 m in thickness, the minimum thickness being the most common. They generally are deeply buried and, with the exception of those near Sardinata (Norte de Santander), can be exploited only by underground mining methods.

DESCRIPTIONS OF PHOSPHORITE DEPOSITS

Department of Norte de Santander

In the Department of Norte de Santander (Figure 2), the Cretaceous sequence is approximately 1800 m thick, consisting of a basal unit of sandstone and limestone overlain by a sequence of black shale, fine-grained sandstone, chert, and dark shale (Richards, 1968). The phosphate beds are in the upper part of the La Luna Formation and in the basal Tres Esquinas Member of the overlying Colon Formation (Figure 3). Three major phosphatic zones are recognized (Figure 2): (1) Tibú-Orú-Las Mercedes, (2) Sardinata, and (3) Sardinata-Lourdes-Gramalate. In the Tibú-Orú-Las Mercedes zone, which extends from Tibú to the town of Las Mercedes, a distance of about 40 km, the phosphorite occurs in the basal Tres Esquinas Member of the Colon Formation. The phosphatic bed is 1-5 m thick and consists of glauconitic sandstone in which are apatite, calcite, and pyrite. The phosphate appears as nodules, pellets, oolites, and phosphatized bone fragments. P_2O_5 content ranges from 8 to 19%. Reserves in this zone are estimated at 13 million tonnes.

The Sardinata zone is along the highway between Astilleros and Sardinata (Figure 2). The phosphatic horizon, which is in the upper part of the La Luna Formation (Figure 4), shows mineralogical and chemical changes caused by weathering. In order of abundance, the fresh phosphate rock consists of calcite, apatite, and quartz, whereas the weathered rock consists of apatite and quartz and only traces of calcite (Zambrano-Ortiz, 1971). The phosphatic material occurs as pellets, oolites, and phosphatized bone fragments. The phosphate mineral is carbonate fluorapatite. The average chemical composition of the weathered phosphorite in this area is (in weight percent):

Humidity (weight loss to 105°)	1.51
Ignition loss (105-1000 °C)	4.98
Insoluble residue	16.55
P_2O_5	27.00
CaO	41.13
Al_2O_3	1.86
Fe_2O_3	0.87
F	3.15
CO_2	2.71

Proven reserves in the Sardinata zone, in a phosphorite unit ranging from 0.5 to 3.5 m in thickness, are as follows:

Under 0 to 12 m of overburden = 4.9 million tonnes
Under more than 12 m of overburden = 4.4 million tonnes

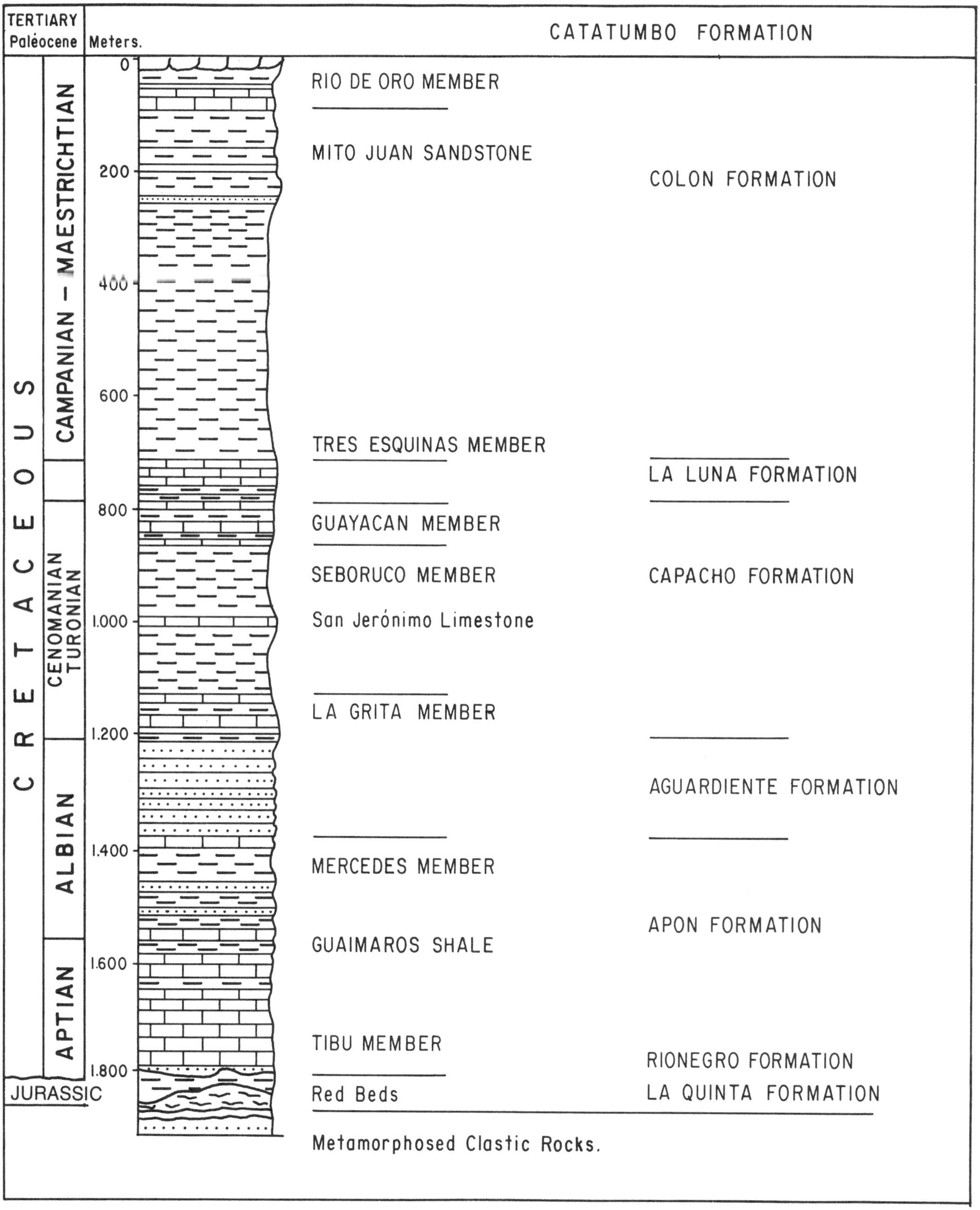

Figure 3. Generalized stratigraphic column of Cretaceous rocks in the area of the Barco concession, Maracaibo Basin. After Richards (1968).

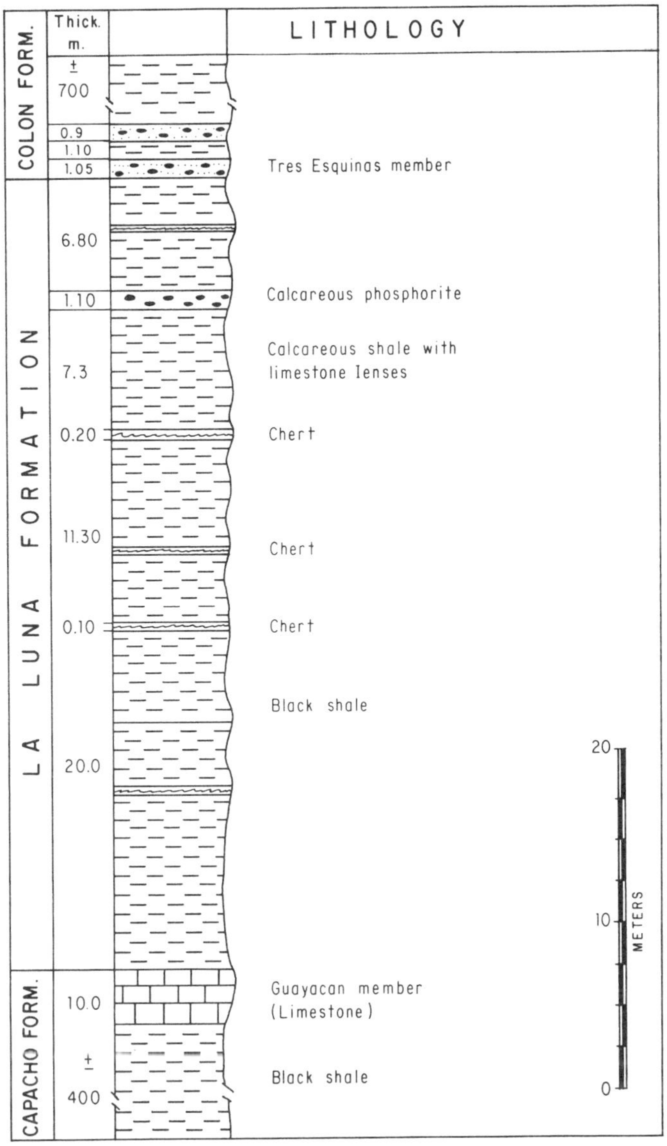

Figure 4. Stratigraphic column of the La Luna Formation near Sardinata, Department of Norte de Santander.

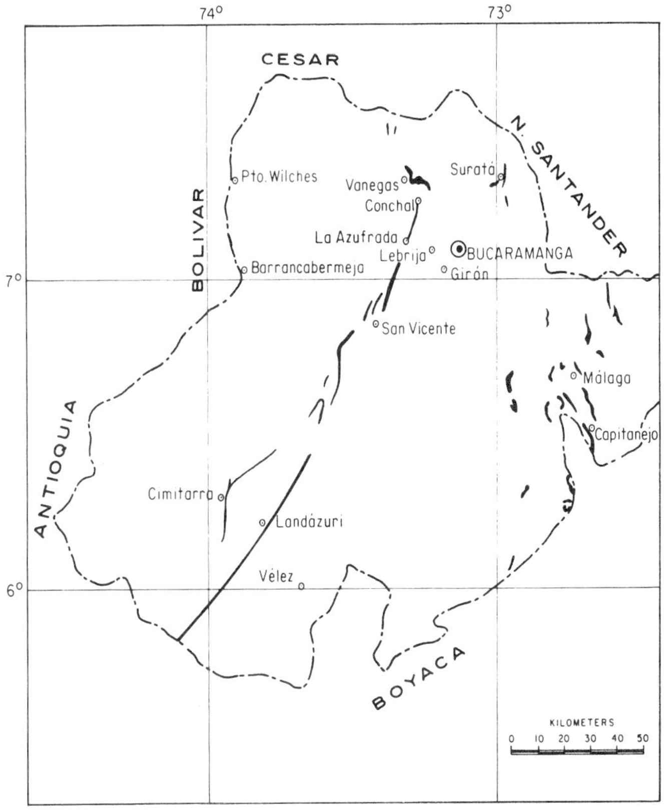

Figure 5. Phosphate rock in the Department of Santander. After Mojica (1978).

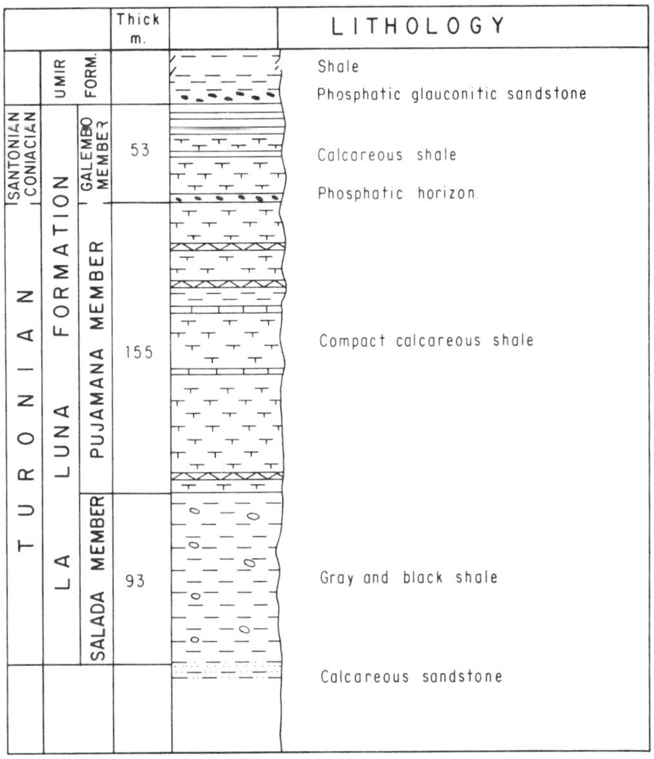

Figure 6. Generalized stratigraphic column of the La Luna Formation in Quebrada La Sorda, near Lebrija, Department of Santander. From Zambrano-Ortiz (1976).

Phosphate rock having less than 12 m of overburden can generally be mined in open cuts, whereas more deeply buried rock requires underground mining methods.

The Sardinata-Lourdes-Gramalate phosphatic zone is about 50 km long, extending from Sardinata to beyond Gramalate in the south (Figure 2). Two phosphatic horizons are present, one at the top of the La Luna Formation (Figure 4) and the other in its lower part (not shown in Figure 4). The latter is made of one to three phosphorite beds having thicknesses of 0.5 to 3.2 m and phosphate contents of 10 to 27% P_2O_5. Potential reserves are approximately 15 million tonnes that can be extracted by underground mining methods.

Department of Santander

The La Luna Formation in the Department of Santander (Figures 5, 6) has been subdivided into three members (Morales, 1958)—the Salada, Pujamana, and Galembo

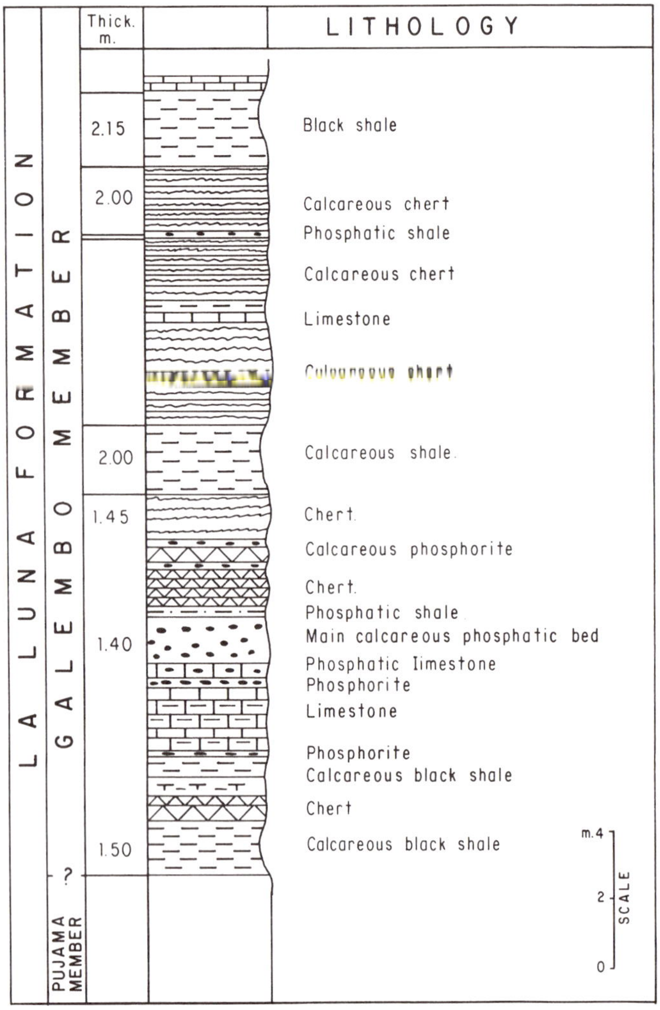

Figure 7. Stratigraphic column of the lower part of the Galembo Member of the La Luna Formation, Quebrada La Sorda, Department of Santander.

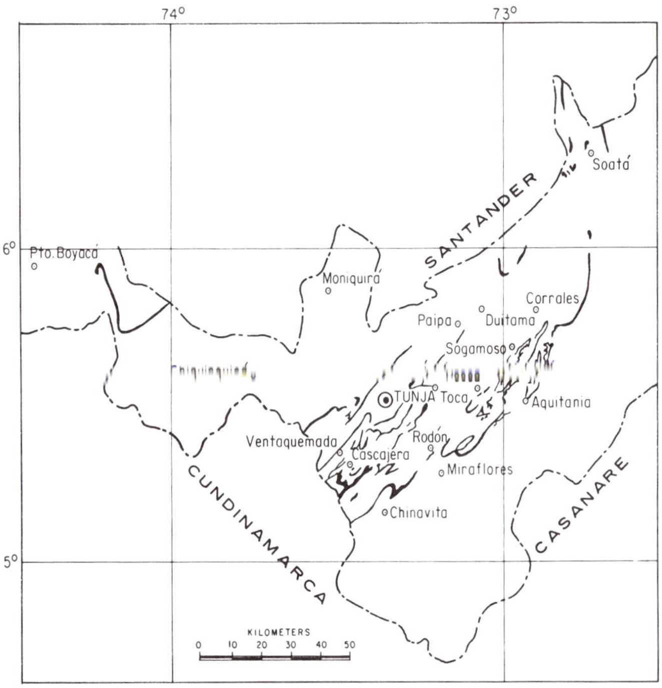

Figure 8. Phosphate rock in the Department of Boyacá. From Mojica (1978).

members—of which only the latter contains phosphorite. The Vanegas-El Conchal-La Azufrada-San Vicente phosphorite zone (Figure 5) extends from Vanegas in the north to San Vicente in the south, a distance of 50 km. The La Luna Formation is well exposed in Quebrada La Sorda (Figures 6, 7), where it consists of a well-stratified, interbedded siliceous limestone and calcareous shale, claystone, chert, and phosphorite. The Galembo Member (Figure 7) has two calcareous phosphorite horizons that contain phosphatic pellets, oolites, and phosphatized foraminifera. The principal phosphorite horizon is 0.7-1.8 m thick and shows differences in chemical composition which are related to the degree of weathering. Fresh rock, which has a high content of calcite, generally contains less than 15% P_2O_5, whereas weathered rock contains from 15 to 27% of P_2O_5. Possible reserves are approximately 30 million tonnes that can be extracted only by underground mining methods.

In the Vanegas-San Vicente area there is also a horizon of glauconitic, phosphatic, and calcareous sandstone at the top of the Galembo Member and at the base of the overlying Umir Formation (Figure 6), which previously was described as the Tres Esquinas Member of the Colon Formation (Richards, 1968) in the Department of Norte de Santander. This upper horizon is persistent, is 1-4 m thick, and contains as much as 18% P_2O_5. It is of particular interest because it contains material that might be utilized for direct application as a soil fertilizer. Preliminary tests by INGEOMINAS show that this material is satisfactory for such use.

Departments of Boyacá and Cundinamarca

In the Departments of Boyacá and Cundinamarca (Figure 1), the Cretaceous shows its maximum development, consisting of about 6000 m of sandstone, black shale, limestone, siltstone, siliceous siltstone, and porcellanite (Etayo-Serna, 1968). The sequence shows a change from a marine environment in early Cretaceous to a continental environment, in which coal beds formed, in late Cretaceous. The phosphate-bearing rocks, which formed toward the end of marine sedimentation, consist of gray claystone and laminar siliceous beds grading laterally into fine-grained sandstone. The distribution of phosphate rock in these two departments is shown in Figures 8 and 9.

Phosphate rock is in the Plaeners Formation (Figure 10) (Renzoni, 1981) of the Guadalupe Group. The principal phosphate deposits, in terms of horizontal extent and thickness, are in the Department of Boyacá, where the Plaeners Formation crops out extensively in many anticlines and synclines over an area of approximately 2600 km^2. Three phosphatic horizons have been identified in the La Conejera and Iza-Cuitiva synclines, of which the most important is near the base of the Plaeners Formation (Figure 10). Another horizon is at the top of this formation (not shown in Figure 10), and a third is in the overlying Labor y Tierna Formation. The main phosphate horizon is 1.8-4.2 m thick, and consists of phosphorite

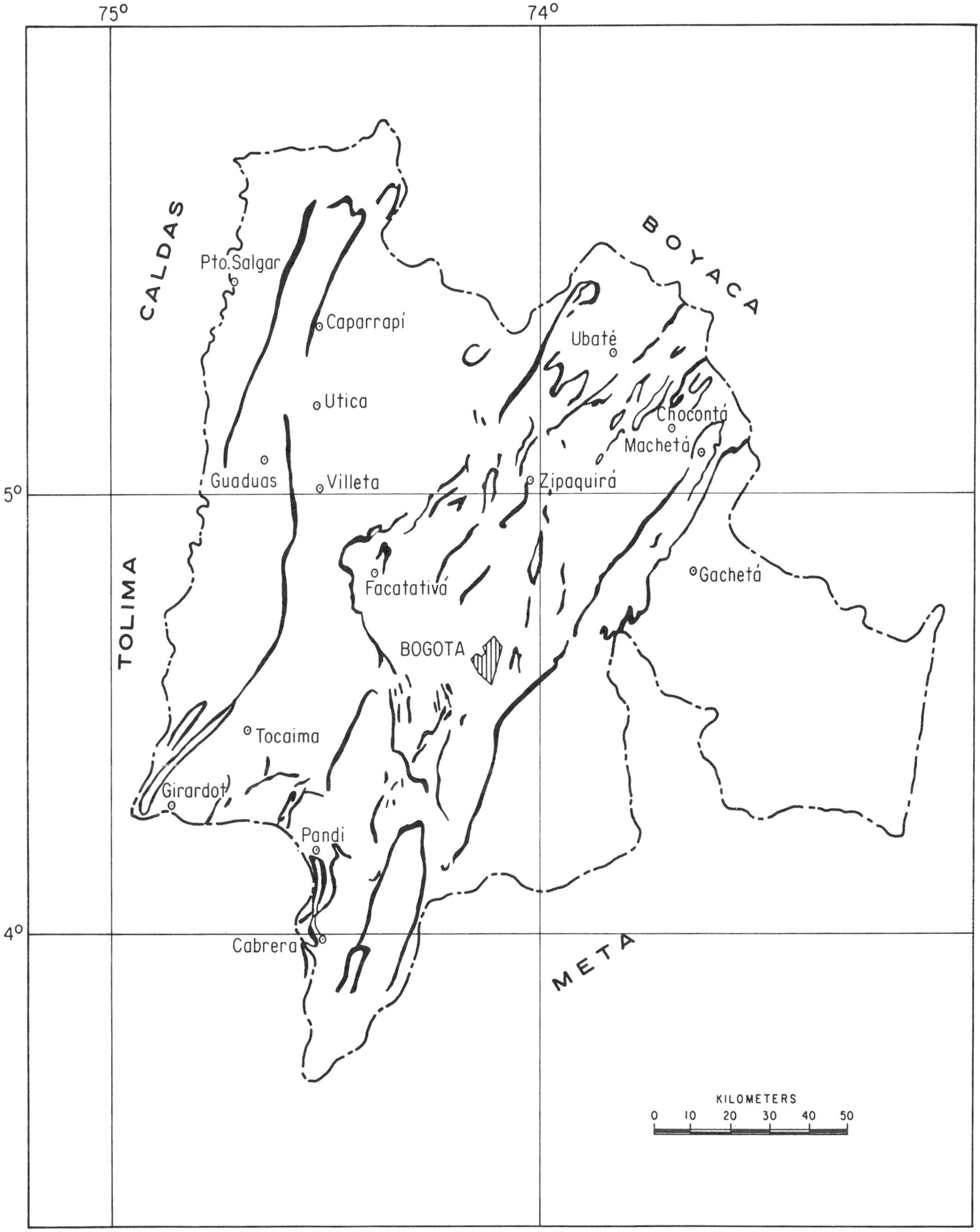

Figure 9. Phosphate rock in the Department of Cundinamarca. After Mojica (1978).

interbedded with siltstone, gray shale, fine-grained claystone, and chert. The phosphorite consists of pellets, nodules, and oolites of fluorapatite, sand, and clay. The average chemical composition is as follows (in weight percent):

Humidity (weight loss to 105 °C)	0.52
Ignition loss (105-1000 °C)	6.00
Insoluble residue	32.00
P_2O_5	22.00
CaO	32.00
MgO	0.09
Al_2O_3	1.55
Fe_2O_3	0.41
F	2.94
CO_2	1.74

Total proven, probable, and possible reserves in these departments are 500 million tonnes. The phosphate deposits of the Department of Boyacá, which have the greatest economic potential, can be extracted only by underground mining methods. The phosphorite in the Department of Cundinamarca contains abundant sandy material and therefore is of less economic importance.

Departments of Tolima and Huila

The Cretaceous sediments in the Departments of Tolima and Huila (Figures 11, 12) show important lithologic differences (Bürgl, 1967). In the Department of Tolima, the lower part of the Cretaceous sequence, the Villeta Group (Figure 13), consists chiefly of black shale in the lower part, giving way to chert, porcellanite, and gray shale in the upper part. The porcellanite units, called Upper Lidita and Lower Lidita, are persistent throughout the Departments of Tolima and Huila. The upper Lidita unit differs from the lower in having a greater thickness of individual porcellanite beds and in having phosphate beds. The overlying Guadalupe and Guaduala formations consist of glauconitic quartz sandstone and red and gray shales, which mark the transition from marine to continental sedimentation.

In the Department of Huila, the Upper Cretaceous, named the Monserrate Formation (Figure 13), is more sandy. Both the upper and lower Lidita porcellanite units contain layers of phosphate rock. The most important phosphate deposits are within a belt about 100 km long, extending from near the northern border of the department, west of Aipe, to Tesalia (Figure 12). The phosphate beds in this belt range from a few centimeters to 2.4 m in thickness. The phosphatic material consists of pellets, oolites, and nodules. P_2O_5 content of the phosphate beds ranges from 10 to 30%, the highest values being for weathered calcareous phosphorite.

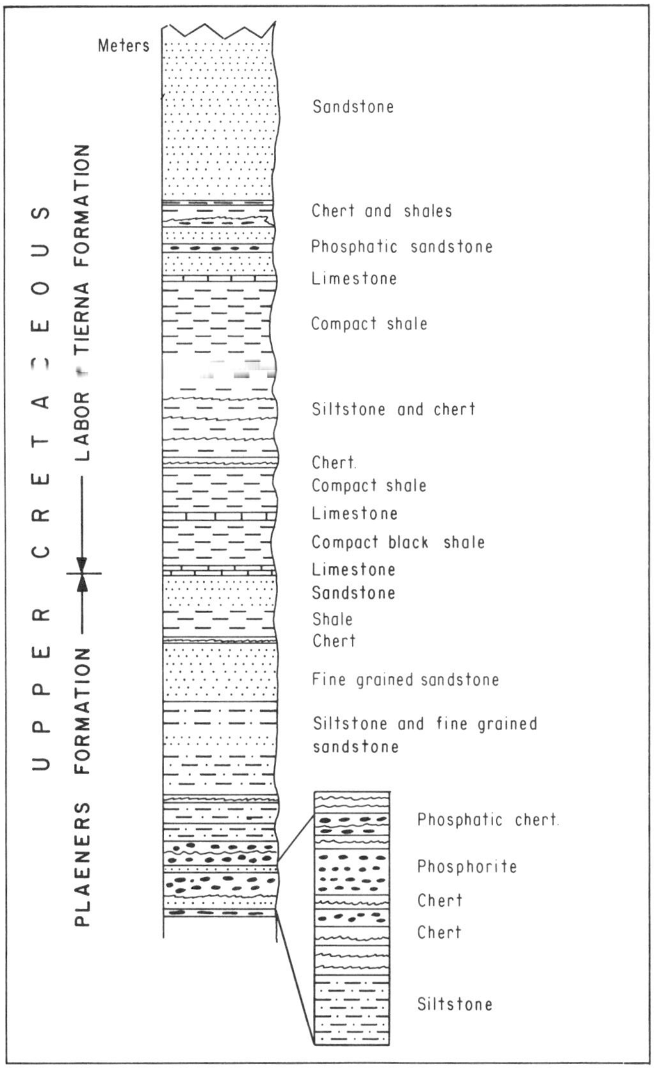

Figure 10. Generalized stratigraphic column for the western limb of the La Conejera Syncline, near Pesca, Department of Boyacá.

REFERENCES CITED

Bürgl, H., 1967, Las capas fosfáticas de la Cordillera Oriental: Instituto Nacional de Investigaciones Geológica-Mineras, Boletín Geológico, v. 15, no. 1-3, p. 7-44.

Cathcart, J. B., and F. Zambrano-Ortiz, 1967, Roca fosfórica en Colombia: Instituto Nacional de Investigaciones Geológica-Mineras, Boletín Geológico, v. 15, no. 1-3, p. 65-162.

Etayo-Serna, F., 1968, El sistema Cretáceo en la Region de Villa de Leiva, y zonas próximas: Geología Colombiana, Universidad Nacional de Colombia, p. 5-74.

Kazakov, A. V., 1937, The phosphorite facies and the genesis of phosphorites: Leningrad, Transactions, Science Institute for Fertilizers and Insect Fungicides, no. 142, p. 95-113.

McKelvey, V. E., R. W. Swanson, and R. P. Sheldon, 1953, The Permian phosphate deposits of western United States: Algiers, Congres Géologique International, Comptes Rendus de la dix-neuviéme session, 1952, p. 54-64.

Morales, L. G., and Colombian Petroleum Industry, 1958, General geology and oil occurrences of middle Magdelena Valley, Colombia: AAPG, Habitat of Oil, Symposium, p. 641-695.

Renzoni, G., 1981, Geología del Cuadrángulo J-12, Tunja: Instituto Nacional de Investigaciones Geológica-Mineras, Boletín Geológico, v. 24, no. 2, p. 31-48.

Richards, H. G., 1968, Cretaceous section in Barco area of northeastern Colombia: AAPG Bulletin, v. 52, p. 2324-2336.

Sheldon, R. P., 1963, Physical stratigraphy and mineral resources of Permian rocks in western Wyoming: U.S. Geological Survey Professional Paper 313-B, p. 50-273.

Slansky, M., 1963, Contribution a l'etude du probléme du phosphate en Colombie: Instituto Nacional de Investigaciones

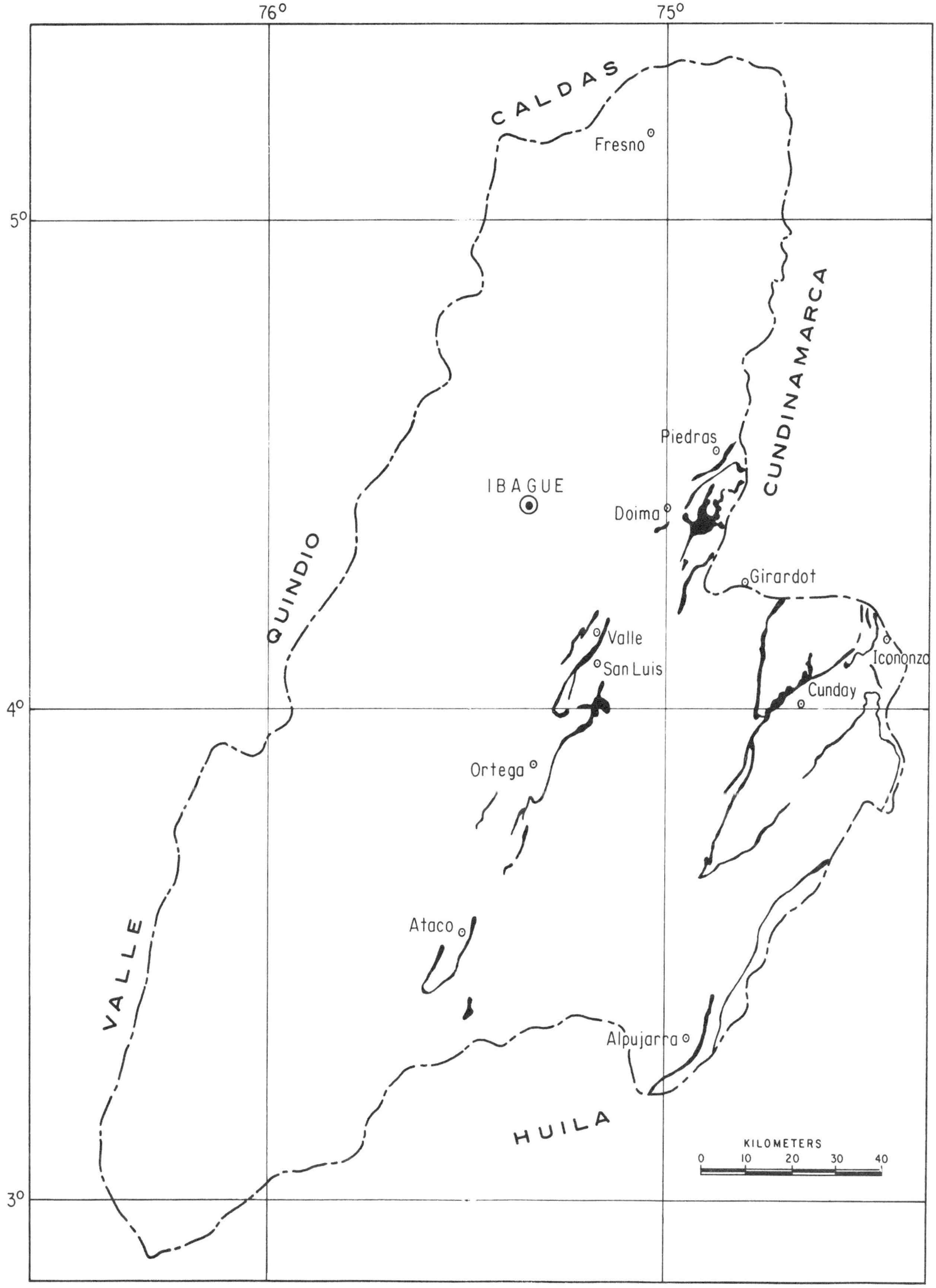

Figure 11. Phosphate rock in the Department of Tolima. After Mojica (1978).

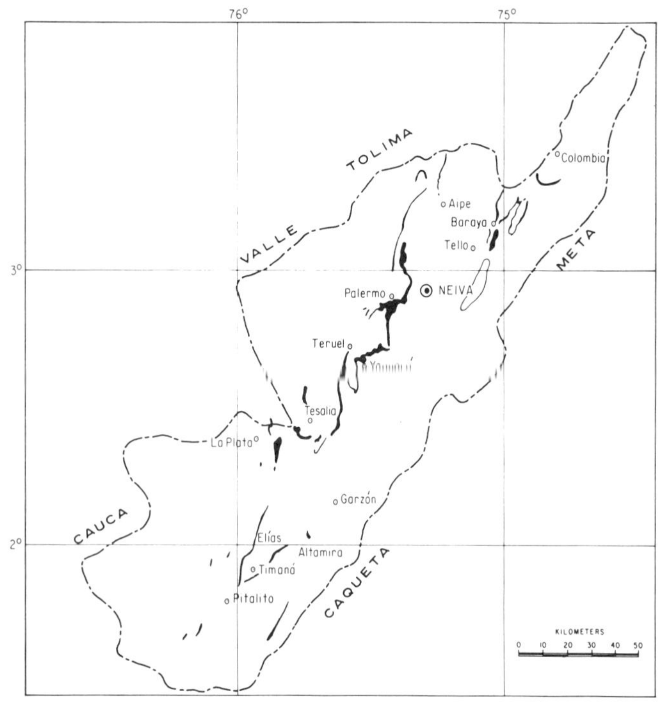

Figure 12. Phosphate rock in the Department of Huila. After Mojica (1978).

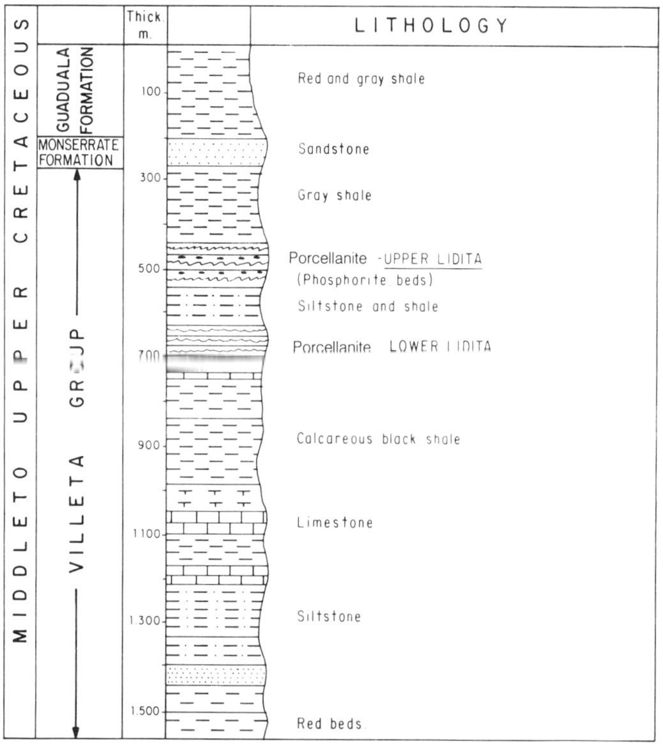

Figure 13. Stratigraphic column for the region of Ortega, Department of Tolima.

Geológica-Mineras, No. 1471, 92 p.

Zambrano-Ortiz, F., 1966, Manifestaciones fosfáticas en el Departamento del Huila y otras localidades del sur del país: Instituto Nacional de Investigaciones Geológica-Mineras, No. 1507, p. 1-23.

Zambrano-Ortiz, F., 1971, Roca fosfórica de Sardinata: Instituto Nacional de Investigaciones Geológica-Mineras, No. 1572, p. 1-44.

Zambrano-Ortiz, F., 1976, Composición de las fosforitas marinas del área La Azufrada, Santander: Memoira del Primer Congreso Colombiano de Geología, p. 325-346.

Zambrano-Ortiz, F., and P. E. Mojica, 1973, Roca Fosfórica en el Alto de La Conejera y Zonas alendañas a Pesca, Boyacá: Instituto Nacional de Investigaciones Geológica-Mineras, No. 1570, p. 1-122.

HYDROCARBON RESOURCES

Ericksen, G. E., Cañas Pinochet, M. T., and
Reinemund, J. A., editors, 1989, Geology of the Andes
and its relation to hydrocarbon and mineral resources:
Houston, Texas, Circum-Pacific Council for Energy
and Mineral Resources Earth Science Series, v. 11.

CHAPTER 25

HYDROCARBON EXPLORATION AND POTENTIAL OF THE PACIFIC COASTAL BASIN OF COLOMBIA

Rafael Bueno Salazar

Empresa Colombiana de Petroleos (Ecopetrol)
Bogotá, Colombia

INTRODUCTION

The Pacific Coastal basin of Colombia (Figure 1) covers an area of approximately 70,000 km^2, of which 14,000 km^2 lie offshore. The structural style of this area corresponds to a convergent plate basin created over folded oceanic sediments and adjacent to the subduction zone. Such a framework could be an attractive exploration target for potential hydrocarbon-bearing traps. The basin is bordered on the east by the Cordillera Occidental and the Atrato Fault and on the west by the so-called Colombian Trough in the Pacific Ocean and the Baudó Range. It extends north into Panama where it abuts the Baudó and Darien ranges (Chucunaque basin), and south into the Esmeraldas and Valdivia-Manta basins of Ecuador.

During the 1974 Circum-Pacific meeting in Honolulu, Hawaii, a summary of the stratigraphy and structure of the Colombian Pacific Coastal basin was given (Bueno and Govea, 1976). The present report focuses primarily on analyzing results obtained by subsequent studies of the region that were aimed at the search for new sources of hydrocarbons. Nine exploratory wells were drilled between 1974 and 1985, of which two are located offshore, which added to previous wells gives a total of 16 exploratory wells in the Pacific basin. Likewise, approximately 16,000 km of seismic profiles have been made of which 13,000 km are in the offshore area. A major geochemical study was made in 1979 to determine the amounts of low-molecular-weight hydrocarbons on the present-day sea floor of the basin. All these investigations were made by the Empresa Colombiana de Petroleos (Ecopetrol), either directly or through its system of association contracts.

The author expresses his appreciation to Mr. Víctor Eduardo Pérez Herrera, a geologist of the Ecopetrol Exploration Division, without whose valuable and enthusiastic contribution this report would not have been possible.

GEOLOGIC FRAMEWORK

The following brief summary of major sedimentary cycles and the common stratigraphic nomenclature (Figure 2) is taken from Bueno and Govea (1976). In addition, geologists from various oil companies also have applied other geologic names that should be considered informal or local in character.

The Cretaceous-Tertiary sedimentary sequences are underlain by a pre-Cretaceous basement complex consisting chiefly of volcanic rocks, including andesitic and basaltic flows, pillow basalts, and basalt dikes and sills. Local felsic intrusions, gneisses, schists, gabbros, and peridotites are poorly exposed in the Cordillera Occidental, the Baudó and Darien cordilleras, the Sautatá Ridge, and Gorgona Island.

Overlying the basement complex is the Groupo Diabasico (Figure 2), a sequence of turbidite and abyssal-pelagic deposits of middle Cretaceous to Paleocene age. This sequence consists of gray siliceous shale, pelagic limestone, claystone and siltstone of pyroclastic origin, black and red radiolarian chert, and interstratified basaltic flows.

The Cretaceous-Paleocene sequence is overlain by sedimentary rocks representing two stages of continuous sedimentation: (1) middle Eocene to the middle upper Miocene, and (2) late Miocene to Pliocene. Lithologic correlations in the basin are difficult owing to numerous facies changes and wedging and pinch-outs of predominantly turbidites interlayered with platform facies rocks. Platform deposition prevailed during the late Tertiary.

Three important regional unconformities have been identified in the Tertiary sedimentary sequence. These are associated with orogenic movement in the Paleocene-middle Eocene, middle-late Miocene, and Pliocene.

BASIN COMPONENTS

The Pacific Coastal basin can be divided into three major morphotectonic units: the Atrato-San Juan trough, the Pacific Frontal basin, and the Urabá basin. The Atrato-San Juan trough is in westernmost Colombia and includes Tertiary and Quaternary sedimentary deposits along the Atrato and San Juan rivers. The Cordillera Occidental and Sautatá-Taumaradó ridge mark its eastern boundary, whereas the western boundary of the basin is marked by the Baudó Range

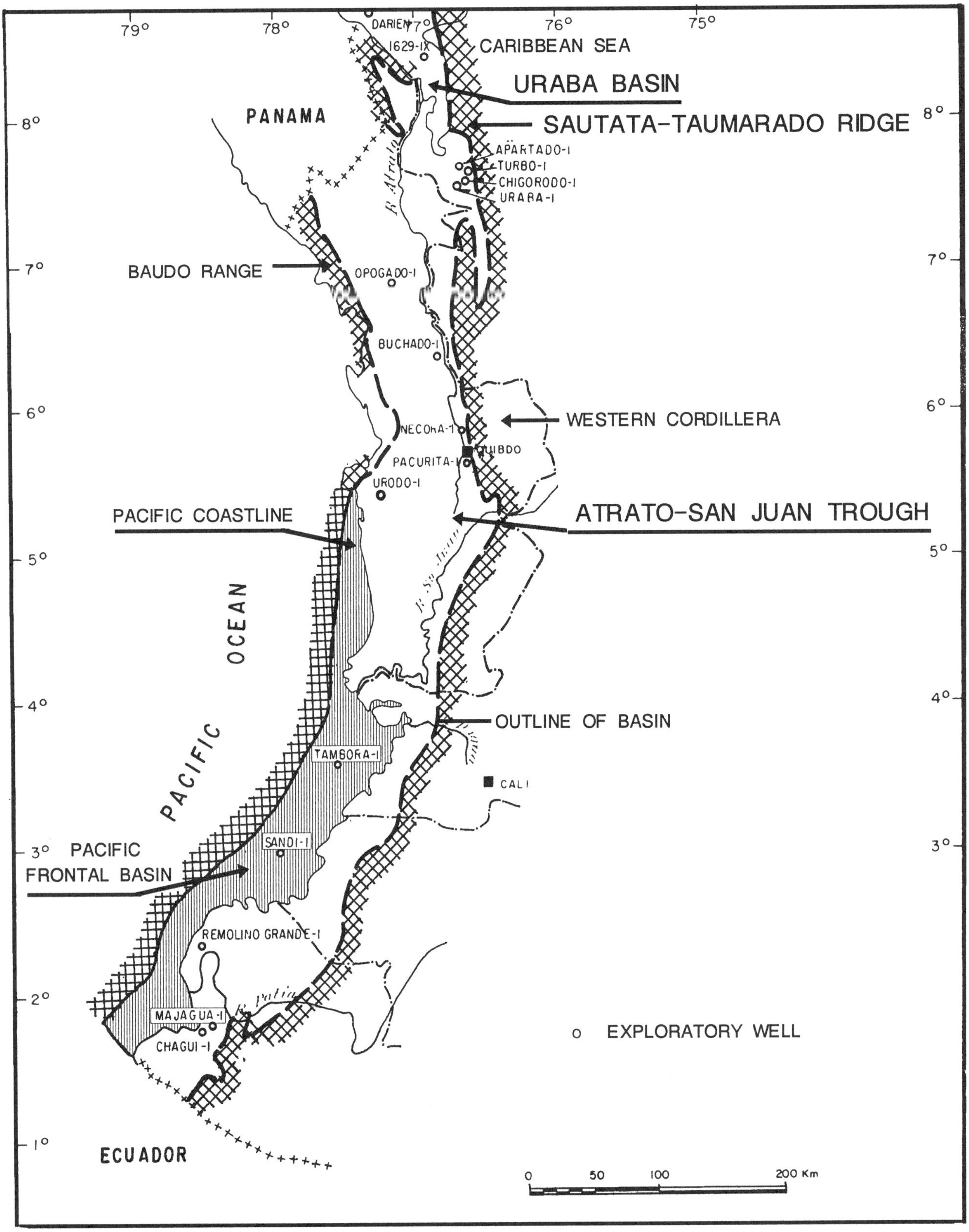

Figure 1—Index map of the Pacific Coastal basin of Colombia.

AGE		FORMATION	LITHOLOGY	DESCRIPTION
Q	PLEISTOCENE		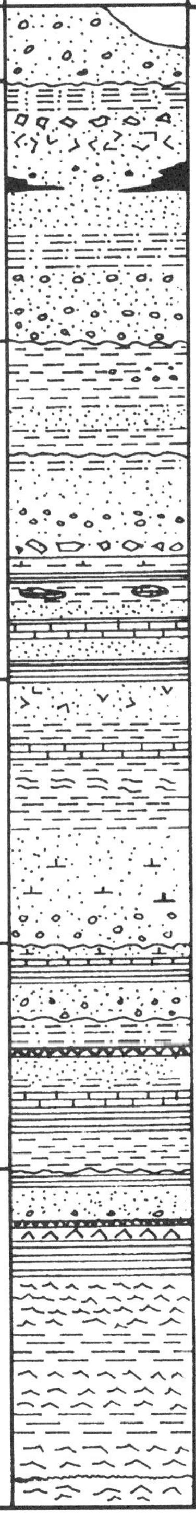	Alluvium
TERTIARY	PLIOCENE	PATO		Conglomerates, sandstones, and siltstones, with carbonaceous lenses, in upper part; Conglomerates and blue and gray sandstone in lower part; 1,300-1,600 m thick. SEISMIC C HORIZON (Aprox.)
TERTIARY	MIOCENE	BOJAYA or TUMACO		Hard claystones and blue-gray siltstones with thin interbeds of brownish sandstones and conglomerate lenses in upper part; calcareous shales, clayey sandstones, and fossiliferous limestones and concretions in lower part; thickness is about 1,400 m.
TERTIARY	OLIGOCENE	UVA or BUENAVENTURA		Predominantly calcareous sandstones anc claystones; conglomerates at base and tuffaceous sandstones at top; thickness is 2,000-3,000 m. SEISMIC B HORIZON (Aprox.)
TERTIARY	EOCENE	SURUCO		Siliceous limestone, black chert, calcareous sandstones and black grayish shale; some conglomerates; thickness is 700-1,200 m. SEISMIC A HORIZON (Aprox.)
CRETACEOUS	MIDDLE - UPPER	GRUPO DIABASICO		Basalts and Diabasic flows with interbeds of pelagic sediments.

Figure 2—Generalized stratigraphic section of the Pacific Coastal basin (from Bueno and Govea, 1974).

in the north and the present coastline in the south. The Atrato-San Juan trough can be divided into three subbasins, which were described in detail by Bueno and Govea (1976). The trough is at the site of the so-called Bolivar geosyncline that extends along the Pacific Coast in the northwestern part of South America. The Pacific Frontal basin is on the Pacific Platform, extending westward from the coastline of Colombia to the Colombia and Panama troughs (Figure 3). Characteristic features of this basin were described by Bueno and Govea (1976). Major geological studies in this area include studies related to drilling of two exploratory wells in 1967, and seismic and geochemical surveys conducted during the late 1970s and early 1980s. The Urabá basin is in the Gulf of Urabá located in the Caribbean between Panama and northwesternmost Colombia. It is a backarc basin (Restrepo and Pérez, 1982) that is bounded on the east by a granodiorite backarc, known as the Sautatá-Taumaradó ridge, and on the west by the Atrato fault system that separates it from the northernmost foothills of the Cordillera Occidental (Figure 3).

STRUCTURAL FRAMEWORK

Different authors have defined the Bolivar geosyncline as consisting of a series of deep structural basins bounded on the east by the Cordillera Occidental and the Sautatá-Taumaradó ridge (Figure 3). These basins originated by accretion of oceanic crust to the continent as the result of plate convergence and subduction during late Cretaceous to Paleocene time (Figure 4).

The predominant tectonic features of the Atrato-San Juan trough trend parallel to the Cordillera Occidental, being about N20°W in the north, changing to N30°E in the south. The structure is characterized by block tectonics; apparent megashears of east-northeast orientation, which cross the trough, are of late Cretaceous age and show lateral movement. Apparently, both vertical and lateral displacement were considerable. Intrusive activity accompanied deformation. Gravity anomalies show no relation to subsurface structures in the trough, which may indicate more intense faulting in Cretaceous time than in the Tertiary or the presence of Cretaceous paleoheights discordantly covered by Tertiary sediments. The most important structural features of the Pacific Frontal basin are diapirs or mud volcanos, which, like salt domes, may have created structural traps for hydrocarbons. These formed during Pliocene-Pleistocene tectonic events. The Urabá basin closes rapidly in the continental sector to the south, where the Atrato fault system joins the Cordillera Occidental. In the north, it extends into the Caribbean toward Panama, forming a type of microplate.

PETROLEUM POTENTIAL AND EXPLORATION

No commercial hydrocarbon deposits have been located in the Pacific Coastal basin of Colombia, but wells drilled in the basin, both on the continent and offshore, indicate the presence of gas with some traces of oil. Geochemical studies of samples from outcrops, wells, and the ocean floor indicate the presence of rocks favorable for generating oil and show the presence of a type of kerogen capable of producing gas. Oil and gas seeps have been identified, and some samples of shales and siltstones have a petroliferous odor when broken. These features indicate adequate thermal maturity for generating hydrocarbons.

The Cretaceous sedimentary rocks are believed to have good hydrocarbon-generating potential, especially in those sectors having less volcanic rocks. The Eocene-Miocene rocks, which are turbidites having a high content of organic matter, are believed to have a good potential for dry gas, particularly in areas where Eocene plutonism increased the geothermal gradient.

Onshore Exploration

Twelve of the 16 exploratory wells in the Pacific basin were drilled onshore, and four of these yielded hydrocarbon shows. In the Urabá basin to the north, Ecopetrol drilled two wells in 1982, Apartado-1 and Chigorodo-1, which produced strong gas shows from a pre-Miocene siltstone sequence. In 1981, the Wainoco Company, working under an association contract with Ecopetrol, drilled Majagua-1 south of the Atrato-San Juan trough. This well produced good shows of gas, with C1, C2, and C3, fine-grained sandstone and siltstone of Oligocene-Miocene age at depths of 2500 to 4200 m. During 1956, Mobil's Chaguí-1 well in the southern part of the trough found strong hydrocarbon shows in Miocene limestone.

Geochemical studies of outcrop samples of Oligocene-Miocene rocks reveal the following characteristics: five of 31 samples gave TOC (total organic carbon) values 2 to 5%, which indicates an organic content within the marginal to very good range. The predominant kerogen is mainly of terrestrial origin, and is a possible generator of gas. The average vitrinite reflectance (R_o) is 0.32, and T_{max} is 426°C, indicating a thermal maturity adequate for the generation of petroleum.

The tectonic-sedimentary framework indicates the presence of structural and combined traps. Future exploratory work should focus on the search for paleodeltas created by ancient rivers draining highlands of crystalline basement rocks that served as sources of clastic sediments. Such highlands would have been located to the east, in a position equivalent to that now occupied by the Cordillera Central. Crystalline basement rocks are considered to be the most favorable sources of permeable sands that are most promising as hydrocarbon reservoirs.

Offshore Exploration

Four wells have been drilled in the offshore sector of the Pacific basin. Two were drilled to depths of more than 3400 m in the Caribbean sector of the Urabá basin with no evidence of hydrocarbons. The other two were drilled in the Pacific Ocean during 1967. Only one of these, Pozo Tambora-1, showed indications of gas and small impregnations of asphalt in lower Tertiary clastic rocks. These wells demonstrated the presence of local torrential clastic sequences having large quantities of conglomerates.

In view of previous exploratory results in the Pacific Frontal basin that yielded evidence as to the partial absence of reservoir rocks, Ecopetrol prepared an exploration program based on the integration of seismic and geochemical studies of the sea floor. A low penetration seismic survey (2 sec) was con-

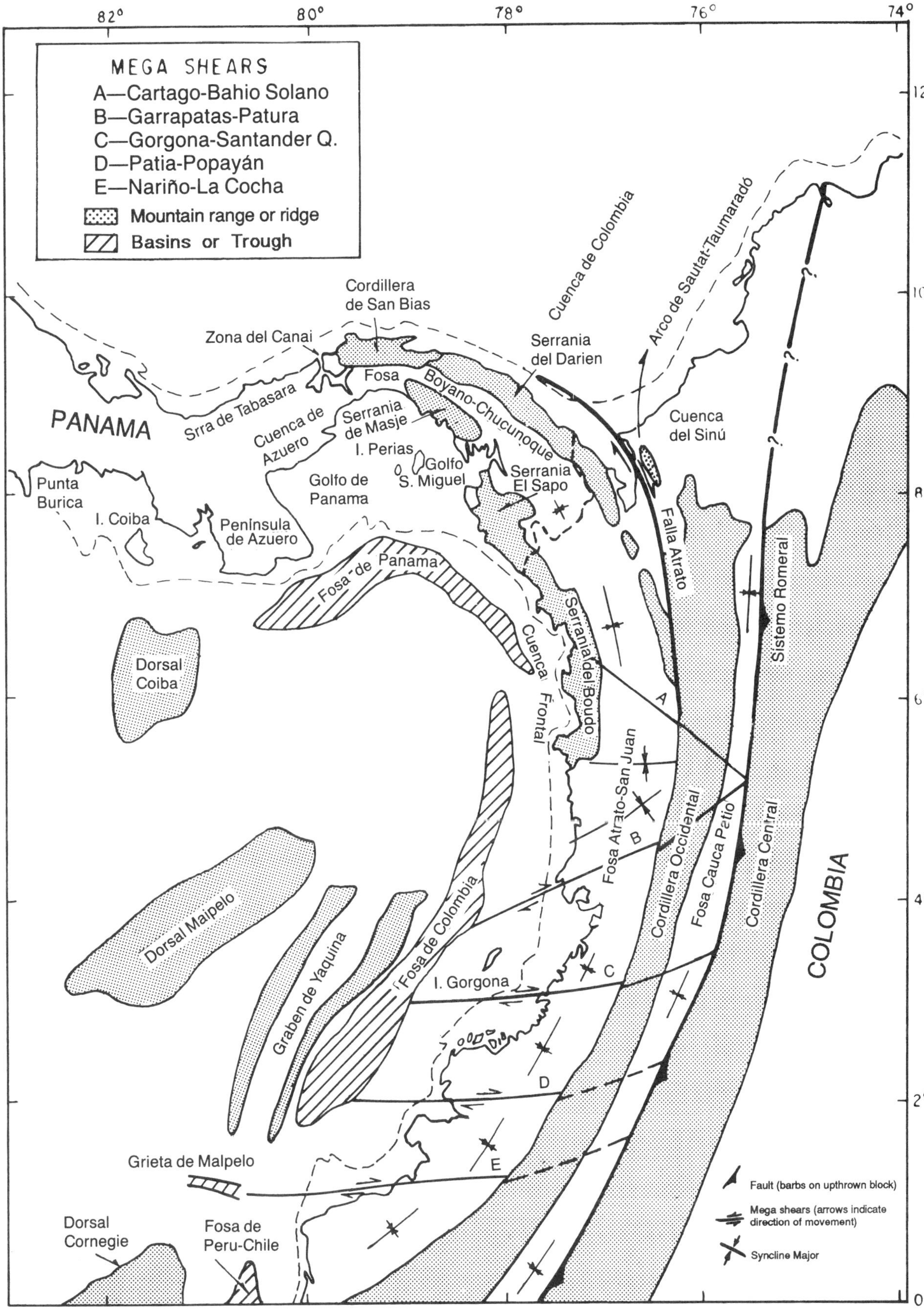

Figure 3—Morphotectonic elements of western Colombia and Panama (from Restrepo and Pérez, 1982).

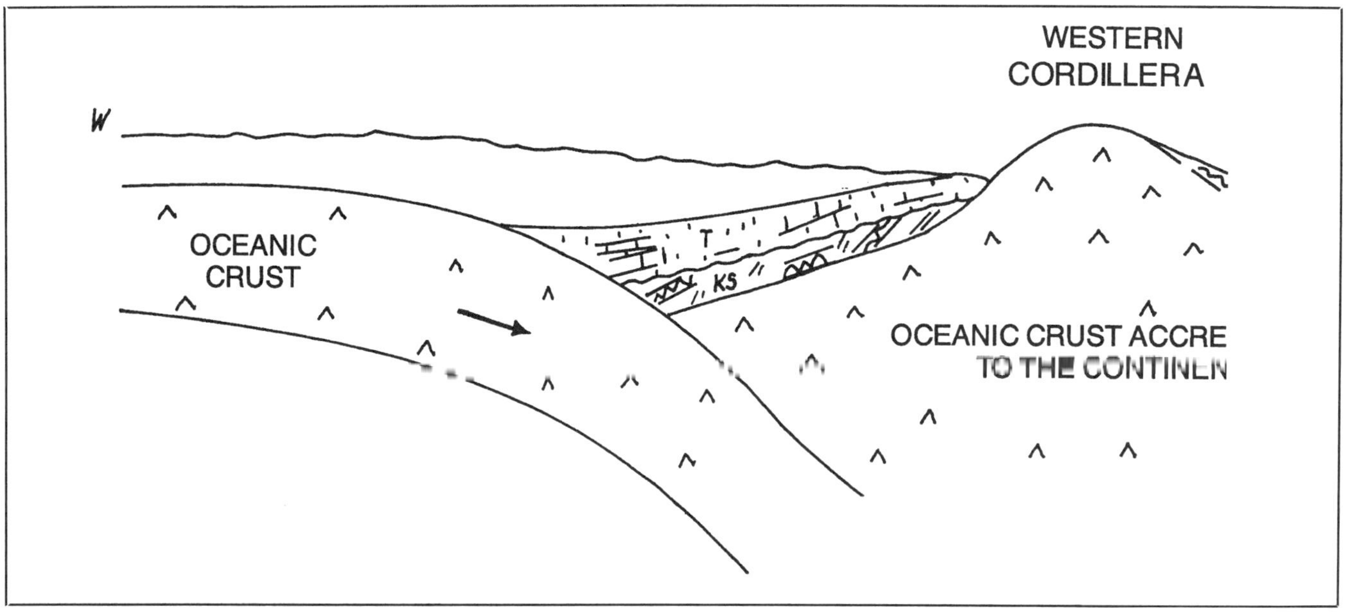

Figure 4—Generalized cross section showing relation of Pacific Coastal basin to oceanic and continental plates (from Govea and Aguilera, 1985).

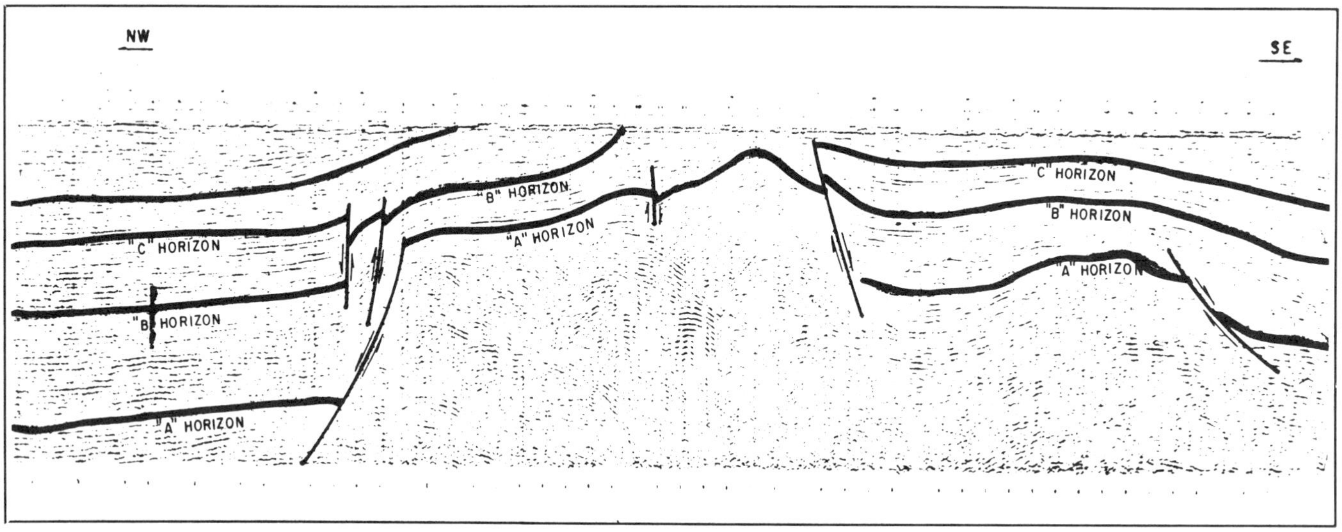

Figure 5—Seismic profile across the southern part of the Pacific Coastal basin showing tentative correlation of Horizons A, B, and C.

ducted in 1978, which complement 4800 km of seismic profiles done from 1971 to 1977. Geochemical samples were collected at seismic shotpoints, which made it possible to prepare maps of anomalous concentrations of low-molecular-weight hydrocarbons that could be compared with amplitude anomalies.

SEISMIC STUDIES

Geophysicists of the Ecopetrol Exploration Division interpreted a 4800-km seismic profile of the Pacific offshore area between the mouths of the San Juan and Patia rivers. This area is approximately 8250 km^2 and includes part of the platform and the slope up to depths of no more than 1000 m. In addition, magnetometer, geochemical, and geological studies were made onshore. All these studies made it possible to differentiate four major "seismic sequences" separated by three unconformities named, from the oldest to youngest, the A, B, and C horizons (Figure 5). Gradual changes were noted in the characteristics of certain reflections in each of these horizons, which indicate lateral facies variations. Structure-closure maps prepared for Horizons A and C (Figures 6 and 7) have served as a basis for locating targets for exploratory drilling.

Horizon A is characterized by a high reflection coefficient, with few variations, broken by several diapirs and by the Pacific Range (Baudó Range) of which Gorgona Island (Figure 6) is a part. The horizon is at the top of the seismic basement and coincides with a regional unconformity marking the base

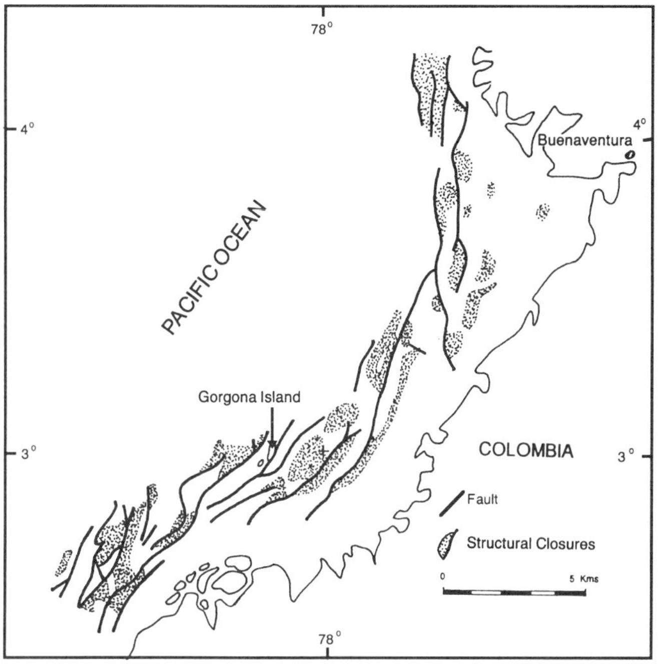

Figure 6—Faults and structural closures as interpreted for the seismic A Horizon in the offshore part of the Pacific Coastal basin (from Garcia and Prince, 1982).

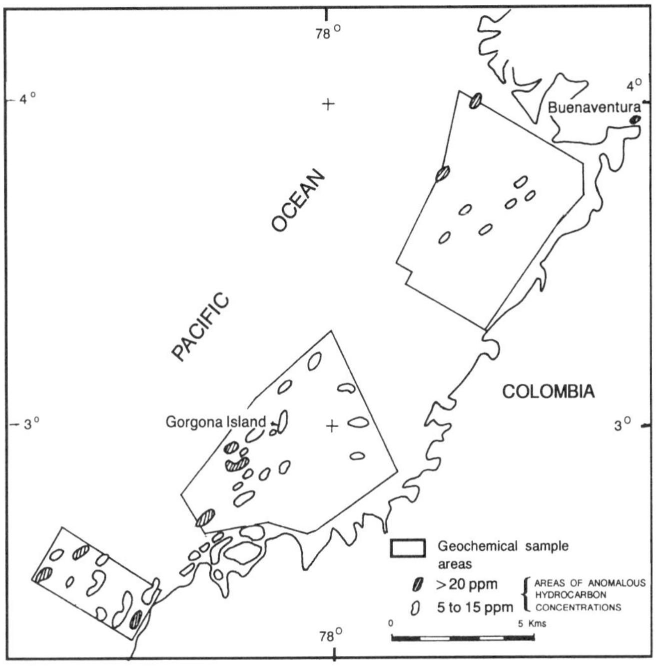

Figure 8—Geochemical study areas and anomalies in the offshore part of the Pacific Coastal basin.

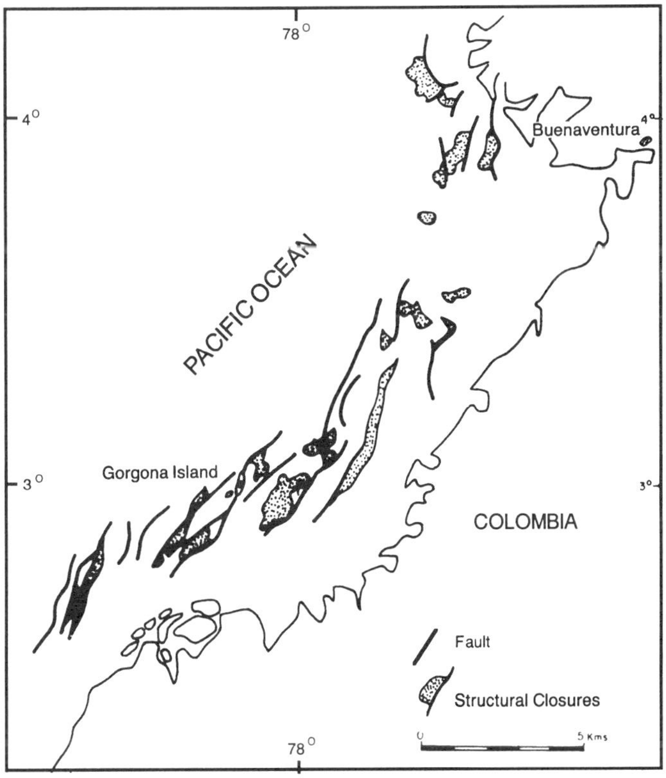

Figure 7—Faults and structural closures as interpreted for the seismic C Horizon in the offshore part of the Pacific Coastal basin (from Garcia and Prince, 1982).

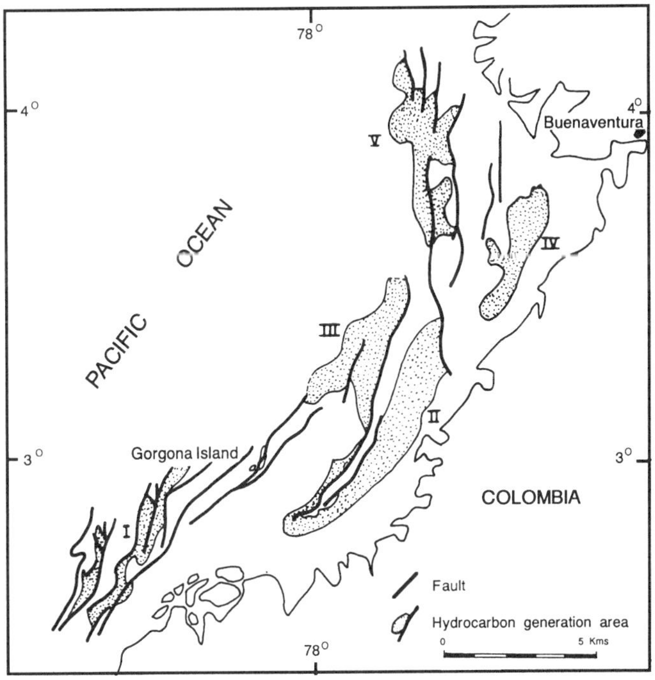

Figure 9—Areas favorable for hydrocarbon generation in the offshore part of the Pacific Coastal basin (from Garcia and Prince, 1982).

of the Tertiary. The nonmigrated isochron map of Horizon A (Figure 6) shows the major faults and structural closures associated with them. In general, these structures are parallel to the coastline. Narrow, elongated structures in the vicinity of Gorgona Island are related to mud diapir zones. Diapir zones also are present toward the coast in the northernmost part of the area. Twenty-two structural highs corresponding to anticlines and 20 closures against faults and/or diapirs have been identified in the seismic profile. Onlap features are abundant in overlying sediments, and reflections in underlying sediments indicate wedging of units and unconformities. These features together with broad continuity of the horizon suggest it is an erosional surface that was subjected to weathering. This surface is probably of middle to upper Miocene age. In general, Horizon C is present throughout the area except where structural highs (Gorgona Island, San Juan ridge) were eroded. It also is interrupted locally by diapir zones. Figure 7 shows faults and structural closures that could be traps. Most of these also are present at the level of Horizon A and are associated with deep basement faults.

It is concluded that Horizons A and C have excellent characteristics in terms of the presence of both structural and combined traps that are favorable sites for oil and gas accumulation. A total of 66 possible traps, including anticlines and closures against faults and diapirs, have been identified.

GEOCHEMICAL STUDIES

The geochemical sea-floor study areas are shown in Figure 8. Maximum depth of the ocean floor in these areas was 150 m. A total of 2400 core samples were collected for these areas and subsequently analyzed by chromatographic methods. Results of the analyses were interpreted statistically in order to identify anomalies and to correlate them with geophysical data. Contents of methane, ethane, propane, butane and isobutane, pentane, and isopentane were measured and evaluated. The sum of hydrocarbons and total quantity of ethane, propane, butane, normal pentane, and isopentane were calculated along with the $C_1/C_2 + C_3$ ratio. Figure 8 shows areas of hydrocarbon concentration of 5 to 15 ppm and greater than 20 ppm. Areas favorable for hydrocarbon generation (Figure 9) in Cretaceous and Eocene rocks were identified by analyzing seismic, geochemical, and drilling data. These areas are considered to have greatest potential for production of oil and gas within the Pacific Coastal basin. A minimum burial depth of 3000 m was established as a limit for possible generating zones. Of the five zones shown in Figure 9, zone II coincides with that part of the basin bordering the continent where the best generating conditions exist in Cretaceous and Tertiary sediments. Depth of the sedimentary column exceeds 7000 m and these sediments have the potential for generating the full range of hydrocarbons. Zone I shows a potential for mature oil in Cretaceous sediments and for gas and mature and immature oil in Tertiary sediments. Generating potentials in zones III, IV, and V are for dry gas in both Cretaceous and Tertiary sequences. Burial depths for Tertiary sequences are between 3000 and 5000 m.

PROSPECT AREAS

The offshore studies in the Pacific Coastal basin enabled identification of 15 prospect areas as having the best potential

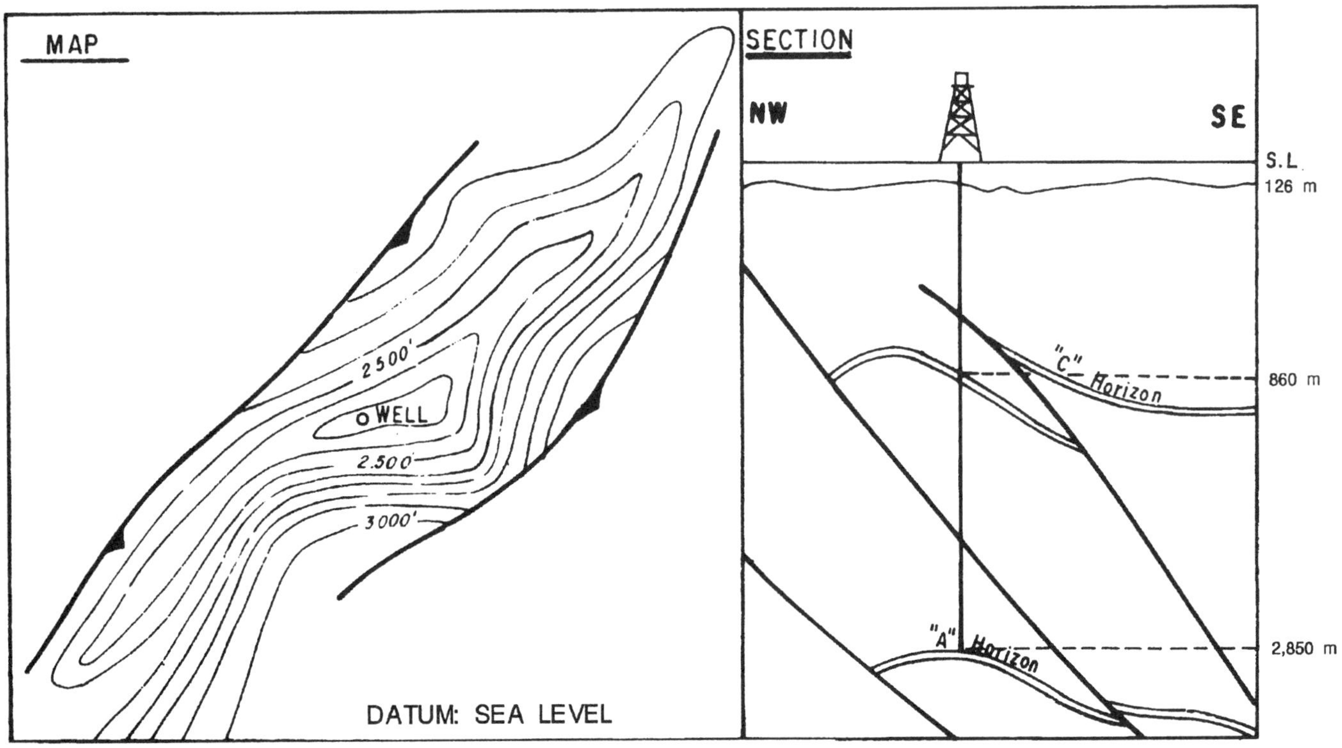

Figure 10—Typical structural pattern of the most favorable exploration targets in the offshore part of the Pacific Coastal basin.

for hydrocarbon accumulation. Prospect area I (Figure 9), which is shown diagrammatically in Figure 10, represents a structural pattern typical of the area. It shows an en echelon pattern of reverse faults resulting from compressive east to west stresses related initially to plate tectonics and sequentially to mud intrusions. The top of Horizon C is slightly above 900 m, and that of the seismic basement (Horizon A) is around 2850 m.

SUMMARY AND CONCLUSIONS

The Pacific Coastal basin of Colombia covers approximately 70,000 km^2 and has a structural style corresponding to a convergent plate basin. Recent geophysical, geochemical, and geological investigations indicate a potential for petroleum accumulation. The geochemical studies indicate the presence of source rocks of Cretaceous and lower Tertiary age having kerogen-rich, gas-prone, organic matter. The existence of oil seeps from Cretaceous outcrops indicate sufficient thermal maturity for oil generation. Adequate reservoirs are present in sandy and calcareous rocks of upper Eocene to Oligocene age. These rocks are predominantly marine and have an estimated thickness in excess of 6000 m.

REFERENCES CITED

Bueno, R., and C. Govea, 1976, Potential for exploration and development of hydrocarbons in Atrato Valley and Pacific Coastal and Shelf Basins of Colombia: American Association of Petroleum Geologists Memoir 25, p. 318-327.

Garcia, C., and R. Prince, 1982, Interpretación geológica de la información sísmica del Pacífico costa afuera: Ecopetrol-Bogota, unpublished report, Inf. Geof. 637, 78 p.

Govea, C., and H. Aguilera, 1985, Cuencas sedimentarias de Colombia: Bogota, II Simposio Bolivariano Exploración Petrolera en las Cuencas Subandinas, v. 2, 93 p.

Restrepo, G., and M. Pérez, 1982, Interpretacíon sísmica regional del area de Urabá: Ecopetrol-Bogota, unpublished report, Inf. Geof. 543, 118 p.

Ericksen, G. E., Cañas Pinochet, M. T., and Reinemund, J. A., editors, 1989, Geology of the Andes and its relation to hydrocarbon and mineral resources: Houston, Texas, Circum-Pacific Council for Energy and Mineral Resources Earth Science Series, v. 11.

CHAPTER 26

HYDROCARBON RESOURCES IN THE SUB-ANDEAN BASINS OF COLOMBIA

Rafael Bueno Salazar
Empresa Colombiana de Petroleos (Ecopetrol)
Bogotá, Colombia

ABSTRACT

Colombia has three sub-Andean sedimentary basins between the Andes Mountains in the west and the Guiana Shield in the east— (1) the Putumayo basin, with oil production averaging 17,000 BOPD; (2) the Llanos Orientales basin to the north, where major petroleum reserves have been discovered recently; and (3) the Amazon basin, which is in its earliest stages of exploration. The Amazon basin is an intracratonic basin which is marginal to, and considered to be a northward extension of, the great Brazilian Amazon basin. Very little subsurface exploration has been carried out in this basin. The Putumayo and Llanos Orientales basins are pericratonic basins. The sedimentary rocks in the Putumayo basin are of Jurassic (Motema Formation) to late Tertiary age. Major hydrocarbon deposits occur in the middle Cretaceous Caballos and Villeta formations as well as in the late Eocene Pepino Formation. In the Llanos Orientales basin, a sequence of Paleozoic rocks, unaffected by metamorphism, is a potential target for hydrocarbon exploration. In addition, Upper Cretaceous and lower Tertiary rocks in this basin have served to generate and trap hydrocarbons.

INTRODUCTION

Colombia has 13 sedimentary basins covering a surface area of 71.6 million hectares (Figure 1). Oil and gas fields have been found in seven of these basins. The three sub-Andean sedimentary basins, two of which have important known reserves, are believed to have major potential for discovery of additional hydrocarbon reserves. The geological conditions and petroleum exploration and development of these three basins—the Amazon, Putumayo, and Llanos Orientales basins—are discussed in this chapter. These are considered to have the greatest oil and gas potential of all the country's sedimentary areas. Hydrocarbon discoveries made in the Putumayo basin in the 1960s and 1970s and in the Llanos Orientales basin in the present decade indicate that this region is Colombia's prime hope for future discoveries.

The region discussed in this chapter covers 350,000 km^2 between the Andes Mountains and the Guiana Shield. It is an area of gentle relief, with grass plains in the north and tropical rain forest in the south. It straddles two topographic basins, the northern draining into the Orinoco River and the southern into the Amazon River.

HISTORY OF HYDROCARBON EXPLORATION AND PRODUCTION IN COLOMBIA

Hydrocarbon development in Colombia began in 1918 with the discovery of the La Cira-Infantas field in the Middle Magdalena Valley basin. Later discoveries in this and other basins provided the country with an oil surplus lasting until 1974, when the nation was obliged to exchange its role as an oil exporter for that of an importer. Discoveries made in the Llanos Orientales basin during 1981-1985 enabled Colombia to resume exports beginning in the second half of 1986. New oil pipelines such as the Caño Limón-Río Zulia, Araguaney-San Cayetano, Apiay-San Cayetano, and San Cayetano-Velásquez lines are being incorporated into the national hydrocarbon transportation network. This expanded network will provide an adequate means for moving crude oil from recently discovered fields in the eastern part of the country.

The discovery and development of petroleum resources in Colombia have occurred in several stages. During the period

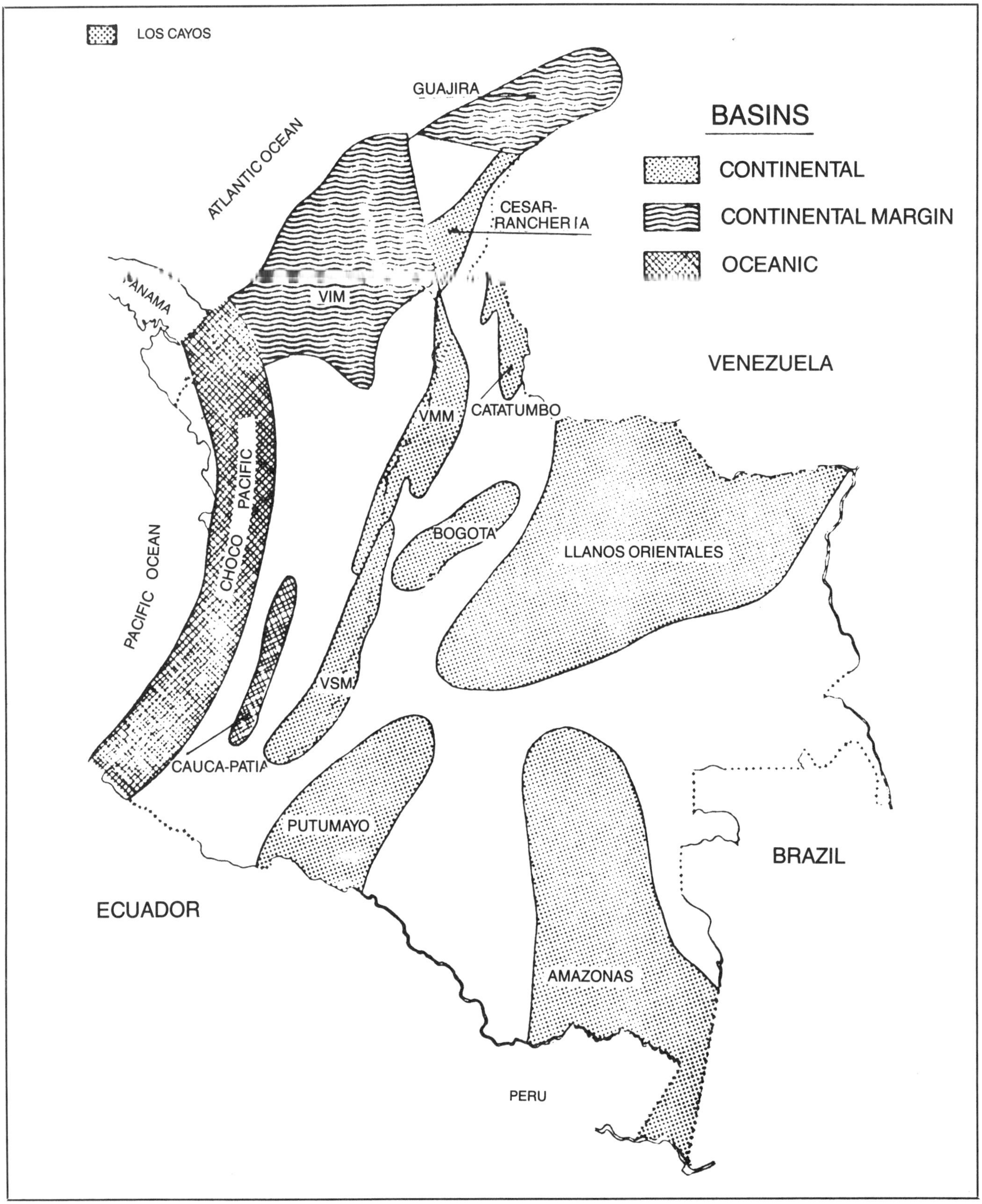

Figure 1. Sedimentary basins of Colombia. (From Govea and Aguilera, 1985.)

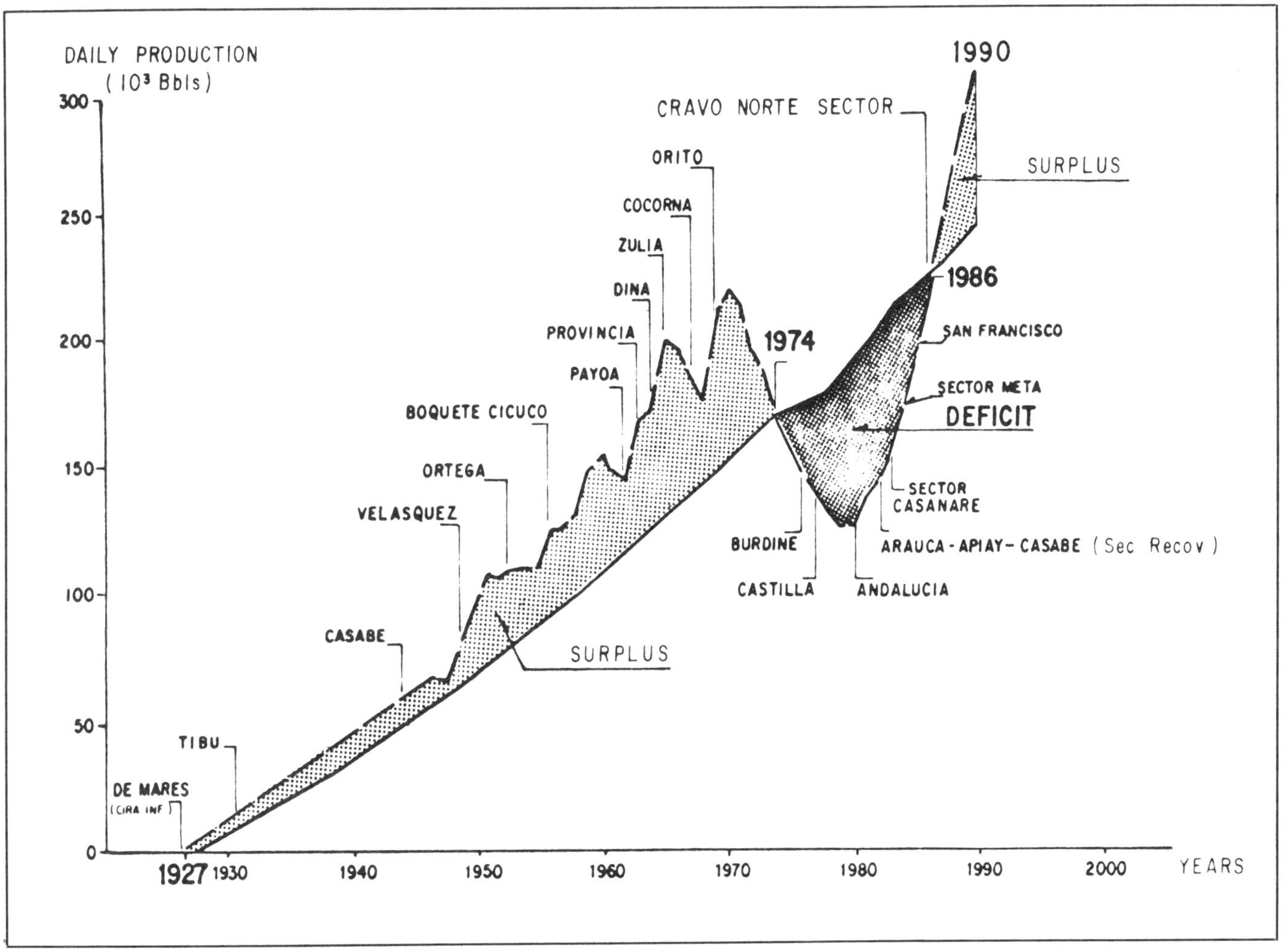

Figure 2. Exploration and production history in Colombia.

between 1917-1927, exploration and development were focused on the giant La Cira-Infantas field in the Middle Magdalena Valley basin (Figure 2), which proved to have approximately 800 million barrels in reserves. Exploitation of this field began in 1927, the same year in which Colombia began exporting crude oil. The next 23 years (1928-1950) witnessed new discoveries, including the Tibú field (reserves totaling 257 million barrels) in the Catatumbo basin, and the Casabe (300 million barrels) and Velásquez (316 million barrels) fields in the Middle Magdalena Valley basin. These increased national reserves to 1200 million barrels. Colombia continued exporting hydrocarbons during this period, and registered production averaging 110,000 barrels per day.

Between 1951 and 1974, fields were discovered at Provincia (Lower Magdalena Valley), Dina-Tello (Upper Magdalena Valley), Orito (Putumayo), and Zulia (Catatumbo). These fields had reserves of 480 million barrels. Production began during this same period. Domestic oil production rose from 130 to 180 thousand BOPD, and reserves quickly fell to a level of 600 million barrels in 1974. However, government projections indicated that by the early 1970s Colombia would undergo a tremendous oil crisis, thereby forcing the country to abandon its export activities and begin importing hydrocarbons. Accordingly, the national government assigned Ecopetrol the difficult task of finding new hydrocarbon resources. The company accepted the challenge and managed to secure enactment of Law 20/1969, which marked a true beginning for the association contract system in Colombia. Based on economic emergency decrees (No. 2223 and No. 2224), this system became the sole means by which oil exploration could be conducted in Colombia as of 1974.

The period 1975-1982 was a time of deficit in domestic oil supply. Reserves declined to 400 million barrels (1978), whereas demand surpassed production by approximately 45,000 BOPD. Starting in 1975, new fields began to be discovered under the Empresa Colombiana de Petroleos (Ecopetrol) association contract system. These included the Burdine field (Putumayo), Andalucia Sur (Upper Magdalena Valley), and the Arauca, Casanare, and Apiay fields (Llanos Orientales). Discovery of this last area, located near the city of Villavicencio, was the result of the direct effort of Ecopetrol through its Exploration Division. These discoveries increased national oil reserves from 400 to 600 million barrels.

The years 1983-1985 set a pattern for oil exploration in Colombia that represents the logical consequence of the exploration policy of the Ecopetrol Vice-Presidency of Exploration. In 1983, this policy led to the discovery (Ecopetrol-Occidental) of fields at Caño Limón, in the Intendency of

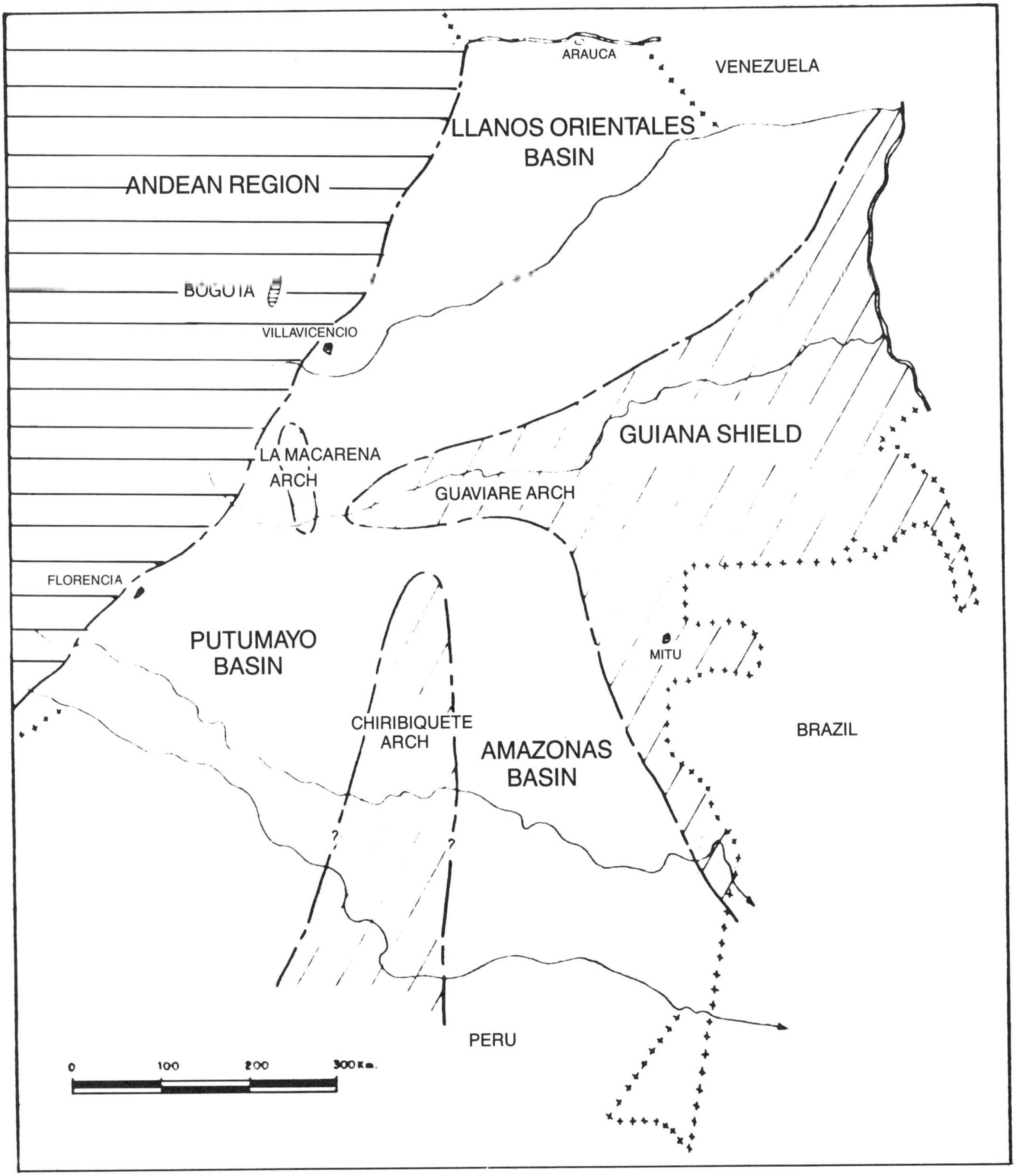

Figure 3. Sedimentary basins in the sub-Andean region.

Arauca, and La Gloria Norte and Caño Garza Norte (Ecopetrol-Aquitaine), in the Intendency of Casanare. These accomplishments were reinforced by discoveries made in 1984, a record year for oil exploration in Colombia. At that time activities developed under the Ecopetrol-Occidental Cravo Norte association contract led to discovery of the Matanegra, La Yuca, and Redondo fields, thereby confirming the presence of a giant field and increasing proven reserves by an additional 600 million barrels. During that same year, the Ecopetrol-Aquitaine association added new reserves to this figure with the Morichal, Sardinas, and Jordán fields located in the Intendency of Casanare. In addition, exploratory work carried out directly by Ecopetrol at the Rancho Hermoso and Entrerrios fields in Casanare and the Guayuriba, Suria, La Reforma, and La Libertad fields in the Meta province, near the city of Villavicencio, also contributed new reserves to the nation's total. Discovery of the San Francisco field (Ecopetrol-Hocol) in the Upper Magdalena Valley basin was a highlight of the 1985 exploratory activities.

Taken as a whole, these new discoveries increased reserves to 1100 million barrels, thus enabling the country to resume exports of crude oil in 1986. Current domestic production amounts to 180,000 BOPD and 521,700 million CFGPD.

PETROLEUM EXPLORATION AND DEVELOPMENT IN SUB-ANDEAN BASINS

Amazon Basin

The Colombian Amazon region is jungle terrain, and is one of the most difficult areas of access in the nation. Practically nothing has been done in the way of hydrocarbon exploration, except for some seismic profiles along the Putumayo River.

The Amazon basin within Colombia is an intracratonic basin considered to be a northward extension of the great Brazilian Amazon basin (Figure 3). It extends over an area of 180,000 km^2, and is limited by the Guiana Shield to the east, the Guaviare and Chiribiquete arches to the north and west, and by Brazil to the south. The basin contains Precambrian crystalline and metasedimentary rocks overlain by sedimentary rocks of early Paleozoic and late Tertiary age, intruded by pre-Tertiary nepheline syenite (Figures 4, 5). In Colombian territory the sedimentary sequence is thought to extend to a maximum depth of 3000 m.

During the 1970s, the Colombian government conducted a multidisciplinary technical-scientific survey of this sector of the Amazon region based on radar images (Amazon Radargrammetric Project). Results indicated that Cretaceous sequences, recognized as major generators of hydrocarbons in adjoining basins, are apparently not present within Colombian territory. However, the presence of Tertiary asphaltic sandstone strata along the Chiribiquete Range suggests that petroleum source rocks may be present in the subsurface farther south, outside Colombian territory. Although no wells have been drilled in the basin, it is still considered an area with important oil potential, possibly containing hydrocarbons which may have migrated from the south where the sedimentary sequence is more complete.

Putumayo Basin

The area underlain by the Putumayo basin is bordered on the west by the Andes, on the east by the Chiribiquete arch, on the north by the Macarena arch and the Guaviare arch, and on the south by Ecuador (Figures 3, 6). The basin is considered to be the northern extension of the eastern basin or Orient basin of Ecuador. The Colombian part extends over an area of 48,000 km^2. Production now averages 18,000 BOPD, ranging between 30° and 34° API. Actual recoverable reserves are estimated at 44 million barrels. This production comes from the Caballos and Villeta formations of Cretaceous age and three zones in conglomerate and sandstone of the Pepino Formation of late Eocene age (Figure 7). Current production comes mainly from the western part of the basin, an area characterized by folding and faulting (Figure 8).

For the past 3 years, Ecopetrol has carried out extensive seismic exploration in the Putumayo basin, as indicated in Table 1. Part of this work focused on the search for oil in previously discovered structures such as Temblón, Hormiga, and San Miguel, whereas other work focused on locating new sources of oil by following a possible continuation of structural-stratigraphic trap conditions revealed by discoveries in the Ecuadorian oil fields.

Reserves discovered in the Putumayo basin are estimated to be equivalent to 800.2 million barrels of oil. There are 14 producing fields. The Orito field, the first to be discovered, was found in 1963 by the Texas Petroleum Company and is now being operated exclusively by Ecopetrol. This field has been the largest producer to date. In an effort to accelerate exploration in the Putumayo basin, Ecopetrol recently offered the oil industry a series of new association contracts covering 32 blocks of 150,000 hectares each, as shown in Figure 9.

Llanos Orientales Basin

The Llanos Orientales basin is a pericratonic basin extending over an area of 190,000 km^2. It is bordered by the Guaviare arch to the south, the Guiana Shield to the east, and the Eastern Cordillera to the west, and extends into Venezuela beyond the Arauca River to the north (Figure 3).

Hydrocarbon production in the basin comes from Upper Cretaceous sandstones in the Guadalupe Formation and lower Tertiary sandstones in the Barco, Mirador, and Carbonera formations. The sedimentary cover includes lower Paleozoic, Upper Cretaceous, and Tertiary sequences (Figure 10).

Structurally, the Llanos Orientales basin may be divided into three parts (Figure 11): (1) a western part characterized by reverse faulting and, along the foothills of the Eastern Cordillera, complex north-trending asymmetrical folds; (2) a central part, or "sub-Andean Trough," having north to northeast-trending antithetic faults of limited displacements, which form structural closures containing hydrocarbon deposits; and (3) an eastern part, which consists of a platform where the sedimentary rocks are less deformed.

For purposes of exploration and development, the Llanos Orientales basin has been divided into three sectors—Meta, Casanare, and Arauca. Except for the Castilla field, all exploratory and production activity in the Meta sector is carried out directly by Ecopetrol. This region is bordered by the Macarena arch and the Vaupés height to the south, by the Eastern Cordillera to the west, the Meta River to the north, and the Guiana Shield to the east. Table 2 gives a summary of seismic exploration in this sector, and oil discoveries are shown in Table 3.

Hydrocarbon accumulations in the Meta sector have been found by Ecopetrol in Upper Cretaceous and lower Tertiary rocks at the Apiay, Guayuriba, Suria, La Reforma, and La Libertad fields (Table 3). The Apiay field consists of a trap

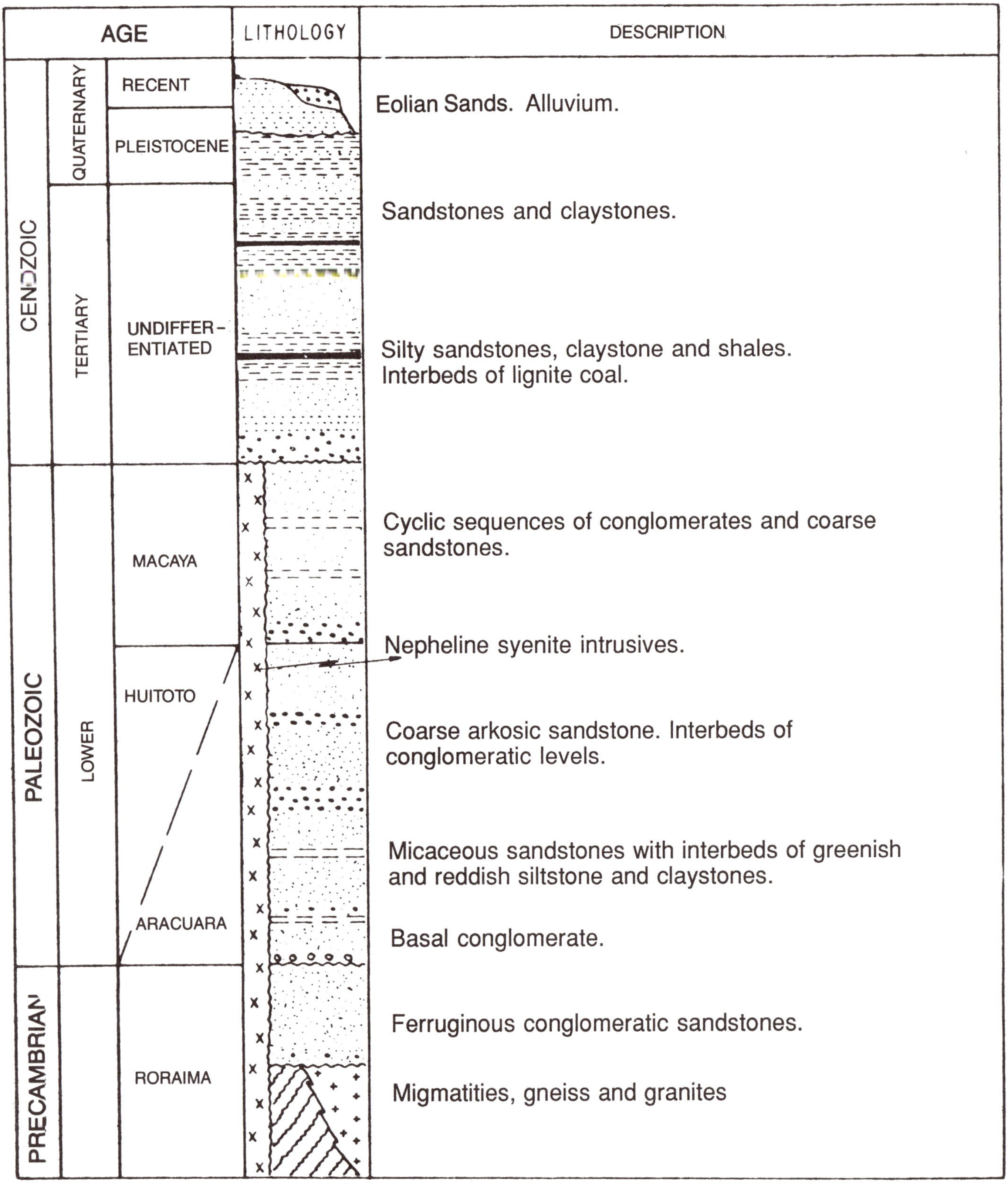

Figure 4. Generalized stratigraphic column, Amazon basin.

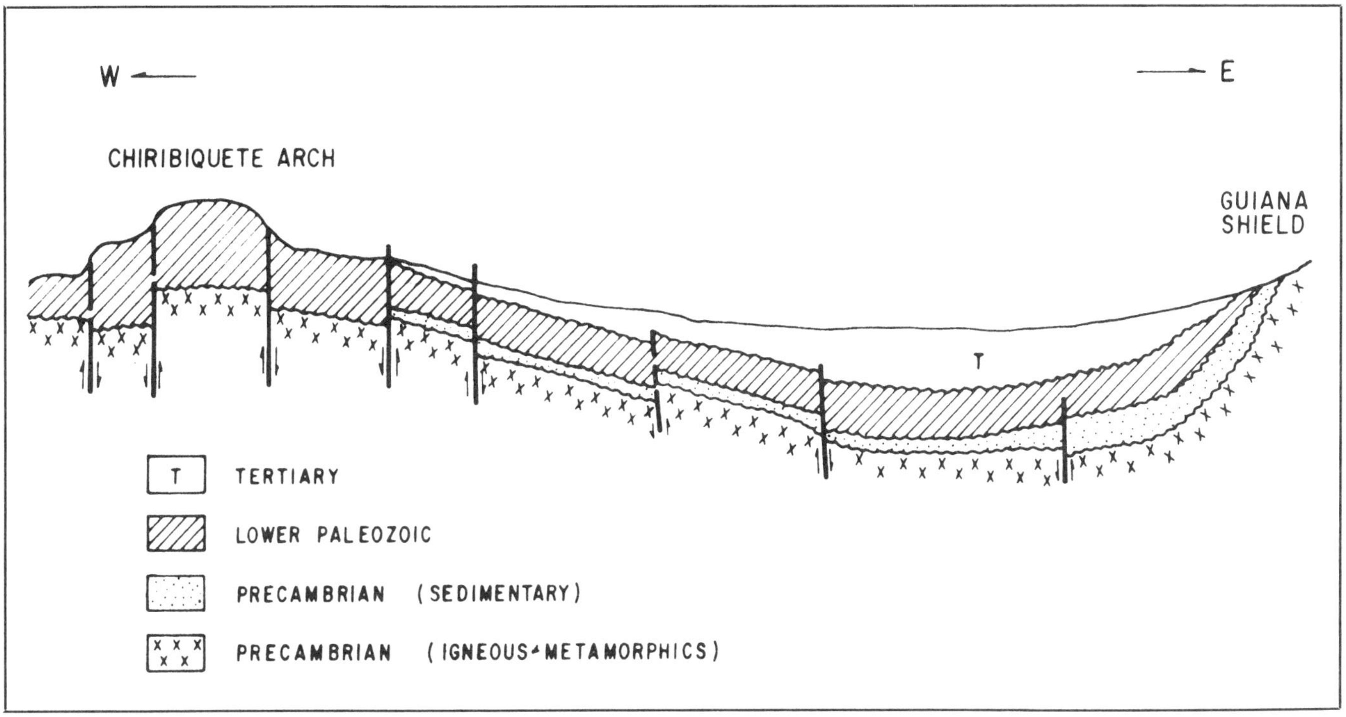

Figure 5. Schematic geological cross section of the Amazon basin.

formed by a reverse fault having an upthrown block to the west (Figure 12), and other fields are characterized by similar, though smaller, structural closures. The geological conditions indicate a depocenter to the west and a continuous wedging of sediments toward the east. Ecopetrol plans to carry out additional drilling and conduct new seismic surveys in the near future, which could lead to additional discoveries (Figure 13).

Current production (1985) of approximately 7000 BOPD at Apiay and 10,000 BOPD at Castilla is transported by tanker trucks to an oil pipeline, where it is pumped across the Magdalena River to the Barrancabermeja refinery. Ecopetrol plans to construct a new pipeline to channel this oil into the national pipeline network, which would decrease transportation costs.

The Casanare sector, in the center of the Llanos Orientales basin, is bordered by the Meta and Casanare rivers to the south-southeast and north, and by the Eastern Cordillera to the west. Most of the exploratory work in this area has been done under the Ecopetrol-Aquitaine association contract, with important direct participation by Ecopetrol. Results of this activity are summarized in Table 4. Trapping is due to structural closures against antithetic faults having upraised blocks on the west (Figure 14). In addition, other companies, including Triton, Union Texas, Union California, Lasmo, Texaco, Leed, Intercol, Alejandra Consultants, Texican, and Creston, have recently begun exploration under association contracts in this sector.

The Arauca sector, bordered by the Casanare and Arauca rivers in the northernmost part of the basin, is the site of the most important discoveries made in Colombia since the giant La Cira-Infantas field was found in 1918. These discoveries were made in the Cravo Norte area, where exploration was carried out under the Ecopetrol-Occidental association contract, and include the fields at Caño Limón, La Yuca, Matanegra, Redondo, and Caño Verde. Corresponding reserves are noted in Table 5, along with those of the Arauca field discovered by Intercol near the Eastern Cordillera to the west. Oil was first discovered at Cravo Norte in 1983 by the No. 1 well at Caño Limón, followed by Caño Limón No. 2. These wells confirmed the existence of 60 million barrels in recoverable reserves. Additional wells drilled at La Yuca, Matanegra, and Redondo (Figure 15) increased reserves to more than 600 million barrels of recoverable crude oil, and continued exploration in the area could eventually increase reserves to more than 1000 million barrels. Oil accumulation is controlled by the Arauca arch, which trends northwestward in the subsurface.

Table 6 gives a summary of drilling in the Llanos Orientales basin since the first well was drilled in 1944.

EXPLORATION PROGRAM AND PROSPECTS

Prospects for Oil in Colombia

Thanks to a comprehensive exploration program on the part of the Colombian government, important discoveries have been made in the Llanos Orientales and Putumayo regions, and the country now ranks as a potential exporter of crude oil in the years to come. Favorable source rocks and structural conditions in the Llanos Orientales and Putumayo basins remain to be explored, and it seems likely that significant new discoveries will be made in this region. Further studies may reveal an oil potential in the Amazon basin to the south, thus making the sub-Andean region of Colombia one of South America's major oil regions.

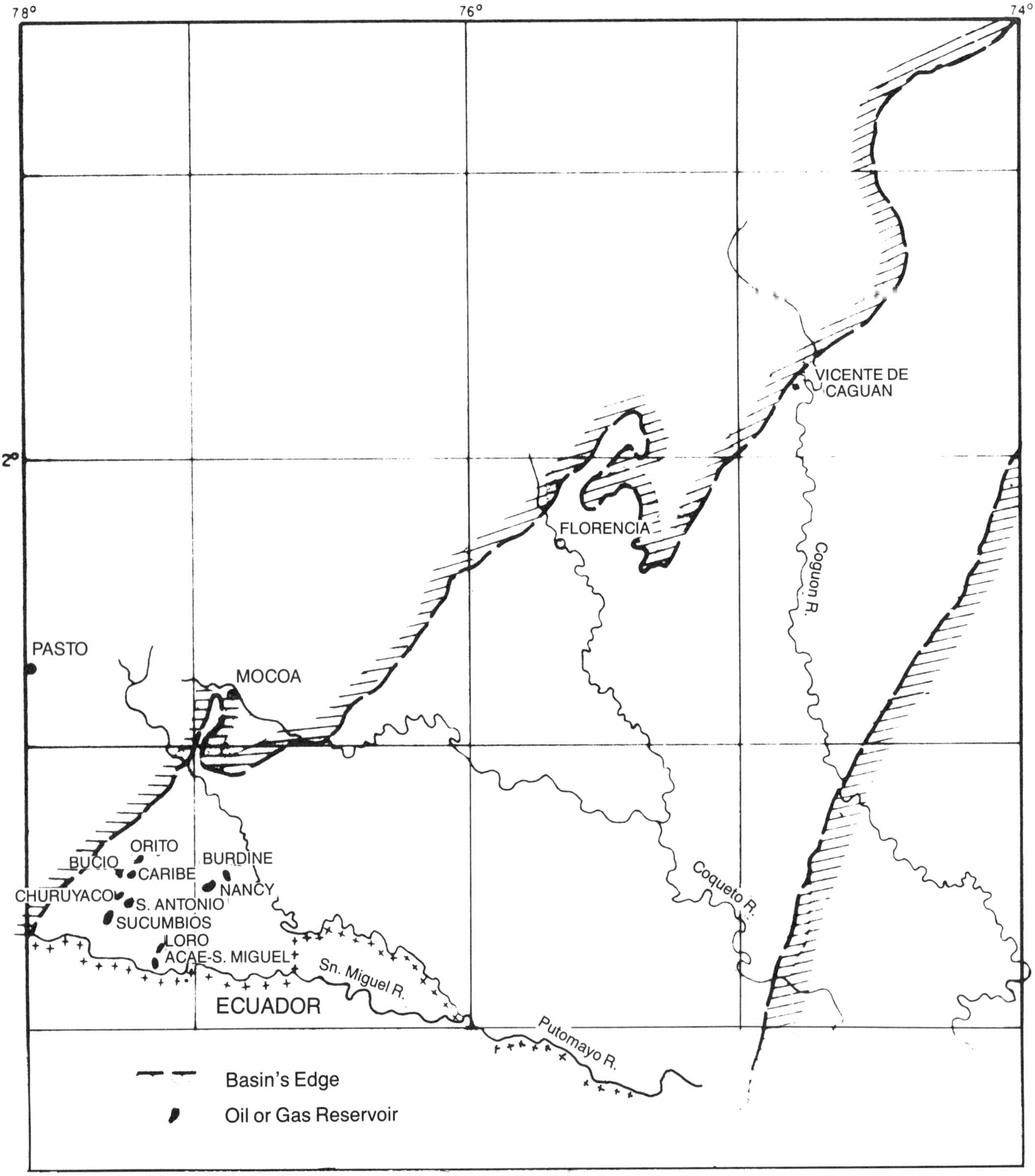

Figure 6. Map showing limits and oil fields of the Putumayo basin.

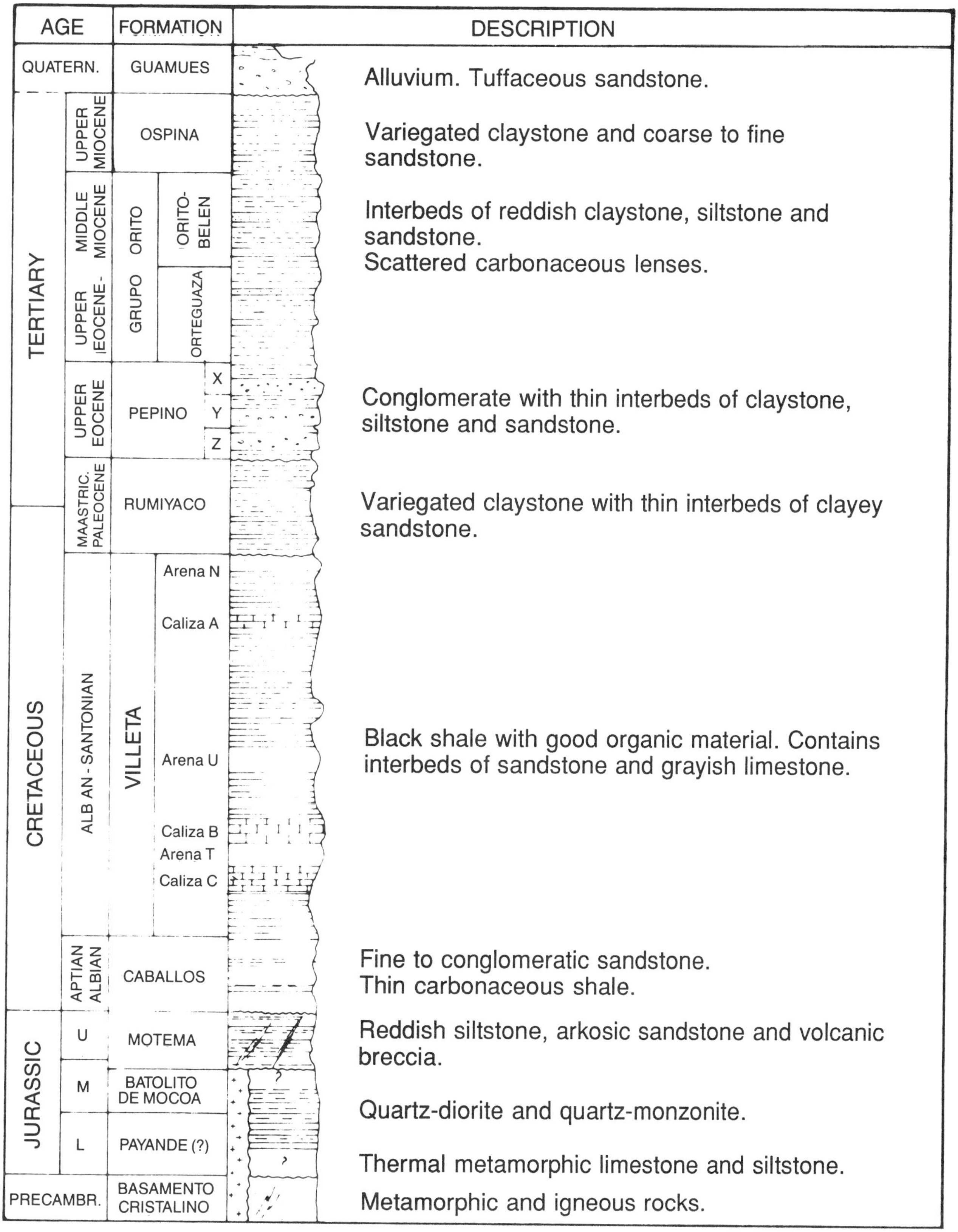

Figure 7. Generalized stratigraphic column for the Putumayo basin.

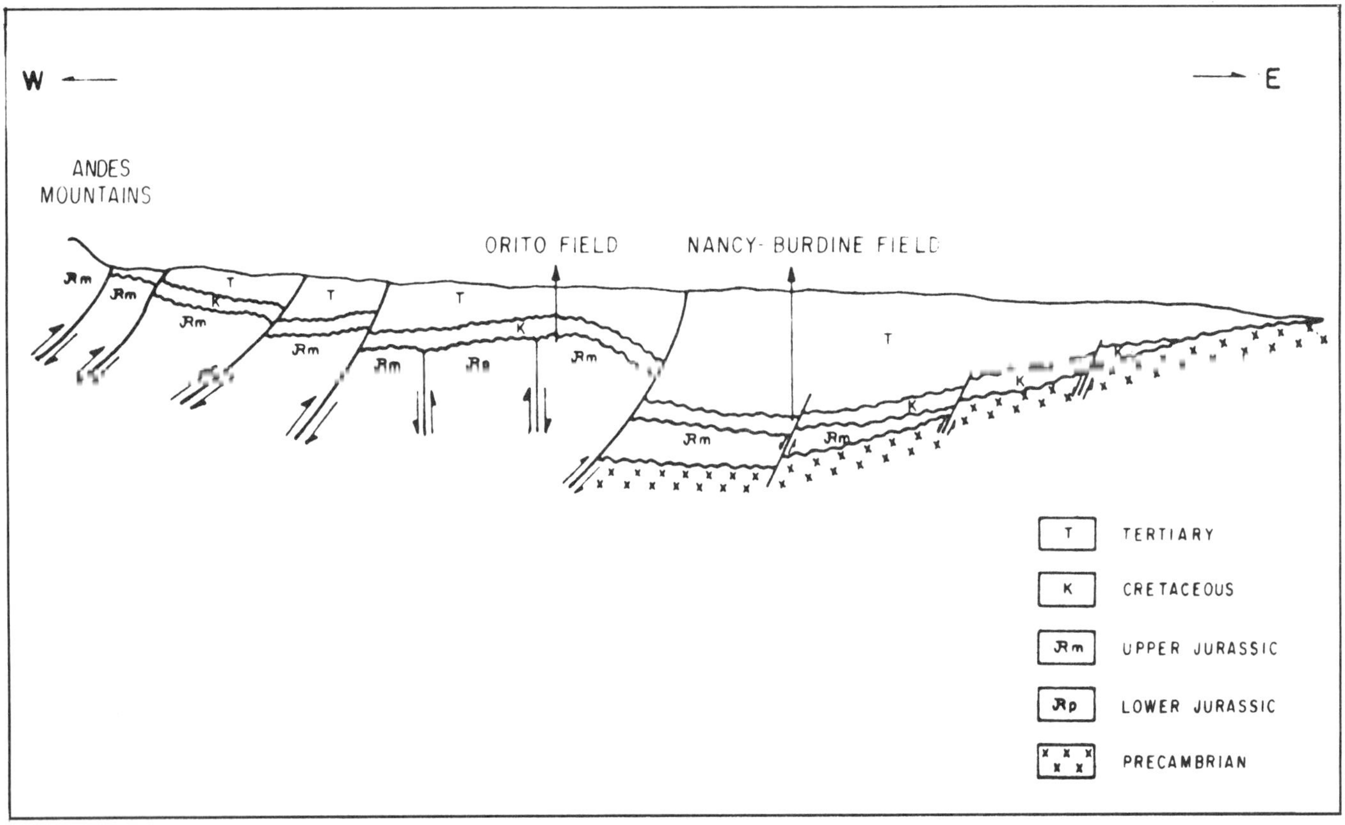

Figure 8. Schematic geological section of the Putumayo basin.

Association Contracts

Colombia has achieved excellent results in the area of hydrocarbon production, thanks to continued and dedicated exploratory efforts carried out directly by Ecopetrol or through the system of association contracts. An association contract is initially signed for a 6-year period, during which time the associate company explores a particular area at its own cost and risk until determining the possible existence of a commercial field. At that moment, Ecopetrol begins to assume responsibility for 50% of all development costs and receives half the production after 10% has been deducted as royalties paid to the various provinces and municipalities. Total duration of the agreement is 28 years. This type of arrangement proved so successful that by September 1985, a total of 187 association contracts had been signed since the system was initiated. Seventy-nine of these are still in effect and cover a surface area of nearly 9 million hectares. Eleven contracts are now in the exploitation phase.

In addition to Ecopetrol, 39 companies have conducted associate operations in the vast Llanos Orientales basin. This contract system was first used in 1971, when nine blocks or sectors were selected in the western part of the basin. Ecopetrol reserved one for its own operations and offered the other eight in a type of international bidding. Later, association contracts became more and more common, increasing to a maximum of 16 in 1975, when the Unete No. 1 well drilled by Ecopetrol revealed the presence of excellent-quality light crude in sandstones of the Mirador Formation. In 1979, with the first commercial fields discovered through association contracts, exploration in the Llanos Orientales increased significantly and eventually was responsible for the discoveries noted previously. These combined discoveries raised national oil reserves to more than 1200 million barrels and place Colombia on the threshold of petroleum self-sufficiency as well as crude oil exportation.

ACKNOWLEDGMENTS

The writer wishes to express his appreciation to Empresa Colombiana de Petroleos geologists Victor Eduardo Pérez

Table 1. Summary of seismic surveys in the Putumayo basin.

Year	Program	Km Registered	Investment (US$)
1981	Putumayo	169	1,048,000
1983	Putumayo	300	715,000
1984	Putumayo	217	543,000

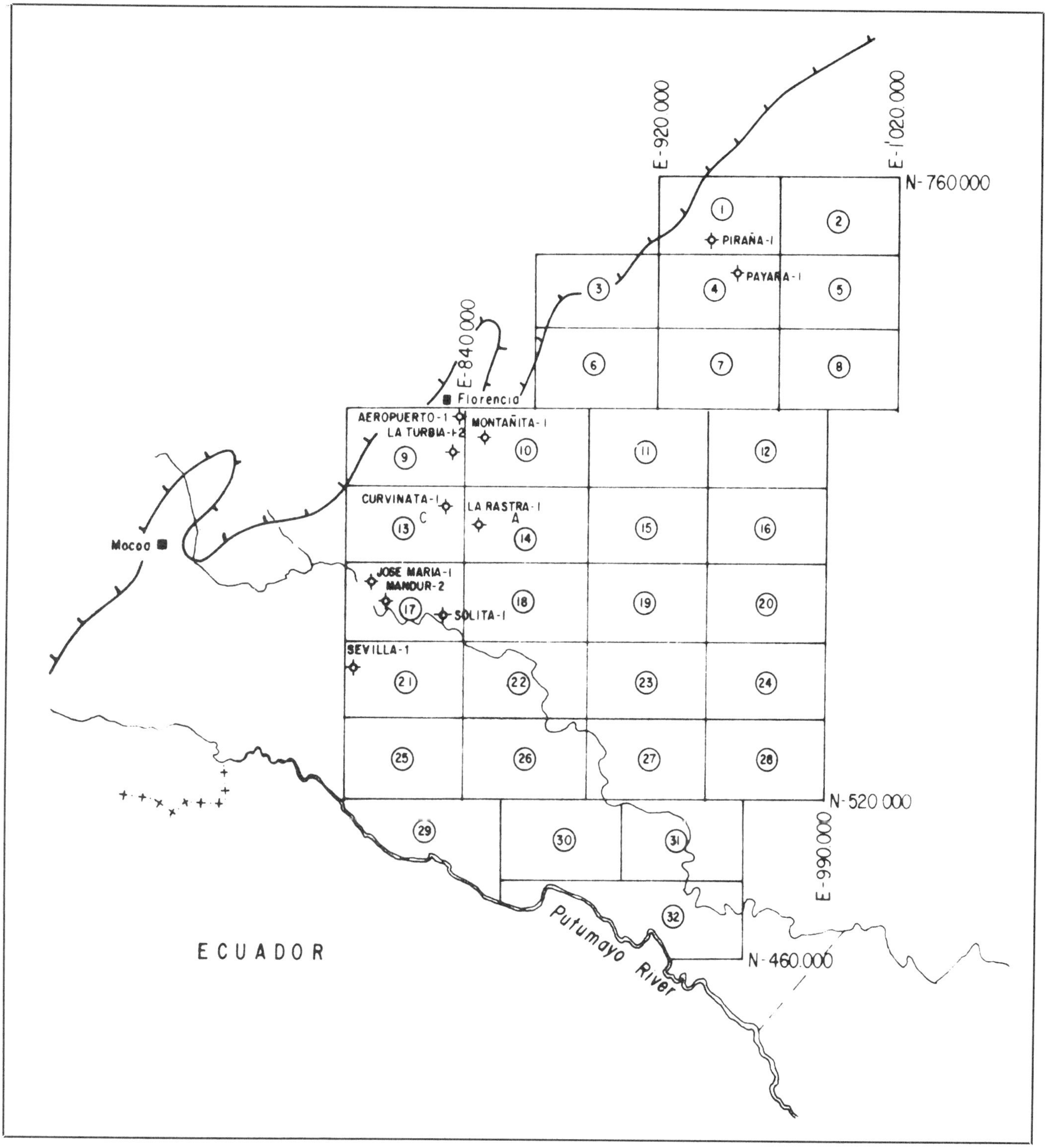

Figure 9. Blocks for proposed association contracts in the Putumayo basin.

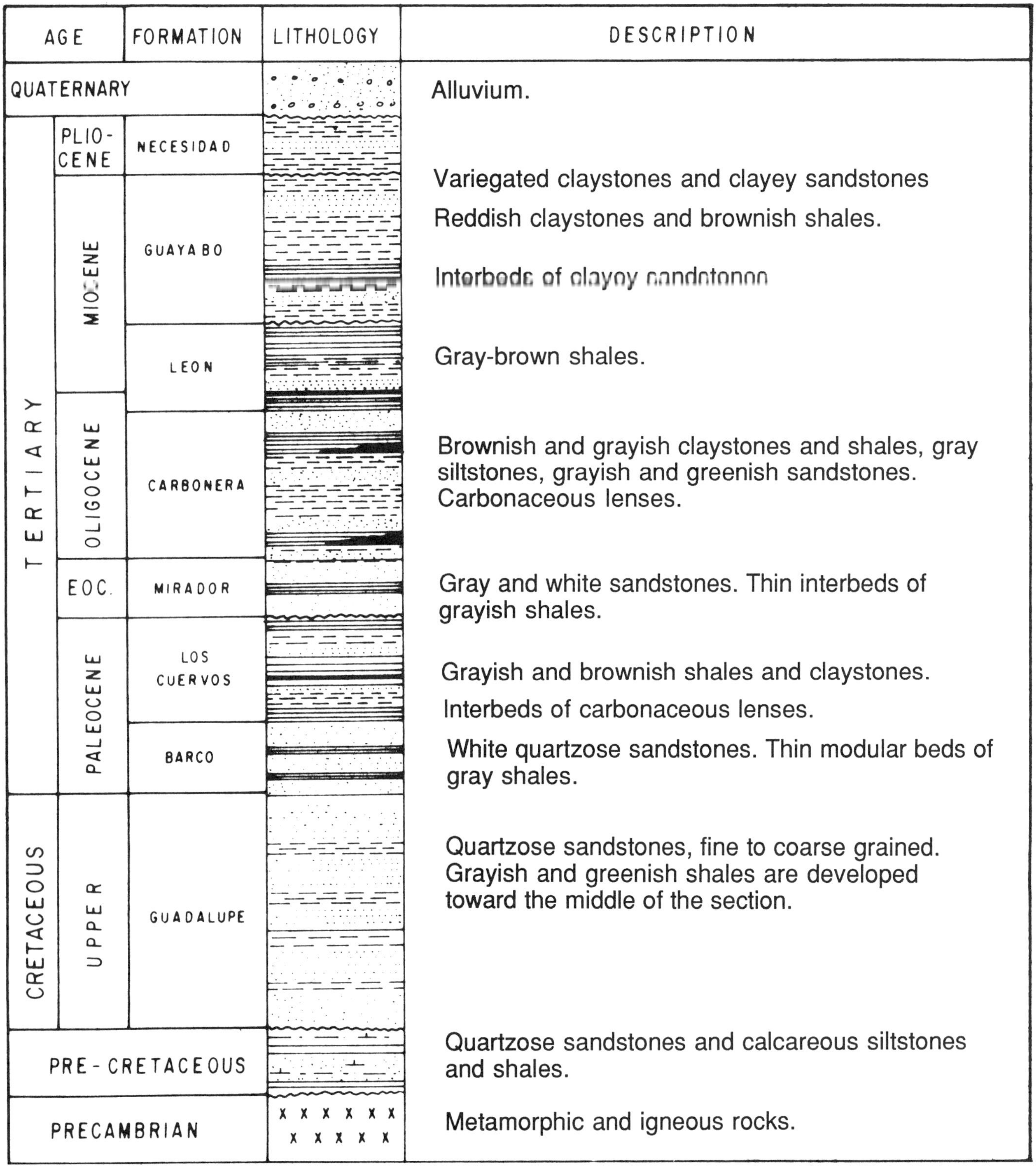

Figure 10. Generalized stratigraphic column for the Llanos Orientales basin.

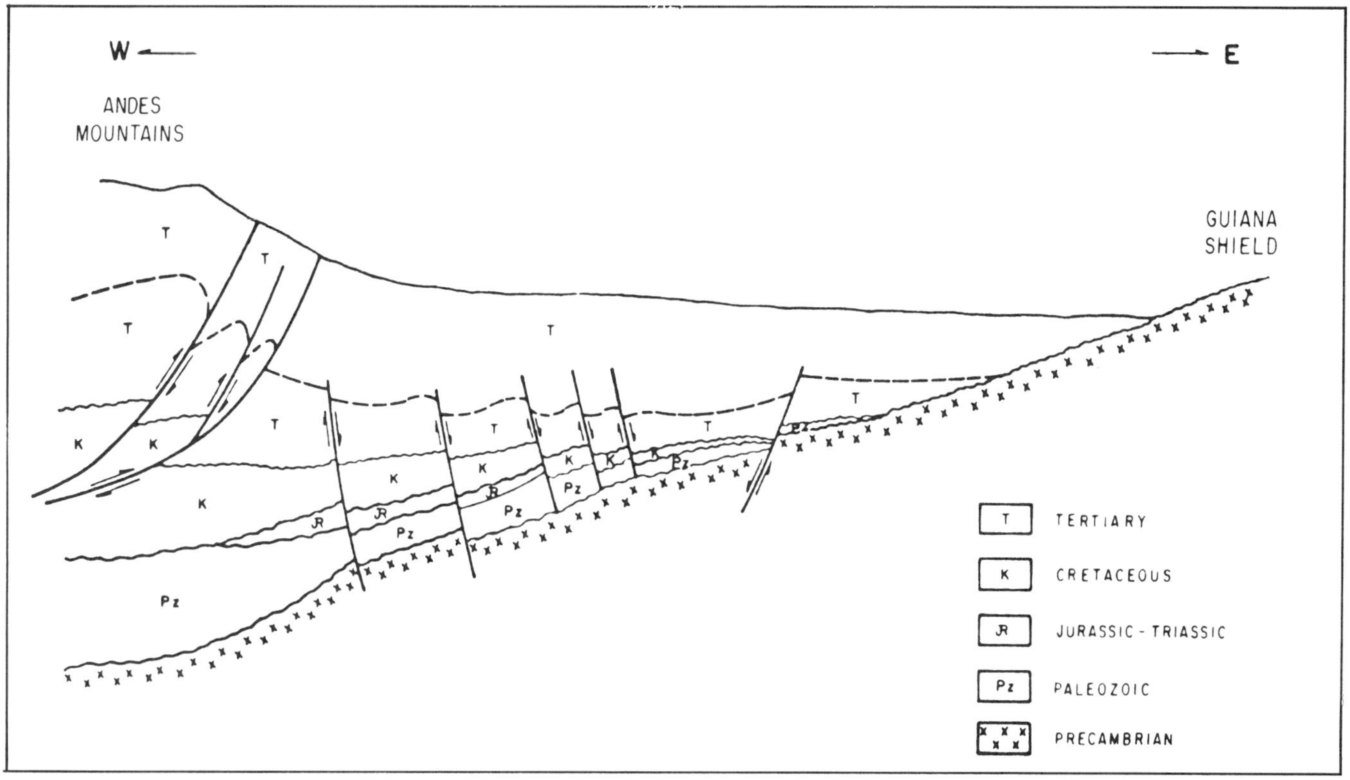

Figure 11. Schematic geological section of the Llanos Orientales basin.

Table 2. Summary of seismic surveys in the Meta sector of the Llanos Orientales basin.

Year	Program	Km Registered	Investment (US$)
1980	Quenane	458	1,250,000
	Chaviva	50	150,000
1981	Quenane	763	2,403,000
	Chaviva	929	2,150,000
1982	Quenane-Chaviva	811	2,890,000
1983	Apiay	189	727,000
	Humadea	119	340,000
1984	Apiay	16	725,000
	Villavicencio	118	341,000
1985	Pachaquiaro	1,000	±3,000,000

Table 3. Oil fields in the Meta sector of the Llanos Orientales basin.

Date of			Reserves	
Discovery	Field	Company	MM Barrels*	MM CFGPD*
1969	Castilla	ECP-Chevron	90.0	
1981	Rubiales	Ecopetrol	22.0	
1981	Apiay	Ecopetrol	30.0	38.000
1984	Guayuriba	Ecopetrol	2.6	13.100
1984	Suria	Ecopetrol	22.0	9.000
1985	La Reforma	Ecopetrol	Under Evaluation	
1985	La Libertad	Ecopetrol	Under Evaluation	

*MM = 10^6 bbl.

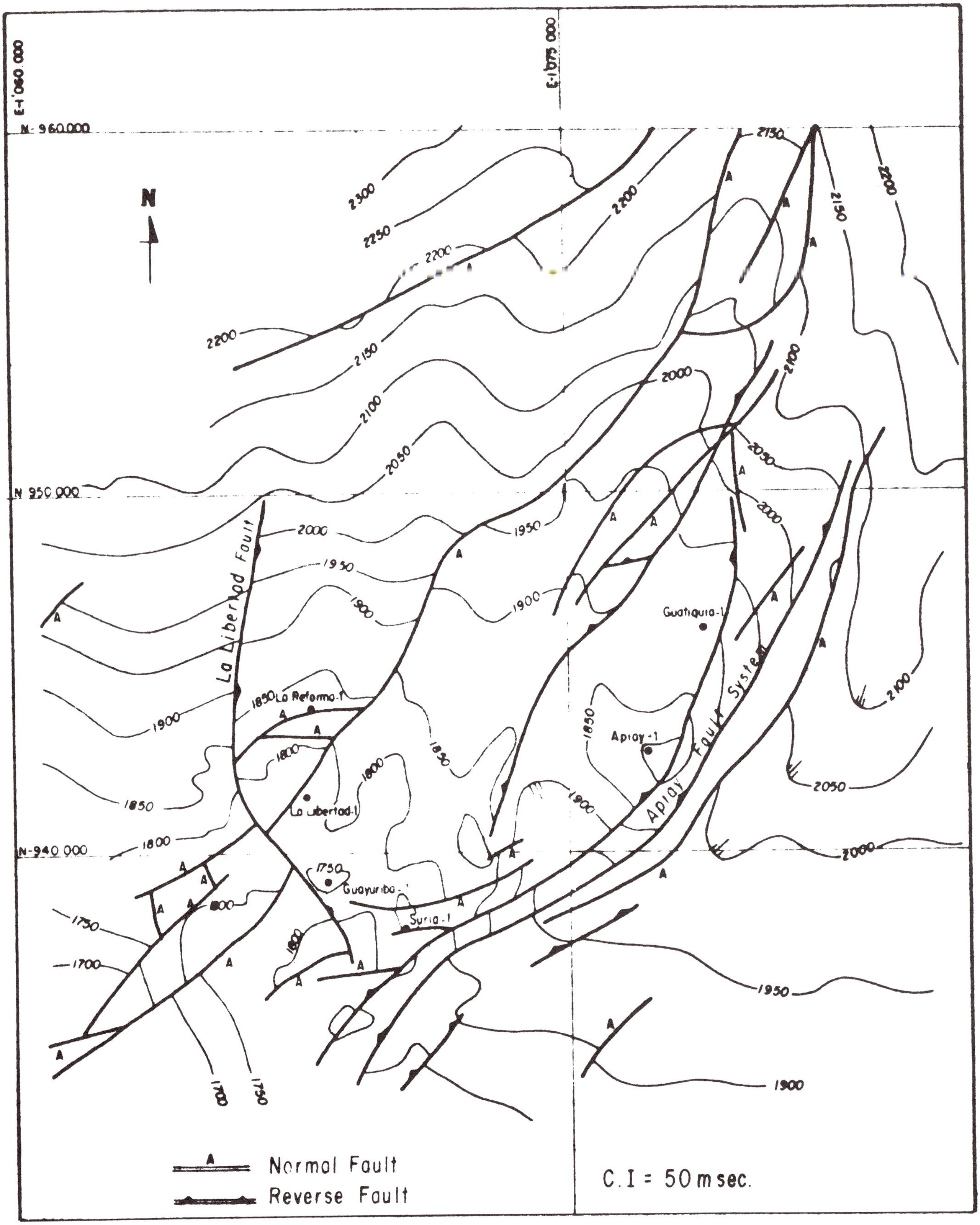

Figure 12. Structure contour map showing top of Cretaceous rocks in Apiay area, Meta sector in Llanos Orientales basin.

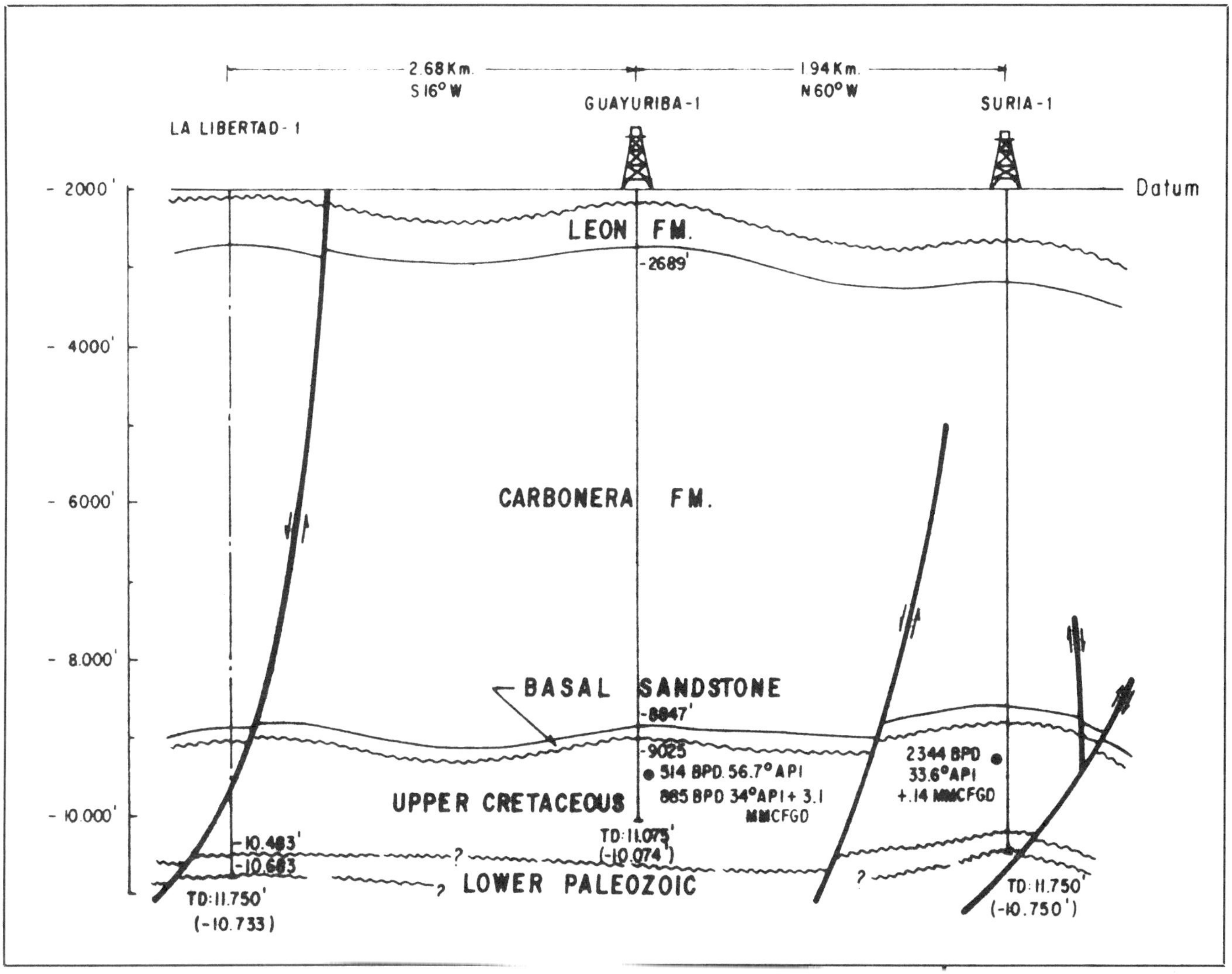

Figure 13. Structural cross section B C D (Figure 12) in Apiay area, Meta sector.

Table 4. Oil fields in the Casanare sector of the Llanos Orientales basin.

Date of Discovery	Field	Company	Reserves MM Barrels*	Reserves MM CFGPD*
1974	Trinidad	ECP-Aquitaine	10.0	
1975	Santiago	ECP-Aquitaine		
1979	Caño Garza	ECP-Aquitaine	14.0	
1980	Tocaria	ECP-Aquitaine	14.0	
1981	El Miedo	ECP-Occidental		
1982	Barquereña	ECP-Aquitaine	4.5	
1982	Sardinas	ECP-Aquitaine	5.0	
1982	Cravo Sur	ECP-Aquitaine	7.3	
1984	La Gloria N	ECP-Aquitaine	6.6	
1984	Entrerrios	Ecopetrol	12.0	
1984	Caño Garza N	ECP-Aquitaine	3.6	
1984	Rancho Hermoso	Ecopetrol	10.5	
1985	Jordan	ECP-Aquitaine	Under Evaluation	
1985	La Flora	ECP-Aquitaine	Under Evaluation	
1985	Caño Duya	ECP-Occidental	Under Evaluation	
1985	Corocora	ECP-Occidental	Under Evaluation	

*MM = 10^6 bbl.

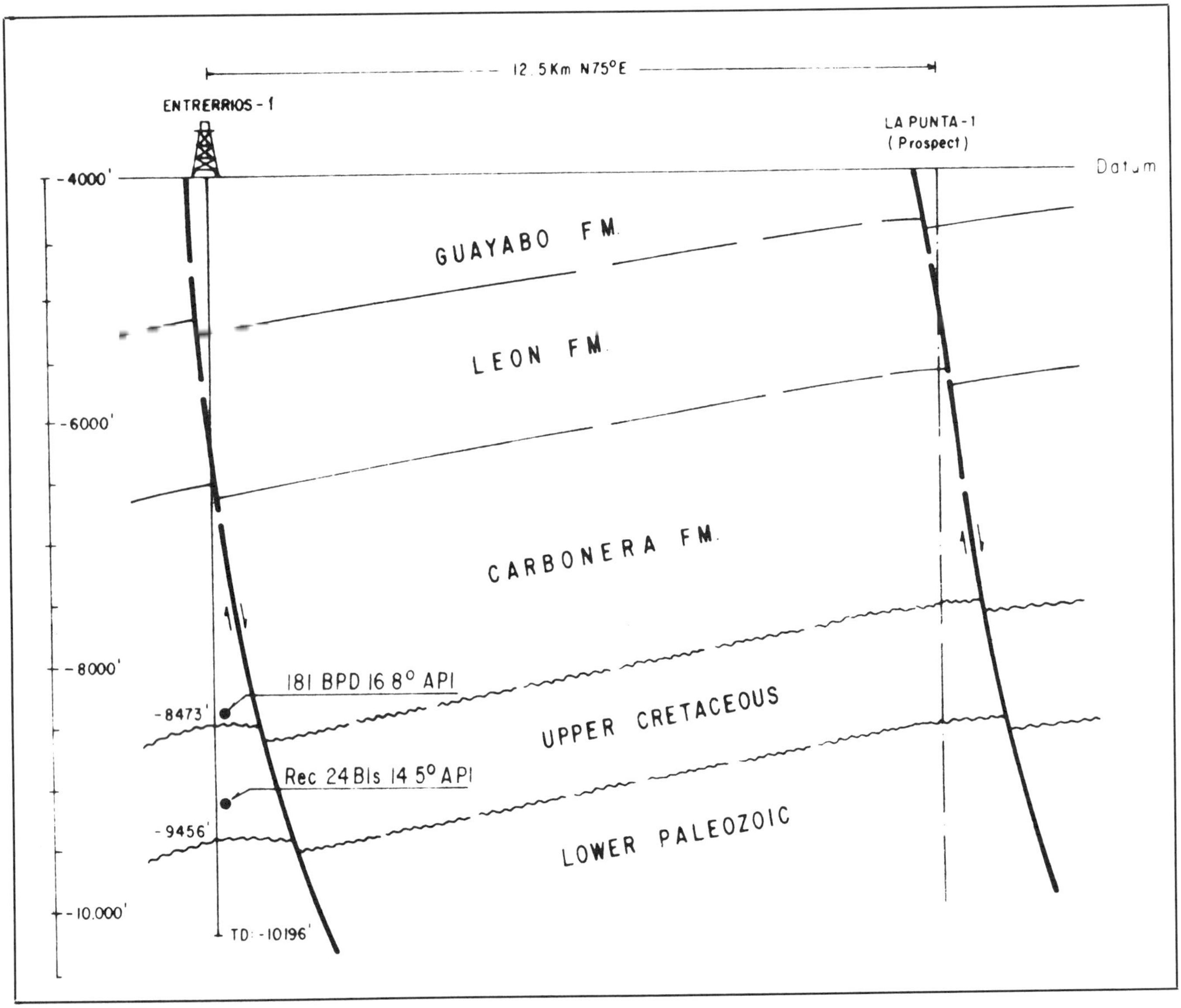

Figure 14. Structural cross section for the Casanare sector, Llanos Orientales basin.

Table 5. Oil fields in the Arauca area of the Llanos Orientales basin.

Date of Discovery	Field	Company	Reserves MM Barrels*	Reserves MM CFGPD*
1980	Arauca	ECP-Intercol	3.0	
1983	Caño Limón	ECP-Occidental	55.0	
1984	La Yuca	ECP-Occidental	299.0	
1984	Matanegra	ECP-Occidental	130.0	
1985	Redondo	ECP-Occidental	100.0	
1985	Caño Verde	ECP-Occidental	Under Evaluation	

*MM = 10^6 bbl.

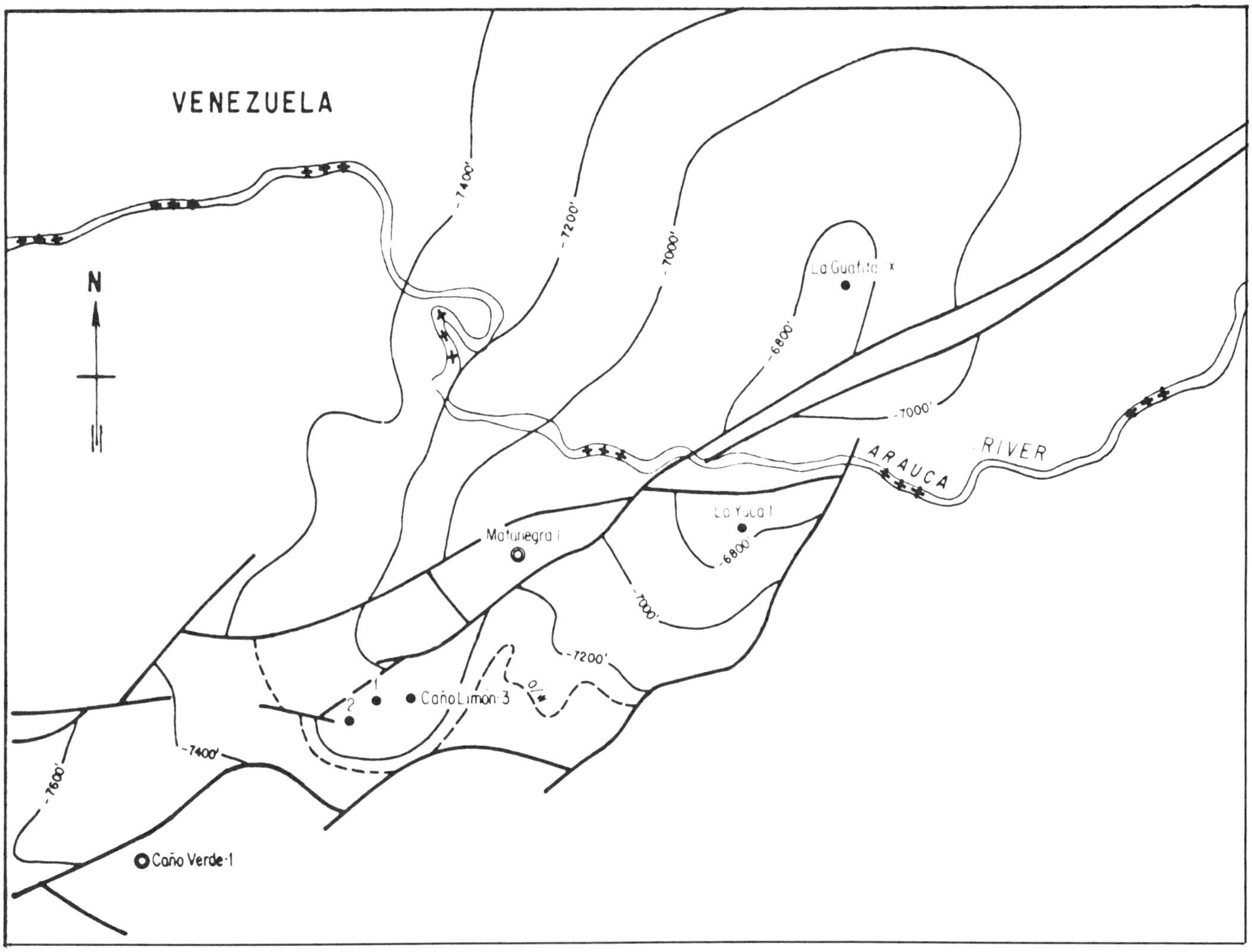

Figure 15. Structure contour map of the top of the Mirador Formation, Caño Limón area, Llanos Orientales basin, Cravo Norte sector.

Table 6. History of drilling in the Llanos Orientales basin.

Period (Years)	Number of Wells Drilled	Lahee Classification A-3	A-2	A-1	Number of Producing Wells	Number of Discovery Wells
1944-1953	9	9	—	—	—	—
1954-1962	5	5	—	—	—	—
1963-1980	35	31	0	4	10	6
1981-1984	71	56	3	12	27	15
1985-(May)	14	13	—	1	8	8

Total Fields Discovered (1981-1985) = 23.
Total A-3 Wells Drilled (1981-1985) = 69.

Herrera and Ernesto Dueñas Jiménez, whose enthusiastic and dedicated cooperation has been essential in the preparation of this report.

REFERENCES

Bolivar, A., and A. Samper, 1983, Interpretación sismica de los programas Quenane y Apiay: Bogotá, Empresa Colombiana de Petroleos Geophysical Report 733, p. 1-42.

Bueno, R., 1985, Presente y futuro de la exploración petrolera en Colombia: Bogotá, II Simposio Bolivariano, Asociación Colombiana de Geólogos y Geofísicos del Petróleo Memoir, p. 1-42.

Caceres, H., and P. Teatin, 1985, Cuenca del Putumayo, provincia petrolífera meridional de Colombia: Bogotá, II Simposio Bolivariano, Asociación Colombiana de Geólogos y Geofísicos del Petróleo Memoir, p. 1-80.

Carvajal, A., 1986, Colombia petroleum seminar, Houston: Empresa Colombiana de Petroleos Geological Report 1266, p. 1-94.

Castro, E., and Y. Puerto, 1985, Desarrollo del Campo Apiay: Bogotá, II Simposio Bolivariano, Asociación Colombiana de Geólogos y Geofísicos del Petróleo Memoir, p. 1-119.

Empresa Colombiana de Petroleos, 1983, Estadísticas de la industria petrolera Colombiana (6th ed.): Bogotá, Empresa Colombiana de Petroleos Geological Report 1103, p. 1-39.

Empresa Colombiana de Petroleos and Occidental, 1986, Joint original oil in place study, Caño Limón field: Bogotá, Empresa Colombiana de Petroleos Geological Report 1281, p. 1-46.

Galvis, J., A. Huguet, and P. Ruge, 1979, Geología de la Amazonía Colombiana: Botogá, Instituto Nacional de Investigaciones Geologico-Mineras Geological Report 1792, p. 1-86.

Govea, C., and H. Aguilera, 1981, Geología de la cuenca del Putumayo: Botogá, Empresa Colombiana de Petroleos Geological Report 789, p. 1-93.

Numpaque, E., 1985, Revisión geológica del bloque Río Ele, Cuenca Llanos Orientales: Bogotá, Empresa Colombiana de Petroleos Geological Report 1236, p. 1-79.

Peron, E., 1985, Exploración petrolifera del precretáceo en Colombia: Qrito, Ecuador, LV Reunión a nivel de expertos Asistencia Recíproca Petrolera Estatal Latinoamericana, Memoir, p. 1-40.

Perez, E., and A. Bolivar, 1985, Exploración petrolera en la subcuenca Apiay-Ariari, Cuenca Llanos Orientales: Bogotá, II Simposio Bolivariano, Asociación Colombiana de Geólogos y Geofísicos del Petróleo Memoir, p. 1-122.

Robertson Research, Inc., 1983, The northern llanos of Colombia hydrocarbon potential and stratigraphic control: Empresa Colombiana de Petroleos Geological Report 1112, p. 1-246.

Ericksen, G. E., Cañas Pinochet, M. T., and Reinemund, J. A., editors, 1989, Geology of the Andes and its relation to hydrocarbon and mineral resources: Houston, Texas, Circum-Pacific Council for Energy and Mineral Resources Earth Science Series, v. 11.

CHAPTER 27

EXPLORATION AND GEOLOGIC FRAMEWORK OF THE CAÑO LIMON OIL FIELD, LLANOS ORIENTALES DE COLOMBIA

Victor H. Gabela
Occidental of Colombia, Inc.
Bogotá, Colombia

ABSTRACT

The discovery of the Caño Limón oil field in July 1983 was the result of 3 years of intensive exploratory activities carried out by Occidental of Colombia in the Llanos Orientales basin, where only minor exploratory success was accomplished during the 40 preceding years. Three structural closures, which constitute the oil field, were defined by conventional seismic surveys. Their origin seems to be related to the lateral movement of a northeast-trending strike-slip fault, called the Caño Limón fault, and associated convergent faults which subdivide the field. The sedimentary sequence ranges in age from pre-Cretaceous to Recent, and includes the fluvial-deltaic sands of Cretaceous, Eocene, and Oligocene age that are the oil-bearing reservoirs. The most important oil reservoir rocks are the deltaic sands of the Mirador Formation (Eocene), which show excellent petrophysical characteristics and contain 80% of the field reserves.

INTRODUCTION

The discovery of the Caño Limón field represents the culmination of 3 years of intensive exploration by Occidental of Colombia in the Llanos basin, where only minor exploration success had been achieved in the 40 preceding years. The field was discovered on July 13, 1983, when the Caño Limón-1 well produced on test 516 M^3/D (3245 B/D) of 30 °API gravity oil from the lower part of the Mirador Formation through a 32/64-in. choke, and 1184 M^3/D (7445 B/D) of 33 °API gravity oil from the upper part of the Mirador Formation through a 1⁷/₈-in. choke, for a total initial combined production of 1699 M^3/D (10,690 B/D). Subsequent to the drilling of these wells and up to the date of this work (July 1985), a total of 11 productive wells and 1 dry hole were drilled. Production rates have averaged 1272 M^3/D (8000 BOPD) per well. Oil production from the Caño Limón field has been limited to small volumes resulting from the production tests of the wells. The field began commercial production in early 1986.

Geographic Location

The Caño Limón field (Figure 1) is located in the Cravo Norte contract, which is one of seven exploration blocks currently operated by Occidental of Colombia in the Llanos region. This is a 260,000-hectare block on the extreme north end of a vast area of low grassy lands of eastern Colombia that extends from the Macarena uplift on the south to the Arauca River, the international boundary between Colombia and Venezuela, on the north. The Caño Limón field is approximately 1/2 km south of the Arauca River and 50 km west of the city of Arauca (Figure 1).

The Llanos region has a tropical climate characterized by a high level of rainfall (1000-2000 mm/yr) during 8 months of the year. This keeps areas such as the northern Llanos almost permanently flooded, making access to the region by land difficult during most of the year. The Llanos summer is a short dry season from mid-December to mid-April.

Geologic Setting

The northern Llanos area is a gently westward-dipping basin flank whose structural deformation consists in general of northeast-trending normal faults. Seismic data indicate that these faults originated during the Paleozoic and have undergone reactivation in the Tertiary. In addition, the Caño Limón area is affected by significant early Tertiary strike-slip faulting oriented in a northeast direction. The origin of these Tertiary faults has not yet been determined, but they could be related to the growth of a subtle basement high, locally known as the Arauca arch, that separates the northern Llanos basin from the Barinas-Apure basin of Venezuela (Figure 2).

Figure 1. Index map showing locations of oil fields and Occidental of Colombia contract blocks.

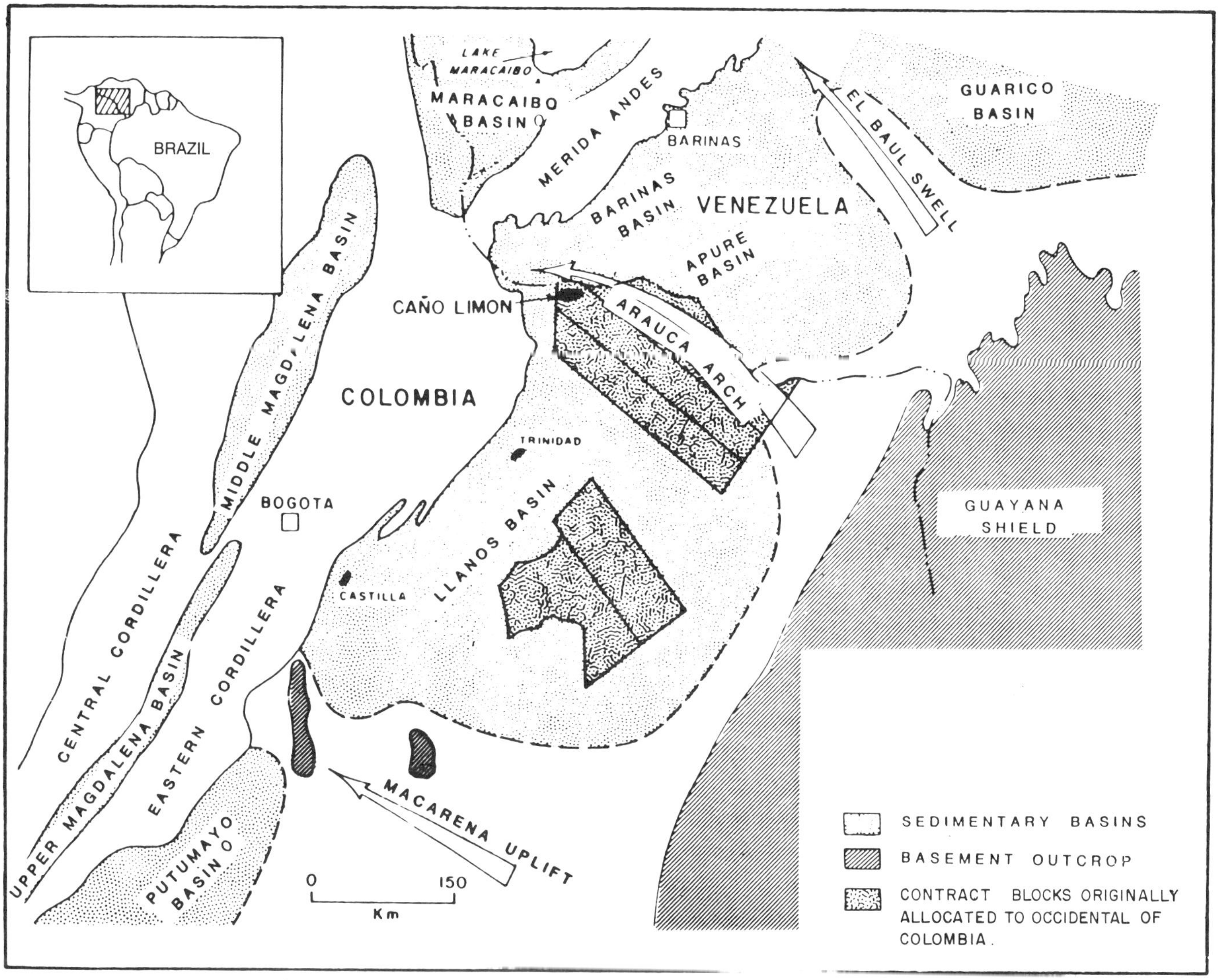

Figure 2. Index map showing sedimentary basins and uplifts in northeastern Colombia and southwestern Venezuela.

The Llanos basin is characterized by a relatively stable tectonic history which favored cyclical sedimentation of predominantly epicontinental facies from Paleozoic to Eocene, followed by transgressive, marine sedimentation during the Oligocene, and by torrential deposition from the Eastern Cordillera during the Miocene (Valderrama, 1982). The regional geology of the Llanos basin has been discussed by Miller and Etayo-Serna (1972), Valderrama (1982), Robertson Research (1983), and McCullough and Padfield (1985), among others, and will not be discussed here.

EXPLORATION HISTORY

Occidental of Colombia began exploration in the Llanos in July 1980 by signing Association Contracts with Ecopetrol over five exploration blocks covering 1 million hectares each. The exploration of the Llanos by Occidental was based on an interpretation of the regional geologic framework established by its successful exploratory efforts in the sub-Andean basins of Perú and Bolivia. Hence, the choice of exploration acreage in the Llanos was influenced by regional geologic concepts developed during the search for oil in those basins. The sub-Andean basins, which are eastward remnants of a Cretaceous miogeosyncline that extended from Venezuela to northern Argentina, share with the Llanos a common genesis and geologic evolution.

Play Concepts

Taylor (1984) summarized some of these concepts as follows:

> The timing of structural development was an important consideration in the choice of acreage because the very young, prominent foothill structures resulting from the Andean orogeny during the Miocene would have missed the main pulse of oil migration, which probably occurred in the Eocene to Oligocene. Therefore, it was expected that these young structures would not be hydrocarbon bearing. This had been found to be true in prominent foothill structures of other sub-Andean basins.

Favorable Cretaceous and Tertiary reservoirs could be anticipated because the Brazilian shield as a source area continued to be prominent during these times. However, uncertainty as to whether shale units were present to seal the reservoirs remained a geological risk, because the paleogeography of the basin and its transgressive-regressive cycles were not well understood.

Older structures away from the foothills would be exploration targets because they were likely to be in place at the time of oil migration. However, they were also likely to be subtle features, because of the relatively stable pre-Andean tectonic history of the basin, and therefore would require extensive and detailed seismic investigation.

Among the old structural features to be explored were the cross-basinal arches such as the Arauca arch that would be of particular importance because they have proved, in other basins, to be foci for hydrocarbon migration.

These concepts were supplemented by preliminary source-rock studies, which indicated that upper to middle Cretaceous shale units buried in the deep part of the basin could have been source rocks for significant amounts of oil and also that the Paleozoic rocks were potential source rocks. If permeable sands were present across the basin, the hydrocarbons originated in these rocks probably migrated toward the periphery of the basin, and filled the pre-existing structures. Thus, with more insight than solid data, and with the general guiding criteria of optimizing the exploration opportunities by acquiring acreage extending from a basin hinge line to a shallow basin edge, Occidental selected five exploratory blocks in the Llanos—Cravo Norte, Cinaruco, Rondon, Orocue, and Guarrojo (see Figure 2). These areas also were considered to have a potential for stratigraphically trapped oil.

Early Results

Once the acreage was contracted, the immediate exploratory challenge was of vast proportions because 5 million hectares of barely explored grassland had to be evaluated in only 12 months. Occidental was contractually committed to relinquish 50% of the original acreage at the end of the first exploration year. Seismic and drilling crews worked frantically during this period to produce 2800 km of seismic data, 19 stratigraphic tests, and 2 exploratory wells. Although this campaign did not result in oil discovery, it did have a positive result as it provided some answers to questions regarding basin stratigraphy, the presence of shale, and possible long-distance oil migration. However, no drillable exploratory structures were identified, and questions remained about structural development within the basin. Thus, the geologic risk remained high, and Occidental began the search for a partner to share the project's risks.

The second year of exploration, from July 1981 to July 1982, also was frantic. Approximately 1000 km of seismic data were recorded and 10 exploratory wells were drilled in the areas retained. Four of these wells found minor accumulations of oil in structural and stratigraphic traps located in the shallow part of the basin (S. Joaquin, El Miedo). These discoveries, although of no commercial value, opened the way for the Caño Limón discovery by confirming that the areas retained by Occidental had the geological ingredients to host an oil field. The task was to find this field.

The restriction on exploration budgets that affected the oil industry during 1982 and 1983, and the exploration risks inherent to the project, led Occidental to continue the search for a partner to the very eve of spudding the Caño Limón-1 well. However, this search failed, and it can be said that Occidental was fortunate in not having found a partner.

Field Definition

Occidental's seismic work in the Cravo Norte "A" and Rondon "A" blocks (Figure 3) began in May 1981. Prior to this work seismic surveys had been carried out by different operators in this part of the Llanos, but they were concentrated mostly to the west and south of Occidental's blocks. No seismic data existed over the Caño Limón area proper. The interpretation of existing seismic data suggested, however, the presence of a northeast-trending major fault in the Rondon "A" block, which motivated the programming of additional seismic lines in the area.

By mid-1982, Occidental had acquired more than 500 km of seismic lines in the Rondon "A" and Cravo Norte "A" blocks, utilizing helicopter-supported portable crews. The source of energy was 2.3-kg dynamite charges placed in single shot holes drilled to a depth of 18 m. The coverage for the most part was 6 fold, and one line was shot with a 12-fold coverage. The data obtained with these parameters were in general of good quality.

The gathering and subsequent processing and interpretation of the data were conducted in stages, with more seismic lines programmed after each interpretation stage. As a result, by mid-1982, a 5 km × 5 km grid of seismic lines had been achieved in the Cravo Norte "A" and Rondon "A" blocks. The interpretation of these data confirmed the presence of an important subregional fault zone, part of which had been recognized by the preliminary interpretation of existing seismic data for the Rondon block. Although the coverage that this seismic data provided was not detailed enough to define a drillable structural closure, it did identify a trend of structural anomalies associated with this fault zone, along which exploratory prospects could be expected to develop.

The exploratory program that culminated with the discovery of Caño Limón started in August 1982 with a detailed seismic survey along this recently delineated fault zone. The preliminary structural interpretation of the seismic data suggested the presence of a fault-bounded anticline, whose critical dips required definition (Figure 3). In search of this critical information, three seismic lines—61.7, 68.3, and 61.6A—were recorded between August 1982 and February 1983. These lines defined the Caño Limón prospect as a fault-bounded anticlinal fold. The location of line 61.7 is shown in Figure 3, and the seismic profile is shown in Figure 4. The dip reversal observed on this line affects the Mirador Formation and older rocks. In addition, a thinning of the overlying stratigraphic sequence was noted over the crest of the fold, indicating that the Caño Limón structure was an old structural feature. The area of closure defined by this structure is approximately 527 hectares. A contour map of this structure is shown on Figure 5.

Additional seismic lines indicated that the Caño Limón fault had little or no vertical throw at the apex of the Caño Limón

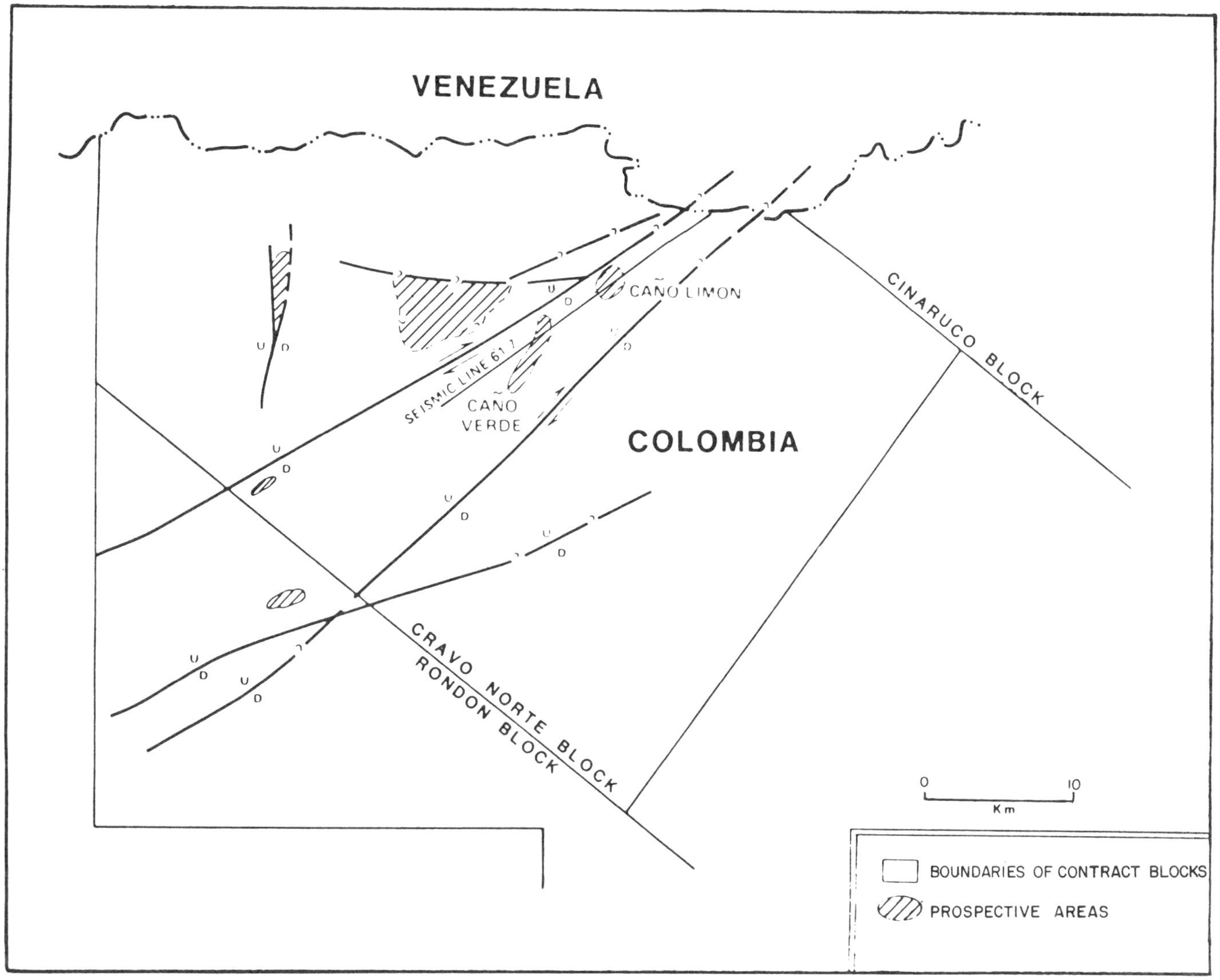

Figure 3. Map showing Cravo Norte, adjacent contract blocks, and known faults as of 1982.

structure. This led to the conclusion that the Caño Limón oil accumulation could extend into an adjacent fault wedge named the Matanegra prospect to the north of the Caño Limón fault (Figure 6).

STRUCTURE

The Caño Limón field was initially thought to be a single structure, a drag fold related to the movement of the Caño Limón fault (Figure 5). Additional seismic control, however, defined two other adjacent structures which either were or were not connected to the Caño Limón field. The drilling of La Yuca-1 and Matanegra-1 wells on those structures, after discovery of Caño Limón, demonstrated that the field actually consists of three structural closures now identified as the Caño Limón, La Yuca, and Matanegra productive areas (Figure 6). These structures appear to have been formed by lateral movement on the Caño Limón fault, which cuts the field from southwest to northeast, and by the associated displacements on the converging Matanegra and La Yuca faults.

Fault Systems

The Caño Limón fault is the longest and most significant fault in the field. It extends to the southwest approximately 40 km outside the field area. The lateral movement of this fault is inferred from its variable vertical displacement, ranging from 0-152 m, and from the apparent offset of local structural features. The magnitude of the lateral displacement is not accurately known because of the lack of regional control data. Locally, the displacement appears to be 1-3 km. Seismic and stratigraphic data indicate that the major strike-slip movement along this fault probably occurred during the late Eocene to early Oligocene. This is suggested by the significant thickening and thinning of the Mirador and overlying Carbonera Formations along the Caño Limón fault; these thickness variations are controlled by paleotopographic lows and

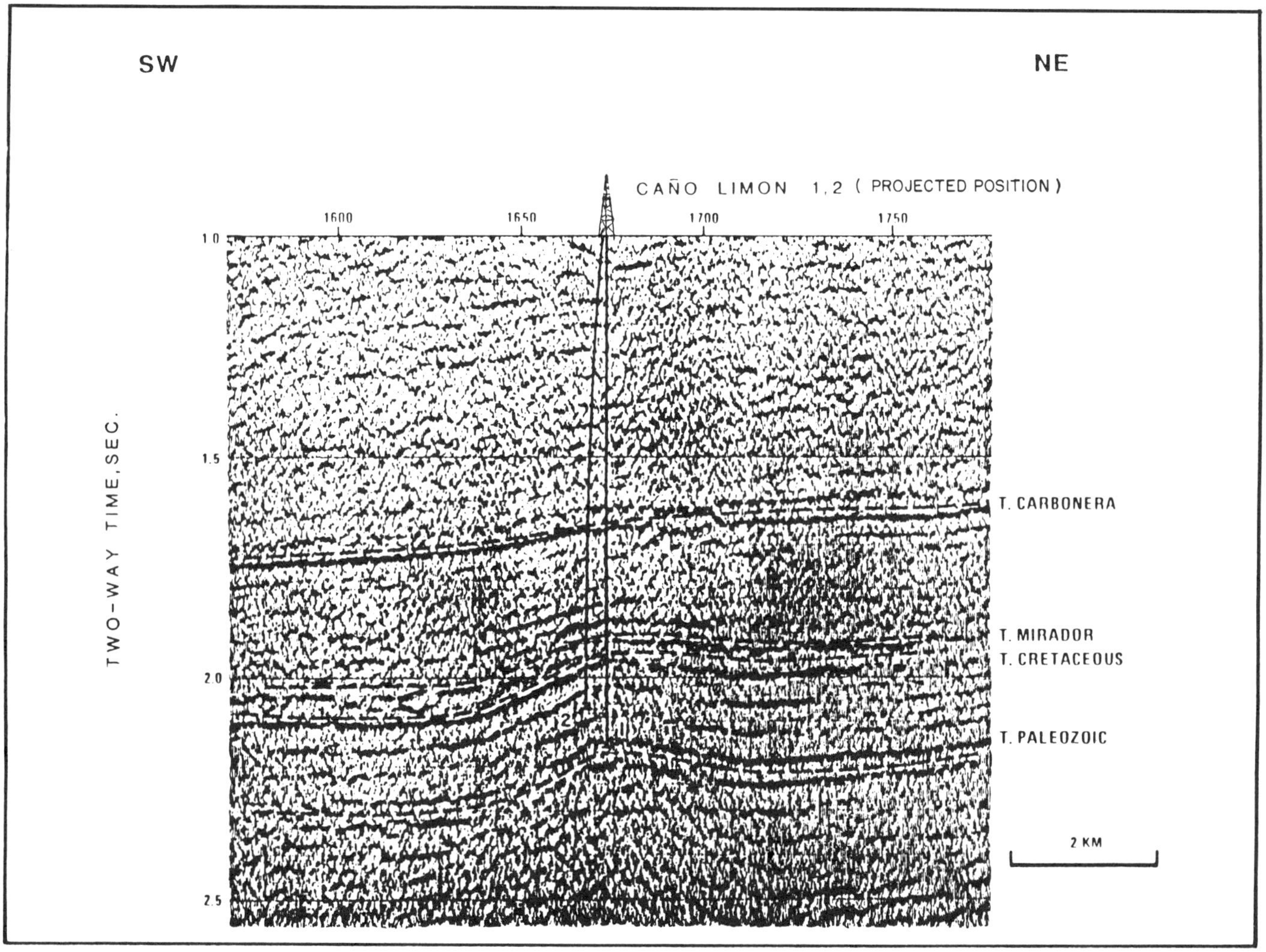

Figure 4. Seismic record along line 61.7 shown on Figure 5, Caño Limón field.

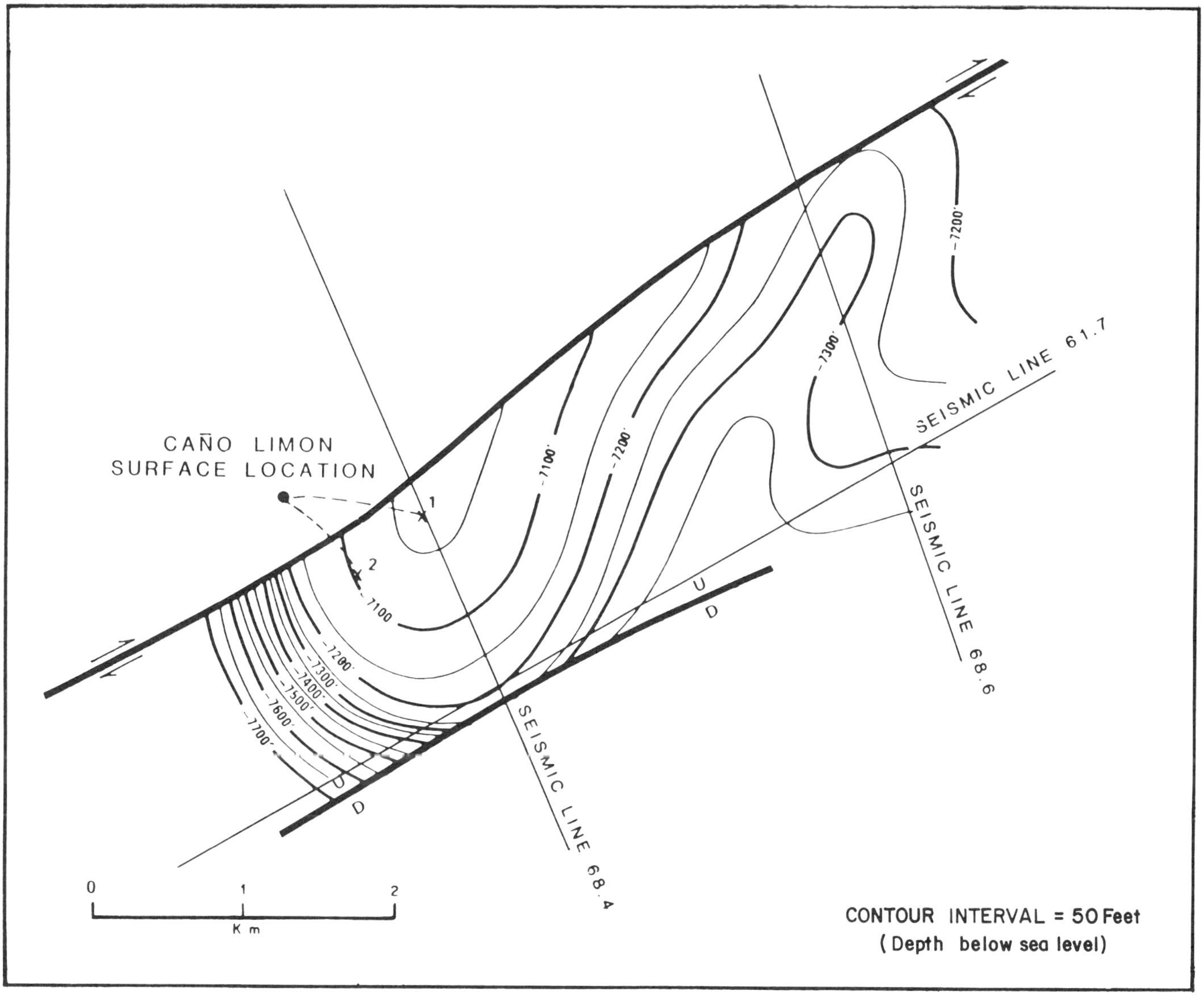

Figure 5. Structure contour map on top of the Mirador Formation, Caño Limón field.

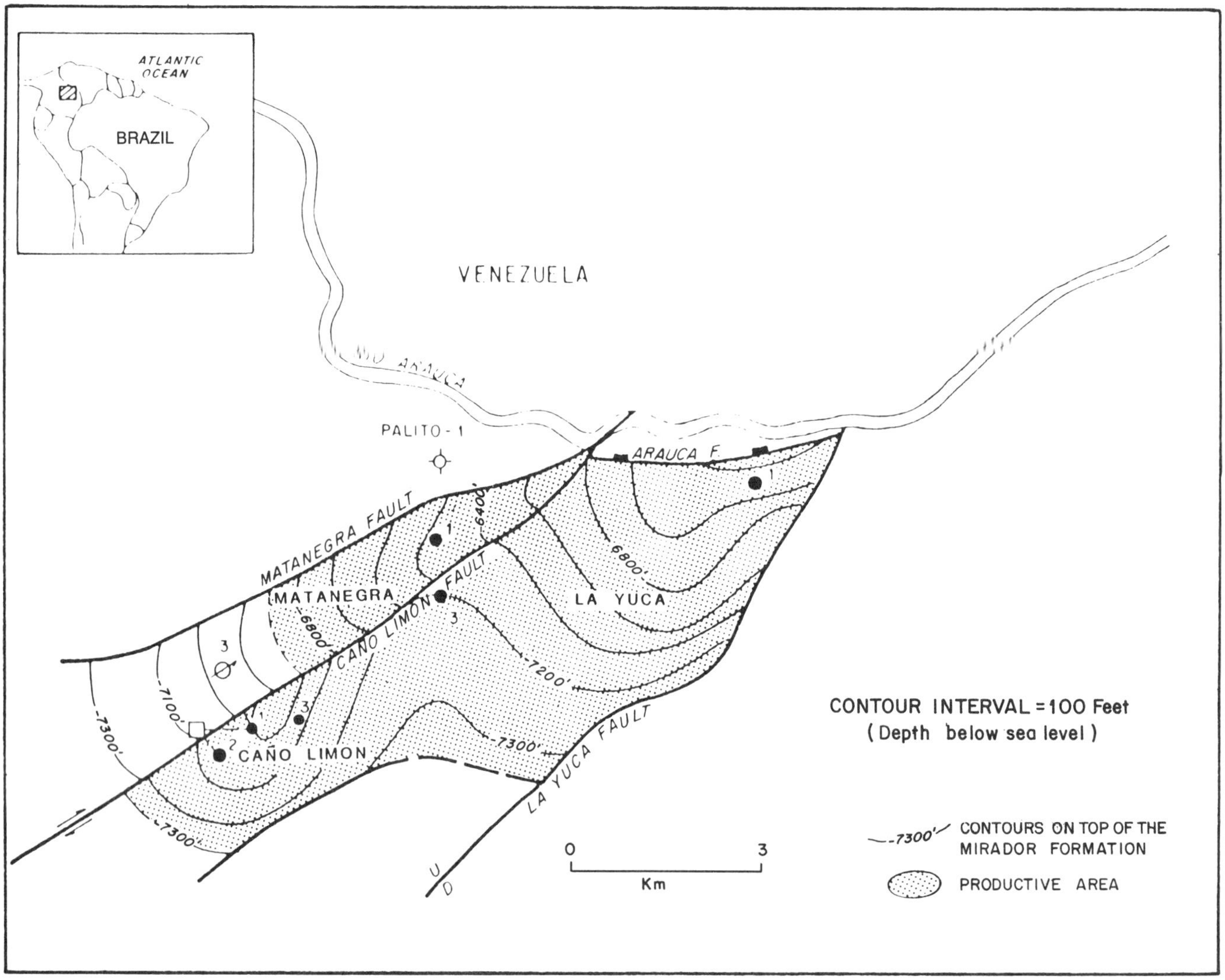

Figure 6. Geologic structure and productive area of the Caño Limón field.

highs produced by the folding that accompanied the fault movement. These data also indicate that the Caño Limón field area was a paleotopographic high at the beginning of Mirador deposition (early Eocene), as demonstrated by the onlap of the basal Mirador over the underlying Cretaceous. The presence of two major erosional unconformities, one of Late Cretaceous to late Eocene age and the other of pre-Albian age, further suggests that the structural activity along the Caño Limón fault can be traced back to pre-Cretaceous orogenic events.

The Matanegra fault branches from the Caño Limón fault just south of the Arauca River (Figure 6) and extends southwesterly for approximately 8 km, where it disappears. This very steep fault shows normal and reverse displacements that suggest a transcurrent movement. The maximum vertical displacement is 61 m.

The La Yuca fault is subparallel to the Caño Limón fault for approximately 35 km, although it could extend farther. It has a reverse displacement of as much as 274 m; it is not yet clear whether this fault also has a component of lateral movement.

The Arauca fault trends east, along a segment of the Arauca River, and joins the Matanegra and La Yuca faults at the north end of the field. It is a normal fault having a downthrown block to the north. Maximum vertical displacement is approximately 46 m.

The Matanegra and Caño Limón faults are considered to be sealing because they separate individual oil accumulations having different oil-water contacts. Additionally, in the case of the Caño Limón fault, interference tests between wells located on opposite sides of the fault confirm that flow does not take place across it. The La Yuca fault is also considered to be sealing because it has not leaked oil in spite of the sand-against-sand relationship across the fault. The sealing nature of the Arauca fault has not yet been determined.

Productive Areas

The Matanegra structure is the northern, wedge-shaped, uplifted block bounded by the Matanegra and Caño Limón faults (Figure 6). Structurally, it is the highest area in the field (Figure 7). It has a total vertical closure of 213 m and contains the largest number of pay zones. The negative results of the Matanegra-3 well, drilled on the flank of this structure,

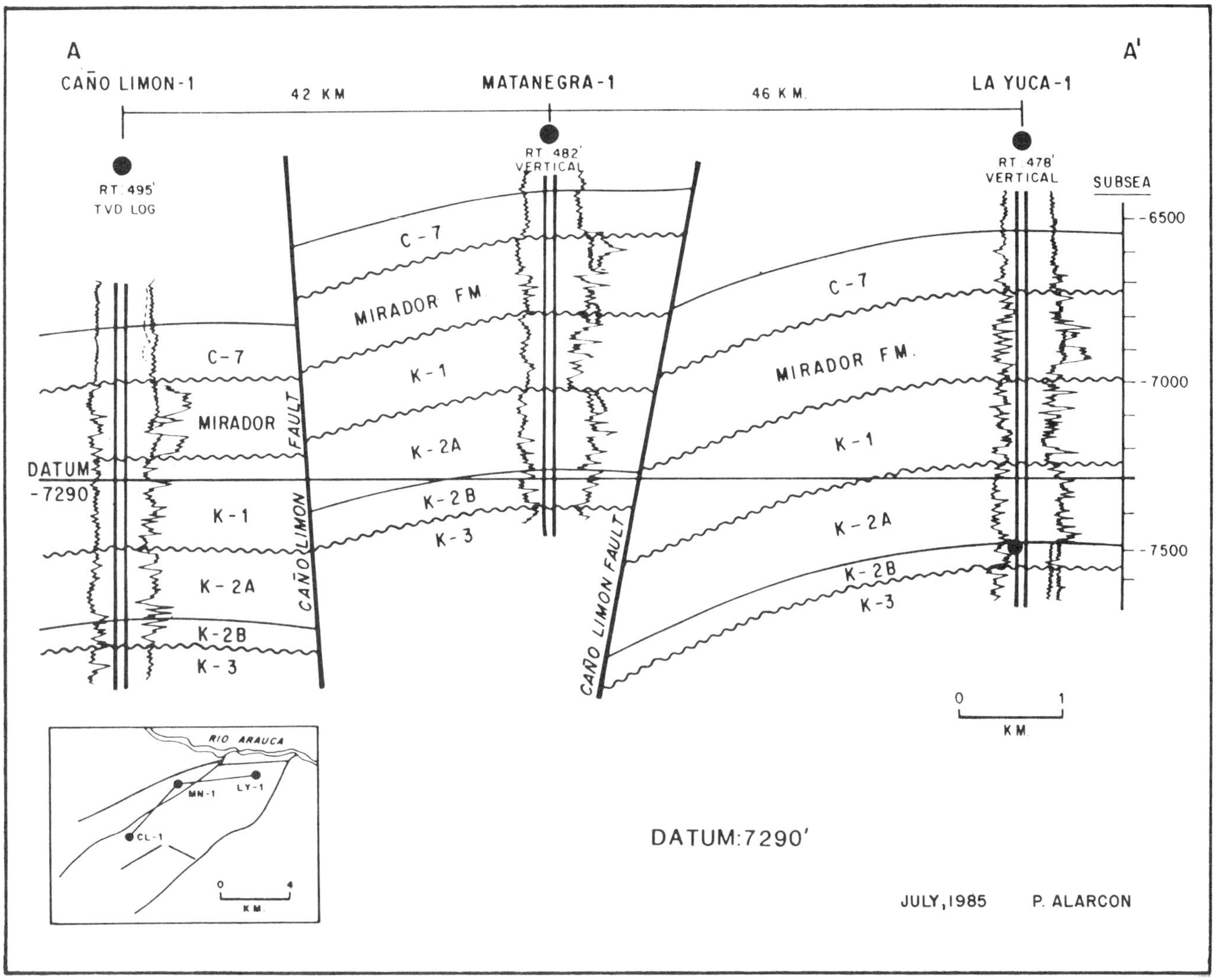

Figure 7. Structural sections in the Caño Limón field.

together with recent interference tests performed between wells located across the Caño Limón fault, indicate that the Matanegra structure is a separate oil accumulation from the Caño Limón-La Yuca structures and therefore has a different oil-water contact, identified at a depth of approximately 2195 m below sea level.

The Caño Limón-La Yuca structures are southwest-plunging anticlinal folds joined by a middle synclinal area. Limbs of these folds dip gently at angles of 5° to 12°, the highest dips being on the southeastern flank of La Yuca structure. The vertical closure of these structures is 91 m at Caño Limón and 180 m at La Yuca. The Caño Limón and La Yuca structures constitute a single oil accumulation and have a common contact at a depth of 2222 m below sea level.

STRATIGRAPHY

The stratigraphic sequence penetrated in the Caño Limón field (Figure 8) ranges from pre-Cretaceous to Miocene and is 2804 m thick. Pre-Cretaceous sedimentary rocks were penetrated only in the Caño Limón-1 well; other wells were drilled only to the basal Cretaceous. Because of absence of both marine fossils and fossil pollen in the short interval penetrated, the age of the pre-Cretaceous rocks could not be determined. Lithostratigraphic correlations with wells drilled outside the Caño Limón field (Joropo-1, La Heliera-1, Chiguiro-1) indicate a tentative Ordovician age for these rocks.

Three major unconformities are recognized: (1) a basal unconformity separating the pre-Cretaceous from the Upper Cretaceous (Senonian); (2) a middle unconformity between the Upper Cretaceous and the Eocene Mirador Formation, which is the main traceable seismic reflector; and (3) an upper unconformity between the Mirador and the Oligocene Carbonera Formation. The Paleozoic-Cretaceous unconformity represents a depositional hiatus during upper Paleozoic and Triassic in the Caño Limón field. Triassic sedimentary rocks were encountered in the Arauquita-1 well located west of the Caño Limón field. The Upper Cretaceous-Eocene unconformity in the Caño Limón field is represented by the Paleocene Barco-Cuervos Formation to the west, in the deeper parts of the basin.

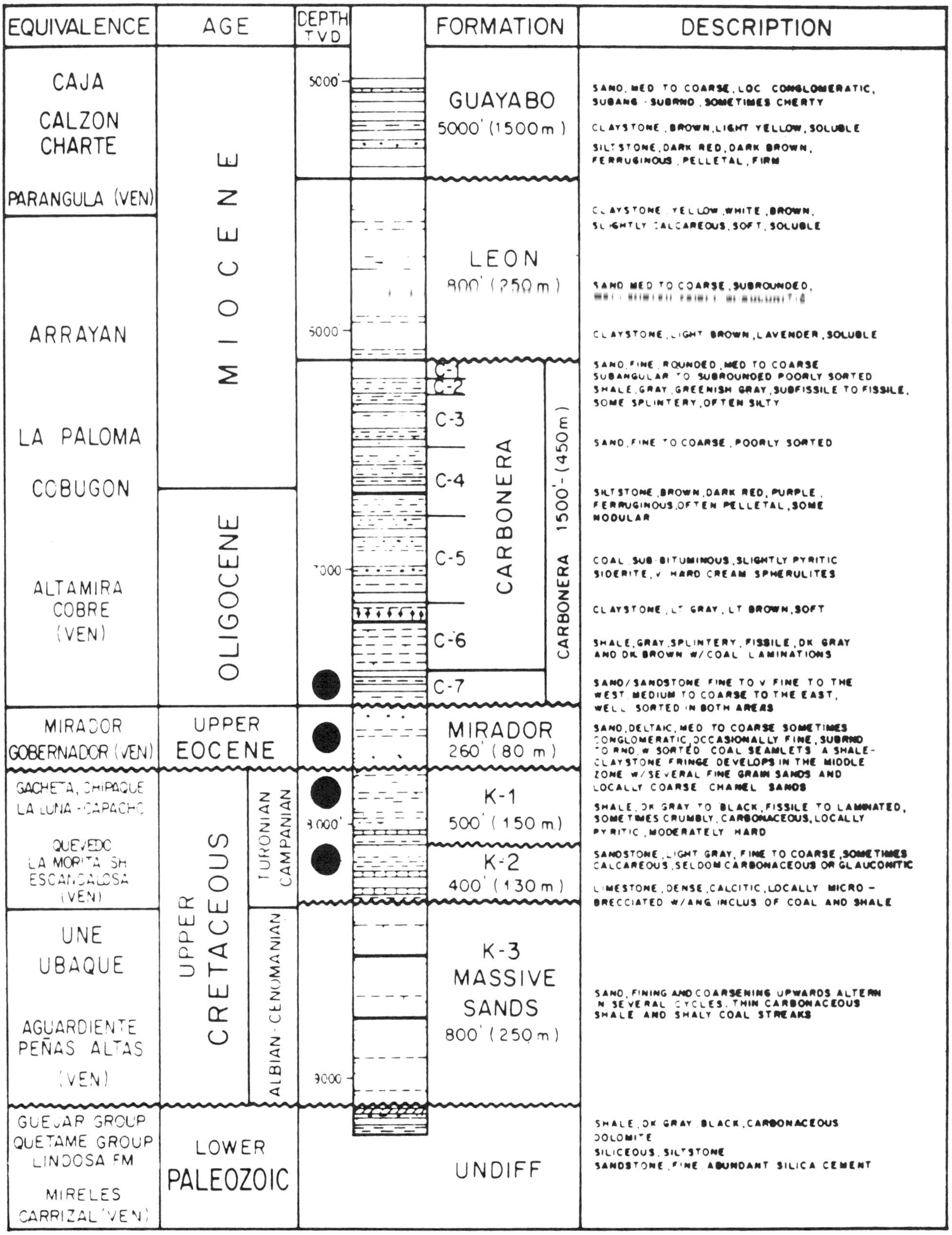

Figure 8. Stratigraphic sequence and productive units in the Caño Limón field.

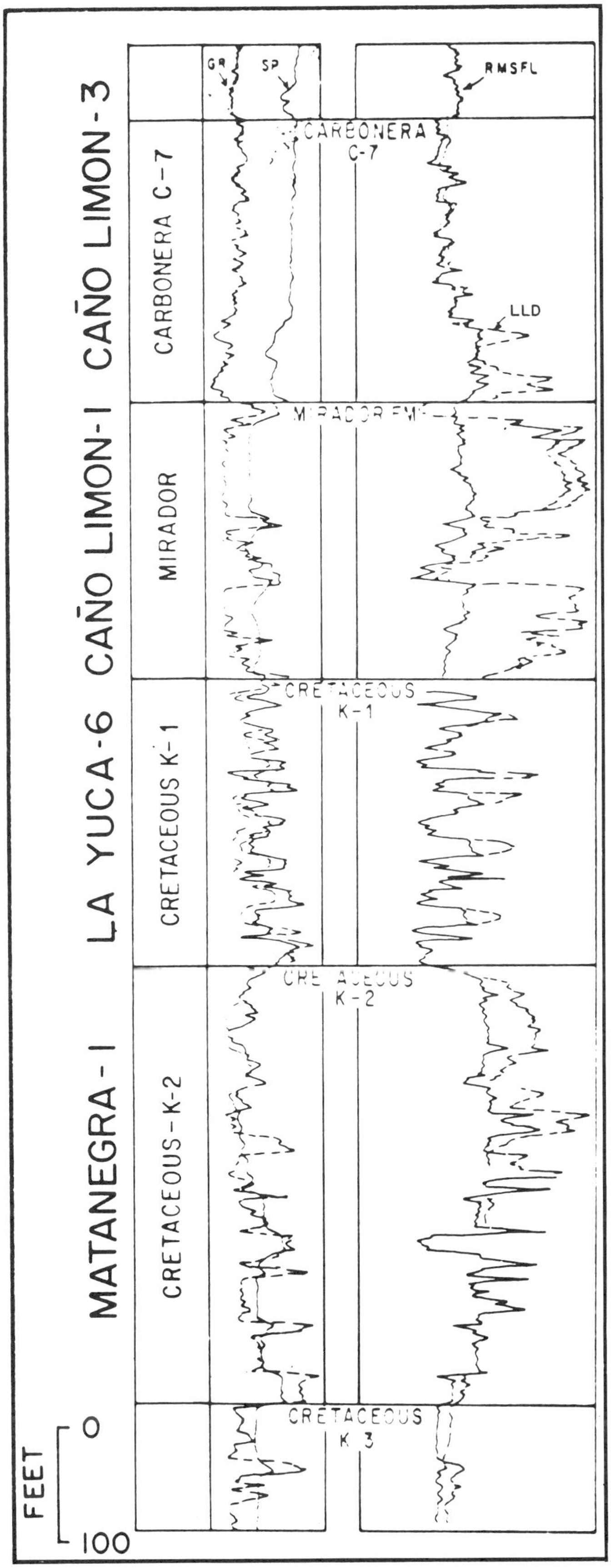

Figure 9. Dual laterolog profiles of productive formations in the Caño Limón field.

The productive beds of the Caño Limón field include fluviodeltaic sands of the Oligocene Carbonera C-7 basal unit, the Eocene Mirador Formation, and Upper Cretaceous K1 and K2 units (Figure 8). A type log for the field is shown in Figure 9. These productive horizons have been subdivided into stratigraphically correlatable units based on identifiable lithological and petrophysical differences (Table 1). Figure 10 shows a stratigraphic correlation of the producing structures of the Caño Limón field.

Deposition of the Carbonera C-7 unit was controlled by structural growth as indicated by its variable thickness and irregular distribution, which makes correlation of individual sand units difficult. These sands filled the paleotopographic low areas that subsided as the Caño Limón field structures grew. As a consequence, these sands are best developed in the synclinal area between the Caño Limón-La Yuca structures and on the flank of the Matanegra structure. They are absent or poorly developed on the crests of these structures. The lower part of the C-7 unit, where it is well developed, appears to be a complex of deltaic channel sands. These are generally medium to coarse grained and fairly well sorted. The upper part of the C-7, on the other hand, is identified as a basal transgressive sand, which is medium to coarse grained and gradually becomes finer grained toward the top, where it grades into siltstone. The sands are 9 m to 53 m thick. Both sands have good reservoir characteristics, and production from the C-7 sands has been obtained in the Caño Limón, La Yuca, and Matanegra structures.

The Mirador Formation, which is present in all wells drilled to date, consists of deltaic sand in stacked distributary channel deposits, with interbedded prodelta clays. It has an average thickness of 75 m. The channel sands comprise two-thirds of the formation. They are medium grained to conglomeratic and show fair to good sorting. The Mirador sands were deposited on an erosion surface and, as previously noted, their thickness, particularly the basal beds, was influenced by paleotopography. For purposes of reservoir management, these sands have been subdivided into four correlatable units (see Table 1). The Mirador sands show excellent reservoir characteristics, are productive in all the Caño Limón structures, and contain 80% of the field's reserves.

The Cretaceous K1 unit consists predominantly of shale with interbedded fine- to medium-grained sand layers deposited in a delta fringe environment. It is 85 m thick. These sands generally exhibit lower permeabilities than the other productive units in the Caño Limón field, and only about 30% of the K1 unit contains reservoir-quality sands. Correlation of individual strata is good over the entire field. The K1 sands are productive in the Matanegra and La Yuca structures. They have a common oil-water contact with the overlying Mirador. The K1 sands are separated from the underlying K2 unit by a shale unit that acts as seal to the oil trapped in the K2A sands, which have a separate and deeper oil-water contact than the K1.

The Cretaceous K2 unit consists of two regressive sand sequences, identified as K2A and K2B, separated by the 6-12 m Palito shale. Only the K2A is productive in the Caño Limón field (Matanegra structure). The K2A sequence consists of five sand layers separated by thin layers of shale or limestone. In general, K2A is homogeneous and exhibits good lateral continuity. Two-thirds of K2A consists of reservoir-

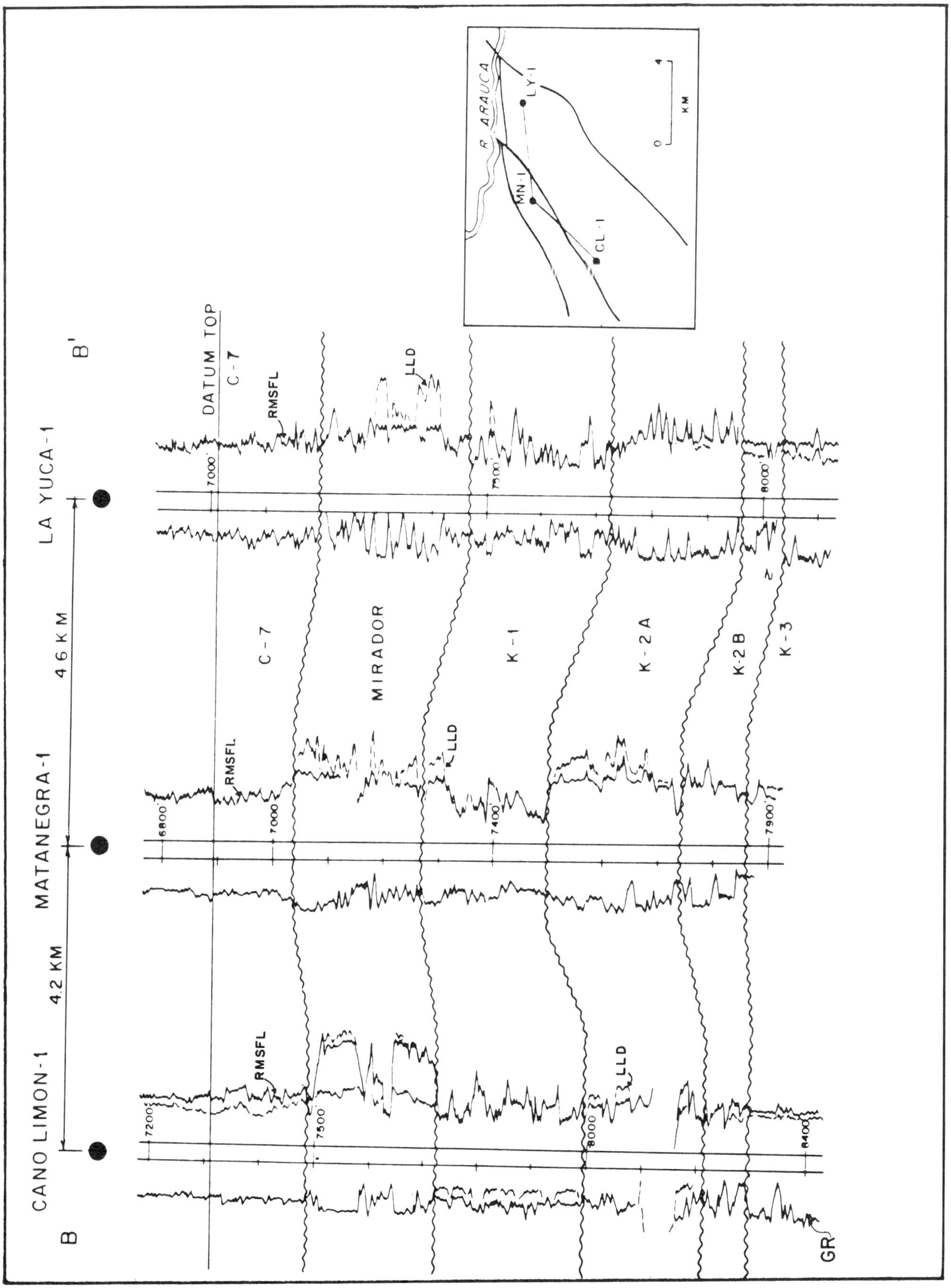

Figure 10. Correlations of productive units in the Caño Limón field based on dual laterolog profiles.

Table 1. Stratigraphic subdivision of productive zones, Caño Limón field.

Formation	Productive Unit	Productive Areas**	Thickness (Meters)
Carbonera	Upper C7	CL, LY, MN	9
	Basal C7	CL, LY	53
	M1	CL, LY, MN	
	M2	CL, LY, MN	
Mirador	M3	CL, LY, MN	75
	M4	CL, LY	
	A	MN, LY	
	B	MN, LY	85
K1	C	MN, LY	
	D	MN, LY	
	E	MN, LY	
	F	—	
	1	MN	
	2	MN	
K2A	3	MN	75
	4	MN	

** CL = Caño Limon; LY = La Yuca; MN = Matanegra.

quality sands. K2A has a thickness of approximately 75 m. K2B consists of laterally continuous, equal proportions, of channel and delta fringe sands. Thin interbedded, discontinuous claystone and limestone layers are common locally. K2B has a thickness of 21-42 m. No oil has been found in K2B in the Caño Limón field. K2A and K2B are considered to be separate reservoirs based on the slightly higher water salinity in K2B.

The Cretaceous K3 unit also has been identified as having excellent reservoir characteristics but as yet has not been found to be productive in the Caño Limón area. It consists of massive fluvio-deltaic sand.

The division of the Cretaceous sequence in the Caño Limón field into units K1, K2 and K3 is based on practical considerations, and these units cannot be correlated with Cretaceous units of similar names assigned by operators elsewhere in the Llanos. The K1 and K2 units in the Caño Limón field have been assigned an age ranging from Cenomanian to Campanian based on palynology (identification of *Araucariacites sp., Dinogymnium sp.,* and *Rugulastiporities caperatus* specimens). Therefore, these units are considered to be equivalent to the Gacheta-Capacho-La Luna-Chipaque formations of the Eastern Cordillera and to the Escandalosa-Quevedo formations and La Morita Shale of Venezuela. The K3 unit in the Caño Limón field has not been dated but, based on its stratigraphic position and lithological similarity, it may correspond to the Albian-Cenomanian sequence in the southern Venezuelan basins (Peñas Altas-Uribante Formation). Also, it is probably equivalent to the Aguardiente, Une, and Ubaque formations of the Llanos basin.

SOURCE ROCKS

It has been generally accepted by the oil industry in Colombia that the Middle and Upper Magdalena basin crude oils, as well as the Llanos basin oils, were probably derived from organic-rich Cretaceous marine shale. This hypothesis was confirmed for the Middle Magdalena basin oils through analyses conducted at the Occidental-Cities Service Research Laboratory on oil samples from Las Monas block oil fields. These analyses show that Las Monas block oil fields are of the same genetic family and that the Cretaceous La Luna and Simiti Formations were the co-sources of these oils. On the other hand, similar geochemical analyses have failed to establish a correlation between these Cretaceous source rocks and the Llanos basin oil, including that from Caño Limón. J. Zumberge (1984), in an intracompany report, states:

"Organic geochemistry parameters suggest that the moderately mature Caño Limón crude oils (produced from the Tertiary Mirador and basal Carbonera formations) were generated from source rocks containing marine derived organic matter. This is in contrast to oils produced from the central Llanos basin (e.g., Caño Garza and Trinidad), which were derived from sediments containing more terrigenous organic matter. All Llanos Basin oils analyzed appear to have varying abundances of terrigenous components however. Although oils from the Middle and Upper Magdalena basins were derived from marine source rocks (i.e., Cretaceous La Luna and Simiti formations), the Caño Limón oils appear to have yet another source with a different marine signature. Thermal maturity and organic richness measurements of a core sample from Caño Limón-1 (9186 ft, Cretaceous) revealed immature (RO = 0.48%) sediments with a poor potential for oil generation due to the predominantly terrigenous nature of the organic matter, indicating that it is not the source of the Caño Limón oils."

The conclusions that may be drawn from these results are that the Llanos basin oils have more than one generating source containing marine and terrigenous organic matter; that the Caño Limón oils were not generated in situ; and that the marine source rocks that generated the Caño Limón oils have not yet been identified within the geographic boundaries of the Llanos basin, and could very well be found in neighboring Venezuela basins.

RESERVOIRS

Petrophysical Properties

All the wells drilled in the Matanegra, La Yuca, and Caño Limón areas were tested with a complete set of electric logs, which allowed among other things the quantitative evalua-

Table 2. Average weighted porosity and residual oil saturation of producing horizons.

Structure/Producing Horizon	Porosity (percent)	Residual Oil Saturation (percent)
CAÑO LIMÓN		
Carbonera	20.6	67.0
Mirador	23.8	76.9
LA YUCA		
Carbonera	21.3	74.9
Mirador	23.8	76.9
K-1	19.9	37.3
MATANEGRA		
Carbonera	25.1	50.0
Mirador	26.8	70.4
K-1	19.9	37.3
K-2	24.6	73.2

tion of the porosity and water saturation of the penetrated formations. In addition to electric and radioactive logs, conventional cores were cut in the Caño Limón-2, Caño Limón-3, and La Yuca-6 wells. Routine tests were made on the available core samples in order to obtain porosity and permeability values of the reservoir sands. Special analyses were made to determine capillary pressure curves, oil-water relative permeabilities, oil residual saturation, cementation factors, saturation exponent, and other parameters.

The porosity values of the Caño Limón reservoirs are very good. When all the producing horizons as a whole are considered, the highest porosities were observed in the Matanegra structure. When each reservoir is considered individually, the highest porosities are those of the Mirador Formation, which range from 23 to 26% in the entire field. The C-7 sands of the Carbonera Formation and K2 Cretaceous unit also have high porosity values, next to those of Mirador. Table 2 summarizes the average weighted porosities per thickness for each productive reservoir of the field. These values are based on the correlation established between the porosities measured in conventional cores and those calculated from electric logs.

Permeability values were determined by analysis of pressures obtained from formation tests and drill stem tests, and by analysis of cores cut to date. Pressure analysis indicates that the Mirador reservoirs have the highest permeabilities, averaging 5 darcys. The permeabilities of the Carbonera Formation C-7 sands range from 3 to 5 darcys, whereas the K1 reservoirs average 1 darcy and the K2 reservoirs 2 darcys. The following are absolute permeabilities of the Mirador Formation determined in one of the cores analyzed: horizontal permeability, 2425 Md; vertical permeability, 1763 Md; variation coefficient, 0.53; Lorenz coefficient, 0.28.

The analysis of fluid samples indicates that the oil-bearing formations are at a pressure greater than the saturation pressure and, therefore, contain only liquid fluids, oil and water. The water and oil saturations were determined by electric log evaluations and by computer programs incorporating parameters of porosity and clay volume of the reservoirs. A residual oil saturation (Sor) of 31.6% was obtained for the Mirador Formation based on relative permeabilities for the oil-water system. The average oil saturation values shown on Table 2 for each field were weighted according to porosity and net productive thickness.

Table 3. Chemical composition of Caño Limón oil (sample at bottom-hole conditions).

Component	Mol %
Hydrogen sulfide	0.00
Carbonic anhydride	0.27
Nitrogen	0.11
Methane	0.05
Ethane	0.03
Propane	0.21
Isobutane	0.33
n-Butane	0.62
Isopentane	0.75
n-pentane	0.72
Hexane	1.61
Neptane +	95.30
	100.00

The properties of the Caño Limón oil (Tables 2 and 3) were determined by analysis of well-bottom samples. These analyses indicated that the Caño Limón oil is subsaturated and that there will be no free gas in the field during its entire productive life. The solution gas measured in the laboratory was lower than 15 standard cubic feet/stock tank barrel for all the analyzed samples. This low gas content has been confirmed by the absence of measurable amounts of gas during the formation tests. The low gas content contributes to the low oil compressibility, approximately CE-6/psi; therefore, because of oil expansion, a minimum energy contribution by the field is expected.

The oil viscosity under reservoir conditions (3200 psi at 200 °F) was measured in the laboratory, obtaining values ranging from 2.4 to 6.9 cp. The API gravity under surface conditions at 16 °C ranges from 26.9° to 32.4°.

Oil In Situ

The oil in situ for the Caño Limón field, based on volumetric calculations and current geologic interpretation, is calculated to be 225 million m^3 (1413 MMbbl), distributed in three productive structures (Table 4).

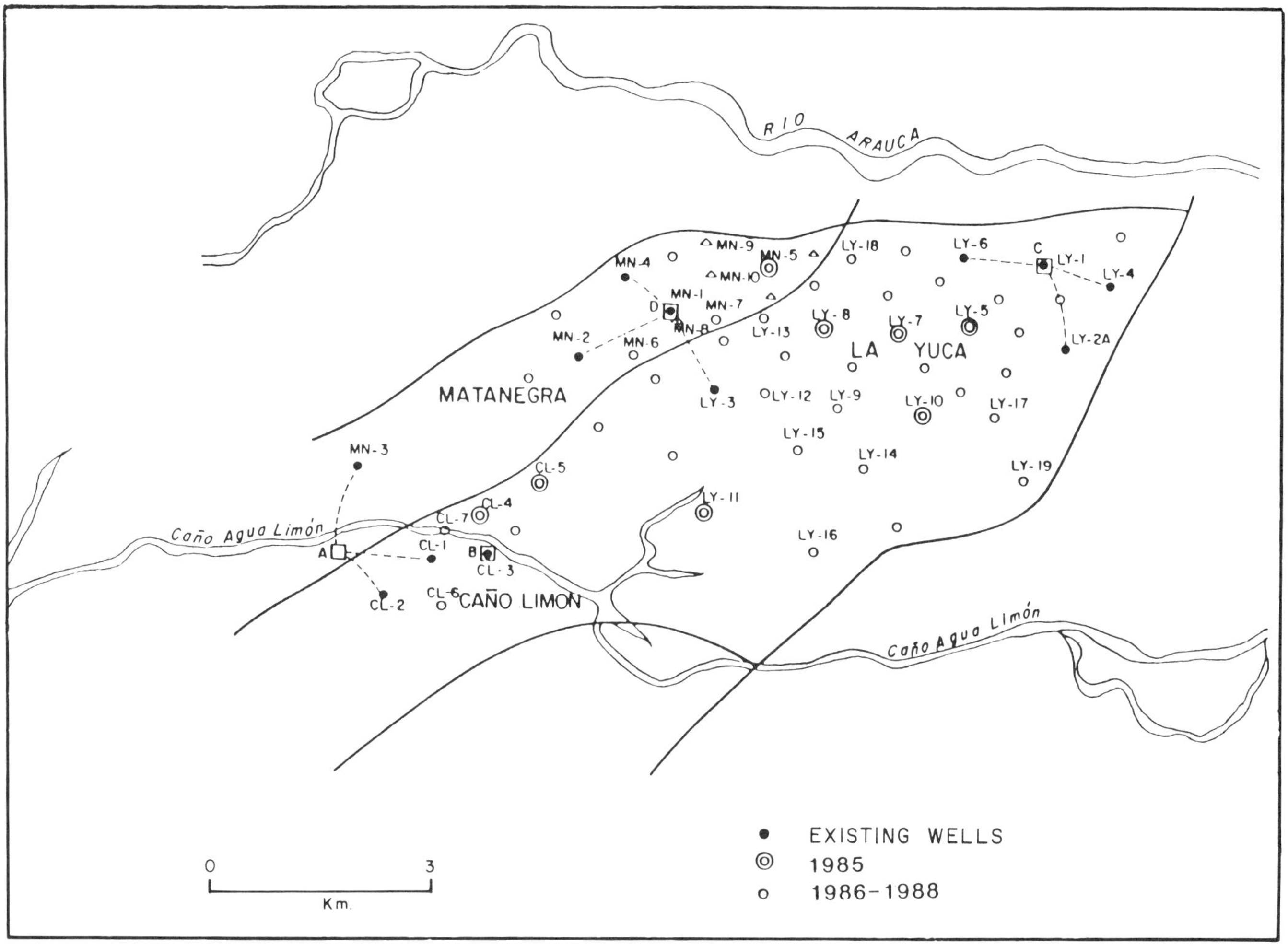

Figure 11. Map showing locations of oil wells for development of the Caño Limón field.

Table 4. Oil in situ, by productive areas (structures).

Structure	Area (Hectares)	Area (Acres)	Volume MM-M^3 (MM-bbl)
Matanegra	1,446	3,570	59 (372)
Caño Limon	2,892	7,140	143 (899)
La Yuca	1,430	3,530	23 (142)
Total	5,768	14,240	225 (1.413)

Water Injection

The size and characteristics of the Caño Limón aquifer are currently unknown. Regional geologic studies suggest, however, the presence of an active aquifer. If the aquifer is not sufficiently active or is limited by the presence of the faults in the field, the natural energy of the field will have to be supplemented with water injection. It is estimated that, in an extreme case, it would be necessary to initiate a water-injection program at an early stage. In this case, the maximum flow to be injected would be 102 M-M^3 (640,000 b/d) by 1991. On the other hand, if the aquifer action is of great magnitude, the need to inject water would be considerably reduced but not eliminated. It is estimated that a 1.5 year injection program would be required after the field starts production, which would reach a maximum volume of 20.7 M-M^3 (130,000 b/d) by 1991.

DEVELOPMENT PLAN

The excellent quality of the field reservoirs justifies a 1000 m spacing between the development wells with good oil recovery efficiency. Initial development of the field will require the drilling of approximately 50 wells, an average of 10 wells per year, up to a total of 65 to 70 wells during total development of the field (Figure 11). The number of injecting wells to be drilled ranges from 8 to 14, depending on the size of the aquifer.

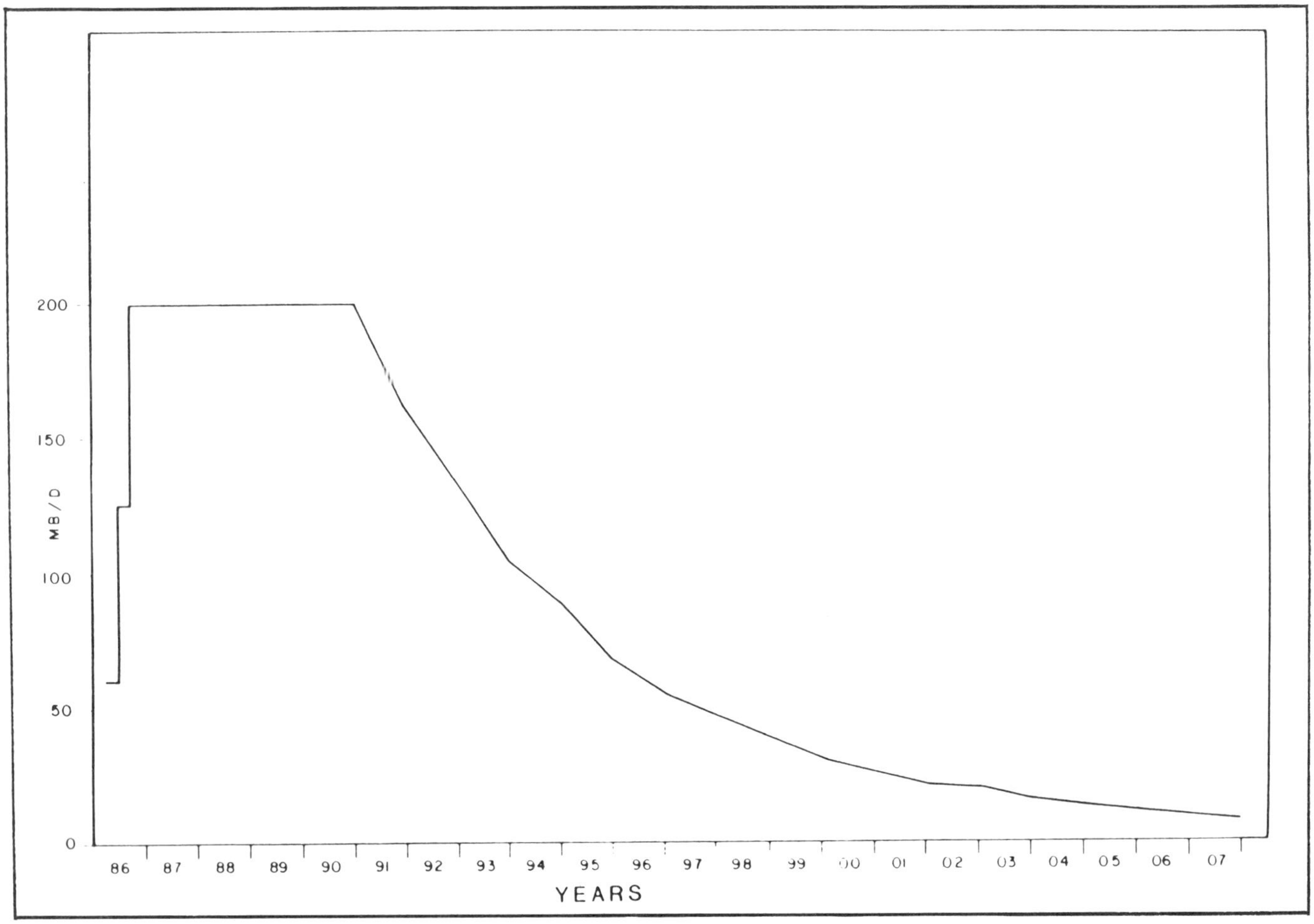

Figure 12. Estimated oil production for the Caño Limón field.

Drilling

Because of the logistical difficulties found in the Llanos region, the initial development plan contemplated the drilling of several directional wells from a common drilling island. Four of these islands were constructed, from which nine deviated wells and three vertical wells were drilled by July 1985. This drilling method resulted in 25° to 35° deviation angles, horizontal displacements as great as 1300 m, and a maximum drilling depth of 2896 m. The resulting spacing was approximately 1000 m. The construction of secondary roads connecting the existing islands and a main access road to the area of the field now allows planning for the drilling of future vertical wells costing less than the wells initially drilled.

Production Forecast

The field development plans contemplated January 1986 as the production commencement date, with an initial volume of 9500 M^3/D (60,000 B/d). The volume was expected to increase to 20,000 M^3/D (125,000 B/d) by the end of the second quarter of 1986, and to reach 32,000 M^3/D (200,000 B/d) by the end of the third quarter of the same year (Figure 12). These plans take into account the initial pipeline limitations, the development well drilling program, and the productivity index of the wells. The maximum production per well is limited to 24,000 M^3/D (15,000 B/d) of fluid.

Artificial Lift

Engineering studies determined that electric submergible pumping is the most appropriate production method for the field's wells. This conclusion was reached after analyzing other possibilities, such as rod pumping and gas lift. The utilization of rod pumping is limited because of the high production rates expected. Gas lift, on the other hand, was not considered to be viable because of the absence of gas in the area. The possibility of utilizing Weir hydraulic pumping is being evaluated at a pilot project level.

Production Facilities

Construction of production and transport facilities was programmed so that operation could begin by December 1985, taking into account the established production forecast as well as pipeline capacity limitations. The initial plans required construction of two crude oil gathering centers designated PF-1 and PF-2 (Figure 13). PF-1 was expected to be operating by December 1985, which coincides with the expected termination date of the Caño Limón-Rio Zulia pipeline. These installations will be capable of processing 22,000 M^3 (140,000 bbl) of fluid per day. PF-1 (see process diagram, Figure 14) has the primary purpose of water separation. The PF-2 production facilities were designed to handle a total of 57,000 M^3 (360,000 bbl) of fluid per day. The PF-2 termination date and commence-

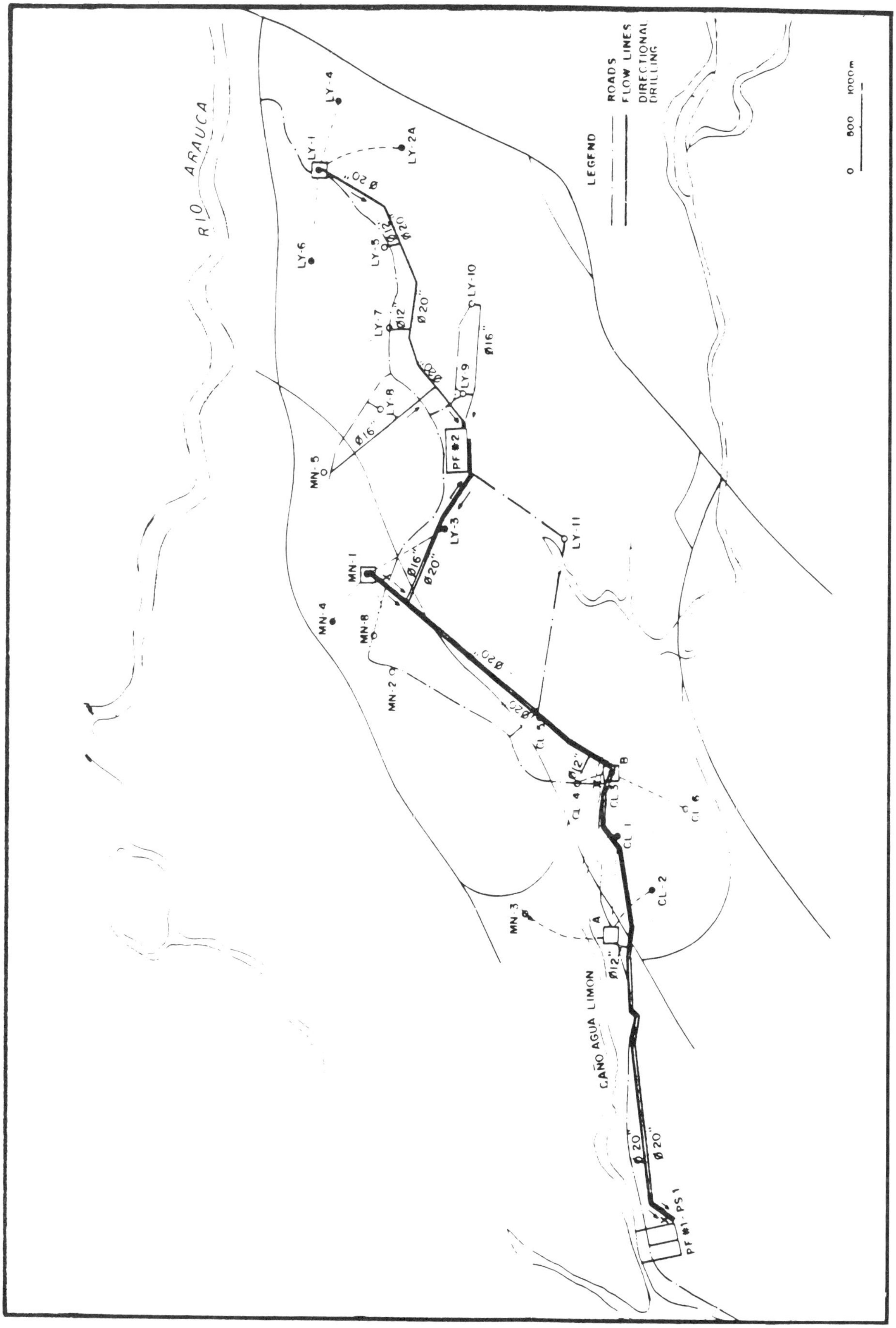

Figure 13. Production and transport facilities in the Caño Limón area.

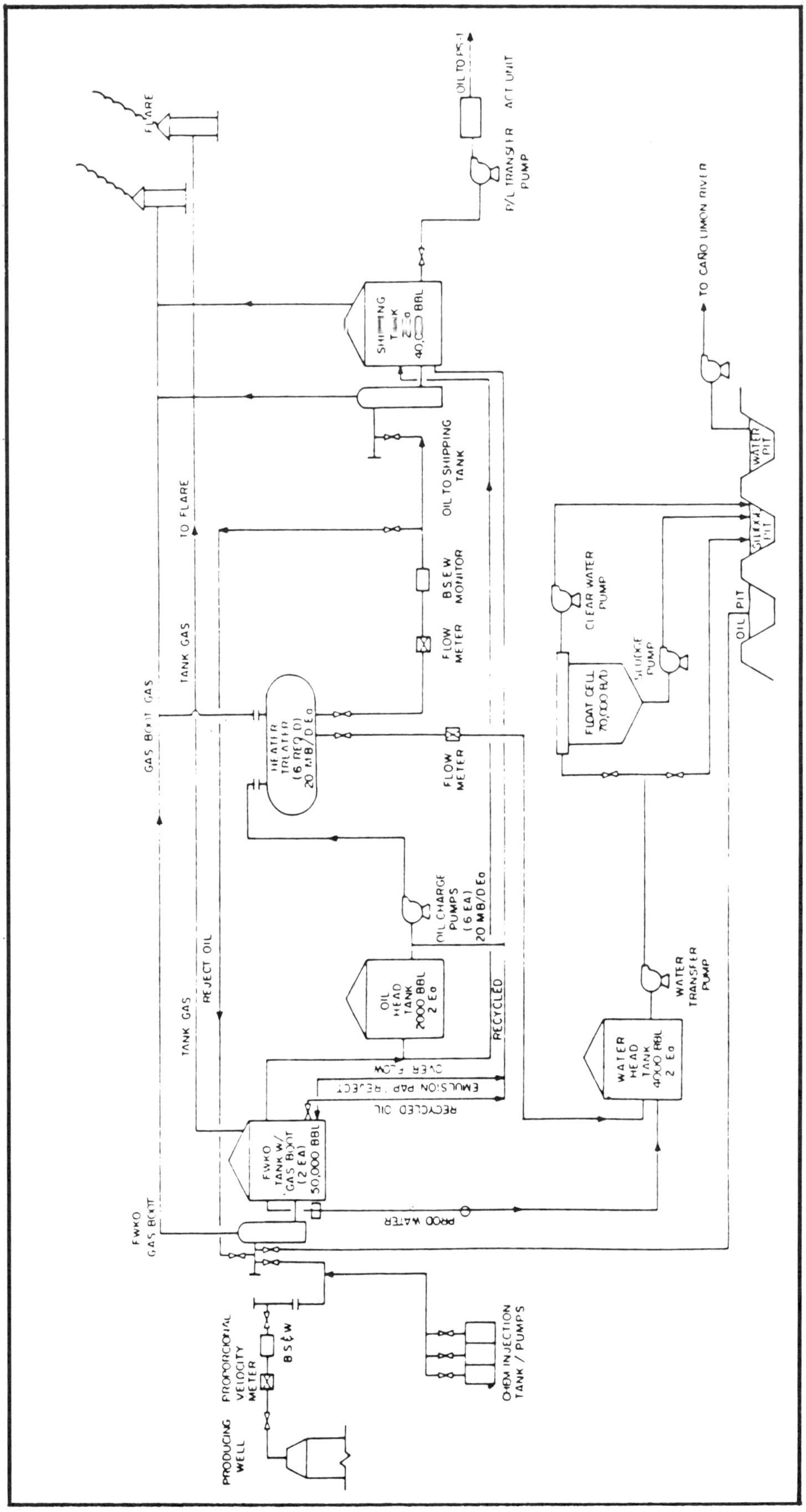

Figure 14. Production facility No. 1, Caño Limón field.

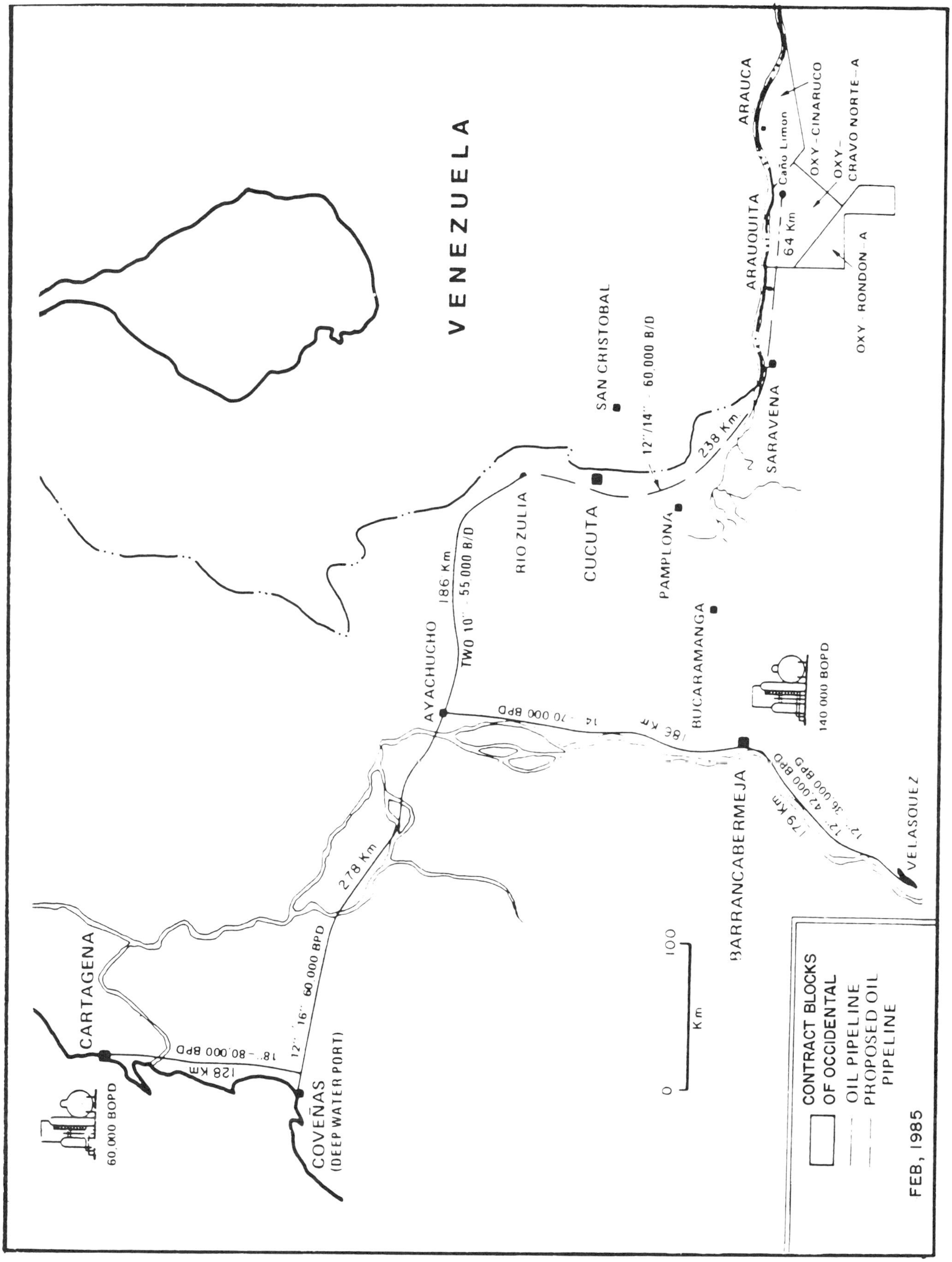

Figure 15. Relationship of Caño Limón field to oil pipeline network in Colombia.

ment of operations was set for June 1986, which likewise coincides with drilling, well completion, and pipeline construction plans.

The construction of two electric plants has been programmed to meet the energy needs of the field. The first plant, to be located at PF-1, will initially have three generators; two or more generators will be added in the future. This plant will provide energy to PF-1, pump station No. 1, and field installations, for a total estimated consumption of 4560 Kw. Each generator will have a capacity of 2400 Kw, which represents an actual available total of 9600 Kw, with an auxiliary generator. The remaining 5000 Kw will be available for the water injection system and the oil artificial lift. Generating plant No. 2 will be located in PF-2 to meet the energy needs of this production station and for the artificial lift of oil from the field's wells. The replacement of these two plants with a central station capable of generating approximately 20-30 megawatts has been anticipated for the future.

Pipeline and Terminal

The Caño Limón-Coveñas pipeline, with a total length of 787 km (Figure 15), was completed on December 19, 1985. Phase 1 of this construction covered the first 286 km from Caño Limón and has four pumping stations (1, 1A, 2, and 3). The pipe diameter ranges as follows: 46 cm between stations 1 and 2; 51 cm in the mountain pass reaching an altitude of 2650 m between stations 2 and 3; and 46 cm from station 3 to kilometer 286. The maximum capacity of this phase 1 is 32,000 M^3/D (200/MB/D). Phase 2 of the pipeline included construction from kilometer 286 to Port Coveñas in the Atlantic Ocean. The pipe diameter is 61 cm and required the initial installation of an intermediate pumping station in Field Six. This pipeline will have a capacity of 32,000 M^3/D (200 MB/D) from kilometer 286 to Ayacucho, and 32,000 M^3 (200 MB/D) from Ayacucho to Coveñas.

A loading terminal, located 10 km offshore, will be constructed in Coveñas. This terminal will have a floating storage unit (FSU), with a capacity of 120,000 dwt, to supply the tankers. The FSU will be connected to the coast by two pipelines of 61 cm and 91 cm diameter, passing through the primary and secondary anchorage buoys, respectively. The terminal operation will require a large infrastructure that includes equipment structure, work and support barges, personnel specialized in maintenance of submarine equipment, fire control, and contamination control.

CONCLUSIONS

The discovery of the Caño Limón field is the most important oil find in a decade because of its giant size and because it opens new exploration frontiers in the Llanos basin of Colombia. This area previously was dismissed by many as marginal, but now emerges as a world-class producing province. The Caño Limón field reservoir parameters are also of world-class quality, a prime reason for which recovery is expected to be at least 50% of the field's currently (November 1986) estimated 2 billion bbl of oil in place. The production and transportation facilities for Caño Limón crude oil has required an investment of approximately 1.5 billion U.S. dollars as well as the latest in engineering technology to meet the challenges of laying a trans-Andean pipeline. The Caño Limón project, with its planned modern infrastructure and unique offshore storage and loading facilities, no doubt will become one of the greatest engineering projects of recent years. Development of the Caño Limón field reserves is expected to significantly affect the economy of Colombia, which will benefit from future oil sales as the country regains its self sufficiency and oil exporting capabilities that were lost in the mid-1970s.

ACKNOWLEDGMENTS

The technical information contained in this chapter was generated by several groups and individuals working for Occidental of Colombia and Occidental Exploration and Production Co. However, the geophysical, development geology, and engineering groups of Occidental of Colombia deserve special recognition; particularly, Rodolfo Anzoleaga, George Kendall, Guimer Dominguez, and Jorge Camacho for their contributions. I would also like to express my gratitude to the executives of Occidental and Ecopetrol, who allowed the presentation of this report.

REFERENCES CITED

McCullough, C. N., and E. P. Padfield, 1985, Petroleum geology of Colombia's Llanos basin: a summary: Oil and Gas Journal, v. 83, no. 15, p. 82-86.

Miller, T. A., and F. Etayo-Serna, 1972, The geology of the Eastern Cordillera between Aguazul-Sogamoso-Villa de Leiva, *in* Geological field trips Colombia 1959-1978: Colombian Society of Petroleum Geologists and Geophysicists, Publication 1979, 13th Annual Field Conference (June 23-24, 1972), p. 349-379.

Robertson Research (US), Inc., 1983, The northern Llanos of Colombia, hydrocarbon potential and stratigraphic control: Special report published for oil company subscribers only, v. 1 Report, p. 1-1 - A-4-20.

Taylor, J. B., 1984, Occidental Caño Limón discovery in the Llanos of Colombia, *in* New ideas, new methods, new developments: New York, Matthew Bender & Company, republication of proceedings of the Southwestern Legal Foundation, Exploration and Economics of the Petroleum Industry, v. 22, p. 279-305.

Valderrama, R., 1982, Development of facies in the basin of the eastern Colombian plains, *in* First symposium of petroleum exploration in the subandean basins of Venezuela, Colombia, Ecuador and Perú, August 1982: Colombian Society of Petroleum Geologists and Geophysicists, v. 2, p. 1-32, (reprint in English, 1984).

Ericksen, G. E., Cañas Pinochet, M. T., and
Reinemund, J. A., editors, 1989, Geology of the Andes
and its relation to hydrocarbon and mineral resources:
Houston, Texas, Circum-Pacific Council for Energy
and Mineral Resources Earth Science Series, v. 11.

CHAPTER 28

HYDROCARBON RESOURCES IN THE COASTAL ZONE OF CHILE

Eduardo González
Empresa Nacional del Petróleo (ENAP)
Santiago, Chile

ABSTRACT

The Chilean continental margin is a plate-convergent boundary marked by an arc-trench system. The Chile-Perú Trench and Andean Cordillera are parallel for about 4300 km, from the border with Perú to the Diego Ramírez Islands in southern Chile. North of Valparaíso, the shelf is extremely narrow—5 to 20 km wide—and has only a thin sedimentary cover, whereas to the south, from Valparaíso to the Diego Ramírez Islands, the shelf is 10 to 100 km wide and has sedimentary basins with relatively thick fills. Twelve forearc sedimentary basins having a total area of more than 130,000 km^2 have been studied in the Valparaíso-Diego Ramírez sector of the continental shelf. More than 4000 m of Late Cretaceous, Tertiary, and Quaternary marine clastic sediments overlie the basement in the depocenters of these basins.

A complex fault and fracture system, characterized by strike-slip faults, dissects the continental shelf, and the sedimentary wedge shows evidence of compressional folding. Several regional unconformities are present—an erosional unconformity at the top of the basement; a low-angle, angular unconformity at the base of the Miocene; an unconformity that separates the Pliocene subhorizontal sedimentary sequence from the faulted and slightly folded Miocene sequence; and an uppermost unconformity that separates Pliocene and Quaternary sediments. The continental slope is abrupt, and is broken by structural terraces having hanging sedimentary basins to ocean depths ranging from 200 to 2000 m. The most remarkable is the Tarapacá basin of northern Chile. Other hanging sedimentary basins are present in the Valparaíso, Madre de Dios, and Diego Ramírez areas. Deep canyons, which are offshore prolongations of main rivers and Pleistocene glacial valleys, cut across the continental shelf and slope.

Between 1962 and 1986, more than 32,500 km of offshore multichannel seismic-reflection lines were recorded on the continental shelf and Gulf of Ancud. As of December 1986, 18 exploratory wells had been drilled to depths of 625 to 3267 m in the continental shelf between Mataquito and Evangelistas Islands.

The basement rocks of the continental shelf and upper slope sedimentary basins are of Paleozoic age, chiefly metamorphic (schists, metasandstones, phyllites, and quartzites) and granitic rocks. The basement of the Madre de Dios basin consists of limestone, sandstone, slate, and basaltic lava of Paleozoic age. The basement rocks probably were deformed and metamorphosed during the late Paleozoic (Permian) to such an extent that they were incapable of generating hydrocarbons when the basins were formed during the Late Cretaceous and Paleogene.

Hydrocarbon-generating rocks, reservoir rocks, and traps are present in the Cretaceous-Tertiary basin fills. The best hydrocarbon-generating rocks are of Late Cretaceous age and are capable of generating gas-condensate and oil. Tertiary rocks are low in organic matter and immature, and tend to generate biogenic gas. As a consequence, in basins throughout the shelf and southern end of the Central Valley, where seismic-reflection lines indicate only a Tertiary sedimentary sequence, the most probable hydrocarbon accumulations are biogenic gas. In basins where pre-Tertiary marine beds are inferred, gas and oil accumulations are possible. The best reservoir rocks observed in the exploratory wells drilled in continental shelf basins and intermontane basins of the south Central Valley are basal sandstones of Late Cretaceous age and basal Oligocene-Miocene sequences. The most common sealing rocks of the sedimentary sequence are Tertiary claystone and shale.

Numerous potential hydrocarbon traps are in basins of the continental shelf and southern part of the Central Valley. These traps consist of beds draped over basement highs or over blocks lifted by faults, fractured or weathered basement rock in uplifted fault blocks, and Cretaceous and early Tertiary sandstone wedges on the flanks of basement highs.

Oil accumulations have not been discovered in the continental shelf or in the southern part of the Central Valley. Gas accumulations of methane, ethane, propane, and butane have been found in sandstones of Late Cretaceous age in the Arauco basin. Accumulations of biogenic gas, having a high content of nitrogen, have been discovered in the Tertiary sedimentary sequence, most notably in the F well of the Valdivia basin. The gas reserves of the Valdivia basin may be of significant size, but additional exploration is required to evaluate their potential.

Additional exploration work is required to completely evaluate the hydrocarbon potential of the continental shelf basins of Chile. Priority for such exploration should be given to areas where Mesozoic source rocks capable of generating hydrocarbons are known to be present and potential reservoirs are anticipated.

INTRODUCTION

Oil exploration began in Chile at the end of the 19th century, with participation by both state-owned and private companies. These early activities were concentrated in the precordillera folded zone of the Magallanes basin, and in the plateau region of Antofagasta and Atacama. Between 1908 and 1917 approximately 10 wells, located near known gas and oil seeps, were drilled in the Magallanes basin. Up to 1926, geological surveys were carried out by private companies holding petroleum concessions. In 1927, Standard Oil Company, Royal Dutch, and Pan America Company requested petroleum concessions in Magallanes, but these concessions were not granted. After that date, exploration activities were carried out by the Superintendencia de Salitre y Minas, and later by the Servicio de Minas del Estado and the Departamento de Minas y Petróleo, and, finally, in 1940 by the Corporación de Fomento de la Producción (CORFO).

Exploration by CORFO culminated with the discovery of Chile's first oil field (Manantiales) in December 1945. Since that date, more than 90 oil and gas fields have been discovered in Magallanes. The Empresa Nacional del Petróleo (ENAP) was created in 1950 for the purpose of exploring and exploiting the hydrocarbon resources of Chile. From its inception until December 1986, ENAP has carried out geological and geophysical surveys, and has drilled more than 2700 onshore and offshore wells throughout the national territory.

The participation of private companies in oil exploration and exploitation activities started with the granting of Operations Contracts in 1977 and 1978 to the consortia Atlantic Richfield Company (ARCO)/Amerada Hess Corporation, and Phillips Petroleum Company/ARCO/Amerada Hess Corporation. Between 1979 and 1982 these consortia carried out geological, seismic, and aeromagnetic surveys, and drilled four offshore wells. In 1982 their contract areas were relinquished, without having discovered commercial oil or gas fields (ENAP, 1982).

To date, the Magallanes basin is the only oil and natural gas producing area in the country. The accumulated production to December 1986 was 54,309.3 Mm^3 of oil (341,064.7 Mbbls), with an annual daily average of 5315 m^3 (33,384 bbls) in 1986 (ENAP, 1986).

ENAP became interested in exploring the Pacific continental shelf as a result of the geophysical studies conducted by the *USNS Charles Davis* (Scholl et al., 1970), which established the existence of offshore forearc sedimentary basins. Exploration activities carried out from 1970 to the present include aeromagnetic, seismic-reflection, seismic-refraction surveys, and drilling of 18 wells (including four drilled by ARCO and Phillips). Cooperation from the United Nations Development Program was received during the first stages of exploration. By the end of 1986, 32,530 km of seismic-reflection lines—both reconnaissance and detail—had been recorded offshore, between Arica and the Diego Ramírez Islands, including surveys made by ARCO and Phillips. The seismic lines have been complemented with gravimetric and magnetic profiles.

ENAP also carried out onshore exploration in the coastal zone, offshore islands, and in the Osorno-Llanquihue area. These exploration activities included geological, aeromagnetic, gravity, seismic-refraction, and seismic-reflection surveys, and drilling of 50 exploratory wells.

REGIONAL GEOLOGY AND MORPHOLOGY

Continental Shelf and Slope

The Chilean continental margin, located in a typical arc-trench system, has an area of more than 4500 km^2. It has probably been subject to active subduction since the Cretaceous, with the oceanic plate moving toward the east beneath the westward-moving continental plate throughout the Chile-Perú Trench (Figure 1) (Mordojovich, 1974, 1981).

From Arica (lat. 18°25′S) to Valparaíso (lat. 33°S), the shelf trends north-northeast to south-southwest and is only 5 to 20 km wide. Forearc sedimentary basins of limited extension are present in this segment, and probably are filled with Cenozoic sediments. The continental slope is abrupt, is interrupted by structural terraces, and has some hanging sedimentary basins. Deep canyons cut the shelf and slope at the outlets of main rivers.

From Valparaíso to the Diego Ramírez Islands (lat. 56°30′S), the shelf has an average width of more than 35 km, and in some sectors of the Itata, Valdivia, Chiloé, and Diego Ramírez basins, it has a width of 60 to 100 km. The general trend is north-northeast to south-southwest from Valparaíso to the Taitao Peninsula, where the trend changes to north-south, and finally a northwest-southeast trend in the southernmost part of the continent. Several large forearc basin segments contain thick sequences of Late Cretaceous and Cenozoic marine clastic sediments. The continental slope is broken by structural terraces on which are hanging basins. The continental shelf and slope are cut by a complex fault system, and are furrowed by deep canyons formed in the outlets of the main rivers and Pleistocene glacial valleys. Offshore from the Taitao Peninsula, the Chile Rise (an active spreading center) intersects the Chile-Perú Trench. The shelf is extremely narrow—no more than 5 km wide—at this point of intersection.

Coastal Zone

The geology of the coastal zone and western part of the Coastal Cordillera has been discussed by Brüggen (1950), Muñoz-Cristi (1956), Zeil (1964), and Ruíz (1965). The Geological Map of Chile, published by the Servicio Nacional de Geología y Minería (1982), gives the latest version of the regional geology of this zone.

North of lat. 21° to 18°20′S, adjacent to the Tarapacá basin, a sequence of marine sedimentary rocks interfingers with Mesozoic and Paleozoic volcanic and continental rocks that unconformably overlie a basement of Paleozoic metamorphic and granitic rocks. Mesozoic and Cenozoic plutonic rocks of wide areal extent intrude this sedimentary sequence in the Coastal Cordillera (Salas et al., 1966; Ferraris and Di Biase, 1978; Maksaev, 1984).

Neritic and coastal plain deposits are exposed in the coastal zone between Valparaíso and the Golfo de Penas (lat. 47°S), where they unconformably overlie Paleozoic metamorphic and granitoid rocks of the crystalline basement of the Coastal Cordillera (Muñoz-Cristi, 1964; González-Bonorino, 1971; Munizaga, Aguirre, and Hervé, 1973; Hervé et al., 1976). The exposed sedimentary sequences are represented by Late Cretaceous and Cenozoic sediments in the offshore forearc basins

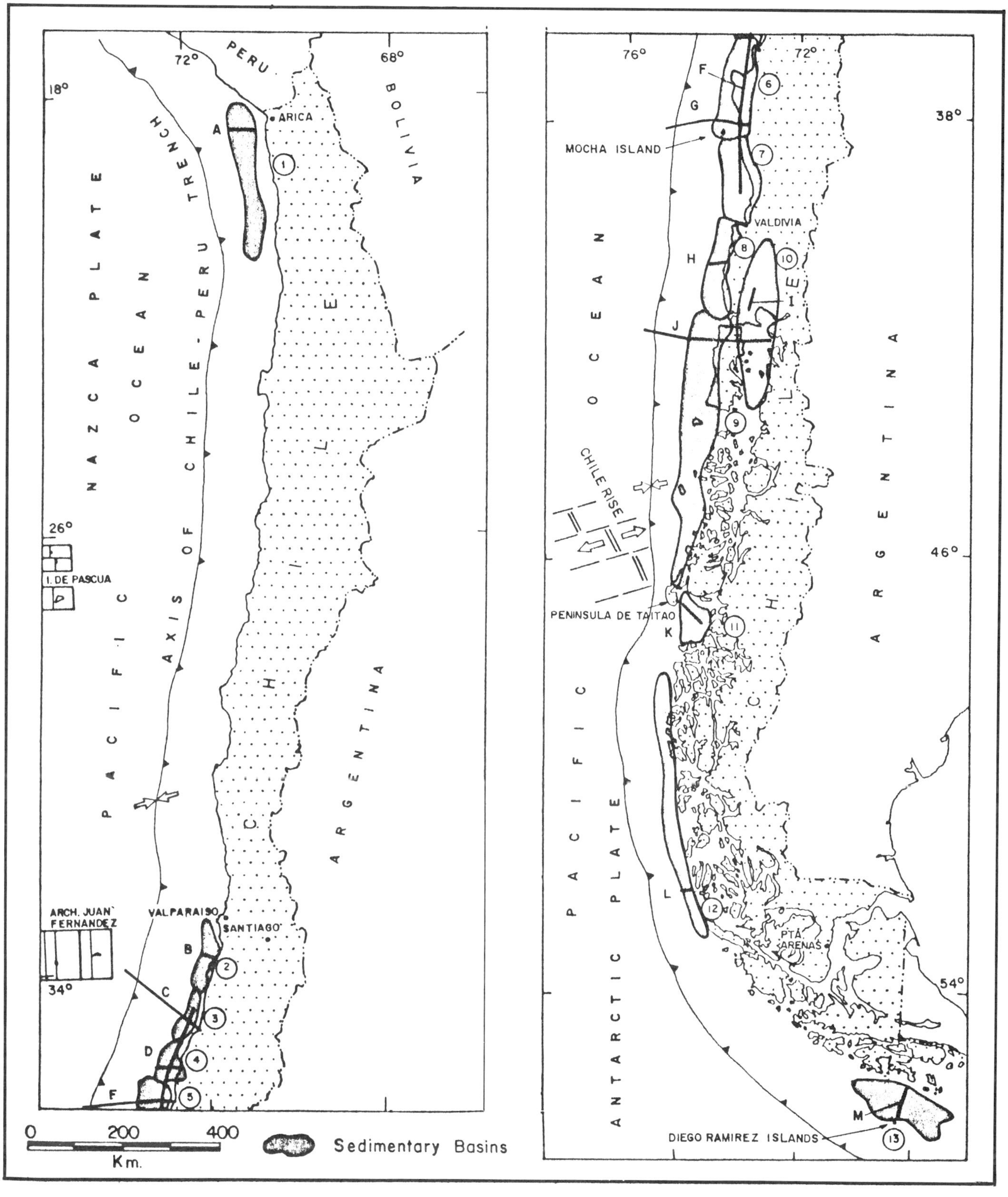

Figure 1. Index map of Chile showing sedimentary basins (shaded areas) along the continental margin, and locations of seismic-reflection profiles and geological cross sections A through L (Figs. 2-14). Sedimentary basins: 1 = Tarapacá; 2 = Navidad; 3 = Mataquito; 4 = Chanco; 5 = Itata; 6 = Arauco; 7 = Valdivia; 8 = Pucatrihue; 9 = Chiloe; 10 = Osorno-Llanquihue; 11 = Golfo de Penas; 12 = Madre de Dios; 13 = Diego Ramirez. Seismic profiles and geological cross sections also shown.

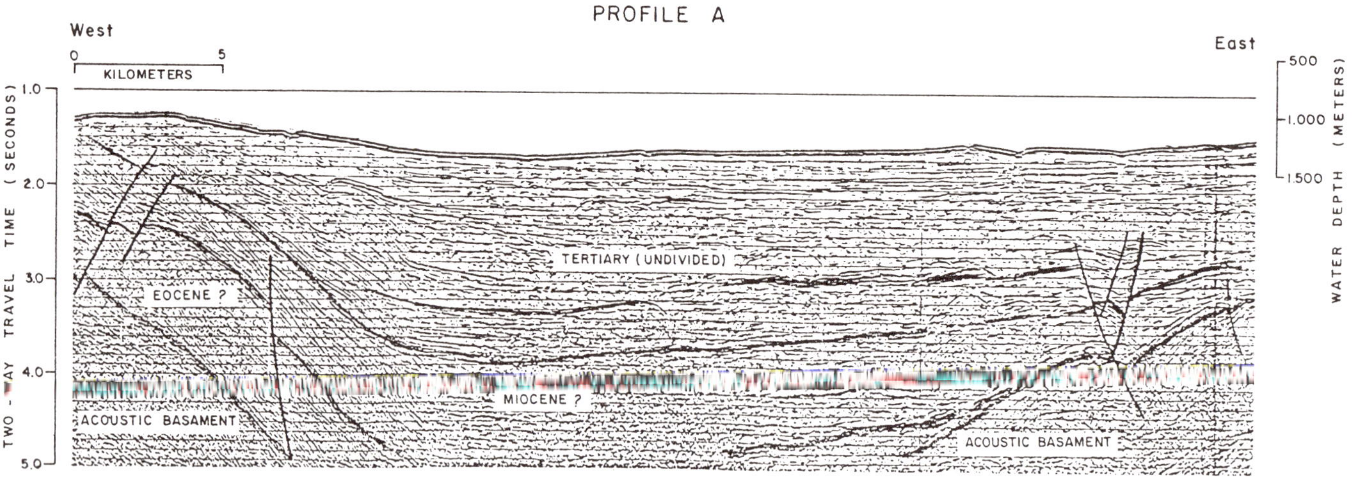

Figure 2. Seismic-reflection profile A, west of Arica, showing seismic stratigraphic characteristics and structural style (wrench faults and flower structure) of the Tarapacá basin. Beds of Miocene(?) age and older onlap the eastern margin of the basin.

of the continental shelf. In general, these beds have gentle dips. Angular unconformities separate some of the Paleogene and Neogene units from the Quaternary units. Tertiary and Late Cretaceous beds (Maestrichtian and Danian) are exposed in the localities of Algarrobo, Navidad-Topocalma, Chanco, Tomé-Concepción-Quiriquina Island, and Arauco Peninsula. Tertiary beds are exposed in the Santa María Island, Mocha Island, Toltén, Parga-Maullín, Chiloé Island, Caleta Manzano, Ayacara, Guafo Island, Guamblin Island, Ipun Island, and Golfo de Penas.

Paleozoic marine beds and doleritic basalts, intruded by Patagonian batholith granitoids of Late Jurassic to Neogene age, crop out in the area from the Golfo de Penas southward to Desolación Island (lat. 53 °S) (Cecioni, 1956; Forsythe and Mpodozis, 1983). These rocks form the basement of the Madre de Dios basin. Granitoid rocks of the Late Jurassic-Neogene Patagonian batholith, Middle Jurassic to Early Cretaceous clastic volcanic beds, Mesozoic-Cenozoic rocks of Ildefonso Island (Suárez and Pettigrew, 1976; Suárez, 1977; Mpodozis, 1980), and metamorphic rocks of the phyllite-blue schist and basaltic lava complex of the Diego Ramírez Islands (De la Rue, 1959; Dalziel et al., 1983) are exposed in the Patagonian Archipelago between Desolación Island and the Diego Ramírez Islands (Figure 1).

In the Coastal Cordillera of the Vichuquén-Tilicura area, between lat. 34 °45 ′ and 35 °S, marine clastic beds of Late Triassic-Early Jurassic age crop out. These have a thickness of more than 1500 m, and fill small graben basins in the Paleozoic basement of granitoid and metamorphic rocks. These basins probably developed in a platform environment west of the Mesozoic volcanic-sedimentary geosyncline and were separated from it by an area of emerged Paleozoic rocks (Corvalán, 1976). Sedimentary sequences of that age have not been observed to date in wells drilled in the continental shelf, but they may be present in graben blocks in the shelf and upper slope. Offshore well data indicate that the crystalline basement of the shelf, between lat. 34 °30 ′ and 40 °S, consists of Paleozoic schist and phyllite. Offshore wells south of Valdivia did not reach the crystalline basement. In some areas of the continental margin, the basement probably consists of rocks of various ages in an accretion complex.

The thickest Late Cretaceous and Early Tertiary sedimentary sequence has been observed in the Itata and Arauco basins (Figure 8), mostly in the western part of the shelf, and probably along the upper slope. The beds in these areas probably accumulated in a neritic to bathyal environment, after the initial folding and intrusion phase that affected the sedimentary sequence of the Andean geosyncline. As a result of this orogeny, most of the Chilean territory probably became an uplifted area (Muñoz-Cristi, 1956; Scholl et al., 1970).

DESCRIPTIONS OF BASINS

The basins are elongate forearc basins in the shelf and upper slope at the front of the South American Plate. Seismic surveys have shown 12 such basins from the border with Perú (lat. 18 °20 ′S) to the Diego Ramírez Islands (lat. 56 °30 ′S) (Figure 1). The basins are related to strike-slip faults caused by oblique convergence of the Pacific oceanic plate and the continental plate. These faults caused displacement and rotation of blocks at the continental margin. Strike-slip faults are particularly prominent in the Madre de Dios and Diego Ramírez basins south of the Chile Rise (Figure 1).

Tarapacá Basin

The Tarapacá basin, in northernmost Chile between lat. 18 °20 ′ and 21 °S, is 20-90 km offshore (Figure 1), and extends over an area of about 8000 km^2 from 900 to 1200 m below sea level. This basin contains more than 4000 m of sedimentary fill, chiefly in a terrace of the upper slope. This fill, which overlies an acoustic basement, is a slightly deformed sequence showing prograded sedimentation. It is inferred to be of Eocene-early Miocene age, and is overlain by a thick sequence of slightly deformed Miocene-Pliocene sediments (Figure 2). Several synthetic and antithetic faults, parallel to the coast, cut the acoustic basement and overlying sedimentary sequence, and delimit the dislocated blocks of the basement. The economic basement probably consists of the same Mesozoic sequence exposed in the coastal zone (Coulbourn, 1977; Ferraris, 1977; Mordojovich, 1974, 1981).

The lack of exploratory wells and insufficient seismic lines

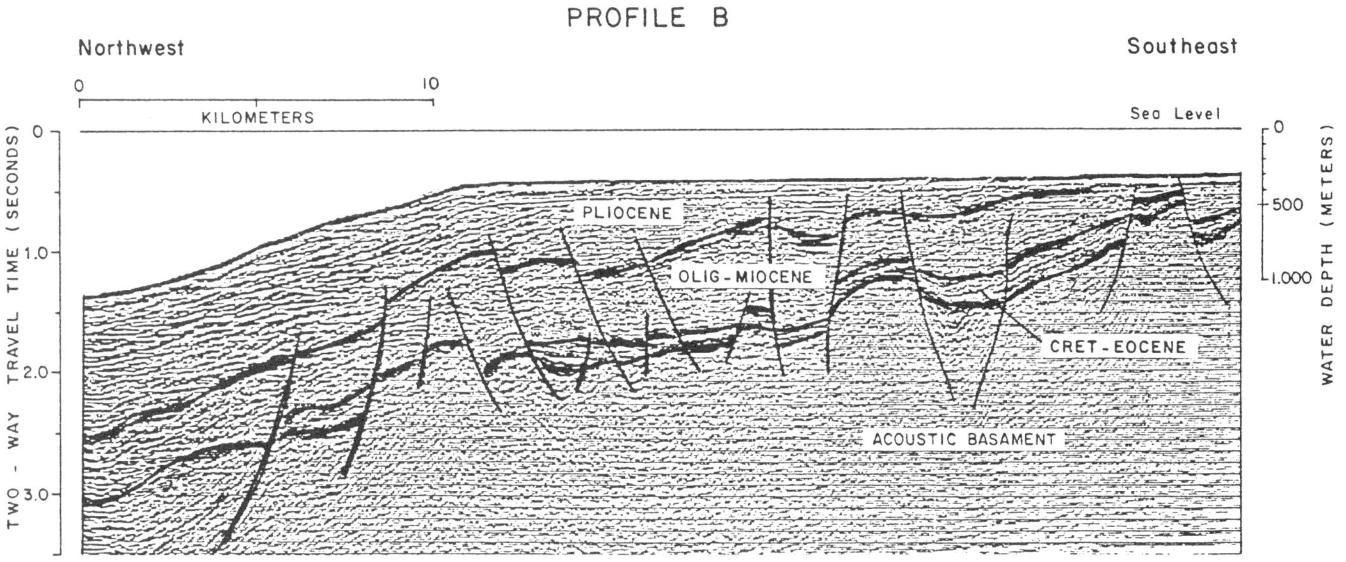

Figure 3. Seismic-reflection profile B, west of Navidad (lat. 33°53′S), showing the sedimentary fill and faulting pattern of the Navidad basin.

does not allow an accurate evaluation of this basin. However, information on the forearc basins of the Peruvian continental margin indicates that the Tarapacá basin probably contains generating, reservoir, and sealing rocks. The best potential for discovery would be in basal clastic reservoirs of traps formed in faulted blocks and other structural and stratigraphic elements.

Navidad-Algarrobo Basin

The Navidad-Algarrobo basin, between lat. 33° and 34°30′S, covers an area of 4000 km^2 and extends to an ocean depth of 200 m. The basin covers the continental shelf and upper slope and extends onshore into the Navidad embayment. The sector under study is the eastern part of the basin where the sedimentary fill consists chiefly of Late Cretaceous, Eocene, Oligocene-Miocene, and Pliocene marine basal conglomerates, sandstones, siltstones, and claystone clastics overlying a Paleozoic basement of granitoid rocks and mica schists. The sedimentary sequence increases gradually in thickness westward from the coast, and reaches a maximum thickness of more than 3000 m in the margin of the shelf. Many steeply dipping synthetic and antithetic faults of pronounced displacement, parallel to the coast, caused formation of horsts and grabens, chiefly in the pre-Miocene part of the sequence. The older beds of the sedimentary sequence, probably of Late Cretaceous and Eocene age, fill the graben depressions in the basement (Figure 3). Unconformities are present at the bases of the Pliocene and Oligocene-Miocene units and at the top of the crystalline basement. The continental shelf and slope are furrowed by a deep canyon at the outlet of the Maipo River, apparently excavated along a slip-strike fault.

Slightly folded sedimentary rocks, as thick as 400 m, are exposed in the coastal zone of the Navidad-Topocalma area (lat. 33°40′-34°10′S) where they overlie granitoid basement rocks (Darwin, 1846, Brüggen, 1950; Leví and Aguirre, 1960; Muñoz-Cristi, 1964; García, 1968; Mordojovich, 1974, 1981; Botinelli, 1975a; Cecioni, 1978; Céspedes, 1978b; Tavera, 1979; Pérez and Reyes, 1980). These consist of Late Cretaceous marine clastic conglomerate and sandstone; Eocene sandstone and claystone containing carbonaceous matter; Oligocene-Miocene siltstone and laminated claystone; and Pliocene marine sandstone and claystone. Navidad is the type locality of the Navidad Formation (Darwin, 1846; Hoffstetter, Fuenzalida, and Cecioni, 1957).

No offshore wells have been drilled in this basin, and consequently the oil and gas potential can be inferred only by comparison with the Mataquito and Chanco basins to the south. In these basins, the hydrocarbon-generating rocks are estimated to be Eocene and Late Cretaceous claystone and shale deposited in grabens. The probable reservoir rocks are quartz sandstone at the base of the sedimentary section. Reservoir traps are closed structural anomalies related to horsts and grabens, structural noses in graben blocks, and stratigraphic traps. The Eocene and Late Cretaceous claystone and shale may be good sealing rocks. Several of the structures outlined by the seismic lines at the southern end of the basin may have considerable closure areas.

Mataquito Basin

The Mataquito basin, between lat. 34°30′ and 35°20′S, covers an area of 2600 km^2 and reaches an ocean depth of 200 m. It extends over the shelf and upper slope. The area under study is the eastern part of the basin. This basin is characterized by blocks separated by high-angle faults of considerable displacement, by downdropped graben blocks to the west, and by strike-slip faults parallel to the coast. The faults chiefly affected the basement and the pre-Miocene sedimentary sequence. The tectonic model suggests the existence of Late Cretaceous or older sedimentary rocks at the continental shelf margin. More than 3000 m of marine clastic sediments were deposited in this basin. From base to top these are: 100-150 m

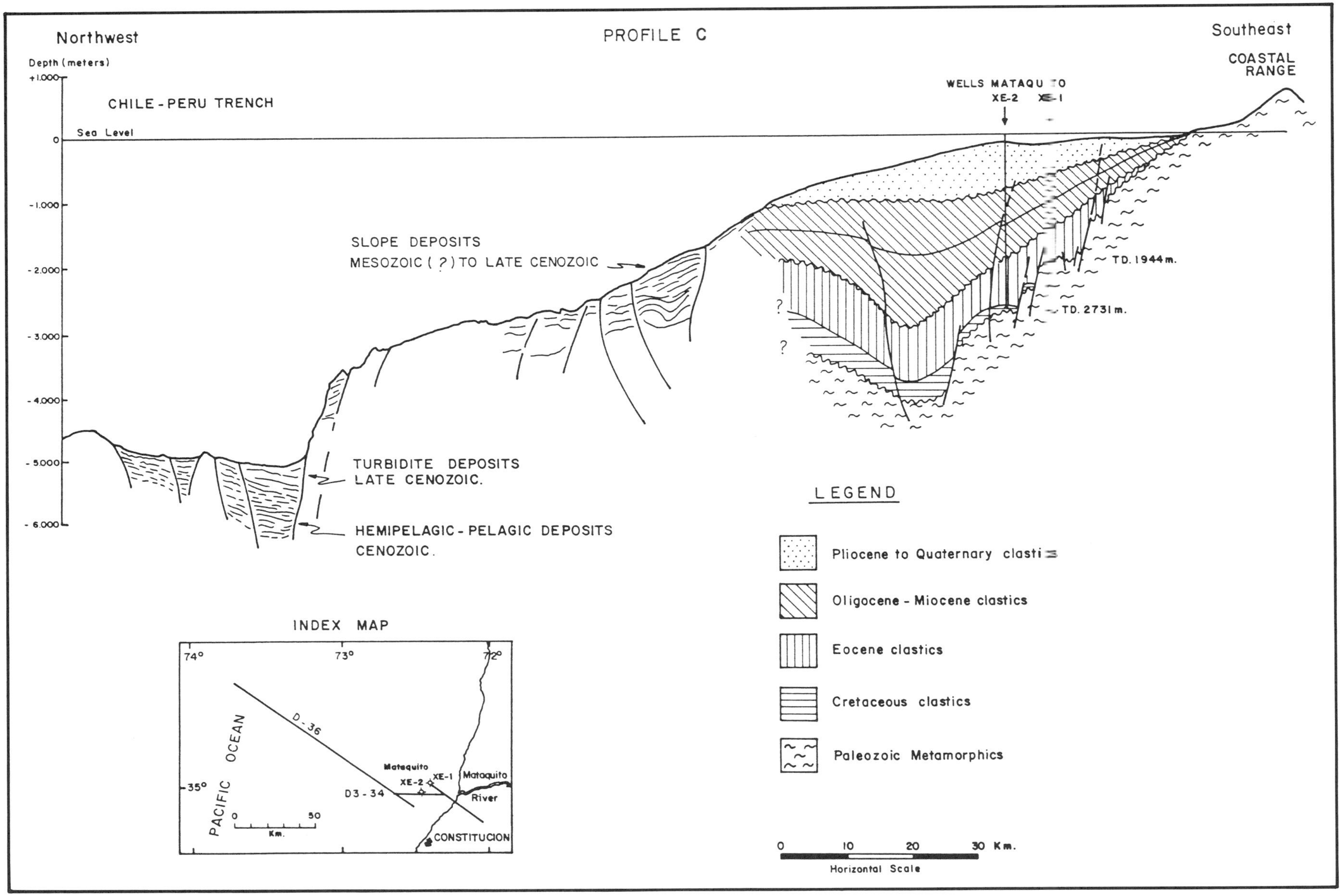

Figure 4. Diagrammatic geologic cross section (profile C), along lat. 35 °S, modified after seismic profile D-36 (Scholl et al., 1970); seismic-reflection profile ENAP D3-34 (Ledezma, 1979b); and exploration well data.

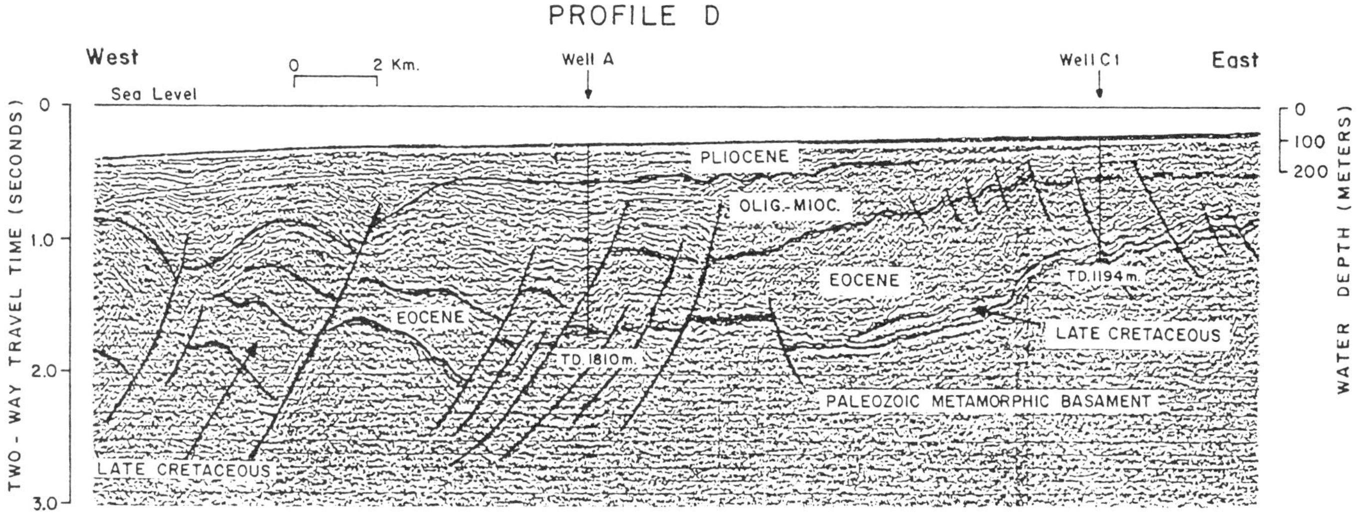

Figure 5. Seismic-reflection profile D, west of Bahia Chanco (lat. 35 °36 'S), showing seismic stratigraphic characteristics and typical structural anomalies (folds and faults) of the Chanco basin. Late Cretaceous beds wedge out downdip at the eastern flank of the basin.

of Late Cretaceous and early Eocene basal sandstone; 300-450 m of Eocene claystone with silty and sandy intercalations; 400-700 m of Oligocene-Miocene laminated claystone with silty and calcareous intercalations; and 800-1000 m of Miocene-Pliocene, locally conglomeratic siltstone and claystone.

The tectonic style of the continental margin and sedimentary fill of the Mataquito basin is shown in Figure 4. Seismic data indicate the presence of positive flower structures produced by strike-slip faults. Important unconformities are at the top of the basement, at the bases of the Oligocene-Miocene and Pliocene sequences.

Possible source rocks are represented by Eocene, Late Cretaceous, and, probably, Jurassic claystone and shale. The most important reservoir rocks are Late Cretaceous and basal Eocene quartz sandstones deposited in a transgressive environment. Their maximum thickness is 130 m, in Mataquito XE-2, where methane showings were registered. Cretaceous and Tertiary claystone and shale overlying potential reservoir sandstones constitute the sealing beds.

Several prospective traps having closed areas of 5 km^2 to more than 20 km^2 and vertical closures of 30 m to more than 40 m have been studied. These traps consist of folds in horst blocks of the basement, positive roll-over and flower structures, and anticlines in graben blocks between faults. The depth of the main reservoirs ranges from 900 m to more than 2700 m. Mataquito XE-1 did not reach the basal reservoir, possibly because of a fault. Mataquito XE-2 reached basal sandstone that apparently was outside the closure area of the trap.

Geochemical analyses of late Eocene to Miocene rocks indicate poor hydrocarbon-generating potential, low content of organic matter (0.65 to 0.70% organic carbon), and immaturity (vitrinite reflectance: 0.48% average Ro). The best prospective areas are in graben blocks or in semigrabens. The hydrocarbons possibly migrated from the upper slope, where Late Cretaceous and Jurassic deposits probably existed (Vichuquén-Tilicura beds of the continental margin).

Chanco Basin

The Chanco basin, between lat. 35 °20 ′ and 36 °S, has an area of 4000 km^2 at the 200-m-depth isobath. The Maule and Itata river canyons furrow the shelf and upper slope, delimiting the basin in the north and south, respectively. The Chanco basin covers the continental shelf and upper slope, and extends onshore in the Chanco embayment (Botinelli, 1974; Mordojovich, 1974, 1981; Young and Valenzuela, 1979; Cecioni, 1983; González, 1983). The eastern part of the basin has been studied.

More than 2500 m of marine clastic sediments were deposited in the depocenter of this basin (Figure 5). From base to top, this section consists of (1) Late Cretaceous and Eocene sandstone, shale, and claystone having a total thickness of as much as 750 m; (2) Oligocene-Miocene claystone with intercalations of siltstone and argillaceous fine-grained sandstone that show prograded stratification, having a maximum thickness of 1270 m; and (3) Miocene claystone as much as 470 m thick.

Northeast-trending antithetic faults, which are most numerous in the pre-Miocene sequence in the eastern part of the basin, have formed structural traps. Normal faults, parallel to the coast and near the margin of the shelf, caused lowering of the Paleozoic basement at the slope.

Geochemical analyses indicate a low content of organic matter (0.39 to 0.91% organic carbon) and immaturity of the Eocene and Miocene generating rocks (vitrinite reflectance: 0.49% average Ro), but show a potential for generation of biogenic gas. The Late Cretaceous rocks have a greater content of organic matter (0.74 to 1.06% organic carbon), with moderate maturity (vitrinite reflectance: 0.65% average Ro), which indicates that these rocks are incipient oil generators. Methane-rich gas recovered from onshore wells confirms the degree of maturity of the Late Cretaceous organic matter, and the best generating rocks are probably the Cretaceous sediments of the upper slope and shelf margin. The best reservoir rocks are Late Cretaceous basal sandstones having thicknesses of 23 to 150 m, porosities of 23 to 28%, and permeabilities of 99 to 1727 md; Eocene sandstones with thicknesses of as much as 50 m, porosities of 21 to 28%, and permeabilities of 37 to 716 md; and Oligocene-Miocene basal sandstones.

Several structural traps have been delineated in the offshore study area. These traps are fault-related blocks over which are closures covering 4 to 10 km^2 horizontally and 30 to 100 m vertically, and which contain good reservoir sandstones. Methane has been observed in the majority of the wells drilled onshore and offshore. Still larger traps are probably present in the western part of the basin, where mature generating rocks are believed to exist, and hydrocarbons probably have migrated to this area from the upper slope.

Itata Basin

The Itata basin, between lat. 36° and 36°46′S, has an area of 5600 km^2 and extends to an ocean depth of 200 m. The Bio Bio River Canyon, which cuts across the shelf and upper slope, marks the southern end of the basin. The basin covers the shelf and upper slope, and extends to the coastal zone between Tomé and Concepción.

The area studied in the Itata basin is characterized by a depression, between a horst block of basement rock at the margin of the shelf and the Coastal Cordillera, having a thick Late Cretaceous and Tertiary fill (Figure 6). More than 400 m of sediments, which overlie the crystalline basement of Paleozoic mica schist and phyllites, are present near the depocenter of the basin (Mordojovich, 1974, 1981; Ledezma, 1979a; González, 1983). From base to top, the sedimentary sequence consists of (1) 1456 m of Late Cretaceous quartz sandstone, siltstone, and shale; (2) as much as 330 m of Eocene claystone, fine sandstone, and siltstone; (3) as much as 1300 m of Oligocene-Miocene silty claystone, siltstone, and argillaceous fine-grained sandstone showing prograded sedimentation; and (4) as much as 1000 m of Pliocene conglomerate, argillaceous sandstone, and claystone. The basement rocks and sedimentary fill of the basin are broken by a complex fault system (Figure 6). Unconformities are present at the bases of the Late Cretaceous, Oligocene-Miocene, and Pliocene sequences. A layer of volcanic rock, encountered at the base of the sedimentary sequence in the Itata XE-3 well, has been interpreted as pre-Cretaceous.

Sedimentary rocks of this basin, exposed in the coastal zone of the Tomé-Concepción area and at Quiriquina Island (lat. 36°35′ to 36°50′S), consist of a 300-m sequence of marine and continental (deltaic) conglomerate, sandstone, siltstone, and shale of Late Cretaceous (the Quiriquina Formation of Maestrichtian age) and early Tertiary age (Darwin, 1846; Hoffstetter, Fuenzalida, and Cecioni, 1957; Chotin, 1969; Céspedes, 1978a; Frutos et al., 1982). This sequence, which shows weak deformation, unconformably overlies the Paleozoic metamorphic and granitic basement.

Geochemical analyses of the J well (Figure 6) indicate poor oil potential for the entire sedimentary sequence. The content of organic carbon in the Tertiary sequence is marginal and only reaches values of 1.3 to 2.6% in the Cretaceous sequence. The organic matter is mostly humic. Vitrinite-reflectance values range from 0.90 to 1.02% average Ro, which indicates that the Late Cretaceous generating rocks are thermally mature at a depth of more than 2500 m, the Eocene rocks are only incipiently mature, and the Oligocene-Miocene rocks are immature. These data suggest a tendency to form gas and condensate in Late Cretaceous generating rocks and biogenic gas in Tertiary rocks.

The main reservoir rocks of the basin consist of (1) as much as 200 m of Late Cretaceous basal quartz sandstone having porosities of 19 to 30% and permeabilities of 5 to 132 md; (2) as much as 70 m of Eocene sandstone having porosities of 26 to 30% and permeabilities of 58 to 74 md; and (3) 50 to 80 m of Oligocene-Miocene argillaceous sandstone. Late Cretaceous and Tertiary shale and claystone overlying the sandstones constitute the sealing rocks.

Reservoir traps consist of folded sedimentary rocks over fault-block horsts, sedimentary wedges on the flanks of basement highs, and diapir structures. Many structural traps showing horizontal closures of 5 to 25 km^2 and vertical closures of 100 m to more than 300 m have been identified.

Mesozoic sedimentary rocks of pre-Cretaceous age may be present on the upper slope and in graben blocks of the shelf, as indicated by some of the seismic profiles. Such rocks would be expected to be a better source of hydrocarbons.

Arauco Basin

The Arauco basin, between lat. 36°46′ and 38°30′S, covers an area of 8000 km^2 and extends to an ocean depth of 200 m. Two noncommercial gas accumulations were discovered in this basin at Lebu and Mocha Island. The Lebu 2 well encountered 117 m of Senonian argillaceous and gasiferous sandstone of low permeability at a depth of 1484 m. Well flow was 82,060 m^3/day of gas (99.6% methane, 0.2% ethane) through a 1/2-inch choke. Two gasiferous sandstones were discovered in the Mocha Island, one at shallow depth at the Oligocene-Miocene basal unconformity, and the other at greater depth, near the base of the Senonian sedimentary sequence. At 1642 m the Mocha Norte 2 well encountered a 9-m unit of Senonian argillaceous and gasiferous sandstone that produced a well flow of 16,500 m^3/day of gas (93.35% methane, 1.37% ethane, 0.44% propane, 0.13% isobutane, 4.53% nitrogen) through a 1/4-inch choke. At 603 m the Mocha Norte 3 well encountered a 12-m unit of Oligocene-Miocene gasiferous quartz sandstone that produced a well flow of 25,300 m^3/day of gas (94.91% methane, 1.21% ethane, 0.25% propane, 3.48% nitrogen) through a 5/16-inch choke.

The Arauco basin covers the shelf and upper slope and extends onshore into the Arauco Peninsula as far as the western front of the Coastal Cordillera. The Arauco Gulf and Arauco Peninsula are the regions where the main early Tertiary subbituminous and bituminous coal mines, currently under exploitation in Chile, are located (Toenges, 1948; Wenzel, 1969, 1982). Because the Arauco Peninsula as well as Mocha and Santa María islands are the uplifted part of the shelf, where much subsurface data are available from many drill holes and extensive underground coal mines, this basin is considered to be a key area for geological study of forearc basins at the Chilean continental margin. However, scanty seismic information and tectonic complexity have inhibited interpretation (Tavera, 1942; Muñoz-Cristi, 1946; Brüggen, 1950; Tavera and Veyl, 1958; Martínez, 1968; Botinelli, 1973; Céspedes, 1976; Céspedes and Johnson, 1985).

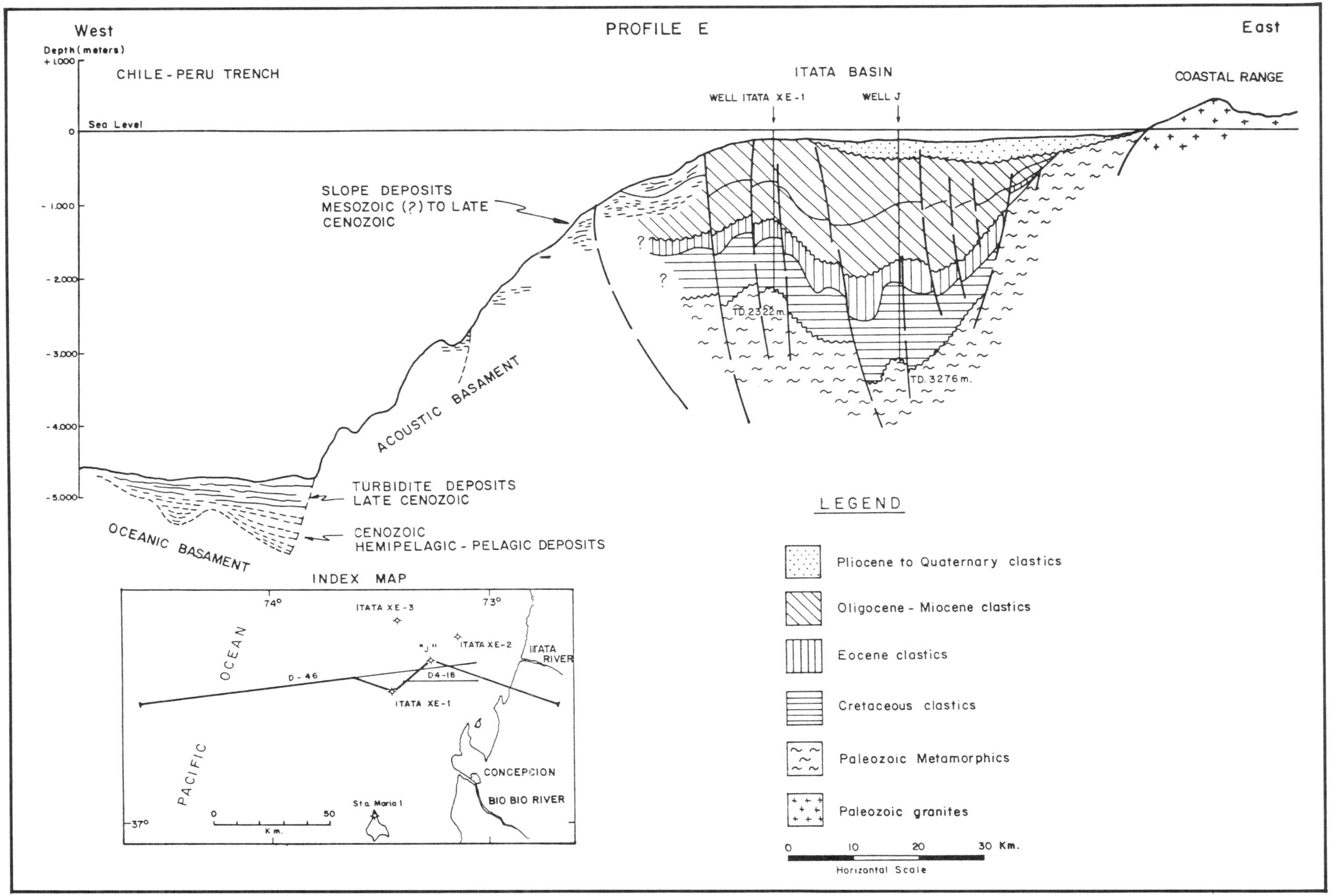

Figure 6. Diagrammatic geologic cross section (profile E), along lat. 36°30′S, modified after seismic profile D-46 (Scholl et al., 1970); seismic-reflection profile ENAP D4-18 (Ledezma, 1981); and exploration well data.

The Arauco basin consists of a depression between the Coastal Cordillera and a horst block of the basement filled with more than 3000 m of Late Cretaceous and Tertiary clastic sediments in the depocenter (Figure 7). During the Late Cretaceous, Eocene, Oligocene, Miocene, and Pliocene the depocenters of the basin migrated eastward. Some seismic lines in the margin of the shelf show a wedging of the basal unit of the sedimentary section, which has been interpreted as pre-Cretaceous. Figure 8 shows the correlation of stratigraphic sedimentary sequences between the Valdivia, Arauco, Itata, Chanco, and Mataquito basins.

From base to top, the sedimentary sequence in the Arauco basin (Figure 7) consists of (1) Senonian basal sandstone, claystone with intercalations of siltstone, and fine-grained sandstone having a maximum thickness of 1150 m; (2) Paleocene and Eocene sandstone with interbedded coal beds, claystone, and siltstone having a maximum thickness of 1600 m, which consists of a deltaic facies in the east and a marine facies in the west over part of the basin; (3) Oligocene-Miocene quartz sandstone and claystone, which contains intercalations of siltstone and fine-grained sandstone, showing prograded stratification to the west and having a maximum thickness of 930 m; and (4) Pliocene claystone and glauconitic quartz sandstone with intercalations of siltstone and fine-grained sandstone, having a maximum thickness of 390 m.

The Paleozoic basement and overlying sedimentary sequence are broken by a complex system of strike-slip and normal faults (Figure 7). Included are predominantly antithetic faults in the Arauco Peninsula, which strike northeast and are parallel to the Coastal Cordillera, and other north-northwest-striking faults associated with elongated anticlinal structures. Unconformities are present at the bases of Late Cretaceous, Oligocene-Miocene (truncating Late Cretaceous, Paleocene, and Eocene sedimentary sequences in the area of the Mocha Este 1 well), and Pliocene sequences.

Geochemical analyses of samples from onshore wells indicate that source rocks in the Tertiary and the uppermost Late Cretaceous are immature or in transition and have only the capacity to generate gas. In contrast, the lower part of the Late Cretaceous sequence, at a depth of more than 1400 m, contains mature rocks having the capacity to generate oil and gas. The composition of the gas from Cretaceous and Tertiary sandstones and gas from Mocha Island indicates adequate maturity of the organic matter in the source rocks.

The main reservoir rocks are (1) Late Cretaceous basal quartz sandstones, having thicknesses of as much as 160 m, porosities of 8 to 18%, and permeabilities of 7 to 444 md; (2) Eocene-Paleocene argillaceous sandstones having thicknesses of 239 to 305 m, porosities of 6 to 15%, and permeabilities of 2 to 3 md; and (3) quartz sandstones above the unconformity at the base of the Oligocene-Miocene sequence on Mocha Island, which have thicknesses of 5 to 35 m, porosities of 11 to 34%, and permeabilities of 2 to 723 md. The main reservoir traps are anticlinal folds associated with lateral faults, fractured basement blocks uplifted by faults, compaction folds above horsts, and stratigraphic wedges.

Because of the geological complexity of the Arauco basin and sparse onshore seismic information, most exploratory wells drilled onshore did not test well-delineated closed traps. However, several structures delineated offshore, near the shelf margin, have horizontal closures of 10 to 40 km^2 and vertical closures of 100 to 200 m (Ledezma, 1981, 1983). These structures are believed to contain Cretaceous and probably Jurassic sedimentary rocks of considerable thickness that are potential source rocks. These rocks probably were deposited in an environment that was more marine than that of the Arauco Peninsula. Their maturity levels are probably higher than those of other rocks in the sedimentary sequence because they are thicker and older. As in the Itata basin, it is possible that the hydrocarbons generated in the Mesozoic rocks of the upper slope migrated toward the structural traps of the shelf margin.

Many gas seeps (93.4% methane, 1.8% ethane, 0.3% propano, 1.3% nitrogen) have been known along the southwestern coast of Mocha Island since the beginning of this century. These gas shows are associated with faulted and fractured claystone and sandy siltstone of Paleocene and Oligocene-Miocene age (Machado, 1903; Tavera and Veyl, 1958; Céspedes, 1980).

Valdivia Basin

The Valdivia basin (Figure 1), between lat. 38°30′ and 39°55′S, covers about 8000 km^2 and extends to an ocean depth of 200 m. The shelf and upper slope in this area are cut by the Toltén and Chaihuin Canyons (Mordojovich and Alvarez, 1977).

The basin extends onshore throughout the Toltén area. It consists of a depression, between the Coastal Cordillera and a basement high in the western part of the continental shelf, filled with a thick sequence of Cenozoic sediments. More than 2500 m of marine clastic sediments have accumulated in the depocenter of the basin. They overlie a Paleozoic mica schist basement.

The F well (Figure 8), which discovered an accumulation of biogenic gas (60.9% methane; 39% nitrogen), was drilled on a small closed structure. The gas accumulation in this well is possibly in a stratigraphic trap that is considerably larger than that of the closed structural anomaly. The reservoir rock is coarse-grained, porous, permeable quartz sandstone, 22 m thick, at the base of the Oligocene-Miocene sequence. It was encountered at a well depth of 1475 m. The gas flow was measured at 225,000 m^3/day through a 5/8-inch choke and a wellhead pressure of 68 Kg/cm^2 (Mordojovich, 1974, 1981; González, 1983).

From base to top, the sedimentary sequence in the Valdivia basin is (1) Eocene siltstone and fine-grained argillaceous sandstone, with intercalations of claystone, having a maximum thickness of 150 m (some seismic profiles show wedging of the Eocene on the flanks of high-relief anomalies); (2) as much as 1100 m of Oligocene-Miocene fine-grained to conglomeratic argillaceous sandstone, claystone, and glauconitic siltstone that overlie the Eocene; the sequence is interrupted by unconformities and erosion channels filled with Pliocene sediments, and shows features of turbidite deposits and prograded sedimentation; and (3) as much as 650 m of Pliocene, fine- to coarse-grained sandstone and claystone containing intercalations of siltstone. North-northwest-trending faults, parallel to the coast, dislocate the basement and sedimentary sequence. Faults having the largest vertical displacements are on the flanks of the basement at the shelf margin. Unconformities appear at bases of the Eocene, Oligocene-Miocene, and Pliocene sequences.

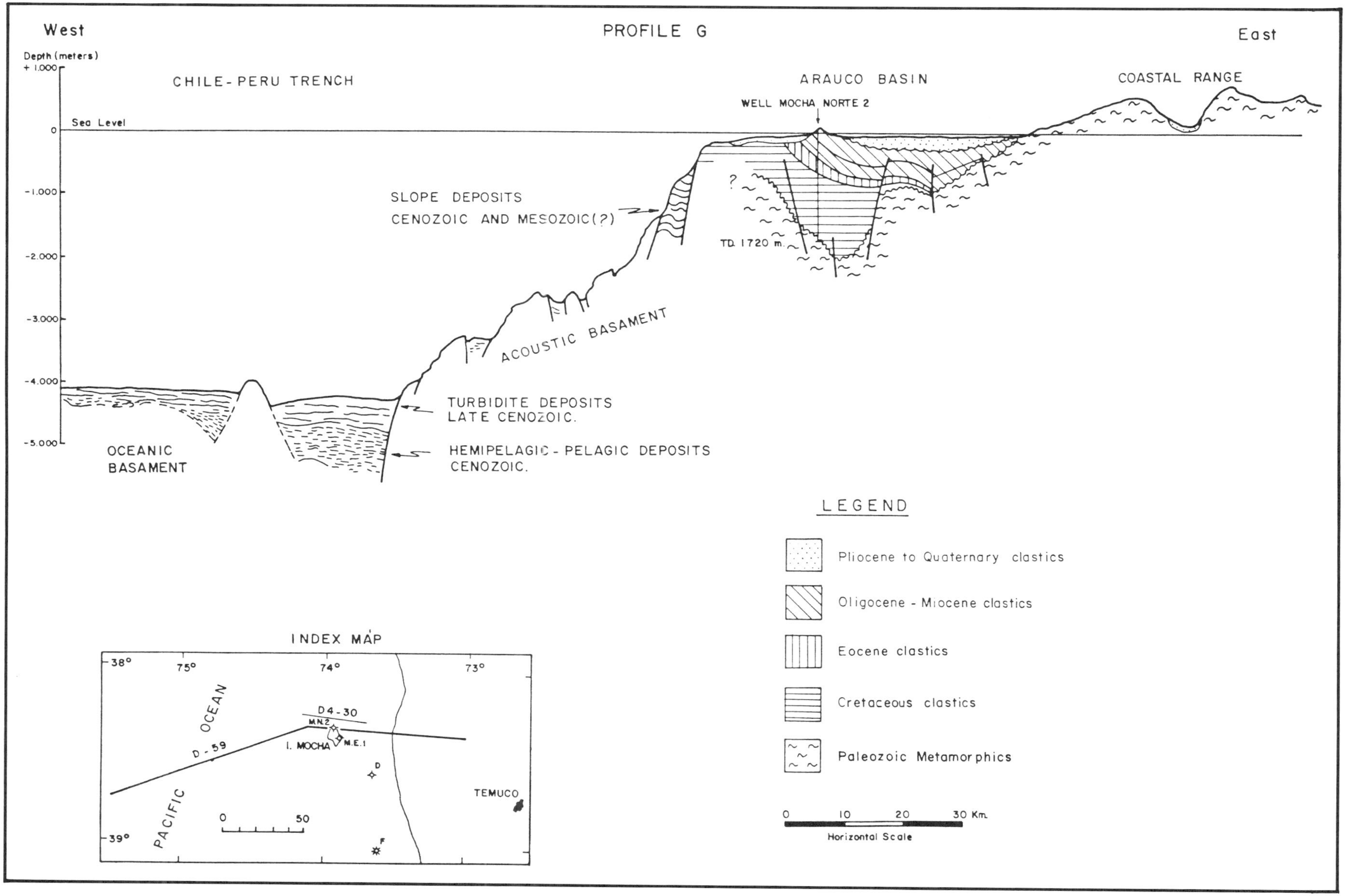

Figure 7. Diagrammatic geologic cross section (profile G), along lat. 38°20′S, modified after seismic profile D-59 (Scholl et al., 1970); seismic-reflection profile ENAP D4-30 (Ledezma, 1981); and exploration well data.

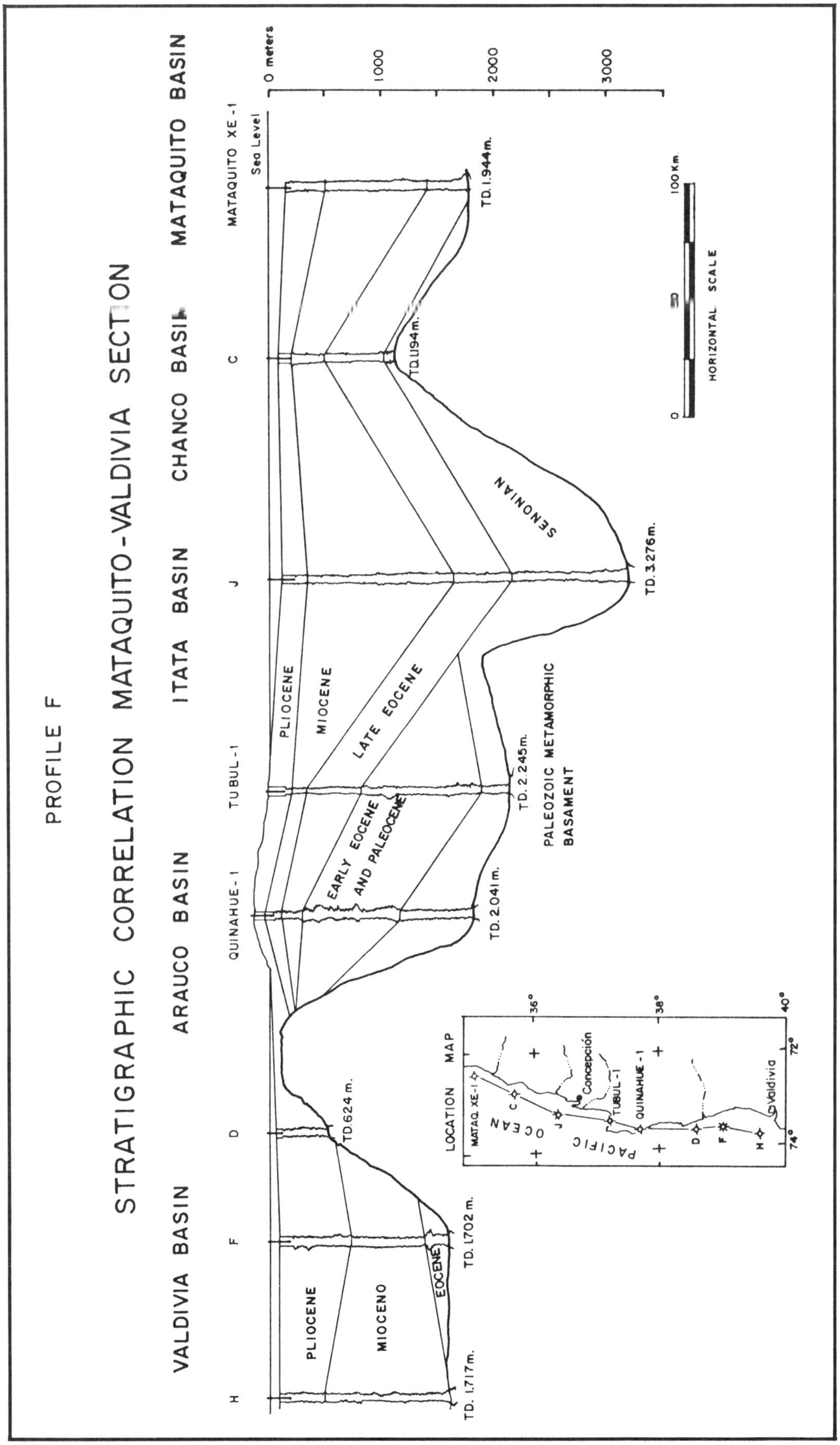

Figure 8. Regional stratigraphic cross section (profile F), showing stratigraphic relationships of Cretaceous and Tertiary units along the Mataquito, Chanco, Itata, Arauco and Valdivia basins (modified after Céspedes, 1976).

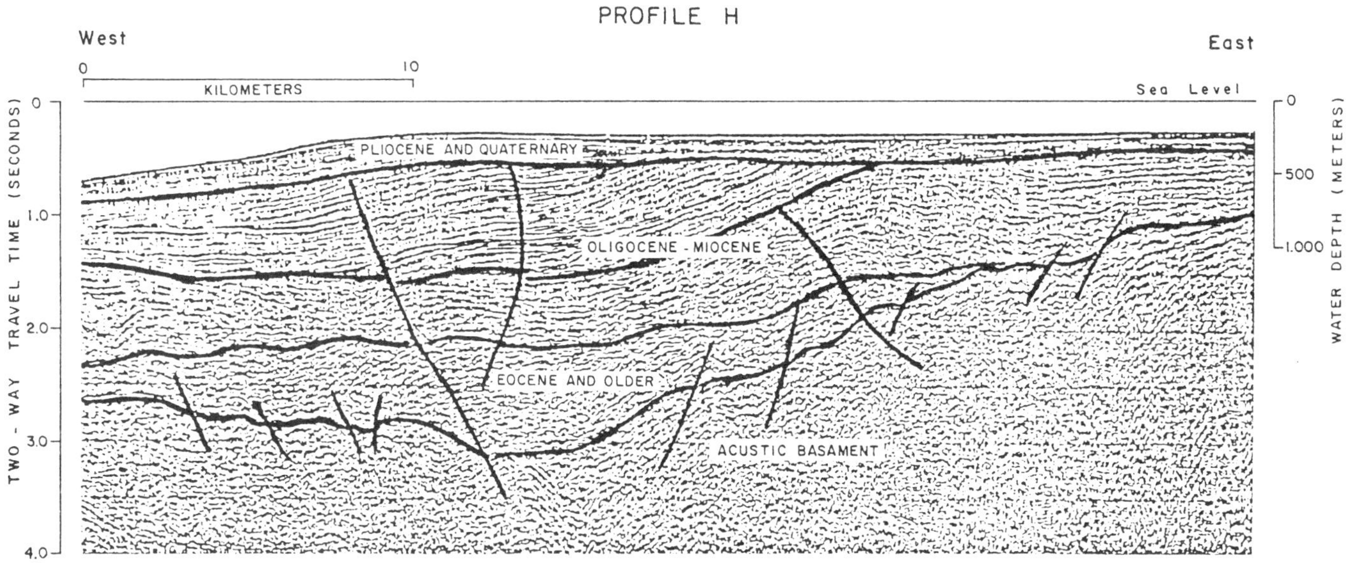

Figure 9. Seismic-reflection profile H, west of Bahia Mansa (lat. 40°35 'S), showing the sedimentary fill and faults in the Pucatrihue basin. Beds of Eocene age and older onlap the eastern margin of the basin.

Geochemical analyses of the F well (Figure 8) indicate that the hydrocarbon potential of the Eocene and Oligocene-Miocene sequences is moderate to poor. The content of organic matter is low, consisting chiefly of kerogen of humic and lignitic origin (types II and III). The Oligocene-Miocene and Eocene sediments are immature and tend to generate biogenic gas only. More mature Late Cretaceous deposits may exist in the deepest parts of the basin. Biogenic gas produced in the sedimentary sequence on the slope probably would migrate toward a Tertiary reservoir sandstone to the west. Several closed structures are present in the Valdivia basin. These show horizontal closures of 2 to 170 km^2 and vertical closures of 20 to 250 m. They might contain commercial gas accumulations in the basal sandstone of Oligocene-Miocene age.

Liquid hydrocarbons, as indicated by ultraviolet fluorescence, are recorded for core samples taken by the *USNS Eltanin* on the Chilean continental shelf, southwest of the F well (Goodell, 1965; González, 1982).

Pucatrihue Basin

The Pucatrihue basin (Figure 1), between lat. 39°55′ and 41°S, covers about 4000 km^2 and extends to an ocean depth of 200 m. The shelf and slope in this area have been furrowed by the Bueno River Canyon.

The basin is on a narrow platform of the shelf and upper slope and has a fill of marine sediments of probable Cenozoic age that is more than 3000 m thick in the depocenter. The sedimentary sequence trends north and the beds dip westward. The basement is probably Paleozoic mica schist and phyllite as exposed on the nearby coast. The lowest sedimentary unit shown in the seismic profile H (Figure 9), which is moderately deformed and unconformably overlies the acoustic basement, has been interpreted as Eocene and older. The slightly deformed middle unit, consisting of two sedimentary sequences (Figure 9) separated by an unconformity, is considered to be Oligocene-Miocene. This unit unconformably overlies the Eocene unit and shows prograded sedimentation to the west. The youngest unit is Pliocene and Quaternary. The Eocene and older sediments and the basement are cut by normal and strike-slip faults striking north-northeast.

The geochemical characteristics of the sediments in the Pucatrihue basin and the petrophysical properties of possible reservoir rocks are unknown. Available seismic information has delineated two prospective structural anomalies near the western end of the shelf having horizontal closures of 20 km^2 and 35 km^2 and vertical closures of 100 m and 200 m. These structures correspond to fault blocks associated with rollover anticlines. As with the other forearc basins along the continental margin of Chile, the hydrocarbons generated in sediments of the shelf margin and upper slope would be expected to migrate into reservoir sandstones in the structural traps in the western part of the shelf.

Chiloé Basin

The Chiloé basin (Figure 10), between lat. 41° and 46°40 'S, covers about 30,000 km^2 and extends to an ocean depth of 200 m over much of the shelf and upper slope. The shelf and slope are furrowed by deep canyons (Mordojovich and Alvarez, 1977; Mordojovich, 1981; Cande and Leslie, 1986). The basin is characterized by a central depression between the coast and a basement high at the margin of the continental shelf (Figure 10). A maximum thickness of more than 3000 m of Cenozoic and possibly Late Cretaceous marine clastic sediments accumulated in the depocenter, which is parallel to the coast. These sediments unconformably overlie crystalline basement, which probably consists of Paleozoic mica schist and phyllite. Because of scanty seismic information and lack of exploratory wells, the nature and morphology of the basement high at the shelf margin are unknown. Subsurface information from test wells indicates that the stratigraphic sequence in the basin consists of two units separated by an unconformity. The upper unit is Oligocene-Miocene, Pliocene, and Pleistocene claystone with intercalations of siltstone,

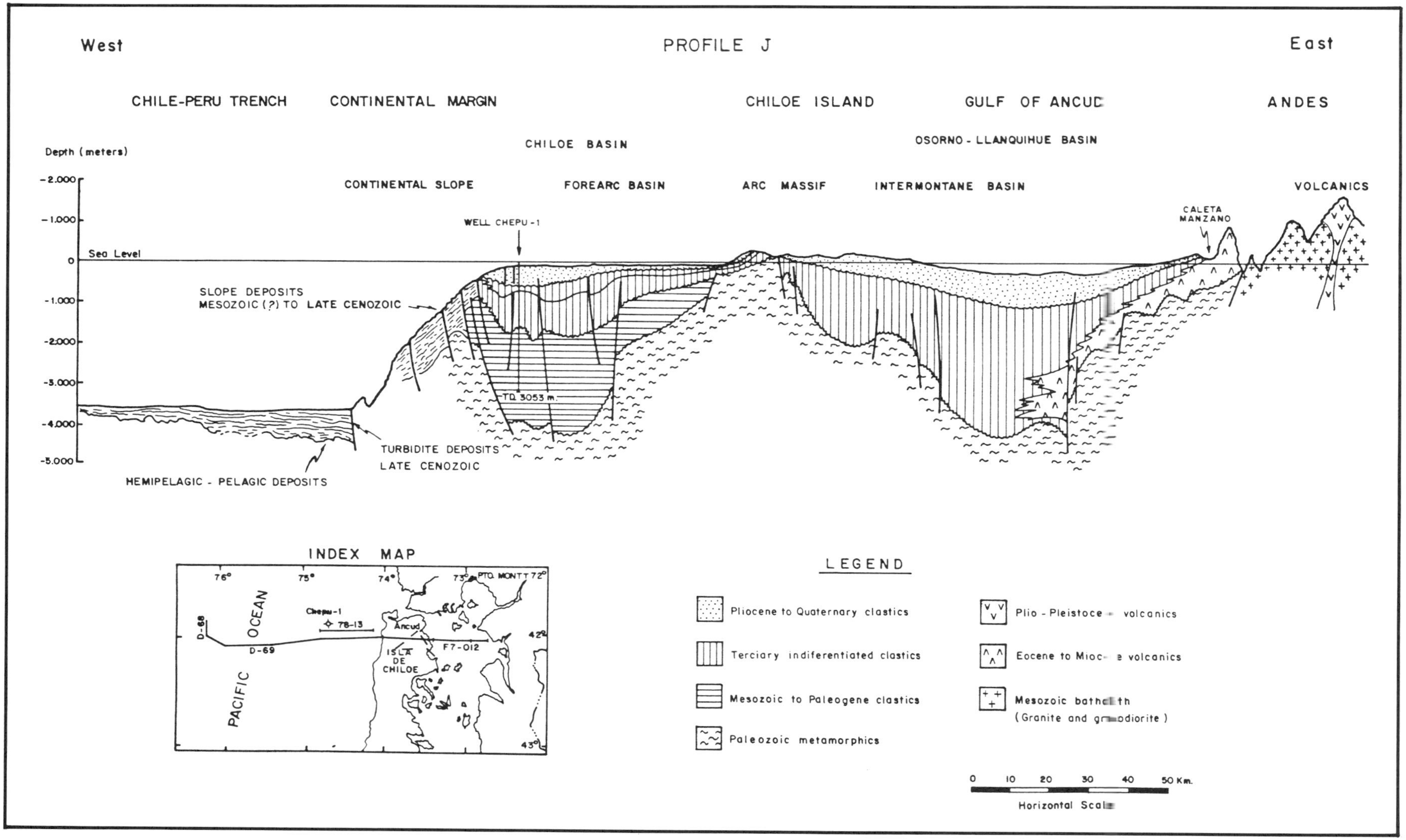

Figure 10. Diagrammatic geologic cross section (profile J) along lat. 42 °S, modified after seismic profile D-59 (Scholl et al., 1970); ENAP seismic-reflection profiles 78-18, F7-012 (Ledezma, 1982); ENAP seismic-refraction profiles. Richie 8, 9, 10, 11, 12, 13 (Botinelli, 1975b, 1976); and exploration well data (Gonzalez, 1983).

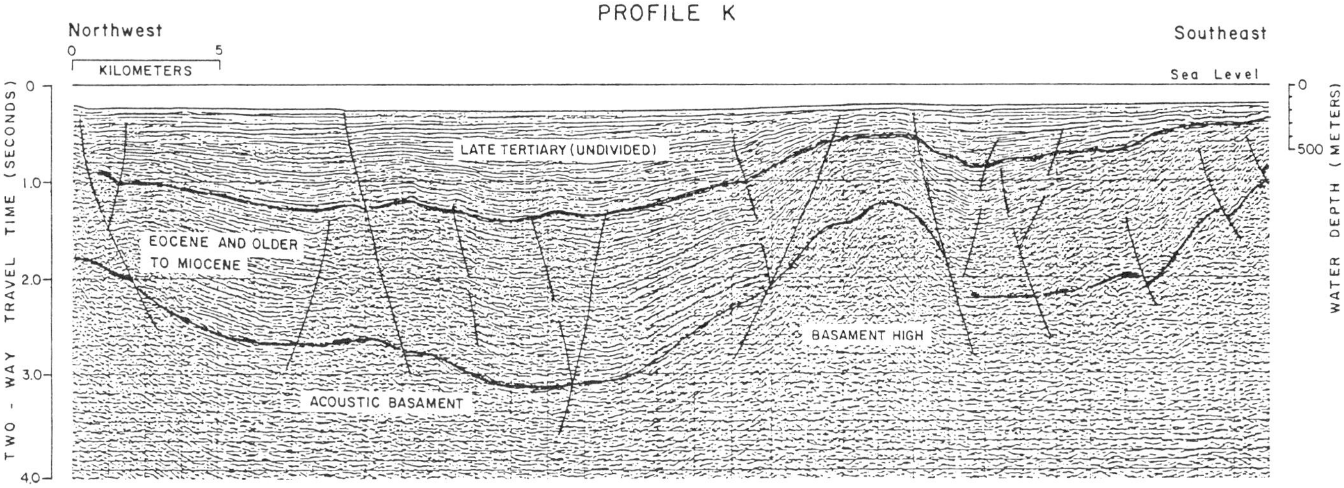

Figure 11. Seismic-reflection profile K, southeast of Peninsula de Taitao, showing seismic characteristics of sediments and typical structural anomalies (wrench faults and flower structures?) of the Golfo de Penas basin. Faults in the southeastern part of the basin are associated with the regional right-lateral Liquiñe-Ofqui strike-slip fault system (Herve, 1984).

sandstone, and conglomerate containing high percentages of volcanic material. The sedimentary paleoenvironment ranged from abyssal to outer neritic but was mainly bathyal. The lower unit, which has a maximum thickness of 1100 m, is predominantly quartzose graywacke sandstone of Paleogene age or older which contains intercalations of hard, fractured shale and siltstone. The sedimentary paleoenvironment of this unit was marginal marine. The entire sedimentary sequence thins rapidly (Figure 10) toward the eastern flank of the basement high at the shelf border (Mordojovich, 1981; González, 1983).

The Chiloé basin is just north of the point where the Chilean rise (a spreading center) intersects the Chile-Perú Trench at the Taitao Peninsula (Herron, Cande, and Hall, 1981; Mpodozis et al., 1985; Cande and Leslie, 1986). The complex tectonic development of the area has resulted in frequent changes of depocenters and has caused several unconformities. The basin is characterized by a series of north-tending anticlines associated with faults. Several normal and strike-slip faults cut the sedimentary sequence and underlying basement rocks. The predominant tectonic style of the Chiloé-Gulf of Ancud area (Figure 10) corresponds to a forearc massif-intermontane basin-volcanic arc fault system (Dickinson and Seeley, 1979).

Geochemical analyses of materials from wells drilled in the Chiloé basin indicate that the sedimentary rocks which tend to generate gas are mature Paleogene beds (vitrinite reflectance values: 0.45 to 0.75% average Ro) at depths of more than 2000 m. These rocks contain type III organic matter in which humic or lignitic kerogen predominate. The Neogene rocks are immature (vitrinite reflectance values are below 0.5% average Ro).

The main reservoir rocks are Oligocene-Miocene basal sandstone and conglomerate as thick as 112 m with porosities of 7 to 20% and average permeabilities of 61 to 77 md; and Paleogene fractured sandstone as thick as 80 m with porosities in the 5 to 20% range and permeabilities restricted to fractures. Reservoir traps are anticlines associated with fault blocks, stratigraphic wedges on flanks of basement highs, and diapiric structures. Several structural traps delineated have horizontal closures of 4 to 90 km^2, and vertical closures of 100 to 150 m. The low content of organic matter, implying poor oil-generating potential, is an unfavorable feature of the sedimentary rocks of the Chiloé basin. However, pre-Paleogene sedimentary rocks with higher content of organic matter and adequate maturity may be present in graben troughs in upper-slope basement rocks.

No oil shows or significant gas shows were encountered in the three wells drilled up to 1986. These results are not totally representative of the potential for oil and gas resources in the Chiloé basin, because the wells are far apart and the entire sedimentary section to the crystalline basement was not penetrated by any of the wells.

Golfo de Penas Basin

The Golfo de Penas basin (Figure 1), between lat. 46°40′ and 47°45′S, extends throughout the continental shelf of Golfo de Penas. It covers 6000 km^2 and extends to an ocean depth of 200 m. This basin has intermontane characteristics and is represented in the coastal region by Tertiary marine rocks exposed in the Tres Montes and Byron Island areas. Its northern, eastern, and southern boundaries are marked by outcrops of Paleozoic mica schist and phyllite, Tertiary volcanic rocks, and Mesozoic granitoid plutons, all exposed in the Taitao Peninsula, western foothills of the Patagonian Andes, and the Guayaneco Archipelago islands. The western boundary is a basement high at the margin of the continental shelf. The basin is isolated and its tectonic history differs from that of the other shelf basins of Chile.

Seismic profile K (Figure 11) shows the sedimentary fill of the basin and its tectonic style. Apparently, the basin originated during the Tertiary through rupture and separation of blocks of continental crust along strike-slip faults. Subsequent subsidence and accumulation of sediments in depressions between faulted blocks was rapid. A northeast-trending fault, related to the Liquiñe-Ofqui fault system (Hervé, 1984), marks the southeastern boundary of the basin, and a northwest-trending fault is at the northern boundary of the basin. A complex system of associated fractures and faults disrupts basement rocks and the overlying sedimentary sequence,

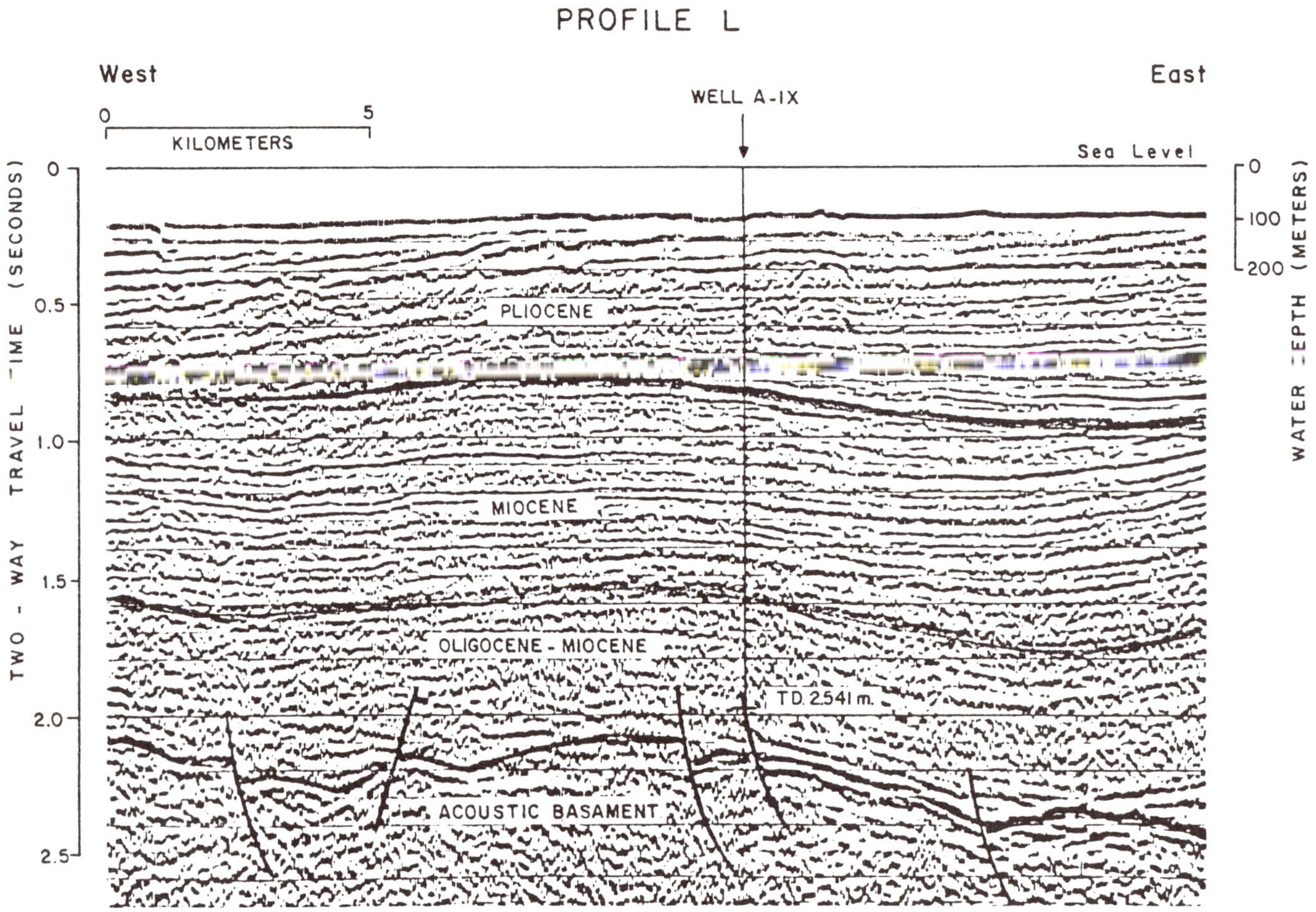

Figure 12. Seismic-reflection profile L, west of Isla Contreras (lat. 52°09′S), showing the sedimentary fill, structural pattern of the Madre de Dios basin, and location of well A-1X.

affecting even recent marine bottom sediments. As a consequence of faulting, the basin actually consists of two superimposed basins; the older trends northeast and the younger, of Neogene age, trends northwest. The depocenter of the basin is at the intersection of the two basins, where Cenozoic and possibly Cretaceous marine clastic rocks attain a thickness of more than 3500 m (López, 1977; Johnson, 1984; Forsythe et al., 1985).

Quartz sandstones in the basin are believed to be satisfactory reservoir rocks, and argillaceous strata rich in organic matter are believed to be capable of generating hydrocarbons. In addition, a high geothermal gradient, as indicated by heat-flow data for the zone of intersection of the Chile rise and trench west of the Taitao Peninsula (Herron et al., 1977; Herron, Cande, and Hall, 1981; Forsythe and Nelson, 1985; Cande and Leslie, 1986), may have caused maturity and enhanced capability of the Tertiary rocks to generate liquid hydrocarbons. The possible traps are structures associated with faults and wedging of stratigraphic units. Tertiary shales overlying the sandstones are sealing rocks.

Geochemical analyses of a sample collected in Eocene beds at Byron Island indicate a vitrinite reflectance of 2.25% average Ro, a value which suggests adequate maturity of the Eocene sediments in the Golfo de Penas.

Madre de Dios Basin

The Madre de Dios basin (Figure 12) is offshore between lat. 47°45′ and 53°S, extending from south of the Golfo de Penas to Desolación Island. It covers about 15,000 km^2 and extends to an ocean depth of 200 m. The basin occupies the shelf and upper slope, and consists of northern and southern sectors separated by a northeast-trending strike-slip fault.

Madre de Dios Norte

The northern basin sector, between lat. 47°45′ and 51°30′S, contains more than 4000 m of Cenozoic and presumably Late Cretaceous clastic sediments overlying acoustic basement in the depocenter. The sedimentary sequence wedges gradually toward basement highs in the west and toward the shelf in the east. The lower part of this sequence, showing prograde stratification, is of Oligocene-Miocene age. It is overlain by a sequence of gently deformed, thick Miocene-Pliocene to Quaternary sediments. Between the Golfo de Penas and approximately lat. 50°S, the shelf is very narrow and appears to be an abrasion terrace having only a thin sedimentary cover, chiefly of Pliocene and Quaternary age. In this zone, the principal basin fill is along the upper slope (Herron, Cande, and Hall, 1981; López, 1983; Cande and

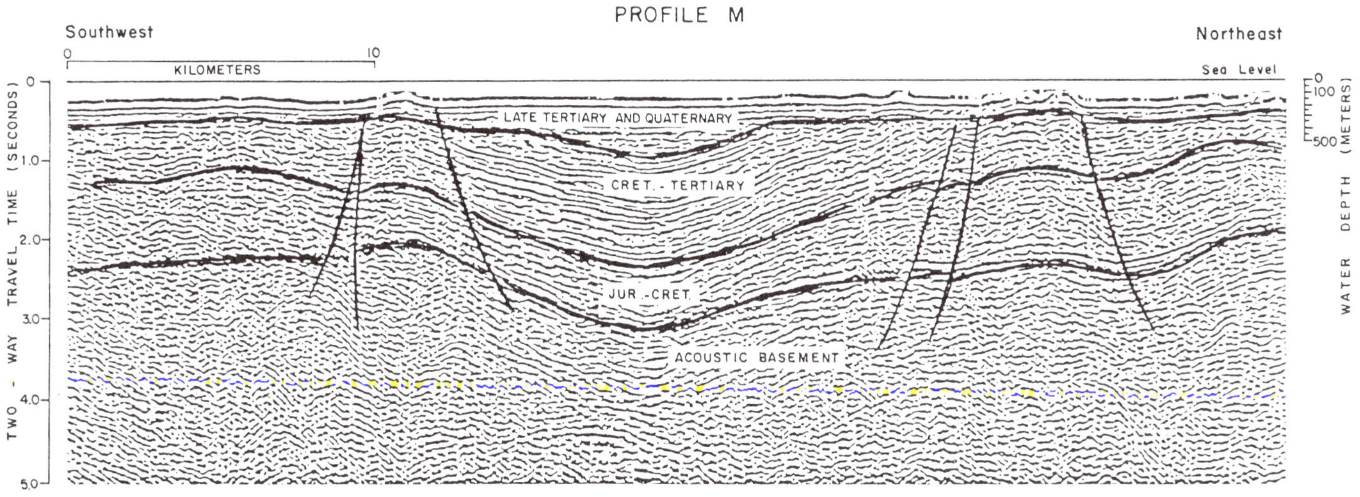

Figure 13. Seismic-reflection profile M, southwest of Peninsula Hardy, showing the sedimentary fill and fault in the Diego Ramirez basin.

Leslie, 1986; Cande et al., 1987).

The hydrocarbon potential of the northern Madre de Dios sector cannot be evaluated because of the lack of information about reservoir rocks and geochemical characteristics. Structural traps are present in the folded sediments above the basement paleorelief, and stratigraphic traps occur along faults at the shelf margin.

Madre de Dios Sur

The southern sector of the Madre de Dios basin, between lat. 51°30′ and 53°S, covers 6000 km^2 and extends to an ocean depth of 200 m in the shelf and upper slope. More than 4000 m of Cenozoic and presumably Late Cretaceous marine clastic sediments overlie basement rocks at the depocenter. Most of this sedimentary fill is in a depression between a north-trending strike-slip fault in the east and an uplifted block of basement rock at the margin of the shelf to the west. The sedimentary fill in the eastern part of the basin has homoclinal dips to the west, and several structural noses in this area dip westward. Several closed structures in the center of the basin are bordered by normal faults (Ledezma, 1982a). Seismic section L (Figure 12) shows the sedimentary fill and fault pattern. The acoustic basement contact is at the top of either the Paleozoic Madre de Dios complex or an overlying Late Cretaceous sequence. The Oligocene-Miocene sequence marks a change which was probably caused by deformation that resulted in slow infill of the trench by fine-grained clastic sediments.

Strike-slip faults are the dominant structural features of this basin sector. One fault system has a northwest trend, and possibly is a continuation of the Straits of Magellan fault (Fuenzalida, 1972). A second fault system that trends northwest may be related to the Shackleton fracture zone (Dalziel and Cortés, 1972; Dalziel, de Witt, and Ridley, 1975; British Antarctic Survey et al., 1985). These two fault systems apparently join near the western side of the Straits of Magellan.

The shelf and slope of this basin segment are furrowed by deep canyons that are the western extensions of the main Patagonian Archipelago canals (Mordojovich, 1981).

The results of the source rock potential analysis of the A-1X well indicate that the Oligocene-Miocene rocks of this basin sector are low in organic carbon, are thermally immature, and contain type III kerogen that tends to generate mostly biogenic gas. Sandstone beds with petrophysical characteristics adequate for reservoir rocks were not observed in the A-1X well. This part of the basin has traps related to the basement paleorelief and to faults, whose horizontal closures range from 5 km^2 to more than 70 km^2 and whose vertical closures range from 70 m to more than 300 m. These traps have not been tested. Although Oligocene-Miocene and older basal sandstones were not penetrated in well A-1X, such rocks may be present deeper in the tested structure and elsewhere in this extensive sedimentary basin, and may be worthy of further exploration.

Diego Ramirez Basin

The Diego Ramírez basin (Figure 1) is a deep depression between the Diego Ramírez Islands (lat. 56°30′S, long. 68°45′W) and the Hardy Peninsula and Patagonian Archipelago islands. It has an area of about 20,000 km^2 on the shelf and upper slope and extends to an ocean depth of 200 m. This basin shows a greater degree of deformation than the Madre de Dios basin, and its history of development differs from that of the other basins at the Chilean continental margin.

The sedimentary fill of the basin, which may be more than 5000 m thick in the depocenter, consists of a slightly deformed upper sequence attributed to the Late Tertiary and Quaternary, disconformably overlying a more intensely deformed lower sequence of high seismic velocity (3000 to 3500 m/sec). This lower sequence consists of two sedimentary units separated by an angular unconformity, which unconformably overlie the acoustic basement (Figure 13). The upper unit is interpreted as Cretaceous and Paleogene flysch deposits, whereas the lower unit is considered to consist of turbidite deposits of the Yahgan Formation (Kranck, 1932; Muñoz-Cristi, 1956; Katz and Watters, 1966) and volcaniclastic rocks of the Hardy Formation (Suárez and Pettigrew, 1976), both of

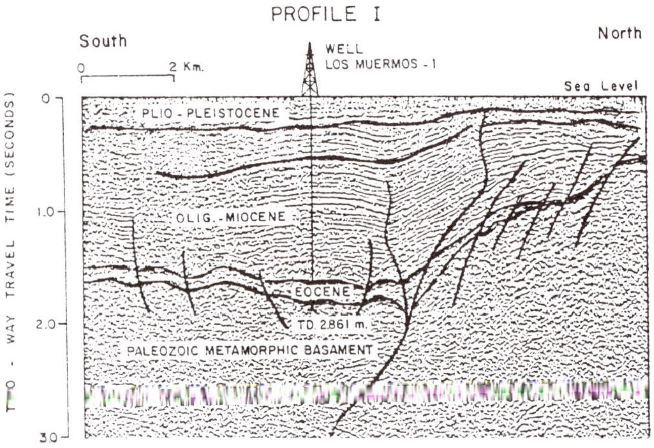

Figure 14. Seismic-reflection profile I, along the western part of Osorno-Llanquihue basin, showing seismic stratigraphic characteristics and typical structural anomalies (wrench faults and flower structure). Beds of Eocene age onlap the southern flank of a basement high.

Late Jurassic-Early Cretaceous age. The acoustic basement probably includes granitoid rocks of the Jurassic-Neogene Patagonian batholith complex (Halpern, 1973; Suárez, 1977) and metamorphic rocks of the Madre de Dios complex and blue schist and phyllites of the Diego Ramírez Islands, both of Paleozoic age (De la Rue, 1959; Mpodozis, 1980; Dalziel et al., 1983).

Northeast-trending horst and graben structures cut by transverse faults predominate, but northwest-trending strike-slip faults, with apparent normal displacement, are found throughout the basin. The northeast margin of the basin is marked by a faulted block of the Patagonian batholith granitoid complex, whereas the southwest margin is at a regional basement high at the shelf margin. This high may represent part of an accreted terrane at the continental margin that shows strong magnetic anomalies.

A relatively thin sedimentary sequence on the continental shelf overlies acoustic basement between the Diego Ramírez basin and southernmost margin of the Madre de Dios basin. The basement in this area is believed to consist of the Madre de Dios Complex or Patagonian batholith. Several hanging basins in the upper slope of this sector are in structural terraces 30 to 40 km wide at 1500 to 2000 m below sea level. These basins contain a sedimentary section of probable Pliocene and Quaternary age showing a 1.5-second two-way travel time.

Lack of data about geochemistry and reservoir rocks make it difficult to estimate the hydrocarbon potential of this basin. However, the following features indicate that the basin merits further study: (1) many fault-controlled closed structures of considerable size and great vertical closure are found in the central part of the basin; horizontal closures on these structures range from 10 km^2 to 50 km^2 and vertical closures range from 30 to 700 m; (2) a thick sedimentary sequence probably contains quartz sandstones, as erosion products of the Patagonian batholith, which are suitable reservoir rocks; and (3) hydrocarbon-generating rocks of Mesozoic age, similar to those of the Magallanes oil basin, may be present.

Bitumen impregnations were observed in a core sample containing foraminifera and bryozoan from the upper slope at the southeasternmost part of the basin, 90 km east-southeast of the Diego Ramírez Islands. This sample, V15-123, was collected from the ocean bottom at a water depth of 287 m, at lat. 56°38.3′S and long. 67°18′W by *R/V Vema* in 1959 (González, 1982).

Osorno-Llanquihue Intermontane Basin

The Osorno-Llanquihue basin (Figures 10, 14) lies between the Coastal Cordillera and the Cordillera de Los Andes, and includes the southern part of the Central Valley from lat. 34°50′S southward and the Gulf of Ancud from lat. 43°10′S northward. It has an area of about 27,000 km^2. This basin is a deep depression bounded on the north and south by Paleozoic basement highs of schist and phyllite. Significant gas shows were encountered in two of the exploratory wells drilled in this basin—Bellavista 1 and Los Muermos 2. The gas in the former contained 69% methane, 15.7% ethane, 2.6% propane, 0.8% isobutane, 1.4% n-butane, and 10.5% nitrogen.

The basin has a thick fill of marine and continental sediments with interbedded volcanic rocks that is broken by many faults. A complex system of north- to northeast-trending normal and strike-slip faults is present. The eastern part of the basin is a north-trending graben extending through the lake district and bounded by steeply dipping faults of great vertical displacement. The Rio Negro high, an uplifted block of crystalline basement apparently bounded by faults, divides the basin into two subbasins.

The western subbasin is characterized by a thick sequence (more than 2300 m) of marine and continental clastic sediments of Tertiary age, containing subbituminous coal beds, and overlying a basement of micaceous schist, phyllite, and granitoid rocks of Paleozoic age. As indicated by seismic data, this subbasin is bounded on the west by a fault system along the eastern side of the Coastal Cordillera (Botinelli, 1980, 1981). The eastern subbasin is characterized by a thick sequence (more than 1000 m) of Quaternary glacial deposits with intercalations of basaltic lava, underlain by more than 3000 m of Tertiary continental and marine clastic sediments, and volcaniclastic dacite rocks. The basement of this basin consists of chlorite schist and phyllite of Paleozoic age (García, 1968; Céspedes, 1979; Céspedes and Johnson, 1984).

The Tertiary rocks in the Osorno-Llanquihue basin are of Eocene to Miocene age. The coal-bearing continental claystones and sandstones at the localities of Riachuelo, Mina Huilma, and Mina Catamutún are Eocene, whereas marine and silty claystones in Rio Hueyusca and Estero Cuz-Cuz are Oligocene-Miocene. Compact, tuffaceous marine shales and sandstones at Caleta Manzano and Ayacara in the southeasternmost part of the basin are Miocene (Martínez, 1965; Leví, Aguilar, and Fuenzalida, 1966; Céspedes, 1979; Céspedes and Johnson, 1984). A major unconformity is at the base of the Oligocene-Miocene sequence.

During the Eocene, the Osorno-Llanquihue basin was probably connected to the Valdivia and Temuco basins, forming a continuous basin between lat. 38°30′ and 43°S, where coal-bearing continental sediments accumulated (Troncoso and Barrera, 1979; Palma-Heldt and Alfaro, 1982; Céspedes and Johnson, 1984). The Eocene sedimentary cycle probably terminated with marine transgression and accumulation of the

Oligocene-Miocene marine sequence now found extending from the continental shelf eastward to the Nirihuau area of Argentina. As a result of Miocene block faulting, the Osorno-Llanquihue basin was separated from the Valdivia and Temuco basins (Ramos, 1982; Céspedes and Johnson, 1984).

Geochemical analyses of well samples indicate that Miocene rocks are low in organic matter and are immature, and, consequently, are poor gas generators. Eocene rocks have more organic matter (types II and III kerogen) and moderate maturity, which place them at the beginning stage of oil- and gas-generating capacity. Older sedimentary rocks that probably exist in the deep eastern subbasin would be expected to have higher contents of organic matter and adequate maturity to generate oil and gas. However, the geothermal gradients measured in the exploratory wells are low, ranging from 18° to 25°C/km.

The best reservoir rocks are basal quartz sandstones of the Oligocene-Miocene sequences, which have thicknesses of 95 to 110 m, porosities of 5 to 32%, and permeabilities of 11 to 168 md. Other potential reservoir rocks are Miocene sandstones having thicknesses of 20 to 150 m and porosities of 2 to 18%, and Eocene continental sandstones having thicknesses of 18 to 200 m and porosities of 6 to 17%. Tertiary claystones overlying sandstone beds constitute adequate sealing rocks. Reservoir traps are related to basement paleorelief, rollover anticlines, and folds associated with strike-slip faults. Traps in the eastern part of the basin, which have not been tested with wells, may contain hydrocarbon accumulations.

SUMMARY AND CONCLUSIONS

Cretaceous and older sedimentary rocks in the coastal basins of Chile have the greatest capacity for oil and gas generation. The older basement rocks are metamorphic and granitic rocks, and the younger sedimentary fill of the basins tends to be low in organic matter and is immature. The basins containing Late Cretaceous beds presumably have undergone continuous subsidence for about 85 m.y., and have an average geothermal gradient of 20° to 30°C/km. Such conditions would establish oil-generating capacity in Cretaceous rocks in sequences more than 2500 m thick. Thicknesses of more than 4000 m have been observed in basins containing Cretaceous beds in graben depressions of the shelf and upper slope. The average geothermal gradients observed in exploratory wells in the basins between lat. 34°30′ and 53°S range from 16° to 37°C/km. The gradients are slightly higher in Late Cretaceous rocks (17° to 37°C/km) than in Tertiary rocks (16° to 27°C/km).

The organic matter in Tertiary marine and continental claystone is thermally immature or in transition. The organic carbon content is 0.26 to 1.23%, and the predominant kerogen is humic to lignitic. The hydrocarbon potential of these rocks is biogenic gas. The organic matter in Cretaceous marine claystone and shale is moderately mature to mature. The organic carbon content is 0.74 to 2.6%, and the predominant kerogen is rich in lipides. These rocks tend to generate gas and oil. Cretaceous rocks in the Arauco and Itata basins are the best generating rocks encountered. Jurassic rocks in the Vichuquén-Tilicura area show an organic carbon content of 1.46% and an average Ro of 2.63%, which indicates advanced maturity and suggests that they are capable of generating gas.

The composition of Cretaceous gas in the Arauco basin indicates that the organic matter in the source rocks is mature. The composition of Tertiary gas in the Valdivia and Osorno-Llanquihue basins indicates low temperature and low maturity diagenesis of the organic matter in source rocks. The Tertiary gas from Mocha Island has a low nitrogen content because it probably migrated from Cretaceous beds.

Lithologies and structures of the sedimentary rocks are favorable for oil and gas accumulations.

Porosity and permeability values of sandstones encountered in offshore and onshore wells are 2 to 31% and 2 to 1727 md, respectively. Reservoir rocks having the best petrophysical properties are Late Cretaceous basal sandstones and Oligocene-Miocene sedimentary sequences. Such Oligocene-Miocene sandstones in the Arauco and Osorno-Llanquihue basins are 5 to 100 m thick, and have porosities of 10 to 34% and permeabilities of 23 to 713 md. Late Cretaceous basal sandstones in the Chanco basin are 23 to 150 m thick, and have porosities of 23 to 28% and permeabilities of 19 to 1727 md. Tertiary and Cretaceous claystone and shale units overlying reservoir sandstones are probably the most widespread seals for the vertical migration of hydrocarbons. Faults and permeability changes due to facies are important lateral seals. Potential hydrocarbon reservoir traps are of four main types: (1) uplifted blocks of the fractured basement; (2) folded sediments above basement-rock horst blocks; (3) anticlines, monoclines, and structural noses associated with faults; and (4) stratigraphic closures.

A total of 18 exploratory wells have been drilled in offshore continental shelf basins, and 50 wells have been drilled in extensions of these basins onshore and in the Osorno-Llanquihue basin. Biogenic gas accumulation was encountered in the offshore Valdivia basin F well, and two accumulations of noncommercial gas were found in onshore wells of the Arauco basin (Mocha and Lebu).

RECOMMENDATIONS

More information is needed to evaluate the oil and gas potential of basins in the coastal region of Chile. In particular, more geological and geophysical data for the shelf margin and upper slope of these basins is needed, chiefly in areas where Cretaceous and Jurassic rocks are believed to be present. The most attractive sector of the continental margin is between lat. 34°30′ and 40°S (Mataquito, Chanco, Itata, Arauco, and Valdivia basins), where additional seismic lines on the shelf and slope are believed to be justifiable. The prospect area of the shelf and upper slope in this sector, to a depth of 1000 m, is 36,000 km^2. In addition, future oil and gas exploration should be aimed at testing possible traps in graben depressions and other closed structures at the western margin of the shelf. Such structures may contain accumulations of hydrocarbons that migrated from Mesozoic sediments and biogenic gas associated with the hydrate layers of the upper slope.

ACKNOWLEDGMENTS

The information contained in this study was obtained from ENAP's Technical Files, and publication was authorized by the Empresa Nacional del Petróleo.

REFERENCES CITED

Botinelli, R., 1973, Interpretación líneas de refracción Lebu-Cañete: Empresa Nacional del Petróleo (Chile), unpublished report.

Botinelli, R., 1974, Prospección sísmica de refracción Zona de Chanco: Empresa Nacional del Petróleo (Chile), unpublished report, 6 p.

Botinelli, R., 1975a, Sísmica de refracción, Navidad: Empresa Nacional del Petróleo (Chile), unpublished report, 4 p.

Botinelli, R., 1975b, Sísmica de refracción, Chiloé: Empresa Nacional del Petróleo (Chile), unpublished report, 26 p.

Botinelli, R., 1976, Plano isobático del basamento Isla de Chiloé: Empresa Nacional del Petróleo (Chile), unpublished report.

Botinelli, R., 1980, Sísmica de reflexión Cuenca Osorno, Ballavista, Huilma: Empresa Nacional del Petróleo (Chile), unpublished report, 2 p.

Botinelli, R., 1981, Interpretación sismica de reflexión, Cuenca de Osorno, Fresia: Empresa Nacional del Petróleo (Chile), unpublished report, 2 p.

British Antarctic Survey et al., 1985, Tectonic map of the Scotia arc.

Brüggen, J., 1950, Fundamentos de la geología de Chile: Santiago, Instituto Geográfico Militar, 374 p.

Cande, S. C., and R. B. Leslie, 1986, Late Cenozoic tectonics of the southern Chile Trench: Journal of Geophysical Research, v. 91, p. 471-496.

Cande, S. C., R. B. Leslie, J. C. Parra, and M. Hobart, 1987, Interaction between the Chile Ridge and Chile Trench: Geophysical and geothermal evidences: Journal of Geophysical Research, v. 92, p. 495-520.

Cecioni, G., 1956, Primeras noticias sobre la existencia del Paleozoico superior en el Archipiélago Patagónico, entre los paralelos 50° y 52° latitud S.: Universidad de Chile, Facultad de Ciencias Físicas y Matemáticas, Departamento de Geología Publicación No. 8, 20 p.

Cecioni, G., 1978, Petroleum possibilities of the Darwin's Navidad Formation near Santiago, Chile: Santiago, Museo Nacional de Historia Natural Publicación Ocasional No. 25, 28 p.

Cecioni, G., 1983, Chanco Formation, a potential Cretaceous reservoir, central Chile: Journal of Petroleum Geology, v. 6, p. 89-94.

Céspedes, S., 1976, Geología y estratigrafía de la Cuenca de Arauco: Empresa Nacional del Petróleo (Chile), unpublished report, 41 p.

Cecioni, G., 1978a, Geología del área de Concepción-Bio Bio: Empresa Nacional del Petróleo (Chile), unpublished report, 9 p.

Cecioni, G., 1978b, Estudio geológico en Navidad, VI Región: Empresa Nacional del Petróleo (Chile), unpublished report, 10 p.

Cecioni, G., 1979, Estratigrafía Cuenca de Osorno y Llanquihue: Empresa Nacional del Petróleo (Chile), unpublished report, 21 p.

Cecioni, G., 1980, Estudio geológico de la Isla Mocha: Empresa Nacional del Petróleo (Chile), unpublished report, 25 p.

Céspedes, S., and C. Johnson, 1984, Cuencas sedimentarias en el Valle Central: Empresa Nacional del Petróleo (Chile), unpublished report, 33 p.

Céspedes, S., 1985, Cuencas sedimentarias en el Pacifico suroriental: Empresa Nacional del Petróleo (Chile), unpublished report, 21 p.

Coulbourn, W. T., 1977, Tectonics and sediments of the Perú-Chile Trench and continental margin at the Arica bight: Honolulu, University of Hawaii, unpublished Ph.D. dissertation, 243 p.

Corvalán, J., 1976, El Triásico y Jurásico de Vichuquén-Tilicura y de Hualañe, Provincia de Curicó, implicaciones paleográficas: Santiago, Primer Congreso Geológico Chileno, Actas, p. A137-A154.

Chotin, P., 1969, Sur la géologie de la Province de Concepción (Chili), la region de Tomé: Paris, Bulletin Societé Géologique Francaise, v. 11, p. 556-560.

Dalziel, I. W. D., and R. Cortés, 1972, The tectonic style of the southernmost Andes and the Antarctandes: Montreal, *in* XXIV International Geological Congress Proceedings, session 3, p. 316-327.

Dalziel, I. W. D., N. J. de Witt, and W. I. Ridley, 1975, Structure and petrology of the Scotia arc and the Patagonian Andes: R/V Hero Cruise 75-A Antarctic Journal U.S., v. 10, p. 307-310.

Dalziel, I. W. D., et al., 1983, First geologic investigation of the Diego Ramírez Islands: Lamont Newsletter, Winter 1983, p. 8.

Darwin, C., 1846, Geological observations on South America: London, Smith Elder and Co., 279 p.

De la Rue, E. A., 1959, Quelques observations faites aux iles Diego Ramírez (Chili): Paris, Bulletin du Muséum 2a, serie XXXI, no. 4, p. 387-391.

Dickinson, W. R., and R. Seeley, 1979, Structure and stratigraphy of forearc regions: American Association of Petroleum Geologists Bulletin, v. 63, p. 2-31.

Empresa Nacional del Petróleo, 1982, Memoria y balance: Empresa Nacional del Petróleo (Chile), p. 12.

Empresa Nacional del Petróleo, 1986, Memoria anual: Empresa Nacional del Petróleo (Chile), p. 30-31.

Ferraris, F., 1977, Antecedentes geológicos de la plataforma continental y fosa Chile-Perú: Santiago, Universidad de Chile, Tésis de prueba para optar al Título de Geólogo, 103 p.

Ferraris, F., and F. Di Biase, 1978, Hoja Antofagasta, Región de Antofagasta, Santiago, Instituto de Investigaciones Geológicas (Chile), Carta Geológica de Chile No. 30, escala 1:250,000.

Forsythe, R. D., and C. Mpodozis, 1983, Geología del basamento pre-Jurásico Superior en el Archipiélago Madre de Dios, Magallanes, Chile: Servicio Nacional de Geología y Minería Boletín No. 39, 61 p.

Forsythe, R. D., and E. Nelson, 1985, Geologic manifestation of ridge collision: evidence from the Golfo de Penas-Taitao basin, southern Chile: Tectonics, v. 4, p. 477-495.

Forsythe, R. D., R. K. Olsson, C. Johnson, and E. P. Nelson, 1985, Stratigraphic and micropaleontologic observations from the Golfo de Penas-Taitao basin, southern Chile: Revista Geológica de Chile, no. 25-26, p. 3-12.

Frutos, J., P. Mencarini, M. Pincheira, Y. Bourret, and G. Alfaro, 1982, Geología de la Isla Quiriquina: Tercer Congreso Geológico Chileno, Actas, p. F307-F338.

Fuenzalida, R., 1972, Geological correlation between the Patagonian Andes and Antarctic Peninsula and some tectonic implications: Palo Alto, California, Stanford University, unpublished Master's thesis, 75 p.

García, F., 1968, Estratigrafía del Terciario de Chile central, *in* Terciario de Chile zona central: Santiago: editorial Andrés Bello, p. 25-77.

González, E., 1982, Demonstraciones de petróleo y gas natural en el margen continental de Chile: Empresa Nacional del Petróleo (Chile), unpublished report, 12 p.

González, E., 1983, Geological and geophysical information for risk contracts: Empresa Nacional del Petróleo (Chile), unpublished report, 88 p.

González-Bonorino, F., 1971, Metamorphism of the crystalline basement of central Chile: Journal of Petrology, v. 17, p. 149-175.

Goodell, H. G., 1965, Marine geology, USNS Eltanin cruises 9-15: Sedimentological Research Laboratory, Florida State University, Contribution 11, 237 p.

Halpern, M., 1973, Regional geochronology of Chile, south of 50° latitude: Geological Society of America Bulletin, v. 84, p. 2407-2422.

Herron, E. M., S. C. Cande, and B. R. Hall, 1981, An active spreading center collides with a subduction zone: A geophysical survey of the Chile margin triple junction: Geological Society of America Memoir 154, p. 683-701.

Herron, E. M., et al., 1977, Post Miocene tectonics of the margin of southern Chile, *in* M. Talwani and W. C. Pittman III, eds., Island arcs, deep sea trenches, and back-arc basins: American Geophysical Union Maurice Ewing Series, v. 1, p. 273-283.

Hervé, F., F. Munizaga, M. Montavini, and M. Hervé, 1976, Edades Rb/St neopaleozoicas del basamento cristalino de la Cordillera de la Costa: Santiago, Primer Congreso Geológico Chileno, Actas, p. F19-F26.

Hervé, M., 1984, La zona de falla Liqueñe-Ofqui en Liquiñe: Santiago, Revista Comunicaciones No. 34, p. 101-105.

Hoffstetter, R., H. Fuenzalida, and G. Cecioni, 1957, Lexique stratigraphique internationale, Amerique Latine, Chili: Paris, v. 5, Fascicule 7, 444 p.

Johnson, C., 1984, Informe geológico preliminar, Golfo de Penas-Península de Taitao: Empresa Nacional del Petróleo (Chile), unpublished report, p. 7.

Katz, H. R., and W. A. Watters, 1966, Geological investigations of the Yahgan Formation (upper Mesozoic) and associated igneous rocks of Navarino Island, southern Chile: New Zealand Journal of Geology and Geophysics, v. 9, p. 323-359.

Kranck, E. H., 1932, Geological investigations in the Cordillera of Tierra del Fuego: Helsinki, Acta Geográfica, v. 4, p. 101-215.

Ledezma, O., 1979a, Isocrónico basamental cuencas Itata a Pucatrihue: Empresa Nacional del Petróleo (Chile), unpublished report.

Ledezma, O., 1979b, Isocrónico basamental plataforma continental Constitución-Algarrobo: Empresa Nacional del Petróleo (Chile), unpublished report.

Ledezma, O., 1981, Isocrónico basamental plataforma continental Constitución-Isla Mocha: Empresa Nacional del Petróleo (Chile), unpublished report.

Ledezma, O., 1982a, Interpretación sísmica Cuenca Maadre de Dios Sud: Empresa Nacional del Petróleo (Chile), unpublished report, 19 p.

Ledezma, O., 1982b, Isocrónico basamental Golfo de Ancud: Empresa Nacional del Petróleo (Chile), unpublished report.

Ledezma, O., 1983, Isocrónico del basamento entre Río Itata y Corral (cuencas de Itata, Arauco y Valdivia): Empresa Nacional del Petróleo (Chile), unpublished report.

Leví, B., and L. Aguirre, 1960, El conglomerado de Algarrobo y su relación con las formaciones del Cretáceo Superior de Chile central: Buenos Aires, Instituto de Investigaciones Geolfgicas and Apartado No. 17 de Actas de la Primera Jornada Geológica Argentina, v. 2, p. 417-431.

Leví, B., A. Aguilar, and R. Fuenzalida, 1966, Reconocimiento geológico en las provincias de Llanquihue y Chiloé: Instituto de Investigaciones Geológicas Boletín No. 19, 45 p.

López, C., 1977, Interpretación sísmica costafuera Golfo de Penas: Empresa Nacional del Petróleo (Chile), unpublished report, 4 p.

López, C., 1983, Interpretación sísmica cuenca Madre de Dios Norte: Empresa Nacional del Petróleo (Chile), unpublished report, 6 p.

Machado, M., 1903, Descripción geológica y paleontológica de Isla Mocha: Santiago, Anales del Museo Nacional de Historia Natural, Chile, p. 36-44.

Maksaev, V., 1984, Mesozoico a Paleogeno de la Región de Antofagasta: Servicio Nacional de Geología y Minería, Seminario Geología de Chile, p. C1-C20.

Martínez, R., 1965, Edad de la Formación Ayacara en Caleta Manzano e Isla Manzano: Sociedad Geológica de Chile, XXI Sesión, Universidad de Chile, p. 19-22.

Martínez, R., 1968, Edad y paleoecología del Piso Millongue in el Terciario de Chile, zona central: Santiago, Editorial Andrés Bello, p. 125-137.

Mordojovich, C., 1974, Geology of a part of the Pacific margin of Chile, *in* C. A. Burk and C. L. Drake, eds., The geology of continental margins: New York, Springer Verlag, p. 591-598.

Mordojovich, C., 1981, Sedimentary basins of Chilean Pacific offshore, *in* Energy resources of the Pacific region: American Association of Petroleum Geologists Studies in Geology, v. 12, p. 63-68.

Mordojovich, C., and J. Alvarez, 1977, Morfología de la plataforma continental de Chile de sector Valdivia-Isla Guafo: Valparaíso, Ciencia y Tecnología del Mar, v. 3, p. 9-22.

Mpodozis, C., 1980, Islas Ildefonos, Tierra del Fuego, depósitos de ante-arco del sistema Megallánico: Revista Geológica de Chile No. 11, p. 91-99.

Mpodozis, C., et al., 1985, El magmatismo Plioceno de Península Tres Montes y su relación con la evolución del Punto Triple de Chile Austral: Revista Geológica de Chile No. 25-26, p. 13-28.

Munizaga, F., L. Aguirre, and F. Hervé, 1973, Rb/Sr ages of rocks from the Chilean metamorphic basement: Earth & Planetary Science Letters, v. 18, p. 87-91.

Muñoz-Cristi, J., 1946, Estado actual del conocimiento sobre la geología de la Provincia de Arauco: Facultad de Ciencias Físicas y Matemáticas, Universidad de Chile, Anales, v. 3, p. 30-53.

Muñoz-Cristi, J., 1956, Chile, *in* W. F. Jenks, ed., Handbook of South American geology: Geological Society of America Memoir 65, p. 191-214.

Muñoz-Cristi, J., 1964, Estudio petrológico sobre el Batolito de la Costa de las Provincias de Santiago y Valparaíso: Santiago, Instituto de Geología, Universidad de Chile, Publicación 25, 95 p.

Palma-Heldt, S., and G. Alfaro, 1982, Antecedentes palinológicos preliminaries para la correlación de los mantos de carbón del Terciario de la Provincia de Valdivia: Concepción, Tercer Congreso Geológico Chileno, Actas, v. 1, p. A207-A233.

Pérez, E., and R. Reyes, 1980, Buchotrigonia (Buchotrigonia) Topocalmensis Sp. Nov. (Trigoniidae, Bivalvia) del Cretáceo Superior de Chile: Revista Geológica de Chile No. 9, p. 37-55.

Ramoz, V. A., 1982, Las ingreciones Pacíficas del Terciario en el norte de la Patagonia (Argentina): Concepción, Tercer Congreso Geológico Chileno, Actas, v. 1, p. A262-A288.

Ruíz, C., 1965, Geología y yacimientos metalíferos de Chile: Instituto de Investigaciones Geológicas (Chile), 305 p.

Salas, R., et al., 1966, Geología y recursos minerales del Departamento de Arica: Instituto de Investigaciones Geológicas (Chile) Boletín No. 21, 114 p.

Scholl, D. W., et al., 1970, Perú-Chile trench sediments and sea floor spreading: Geological Society of America Bulletin, v. 81, p. 1339-1360.

Servicio Nacional de Geología y Minería, 1982, Mapa geológico de Chile, escala 1:1,000,000.

Suárez, M., 1977, Geología del área sur del Canal Beagle, Región Magallanes y Antártica chilena: Servicio Nacional de Geología y Minería, Carta Geológica de Chile, escala 1:250,000.

Suárez, M., and T. H. Pettigrew, 1976, An upper Mesozoic island arc-back system in the southern Andes and South Georgia: Geological Magazine, v. 113, p. 305-310.

Tavera, J., 1942, Contribución al estudio de la estratigrafia y paleontología del terciario de Arauco: Primer Congreso Panamericano de Ingeniería de Minas y Geología, Anales, v. 2, p. 580-632.

Tavera, J., 1979, Estratigrafía y paleontología de la Formación Navidad, Provincia de Colchaqua, Chile: Santiago, Museo Nacional de Historia Natural Boletín No. 36, 176 p.

Tavera, J., and C. Veyl, 1958, Reconocimiento geológico de la Isla Mocha: Departamento de Geología Universidad de Chile Publicación No. 12, 30 p.

Toenges, A. L., 1948, Coals of Chile: U. S. Bureau of Mines Bulletin 474, 106 p.

Troncoso, A., and E. Barrera, 1979, Estudio palinológico de tres testigos del pozo Rio Blanco No. 1 (Osorno): Arica, Segundo Congreso Geológico Chileno, Actas, v. 3, p. H1-H14.

Wenzel, O., 1969, Geología, reservas y petrofrafia de los carbones chilenos, *in* El Carbón la Siderurgia Latinoamericana: Santiago, Seminario ILAFA, p. 173-193.

Wenzel, O., 1982, Estratigrafía de las series carboníferas de Arauco y Concepción: Tercer Congreso Geológico Chileno, v. 3, p. F256-F279.

Young, J., and H. Valenzuela, 1979, Sismica marina y geología de la cuenca de Chanco: Empresa Nacional del Petróleo (Chile), unpublished report, 12 p.

Ziel, W., 1964, Geologie von Chile: Berlin, Gerbruder Berntraeger, 233 p.

Ericksen, G. E., Cañas Pinochet, M. T., and
Reinemund, J. A., editors, 1989, Geology of the Andes
and its relation to hydrocarbon and mineral resources:
Houston, Texas, Circum-Pacific Council for Energy
and Mineral Resources Earth Science Series, v. 11.

CHAPTER 29

HYDROCARBON RESOURCES OF VENEZUELA—THEIR SOURCE ROCKS AND STRUCTURAL HABITAT

Hans H. Krause and Keith H. James
Maraven, S.A.
Caracas, Venezuela

ABSTRACT

Venezuela has three major hydrocarbon provinces—the Maracaibo, the Barinas-Apure, and the Eastern Venezuela basins (the latter including the supergiant Orinoco oil belt). Their hydrocarbons originated mainly from rich and ubiquitous, kerogenous source rocks laid down in the Late Cretaceous over a vast, restricted (euxinic) platform on the northern margin of the Guayana Shield. Maturation occurred below Eocene to Recent depocenters generated by Caribbean, South American, and Pacific Plate interactions. The latest tectonic phase created the present-day basins and their confining, transpressional Andean and Coastal/Interior Range uplifts.

In the Maracaibo basin oil is produced mainly from fractured igneous basement, Cretaceous limestone, and Tertiary sandstone. The traps are structural and stratigraphic and are associated with north- to northeast-trending sinistral faults. Elsewhere in Venezuela, production is predominantly from Tertiary sandstone. In the Orinoco oil belt, trapping is essentially stratigraphic. In the flexure zones of the Eastern Venezuela and Barinas-Apure basins, normal faults provide traps. In the northern part of the Eastern Venezuela basin, northeast-trending compressional structures are associated with northwest-trending dextral faults.

Accumulated production to the end of 1986 was 40×10^9 barrels of oil, most of which had come from the Maracaibo basin. Official oil reserves as of 1986 were 55.5×10^9 barrels, slightly less than half of which are located in the same basin, and gas reserves are 92×10^{12} cubic feet, mostly located in the Eastern Venezuela basin. Current production is 660×10^6 barrels of oil and 1×10^{12} cubic feet of gas per annum. The Orinoco oil belt has proven in-place resources of 1.2×10^{12} barrels, only partly included in the official reserves. Production in the next century will shift increasingly to the Eastern Venezuela basin.

INTRODUCTION

Purpose and Scope

So much has been written about the hydrocarbon accumulations and source rocks of Venezuela that it becomes difficult not to merely repeat previous reports. There remains, however, a variation in the quality of knowledge of the various provinces which also tends to be reported independently. In this chapter we refer to the latest information available and attempt to provide a regional synthesis. In this way we present some regional concepts concerning the paleogeography of Cretaceous source rocks, their present disposition, and their total original generative capacity. In addition, we describe the plate tectonic setting and the structural development of Mesozoic-Cenozoic Venezuela as these relate to the various hydrocarbon provinces.

Venezuela has three major onshore hydrocarbon provinces: the Maracaibo, Barinas-Apure, and Eastern Venezuela basins (Figure 1). The latter includes the supergiant accumulation of the Orinoco oil belt. In addition, in the offshore, there is gas (biogenic) in the Carúpano basin, some (at present uncommercial) oil in the Cariaco Trough, and oil and gas in La Vela Bay. The Gulf of Venezuela has not been drilled at this time. These latter basins will not be discussed, as they are somewhat unrelated to the much more important onshore basins.

Production of Oil and Gas

In the Maracaibo basin, production comes from igneous basement, Cretaceous limestone, and Tertiary sandstone. Sealing is by interbedded shale or clay and, in some cases, by faults or by heavy-oil plugs. In the other basins, production is mainly from Tertiary sandstone sealed by interbedded shale

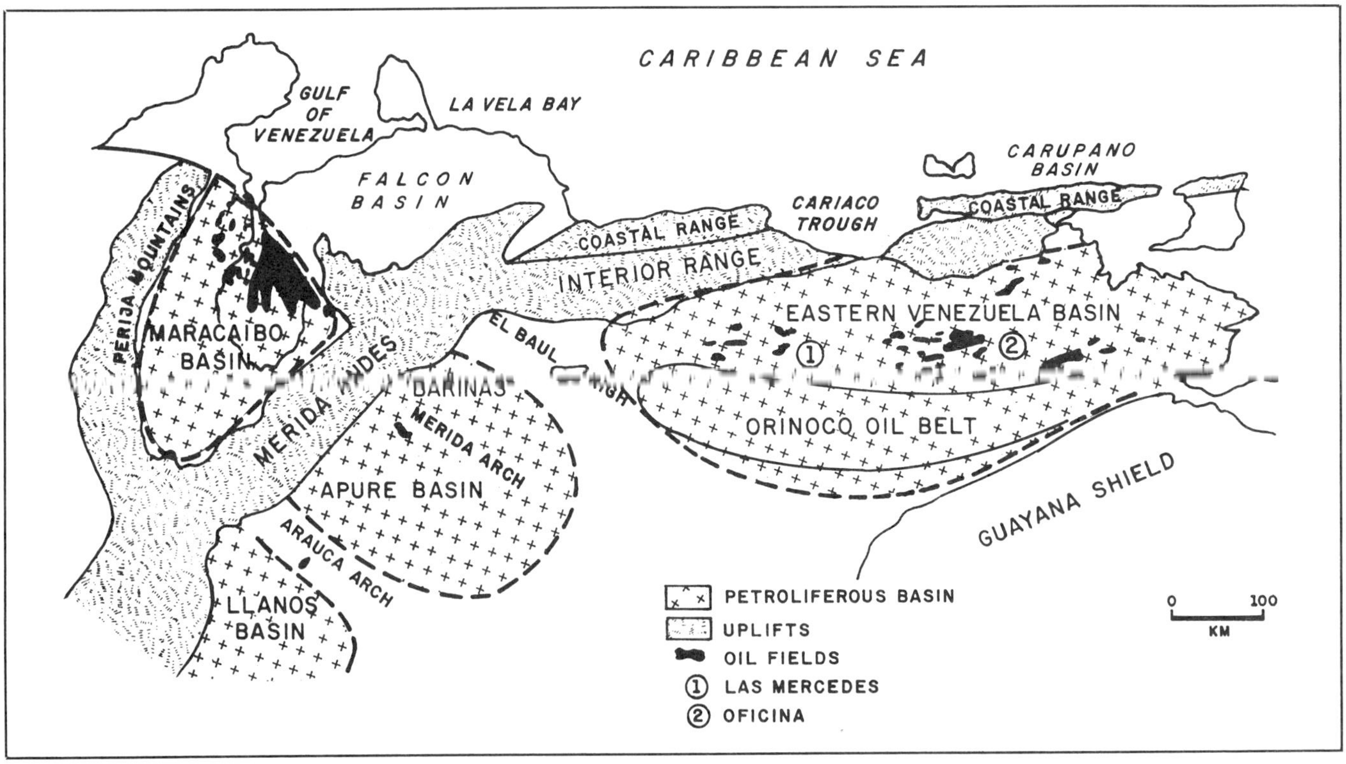

Figure 1. Main hydrocarbon provinces in northern Venezuela.

or clay, by faults, and, in the Orinoco oil belt, by pinchouts and by heavy-oil plugs.

By the end of 1986, 40×10^9 barrels of oil had been produced in Venezuela; most of this had come from the Maracaibo basin. Venezuela's official reserves at the same date were 55.5×10^9 barrels, of which 19.2×10^9 barrels were in the Maracaibo basin, 35.6×10^9 barrels in the Eastern Venezuela basin, and 700×10^6 barrels in the Barinas-Apure basin. In addition, there are (official) unproven expectations of 2.5×10^9 barrels of light- and medium-gravity oil. Only partly included in these figures are the recoverable reserves corresponding to the proven 1.2×10^{12} barrels of in-place resources in the Orinoco oil belt (Fiorillo, 1985). Official gas reserves are 92.10^{12} cubic feet, of which 66×10^{12} cubic feet occur in the Eastern Venezuela basin. Oil production in 1986 was 650×10^6 barrels. Gas production is just over 1 trillion cubic feet per annum.

SUMMARY OF MESOZOIC-RECENT GEOLOGIC HISTORY

The sedimentary section of interest to the hydrocarbon geologist in Venezuela ranges from the Jurassic to the upper Tertiary.

The basement is built of Precambrian to Mesozoic igneous rocks and Paleozoic to Mesozoic metamorphic rocks. Jurassic red beds and volcanic rocks accumulated in northeast-southwest-trending grabens which formed in response to opening of the North Atlantic. In the Early Cretaceous a broad peneplain formed along the northern flank of the Guayana Shield. During the Late Cretaceous, it became the site of sandy deposition in the south, and lagoonal or marine, carbonate and shale deposition in the north. Venezuela's most important source rocks were formed at this time. Farther to the north an island arc formed in response to convergence between the Caribbean and South American Plates. At the end of the Cretaceous this arc collided with the continent, and arc and oceanic material were thrust southward. The northernmost Mesozoic sediments became metamorphosed, and a flysch trough formed south of the thrust front. Island-arc activity ceased at the end of the Cretaceous, but the general scenario of compression continued along the northeastern margin of the continent throughout the Tertiary and still persists today. Thus, Tertiary sedimentation in this area followed a general pattern of paralic or continental sand deposition in the south and flysch or molasse deposition in the north.

In contrast, the northwestern part of the continent is being thrust northward onto the Caribbean Plate. Since the Caribbean-South American plate boundary is one of dextral transcurrency, this northwestern area is characterized by major pull-apart basins where lacustrine to open-marine clastics and local carbonates have accumulated from Oligocene to Recent.

CRETACEOUS SOURCE ROCKS

Stratigraphic Units of Major Hydrocarbon Provinces

The most researched and best documented source rock of Venezuela is the Upper Cretaceous (Cenomanian-Turonian-Coniacian) La Luna Formation of the Maracaibo basin (Table

Table 1—Cretaceous source rock and adjacent lithounits

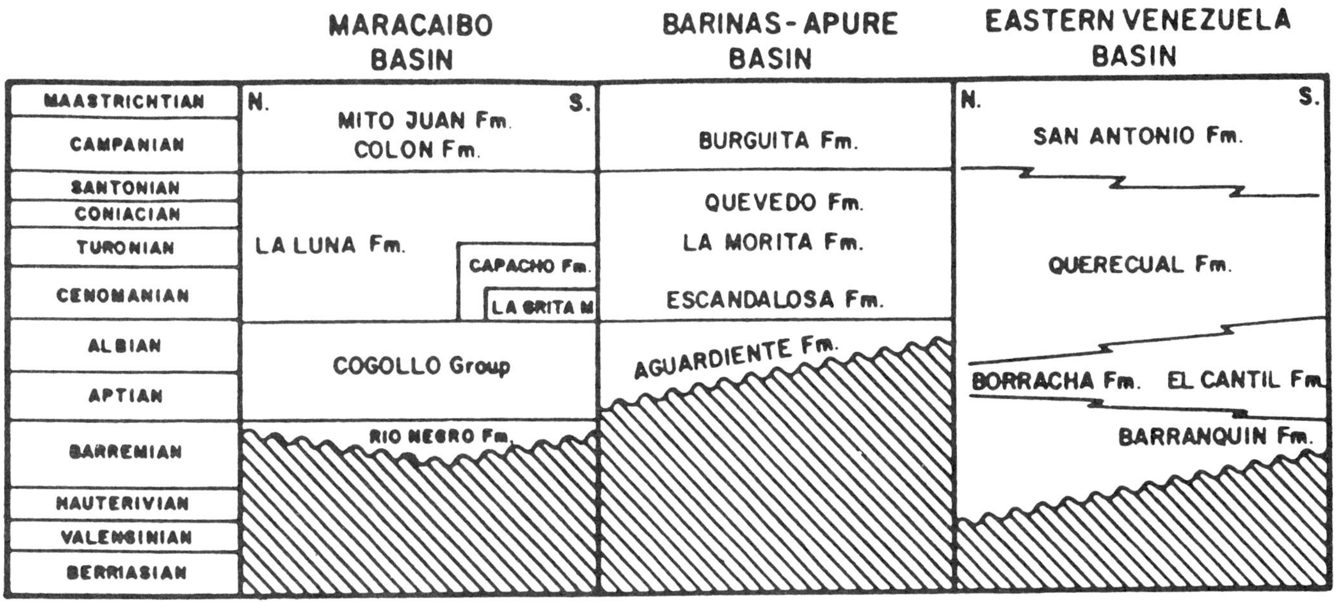

1). Latest reports were presented by Talukdar, Gallango, and Ruggiero (1985), Talukdar, Gallango, and Chin-A-Lien (1985), and Vierma (1985a, 1985b).

The unit consists of dark-gray to black, organic-rich, finely laminated detrital limestones and calcareous shales. Nodules are common. Planktonic and pelagic fossils are abundant; benthonic fossils are absent except in some sections in the Mérida Andes (González de Juana, Iturralde de Arozena, and Picard, 1980). Chert is common in the section, becoming more abundant upward and northward. The unit becomes sandy to the south. The average thickness in the Maracaibo basin is around 110 m. High organic content of kerogen type II, absence of benthonic fossils, and abundance of planktonic fossils are generally argued as indicating a relatively deep marine, euxinic environment of deposition similar to the Cariaco Trough or the Black Sea. There are, however, no faunal depth indicators. The underlying Cogollo Group, on the other hand, is demonstrably the product of a shallow carbonate environment. It contains grainstone and ooid offshore bars and lagoonal, organic-rich sediments which are geochemically indistinguishable from the La Luna.

The net source thickness of the La Luna is some 66% of the gross (from core and electric-log studies) and the oil yield has been estimated at 290×10^6 bbl/km^3 (Talukdar, Gallango, and Ruggiero, 1985). The total hydrocarbon drainage area of the La Luna below the Maracaibo basin is some 50,000 km^2.

The potential of the Cogollo Group source rocks has not yet been investigated in such detail, but in one well geochemical logs indicated source-rock quality as good as the La Luna over a 130-m interval. The geographic extent of this material remains to be determined.

In the south of the Maracaibo basin the lateral equivalent of the lower part of the La Luna is the Capacho Formation, whose La Grita Member (as much as 50 m thick) is of the same facies as the La Luna. Like the Cogollo Group, little information exists on the extent of this unit.

In the Eastern Venezuela basin, the Querecual Formation, age equivalent of the La Luna, also consists of black limestone and shale rich in planktonic forams and containing abundant nodules. Chert is present and increases in abundance upward and northward. Toward the south the unit becomes sandy. Thickness ranges from 353 to 700 m. The unit thins to the south as a function of facies change to the overlying, prograding, sandy San Antonio Formation. Campos, de Cabrera, and Lander (1985) describe the environment of deposition as deep marine in the north and shallow marine in the south. The Querecual lies above the El Cantil (coastal sandstone and limestone) and Borracha (reefal limestone) Formations (cf. the Cogollo Group of the Maracaibo basin). Talukdar, Gallango, and Ruggiero (1985) determined that this source rock (again, kerogen type II) has a generative capacity of 56 to 152×10^6 bbl/m^3. Arnstein et al. (1982) indicate a kitchen area of some 28×10^3 km^2.

In the Barinas-Apure basin, the Upper Cretaceous Quevedo, La Morita, and Escandalosa Formations have been identified as source-rock-bearing units (Chigne, 1985), which are sandy equivalents of the La Luna. The are coastal deposits in the south and of more open marine environments in the north. Again, the organic material is kerogen type II in the northwest, but kerogen type III becomes important southward toward the craton and over the areas today occupied by the Mérida and Arauca swells (Figure 1). There are no available data on net thicknesses or generative capacities of these units. Chigne's map indicates a kitchen area of some 40×10^3 km^2.

Paleogeographic Environment

From a paleogeographic point of view, it is logical to link the source rocks of today's basins into a single, large depositional province which existed along the northern flank of South America in the Late Cretaceous. The Cogollo Group in the Maracaibo basin clearly documents the development of a

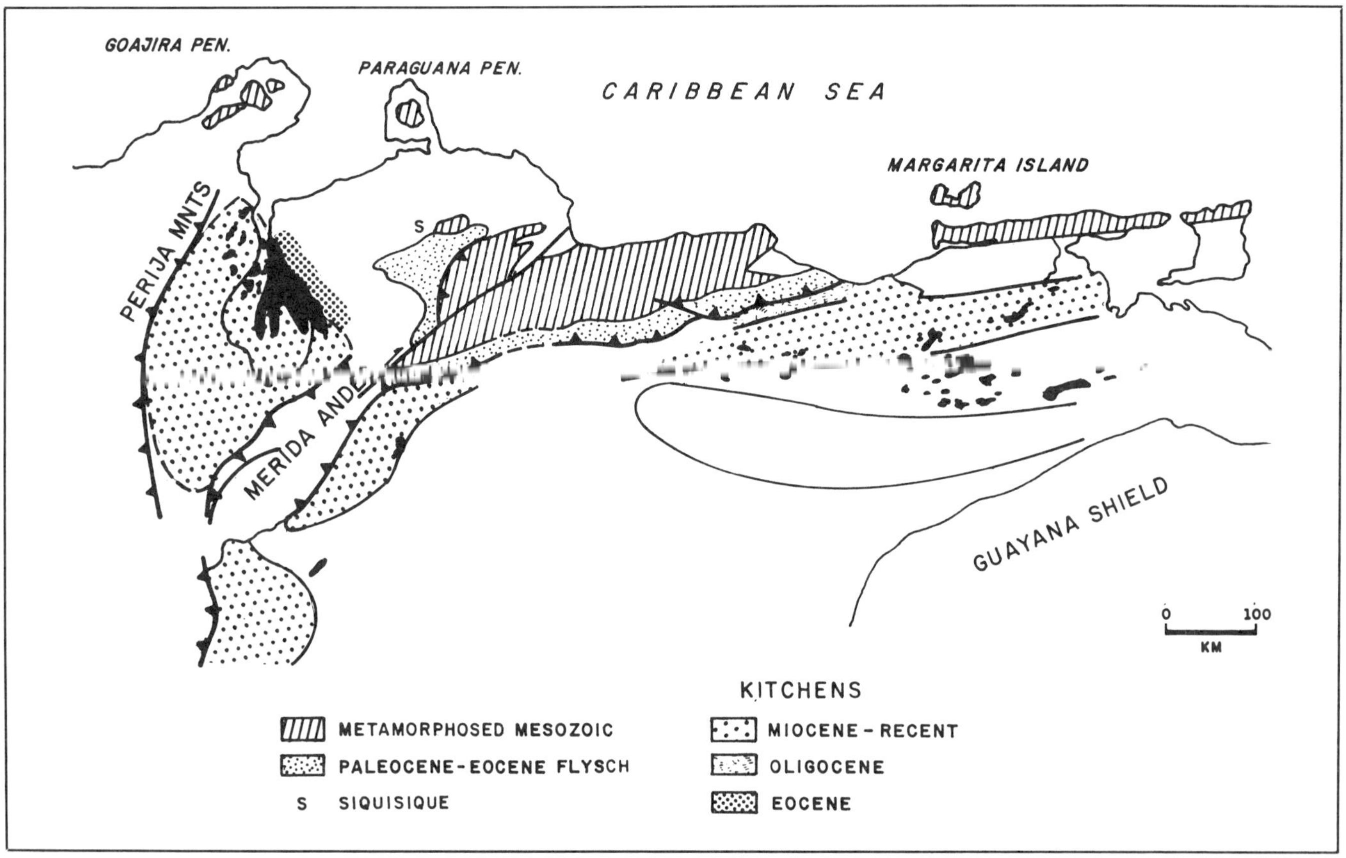

Figure 2. Documented kitchens for Cretaceous source rocks in northern Venezuela.

shallow carbonate platform, above wavebase, characterized by grainstone bars and intervening lagoons with organic-rich sedimentation. Expansion of this latter (euxinic) environment led to widespread deposition of source rocks of the La Luna, Querecual, Quevedo, La Morita, and Escandalosa Formations. What caused this to occur remains speculative. A deepening of water may have been involved, but what was essential was widespread and prolonged existence of anoxic bottom waters. Such conditions may have resulted from the development of a very broad lagoon behind a regional, seaward barrier (a reef?).

The original extent of this Cretaceous facies is unknown. Outcrops in the Perijá and Mérida Andes show that it extended across these areas. Gabela (1985) mentions that the La Luna is a source rock in the middle Magdalena basin of Colombia. The Cenomanian metasedimentary Los Robles Group of Margarita Island contains thin lenses of black limestone, black marble, and graphitic schist (González de Juana and Vignali, 1972). In the Interior Ranges of north-central Venezuela and in the south of the Falcón basin, there are Paleocene-Eocene flysch sediments (the Guárico Formation, the Matatere Flysch) which contain northerly derived exotic blocks of La Luna or Querecual facies (Peirson, Salvador, and Stainforth, 1966). Clearly, the same source rock facies extended northward into an area which suffered uplift and erosion in the early Tertiary.

Further evidence of organic-rich sediments of Cretaceous age is found throughout the northern coastal ranges (Dengo, 1953). Castro and Mederos (1985) describe the Barremian-Santonian Mejillones Group in the offshore Carúpano basin, at least 1,100 m thick, which contains cherty brown limestone of euxinic origin interbedded with radiolarite and basalt.

Maturity of Cretaceous Source Rocks

The La Luna reached maturity for oil generation in the northeast of the Maracaibo basin (Figure 2) during the Eocene (Talukdar, Gallango, and Chin-A-Lien, 1985). It is now mature for gas generation in this area (note that the extension of this Eocene kitchen to the northeast below the Falcón basin is unknown).

In the west and south of the basin, maturity for oil generation was attained in the Miocene to Recent (Talukdar, Gallango, and Chin-A-Lien, 1985). The La Luna is at the threshold of generation (Talukdar, Gallango, and Chin-A-Lien, 1985), and concretions within the unit contain oil at its type locality int he Perijá Mountains (Hedberg, 1931). At outcrop in the Mérida Andes the La Luna is overmature, whereas the La Grita Member, A lateral equivalent, is mature in the south of the Maracaibo basin.

The Querecual Formation of the Eastern basin generated oil in the middle Oligocene-middle Miocene in the northern part of the basin, where it is now producing gas; it has been generating oil farther south since the Miocene (Arnstein et al., 1982). Thus the Eastern basin is now characterized by gas in the north and heavy oil (mostly degraded; Fiorillo, 1985) in the

Table 2—Tertiary source rock and adjacent lithounits

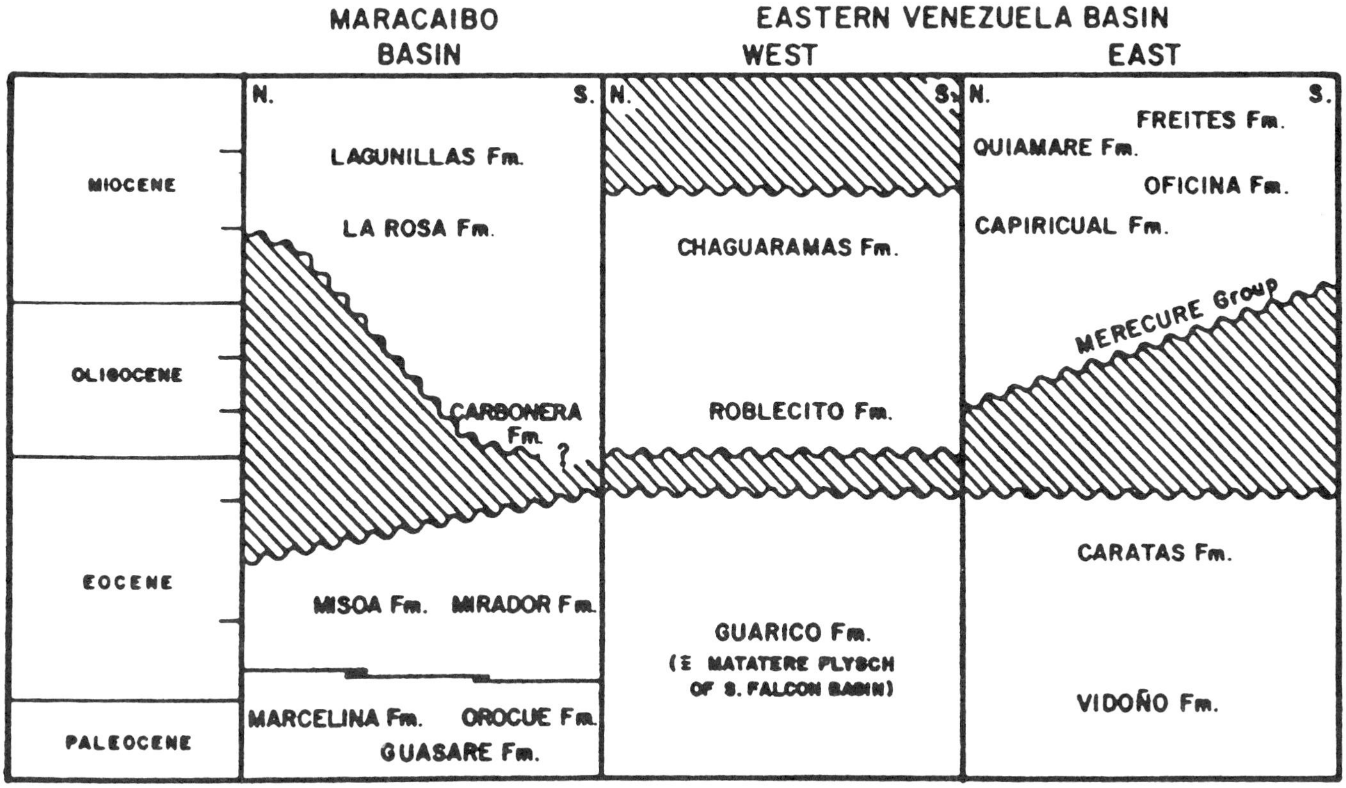

south. In the Barinas-Apure basin, generation began in the middle Miocene (Chigne, 1985) in the depocenters adjacent to the Andes. Farther south, the source rocks are immature.

Summarizing the foregoing, the general picture is that Cretaceous source rocks are immature toward the craton and are mature or overmature below Tertiary depocenters. The oldest known depocenter (Eocene) was below the northeastern part of the Maracaibo basin and presumably extended some distance below the Falcón basin.

If one extrapolates the trend of increasing maturity northward to the graphitic schists of the northern ranges, then one arrives at a regional model of earlier maturation in more northern basins, the oldest of which became metamorphosed and inverted. This inversion is shown by the presence of La Luna/Querecual olistoliths within Paleocene-Eocene flysch to the south. Moreover, mature or overmature La Luna at outcrop in the Perijá Mountains and the Mérida Andes show that these areas also were kitchens at some time in the past. We conclude that Cretaceous source rocks originally were distributed over an extensive area of northern South America. Likewise, kitchen areas were more extensive than at present, and maturation/maturity migrated progressively southward.

TERTIARY SOURCE ROCKS

Blaser and White (1984) reported that Tertiary source rocks, characterized by land-plant material, are present in the south and west of the Maracaibo basin. They occur in the Paleocene Orocué and Marcelina Formations, the Eocene Mirador and Oligocene Carbonera Formations, and the Miocene Lagunillas and La Rosa Formations (Table 2). They could be mature in the north Andean foredeep, in the south of the basin, where little exploration has been carried out. Talukdar, Gallango, and Chin-A-Lien (1985) reported that three fields in the southwest of the basin produce crude oils derived from terrestrial source rocks.

In the Eastern Venezuela basin there are source rocks in the Merecure Group and Roblecito Formation of Oligocene age, and in the Oficina and Chaguaramos Formations of Miocene age (Arnstein et al., 1982). The organic material is mainly terrestrial, and some waxy crude oils in the Las Mercedes and Oficina areas (Figure 1) are probably derived from these units.

Quantitative data on generative capacities, net thicknesses, and areal extents of Tertiary source rocks are not yet available.

STRUCTURAL CONSIDERATIONS

The structural development of northern Venezuela—the rise of the Perijá, Andean, and Coastal and Interior Range uplifts; the development of complementary depocenters and kitchen areas; and the growth of structural traps—resulted from interaction between the Caribbean and South American Plates. Activity of a Pacific plate or plates has influenced the development of northwestern Venezuela.

Present-Day Structures

Figure 3 shows the main faults and folds of northern Venezuela. In the northeastern and north-central areas the dominant structures are east-northeast-trending folds and thrusts,

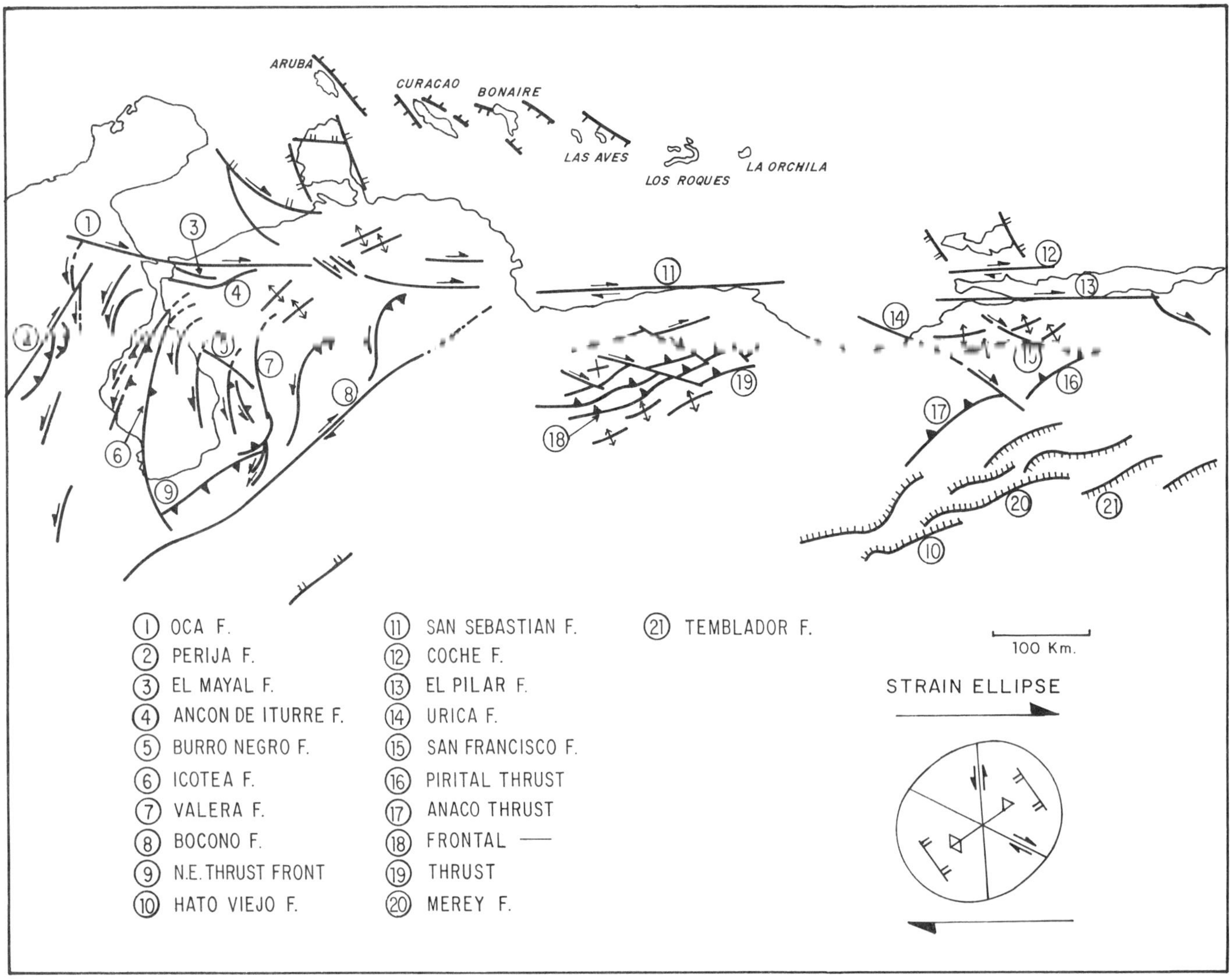

Figure 3. Main structural features in northern Venezuela.

and dextral faults which trend northwest in the north and which swing to east-west in the south. Many of the oil and gas fields in the Eastern Venezuela basin are located in the above folds. The northern part of the Falcón basin has similar structural features.

In the Maracaibo basin, in contrast, there is a fan, opening to the south, of sinistral faults trending northeast to north. The main folds in the area are subparallel to, and were generated by compression along, these faults. The folds are most prominent in the north, where compression is greatest. These sinistral faults are of great importance to the basin's hydrocarbon accumulations. They not only generated the structural traps of the basin, but they also created the fractures essential for production from igneous basement and tight Cretaceous limestone. In addition, synsedimentary structural growth resulted in the development of pinchout traps. Finally, extensional strain elements of the sinistral systems may well provide vertical migration paths for hydrocarbons ascending from Cretaceous source rocks to Tertiary reservoirs.

Lastly, normal faults are abundant above the southern hingelines of the Eastern Venezuela and Barinas-Apure basins. These, too, are important because they form fault-sealed traps for hydrocarbons.

Origins of Structures

The structures of northeastern and north-central Venezuela, and of the northern part of the Falcón basin, resulted from dextral transpression along the east-trending, Caribbean-South America Plate boundary and along dextral synthetics to that system. The relatively straight, east-trending coastlines of the northeast and north-central areas are manifestations of the plate-to-plate interaction and the associated earthquake activity. Transpression here has caused the uplift of the Coastal and Interior Ranges, which form the northern limit of the Eastern Venezuela basin.

In the northwest, in contrast, the coastline bulges convex to the north and a simple, east-trending, plate-to-plate boundary is not apparent. Nevertheless, in the northern half of the Falcón basin there is a well-developed fault/fold system which results from east-west dextral deformation. The Oca, El Mayal, and Ancón de Iturre faults, for example, are part of a dextral system which continues eastward across the basin, with right side-step, to link up with the San Sebastian fault which defines the straight coastline mentioned above. In addition, the northwest-trending coastal segments of the Gulf of Venezuela, the Maracaibo basin, the Paraguaná Peninsula, and the Falcón basin are extensional strain elements of the east-west, dextral system. Thus, the east-trending Caribbean-South America Plate boundary does have expression in the northwest, but it seems to be somewhat displaced to the north, relative to the eastern areas, and transpressive uplifts are not present.

Most modern earthquake epicenters in the northwest are aligned parallel to the Bonocó fault, and some authors conclude that this fault marks the Caribbean-South America Plate boundary in this area. Dextral movement along the fault is generally accepted, though the amount of such movement is debated (see, for example, Schubert, 1984). Pindell (1985) indicated as much as 290 km of lateral displacement and 15-20 km of compressive shortening associated with the Bonocó system. Transpression along the fault has caused the uplift of the Mérida Andes in flower-structure form, lifting basement as much as 5,000 m above sea level. In the adjacent North Andean foredeep, basement lies at 11,000 m below sea level. Similar dextral transpression along the Perijá fault, acting somewhat earlier, generated the Perijá Mountains. The Maracaibo basin is thus defined by two transpressional uplifts.

According to this paper, the result of dextral movement along the Bonocó fault is that the area northwest of the fault is approaching and overriding the Caribbean Plate. This explains the northward convexity of the northeastern Venezuela coastline. It also explains the east-west dextral strain elements, described above, of the Falcón basin, the Maracaibo basin, and the Gulf of Venezuela, for progressive movement of northwestern South America onto the Caribbean Plate means that the former must progressively assume the movement of the latter.

The compressional element of the interaction of northwestern Venezuela with the Caribbean area is responsible for enhanced activity along antithetics to the east-west dextral system, and this explains why the Maracaibo basin is dominated by sinistral faults (Figure 4).

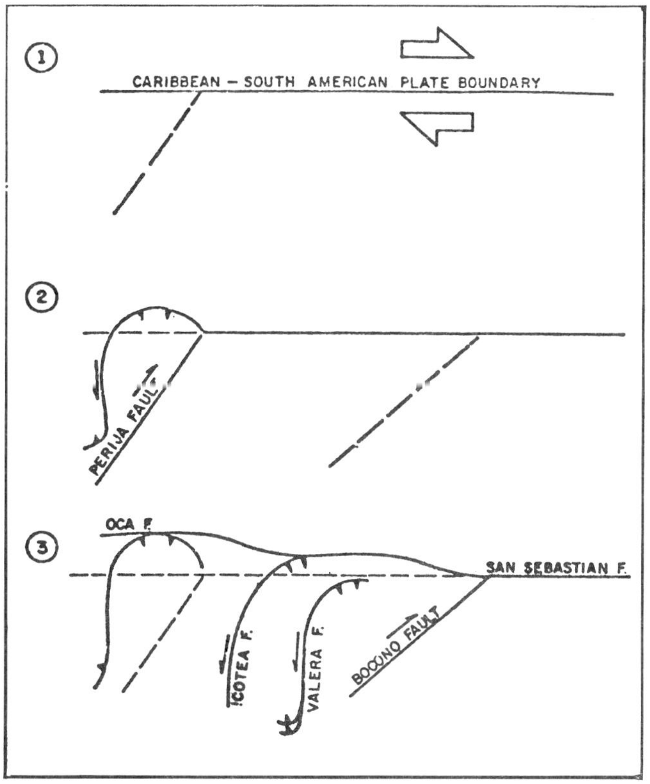

Figure 4. Schematic representation of interaction between N.W. Venezuela and the Caribbean Plate.

Cause and Timing of Dextral Movements Along the Perijá and Bonocó Faults

The driving force for these movements lies in the Pacific where the Cocos Plate, 25 Ma, according to Hey (1977), or 28 Ma, according to Lonsdale and Klitgord (1978), was moving N70°E as the result of spreading along both the East Pacific rise and the Cocos-Nazca Plate boundary. This movement was at right angles to, and generated, the Middle America thrust. It resulted in compression in that direction against northwestern South America, and this reactivated northeast-trending faults originally formed in the Late Jurassic-Early Cretaceous (James, 1985). Dating of the beginning of northeasterly Cocos Plate movement depends upon identification of the oldest magnetic anomalies generated by the Nazca-Cocos spreading. Literature puts this at 25 or 28 Ma, as already mentioned. However, the oldest anomalies may have been subducted below Central America, and thus no longer be detectable. West of the Cocos Ridge there is a set of northwest-trending anomalies (anomalies 12, 13, etc., Corvalán et al., 1981). These imply northeastward movement of Pacific crust in that area and their ages, at youngest, are Oligocene. Northwestward movement of northwestern Venezuela, therefore, could have begun in the Eocene. Thus it is interesting to point out that a regional unconformity, which developed in the late Eocene, is present in the Maracaibo basin. This unconformity, which has its greatest erosional vacuity in the north, could mark the beginning of the collision between northwestern Venezuela and the Caribbean. Timewise, it coincides with the shift of the relative motion of the Caribbean Plate from northeast to east (Mattson, 1984).

GENERATIVE CAPACITY

By restoring northwestern Venezuela to its supposed original position to the southwest (Figure 5), so that the metamorphosed Mesozoic rocks in the north form a continuous belt, south of which lie the Paleocene-Eocene flysch deposits, one sees that the oil fields and kitchen areas also line up in an approximately east-trending zone. Knowing that the flysch deposits contain allochthonous source rocks derived from the north, that graphitic metasediments of Late Cretaceous age lie in that direction also, and that much crustal shortening has taken place in the coastal and interior ranges, we suggest that the original areal extent of the Upper Cretaceous source rocks may have been in the order of some half a million square kilometers (Figure 6).

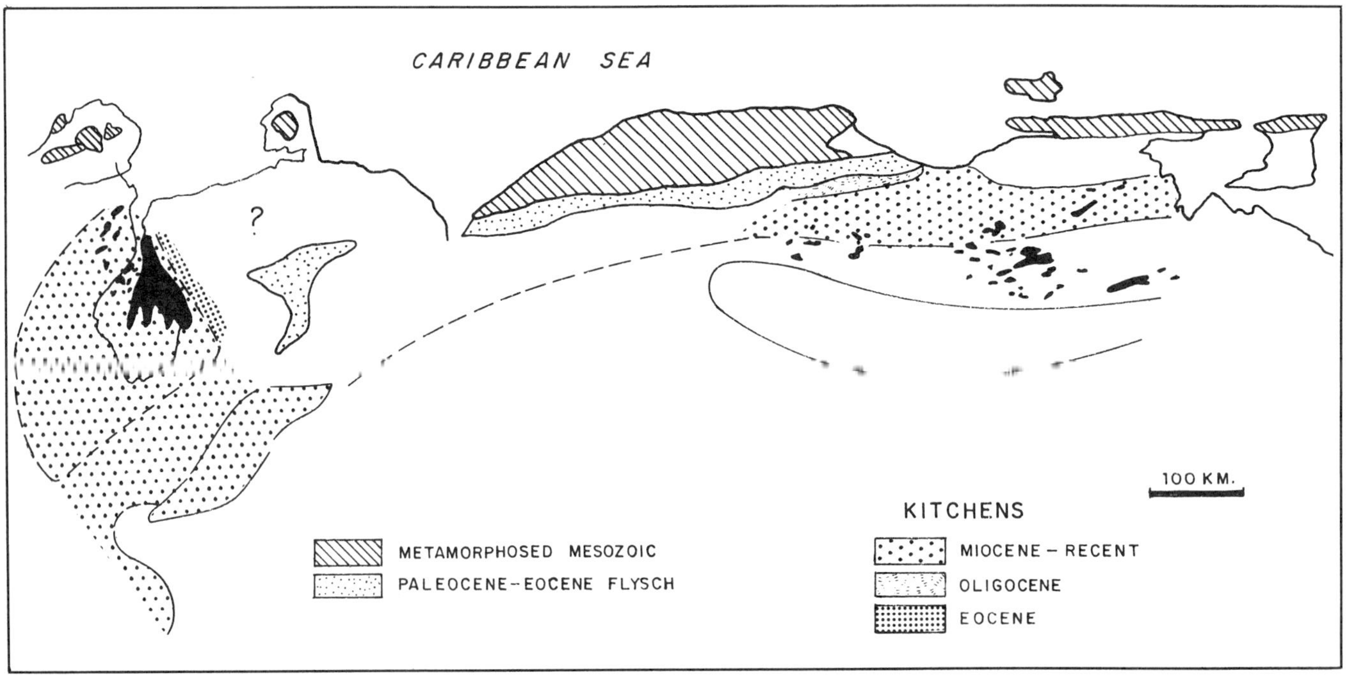

Figure 5. Displacements along the Perijá and Bonocó faults restored, northern Venezuela.

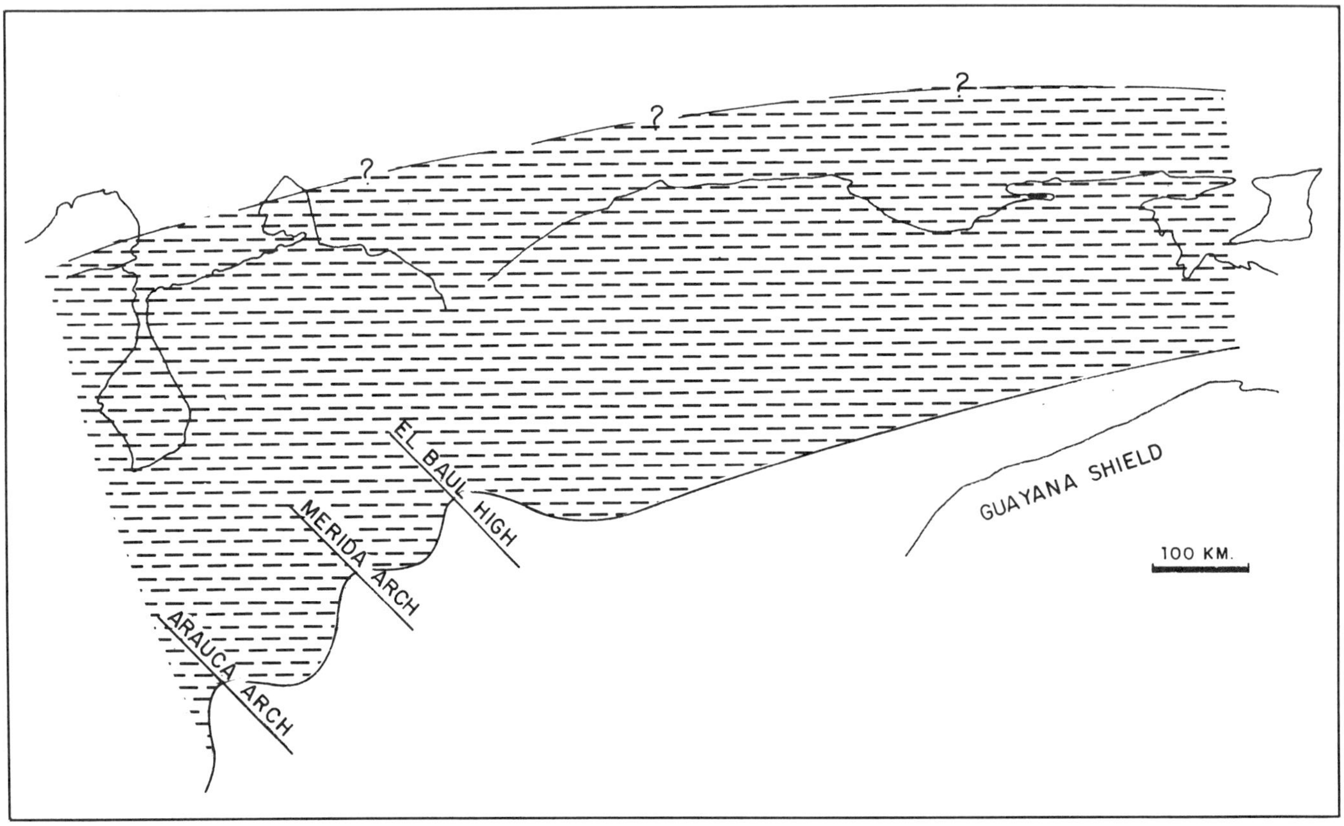

Figure 6. Possible original extent of Cretaceous source rocks in northern Venezuela.

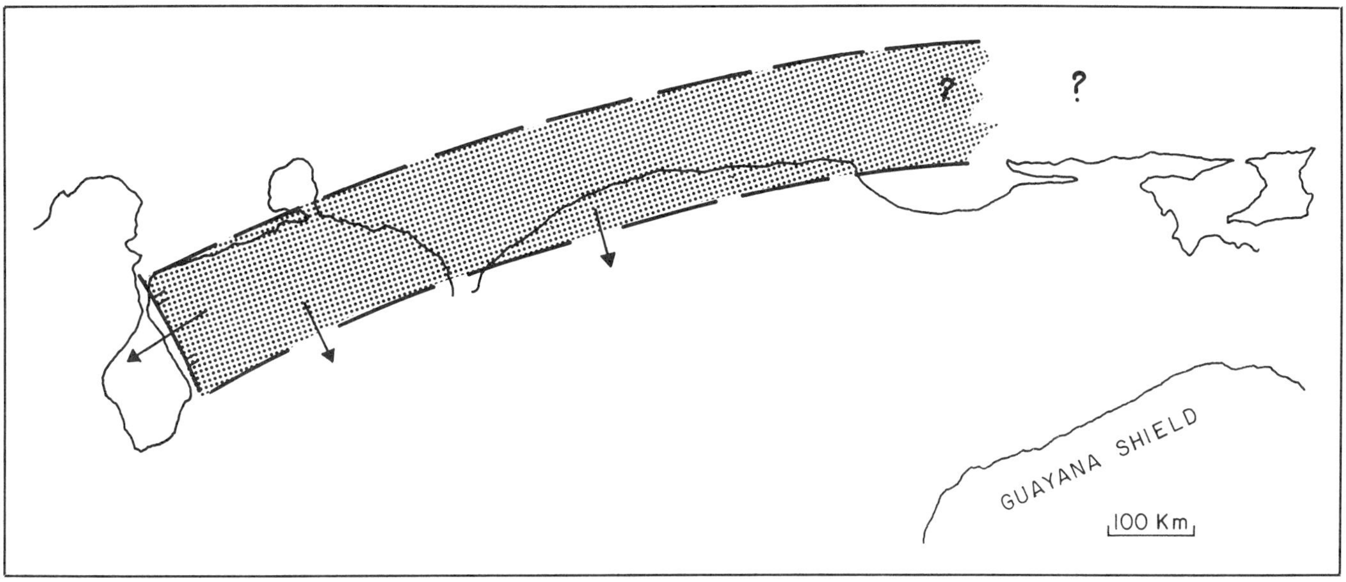

Figure 7. Possible extent of Eocene kitchen in northern Venezuela.

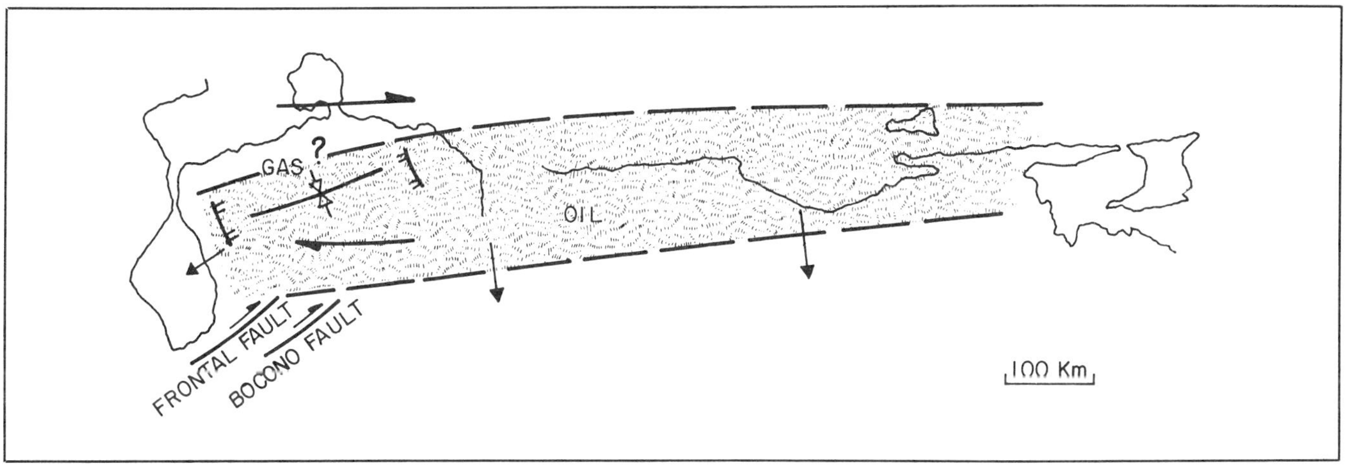

Figure 8. Possible extent of Oligocene kitchen in northern Venezuela.

Maturation occurred below a series of depocenters aligned roughly east-west, which migrated southward from Eocene (possibly Late Cretaceous) to Recent (Figures 2, 7, 8). Of the original source-rock area, some 400,000 km^2 could have passed into the maturity window (cf. total of published kitchen areas 118,000 km^2).

If we assume a generative capacity somewhere between that of the La Luna (290×10^6 bbl/km^3) and the Querecual (56 to 152×10^6 bbl/km^3)—say 200×10^6 bbl/km^3—and a *conservative*, regional average net source-rock thickness of 75 m, the amount of oil which could have been generated is in the order of 6 trillion barrels. This figure does not include the products of other source rocks, of both Cretaceous and Tertiary age, for which quantitative data are not yet available. McDowell (1975), summarizing the data from several basins, suggested entrapment in reservoirs of between one-quarter and one-third of the oil generated. For Venezuela, this gives an estimate of 1.5 to 2 trillion barrels of entrapped oil, which compares well with the estimated total known original in-place resources of 1.5 trillion barrels. Of this amount, 40×10^9 barrels have been produced and 295×10^9 barrels are recoverable reserves at today's economics and technology. These considerations are encouraging for continued exploration and appraisal in Venezuela.

Whereas production at present comes mainly from the Maracaibo basin (with currently known reserves of 19.2×10^9 barrels), in the next century attention will turn increasingly to the massive resources of the heavy-oil belt of the Eastern Venezuela basin. Of these, 700 billion barrels of oil in-place, 217 billion recoverable, lie in designated priority areas where recovery will be reasonably economical (Fiorillo, 1985). In nonpriority areas there is an additional 50×10^9 recoverable barrels from 500×10^9 barrels of in-place resource.

ACKNOWLEDGMENTS

This information is published with the kind permission of Maraven, S.A. and Petroleos de Venezuela, S.A. The authors acknowledge that reference has been made to internal company reports.

REFERENCES CITED

Arnstein, R., C. Betoret, E. Molina, L. Mompart, J. Ortega, F. Russomano, and H. Sánchez, 1982, Geología petrolera de la Cuenca de Venezuela Oriental: Mexico, XLV Reunión a Nivel de Expertos, Asistencia Recíproca Petrolera Estatal Latino-Americana, 33 p.

Blaser, R., and C. White, 1984, Source-rock and carbonization study, Maracaibo basin, Venezuela, *in* G. Demaison and R.J. Murris, eds., Petroleum geochemistry and basin evaluation: American Association of Petroleum Geologists Memoir 35, 426 p.

Campos, V., S. de Cabrera, and R. Lander, 1985, Estratigrafía del Noreste de Anzoátegui: Caracas, VI Congreso Geológico Venezolano, Memoria, Tomo I, p. 156-200.

Castro, M., and A. Mederos, 1985, Litoestratigrafía de la Cuenca de Carúpano: Caracas, VI Congreso Geológico Venezolano, Memoria, Tomo I, p. 201-225.

Corvalán, J., et al., 1981, Plate-tectonic map of the Circum-Pacific region, southeast quadrant, scale 1:10,000,000: American Association of Petroleum Geologists, Circum-Pacific Map Project.

Chigne, N.C., 1985, Aspectos relevantes en la exploración de Apure: Caracas, VI Congreso Geológico Venezolano, Memoria, Tomo V, p. 2891-2929.

Dengo, G., 1953, Geology of the Caracas region: American Association of Petroleum Geologists Bulletin, v. 15, p. 7-40.

Fiorillo, G.J., 1985, Exploration and evaluation of the Orinoco Oil Belt, final results: Long Beach, California, Third International Conference on Heavy Crudes and Tar Sands (UNITAR), v. 1, p. 332-361.

Gabela, V.H., 1985, Campo Caño Limón, Llanos Orientales de Colombia: II Simposio Bolivariano, Exploración Petrolera en las cuencas sub-Andinas, Publicación I, 28 p.

González de Juana, C., J. Iturralde de Arozena, and X. Picard, 1980, Geología de Venezuela y sus cuencas petrolíferas: Caracas, Ediciones Foninves, Tomo II, 1030 p.

Gonzáles de Juana, C., and M.C. Vignali, 1972, Rocas metamórficas e igneas en la Península de Macanao, Margarita, Venezuela: Porlamar, VI Conferencia Geológia del Caribe, Memoria, p. 63-68.

Hedberg, H.D., 1931, Cretaceous limestones as petroleum source-rocks in northwestern Venezuela: American Association of Petroleum Geologists Bulletin, v. 15, p. 229-244.

Hey, R., 1977, Tectonic evolution of the Cocos-Nazca spreading center: Geological Society of America Bulletin, v. 88, p. 1404-1420.

James, K.H., 1985, Marco tectónico, estilos estructurales y habitat de hidrocarburos Cretacicos, Venezuela: Caracas, VI Congreso Geológico Venezolano, Memoria, Tomo IV, p. 2452-2469.

Lonsdale, P., and K.D. Klitgord, 1978, Structure and tectonic history of the eastern Panamá Basin: Geological Society of America Bulletin, v. 89, p. 981-999.

Mattson, P.H., 1984, Caribbean structural breaks and plate movements: Geological Society of America Memoir 162, p. 131-152.

McDowell, A.N., 1975, What are the problems in estimating the oil potential of a basin?: Oil and Gas Journal, June 9, p. 85-90.

Peirson, A.L., A. Salvador, and R.M. Stainforth, 1966, The Guárico Formation of north-central Venezuela: Asociación Venezolana de Geológia, Minería y Petróleo, Boletín Informativo, v. 9, p. 183-224.

Pindell, J.L., 1985, Alleghenian reconstruction and subsequent evolution of the Gulf of Mexico, Bahamas and proto-Caribbean: Tectonics, v. 4, p. 1-39.

Schubert, C., 1984, Basin formation along the Bonocó-Morón-El Pilar Fault system, Venezuela: Journal of Geophysical Research, v. 89, p. 5711-5718.

Talukdar, S., O. Gallango, and A. Ruggiero, 1985, Formaciones La Luna y Querecual de Venezuela: Rocas madres de petróleo: Caracas, VI Congreso Geológico Venezolano, Memoria, Tomo VI, p. 3606-3642.

Talukdar, S., O. Gallango, and M. Chin-A-Lien, 1985, Generation and migration of hydrocarbons in the Maracaibo Basin, Venezuela (abs.): Munich, 12th International Meeting on Organic Geochemistry, Abstracts, p. 28.

Vierma, L., 1985a, Caracterización geoquímica de kerógeno en rocas carbonáticas de la Formación La Luna: Caracas, VI Congreso Geológico Venezolano, Memoria, Tomo IV, p. 2248-2262.

Vierma, L, 1985b, Correlación de crudos y rocas madres en una porción de la Cuenca de Maracaibo, Venezuela: Caracas, VI Congreso Geológico Venezolano, Memoria, Tomo IV, p. 2263-2280.

Ericksen, G. E., Cañas Pinochet, M. T., and
Reinemund, J. A., editors, 1989, Geology of the Andes
and its relation to hydrocarbon and mineral resources:
Houston, Texas, Circum-Pacific Council for Energy
and Mineral Resources Earth Science Series, v. 11.

CHAPTER 30

PETROLEUM PROSPECTS OF THE SEDIMENTARY BASINS OF ECUADOR

Giovanni Rosanía S.
Texaco Petroleum Inc.
Quito, Ecuador

ABSTRACT

Ecuador has four major sedimentary basins that are prospective for oil, three in the Coastal region and one in the Oriente region. In the Coastal region the stratigraphic sequence starts in the Cretaceous with the deposition of the Piñón Formation, which is the substratum for all the area. Rocks ranging from Cretaceous age to Recent are present. A Paleocene-lower Eocene hiatus is noticeable in the Borbón and Progreso basins, whereas an upper Eocene, upper Oligocene hiatus is present both in the Manabí and Progreso basins.

The Coastal basins are separated from the Andean Range by a major fault zone which extends mostly under Recent deposits. Its extension is from south of Guayaquil to the north toward the Colombian border. The main structural features separating the basins are the Chongón-Colonche Hills; and the Jama-Quinindé, Esmeraldas-Pinchincha, and La Cruz faults.

The Oriente basin is filled with sediments ranging from Early Paleozoic age to Recent, which have been deposited in marine and continental environments. From the Cretaceous onward, a nonmetamorphosed sedimentary sequence has been deposited. More than 11,000 m of sediments have accumulated in the Oriente basin, which is asymmetric and plunges to the south. To the west, two major north-trending uplifts are present: the Cutucú north of the Peruvian border and the Napo near Colombia.

Oil is produced from Tertiary rocks in the Progreso basin and from Cretaceous rocks in the Oriente basin. More than 100 million barrels of oil have been produced from the Progreso basin and 1,100 million barrels have been produced from the Oriente basin. The major reservoir rocks in the Oriente basin are the Hollín and Napo formations.

INTRODUCTION

Extent and Distribution of Basins

Ecuador is divided into three geographic provinces: the western coastal lowlands, the Andean Mountains which run north through the center of the country, and the low-lying eastern or Oriente region. This chapter will discuss the geology and hydrocarbon potential of the sedimentary basins in the Coastal and Oriente regions.

The Coastal region of Ecuador is generally flat but contains low mountain ranges, including the Chongón-Colonche Hills which trend northwesterly from Guayaquil, and the Jama Hills which parallel the coast along northwestern Ecuador. These ranges divide the coastal province into four separate basins which are, from north to south, the Borbón, Esmeraldas-Caraquez, Manabí, and Progreso basins (Figure 1). The Esmeraldas-Caraquez basin will not be discussed separately. The northern part (Esmeraldas) is geologically similar to the Borbón basin, and the Caraquez part is similar to the Manabí basin.

The eastern or Oriente region covers approximately 100,000 km^2 and is part of the foreland zone between the Andean Mountains and the Guayana Shield. Near the Andean Mountains a foothill zone is present where outcrops can be studied, but generally the Oriente region is flat, low lying, and jungle covered. Only one basin, the Oriente basin, is present and the geology is remarkably similar from the northern border with Colombia to the southern border with Perú. Most of the known oil reserves of Ecuador are in the northern part of the Oriente basin (Figure 1).

COASTAL REGION

The Coastal region of Ecuador extends from the Andean Mountains west to the Pacific, and from the Colombian border south to the border with Perú (Figure 2). The region contains a thick sequence of Mesozoic and Cenozoic sediments, mostly deposited in a marine environment overlying a volcanic substratum of probable Early Cretaceous age. Folding and faulting during the Tertiary resulted in many facies changes within

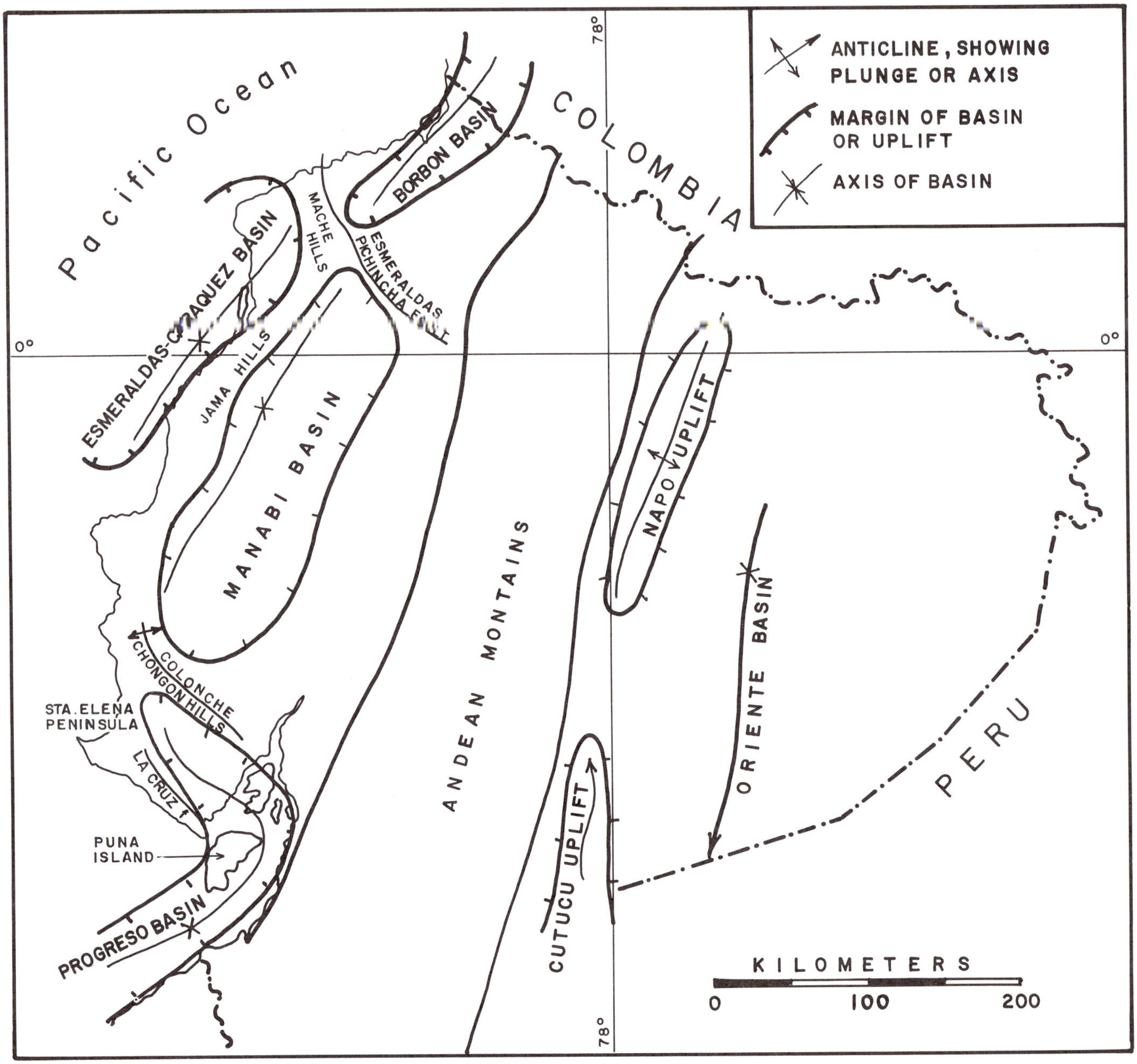

Figure 1. Index map of Ecuador showing principal sedimentary basins.

the sedimentary units. At the end of the Neogene, or later, the region was deformed into four separate basins.

Stratigraphy

The thickest accumulation of sediments in the Coastal region is apparently in the Gulf of Guayaquil (Progreso basin), where the Tertiary deposits alone may be more than 10,000 m thick. The thickness of Cretaceous sediments is unknown but is estimated to be more than 3,000 m. Stratigraphic units in the Coastal region are shown in Figure 3. Stratigraphic formations of interest for petroleum exploration are discussed below for each basin in this region.

Borbón Basin

Two wildcat oil and gas test holes were drilled in the Borbón basin, Borbón-1 and Telembi-1 (Figure 2), and one in the Esmeraldas basin, Camarones-1 (Figure 2), during the mid-1940s. Very little exploratory work has been done in these areas since those holes were drilled, probably because none of the holes had hydrocarbon shows. The following discussion of the stratigraphy is based largely on studies made prior to drilling and on data from those drill holes.

Cretaceous

Piñón Formation—The volcanic substratum underlying the marine sediments throughout the Coastal region of

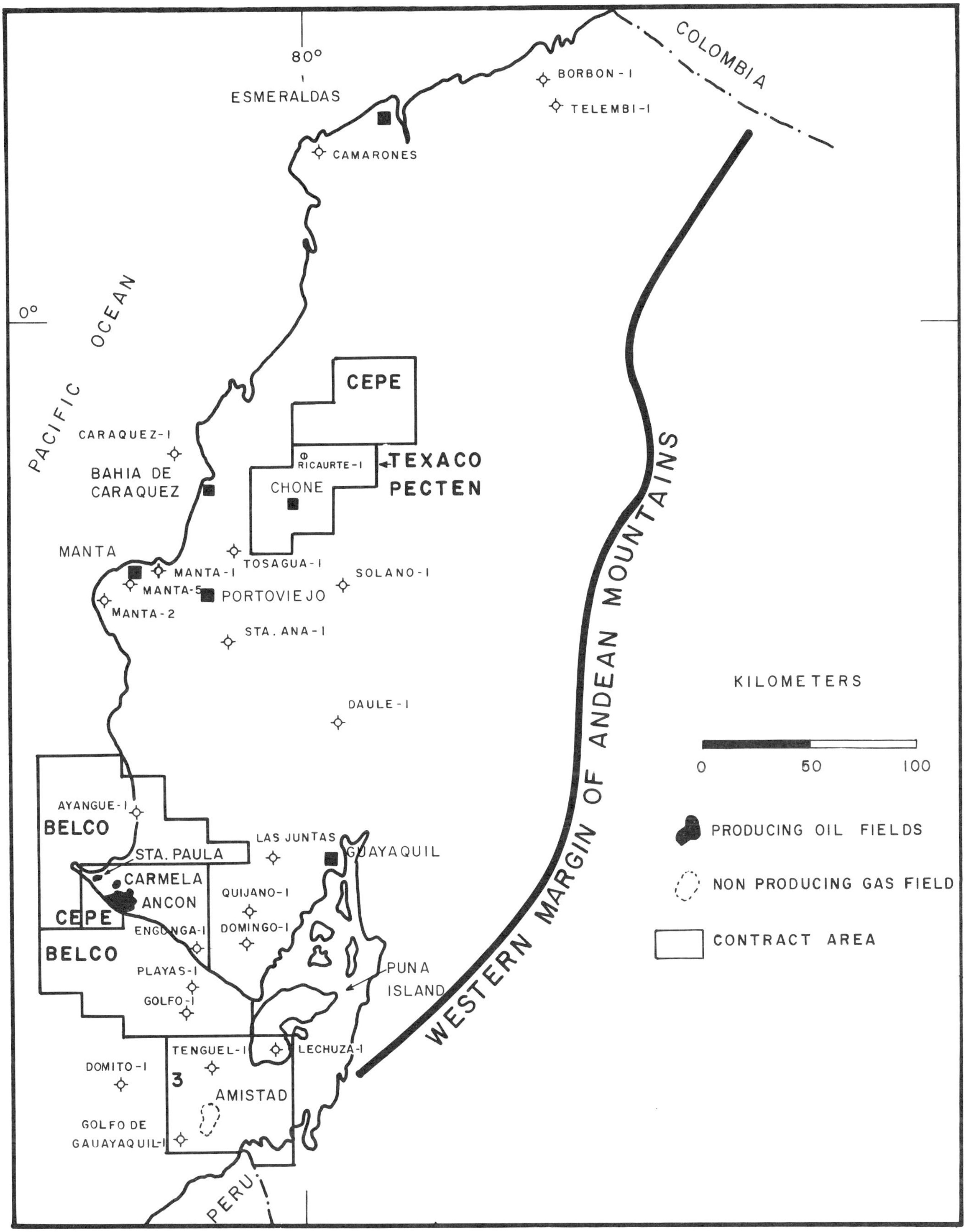

Figure 2. Contract areas oil and gas fields and important exploratory wells in Coastal Ecuador.

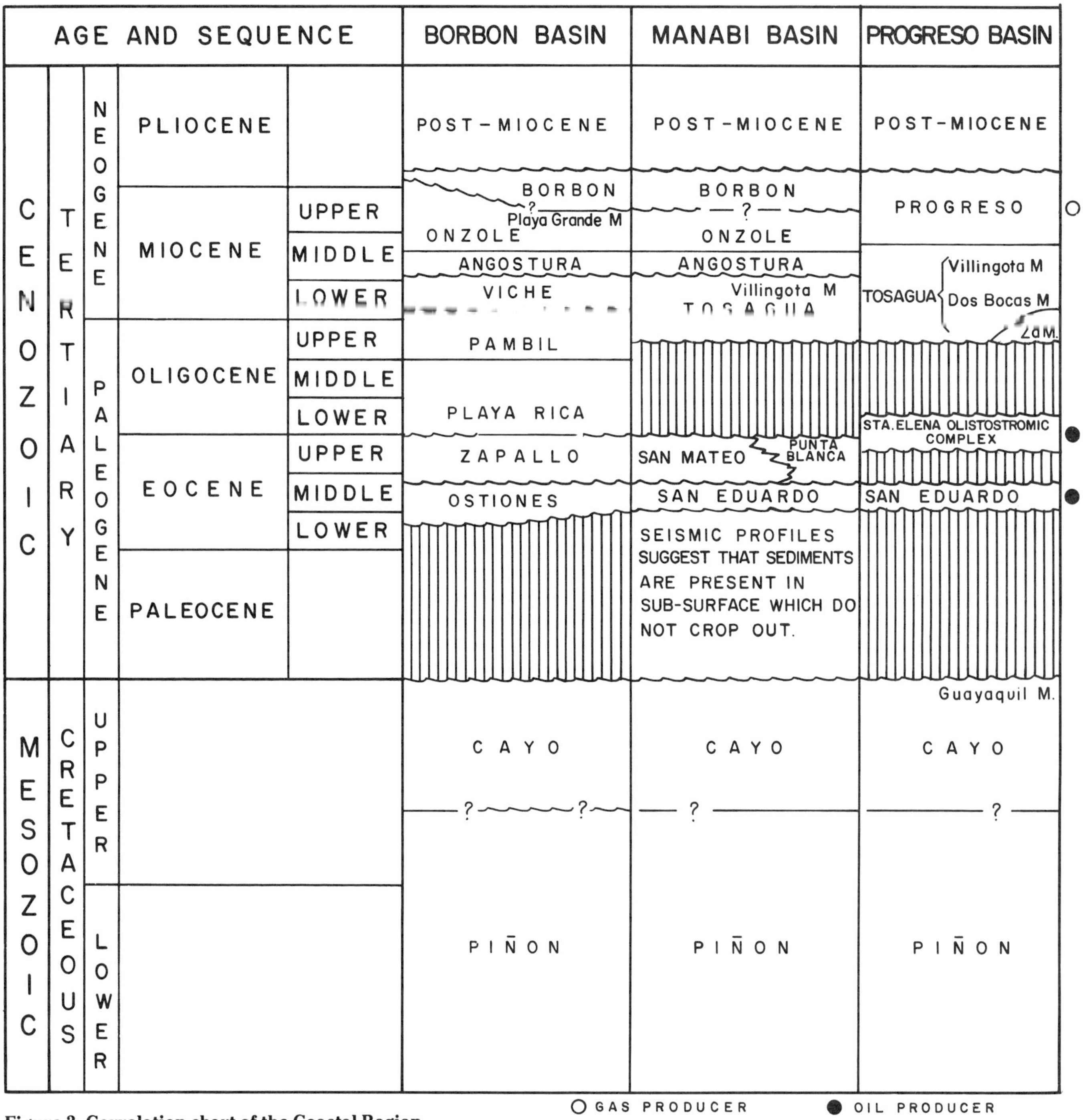

Figure 3. Correlation chart of the Coastal Region.

Ecuador is called the Piñón Formation, which consists largely of basalt and diabase flows, breccia and agglomerate, mostly of submarine origin. Green tuff and porphyritic lava are present in some places, and shale has also been reported, but these reports may result from confusion with the overlying Cayo Formation. The base of the Piñón Formation has never been encountered and a thickness cannot be given. In the Camarones well about 210 m of the formation were penetrated. The contact with the overlying Cayo Formation is apparently conformable in the Manabí and Progreso basins, but may be unconformable in the northern Borbón and Esmeraldas basins. An Early Cretaceous age has been assigned to the Piñón Formation on the basis of its stratigraphic position below the upper Cretaceous Cayo Formation. However, the uppermost part of the Piñón may be of Late Cretaceous age and the lowermost part older than Cretaceous.

Cayo Formation—The Cayo Formation has not been separated from the Piñón in the Borbón basin because outcrops are generally quite poor and exploratory wells did not encounter the Cayo, having penetrated Piñón volcanics directly below the Tertiary sediments. However, the Cayo may be present in the exploratory well Camarones-1 in the Esmeraldas basin. The strata overlying the Piñón in that well consist of dark-greenish-gray to bluish-gray, hard, tuffaceous, silicified, slaty shale having minor layers of ash and ashy sandstone and a few layers of cherty limestone. Microfossils were reported throughout this sequence, but results of the study of these fossils are not available to the writer. The sediments tentatively assigned to the Cayo are about 365 m thick in the Camarones well. They are underlain by Piñón volcanics which were reported to have a weathered zone, suggesting a hiatus. The upper contact of the Cayo is unconformable.

Tertiary

Ostiones Formation—Light-gray to white, brecciated, silicified and strongly folded beds of limestone containing Discocyclina are present in the Rio Verde (Figure 2), where they unconformably overlie the Piñón volcanics. These limestone beds, called the Ostiones Formation, are apparently found only in patches in the northern part of the Borbón basin. They are absent in all exploratory wells. Similar limestone of equivalent age, the San Eduardo, in the Progreso basin is much more continuous. Fossils indicate that the Ostiones Formation is early to middle Eocene in age. The limestone beds probably do not exceed 105 m in thickness in the Borbón basin.

Zapallo Formation—The Zapallo Formation is present throughout the Borbón (and Esmeraldas) basins and can generally be separated into two members. The lower member, sometimes called the Santiago limestones or Santiago formation, generally has a calcareous basal conglomerate overlain by layers of calcarenite and calcareous sandstone, interbedded with shale. Limestone is absent in the upper member, but locally both shale and sandstone units are calcareous. Elsewhere, the upper member is noncalcareous, consisting chiefly of shale with interbeds of sandstone, tuff, and chert. The shale is commonly dark-brown to gray and fairly hard, and grades upward to bluish-gray and greenish gray, hard, argillaceous siltstone.

The Zapallo generally has an abundant fauna, which indicates that the formation ranges from middle to late Eocene in age. The lower contact of the Zapallo is unconformable. The upper contact also is unconformable at the type section of the overlying Playa Rica (Canfield, 1966) but may be conformable elsewhere in the basin. The thickness of the Zapallo apparently changes considerably in the area. Estimates of about 500 m for the lower member and possibly 1,000 m for the upper member have been made (Baldock, 1982), but thicknesses in drill holes do not exceed 300 m.

Playa Rica Formation—At its type locality on the Santiago River (Figure 2) the Playa Rica Formation consists of a basal conglomerate overlain by dark-brown to gray and black foraminiferal shale with some interbedded sandstone. However, in the two holes that have been drilled in the basin, no basal conglomerate was encountered. The lower part of the Playa Rica in these holes consists of green to brown and black, locally tuffaceous shale with very minor amounts of sandstone and siltstone. The upper part of the Playa Rica consists of poorly sorted pebble conglomerate having interbeds of ashy sandstone in one drill hole (Telembi-1), and a brown to gray silty shale with interbedded coarse-grained sandstone and conglomerate in the other drill hole (Borbón-1). Microfauna indicate early to middle Oligocene age for the Playa Rica Formation.

As stated above, at its type section the Playa Rica unconformably overlies the Zapallo Formation but elsewhere may be conformable. It grades into the overlying Pambil Formation. Baldock (1982) indicates a maximum thickness of 800 m for the Playa Rica, but in the drill holes the formation is not more than about half that thickness.

Pambil Formation—Overlying the Playa Rica is the Pambil Formation, which consists of greenish-gray brown, fairly hard silty shale with thin beds of gray ash and gray, fine- to medium-grained sandstones. The shale contains an abundant fauna, indicative of a late Oligocene to early Miocene age. The Pambil has a maximum thickness of 750 m (Canfield, 1966), but the thickness changes laterally because of the gradational lower contact and the unconformity at the top. In drill hole Borbón-1 strata assigned to the Pambil are about 190 m thick, whereas in drill hole Telembi-1 they are more than 450 m thick.

Viche Formation—Though the uppermost part of the Pambil Formation may be of Miocene age, sediments of the Viche Formation are essentially the first true Miocene rocks encountered in the Borbón basin. In the outcrop the Viche Formation consists of light- to dark-green and brown silty claystone with a few sandy interbeds. Locally tuff and tuffaceous sandstone are present. The claystone is generally rich in foraminifera and lamellibranchs. The rich fauna (mostly benthonic) indicates an early to possibly middle Miocene age for the Viche Formation. It is possible that in local areas of the basin the basal Viche could be late Oligocene in age, but the bulk of the formation is early Miocene.

Drill hole Borbón-1 penetrated about 900 m of strata assigned to the Viche Formation, of which the lower 175 m is mostly conglomerate and sandstone with interbeds of brown shale. A few of the sandstone layers are quite porous. Overlying these coarse clastics is about 725 m of brown to gray, generally silty, calcareous shale with very minor thin beds of gray to green siltstone and fine-grained, locally calcareous sandstone. Glauconite is common in the uppermost 200 m.

The Viche Formation unconformably overlies the Pambil and is, in turn, overlain unconformably by the Angostura Formation. Because of these unconformities the Viche has a wide range of thickness. As stated above, the thickness is about 900 m in drill hole Borbón-1, and a similar thickness has been measured on the outcrop along the Rio Verde (Faucher et al., 1971). However, the Viche is not present in drill hole Telembi-1 and apparently has been removed by erosion prior to deposition of the Angostura Formation.

Angostura Formation—The transgressive Angostura Formation crops out over large areas in the Borbón basin, overlying the Playa Rica, Pambil, and Viche formations. The basal Angostura is conglomeratic but becomes progressively finer upward, grading into siltstone and claystone which predominate in the overlying Onzole Formation. Molluscs are widespread in the Angostura, and microfauna is abundant in the finer clastics. The fauna indicates an early middle Miocene age.

The Angostura Formation is as much as 600 m thick, but in drill hole Telembi-1 it consists of about 100 m of interbedded conglomerate and sandstone with minor gray to brown shale. In drill hole Borbón-1 the Angostura consists of about 290 m of greenish-gray to brown siltstone and slightly tuffaceous shale having a few interbeds of fine- to coarse-grained ashy sandstone.

Onzole Formation—The Onzole Formation is composed mostly of greenish- to bluish-gray siltstone and tuffaceous shale with minor amounts of sandstone. At the type locality, along the Cayapas and Onzole rivers (Figure 2), the lower part of the formation contains thin white ash beds that give it a banded appearance. The Onzole contains abundant megafossils and microfossils. Though the megafossils are not very indicative of a precise age, the microfossils suggest that the formation is of middle Miocene to early Pliocene in age. In the eastern part of the Borbón basin beds of tuffaceous, thin-bedded argillaceous sandstone, siltstone, and shale may be equivalent to the upper part of the Onzole Formation. This sequence is about 500 m thick in the type locality; it is 425 m thick in drill hole Borbón-1. An upper member of the Onzone, called the Playa Grande, has been reported by Canfield (1966). Fossil evidence, though scant, suggests a late Miocene age (Bristow and Hoffstetter, 1977).

Because of its gradational contact with the underlying Angostura Formation and the unconformity which is present in many areas at its top, the Onzone Formation is not uniform in thicknesses. In drill hole Borbón-1, the Onzole Formation (below the Playa Grande unit) is more than 380 m thick.

Borbón Formation—In the northwestern part of Ecuador, the Borbón Formation generally overlies the Onzole unconformably. However, in some areas it appears that the Borbón may be gradational with the Onzole. The Borbón consists of bluish-gray, fine- to coarse-grained sandstone containing a few interbeds of blue-gray or tan tuff. A basal conglomerate is generally present. The amount of tuff increases in the upper part of the Borbón and the sandstone becomes finer, grading into siltstone. In general the Borbón also becomes more shaly and tuffaceous to the west. The Borbón is richly fossiliferous, containing both molluscs and microfauna, which indicate an age ranging from late Miocene to Pliocene. The thickness of the Borbón is uncertain but probably exceeds 750 m in the eastern part of the basin.

Manabi Basin

Three wells were drilled by International Ecuadorian Petroleum Company (IEPC) in the Manabí basin in the mid-1940s, one of which (Solano-1) had slight shows of oil and gas. Interest lagged in the area until the Ecuadorian government oil company, Corporacion Estatal Petrolera Ecuatoriana (CEPE), initiated seismic and surface studies in the early 1980s. Seismic investigations defined a depositional basin and indicated that the structure of the basin is favorable for accumulation of hydrocarbons. Recently, a consortium of Texaco and Pecten companies signed a contract for 200,000 hectares in the northern part of the basin (Figure 2) and began exploration. The consortium planned to drill its first exploratory well (Ricarute-1) in 1987.

Data from the recent studies are not available, and the following discussion of the stratigraphy is based on earlier work. The Cretaceous Piñón volcanics will not be discussed, nor will the post-Miocene sediments.

Cretaceous

Cayo Formation—The Cayo Formation consists of a thick sequence of well-indurated erosion-resistant sediments and volcanics including units of green to gray slate, silicified tuff, tuffaceous hard sandstone, graywacke, volcanic breccia, and, locally, siliceous marl. Volcanics predominate in the lower part of the Cayo, whereas sediments predominate in the upper part The formation shows a range in thickness, but is believed to be at least 3,000 m throughout the Manabí basin. The Cayo apparently grades into the Piñón Formation, though in some areas the contact may be disconformable. The Piñón has been dated radiometrically as Cenomanian or Senonian. Radiolaria and foraminifera indicate that the Cayo is Senonian to Maestrichtian and possibly Danian (Bristow and Hoffstetter, 1977).

Tertiary

San Eduardo Formation—The San Eduardo Formation consists chiefly of limestone similar to the Ostiones Formation of the Borbón basin. This limestone is better exposed and developed in the Progreso basin to the south, and a more complete description will be given in the discussion of that basin. In the Manabí basin exposures are generally poor. In the Manta area, small exposures of the San Eduardo show it to be a calcareous microbreccia composed of grains from the Cayo Formation and weathered dolerite. Another exposure of the San Eduardo is along the Pacific coast north of Jama where a small,40-m-thick biotherm occurs in the formation. This exposure also shows a detrital marl back-reef facies (Faucher et al., 1971). Fauna indicate that the San Eduardo is of middle Eocene age in the Manabí basin.

San Mateo Formation—Outcrops of sediments assigned to the San Mateo Formation are limited to the southern part of the Manabí basin. The lower part of the sequence consists of interbedded conglomerate and sandstone, and includes minor beds of gray tuffaceous shale. The conglomerate and sandstone are composed chiefly of boulders, pebbles, and grains of porphyritic lavas. The cementing material is generally calcareous. Overlying this lower unit is a sequence of finer grained rocks, including greenish-gray tuffaceous shale and siltstone and lesser amounts of argillaceous sandstone, some of which is porous and permeable. Locally, argillaceous or clayey limestone is present. The San Mateo Formation unconformably overlies the San Eduardo or the Cayo Formation and is, in turn, unconformably overlain by the Tosagua Formation. The thickness is estimated to be about 800 m, but a sequence about 1,300 m thick was measured at Pacoche, southwest of Manta (field section 26, Faucher et al., 1971).

Fossils in the San Mateo indicate a late middle to late Eocene age and suggest that the formation accumulated in a nearshore environment (Bristow and Hoffstetter, 1977).

Punta Blanca Formation—The Punta Blanca Formation in the northern part of the Manabí basin consists chiefly of siliceous claystone and tuff. A clastic calcareous unit, which is present in places at the base of the formation, may be either debris from the eroded San Eduardo Formation or part of the San Eduardo Formation. The Punta Blanca is believed to lie unconformably on the San Eduardo or Cayo Formation and is unconformably overlain by the Tosagua Formation. The thickness is about 1,000 m. Radiolaria are common throughout the Punta Blanca but foraminifera are rare. Those collected from various outcrops indicate a middle to late Eocene age. Apparently the Punta Blanca in the northern part of the Manabí basin correlates with the more clastic San Mateo Formation of southern Manabí.

Tosagua Formation—The Tosagua Formation consists chiefly of massive cream to brown and chocolate-brown claystone and shale. The shale is commonly tuffaceous. Argillaceous sandstone and siltstone are generally present. Locally, a conglomeratic sequence at the base of the Tosagua contains pebbles from the underlying formation. At the top of the Tosagua near Charapoto, about 25 km south of Bahía de Caraquez, is a sequence of thin-bedded, soft to hard, tuffaceous, generally white to light-olive-green shale units containing dolomitic lenses in some interbeds of siltstone and sandstone, which were formerly called the Charapoto Formation. These shale units are now considered to be an upper member (Villingota member) of the Tosagua (Bristow and Hoffstetter, 1977). More than 1,300 m of strata in the Tosagua Formation have been measured in the northern part of the Manabí basin, and at least 650 m in the Villingota member have been measured in the southern part of the basin. As stated earlier, the Tosagua unconformably overlies older sediments. The upper contact is also apparently unconformable.

The Tosagua Formation of the Manabí basin is fairly rich in foraminifera. The lower part of the formation is late Oligocene in age, but the Villingota member is middle Miocene. The Tosagua apparently correlates with the Viche Formation of the Borbón basin but may be slightly older in the basal part (Figure 3).

Angostura and Onzole Formations—The Angostura and Onzole formations, which were described in the Borbón basin, also are widespread in the Manabí basin. The Angostura unconformably overlies rocks ranging from Cretaceous (Piñón Formation) to middle Miocene (Tosagua Formation). It attains a thickness of 600 m in the Manabí basin. The overlying Onzole Formation may be about 550 m thick in drill hole Solano-1 in the eastern part of the Manabí basin (Bristow and Hoffstetter, 1977).

Borbón Formation—The upper Miocene to Pliocene Borbón Formation described in the discussion of the Borbón basin is of similar lithology in Manabí. Near the coast west of the Jama Hills, the Borbón Formation consists of medium-grained sandstone to siltstone containing interbeds of tuffaceous shale. Locally, the sandstone is friable and has fair to good porosity. A maximum of about 750 m of Borbón may be present in outcrop. East of the Jama Mountains the sandstone is generally coarser, is calcareous, and contains molluscs.

Progreso Basin

The Progreso basin and the neighboring Santa Elena Peninsula (Figure 1) have been the most extensively explored areas in the Coastal region of Ecuador. The first well was successfully completed on the Santa Elena Peninsula in 1911, and drilling was nearly continuous from 1919 to the late 1960s. During this time, geological (and minor geophysical) investigations were also carried out in the Progreso basin. Recently, considerable geophysical work has been done in the Gulf of Guayaquil by CEPE. Belco is currently evaluating two large contract areas (400,00 hectares each) mostly offshore north of the Santa Elena Peninsula and in the Gulf of Guayaquil (Figure 2).

Cretaceous

Piñón Formation—The Piñón Formation is best exposed along the north flank of the Colonche-Chongón Hills, which form the northern limit of the Progreso basin. Here the formation consists chiefly of volcanic rocks, including flows, breccia, and agglomerate, but near the top, thin beds of tuffaceous shale, siltstone, and sandstone are present. The base of the Piñón has not been encountered, but its upper contact with the Cayo appears to be gradational. Faults are present at many places along the upper contact.

Cayo Formation—The Cayo Formation also is well exposed in the Colonche-Chongón Hills, where it consists of brown to greenish-gray sandstone and conglomerate, light-tan to olive-green claystone, siliceous shale, and locally hard, brown, partly siliceous limestone. The basal Cayo commonly contains volcanic agglomerate, and beds of tuff are present throughout the formation. The upper part of the Cayo is a sequence of chert beds known as the Guayaquil cherts or Guayaquil member. This sequence consists of generally thin-bedded, varicolored chert with interbeds of siliceous shale, tuff, and claystone. Locally, the uppermost part is calcareous, apparently the result of deposition of calcium carbonate leached from the overlying San Eduardo limestone by ground water (Canfield, 1966). Much of the silica also appears to have been deposited from ground water.

The Cayo changes laterally in thickness because of apparent gradational lower contact with the Piñón Formation. Faults along the contact and within the formation have caused apparent differences in thickness. It is estimated, however, that the formation is at least 3,250 m thick in the Progreso basin. Microfauna indicate a Senonian to Maestrichtian age. It is possible that the Guayaquil chert sequence may be as young as Danian.

Tertiary

San Eduardo Formation—The San Eduardo Formation also is exposed along the Colonche-Chongón Hills, where it disconformably overlies the Cayo Formation. The San Eduardo consists of generally well-bedded calcarenites, which possibly are the erosion products of former reefs. Argillaceous limestone units (calcilutites) also are generally present near the top of the formation. The limestone is light gray to tan and black and is locally petroliferous. Minor interbeds of siliceous and slightly calcareous shale are present locally. Near the base of the formation are cobbles of black chert which were probably derived from the underlying Guayaquil member of the Cayo Formation. Fossils indicate that the San Eduardo is of middle Eocene age. The formation apparently attains a thickness of 200 m in the Rocafuerte cement quarry near Guayaquil (Faucher et al., 1971), but elsewhere it generally is thinner. The lower contact with the Cayo is disconformable but locally appears to be an angular unconformity. The upper contact is marked by unconformity.

Santa Elena Olistostromic Complex—On the Santa Elena

Peninsula the rocks present between "basement" (Cretaceous Piñón or Cayo) and the Quaternary are allochthonous and form the Santa Elena Olistostromic Complex. The complex is very chaotic, but geologists from Anglo-Ecuadorian Oilfields Ltd. separated it into three olistostromes and three slides in the Ancon area, where many holes have been drilled. Away from this area, this unit has not yet been recognized (Colman, 1970). The Santa Elena Complex consists of clay to turbidite sandstone, shale, and sandy clay that contains clay pebbles. The olistoliths in the lower part of the complex include large blocks of chert and igneous rocks, together with sandstone and siliceous mudstone. Higher in the complex the igneous and chert blocks disappear, but sandstone, tuffaceous sandstone, conglomerate, and minor amounts of limestone are present as olistoliths. The oil fields on the Santa Elena Peninsula produce from very large fragmented sandstone blocks. Locally, oil is obtained from fragmented chert blocks within the olistostromes. Some have regarded the Elena Peninsula geologically as a single basin, called the Daule basin (Drummond, 1986); however, the oil production does not come from a continuous mappable reservoir.

The Santa Elena Complex was emplaced probably during the late Eocene to Oligocene. Foraminifera of late Eocene age are found from the lowermost olistostrome to the uppermost slide. In the Progreso basin east of the peninsula the complex is unconformably overlain by upper Oligocene sediments.

Tosagua Formation—The Tosagua Formation in the Progreso basin has been divided into the Zapotal, Dos Bocas, and Villingota members. The lowermost Zapotal member has basal conglomerate that grades upward into sandstone and then into shale. It was deposited at the margins of the basin, whereas in the deeper part of the basin, the much finer grained clastic material of the Dos Bocas member was deposited. The Zapotal grades both laterally and also upward into the Dos Bocas member. The Dos Bocas, which is very similar to the Tosagua Formation in the Manabí basin, consists mostly of brown to chocolate-brown shale with a few interbeds of siltstone, sandstone, and argillaceous limestone. A few beds of lignite have been reported, and bentonite also is locally present. The brownish shale of the Dos Bocas member grades upward into the diatomaceous white shale in the lower part of the Villingota member. The upper part of the Villingota member is coarser and contains interbedded siltstone and sandstone. Fossils (principally molluscs) in the lowermost part of the Tosagua indicate a late Oligocene to early Miocene age. The Dos Bocas member probably extends from early Oligocene (in the central part of the basin, where it laterally grades into the Zapotal member) to lower middle Miocene. The Villingota member apparently is mostly middle Miocene.

The Tosagua Formation unconformably overlies the Santa Elena Olistostromic Complex. The upper contact with the Progreso Formation is commonly gradational, but in the northeast part of the basin the contact is apparently unconformable. The thickness of the Tosagua changes across the Progreso basin because of the unconformity at its base and the gradational upper contact, but probably is more than 3,500 m in the deeper parts of the basin.

Progreso Formation—The Progreso Formation consists of a thick sequence (3,300 m) of sandstone, siltstone, and shale. The lower part was formerly called the Subibaja Formation, but recent publications do not recognize this unit as a separate formation. The lower Progreso apparently is transitional with the upper member of the Tosagua Formation and consists of dark-gray to locally brownish silty shale containing interbeds of siltstone and fine-grained sandstone. Locally, the contact is believed to be unconformable. Calcareous concretions are present in places. Overlying the sequence is greenish-gray siltstone which grades upward into fine-grained sandstone. Glauconite is common and carbonaceous material is present locally. The upper part of the Progreso consists of sandstone, siltstone, shale, and claystone. Some of the sandstone contains abundant shell fragments and grades into a coquina. Generally, the sandstone is fine grained, fairly soft and poorly cemented. The northern part of the Progreso basin contains light-gray to tan algal limestone and locally bentonite, bentonitic clay, tuff, and pebble sandstone (Bristow and Hoffstetter, 1977). The fossils in the Progreso Formation indicate a late middle Miocene to possible Pliocene age. In the Amistad area of the Gulf of Guayaquil, gas has been found in the lower part of the formation.

Structure

The Coastal basins are separated from the Andean Mountains by a major fault zone, the Guayaquil-Babahoyo-Santo Domingo fault zone (Figure 4), which apparently extends mostly under Recent deposits from south of Guayaquil northward to the border with Colombia. Other faults in the Coastal region of Ecuador were also important in the formation of the basins (Figure 4).

Borbón Basin

The Borbón basin is limited on the southeast by outcrops of the Piñón Formation, on the west by the Esmeraldas-Pichincha fault (Figure 1), and on the northwest, apparently, by the Valdez fault (Figure 4). The basin plunges northeasterly into the offshore area of Colombia, where more than 4,000 m of Tertiary sediments are probably present. The Borbón Basin has been subjected to considerable compression, and it is likely that faults have played an important role in the basin's development (Canfield, 1966). The earliest faults strike northeasterly, essentially paralleling the coast and the Cretaceous outcrops. Later faults strike northwesterly and offset the earlier group. The Esmeraldas-Pichincha fault, which apparently has major lateral movement, is an example of a later fault.

The Esmeraldas-Caraquez basin is separated from the Borbón basin by the Esmeraldas-Pichincha fault and the Mache Hills (Figure 1). The northern part of the basin, called the Esmeraldas basin, is structurally similar to the Borbón basin and may have become a separate basin as recently as the Pliocene.

Manabí Basin

The Manabí Basin lies east of the Jama-Quinindé and Portoviejo-Jipijapa faults (Figure 4). The Esmeraldas-Pichincha fault apparently limits the basin on the north, the Colonche-Chongon Hills limits the basin on the south, and the Guayaquil-Babahoyo-Santo Domingo fault zone limits the basin on the east. Most of the major faults in the Manabí area strike northeasterly as shown by the Jama Hills, an

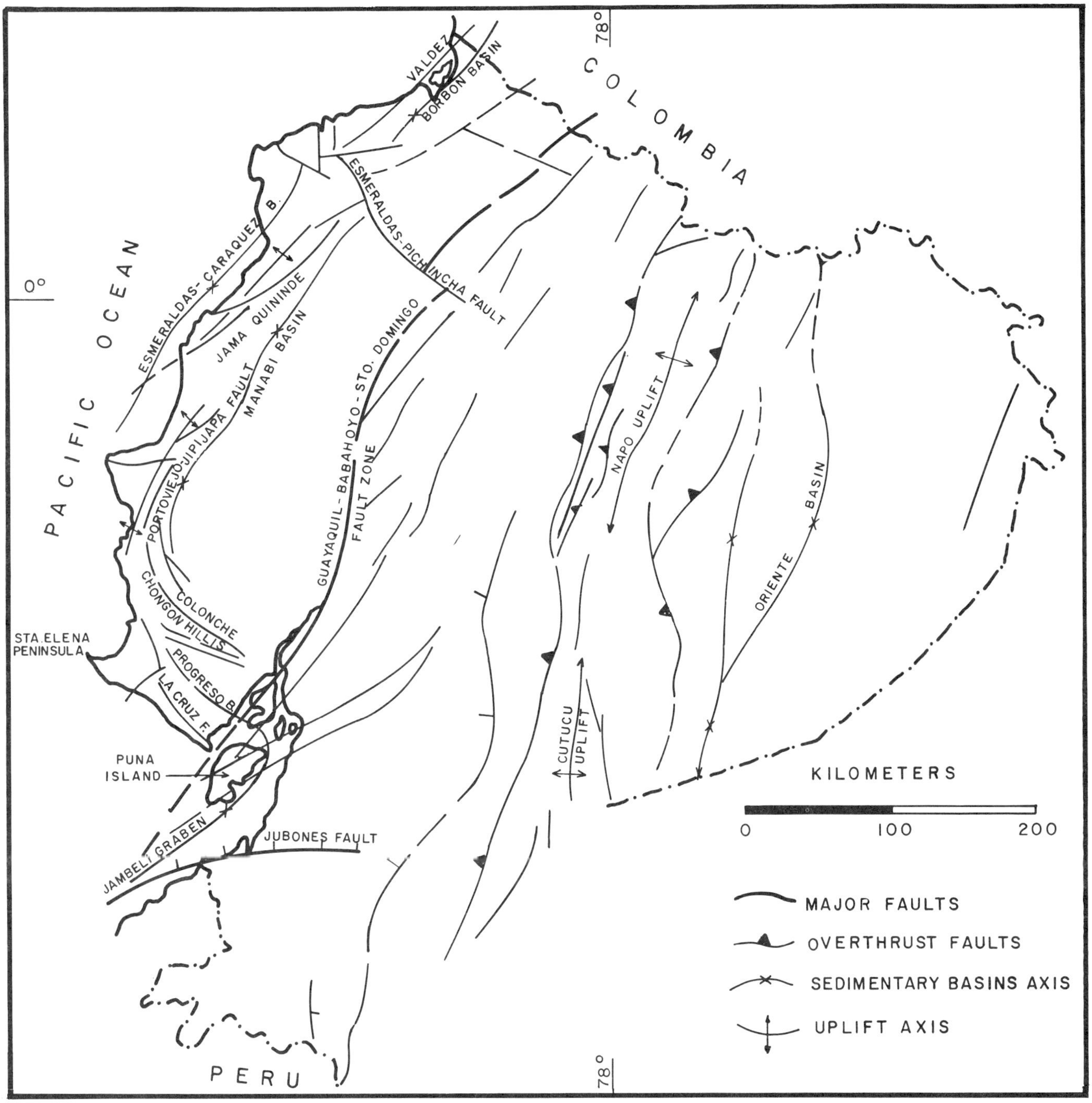

Figure 4. Principal structural features of Ecuador (after J.W. Baldock, 1982).

uplifted area west of the Jama-Quinindé fault. Some irregular block faulting is present in the hills, and, in general, smaller faults tend to strike northwesterly. A major east-trending fault has been mapped on the Manta Peninsula, and faults having this strike also may be present farther east. Lateral movement evidently took place along major faults.

Seismic studies indicate that the northern part of the Manabí basin is deeper than the southern part, probably because of faulting, and that more than 6,000 m of Tertiary sediments may be present in that area.

Progreso Basin

The Progreso basin is bounded on the northeast by the Colonche-Chongón Hills and on the southwest by the La Cruz fault zone (Figure 1). The axis of the basin plunges southwesterly, wraps around the Puna Island, and plunges southwesterly along the Jambeli graben into the Gulf of Guayaquil. The La Cruz fault zone separates the Progreso basin from the Santa Elena Peninsula. All oil fields of the coastal region of Ecuador are located in this peninsula area. On the peninsula, Miocene rocks are absent, probably having been removed by erosion after the area was uplifted. The La Cruz fault also apparently had considerable lateral movement. Later faults striking northeasterly cut the La Cruz fault at nearly right angles. The Colonche-Chongón Hills may be faulted on their northeast side, but it appears that they are the result of simple uplift. Faults paralleling the Colonche-Chongón Hills are present on the southwest flank. Other faults in the area trend east-northeast and northwest.

The Jambeli graben is bounded by major faults having both right-lateral and vertical components of movement. South of Puna Island more than 9,000 m of Tertiary sediments may be present.

Oil and Gas Possibilities

Except for the Santa Elena Peninsula, very little exploratory drilling has been carried out in the coastal region of Ecuador. The results of this drilling are discouraging. A few of the drill holes had minor oil and/or gas shows, but successful completions were nil except for three gas wells completed in the early 1970s in the Amistad field in the Gulf of Guayaquil. Onshore reflection seismic surveys were practically nonexistent in the coastal region until CEPE initiated work in 1980. Exploration now (1987) under way by Belco and the Texaco-Pecten Consortium is very important in determining the hydrocarbon potential of coastal Ecuador.

Borbón Basin

No oil or gas shows were encountered in the three holes drilled in the northern coastal area. The reservoir characteristics shown by the hole drilled in the Esmeraldas basin (Camarones-1) are poor, but a few of the sandstone units in the two Borbón basin drill holes appear to have fair porosities (no measured porosities are available). Some of the sandstone units in outcrops of the Zapallo and Viche formations also appear to have fair porosities and are considered as potential reservoir rocks.

Although geochemical data are not available, it appears that the finer grained sediments in the Borbón basin do not have good oil-generating potential. However, some of the sediments from the deeper part of the basin may have sufficient depth of burial to have potential.

Manabí Basin

Several holes drilled in the Manabí basin had slight to good oil and gas shows. Minor gas seeps (probably methane) have been reported from outcrops. Potential reservoirs are present in the San Mateo Formation and locally in the sandstone beds of the Angostura and Borbón formations. The reefal facies of the San Eduardo Formation may be prospective in local areas.

Geochemical studies of surface samples of Tertiary rocks in the Manabí basin have indicated TAI (thermal alteration index) readings to be no higher than 1.7. Probably the outcropping rocks have never been subjected to deep burial. It is possible, however, that generation of petroleum may have occurred within the 6,000 m of Tertiary sediments thought to be present in the northern part of the Manabí basin. Seismic records suggest that Tertiary sediments older than the upper Eocene San Eduardo Formation may be present in the deeper parts of the basin (Santos, 1983). If the correlation is correct, these sediments could provide both source and reservoir beds that have not been seen in outcrop, and have been encountered in subsurface, in the drill hole Ricaurte-1 (August 1987).

Progreso Basin

Although no oil has been discovered within the Progreso basin, about 115 million barrels of oil have been produced from the Olistostromic Complex of the Santa Elena Peninsula. Strata equivalent to the oil-bearing units of the Santa Elena Peninsula, and believed to have the same geologic history, are present in the Progreso basin, and therefore the potential of the basin must be considered good. Because of the irregular distribution of the olistostromes, however, the mapping of prospective areas is very difficult. Also, although considerable oil has been produced on the peninsula, the producing rates from individual wells have never been very high and now, near the end of the producing life of the fields, are averaging less than 4 barrels of oil per day. The beds of sand and conglomerate of the Zapotal Formation appear locally to have good porosities and permeabilities and should be considered as potential reservoirs in the Progreso basin. Also, favorable Miocene sand strata are present in the Gulf of Guayaquil, which contains gas tested at high rates (Amistad field).

Some of the Tertiary rocks are apparently immature and are generating only gas in the Gulf of Guayaquil. Other sediments, however, generate petroleum, as proved by the oil at Santa Elena and a small amount that was encountered in CEPE's offshore Golfo de Guayaquil-1 exploratory well drilled south of the Amistad field near the border with Perú.

ORIENTE REGION

The Ecuadorian Oriente region, or the Oriente basin, is a jungle-covered area of about 100,000 km^2, limited on the north by Colombia and on the south and east by Perú. The basin is filled with sediments ranging in age from pre-Cretaceous (early Paleozoic) to Recent, which have been deposited in both marine and continental environments (Figures 5, 6). The Oriente basin forms part of an extensive sub-Andean basin which separates the Andes from the Guyana and Brazilian Shield areas. The western part of the Ecuadorian Oriente basin contains foothills including the Napo uplift to the north and the Cutucú uplift to the south (Figure 1). Cretaceous sedimentary rocks, the principal hydrocarbon objectives in the Oriente, are exposed in these uplifts. Still older rocks are exposed in the more deeply eroded central parts of the uplifts. East of the foothill zone, in the basin proper, outcrops of any importance are lacking.

Stratigraphy

Although more than 11,000 m of sediments have accumulated in the Oriente basin, this thickness is probably not present in any one location. Many major unconformities limit thickness and extent, especially of pre-Cretaceous rocks.

Pre-Cretaceous

Four pre-Cretaceous formations have been recognized in the Oriente basin. The Pumbuiza Formation includes Upper Silurian (or Lower Devonian) to Mississippian sediments. The

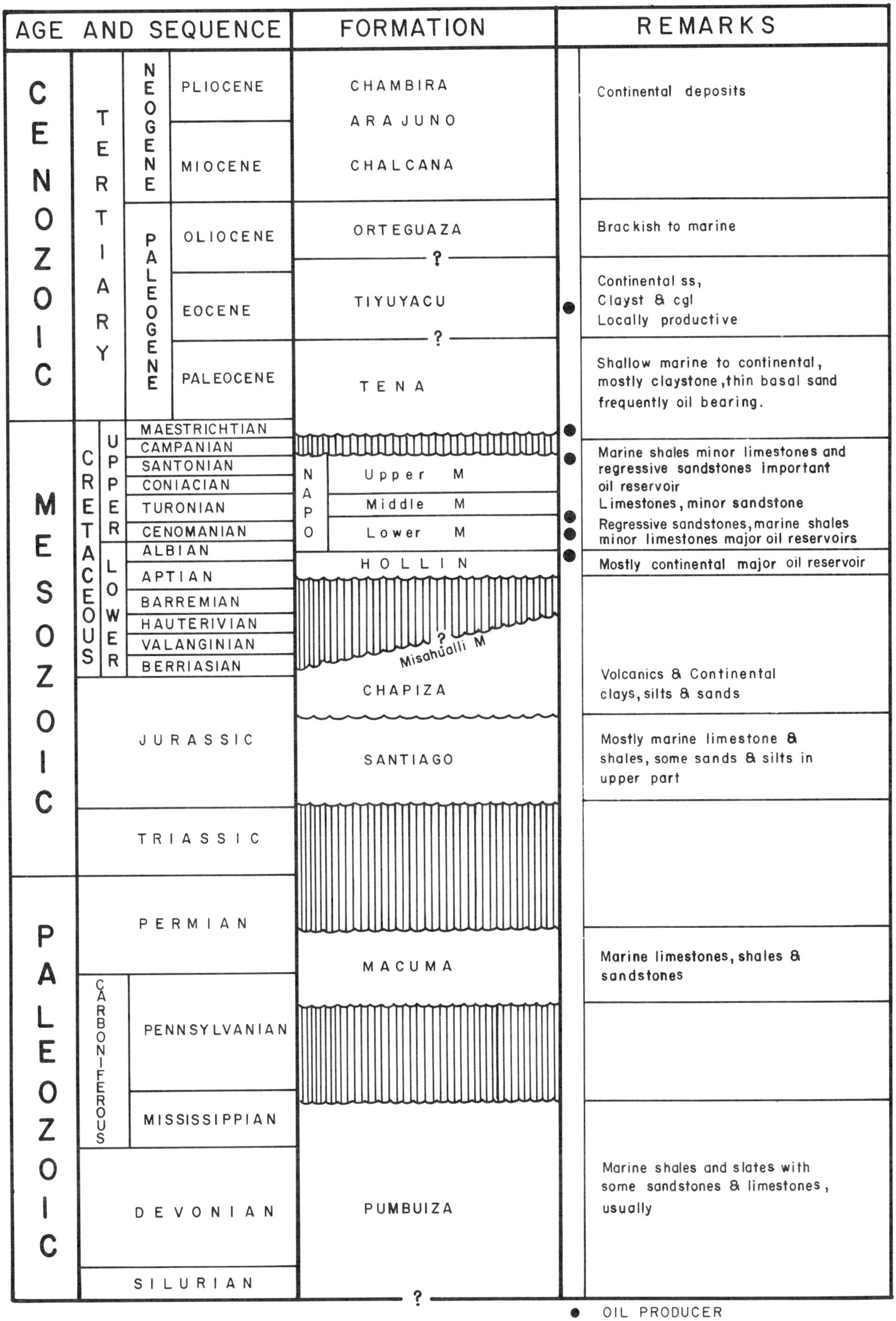

Figure 5. Stratigraphic column of the Oriente basin.

base of the formation has never been observed in outcrop or in the subsurface. The Pumbuiza is marine and consists mainly of phyllitic black to gray shale and slate with interbeds of quartzitic sandstone and very hard limestone and dolostone. The unconformably overlying Permian-Pennsylvanian Macuma Formation consists mostly of marine limestones with less important shales, dolostones, siltstones, and sandstones. The Macuma is more fossiliferous than the underlying Pumbuiza. The third pre-Cretaceous formation, the Lower Jurassic Santiago, consists of mostly limestone and locally bituminous shale with some tan and gray sandstone and siltstone. The sequence is marine, as proved by the presence of ammonites, but found is generally scarce. The base of the formation has not been observed but is believed to be unconformable. The youngest pre-Cretaceous formation is a continental red-bed series, Chapiza, which unconformably overlies the older rocks. The upper part of the Chapiza contains volcanics and has been named the Misahualli member. The upper part of the Chapiza probably is Early Cretaceous (pre-Aptian), but the bulk of the formation is believed to be Jurassic (Figure 5).

Cretaceous

Hollín Formation

The Hollín Formation is present throughout the Oriente basin, except possibly in the northeasternmost part where it either pinches out or becomes indistinguishable from other Cretaceous rocks. The Hollín is predominantly quartzose sandstone, which is thick bedded to massive, coarse to very fine grained, white to tan, and commonly crossbedded. Irregular thin, locally micaceous layers of shale are present in places. In the upper part of the formation the shale content (locally) increases, some of the sandstone units are glauconitic and less porous, sandy limestone units are present, and few marine fossils have been reported. Although the bulk of the Hollín Formation was apparently deposited in a continental, braided-stream environment, the uppermost part was at least locally deposited under marine conditions. Over most of the Oriente basin the thickness of the Hollín ranges from about 100 to 150 m, but in the extreme southwest, near Perú, the thickness is estimated to be more than 300 m (Canfield, Bonilla, and Robbins, 1982). The formation generally thins in a northeasterly direction and, as stated above, may pinch out in the northeastern Oriente.

The Hollín overlies older sediments with an abrupt angular unconformity, generally the Chapiza volcanics of Mesozoic age, but Paleozoic and "shield" rocks have been reported from drill holes.

Napo Formation—Tschopp (1953) divided the Napo Formation into three members based on field studies along the western foothills and five holes drilled in the southern part of the area. The lower Napo of Tschopp had three sand units, but the lowermost zone is now included in the upper part of the Hollín Formation because drilling has shown that separation of the two sand bodies is not warranted. The lower Napo, as redefined, contains two major regressive sand zones—the "T" sand and "U" sand, which are important oil reservoirs in the Oriente basin. The lower Napo also contains abundant shale and minor limestone. One limestone unit lying above the "T" sand zone, the "B" limestone, is locally fairly thick. Both sand zones are quartzose, coarse to fine grained, porous, and permeable in the eastern part of the basin. In the western part of the basin the sand zones are thinner and glauconitic, and the reservoir characteristics are less favorable for petroleum.

The middle Napo of Tschopp, or "main limestone," is very well developed along the western side of the Oriente basin. The limestone, which is also named the "A" limestone, is massive, dense, and finely crystalline along the foothill zone but becomes sandy and much thinner toward the east. The middle Napo also includes a sandstone unit which lies above the "A" limestone in the northeastern part of the basin. This unit, designated the "M-2" sandstone, is also a regressive deposit similar to the "T" and "U" sands of the lower Napo. Although oil was encountered in this sandstone, none has yet been produced.

The upper Napo over most of the Oriente basin is composed of dark-gray shale and minor thin beds of limestone. In the eastern and southeastern parts of the basin, however, there is an important sand at the top of the member which has been named the "M-1" sandstone or the Vivian sandstone. This also is a regressive unit consisting of medium- to fine-grained sand which is locally microconglomeratic. It is an important reservoir in some fields of the northeastern Oriente.

The Napo, like the underlying Hollín, attains its greatest thickness of about 925 m in the extreme southwestern Oriente and thins northeasterly. The lower and middle members become very sandy toward a northeastern source and become indistinguishable from the Hollín (Canfield, Bonilla, and Robbins, 1982).

The Napo Formation is conformable with the Hollín and ranges in age from late Albian (lower Napo) to Senonian (upper Napo). The middle Napo has been dated as Turonian by ammonites (Tschopp, 1953). The Tena Formation unconformably overlies the Napo.

Tena Formation

The Tena Formation consists of multicolored claystone, siltstone, and fine-grained argillaceous sandstone. The predominant colors are various shades of red, but greenish rocks are present in some areas. In places the formation contains a basal sand which is oil bearing in several Oriente fields.

The Tena Formation represents a transition stage between the marine Napo Formation and the brackish water to continental facies of the Tertiary. A few dwarf foraminifera collected near the base of the Tena indicate a Late Cretaceous (Maestrichtian) age (Canfield, Bonilla, and Robbins, 1982). The upper part of the formation is believed to be Paleocene and possibly early Eocene. The upper contact with the overlying Tiyuyacu Formation is quite abrupt, marked by red claystone underlying a cherty conglomerate. One would suspect that the contact is at least disconformable, but no field evidence supports this view. It is believed that the contact is conformable and that the change to conglomerate was caused by a rapid uplift of the source area to the west. Also, locally, there are a few coarse-grained to conglomeratic cherty sandstone units present in the upper part of the Tena which suggest a transitional contact. The Tena is thickest (about 1,000 m) along the mountain front. It thins easterly and in the easternmost Oriente may be only 75 to 90 m thick.

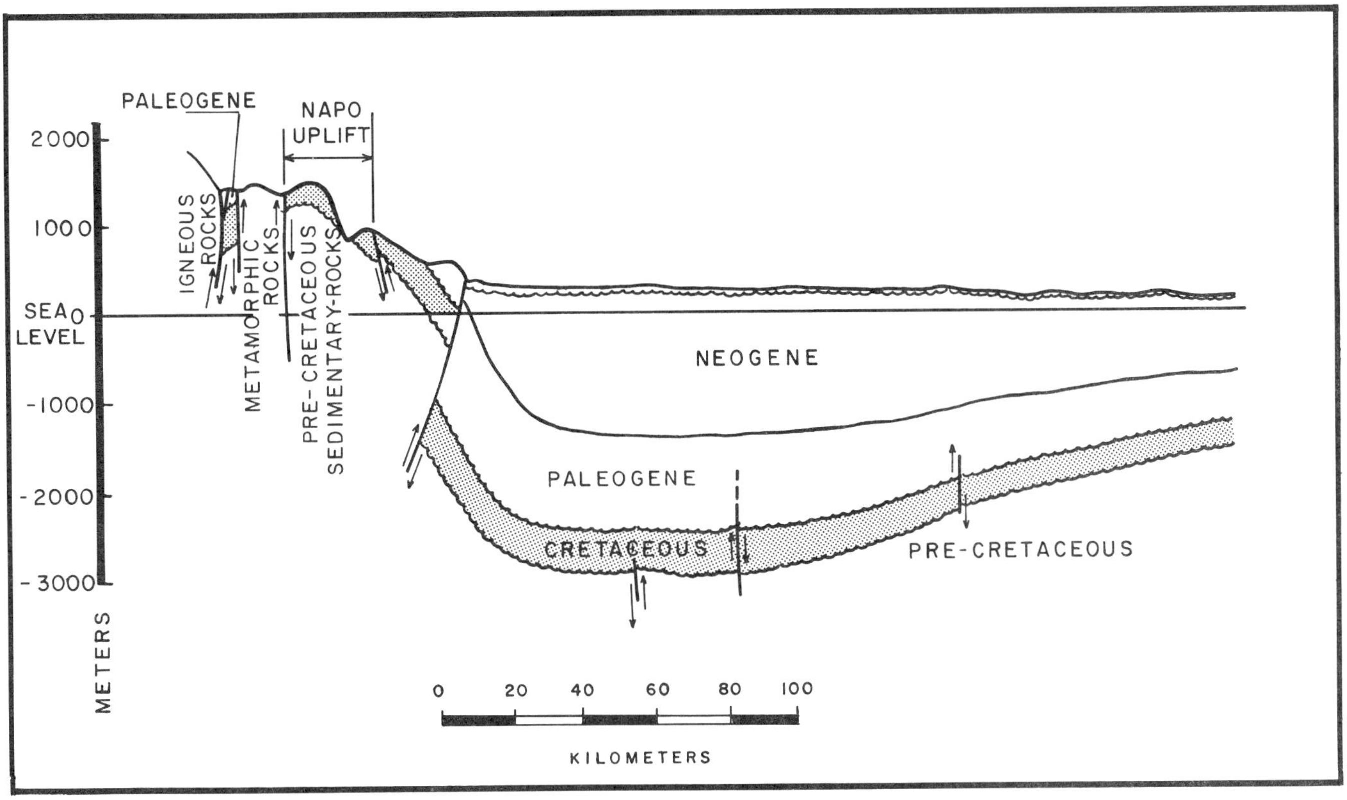

Figure 6. Schematic geological cross section of the Northern Oriente basin of Ecuador.

Tertiary

Tertiary rocks in the Oriente basin which are known to have oil are the Tiyuyacu and Orteguaza formations. The Tiyuyacu Formation of probable Eocene age consists of three separate units: (1) a thick basal cherty conglomerate and coarse-grained sandstone including interbeds of claystone, (2) a middle unit composed of multicolored claystone, siltstone, and shale, with only minor sandstone, and (3) an upper conglomerate very similar to the basal unit. Glauconite and shale have been reported from the upper unit. The upper unit grades into a gray to bluish-gray shale containing minor fine-grained, locally glauconitic sandstone of the Orteguaza Formation of Oligocene age. Inasmuch as the contact with the Tiyuyacu Formation is gradational, it is likely that the upper part of the Tiyuyacu also is Oligocene, but most of the formation is believed to be Eocene. Overlying the Orteguaza are several thousand feet of mostly fine to coarse clastic sedimentary rocks of continental origin (Figure 5).

Oil shows have been reported from the Oreteguaza and Tiyuyacu formations, and a small amount of oil has been tested from the Tiyuyacu. At the Orito field, Colombia, the Tiyuyacu equivalent (Pepino Formation) is an important producer.

Structure

The Oriente basin is asymmetric and plunges to the south, where it is much deeper than in the north. The western margin of the basin has two major north-trending uplifts—the Cutucú just north of the Peruvian border and the Napo near Colombia. East of these two uplifts the stratigraphy of the broad syncline shown in Figure 6 is taken from well logs. In the northern half of the basin the syncline contains several low-relief anticlines (Canfield, Bonilla, and Robbins, 1982). A subtle cross-basin arch in this area may account for the broad axis of the basin. In the southern half of the basin the synclinal axis is better defined and plunges steeply toward Perú (Figure 4).

Most of the oil that has been discovered in the Oriente is trapped in north-trending anticlines which exhibit early structural growth. Some of the Napo sand accumulations may be in part stratigraphically controlled.

Oil and Gas Resources and Possibilities

Prior to the recent OPEC decision to reduce production, the Oriente region produced more than 300,000 barrels of oil per day, but production has now been reduced to about 260,000 barrels per day of oil ranging from about 18° to 35° API. Production is from the north-central part of the basin, where most of the exploratory activity has taken place. Somewhat heavier oil, ranging from 15° to 25° API, has been tested in wells in the eastern and southeastern Oriente and even heavier oil, 10° to 12° API, in the southwestern region. More than 1.1 billion barrels has already been produced from the north-central oil fields, and the remaining recoverable reserves in these fields are about 1.5 billion barrels. In addition to the producing fields there are several important discoveries

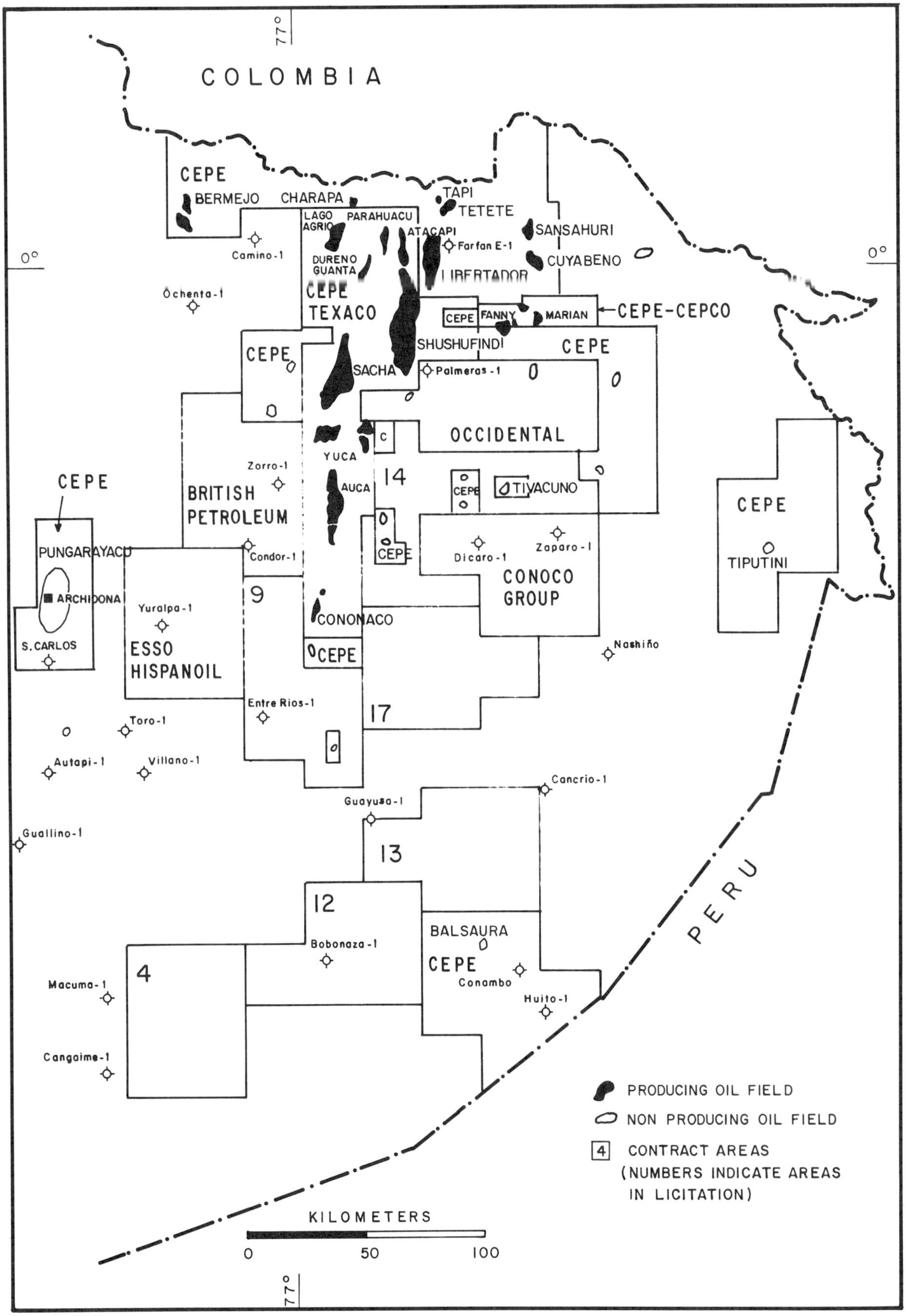

Figure 7. Contract areas, oil fields, and important exploratory wells in the Oriente of Ecuador.

that are not yet tied into the gathering system and therefore are not in production (Figure 7).

Source Rocks

The organically rich dark-gray to black marine shale and limestone in the Napo Formation are believed to have been the major sources for the oil produced in the Oriente. These sediments, which are present throughout the region, exhibit TAI values of 2.5 to 3.5. In the western part of the basin there is an increase of dark shale in the Hollín Formation, which can also be considered as a source rock. The bituminous shale and limestone of the Lower Jurassic Santiago Formation may also be important source rocks in the southern Oriente.

Reservoir Rocks

Hollín Formation Reservoirs

The coarse-grained to conglomeratic sandstone in the lower part of the Hollín Formation generally contains salt water, the fine- to medium-grained sandstone in the middle of the formation contains fresh water, and the fine- to medium-grained sandstone in the upper part of the formation contains salt water, meaning an interfingering of the sediments. The middle or main Hollín sand is the most important reservoir rock in the Bermejo, Lago Agrio, Sacha, Auca, and Cononaco oil fields (Figure 8). The porosities average about 16%; permeabilities range from about 200 to more than 1,000 md (millidarcys) and average about 650 md based on a few conventional cores (Canfield, Rosanía, and San Martin, 1982). The upper part of the Hollín is also productive but is less porous and permeable than the main sand.

Napo Formation Reservoirs

Two sand units in the lower part of the Napo Formation, the "T" and the "U" sands, also are very important reservoirs in the Oriente. These are the principal oil-producing sands in the large Shushufindi and Libertador fields and are also important oil producers elsewhere. The "U" sand has been further divided into two subunits, the upper and lower "U" sands, in the more easterly Libertador field. Apparently, there was short-lived marine transgression in the interval between the deposition of these "U" sand bodies. Both sand units have a lower "main" sand and thinner overlying sands. The main sands are apparently deltaic, and partial cores indicate that they have porosities of 16 to 20% and average permeabilities of about 400 md. The overlying thinner sands are generally glauconitic and have lower porosities and permeabilities. Both sand units are thinner and less important reservoirs toward the west.

Other Reservoirs—East of the Shushufindi field the upper Napo "M-1" sand is a major objective for drilling. This sand is generally fine to medium grained, quartzose, and has porosities of more than 20%. It produces oil of about 21° API in several smaller fields.

The basal sand in the Tena Formation is a secondary objective in the Oriente. This sand produces 20° to 25° API oil in several of the fields, and has porosities ranging from 15 to 20% where productive (Canfield, Rosanía, and San Martin, 1982).

INVESTIGATIONS IN THE PUNGARAYACU AREA

In 1975, the writer initiated surface geological studies for CEPE in the Pungarayacu area, located along the south plunge of the Napo uplift. This area has long been known to have important tar impregnations and oil seeps, but CEPE's efforts are the first to determine the potential of the area. After the initial surface work and aerial photography interpretation, a drilling program was started in October 1980. Twenty-four core holes were drilled, ranging in depth from about 120 to 460 m. All of the holes were continuously cored and penetrated the entire Hollín Formation, which is the most important reservoir. Depth to the Hollín ranged from about 50 to more than 330 m. Additional exploration is planned for the area.

It appears that the anticlinal structure in the Pungarayacu area is closed to the west and north by a major fault which places granite against Tertiary and Cretaceous sediments. Along this fault heavy oil seeps are reported. Also tar and heavy oil are present in Cretaceous outcrops that are exposed in the streams which dissect the area. The structure is closed to the east by faulting and dip of the beds, and to the south by dip of the beds. An apparent interface between water and the heavy oil or tar was noted in the easternmost drill holes.

The Hollín Formation is about 90 m thick in the Pungarayacu area. The upper part of the formation is generally glauconitic and has a calcareous cement. The porosity is generally less than in the main sand, and the oil saturation is also much less, being about 16 to 20%. In the northern part of the area (north of Archidona) the main Hollín sand has porosities of about 25% and oil saturations of about 48%. South of Archidona, oil saturation is less, being from 25 to 37%. Cross faulting has affected the structure, and oil saturations appear to be higher in elevated fault blocks.

The "T" sand of the lower Napo Formation and the "A" limestone of the middle Napo in the Pungarayacu area also contain heavy oil, but neither of these units is as important as the Hollín. The "T" sand has porosities and oil saturations similar to the uppermost glauconitic sand of the Hollín. The limestone locally has high oil saturation, but porosities are much lower than in the main Hollín reservoir. The "U" sand, an important productive zone in the oil to the northeast, is locally present but very poorly developed at Pungarayacu.

A conservative preliminary estimate of reserves in place in the Hollín reservoir of the Pungarayacu area amounts to 7.2 billion barrels. Oil gravity ranges from 8° to 16° API. It is interesting to note that a few wells had gas shows. At present, CEPE is evaluating possible methods of extracting the Pungarayacu reserves.

ACKNOWLEDGMENTS

The author expresses his gratitude to R.W. Canfield of Texaco Petroleum Company for his suggestions and assistance, to Francisco Silva V. of the National Hydrocarbon Directorate, and to Miss Adriana Naranjo S., and the Drafting Department of the CEPE-TEXACO Consortium.

REFERENCES

Baldock, J. W., 1982, Geology of Ecuador, explanatory bulletin of the National Geological Map of the Republic of Ecuador: Ministerio de Recursos Naturales y Energëticos, Direcciön General de Geologia y Minas, Ecuador, p. 45-54.

Bristow, C. R., and R. Hoffstetter, 1977, Lexique stratigraphique international: Paris, Centre National de la Recherche Scientifique, 2nd. ed., p. 66, 79, 211-224, 252, 328-332.

Canfield, R. W., 1966, Geological report of the coast of Ecuador: Ecuador, Ministerio de Industrias y Comercio, p. 1-90.

Canfield, R. W., G. Bonilla, and R. K. Robbins, 1982, Sacha oil field of Ecuadorian Oriente: American Association of Petroleum Geologists Bulletin, v. 66, p. 1076-1090.

Canfield, R. W., G. E. Rosanía, and H. F. San Martin, 1982, Geología de los compos gigantes del Oriente Ecuatoriano: Bogota, Symposium on petroleum exploration in the subandean basins of Venezuela, Colombia, Ecuador and Perú, v. 2, p. 2-14.

Colman, J. A. R., 1970, Guidebook to the geology of the Santa Elena Peninsula: Ecuadoran Geological and Geophysical Society, p. 3-34.

Drummond, K. J., 1986, Explanatory notes for the energy-resources map of the Circum-Pacific region, northeast quadrant: American Association of Petroleum Geologists, Circum-Pacific Council for Energy and Mineral Resources, p. 13.

Faucher, B., R. Vernet, G. Bizon, J. J. Bizon, N. Grekoff, M. Lys, and J. Sigal, 1971, Sedimentary formations in Ecuador: Bureau D'etudes Industrielles de Cooperation de L'Institute Francais du Petrole (BEICEP), p. 25-29, 105.

Santos, M., 1983, La subcuenca Manabí y su contexto regional: Guayaquil, Ecuador, III Ecuadorian Congress of Geological Engineers in Mining and Petroleum, p. 2-35.

Tschopp, H. J., 1953, Oil exploration in the Oriente of Ecuador: American Association of Petroleum Geologists Bulletin, v. 73, p. 2303-2347.

Ericksen, G. E., Cañas Pinochet, M. T., and Reinemund, J. A., editors, 1989, Geology of the Andes and its relation to hydrocarbon and mineral resources: Houston, Texas, Circum-Pacific Council for Energy and Mineral Resources Earth Science Series, v. 11.

CHAPTER 31

ANDEAN HYDROCARBON RESOURCES—AN OVERVIEW

Colin L. Stabler
Pecten International Company
Houston, Texas, U.S.A.

ABSTRACT

The Andean province is among the most prolific oil-producing areas of the world, with some 82 billion barrels of discovered recoverable oil. The distribution of these reserves is mainly controlled by the evolution of the Andean orogenic belt. Andean basins are described by their relationship to the Mesozoic Andean volcanic arc, as forearc and backarc basins. The principal source rocks were developed in backarc basin settings during the Mesozoic, especially during the Cretaceous. Most of the reservoir sandstones were deposited in backarc basins, but some were deposited in Cenozoic foredeeps developed over the backarc areas and in trenches in the forearc areas. Unmetamorphosed Paleozoic sediments are preserved in the Tarija basin of Bolivia and northern Argentina and have generated relatively minor volumes of oil and gas.

Much of the oil is contained in structural traps. In some areas, especially in the northern Andes, the oil is in closures related to wrench faults. Andean thrust-fault anticlines have trapped oil in places all along the Andean foothills belt. Drape structures and stratigraphic pinch-outs also form important traps, especially on the undeformed foreland flank of the basins.

Some basins are very rich in oil; some basins appear to be lean. In other basins, oil has not yet been discovered. Significant reserves of oil are expected to be added to these provinces in the future.

INTRODUCTION

Hydrocarbons are present along the extent of the Andes from Venezuela to Tierra del Fuego. Recoverable oil from discoveries to date total 82 billion barrels of which 44 billion barrels have already been produced. Clearly, the Andean province ranks among the world's most prolific oil-producing areas.

BASINS

The oil-producing areas of the Andes are within structural depressions filled with sedimentary rocks to form basins. Depending on their present position with respect to the physiography of the Andean chain, these basins may be classed as sub-Andean, Intermontane, and Coastal.

The basins can also be classified on the basis of their paleogeographic position with respect to the Mesozoic volcanic arc. Figure 1 shows the locations of these basins, their most commonly used names, and their estimated discovery oil volumes (Petroconsultants, 1984). Many of the so-called sub-Andean basins represent Mesozoic backarc settings that became foredeeps in the Tertiary. Examples are the Barinas-Apure-Meta and the Putumayo-Maranon basins. Other sub-Andean basins, for example the Cuyo basin, appear to have had a rift origin, probably related to breakup of Gondwana. They were influenced by backarc sedimentation in the late Mesozoic, and later became foredeeps in their western parts. Intermontane basins such as the Upper and Middle Magdalena seem to have had a Mesozoic backarc history but became separated from their sub-Andean counterparts to the east in the late Tertiary. The coastal basins represent forearc basins which received various thicknesses of clastic deposits throughout the Tertiary.

Known oil reserves are not equally distributed between each type of basin. As can be seen from Figure 1, the backarc basins contain by far the larger percentage of reserves compared with the forearc basins. Figure 2 shows the distribution of oil reserves among the separate basins. The Talara-Progresso area is by far the most prolific of the forearc basins; the Maracaibo is the richest of the backarc basins. If we consider only those basins outside Venezuela, the Putumayo-Maranon is the most prolific.

Although the reasons for the uneven distribution of oil reserves are far too complex to address in this brief summary, the key factors that are involved will be summarized by reviewing firstly, the paleogeographic evolution of the area; secondly, the stratigraphic sequence; and thirdly, the structural style.

PALEOGEOGRAPHY

In the Devonian, intercontinental and epicontinental basins flanked the Guyanan and Brazilian shield areas. Thick sequences of sandstone interbedded with shale were depos-

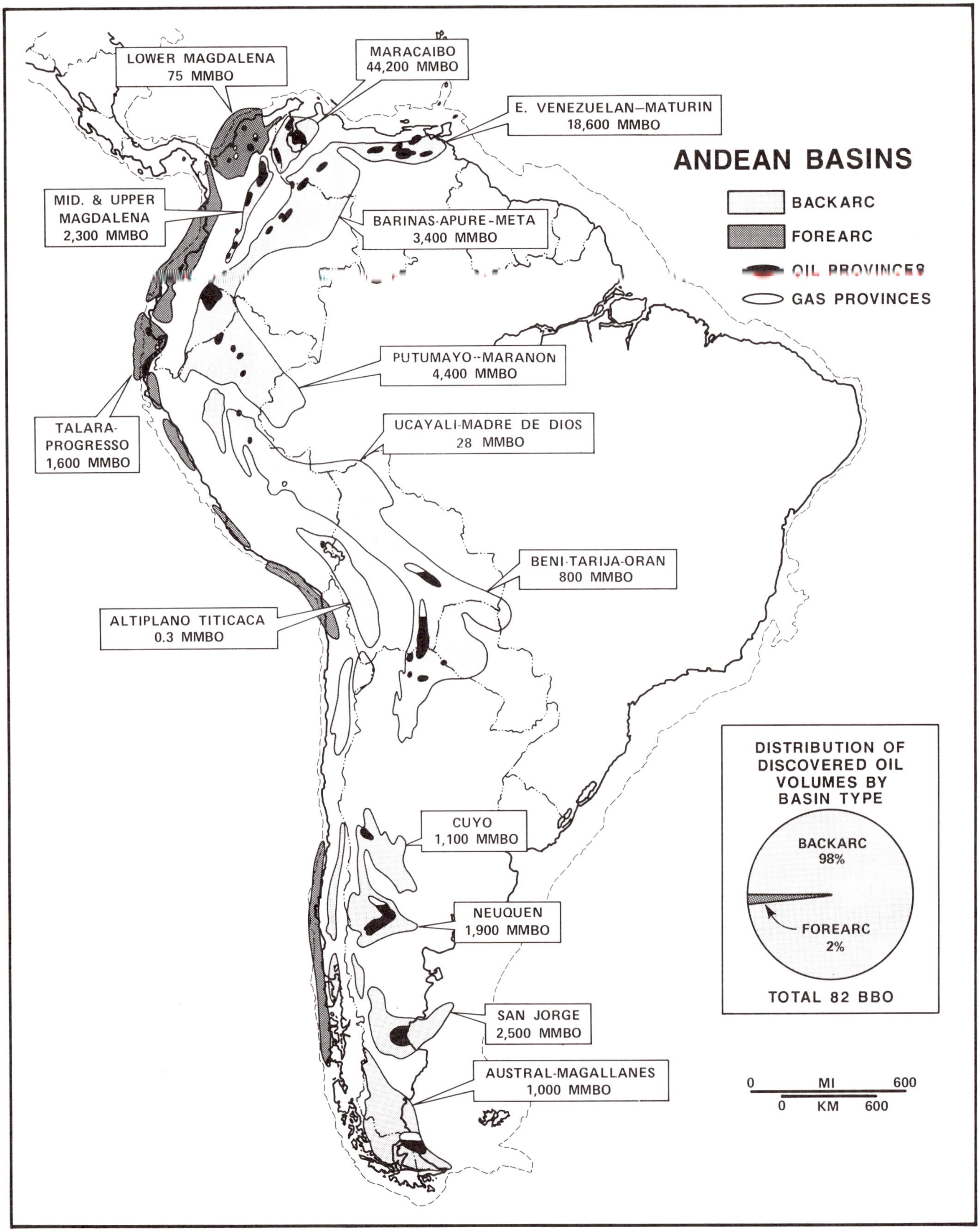

Figure 1. Locations of Andean basins, showing most commonly used names and estimated discovery oil volumes. (Data from Petroconsultants)

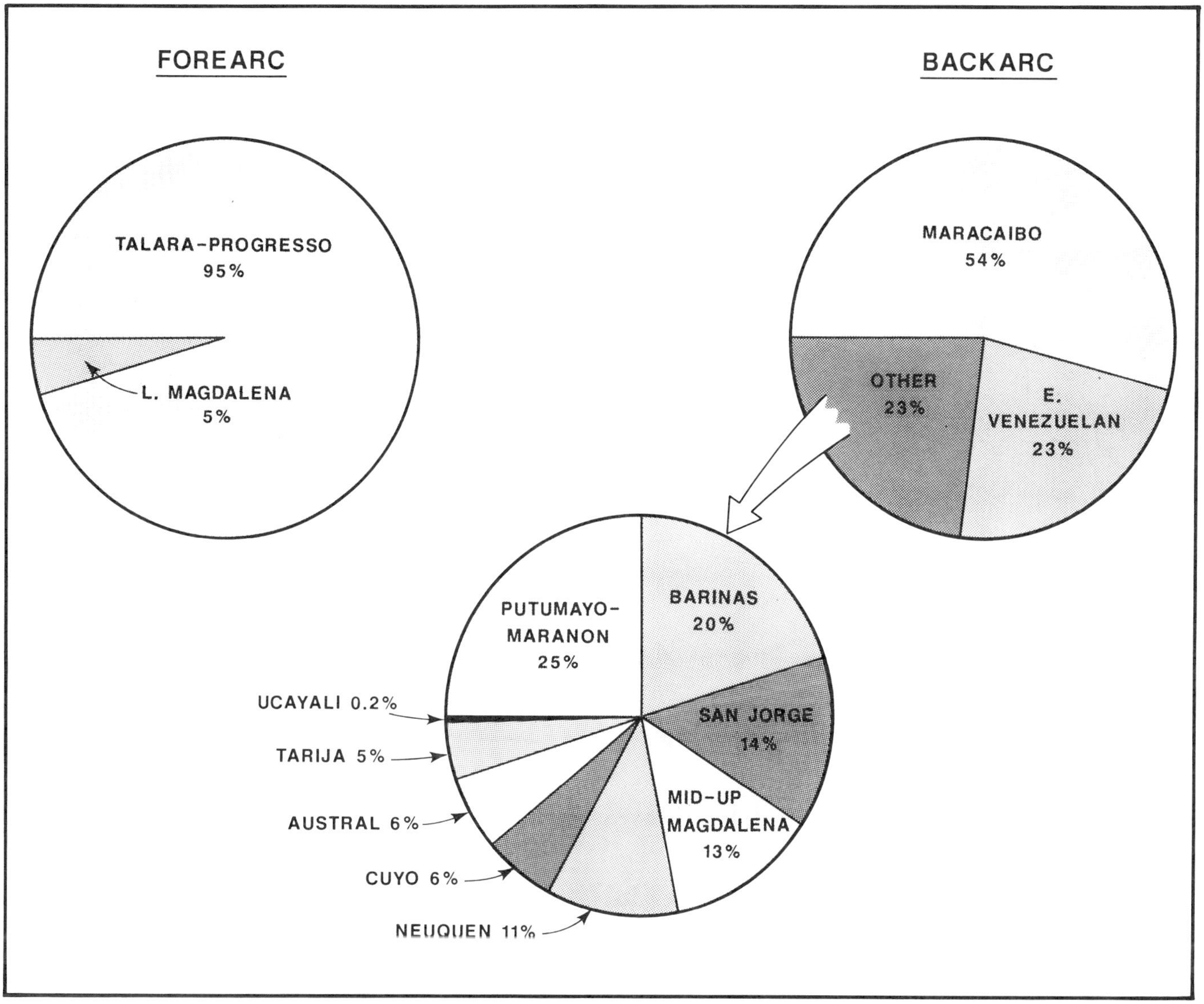

Figure 2. Distribution of discovered oil volumes by basin. (Data from Petroconsultants)

ited; some of the shale is rich in organic matter. These rocks were metamorphosed and deformed, except in the central Andes where they contain important oil and gas reserves (Tarija basin). The Carboniferous of southeastern Perú and northeastern Argentina is represented by sandstone and shale, periglacial to marine in nature, which may contain important reservoirs and source rocks.

By Cretaceous time a chain of island arcs had formed (Figure 3A). East of this chain, source rocks were deposited in marine, perhaps restricted, conditions. The shield areas provided good-quality sands that now serve as some of the best reservoir rocks. Oceanic sediments and volcanic rocks occur in the forearc setting to the west.

The Tertiary saw the main development of the Andes (Figure 3B). Marine connections with the Pacific were cut off by uplift of the island arcs. Thick wedges of immature sediment were shed into developing foredeeps and deposited in an increasingly continental environment as the seas retreated to north and south.

STRATIGRAPHY

Considering the distribution of oil discoveries with respect to the age of the reservoir rocks, by far the largest volumes of oil appear to be in Tertiary sandstones of Venezuela and Colombia, whereas Mesozoic strata are the principal reservoirs throughout most of the rest of the Andean trend.

To understand this distribution it is useful to refer to a time-stratigraphic chart, such as that in Figure 4, which is constructed with geologic time on the vertical axis. The backarc basins in the north from Venezuela to Perú were transgressed in the Cretaceous when the principal source rocks were deposited. The basins in Argentina had different sedimentary histories; the Neuquen and Austral had some marine influence throughout much of the time. Their source rocks are marine Jurassic and Lower Cretaceous shales. On the other hand, the Cuyo and San Jorge basins were predominantly continental. Their source rocks are lacustrine shales.

The early Andean uplift resulted in rapid marine regression

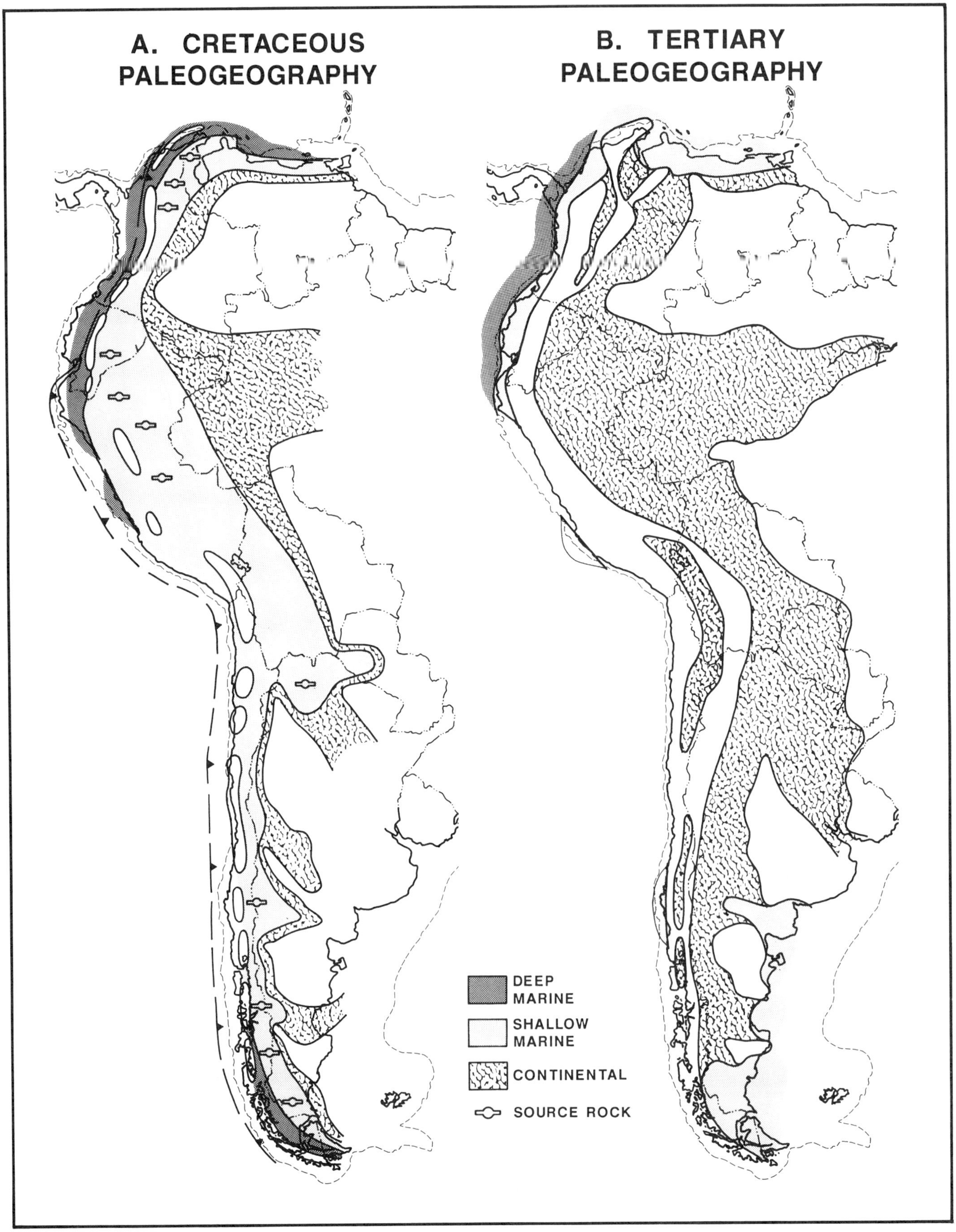

Figure 3. Paleogeography of the Andean province.

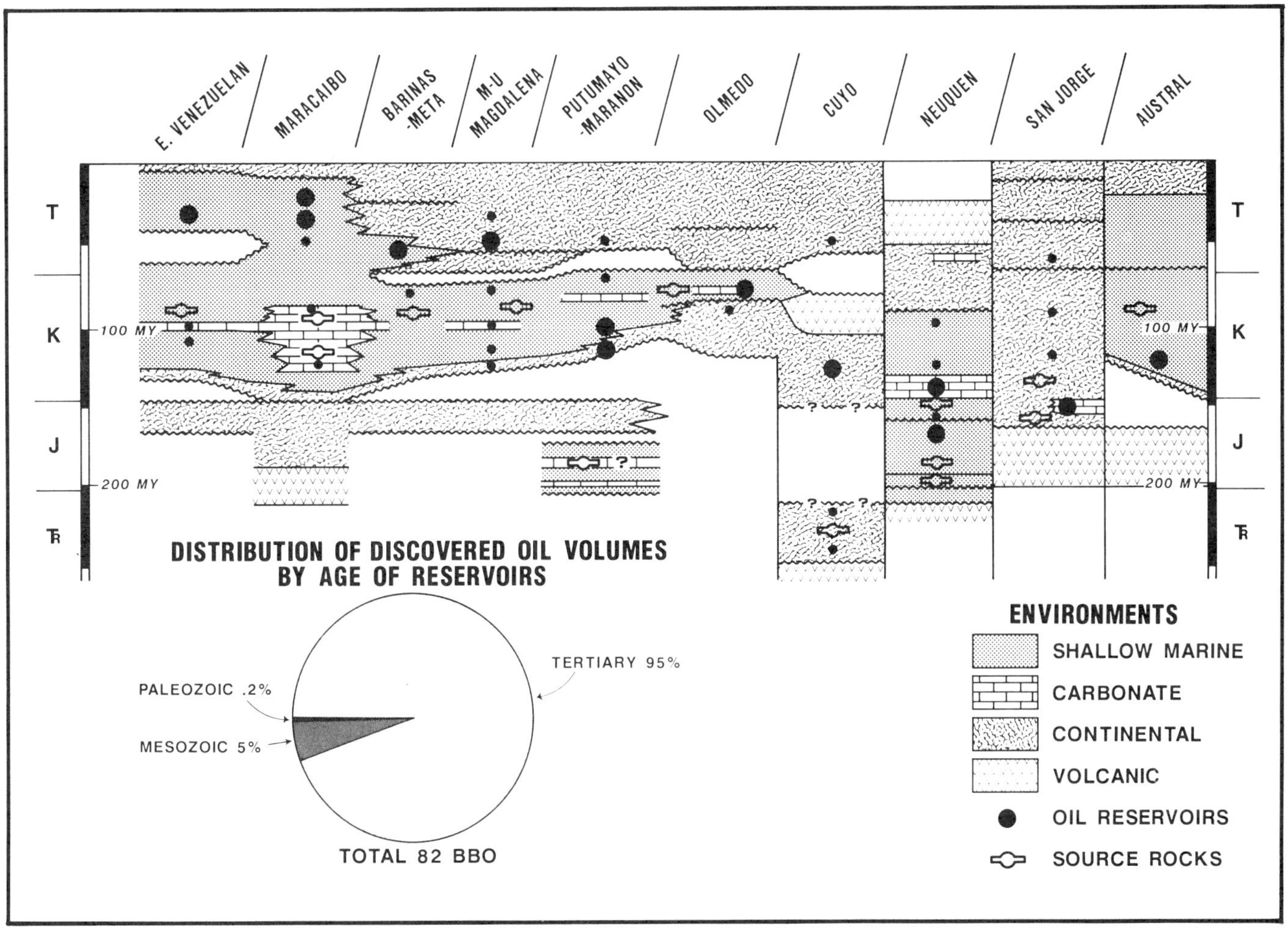

Figure 4. Time-stratigraphic framework of Mesozoic Andean backarc basins. (Data for distribution of estimated discovered oil volumes by age of reservoir from Petroconsultants.)

over many of the basins at the beginning of the Tertiary. Only in the far north and south did marine conditions persist: Caribbean influence in the Maracaibo and East Venezuelan basins, South Atlantic influence in the Austral basin. Figure 4 also illustrates where the oil reservoirs lie in relation to stratigraphy. Much of the oil is produced from the Tertiary, especially from the marine Tertiary in the north, where cycles of reworked sand and marine shale provide multiple interbedded reservoir-seal pairs. If the mature source rocks are in the Mesozoic, then vertical migration must be important. In most of the rest of the basins, Mesozoic clastic and carbonate reservoirs, stratigraphically close to mature source rocks, contribute significant volumes of oil.

STRUCTURE

Structurally, the basins differ significantly in size and shape. Some basins are not more than 5 km deep; others are more than 8 km deep (Figure 5). Many basins are asymmetrical. The northern sub-Andean basins have their deepest part in the west, whereas the northern Intermontane basins have their deepest part in the east. This suggests that some of these basins are remnants of larger basins split in two by uplift of the Eastern Cordillera; for example, the Maracaibo and Barinas basins in Venezuela, and the Middle Magdalena and Meta-Llanos basins in Colombia. In contrast, the Argentinian basins appear more symmetrical and elongate, suggesting that they each had separate extensional-rift histories.

Structural style within the basins changes away from the Andean deformation front and dictates the type of hydrocarbon traps (Figure 5). In the former backarc basins thrust-faulted anticlines occur in a zone subparallel to the Andean trend. They are hydrocarbon-bearing in some basins, such as the Putumayo, Tarija, and Cuyo basins, but barren in others, perhaps because of factors such as timing of deformation with respect to hydrocarbon migration and rupture of sealing strata. Strike-slip faults are associated with and in front of the thrust zone and cross basins at acute angles to the Andean front. This structural style is especially important in the northern Andes, where many of the oil fields show trap styles that are associated with strike-slip faults. This phenomenon could have permitted oil to migrate by upward-spiralling from Cretaceous source rocks into Tertiary reservoirs. Elsewhere, as for example in the Putumayo basin, subtle low-relief anticlinal structures occur in the axial parts of the backarc basins. They probably formed by rejuvenation of basement features, and

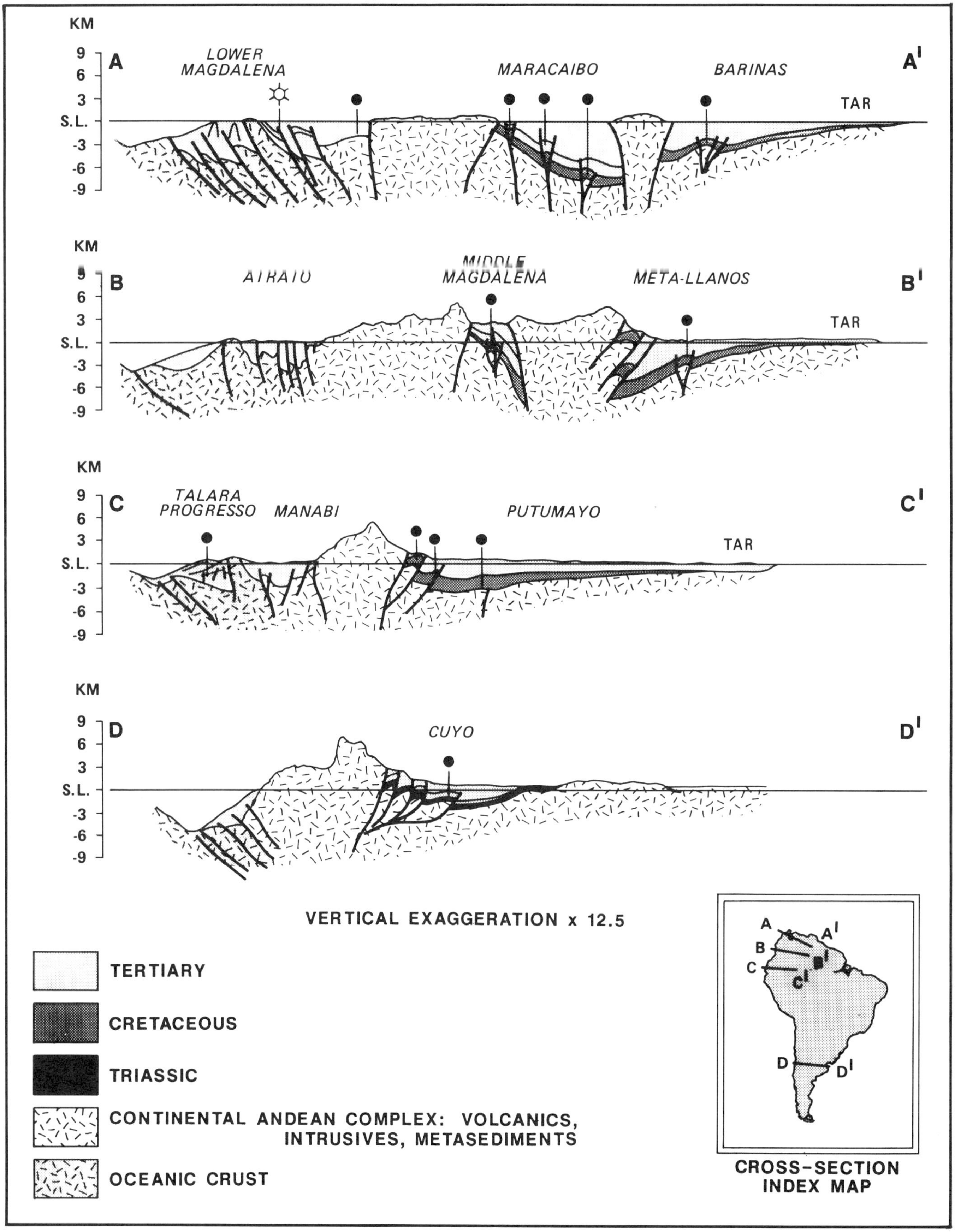

Figure 5. Structural cross-sections showing Andean basins.

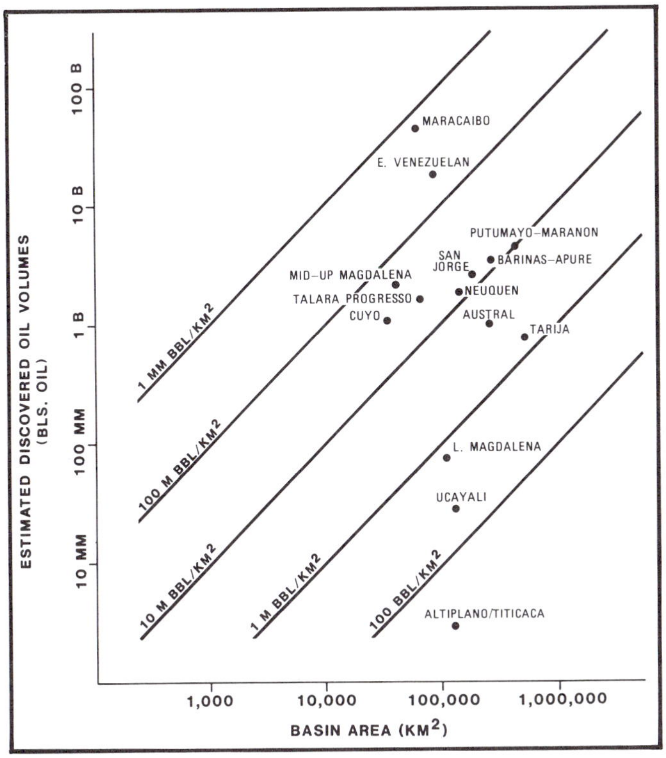

Figure 6. Oil richness of Andean basins. (Data from Petroconsultants).

contain oil in Cretaceous sand reservoirs. Farther away from the thrust zone, the eastern flanks of the basins are essentially undeformed and traps tend to be stratigraphic. Vast quantities of oil have migrated to these flanks and are now found in tar belts. The Orinoco tar belt is believed to contain 1 to 2 trillion barrels of heavy oil and tar.

The forearc basins are commonly divided into two by a zone of upthrusted ophiolitic material. The inner forearc basins show strike-slip faulting and folds. Minor oil and gas are present in the inner part of the Lower Magdalena basin of Colombia. The outer forearc basins are best represented in the Talara-Progresso basin of Ecuador-Perú. There oil is trapped by extensional structures such as growth faults, shale diapirs, normal faults, and olistostromes, mainly in the Tertiary sediments. Mud diapirs in the outer Lower Magdalena basin contain shows of oil and gas. Farther south into Perú and Chile, narrow, relatively thin wedges of Tertiary sediment are perched in a trench-slope setting. Although not as heavily explored as the sub-Andean basins, most of the forearc basins have not yielded commercial hydrocarbons, perhaps because of low thermal gradients, insufficient sediment thickness, and lack of good source rocks.

CONCLUSION

Presently known hydrocarbon reserves are restricted to the sedimentary basins. The accumulations are in reservoirs of various ages and lithologies, but mainly in Tertiary sandstone. The traps are of various styles, mainly structural. The different basins can be compared on a basin richness graph (Figure 6) showing discovered volumes versus basinal area. The Maracaibo and East Venezuelan are clearly the richest basins. A possible reason is vertical migration from good marine source rock into multiple reservoirs. Other basins appear less rich. The Cuyo and San Jorge basins may be less rich because of their more continental stratigraphy, whereas the Talara-Progresso basin may be less rich because of its forearc setting. In other basins the apparent richness may merely be a function of exploration intensity. There is potential for the Barinas-Apure and Putumayo-Maranon to become more prolific; likewise, the Lower Magdalena, Ucayali,[1] and the virtually untested Altiplano basins.

There are many basins in which commercial hydrocarbons have not been found. It becomes important to answer this question, "Is the reason for lack of success due to geologic phenomena, such as lack of good source rocks, lack of adequate seals, and no migration paths to reservoirs, or is the reason one of logistics, economic incentives and low intensity of exploration?"

Recent important discoveries (for example: Yacimientos Petroliferos Fiscales' discoveries at Palmar Largo, Argentina; Occidental's discovery at Caño Limón, Colombia; British Petroleum's discovery at Payamino in Ecuador; Lagoven's discovery at Furrial in the East Venezuelan basin[1]) indicate that significant oil reserves will be added by the future exploration efforts of private and national oil companies.

ACKNOWLEDGMENTS

The author wishes to thank the management of Pecten International Company for permission to publish this information; the staff of the New Ventures, Latin America group, especially L. Gordon and D. Malouta, for helping in its preparation; and Pecten's drafting department for making the illustrations.

REFERENCES

Bueno Salazar, R., 1985, Present and future petroleum exploration in the sub-Andean basin (Colombia), *in* Petroleum exploration in sub-Andean basins: Asociacion Colombiana de Geologos y Geofisicas del Petroleo, II Symposium Bolivariano Bogotá, Ago 1985, 42 p. (in Spanish).

Govea, R.C., and B.H. Aguilera, 1985, Sedimentary basins of Colombia, *in* Petroleum exploration in sub-Andean basins: Asociacion Colombiana de Geologos y Geofisicas del Petroleo, II Symposium Bolivariano Bogotá, Ago 1985, 93 p. (in Spanish).

Lesta, P.J., J. Digregorio, and M.E. Mozetic, 1985, Present and future petroleum exploration in sub-Andean basins of Argentina, *in* Petroleum exploration in sub-Andean basins: Asociacion Colombiana de Geologos y Geofisicas del Petroleo, II Symposium Bolivariano Bogotá, Ago 1985, 34 p. (in Spanish).

Lonsdale, P., 1978, Ecuadorian subduction system: American Association of Petroleum Geologists Bulletin, v. 62, p. 2454-2477.

Petroconsultants, 1984, Basin data sheet file.

Rodriguez, A., 1985, Petroleum exploration in sub-Andean basins, *in* Petroleum exploration in sub-Andean basins: Asociacion Colombiana de Geologos y Geofisicas del Petroleo, II Symposium Bolivariano Bogotá, Ago 1985 (in Spanish).

[1]Shell's recent discovery of 7 TCF gas and 200 MMB condensate in the Ucayali basin of Perú is also significant in this regard but was announced too late to include in the reserve numbers in this chapter.

Touzett, P.J., and V.R. Sanz, 1985, Present and future petroleum exploration in the sub-Andean basins of Perú, *in* Petroleum exploration in sub-Andean basins: Asociacion Colombiana de Geologos y Geofisicas del Petroleo, II Symposium Bolivariano Bogotá, Ago 1985, p. 1-93 (in Spanish).

Urien, C., and J. Zambrano, 1985, sub-Andean basins from Venezuela to Malvinas Plateau, *in* Petroleum exploration in sub-Andean basins: Asociacion Colombiana de Geologos y Geofisicas del Petroleo, II Symposium Bolivariano Bogotá, Ago 1985 (in Spanish).

Ericksen, G. E., Cañas Pinochet, M. T., and Reinemund, J. A., editors, 1989, Geology of the Andes and its relation to hydrocarbon and mineral resources: Houston, Texas, Circum-Pacific Council for Energy and Mineral Resources Earth Science Series, v. 11.

CHAPTER 32

SUB-ANDEAN HYDROCARBON RESOURCES OF ARGENTINA

Marcelo Reinaldo Yrigoyen
Esso Exploradora y Productoro Argentina Inc.
Buenos Aires, Argentina

ABSTRACT

The total sedimentary basinal area in Argentina is about 1.4 million km^2 onshore and 400,000 km^2 offshore down to 200 m water depth. Six of the onshore pericratonic basins have been productive for both oil and gas, and all of them belong to the sub-Andean domain. Their current daily production amounts to 480,000 BOPD and 1800 MCFGPD. Present proven plus probable/possible reserves are nearly 87 billion barrels of oil and 77 trillion ft^3 gas. Cumulative production to December 31, 1984, was 4.2 billion barrels of oil and 8.8 trillion ft^3 of gas, 93% of which was associated gas.

The most important basins from the standpoint of current oil production are San Jorge (yielding 41% of Argentina's total production), Neuquen (27%), Cuyo (23%), Magallanes (6.5%), and Tarija (2%). Yacimientos Petroliferos Fiscales (YPF), the Argentine state oil agency, is the largest single operator in the country, producing about 64% of the total crude oil. The remainder is produced by private Argentine and international companies.

By the end of 1984, 31,282 wells had been drilled in the country, from which 10,555 were oil producers and 1010 gas producers. The exploratory drilling effort amounted to 4132 wildcats or 13% of the total, the balance being mostly development wells.

INTRODUCTION

Located at the southern end of Latin America (Figure 1), Argentina's surface area is 2,791,810 km^2 on land, and 1,235,214 km^2 offshore. Population is 30 million inhabitants. Energy requirements during 1984 were on the order of 52 million m^3 of oil equivalent, of which 54% came from oil, 29% from natural gas, 13% from hydroelectricity, 3% from nuclear energy, and 1% from coal.

Argentina's geological configuration shows about 20 major sedimentary basins on land with a surface area of about 1,400,000 km^2 and 9 basins in the South Atlantic shelf with an area of about 400,000 km^2. In other words, 23% of the total area occupied by sedimentary basins is offshore and 77% is onshore. As of 1985, six basins have produced commercial volumes of oil and natural gas.

The presence of commercial hydrocarbons in Argentina has been known since the 18th century. Preliminary studies of some shallow oil seeps were carried out in 1865, when the first oil concessions were granted, but it was not until 1881 that hydrocarbon deposits discovered in the Tarija and Oran sedimentary basins in northern Argentina (Figure 2) were first investigated. The first exploratory well in the search for oil in Argentina, drilled by a private businessman from Mendoza in 1980, was in the vicinity of the Cacheuta natural oil seepages in the northern Cuyo basin. The area was later developed and became an important commercial oil field, where 30 producing wells were drilled before the end of the 19th century. This early exploration, development drilling, and exploitation of oil resources in Argentina was started by a group of private investors who realized the value of hydrocarbons and their significance in the future economic development of the country.

OIL PRODUCING BASINS

Figure 2 shows the distribution of the main Argentina sedimentary basins, including the areas where hydrocarbons are being produced at present. The six basins that have been productive are the Tarija, Oran, Cuyo, Neuquen, San Jorge, and Magallanes (or Austral). The first two basins are independent tectono-sedimentary units which, because they are contiguous, are generally referred to as a single basin. the "Northwest basin." The Tarija and Oran basins have been recently described in several published reports, including those of Turic et al. (1981), Cáceres and Montangero (1985), and Lesta, Digregorio, and Mozetic (1985). A geologic summary of these and other basins in Argentina is given in Table 1, and information about their reservoirs, production, and reserves is presented in Table 2. These tables give the reader a general idea of basin size and characteristics. Supplementary information is provided in the following discussions of each individual pro-

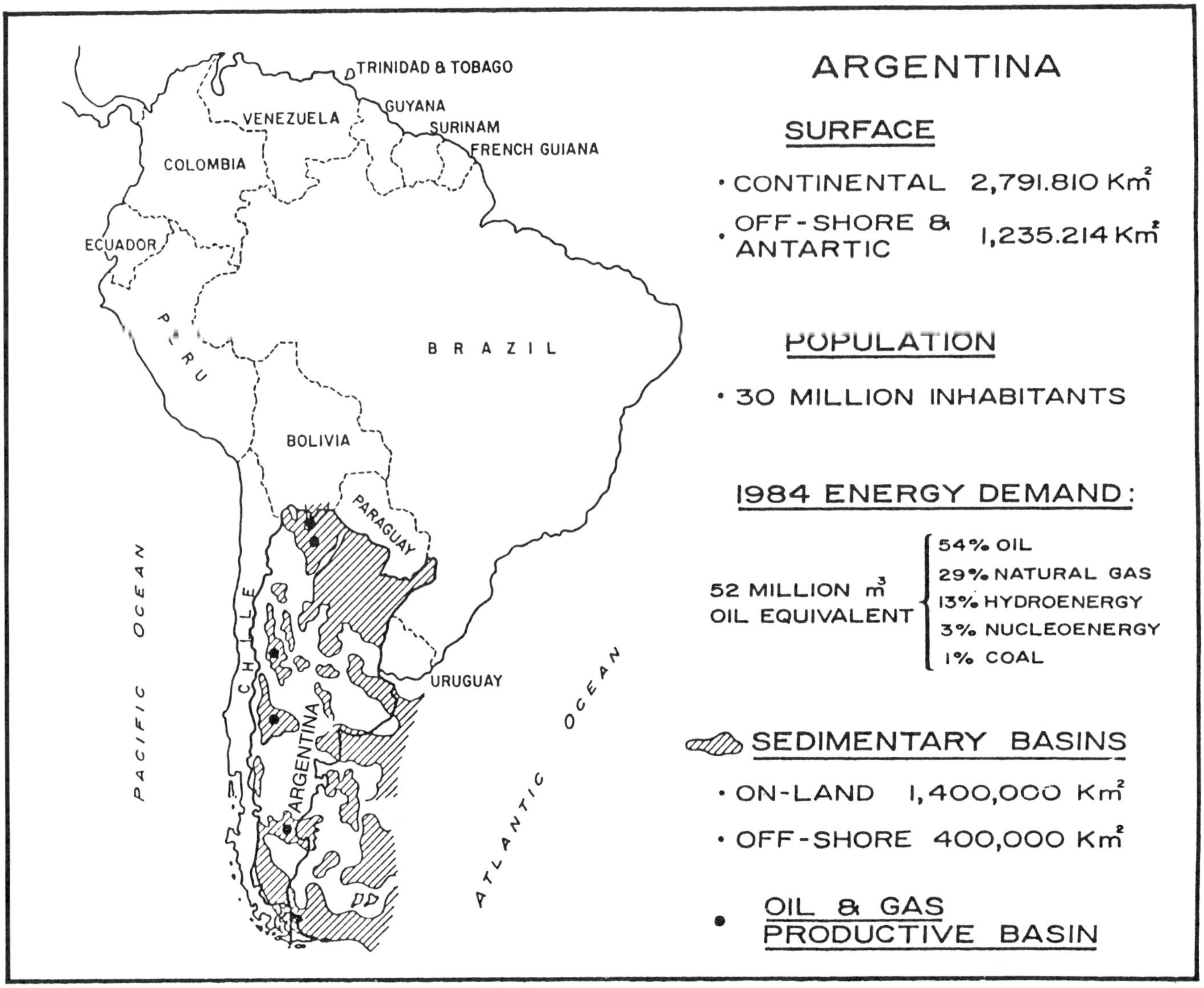

Figure 1. Index map of South America showing sedimentary basins and energy requirements of Argentina.

ductive basin. In addition, typical transverse sections of the basins are shown in Figures 3-5.

Tarija Basin

The Tarija basin is a segment of a large pericratonic basin that extends from central Bolivia to northern Argentina. Like many other basins of the sub-Andean region, the Tarija basin may be subdivided into two areas: (1) a structurally complex area along the Andean front in the west, which is characterized by persistent, longitudinal, overthrust asymmetrical anticlines that have eroded to a distinctive topography of parallel ridges and valleys, and (2) a platform area in the east, which is characterized by more gently folded rocks.

This basin has a Precambrian metamorphic basement covered by several thousand meters of Cambrian, Ordovician, Silurian, and Devonian marine sediments. These early basin sediments are unconformably overlain by thick nonmarine deposits (partly fluvioglacial) of Carboniferous age, by marine and nonmarine sediments of Triassic age, and finally by a thick sequence of Tertiary red beds.

The main source rock for hydrocarbons appears to be the widespread black shales of the Los Monos Formation of Middle Devonian age. The reservoir rocks consist of several sandstone units within the Silurian and Devonian to Pennsylvanian sedimentary sequence. The most prolific units have been the Early Devonian Santa Rosa Formation and the Mississippian Tupambi sands.

The present tectonic configuration of the Tarija basin (Figure 3) is the result of the superposition of several phases of deformation, presumably controlled by tectonic alignments delineated during the early Paleozoic. The main periods of deformation that affected the area occurred in the late Paleozoic and during the Andean orogeny in the late Mesozoic and Tertiary. Compressional movements from the west during the Andean orogeny caused imbricate overthrusting in some areas, as shown in Figure 3.

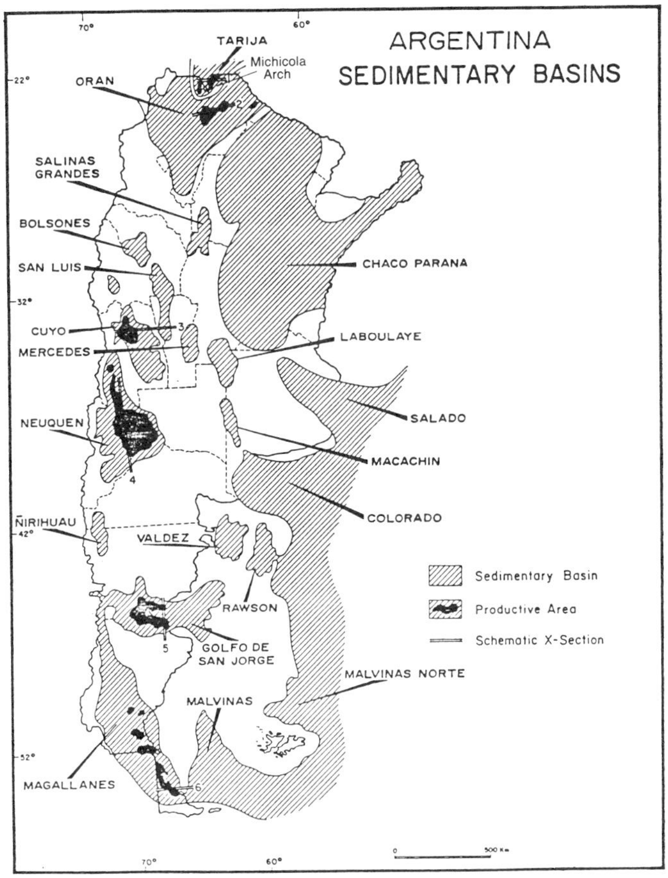

Figure 2. Productive areas in relation to sedimentary basins. Cross sections shown on Figures 3-5.

Oran Basin

The Oran basin, located east of the Andes in northernmost Argentina, is another structural remnant of a much larger Cretaceous-Tertiary basin. In the west it probably was open to the Pacific Ocean for a short period. One arm of the basin appears to have also extended to the north-northwest and connected with the Altiplano region of Bolivia and Southern Perú. Andean uplift during the middle to late Tertiary destroyed most of the western part of the basin and left a long northeast-trending trough that extends across northern Argentina into the plains of western Paraguay (Figure 3). A second smaller trough, known as Metán subbasin, extends southward along the sub-Andean front. The Michicola arch, near the Bolivian border, forms the northern boundary of the Oran basin. The western margin of the basin ends abruptly against the eastern side of the Andes.

The basin floor, which is a block-faulted peneplane surface on Precambrian to Early Devonian metasediments, forms an extensive graben having a northeast-trending axis. The graben began to fill with nonmarine clastic sediments during the early to middle Cretaceous. During the Maestrichtian-Paleocene the carbonate/sand/shale sequence of the Yacoraite Formation was deposited during an extensive shallow-marine transgression. This sequence is the main producer of hydrocarbons in the Oran basin. During the Eocene-Oligocene, widespread red beds were deposited, and from the Miocene to the present the erosional products of the uplifting Andes formed huge synorogenic and postorogenic clastic deposits that spread eastward over large areas of South America. These clastic deposits are more than 5 km thick in the Oran basin. As the proto-Andean mountains rose, the clastic deposits, which were coarse near the mountain front, became progressively finer eastward across the Chaco plains.

The difference between the western and eastern parts of the Oran basin is mainly one of tectonic style. Prominent longitudinal structural trends resulting from compression and thrusting in the west give way to gentle folding in the east. The Lomas de Olmedo structural trend (Figure 3), the easternmost zone of deformation in the Oran basin, separates the eastern and western parts of the basin. Some oil production was recently discovered at the eastern margin of this isolated structure.

Table 1. Producing basins of Argentina: description of basin geology.

Geologic Basin Description	Tarija	Oran	Cuyo	Neuquen	San Jorge	Magallanes
Basinal Area ($\times$ 1000 Km^2)	76	142	40	120	88	120
Sedimentary Volume ($\times$ 1000 km^3)	286	350	112	290	222	320
Average Sedimentary Thickness (m)	3,800	2,600	2,500	2,600	2,500	2,500
Max. Depth to Economic Basement (m)	4,200	5,700	6,000	6,500	6,000	9,500
Basin Classification	Pericr. asymetr. foreland	Pericrat. interm. fault through	Pericrat. interm. fault through	Pericr. foreland embayment	Intra/pericrat. fault through	Pericr. foreland asym. through
Basin Shape	Elongated	Elongated	Elongated	Triangular	Oval	Elongated
Azimuth of Depositional Axis	S-N	SW-NE	S-N	S-N	W-E	SSE-NNW
Sedimentary Fill	Pal + Tr + K + T	K + T	Tr + K + T	(Tr)J + K + T	J_L + K + T	K + T
Basinal Basement	Paleozoic	Paleozoic	Pr€ + Paleoz.	Pal. + Triassic	Pal./J. volc.	Pal./J. volc.
Main Source Rock	Mid. Devonian	Late Cretac.	Late Triassic	Late Jurassic	Early Cretac.	Early Cretac.
Main Reservoir Units & Lithology	Tupambi SS/ Sta. Rosa SS	Yacoraite Lms	Barrancas SS Victor-Porter SS	Several Lms & SS Units	Chubut SS	Springhill SS
Age of Reservoir	Carboniferous Early Devon.	Late Cretac.	Early Cretac./ Late Triassic	Cretac. to Jurass.	Late Cret./Pal.	Early Cretac.

Table 2. Production and reserve data for Argentina producing basins.

Production and Reserve Data	Tarija	Oran	Cuyo	Neuquen	San Jorge (on land)	Magallanes (on land)
Average Depth to Reservoirs (m.b KB)	3800/5500	4000	2000/2900	1700/2500	1700/2300	2050
Average Oil Gravity (°API)	54/40	44	29.4	29.5/33.2	23.5	40.8
Number of Wells Drilled (approximate)	920	80	2,560	5,880	20,140	1,520
Oil Recovery per Well ($\times$ 1000 m^3)	135	80	110	26/70	22	60
Oil Recovery per hectare (m^3/ha)	3,320	4,700	4,500	2,750	1,100	2,250
Average Daily Production per well (m^3/day)	7/90	90	22	9	8	24
Number of Giant Fields (> 80 million m^3)	—	—	1 oil	1 gas, 1 oil + cond.	2 oil	—
Original Proven Oil Reserves (million m^3)	40.7	10.3	176.9	315.7	458.3	58.4
Cumulative Oil Production (million m^3)	17.1	9.5	141.8	153.3	316.7	27.2
Remaining Proven Oil Reserves (million m^3)	23.6	.8	35.1	162.4	141.6	31.2
Probable Oil Reserves (million m^3)	16.5	.1	7.6	83.8	63.2	28.8
Original Proven + Probably Oil Reserves (million m^3)	57.2	10.4	184.5	399.5	521.5	87.2
1984 Oil Production ($\times$ 1000 m^3)	677	166	4,429	9,185	11,600	1,798
1984 Average Daily Oil Production (m^3/day)	1850	454	12,101	25,096	31,694	4,912
Original Proven Gas Reserves (million m^3)	138,842	3,984	4,963	516,046	100,627	150,365
Cumulative Gas Production (million m^3)	42,041	2,597	4,055	77,452	62,993	46,801
Remaining Proven Gas Reserves (million m^3)	96,801	1,387	908	438,594	37,634	103,564
Probable Gas Reserves (million m^3)	44,937	1,000	—	88,240	3,256	71,588
Original Proven + Probable Gas Reserves (million m^3)	183,779	4,984	4,963	604,286	103,883	221,953
1984 Gross Gas Production (million m^3)	2,070	41	764	8,408	4,007	3,463
1984 Average Daily Gas Production ($\times$ 1000 m^3)	5,656	112	2,086	22,973	10,948	9,462

Cuyo Basin

The most productive part of the Cuyo basin, which covers an area of 40,000 km^2, lies at the eastern foothills of the Andes (Figure 2). Cuyo is a typical pericratonic landlocked intermontane basin of a transtensional taphrogenic style. It includes several fault-bounded subbasins filled with a nonmarine sedimentary sequence as much as 12,000 m thick, mainly of Triassic and Tertiary ages, which overlies a complex basement of Precambrian and early Paleozoic metasedimentary rocks and Permo-Triassic volcanic rocks. The Jurassic(?) and the Cretaceous are poorly represented, but Neocomian clastic sediments and basalts constitute the principal reservoirs and caprocks, respectively, in the main productive area (Figure 4).

The basin fill, chiefly of middle-late Triassic age, consists of alluvial-fan and fluvial-plain deposits containing tuffaceous material in the upper part. An intermediate unit of lacustrine shale, the Cacheuta Formation, is generally accepted as being the main source of the oil produced from the Cuyo basin. The uppermost Triassic Rio Blanco Formation is overlain with slight unconformity by nonmarine and volcanic deposits of Neocomian age. Late Cretaceous to early Tertiary movements started a new erosional and depositional cycle leading to accumulation of thick Tertiary terrestrial deposits. This Cenozoic sequence begins with the oil-productive Papagayos Formation at the base, overlain by several thousand meters of middle to late Tertiary red beds.

The folds and faults in the Cuyo basin were formed largely during the Tertiary-Quaternary, but some structures were formed during the Mesozoic. Structural traps are the most important sources of oil (Figure 4), but several oil pools are also in stratigraphic traps. In the western, productive, part of the basin the structures are faulted, asymmetrical closed anticlines along two (or three) almost parallel trends. The asymmetrical folds and thrust faults, which generally strike north, decrease in intensity from west to east, giving way to fault-controlled broad, flat, intermontane valleys. Oil reservoir sands lie within the Triassic and basal Tertiary sequences, but the Early Cretaceous Barrancas Formation is by far the main oil producing unit (Figure 9).

Neuquen Basin

The Andean segment of the Neuquen basin is in the backarc sub-Andean domain, but the so-called "Neuquen embayment" basin segment shows an aulocogenic pericratonic character that extends into the eastern foreland. As a whole, the Neuquen basin is a predominantly Jurassic-Cretaceous depositional basin. It is roughly triangular (Figure 2), and its western margin terminates against the north-trending Tertiary structures of the Andean Front. From a tectonic standpoint, the Neuquen basin lies on the continental interplate at the margin of the Andean belt.

Neuquen basin sedimentation began as Early Jurassic to Kimmeridgian fill in a rift system that opened in the Late Triassic. This fill consists of marine and nonmarine clastic sediments and subordinate carbonates and evaporites. During the Late Jurassic and Early Cretaceous the basin evolution shifted into an early stage of subsidence, and a thick series of carbonates, shales, and sandstones was deposited. In the Late Creta-

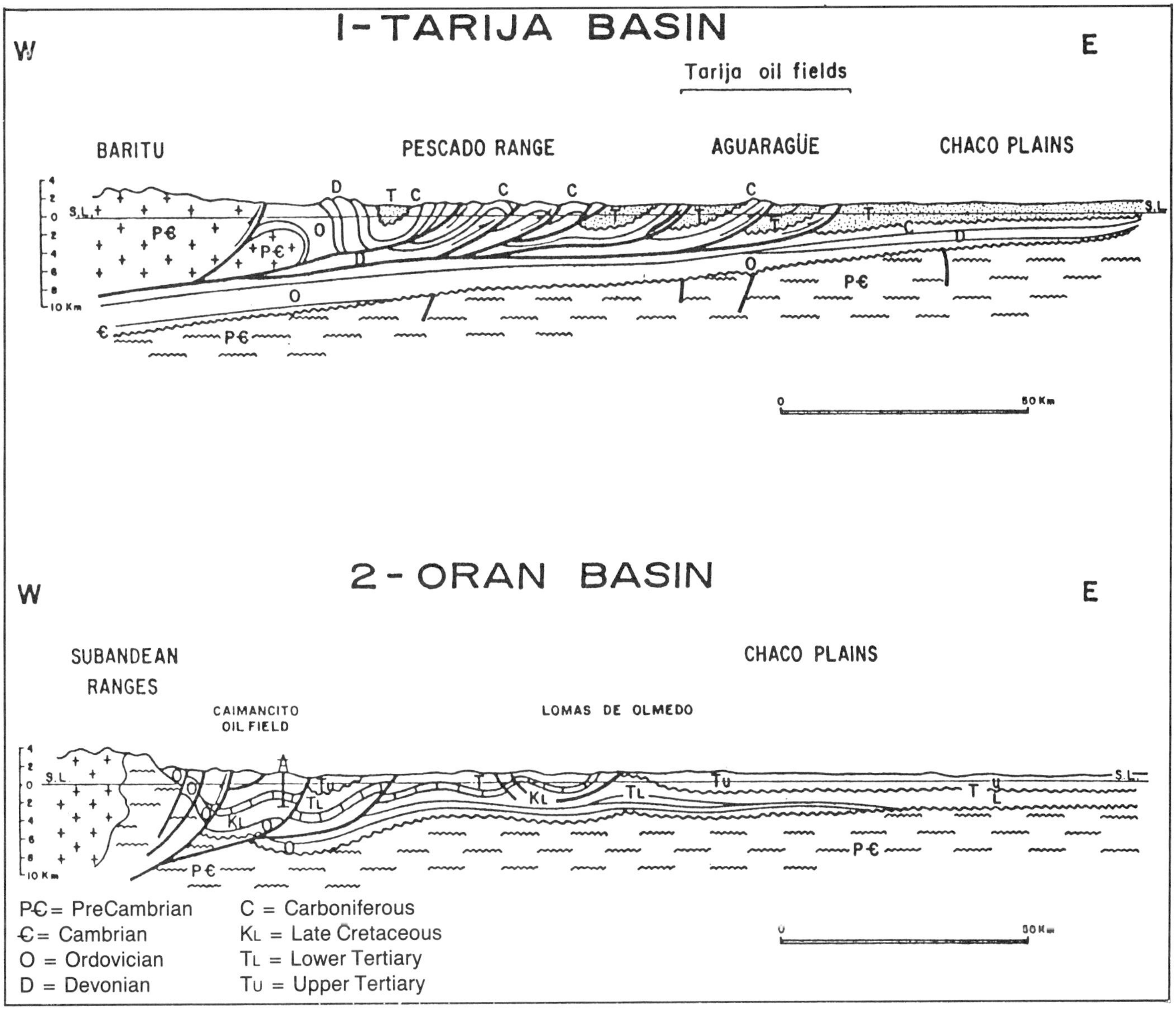

Figure 3. Schematic cross sections of the Tarija and Oran basins. Locations of cross sections shown on Figure 2.

ceous, after the Subhercynian orogeny, a thick postorogenic unit of red beds was deposited in an extended area during a late phase of subsidence. These sedimentary rocks are overlain by a discontinuous sequence of Cenozoic rocks, coeval with the Andean deformation to the west (Figure 4).

The Neuquen basin is the second most prolific hydrocarbon producer of Argentina. Nearly all potential reservoirs have been productive somewhere in the basin, but the Jurassic Lotena-Punta Rosada and Tordillo-Sierras Blancas sands and the Late Jurassic-Early Cretaceous Quintuco-Loma Montosa carbonates are the major commercial reservoir rocks. Los Molles and Vaca Muerta black shales, which are widespread in the basin, are rich oil and gas source rocks. Block faulting, wrenching, and drape folding during the Jurassic, and in subsequent Andean orogenic deformation, formed the many structural traps which, together with stratigraphic traps, controlled commercial accumulations in the Neuquen basin.

San Jorge Basin

Although it is one of the smaller basins (roughly 88,000 km^2), the San Jorge basin is the most important hydrocarbon-producing basin of Argentina. Its sedimentary fill accumulated during several episodes. Northwest-trending extensional graben systems were formed and filled during a Triassic and Early Jurassic volcanic episode (Lonco Trapial and Bahia Laura Groups). The basin continued to subside and fill during Late Jurassic-earliest Cretaceous rifting coincident with the opening of the Atlantic Ocean, and during the following Cretaceous phase of subsidence. Finally, a latest Cretaceous-early Tertiary tensional deformation gave rise to the structural alignments now prevalent in the basin. Many structural traps along normal faults were formed at this time.

A middle Tertiary compressional event produced the narrow north-trending San Bernardo fold belt, which exhibits

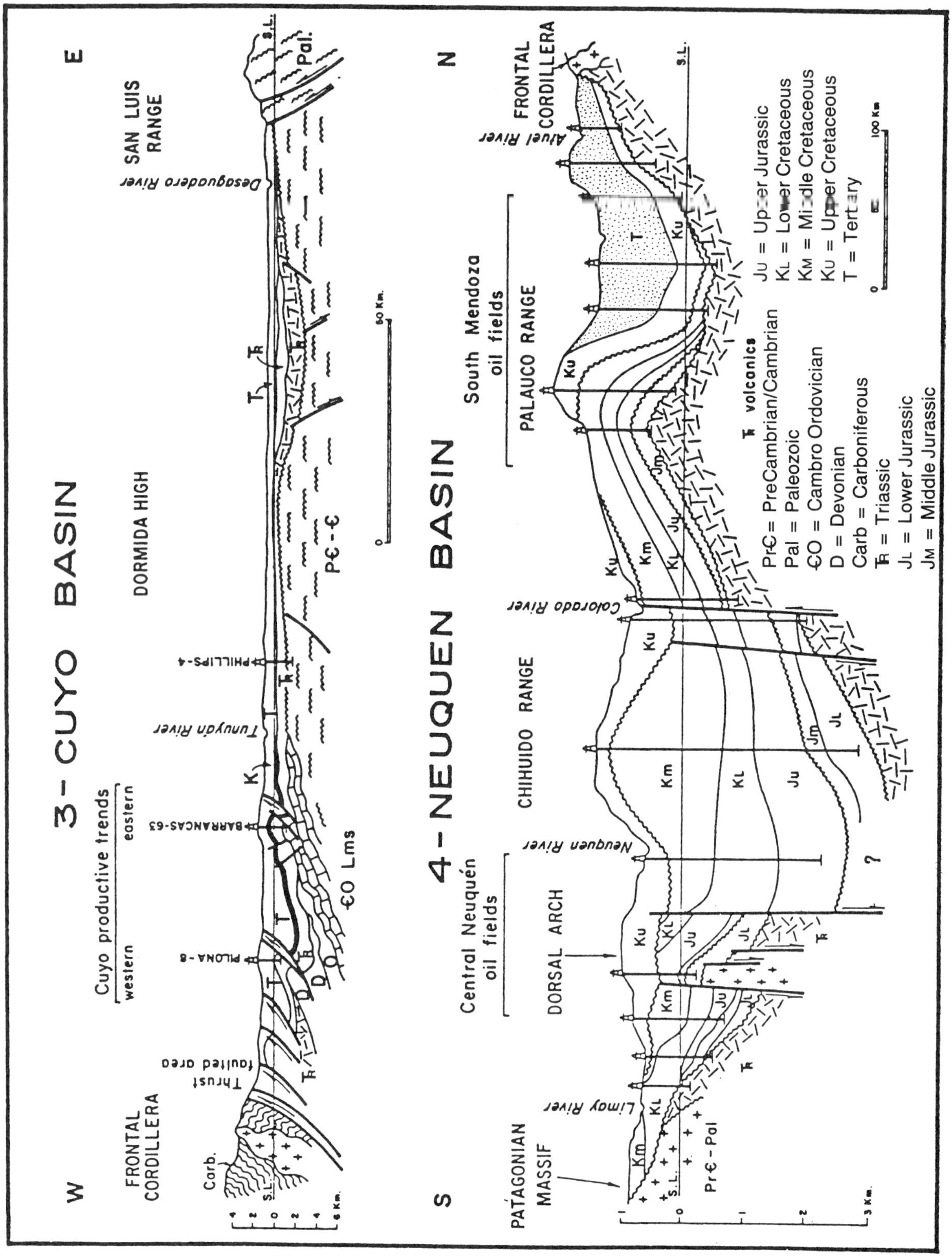

Figure 4. Schematic cross sections of the Cuyo and Neuguen basins. Locations of cross sections shown on Figure 2.

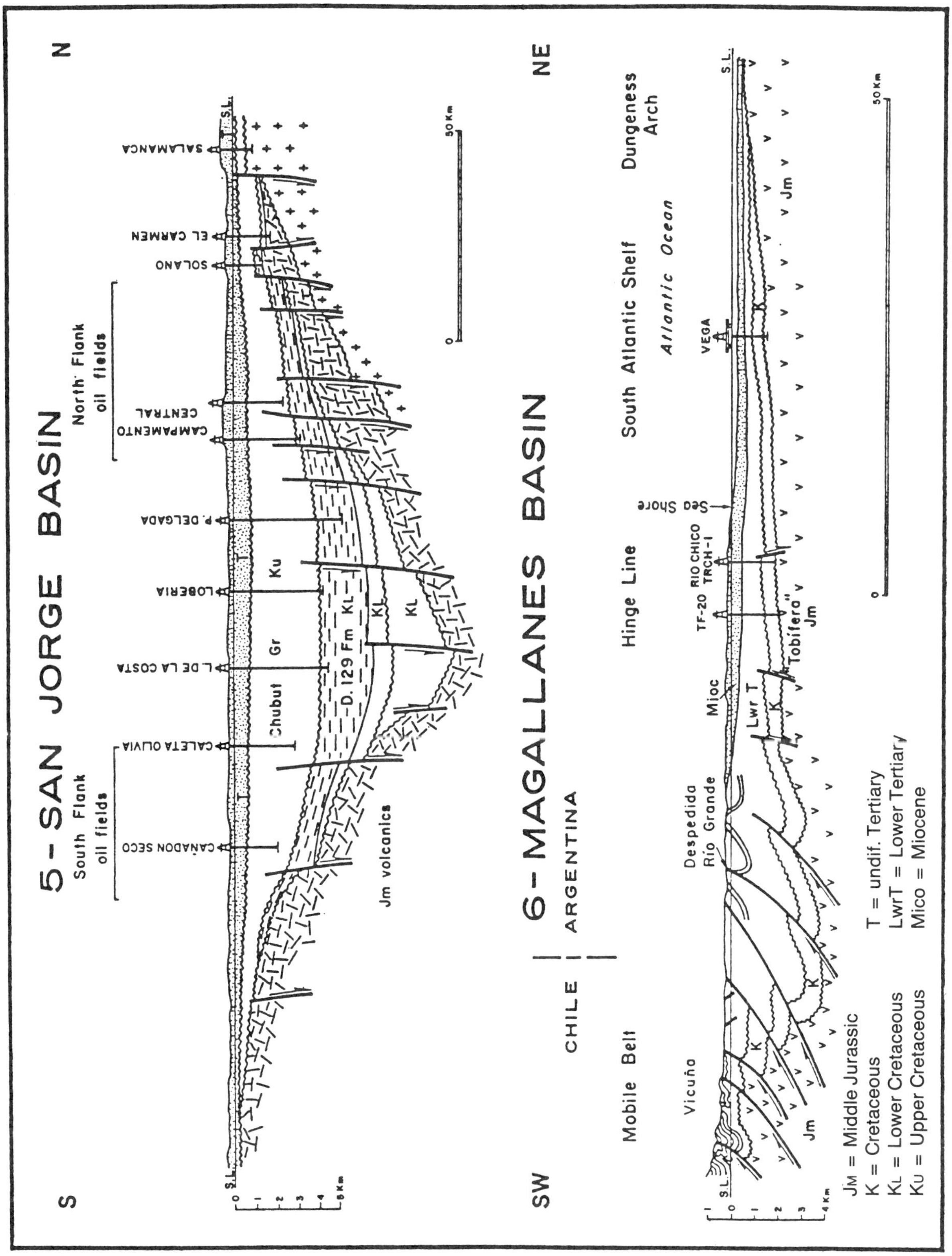

Figure 5. Schematic cross sections of the San Jorge and Magallanes basins. Locations of cross sections shown on Figure 2.

Table 3. Oil and gas statistics for Argentina, 1984

OIL		
Total demand	26,453,504 m^3/year	72,277 m^3/day
Local production	27,854.600 m^3/year	76,105 m^3/day
Importation (Crude & Prod.)	156.600 m^3/year	428 m^3/day
Exportation	1,557.696 m^3/year	4,256 m^3/day
PROVEN OIL RESERVES (Dec. 31, 1984)	394.7 million m^3 (≈ 14 years)	
NATURAL GAS		
Total Production	18,753 million m^3/year	51.237 million m^3/day
Delivered to trunk-lines	12,772 million m^3/year	34.896 million m^3/day
Importation from Bolivia	2,210 million m^3/year	6.040 million m^3/day
Flared/vented	3,550 million m^3/year	9.700 million m^3/day
PROVEN GAS RESERVES	[illegible] million m^3 (≈ [illegible]) (≈ 41 years)	

DRILLING				
Wells drilled during 1984	Wildcats	136 wells	Total wells 1031	785 oil producers
	Outpost	168 wells		35 gas producers
	Development	727 wells		211 dry & abandoned
1984	on oil & oil/gas production	= 8553 wells		
	on gas production	= 312 wells		

reverse movements on older, normal graben-bounding faults, and local low-angle thrust faults. A prominent early to middle Tertiary unconformity resulted from uplift and erosion of Cretaceous sediments around the basin margins and the accumulation of continental tuffaceous and shallow-marine sediments within the basin during the Cenozoic.

The Lower Cretaceous lacustrine dark shales of the "D-129" Formation (Figure 5) are the most favorable source rocks yet identified. Most hydrocarbon production has come from middle to upper Cretaceous sands above this source rock, and from Paleocene sands along the north and south flanks of the basin and their distal extensions. This pattern suggests that hydrocarbon trapping may have been related to vertical migration along reactivated graben faults. Volcanic tuffs, tuffaceous sands, and other evidence of igneous activity are present throughout the basin fill, and reservoir quality is generally low because of the highly tuffaceous sand lenses. Sills, dikes, and other intrusive bodies are widespread in the basin.

Magallanes Basin

The Magallanes basin is located in the southernmost tip of South America (Figure 2). It is bounded on the west by the Patagonian batholith and the Southern Andes deformed belt. In the north and east, it is bounded by the Dungeness arch, a large long-lived basement high. In the northwest, a shallow embayment apparently connects the Magallanes basin to the western side of the San Jorge basin. In the southeast, another embayment, around the southern tip of the Dungeness arch, connects the Magallanes basin with the offshore Malvinas basin.

The Magallanes basin lies on the southernmost part of the South American Plate, where it has been subjected to complicated compressional and transcurrent deformation along the convergent margin between the South American and Scotia plates. To the west, the southern extension of the Chile Trench forms a plate boundary between the Antarctic (and other microplates) and the South American Plate. South of the Chile Rise, this boundary is the site of strike-slip motion and slow subduction, whereas to the north of the Chile Rise, subduction of the Nazca Plate beneath South America is considerably faster. In the past, plate interaction along the western margin of South America has controlled much of the deformation and deposition seen today in the Magallanes basin.

The Magallanes basin underwent a two-phase development history. The basin first formed by rifting during the Middle Jurassic to Early Cretaceous, providing a substratum of normal fault blocks. After this phase of rifting a foredeep basin formed in front of advancing Andean thrust sheets. This phase of basin development was long-lived and diachronous. It began during the Late Cretaceous and continued well into the Tertiary. Of the two phases of basin development, the first was caused by thinning and thermal cooling of the lithosphere, whereas the second was caused by flexural loading.

All commercial pools of the Magallanes basin are located in the platform area near the Atlantic coast and in the area immediately offshore (Figure 5). Major hydrocarbon accumulations are in the Late Jurassic-Early Cretaceous Springhill transgressive basal sand, although some production has come from the weathered and fractured top of the Middle to Late Jurassic volcanic basement. The main source rock is the Early Cretaceous black shale that conformably overlies the Springhill sands. The most common traps are positive topographic features on the eroded surface of the volcanic basement, as well as elongated anticlines flanked by faults. Stratigraphic control in some fields is indicated by the presence of barriers and local pinching-out of the Springhill sand on top of the so-called "bald-highs."

PRODUCTION AND RESERVES OF OIL AND GAS

Argentina's total oil requirements in 1984 were about 26.5 million m^3, or 72,277 m^3/day, distributed as shown on Table 3.

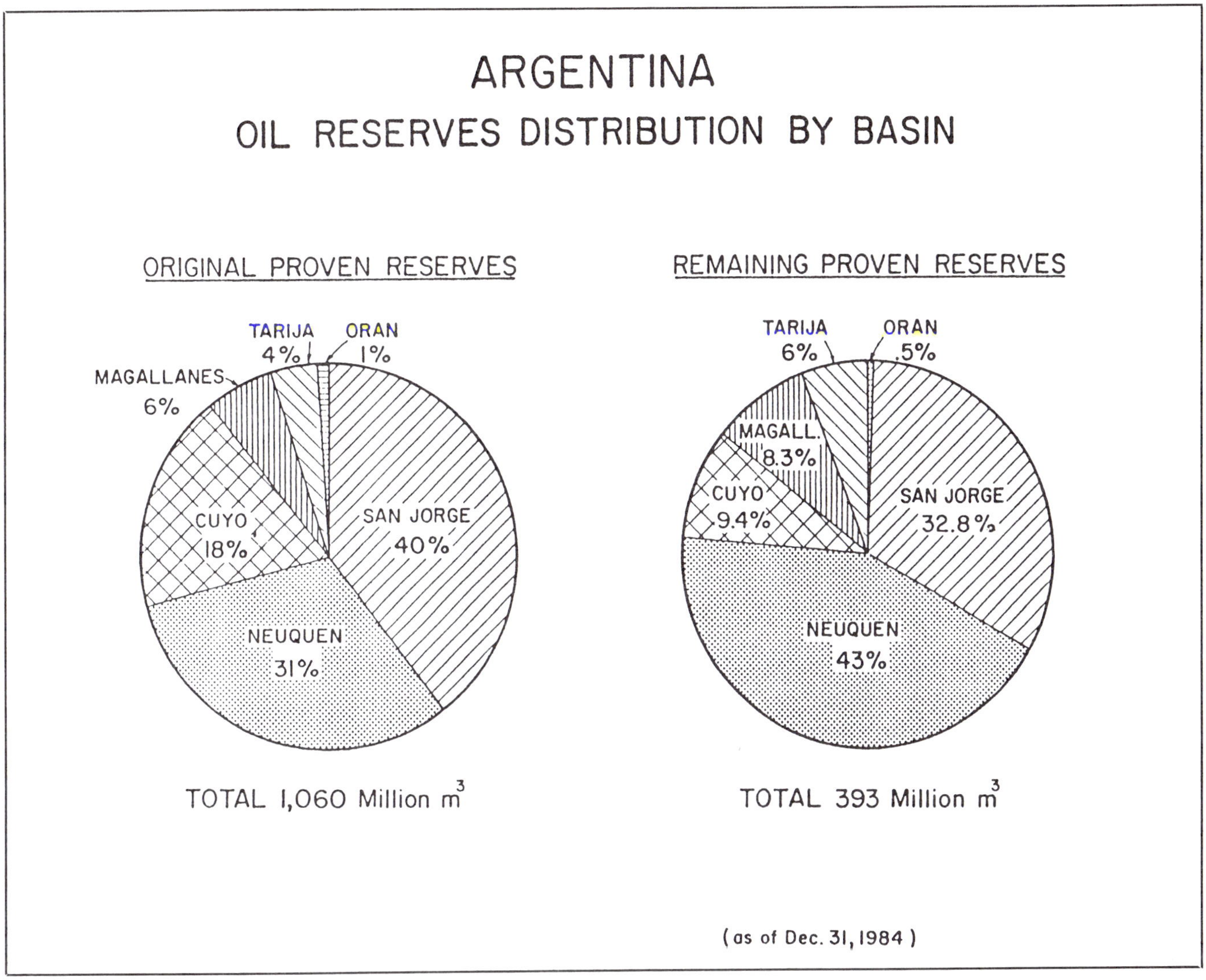

Figure 6. Percentage distribution of original and remaining proved reserves of oil in basins of Argentina.

Total oil produced (27,854,600 m^3) during the same year exceeded demand for the period and allowed exportation of 1.6 million m^3 to neighboring countries. Imports consisted of small amounts (156,000 m^3) of special crude oil.

At the end of 1984, proven oil reserves in YPF and contractor's fields, plus those corresponding to old concessions, totaled 394.7 million m^3. In other words, proven reserves were adequate to meet the country's needs for more than 14 years at the 1984 production rate.

The natural gas reserves of Argentina are larger and are sufficient for a longer period than the oil. Gas production in 1984 was 18,753 million m^3 (51,237,000 m^3/day), 68% of which was delivered to trunk lines. Of the remainder, 13% was reinjected and used in field and transport operations, and 19% unspent (flared or vented due to lack of gas lines in isolated areas or because of undesirable contaminants such as sulphur or carbon dioxide). Argentina's present proven reserves of natural gas amount to 680,000 million m^3 (around 24 trillion ft^3) or enough to meet present demand for 41 years. It should be noted that in 92% of the fields, natural gas in Argentina is associated with liquid hydrocarbons.

Oil and gas production requires a significant drilling effort. Total drilling in the country in 1984 amounted to 1031 wells, 136 of which were wildcats, 168 outpost wells, and 727 development wells. As a result of this activity, 783 new oil wells and 35 gas wells were put in production, whereas 211 wells were dry. At the end of 1984, there were 8553 producing oil wells and 312 gas wells.

Statistics show that present depletion of oil reserves in Argentina is almost twice that of the natural gas reserves. In other words, the smaller resource is being exploited at a rate faster than the larger resource. Several measures should be implemented to change this situation. For example, an intensive exploration drilling program should be carried out, and development drilling completed in the production areas of known reserves. Of the total area of known sedimentary basins, covering more than 1,800,000 km^2, only 28% has been studied in some detail. Thus, more than a million km^2

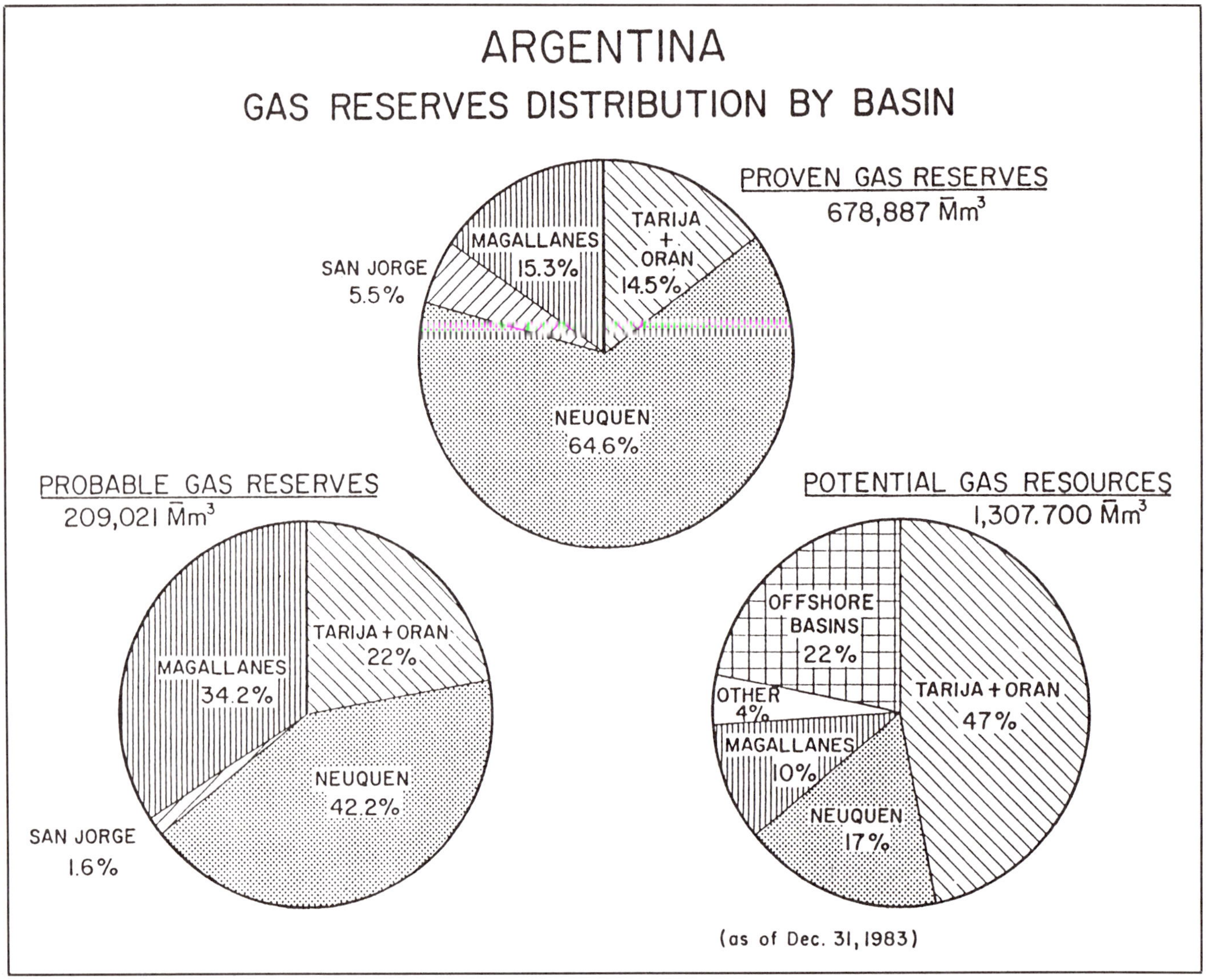

Figure 7. Percentage distribution of original and remaining proved reserves of gas in basins of Argentina.

remain to be explored in search for commercial hydrocarbons. Exploration effort to date is equivalent to barely one exploratory well for every 470 km^2, an indication of the large amount of exploration drilling yet to be done in Argentina. Extensive areas of onshore basins and significant areas of offshore basins require immediate exploration if depletion rates are to be changed.

DISTRIBUTION OF OIL AND GAS RESERVES

The distribution of oil and gas reserves within the six sedimentary basins is shown in Figures 6 and 7. Figure 6 shows the percentage distribution of the 1060 million m^3 of original proven oil reserves and the 1984 proven reserves (393 million m^3). As can be seen, the San Jorge and Neuquen basins have three-fourths of the oil discovered in Argentina. The remaining reserves, in order of magnitude, are in the Cuyo, Magallanes, Tarija, and Oran basins. The largest volumes of gas reserves (Figure 7), both free and associated, are in the Neuquen basin, followed by the Magallanes, Tarija, and Oran basins. In respect to yet-to-be-discovered potential gas reserves, great significance is given to the basins in the northwestern part of the country and the offshore basins being developed in the South Atlantic.

The distribution of proven oil reserves according to the age of the reservoirs, is shown in Figure 8. As can be seen, 70% of the total proven oil reserves are in Cretaceous reservoirs, especially within Late Cretaceous sediments. Jurassic reservoirs rank second. The remaining proven reserves are in Triassic, Tertiary, and Paleozoic sediments.

Figure 9 shows the proven oil reserves in each reservoir formation within its respective basin. The very large reserves within the Chubut Group (Upper Cretaceous) in the San Jorge basin are evident, as are the somewhat smaller reserves within Lower Cretaceous rocks of the Neuquen basin. Reserves are found in Paleozoic reservoirs only in the Tarija basin, whereas reserves are in Triassic, also Cretaceous and Tertiary reservoirs in the Cuyo basin. Proven reserves of the

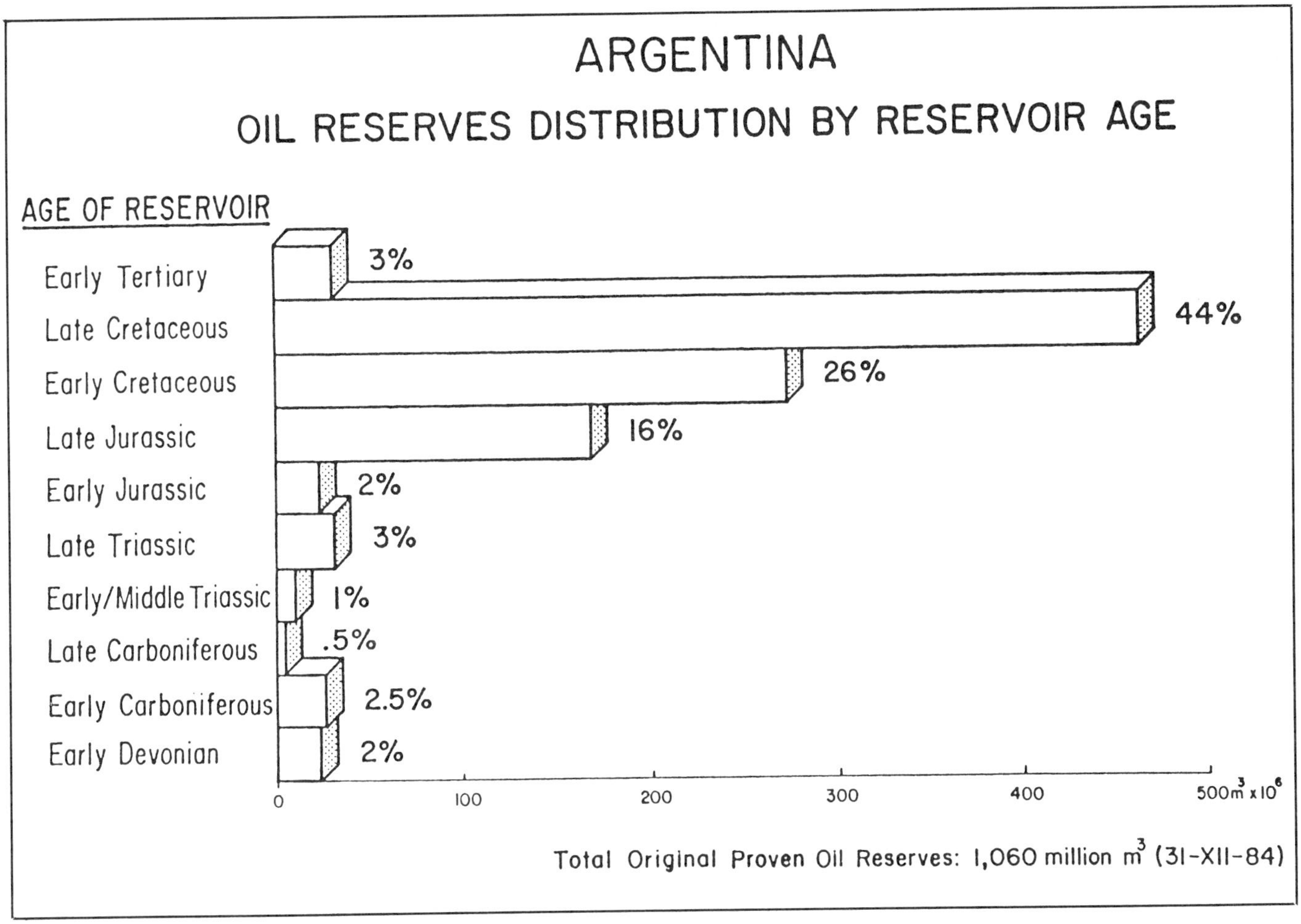

Figure 8. Distribution of oil reserves in Argentina by reservoir age.

Magallanes basin are almost entirely in Lower Cretaceous (and Upper Jurassic) sandstone of the Springhill Formation, and to a much lesser degree in volcanic rocks of early to middle Jurassic age, informally known as "Serie Tobífera." In the Oran basin, the largest reservoir is in Upper Cretaceous fractured limestone of the Yacoraite Formation. Smaller reserves are in sandstones of the overlying lower Tertiary Santa Barbara Group.

In several fields, significant hydrocarbon production is being obtained from igneous reservoirs, both extrusive and intrusive rocks. For example, a field recently discovered in the western part of the Oran basin produces from early Tertiary weathered basalt. In the Neuquen basin, commercial production was obtained from the top of a sequence of weathered volcanic flows of the Choiyoi Group (Triassic) and from andesitic dikes and sills (early Tertiary) with good fracture porosity. The intrusions are particularly prevalent in the northern segment of the basin that extends south of Mendoza province.

SIZE AND NUMBER OF OIL FIELDS

The distribution of Argentine oil reserves according to the size and number of fields is shown by a tridimensional chart (Figure 10). The vertical axis shows fields of different size, according to their proven reserves, ranging from a minimum of less than 100,000 m^3 to a maximum of more than 16 million m^3. The other two orthogonal axes show the number of fields in each category. Original proven oil reserves amounting to 1060 million m^3 (Figure 6) are distributed in 284 oil fields of different size. More than half of these reserves (540 million m^3) are in 15 major fields having more than 100 million barrels, whereas 37% (394 million m^3) is found in 52 fields of the next two lower categories, which range from 4 to 16 million m^3 (from 25 to 100 million barrels). The remaining proven oil reserves (126 million m^3) are in more than 200 small fields of the lowest four categories.

If we accept the classification by Halbouty et al. (1970), that "giant" oil fields are those having original reserves of more than 500 million barrels (about 80 million m^3), in Argentina four giant oil fields have been discovered to date. Two of these giant fields—the North Flank Group with 131.5 million m^3 of reserves and the South Flank Group with 169 million m^3 of reserves—are in the San Jorge basin. A third giant field, Loma de la Lata, with 122 million m^3 of reserves of oil and condensate, is in the Neuquen basin; the fourth, the Punta Bardas-Vacas Muertas field, with more than 80 million m^3 of reserves, is in the Cuyo basin.

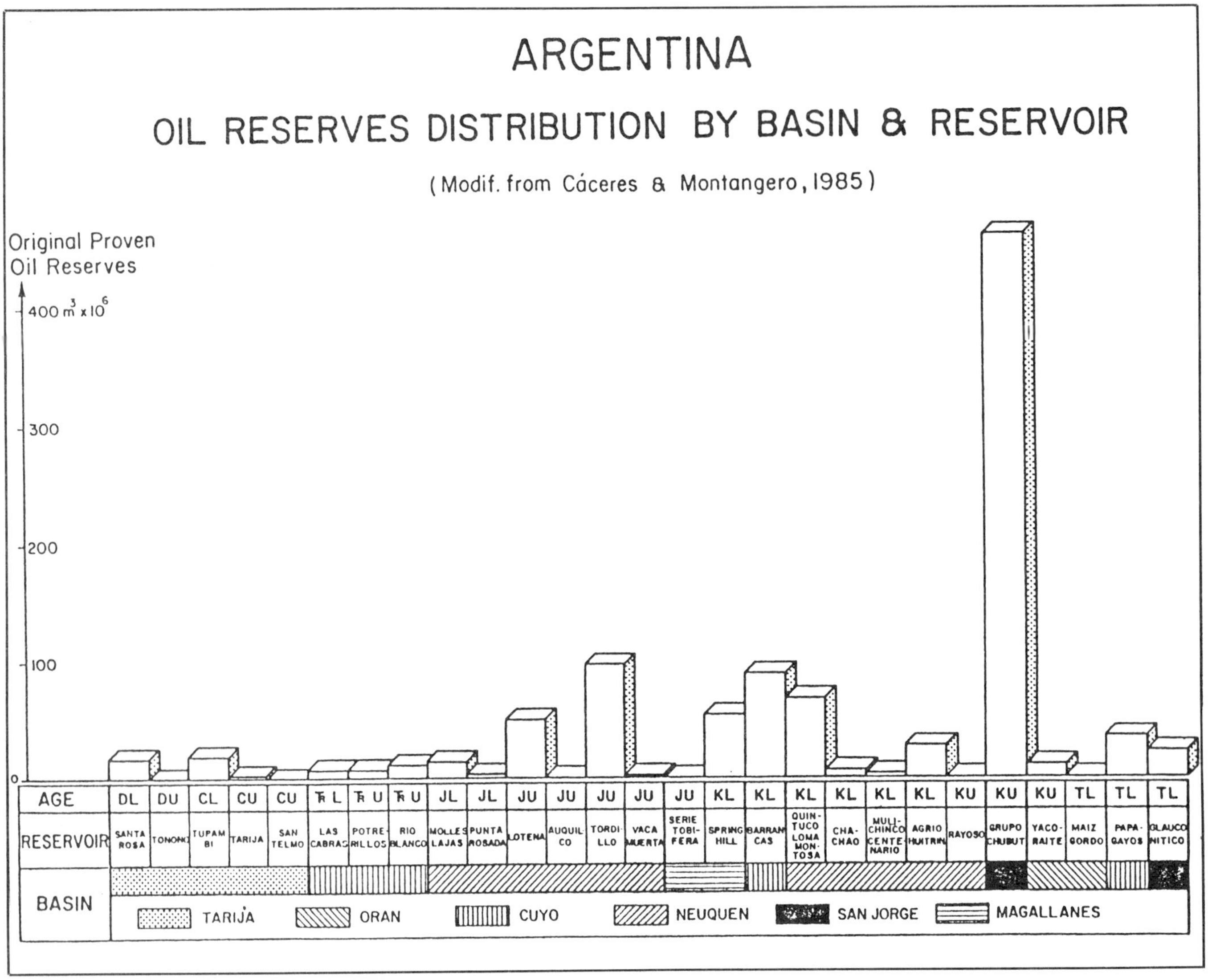

Figure 9. Distribution of oil reserves in Argentina by basin and reservoir.

EVOLUTION OF HYDROCARBON RESERVES IN ARGENTINA

The rate of increase in oil and gas reserves is closely related to drilling rates. Figure 11 compares drilling rates and oil and gas reserves from 1950 until the end of 1984. As can be observed, oil reserves in Argentina reached their highest peak in 1961, with a total proven reserve of 598 million m^3. After 1961, reserves gradually decreased to 250 million m^3 in 1967. From 1970 to 1984, proven oil reserves remained at the nearly constant level of 393 million m^3.

The evolution of proven reserves of natural gas differs substantially from that of proven oil reserves. Since 1978, gas reserves have been considerably higher than oil reserves, chiefly because of the discovery of the giant field Loma de la Lata in the central part of the Neuquen basin. This field has reserves of 396,000 million m^3 of natural gas plus 52.1 million m^3 of condensate in the Sierras Blancas reservoir of Callovian age, and 70 million m^3 of oil in the Quintuco (Berriasian limestones) reservoir. The magnitude of Argentine gas reserves has increased in recent years with new gas discoveries in the Tarija basin, which reached a maximum of 691,000 million m^3 in 1981/1982 and then decreased to 680,000 million m^3 in 1984.

The drilling rate curves in Figure 11 show that the most intensive drilling was in 1961, when a total of 1620 wells were completed. The drilling rate declined to a minimum of 422 wells in 1969 and then increased gradually to 1031 wells in 1984. Figure 11 also shows that the yearly rate of exploratory drilling also was variable, reaching a maximum of 155 wildcats in 1971.

An evaluation of required drilling rates until the year 2000, a period of about 14 years, to maintain both self-sufficiency and the current reserves/demand relationship, was made recently by the Chamber of Argentine Oil Companies. This study reveals that new oil reserves tantamount to the oil dis-

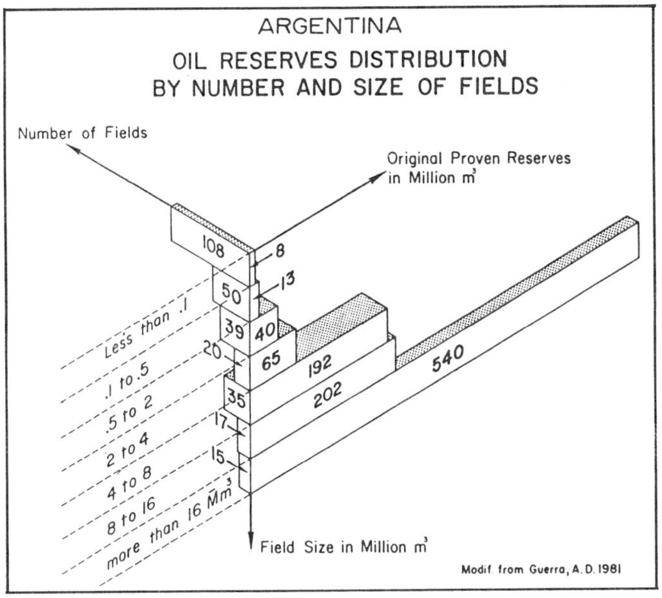

Figure 10. Distribution of oil reserves in Argentina by number and size of fields.

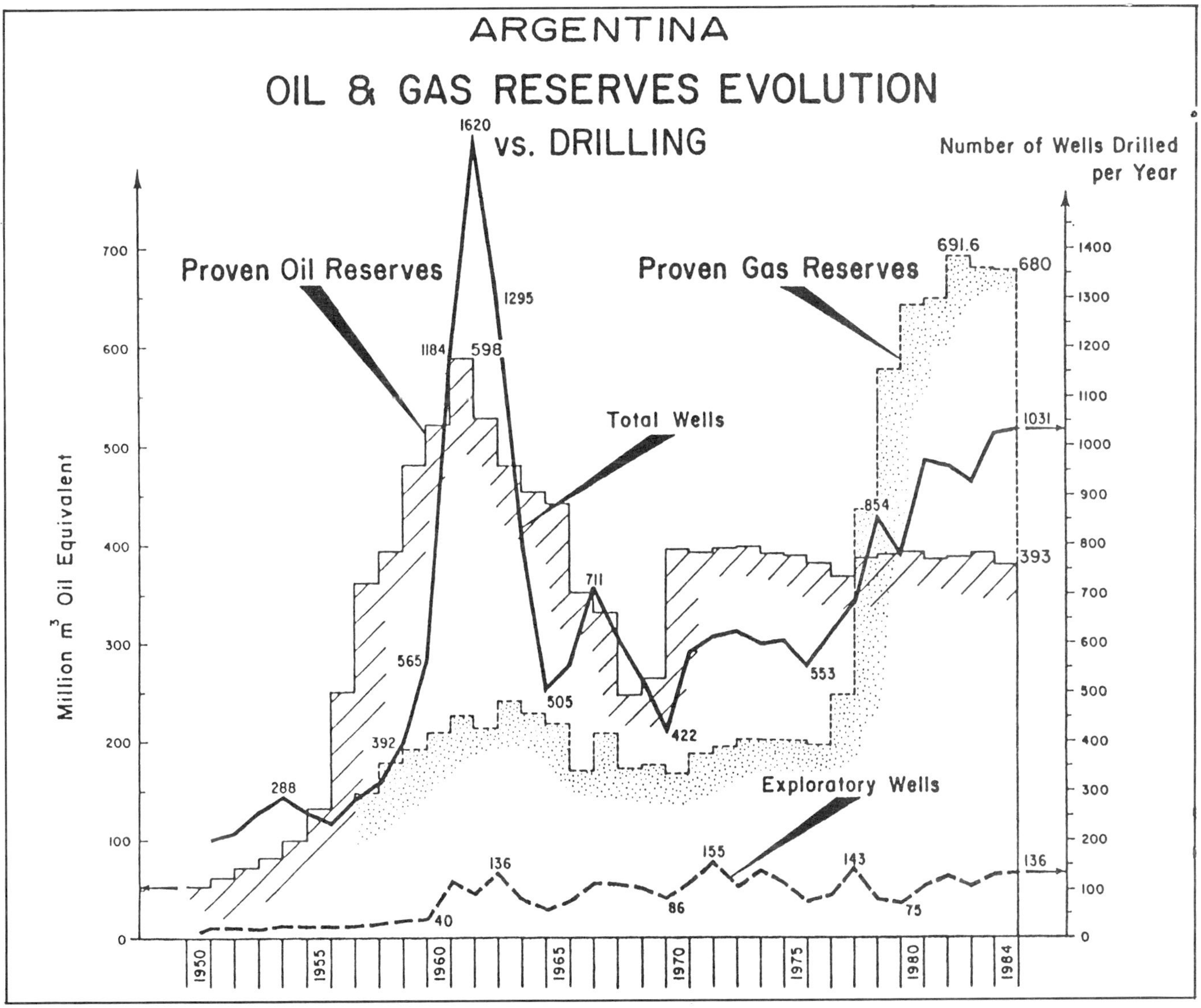

Figure 11. Proven oil and gas reserves in Argentina resulting from drilling each year from 1950 to 1984.

covered and proven in Argentina from 1907 to 1984, or around 1000 million m^3, will have to be found. On the basis of historical experience and considering a cummulative annual consumption growth of 3 to 4% per year, it was estimated that about 25,000 development wells (1450 wells/yr), and a total of 3300 wildcat wells (200 exploratory wells/yr) will be required to achieve this target. These drilling rates are far greater than the 1984 rates.

REFERENCES CITED

Cáceres, A.E., and G. Montangero, 1985, Reservorios precretácicos argentinos: Asistencia Recíproca de Empresas Petroleras Latinoamericanas, LV Reunión a nivel de expertos, p. 1-31, Quito, Ecuador.

Halbouty, M.T., A.A. Meyerhoff, R.E. King, R.H. Dott, Sr., D. Klemme, and T. Shabad, 1970, World's giant oil and gas fields, geologic factor affecting their formation and basin classification. Geology of giant petroleum fields: American Association of Petroleum Geologists Memoir 14, p. 502-575.

Lesta, P., J. Digregorio, and M.E. Mozetic, 1985, Presente y futuro de la exploración de petroleo en las cuencas subandinas argentinas: II Simposio Bolivariano-Exploración Petrolera de las Cuencas Subandinas, v. 3, p. 1-34, Bogotá, Colombia.

Turic, M., C. Fernández-Garrasino, R. Pombo, J.L. Bianchi, and H. Dibenedetto, 1981, Cuencas sedimentarias en la Argentina: Published communication of Yacimientos Petrolíferos Fiscales, p. 3-40, Buenos Aires.